Energy, Environment, and Sustainability

Series Editor

Avinash Kumar Agarwal, Department of Mechanical Engineering, Indian Institute of Technology Kanpur, Kanpur, Uttar Pradesh, India

AIMS AND SCOPE

This books series publishes cutting edge monographs and professional books focused on all aspects of energy and environmental sustainability, especially as it relates to energy concerns. The Series is published in partnership with the International Society for Energy, Environment, and Sustainability. The books in these series are edited or authored by top researchers and professional across the globe. The series aims at publishing state-of-the-art research and development in areas including, but not limited to:

- Renewable Energy
- Alternative Fuels
- Engines and Locomotives
- Combustion and Propulsion
- Fossil Fuels
- Carbon Capture
- Control and Automation for Energy
- Environmental Pollution
- Waste Management
- Transportation Sustainability

Review Process

The proposal for each volume is reviewed by the main editor and/or the advisory board. The chapters in each volume are individually reviewed single blind by expert reviewers (at least four reviews per chapter) and the main editor.

Ethics Statement for this series can be found in the Springer standard guidelines here https://www.springer.com/us/authors-editors/journal-author/journal-author-helpdesk/before-you-start/before-you-start/1330#c14214

More information about this series at https://link.springer.com/bookseries/15901

Yogalakshmi Kadapakkam Nandabalan · Vinod Kumar Garg · Nitin K. Labhsetwar · Anita Singh
Editors

Zero Waste Biorefinery

Springer

Editors
Yogalakshmi Kadapakkam Nandabalan
Department of Environmental Science
and Technology
Central University of Punjab
Bathinda, Punjab, India

Vinod Kumar Garg
Department of Environmental Science
and Technology
Central University of Punjab
Bathinda, Punjab, India

Nitin K. Labhsetwar
ERMD Division
NEERI
Nagpur, India

Anita Singh
Department of Environmental Sciences
Central University of Jammu
Samba, India

ISSN 2522-8366 ISSN 2522-8374 (electronic)
Energy, Environment, and Sustainability
ISBN 978-981-16-8681-8 ISBN 978-981-16-8682-5 (eBook)
https://doi.org/10.1007/978-981-16-8682-5

This Springer imprint is published by the registered company Springer Nature Singapore Pte Ltd.
The registered company address is: 152 Beach Road, #21-01/04 Gateway East, Singapore 189721, Singapore

Preface

Energy demand has been rising remarkably due to the increasing population and urbanization. The global economy and society significantly depend on energy availability because it touches every facet of human life and activities. Transportation and power generation are the two major examples. Without millions of personalized and mass transport vehicles and the availability of 24×7 power, human civilization would not have reached contemporary living standards.

The International Society for Energy, Environment and Sustainability (ISEES) was founded at the Indian Institute of Technology Kanpur (IIT Kanpur), India, in January 2014 to spread knowledge/awareness and catalyze research activities in the fields of energy, environment, sustainability, and combustion. Society's goal is to contribute to the development of clean, affordable, and secure energy resources and a sustainable environment for society and spread knowledge in the areas mentioned above and create awareness about the environmental challenges the world is facing today. The unique way adopted by ISEES was to break the conventional silos of specializations (engineering, science, environment, agriculture, biotechnology, materials, fuels, etc.) to tackle the problems related to energy, environment, and sustainability in a holistic manner. This is quite evident by the participation of experts from all fields to resolve these issues. ISEES is involved in various activities such as conducting workshops, seminars, conferences in the domains of its interests. Society also recognizes the outstanding works of young scientists, professionals, and engineers for their contributions in these fields by conferring them awards under various categories.

The Fifth International Conference on 'Sustainable Energy and Environmental Challenges' (V-SEEC) was organized under the auspices of ISEES from December 19 to 21, 2020, in virtual mode due to restrictions on travel because of the ongoing COVID-19 pandemic situation. This conference provided a platform for discussions between eminent scientists and engineers from various countries, including India, Spain, Austria, Bangladesh, Mexico, USA, Malaysia, China, UK, Netherlands, Germany, Israel, and Saudi Arabia. At this conference, the eminent international speakers presented their views on energy, combustion, emissions, and alternative energy resources for sustainable development and a cleaner environment. The conference presented two high-voltage plenary talks by Dr. VK Saraswat, Honorable Member, NITI Ayog, on 'Technologies for Energy Security and Sustainability' and Prof. Sandeep Verma, Secretary, SERB, on 'New and Equitable R&D Funding Opportunities at SERB.'

The conference included nine technical sessions on topics related to energy and environmental sustainability. Each session had 6–7 eminent scientists from all over the world, who shared their opinion and discussed the trends for the future. The technical sessions in the conference included emerging contaminants: monitoring and degradation challenges; advanced engine technologies and alternative transportation fuels; future fuels for sustainable transport; sustainable bioprocessing for biofuel/non-biofuel production by carbon emission reduction; future of solar energy; desalination and wastewater treatment by membrane technology; biotechnology in sustainable development; emerging solutions for environmental applications' and challenges and opportunities for electric vehicle adoption. Five hundred plus participants and speakers from all over the world attended this three days conference.

The conference concluded with a high-voltage panel discussion on 'Challenges and Opportunities for Electric Vehicle Adoption,' where the panelists were Prof. Gautam Kalghatgi (University of Oxford), Prof. Ashok Jhunjhunwala (IIT Madras), Dr. Kelly Senecal (Convergent Science), Dr. Amir Abdul Manan (Saudi Aramco), and Dr. Sayan Biswas (University of Minnesota, USA). Prof. Avinash K Agarwal, ISEES, moderated the panel discussion. This conference laid out the roadmap for technology development, opportunities, and challenges in energy, environment, and sustainability domain. All these topics are very relevant for the country and the world in the present context. We acknowledge the support received from various agencies and organizations for the successful conduct of the fifth ISEES conference V-SEEC, where these books germinated. We want to acknowledge SERB (special thanks to Dr. Sandeep Verma, Secretary) and our publishing partner Springer (special thanks to Ms. Swati Meherishi).

The editors would like to express their sincere gratitude to a large number of authors from all over the world for submitting their high-quality work on time and revising it appropriately at short notice. We would like to express our special gratitude to our prolific set of reviewers, Dr. Sakar Mohan, Dr. G. N. Nikhil, Dr. Anjireddy, Dr. Omprakash Sarkar, Dr. Sumer Singh Meena, Dr. Kashyap Dubey, Dr. Abhishek Chandra, Prof. A. K. Jain, Dr. Krishna Kadirvelu, Dr. M. D. warakanath, Dr. D. Prabha, Dr. Sri Shalini, Dr. R. Selvakumar, Dr. Godvin Sharmila, Dr. R. Yukesh Kannah, Dr. M. Dinesh Kumar, Dr. P. Balasubramanian, Dr. Dhanya M. S., Dr. Khetan Shevkani, Dr. S. Shantha Kumar, Dr. P. Senthil Kumar, Dr. Rajeev Pratap Singh, Dr. C. Sivaraman, Dr. V. Preethi, Dr. Divya Subhash, Dr. Ritu Singh, Dr. Rashmi Kataria, Dr. K. Sivagami, Dr. R. Arthur James, Dr. S. Sudalai, Dr. Mukesh Awasthi, Dr. Sushma Yadav, Dr. Lakhvinder Singh, Dr. Neelam Yadav, Dr. Sartaj Bhat, Dr. Hardeep Rai Sharma, Dr. V. V. Tyagi, Dr. S. Adish Kumar, Dr. Manpreet Singh Bhatti, Dr. Divya Nair, Dr. Anoop Yadav, Dr. Kiran Bala, Dr. Anju Malik, Dr. Somvir Bajar, Dr. Minakshi Suhag, Dr. Naresh Rawat, Dr. Navish Kataria, Dr. Anita Singh Kirrolia, Dr. Mona Sharma, Dr. Asheesh Kumar Yadav, Dr. Poonam Yadav, Dr. Richa Kothari, Dr. Kashif Kidwai, Dr. Kavita Sharma, Dr. K. V. Yatish,

Dr. G. Velvizhi, Dr. Simranjeet Singh, Dr. Kulvinder Bajwa, Dr. Renu Singh, Dr. Atin Kumar, Dr. Arti Devi, and Dr. Harmohan Singh who reviewed various chapters of this monograph and provided their valuable suggestions to improve the manuscripts.

This book is a compilation of process, technologies, and value-added products such as high-value biochemicals and biofuels produced from different waste biorefineries. The book is sectioned into four categories where few chapters in Part I provide a comprehensive outlook about zero-waste biorefinery and technologies associated with it. The emerging technologies that potentially put back the lignocellulosic waste, municipal solid waste, and food waste into intrinsic recycling for production of high-value biochemicals and bioenergy, along with associated challenges and opportunities, are compiled up in Parts II and IV. Algal biorefineries leading to sustainable circular economy through production of broad spectrum of bioactive compounds, bioethanol, biobutanol, biohydrogen, biodiesel through integrated biorefinery approach are discussed in Part III. The book includes few chapters on conversion technologies and mathematical models applied for process optimization. Chapters include recent results and are focused on current trends of waste biorefineries. In this book, readers will get a sound foundation about the underlying principles of biorefineries and a up-to-date state-of-the-art-based overview on the latest advances in terms of scientific knowledge, techno-economic developments, and life cycle assessment methodologies of integrated waste biorefinery. The book is envisioned for a broader audience, and the editors hope that the book would greatly interest the professionals, postgraduate students, and policy makers involved in waste management, biorefineries, circular economy, and sustainable development.

Bathinda, India — Yogalakshmi Kadapakkam Nandabalan
Bathinda, India — Vinod Kumar Garg
Nagpur, India — Nitin K. Labhsetwar
Samba, India — Anita Singh

Contents

Editors and Contributors

About the Editors

Dr. Yogalakshmi Kadapakkam Nandabalan is currently working at Central University of Punjab, as Associate Professor in the Department of Environmental Science and Technology. She holds a Master's and Ph.D. degree in Environmental Science from Anna University, Chennai, India. She has about 12 years of teaching and research experience. Her areas of research interest include water and wastewater management, energy generation from waste and wastewater, solid waste management, and nanobioremediation. She has authored around 15 book chapters and 40 research and review articles in peer-reviewed, high-impact factor journals. She has developed a number of e-content modules. She has completed three sponsored research projects funded by various agencies as Principal Investigator. She is a member of several professional and scientific bodies.

Prof. Vinod Kumar Garg is presently working as Professor in the Department of Environmental Science and Technology, Central University of Punjab, India. He is a well-rounded researcher with over 30 years of experience in leading, supervising, and undertaking research in the broad field of solid and hazardous waste management. His research group is working on water and wastewater pollution monitoring and abatement, solid waste management, pesticide degradation, and heavy metal detoxification. He has published more than 200 research and review articles, 22 proceedings, and 6 editorials in peer-reviewed journals of international and national repute with more than citations 12000 and h-index 57. In addition, he has published 2 books and 12 book chapters and completed 10 sponsored research projects as Principal Investigator funded by various agencies and departments. He is an active member of various scientific societies and organizations including, the Biotech Research Society of India, the Indian Nuclear Society, etc.

Dr. Nitin K. Labhsetwar is a Ph.D. with over 34 years of research experience in environmental and energy-related research. He has been actively working on cleaner energy generation including photocatalytic hydrogen generation, bioenergy, carbon capture, chemical looping combustion, CO_2 reforming, automobile emission control and rural energy (development and deployment of improved biomass cook stoves), etc. He has worked as STA/JSPS Fellow and Visiting Overseas Researcher at NIMS, Tsukuba, Japan, and as Visiting Professor at Kyushu University, Japan, under the Global COE Program on Novel Carbon Resource Sciences. He has over 170 research publications with over 6800 citations and 22 international patents and 8 design registrations in addition to a few contributions in books. He has so far contributed to more than 75 R&D projects.

Dr. Anita Singh is working as Assistant Professor, Department of Environmental Sciences, Central University of Jammu, India. She has over 8 years of teaching and research experience. Dr. Singh pursued her Ph.D. on bioethanol production from lignocellulosic biomass from Guru Jambheshwar University of Science & Technology, India. Her post-doctoral research was at the University of Georgia, USA, on selective removal of inhibitors from lignocellulosic biomass. She has published 40 research papers in international and national journals and 6 book chapters in various edited books. She has guided 33 master's dissertation and is currently supervising 4 Ph.D. students. She was awarded CSIR-SRF and CSIR-RA fellowships for research during Ph.D. and UGC Raman Post-doctoral Fellowship for doing research in the USA. She is working in the field of biofuel production and value-added product production from lignocellulosic biomass.

Contributors

Abdullah Department of Plant Sciences, Quaid-i-Azam University Islamabad, Islamabad, Pakistan

Zeeshan Ahmad Department of Plant Sciences, Quaid-i-Azam University Islamabad, Islamabad, Pakistan

S. Aishwarya School of Chemical Engineering, Vellore Institute of Technology, Chennai, Tamil Nadu, India

Muthahar Fazila Amanullah Centre for Food Technology, Department of Biotechnology, Anna University, Chennai, India

E. Anagha School of Chemical Engineering, Vellore Institute of Technology, Chennai, Tamil Nadu, India

R. Aanandhalakshmi Department of Biotechnology, School of Bio and Chemical Engineering, Kalasalingam Academy of Research and Education, Krishnankoil, Tamil Nadu, India

Venkata Naga Surya Gunasri Appala Department of Chemical Engineering, Dr. B. R. Ambedkar National Institute of Technology, Jalandhar, Punjab, India

Somvir Bajar Department of Environmental Science and Engineering, J.C. Bose University of Science and Technology, YMCA, Faridabad, Haryana, India

Shailendra Bajpai Department of Chemical Engineering, Dr. B. R. Ambedkar National Institute of Technology, Jalandhar, Punjab, India

Narsi R. Bishnoi Department of Environmental Science and Engineering, GJUS&T, Hisar, India

Samarshi Chakraborty School of Chemical Engineering, Vellore Institute of Technology, Chennai, Tamil Nadu, India

Rohit Dalal Energy and Resource Management Division, CSIR-National Environmental Engineering Research Institute, CSIR-NEERI, Nagpur, Maharashtra, India

Anirban Dey Department of Chemical Engineering, Pandit Deendayal Energy University, Gandhinagar, Gujarat, India

R. Divyabharathi Department of Renewable Energy Engineering, Tamil Nadu Agricultural University, Coimbatore, India

M. S. Dhanya Department of Environmental Science and Technology, Central University of Punjab, Bathinda, Punjab, India

P. Ezhilarasi Department of Biotechnology, School of Bio and Chemical Engineering, Kalasalingam Academy of Research and Education, Krishnankoil, Tamil Nadu, India

Nandhini Devi Ganesan Centre for Food Technology, Department of Biotechnology, Anna University, Chennai, India

Vinod K. Garg Department of Environmental Science and Technology, School of Environment and Earth Sciences, Central University of Punjab, Bathinda, India

P. S. Gowtham Department of Biotechnology, School of Bio and Chemical Engineering, Kalasalingam Academy of Research and Education, Krishnankoil, Tamil Nadu, India

Iram Gul Department of Plant Sciences, Quaid-i-Azam University Islamabad, Islamabad, Pakistan

Srisowmeya Guruchandran Centre for Food Technology, Department of Biotechnology, Anna University, Chennai, India

Zahoor Ul Haq Department of Plant Sciences, Quaid-i-Azam University Islamabad, Islamabad, Pakistan;
Department of Botany, Shaheed Benazir Bhutto University, Wari Campus, Wari, Dir Upper, Pakistan

Majid Iqbal Department of Plant Sciences, Quaid-i-Azam University Islamabad, Islamabad, Pakistan

V. M. Jaganathan Department of Energy and Environment, National Institute of Technology, Tiruchirappalli, Tamil Nadu, India

Justin Joy Department of Chemical Engineering, Amal Jyothi College of Engineering Kanjirappally, Kottayam, India

P. Jyothirmai School of Chemical Engineering, Vellore Institute of Technology, Chennai, Tamil Nadu, India

Sonika Kag Delhi Technological University (DTU), Delhi, India

Rashmi Kataria Delhi Technological University (DTU), Delhi, India

Ravneet Kaur Sant Longowal Institute of Engineering and Technology, Sangrur, Longowal, India

Shujaul Mulk Khan Department of Plant Sciences, Quaid-i-Azam University Islamabad, Islamabad, Pakistan

Bala Kiran Department of Biosciences and Biomedical Engineering, Indian Institute of Technology Indore, Indore, Madhya Pradesh, India

Lukas Kratky Faculty of Mechanical Engineering, Department of Process Engineering, Czech Technical University in Prague, Prague 6, Czech Republic

Neha Kukreti Delhi Technological University (DTU), Delhi, India

Nitin K. Labhsetwar Energy and Resource Management Division, National Environmental Engineering Research Institute, Nagpur, India

Vishnu Manirethan Department of Chemical Engineering, Amal Jyothi College of Engineering Kanjirappally, Kottayam, India

Sweeti Mann Delhi Technological University (DTU), Delhi, India

Parth G. Marakana Department of Chemical Engineering, Pandit Deendayal Energy University, Gandhinagar, Gujarat, India

Chakravarthy Muninathan Centre for Food Technology, Department of Biotechnology, Anna University, Chennai, India

Abhilash T. Nair Department of Applied Sciences and Humanities, National Institute of Foundry and Forge Technology (NIFFT), Hatia, Ranchi, Jharkhand, India

Umar Nawaz Department of Nanoscience and Technology, NCP, Quaid-i-Azam University Campus, Islamabad, Pakistan

Regina Nogueira Institute for Sanitary Engineering and Waste Management, Leibniz Universitat Hannover, Hannover, Germany

K. Palanivelu Centre for Climate Change and Disaster Management, Department of Civil Engineering, College of Engineering Guindy Campus, Anna University, Chennai, Tamil Nadu, India

Nitin Naresh Pandhare Department of Chemical Engineering, Dr. B. R. Ambedkar National Institute of Technology, Jalandhar, Punjab, India

A. Ramachandran Centre for Climate Change and Disaster Management, Department of Civil Engineering, College of Engineering Guindy Campus, Anna University, Chennai, Tamil Nadu, India

Pawan Kumar Rose Department of Energy and Environmental Sciences, Chaudhary Devi Lal University, Sirsa, Haryana, India

Rohit Ruhal Regional Centre of Biotechnology, Faridabad, India

Saloni Sachdeva Department of Biotechnology, Jaypee Institute of Information Technology, Noida, India

Bharti Saini Department of Chemical Engineering, Pandit Deendayal Energy University, Gandhinagar, Gujarat, India

Rimjhim Sangtani Department of Biosciences and Biomedical Engineering, Indian Institute of Technology Indore, Indore, Madhya Pradesh, India

Jaigopal Sharma Delhi Technological University (DTU), Delhi, India

Anita Singh Department of Environmental Sciences, Central University of Jammu, Jammu, Jammu and Kashmir, India

Simar Preet Singh Bennett University, Greater Noida, Uttar Pradesh, India

K. Sivagami School of Chemical Engineering, Vellore Institute of Technology, Chennai, Tamil Nadu, India

Smrity Sonbhadra Department of Biosciences and Biomedical Engineering, Indian Institute of Technology Indore, Indore, Madhya Pradesh, India

S. Sri Shalini Centre for Climate Change and Disaster Management, Department of Civil Engineering, College of Engineering Guindy Campus, Anna University, Chennai, Tamil Nadu, India

V. Sruthi School of Chemical Engineering, Vellore Institute of Technology, Chennai, Tamil Nadu, India

P. Subramanian Department of Renewable Energy Engineering, Tamil Nadu Agricultural University, Coimbatore, India

Priyanka Uddandarao School of Nature Sciences, National University of Ireland, Galway, Republic of Ireland

B. Vanavil Department of Biotechnology, School of Bio and Chemical Engineering, Kalasalingam Academy of Research and Education, Krishnankoil, Tamil Nadu, India

Rijin Thomas Varghese Department of Chemical Engineering, Amal Jyothi College of Engineering Kanjirappally, Kottayam, India

S. Varunkumar Department of Mechanical Engineering, Indian Institute of Technology Madras, Chennai, Tamil Nadu, India

Roshan Wathore Energy and Resource Management Division, CSIR-National Environmental Engineering Research Institute, CSIR-NEERI, Nagpur, Maharashtra, India

Deepak Kumar Yadav Department of Environmental Science and Engineering, GJUS&T, Hisar, India

K. N. Yogalakshmi Department of Environmental Science and Technology, School of Environment and Earth Sciences, Central University of Punjab, Bathinda, India

Part I
General

Chapter 1
Zero Waste Biorefinery: A Comprehensive Outlook

Saloni Sachdeva, Vinod K. Garg, Nitin K. Labhsetwar, Anita Singh, and K. N. Yogalakshmi

S. Sachdeva
Department of Biotechnology, Jaypee Institute of Information Technology, A-10 Sector 62, Noida 201309, India

V. K. Garg · K. N. Yogalakshmi (✉)
Department of Environmental Science and Technology, School of Environment and Earth Sciences, Central University of Punjab, Bathinda 151001, India

N. K. Labhsetwar
Energy and Resource Management Division, National Environmental Engineering Research Institute, Nagpur 440020, India

A. Singh
Department of Environmental Sciences, Central University of Jammu, Jammu, Jammu and Kashmir 180011, India

Y. K. Nandabalan et al. (eds.), *Zero Waste Biorefinery*, Energy, Environment, and Sustainability, https://doi.org/10.1007/978-981-16-8682-5_1

GRAPHICAL ABSTRACT

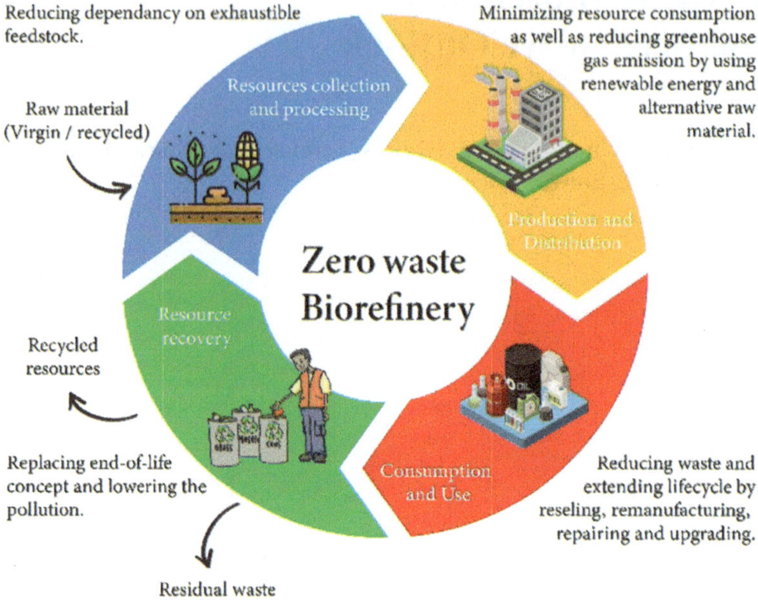

1.1 Introduction

The contemporary world is undergoing prodigious expansion in human population. On an average 81 million people are added annually to the existing 7.9 billion people (Worldometers.info 2021). In such circumstances, the availability of resources such as water, food, energy, forest, and biodiversity that are necessary for human survival has become scarce in addition to exacerbation in energy demands, pollution, waste generation, and climate change leading to an unsustainable habitat (Pimentel and Pimentel 2003). To cope with this ever-increasing trend in population and limited supply of resources, countries across the world have proclaimed their focus on strengthening sustainable and carbon-free alternatives to the conventional production mechanisms. One such substitute is “Biorefineries” which are infrastructural establishments that aim to convert biomass into numerous commercial commodities or energy in a sustainable manner (Kamm and Kamm 2004; Fernando et al. 2006; Cherubini 2010; Jong and Jungmeier 2015; Mohan et al. 2016). Biorefinery is not a modern concept; biomass has been exploited in the pulp and paper industry, fuel industry, and food industry for a considerable time. Biomass includes all sorts of organic materials from renewable sources. However, the efficacy and on-field existence vary significantly for different classes of biorefineries (as listed in Table 1.1). Towards future biorefineries, the scientific community seeks alternate substrates with

Table 1.1 Types of biorefineries based on the type of feedstock and their significance

Substrate	Type of biorefineries	Products	Significance
Sugar/Starch crops	One-platform (C6 sugar) biorefinery	Bioethanol Animal feed	+ The primitive attempt replaces non-renewable resources – Compete with food and feed supply
Oil crops	One-platform (oil) biorefinery	Biodiesel Animal feed Glycerin	
Lignocellulosic biomass	One-platform (syngas) biorefinery	Biofuels Chemicals (alcohols)	+ Lignocellulosic biomass is the most abundant available substrate in the environment with no competition for food and feed production + Wide range of established pre-treatment methods – Complex structure/chemical composition
	Four-platform (lignin/syngas, C5/C6 sugar) biorefinery	Biofuels Animal feed Bioethanol	
Grasses/Algae	Green Biorefinery	Biomethane Chemical building blocks (lactic acid and amino acids) Biomaterials Fertilizer	+ Grasses/algal biomass is the second most abundant available substrate that naturally occur in environment + Reduction in greenhouse gas emissions – Difficult to maintain cultivation requirements

cleaner conversion techniques to transform biomass into a range of value-added products. Building such an eco-efficient technology will not only pave way for a sustainable living but also a way out for current concerns such as waste management, and increasing greenhouse gas emissions (Angouria-Tsorochidou et al. 2021).

The concept of the biorefinery is analogous to traditional oil refineries except for the fact petroleum-based refineries make use of non-renewable (fossil fuels) feedstocks which liberate greenhouse gases imposing detrimental impacts on the environment. On contrary, an enormous variety of biomass (such as edible crops, algae, and residues from industries) is available that can be transformed using physical, chemical, or biological processes for generating a wide range of products or energy to suffice the demands of mankind (Uellendahl and Ahring 2010; Kumar et al. 2013; Hernández et al. 2014; Salvachúa et al. 2016; Gnansounou and Pandey 2017; Ichikawa et al. 2017; Hingsamer et al. 2019; Bittencourt et al. 2019; Monte et al. 2020). Besides environmental and economic benefits of organic energy sources, food security menace and recurring shortfall of input resources limit the implementation of the biorefineries in reality (Espinoza Pérez et al. 2017; Gírio et al. 2017; Byun

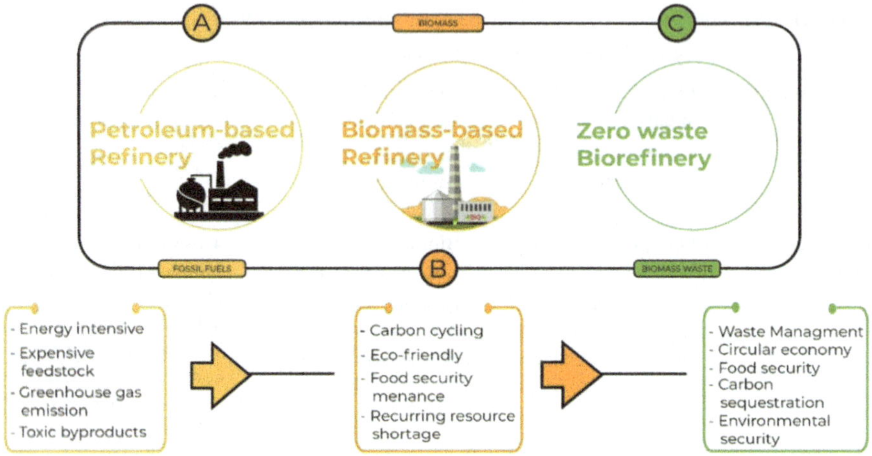

Fig. 1.1 Transition from petroleum biorefineries to zero waste biorefineries

and Han 2020). In this regard, residues from industries/biorefineries are often overlooked but they are the quintessential alternative to the aforementioned biorefineries. Wastes in solid, liquid, or gaseous form have net positive energy incorporated, which holds propitious capabilities to be transformed into bioproducts and biofuels (Oliveira and Navia 2017; Zheng et al. 2020; Venslauskas et al. 2021). The waste utilization approach attributes to repeated use of the substrates which further brings down the limitation of finite resources and emphasizes zero waste generation. Moreover, the concept adds auxiliary value along with improved resources utilization efficiency (Byun and Han 2020). These zero waste bio-based refineries open the door for the paradigm shift from linear mass production involving extraction, manufacturing, utilization, disposal to a closed-loop system and thus evolving to a sustainable bioeconomy as illustrated in Fig. 1.1 (Nizami et al. 2017; Venkata Mohan et al. 2019; Ubando et al. 2020). The rational idea behind sustainable biorefineries solely depend on reusing and recycling the residues, which require a thorough understanding of gaps in technical and biological proceedings. On that account, the present chapter focuses on providing a comprehensive overview of carbon-neutral biorefinery systems using feedstocks such as lignocellulose, algae (micro-, and macro-algae), residue (solid waste, paper, food, manure), and improvised treatment processes.

1.2 The Zero-Waste Biorefinery Concept

Biorefineries are the consolidated system of physical, chemical, and biological processes in well-defined order which can convert biomass or residues to marketable services viz., chemicals, feed, fuels, energy, etc. Biomass is regarded as a renewable source of energy that can effectively contribute to global energy demand through fuels and economic prosperity by producing high-value products (Zetterholm et al. 2020). Analogous to petroleum refineries, biomass-based refineries also account for certain downsides. The major pitfall are limited space and added waste generation. To overpower these foregoing issues, residues or wastes from biorefineries unfold the possibility to reuse and recycle using effectual mechanisms which can be termed as zero waste biorefineries (Brunklaus et al. 2018). Zero waste biorefineries are expected to develop an effective circular bioeconomy by directly benefitting society and the economy (Venkata Mohan et al. 2019; Ubando et al. 2020). Figure 1.2 illustrates how biological-based products revolve in a closed and re-circulating loop system. According to Ellen MacArthur Foundation (2013), a compact circle establishes maximum savings on raw material, capital, and energy (Ellen MacArthur Foundation 2013). The circular model of consumption and production epitomizes multiple re-use of initial feedstock material across the loop, and diminishing the need for fresh feedstock. Aside from economic benefits, the zero-waste biorefinery can also strengthen climate change control by utilizing carbon dioxide into efficacious bio-feedstock.

Fig. 1.2 Zero-waste biorefinery—Circular and carbon-neutral system of production and consumption

Bioeconomy and circular bioeconomy are often interchangeably used; however, the two terms differ slightly in their principles. The term bioeconomy was originally defined in 1998 with a target of developing bio-commodities using renewable sources of input and replacing oil-based commodities (Enriquez 1998). The published literature so far promises to curb climate change effects by providing renewable biomass together with generating opportunities and employment (Ubando et al. 2020). Whereas circular bioeconomy endorses circular economy framework, wherein bio-based products are generated using a variety of biomass. The circular loop ensures economic viability, waste reduction, and sustainability (Carus and Dammer 2018).

1.2.1 Lignocellulosic Biorefinery

Lignocellulosic biomass attribute to the plant dry mass which consists of cellulose (35–48%), hemicellulose (22–30%), and lignin (15–27%) (Yousuf et al. 2020). This is the most abundant and cost-effective raw material present, which includes agricultural and forestry residues. For a long time, lignocellulosic biomass has been transformed into marketable commodities via physical, chemical, and biological processes. However, the prime focus has been extracting and exploiting polysaccharides to manufacture alcohol, methane, phenolic compounds, furans, etc., and on the other hand, lignin is either transformed into low-quality aromatic products or burnt to generate heat. On account of ineffective handling of the lignin component limit the functioning of biorefineries. Nonetheless, several studies affirm that lignin with little or no degradation can yield high-value chemicals and fuels. The development of eco-friendly and milder procedures for structural degradation of lignin into beneficial products is the need of the hour. Garlapati et al. (2020) scrutinized biological and chemo-catalytic processes for the effective valorization of the lignin component. It was concluded that bioconversion of lignin using peroxidases, laccases, and other auxiliary enzymes gained a victory over chemical processes. Keeping in mind the concept of "Zero waste biorefinery" in lignocellulose biomass-based refinery, the scientific community explored limited bottlenecks (as depicted in Fig. 1.3). Qiu et al. (2021) peeked into alternatives of traditional oil plants, such that the efficacy of production-consumption can be elevated. The study found *Litsea cubeba* as an ideal substitute as it holds the potential to produce oil in addition to value-added products. Citral, being the dominant component, the applicability of *Litsea cubeba* can be extended to the perfume industry, pharmaceutical industry (anti-oxidants, anti-microbial, inflammatory), and food industry (food flavoring). Another innovative approach explored the efficacy of adsorbents fabricated from renewable polysaccharides residues (Zuin et al. 2017). Mesoporous carbonaceous sorbents formulated using starch (Starbon), alginic acid (Algibon), and pectin (Pecbon) were manifested as suitable substrates due to their functional and structural flexibility. Among three sorbents, starbon overpowered the other two as its fabrication required limited steps and had an improved environmental footprint.

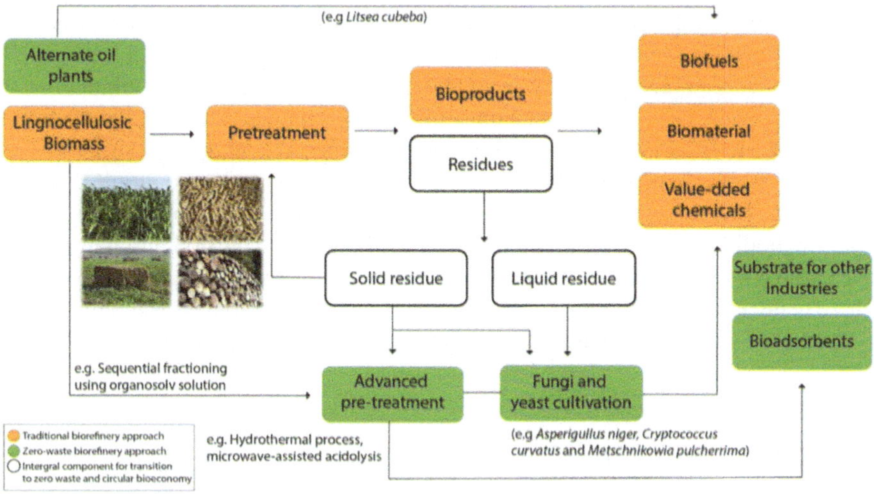

Fig. 1.3 Zero-waste biorefinery processes for lignocellulosic biomass

Another major bottleneck in lignocellulose biorefinery is the selection of a meticulous pretreatment method (Liguori and Faraco 2016; Arevalo-Gallegos et al. 2017; Galbe and Wallberg 2019). An effective treatment technology is crucial for lowering down the production cost involved and making the process more sustainable. Recently, a study conducted by Squinca et al. (2018) opened the doors for utilizing liquefied materials in cultivations of *Aspergillus niger* using enzymatic pretreatment. The findings of the study concluded that using the above-mentioned combination can cut down the cost of enzyme production and attain a closed-loop within the biorefinery. Following the same trend, waste pomegranate peels were digested via the hydrothermal procedure and the residue was further treated with enzymes aiming to increase the overall efficiency of the system (Talekar et al. 2018). During the initial phase of the setup, a substantial yield was achieved with 18.8–20.9% of food-grade quality pectin and 10.6–11.8% of phenolic compounds. The residual biomass was subjected to enzymes which in turn generated 95% of glucose. A primitive study focusing on unconventional pre-treatment technology was conducted by Abdelaziz et al. in 2015 where an organosolv aqueous solution was investigated for sequential fractioning of the forest residues (Abdelaziz et al. 2015). The organosolv consists of 2-methyltetrahydrofuran and oxalic acid as organic solvent and catalysts respectively. At the end of the experiment, the researchers successfully fractionalized the components in their compact form. A new method of pretreatment i.e., rapid microwave-assisted acidolysis was scrutinized and further residual biomass was investigated for potential as yeast substrate (Zhou et al. 2017). Three types of lignocellulosic biomass were considered including herbaceous, hard, and softwood, out of which softwood on treatment liberated lignin component with 93% purity. Furthermore, two yeast strains (*Cryptococcus curvatus* and *Metschnikowia pulcherrima*) were able to proliferate on the residual biomass from the first proceeding. In recent years, it is observed that the association of two or more treatments methods has been a preferred choice to integrate into the circular economy. Dávila et al. (2017, 2019) proposed a combination of hydrothermal treatment followed by delignification and enzymatic saccharification to segment different components of vine shoots as the lignocellulosic biomass. Under optimized conditions, the study was successful in extracting 82 wt. % of lignin along with deceased cellulose degradation (35 wt %) (Dávila et al. 2017, 2019). Another progressive study made effort to omit sulfur from the traditional fractioning methods by using steam explosion and hydrotropic extraction (Olsson et al. 2019). Hardwood was first subjected to steam pre-treatment where hemicellulose was solubilized and subsequently, the biomass was subjected to hydrotropic extraction to extract lignin, leaving behind a solid mass rich in cellulose.

1.2.2 *Algal Biorefinery*

Algae are unicellular aquatic photosynthetic organisms that lack root, shoot and stem characteristics. Based on morphology, algal species can be macroalgae (rich in carbohydrates) and microalgae (rich in lipid content). Owing to the macromolecule abundance and relatively cheaper cost to edible crops, algae became an emerging feedstock for biorefineries. However, the conventional biorefineries were pronounced to manufacture one class of commodities, but with progression in technology, a wide range of high-value products are developed (Rajak et al. 2020). Incorporating the concept of zero waste biorefinery, algal residues (solid or liquid) serve as a prime asset as illustrated in Fig. 1.4 (Mitra and Mishra 2019). Solid residual mass can be utilized as a feedstock to other industries or as a sustainable adsorbent to remediate a polluted

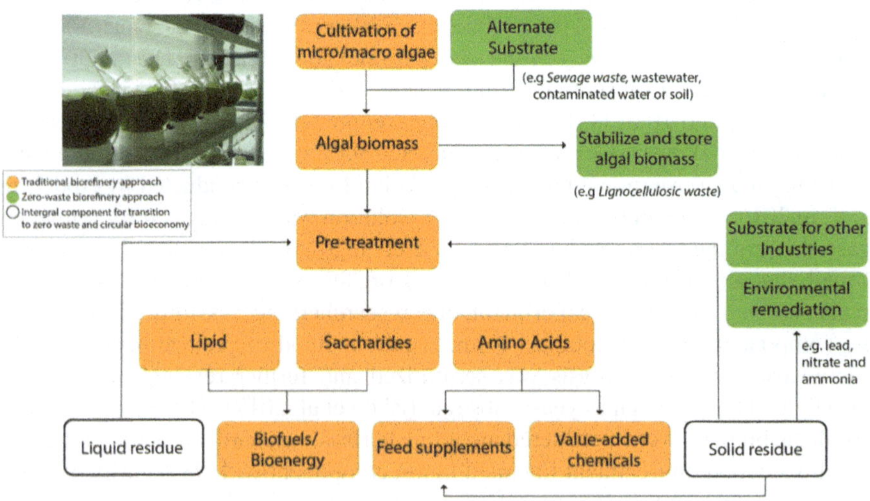

Fig. 1.4 Zero-waste biorefinery processes for algal biomass

environment. Additionally, biomolecules and nutrients present in the residues can be reused or recycled to generate biofuels and valuable co-products (Gifuni et al. 2019). In 2016, Tedesco and Stokes evaluated the biogas production of six indigenous algal-treated biomass for the production of biogas. And it was observed that *Laminaria* and *Fucus* spp can produce 187–195 mL CH_4 gVS^{-1} and 100 mL CH_4 gVS^{-1}, respectively (Tedesco and Stokes 2016). Recently, a study proposed a mechanism using *Fucus spiralis* as a feedstock in the circular bioeconomy. Algal biomass was put to use for the generation of economically benefiting services and the leftover residue was utilized to remediate lead (Pb(II)) contaminated environment (Filote et al. 2019). For a long time, algae have been used to produce marketable commodities, therefore the focus of the study was to evaluate the efficiency of contaminant removal and further using the biomass as biochar, thus fulfilling the "zero waste biorefinery" concept. The optimistic results of the studies expand our horizon for biofuel and biochar co-production in an environment-friendly manner (Filote et al. 2019; Bhowmick et al. 2019). Also, encourage industries to opt for green resources to make the transition from the traditional linear economy. Wuang et al. (2016) exploited the remediation capability of *Spirulina platensis* as a feedstock in aquaculture and further facilitated the residue as fertilizers for leafy vegetables. The study depicted optimistic results, indicating the high removal capacities of ammonia and nitrate in water from aquaculture. Moreover, the residual biomass also enhanced the growth of vegetables, in comparison to the controls.

A prominent economic barrier in algal biorefinery is the seasonal variation in the production of biomass. Wendt et al. (2017), Jarvis et al. (2018), and Wahlen et al. (2020) collectively, explored the methods to stabilize and store algal biomass for off-season requirements. The primary study conducted by Wendt et al. (2017) and Wahlen et al. (2020) blended corn stover and yard waste respectively with algal strains. The results showed effective preservation of the blended biomass for upto 30 days. Also, the organic acid produced during the storage remarkably increased its proportion in the biomass that can be extracted later. Jarvis et al. (2018) established a synergistic effect of blended biomass of algae and lignocellulose. Usually, certain oils are used as a buffer to stabilize the blend, however in this case organic acids from algae substituted the need for an external buffer source. The results depicted an increased yield of biocrude that can further be processed for value-added products.

For sustainable production using algal biomass as feedstock, the primary focus depends on cultivation and harvesting techniques. In this regard, a couple of researchers have made an effort to use inexpensive substrates for the growth of algae and subsequently producing high- value products. A study in 2019 evaluated the algal residue for the absorption of contaminants in the environment (Sadhukhan et al. 2019). The experiment was designed in a way, where the growth of algae was carried out using the contaminated water/soil inputs which could effectively remove toxic components via phytoremediation, and then algal biomass can be utilized further for the generation of energy, chemicals, and other beneficial products. Another study investigated sewage waste as its growth medium for algae to substitute freshwater dependency. The results affirmed positive outcomes to generate high-value services which further make biorefinery more cost-effective and self-sustainable (Wuang et al. 2016; Mishra and Mohanty 2019). To reduce the cultivation cost, Sachdeva et al. (2018) estimated nitrite as an alternative to nitrate in the culture medium of *Arthrospira* sp in photosynthetic biorefineries using wastewater as input material. The algal strain could assimilate up to 120 mM nitrate in the form of urea in the biorefinery without any ill effects on the products. Moreover, utilizing wastewater can effectively sequester carbon dioxide along with cost reduction by 35–86%. In terms of the economic prosperity of utilizing inexpensive feedstock, Judd et al. (2017) evaluated the relationship between algal productivity and production cost. The analyses reveal that production cost can be minimized to $1 per liter, in addition, to harnessing beneficial components from the waste sources under optimized conditions. The aforementioned processes manifest a win-win situation with economic and environmental stability. However, these approaches are still in their infancy stage and require elaborative analysis at the pilot scale to substantiate feasible zero waste algal biorefinery.

1.2.3 Integrated Biorefinery

So far, the studies mentioned capsulate the advancements in the biorefineries that drive the traditional economy to circular bioeconomy and zero-waste concept. This section of the chapter includes the integration of lignocellulosic biomass and algal biomass to compensate for each other's downsides. De Bhowmick et al. (2018a) were pioneers who researched an innovative solution where fermentation (dark/photo) of lignocellulosic biomass was combined with cultivation of microalgae (Bhowmick et al. 2018a). The results of the study established a cost-effective cycle where lignocellulose and algae were completely degraded to produce bioenergy (hydrogen, gasoline, diesel, and ethanol), and carbon dioxide emission was reduced. Based on Taguchi's approach of experimental design, rice husk, pinewood, and *Sargassum* sp. were selected to produce biochar and the process parameters were further evaluated (De Bhowmick et al. 2018b). Temperature (500 °C) and algal ratio (70%) were critical parameters in enhancing the yield, surface area, ash content, and thermal stability of biochar produced. Another integral biorefinery involved oil palm residues and microalgae cultivation for bioenergy, bioproducts, bioactive compounds, and biopharmaceutical production (Abdullah and Hussein 2021). The study was recognized as an environmentally friendly technique for reducing the emission of greenhouse gases. The integrated system enhanced the economic competitiveness and paved the way to achieve a zero-waste biorefinery approach.

1.2.4 Residue Biorefinery

Residue or waste-based biorefineries are a consolidated system of cascading processes to produce biochemicals, bioactive agents, biopolymers, biofuels, and many more (summarized in Table 1.2). With rapid expansion in economic activities and urbanization, waste has been recognized as a resource that will never lessen. Therefore, the production of commercial commodities from wastes bring forth advantages such as reduction in greenhouse gas emissions, waste management, decrease landfill costs, and other economic benefits (Caldeira et al. 2020). Additionally, residual biomass as substrate can impart solution to food-energy indecision, and thus advocate sustainable design in comparison with traditional biorefineries. Broadly, wastes can be characterized as agricultural, industrial, food, biomass, and municipal wastes, however, the composition is highly susceptive of geographical location.

Globally, it is estimated that 2.01 billion tons of municipal waste are generated by the nations, out of which 33% of the wastes remain as such and compound modern issues related to sustainable living (Datatopics.worldbank.org 2021). Therefore, waste biorefineries are advocated as an integral component to extend its boundary into circular bioeconomy (Oliveira and Navia 2017; Dahiya et al. 2018; Zabaniotou and Kamaterou 2019; Lappa et al. 2019; Yadav et al. 2020; Nawaz et al. 2020; Alfio et al. 2021). To counterbalance the ominous effect, the extraction process limits the development of zero waste biorefineries (Villacís-Chiriboga et al. 2020). Li et al. (2019) explored an innovative transformation for the extraction process that accounts for the major expenditure in the production of bioactive compounds from the biomass. The authors utilized vegetable oils and their derivatives to effectively extract volatile, as well as non-volatile compounds from rosemary leaves with keeping in mind the concept of zero waste biorefinery. The study concluded that oleo-extraction using soybean oil reduced the need for separation processes, which hold the potential to substitute traditional methods. Fish waste from aquaculture is also considered a potent hurdle. With the expansion in the market for salmon derived products, the co-stream and fish waste have been becoming more evident, as the residue generation is not seasonal but produced throughout the year. Considering the issues, a study evaluated a two-way approach for the sequential extraction of profitable commodities (Venslauskas et al. 2021). Firstly, the residual biomass was subjected to mild pre-treatment to extract high-value oil and the further enzymatic process was applied to efficiently separate oil and protein content. Once all the chemicals are extracted, the solidified biomass can be used as feed to animals. Economic and environmental assessment of the approach concluded that the two-way method can increase the return on investment along with decreasing the environmental footprint.

Food waste generated from the agro-industrial processes is of major concern and requires adequate steps for sustainable management. The cascading approach has emerged to be one of the beneficial approaches where primary products such as bioethanol, biodiesel, tannins, phenolic compounds along with secondary products as feedstock for energy generation, biochar, and carbon material can be obtained (Dahiya et al. 2018). This approach not only suffices the maximum resource utilization but also integrates biorefinery into a closed-loop pattern of the economy. Jin et al. (2020) explored the efficiency of grape pomace in lead remediation. In conventional biorefinery with the grape as substrate, a larger portion of the residues persists in the landfills, thus reducing the efficiency of the system. Therefore, the authors

Table 1.2 Bioproducts produced from wastes

Type of waste	Extraction process	Bio-products	References
Food			
Fish waste	Chemical/Biological extraction	Biopolymers, Biofuels	Lionetto and Esposito Corcione (2021), Paone et al. (2021)
	Enzymatic extraction	Polyunsaturated fatty acids, Feed, Food supplements	Alfio et al. (2021), Melgosa et al. (2021)
	Hydrolysis (Chemical or enzymatic)	Collagen, Bioactive peptides, Fish protein hydrolysate, Biopolymer	Coppola et al. (2021)
	Co-stream processing	Biogas, fertilizers, Fish protein hydrolysate	Venslauskas et al. (2021)
Animal waste	Fermentation	Biogas, Enzymes, bioactive agents	Schwede et al. (2016), Shahzad et al. (2017)
	Enzymatic hydrolysis and other conventional Methods	Bioactive peptides, Protein hydrolysate, Lipid hydrolysate	Shahzad et al. (2017)
Oil-crops	Microwave and ultrasound-assisted Method	Bioactive compounds	Li et al. (2019)
	Conventional Methods	Animal feed, Biofuel, Biolubricant	Qiu et al. (2021)
Vegetables and fruits	Enzymatic Methods	Animal feed, Pectin	López et al. (2010)
	Microwave and ultrasound-assisted Method	Animal feed, bioactive compounds, nanoparticles, biofuels	Joglekar et al. (2019)
	Conventional Methods	Hydrochar, Microfibrillated cellulose, Phenolics, Bioplastics	Joglekar et al. (2019), Getachew and Woldesenbet (2016), Gao et al. (2021)
	Chemical/Biological Methods	Biofuels, Enzymes, Carotenoids, Biofertilizers, Essential oils	Patsalou et al. (2019), Patsalou et al. (2020), López et al. (2010), Pathak et al. (2017), Rodríguez-Valderrama et al. (2020), Qin et al. (2021)
Biomass			
Lignocellulosic	Extraction Methods	Biopesticides, Bioproducts	Dutta et al. (2020)

(continued)

Table 1.2 (continued)

Type of waste	Extraction process	Bio-products	References
	Fermentation	Biofuel	Liu et al. (2016)
	Chemical/Biological Methods	Biofuels, Biopolymers, Bioadsorbent,	Liguori and Faraco (2016), Liu et al. (2016), Surra et al. (2019), Islam et al. (2020), Zhai et al. (2020)
Algal	Fermentation	Biofuels, Bioproducts	Overbeck et al. (2016), Chen et al. (2018)
	Conventional Methods	Biochar, Bioadsorbent, Biofuels	Filote et al. (2019), Shahid et al. (2019)
	Chemical/Biological Methods	Bioadsorbent, Biofuel, Biopolymers	Shukla et al. (2016), Naresh Kumar et al. (2020), Fawzy and Gomaa (2020)
Industrial waste			
	Biological Methods	Substrate for algae production, Bioactive compounds, Acids	Nawaz et al. (2020), Yadav et al. (2019)
	Fermentation	Biohydrogen, Biopolymers	Koutinas et al. (2014), Rajesh Banu et al. (2020)
Municipal waste			
	Fermentation	Bionanoparticles,	Romero-Cedillo et al. (2020)
	Hydrothermal liquefaction	Biochar, Biogas, Biocrude	Mishra et al. (2019)
	Conventional Methods	Biopolymers, Biofuel, Substrate for algae production	Moretto et al. (2020), Nanda and Berruti (2021)

processed the residues for biochar and bioremediation via pyrolysis of lignin components to achieve the concept of zero waste biorefinery. Management of citrus fruits by-products from agro-industries is another challenge faced by the industries because converting it into feed for animals is a costly procedure. Consecutively, low pH, high organic content, and moisture attribute to the citrus residues making it inappropriate for landfills as well. In regards to this, Patsalou et al. (2019) investigated the combined pretreatment method (acid and enzymatic hydrolysis) for ethanol and methane production in the batch fermentation process. The results of the study depicted that 30.7 g L^{-1} of ethanol and 342 mL gVS^{-1} of methane were produced as byproducts which were equivalent to the amount of alcohol obtained from the fresh feedstock. Furthermore, the authors explored the fermentation process using *Actinobacillus succinogenes* to produce other commodities such as fertilizers, essential oils, pectin, and succinic acid (Patsalou et al. 2019). The studies successfully established a foundation for utilizing futile substrates into valuable energy sources and following the zero-waste approach.

Another potential renewable source of energy is Municipal Solid Waste (MSW). MSW is defined as solid/semi-solid material which includes plastics, food waste, paper waste, yard waste, construction and demolition wastes from urban as well as rural areas. The composition of MSW makes it an ideal substrate for generating an extensive range of economically benefiting products or biofuels (Nanda and Berruti 2021). Additionally, the application of pre-treatments can substantially decrease the greenhouse gas emissions from wastes/landfills. Among numerous conversion technologies available for waste valorization, pyrolysis is the most suitable process for MSW to produce biochar (Romero-Cedillo et al. 2020; Gopu et al. 2018). Recently, a study conducted by AlDayyat et al. (2021) assessed the conversion capacity of MSW into commodities via pyrolysis. With optimum conditions (temperature 500 °C and 20 °C/min heating rate), biochar and bio-oil were obtained. On further analysis, it was concluded that bio-oil contained enough amount of alkanes and alkenes which indicate a promising potential for alternative fuel production. On the other hand, biochar produced was of low calorific value (11.5 MJ//kg) and can be utilized for soil amendment. Pyrolysis on combining with liquefaction and gasification can produce dense bio-oil and biochar, respectively. Gasification of MSW produces hydrogen-rich syngas. However, the quantity and quality of the products formed are determined by the operational conditions. Enough studies have been published so far to substantiate the production of value-added products but literature related to the environmental study and the effect of operational parameters on quality is insufficient.

1.3 Conclusions

So far, carbon-based and non-renewable sources have been the pillar for the economy. But with expansion in population, the production-consumption equilibrium has become erratic. Moreover, the waste generated by the urban area and agriculture aggravates the inefficiency of the economic system. In that regard, waste generated across the nations can become a remarkable feedstock such that its components can be recycled or reused into energy, chemicals, and other marketable materials via numerous biorefinery technologies, called zero-waste biorefinery. The concept of zero waste biorefinery is not just limited to one sector, rather it calls for exploration from different segments of sciences to make the biorefinery more self-sustainable and carbon–neutral. The recurring movement of biomass across the production-consumption chain opens a new dimension to industries that can cost-effectively produce commodities and resolve environmental issues in a unified approach. Based on the extensive exploration, the notion of "Zero waste biorefinery" is in its early stage therefore more vigorous and proactive studies need to be carried out with an eye on producing sustainable and eco-efficient products to compete with petroleum-based products. Also, shortage of economic and environmental assessment studies of zero waste biorefineries, it was difficult to draw a decisive conclusion, although the chapter highlights all the recent advancements in the zero-waste biorefinery.

References

Abdelaziz O, Gadalla M, El-Halwagi M, Ashour F (2015) A hierarchical approach for the design improvements of an Organocat biorefinery. Biores Technol 181:321–329

Abdullah M, Hussein H (2021) Integrated algal and oil palm biorefinery as a model system for bioenergy co-generation with bioproducts and biopharmaceuticals. Bioresour Bioprocess 8(1)

AlDayyat E, Saidan M, Al-Hamamre Z, Al-Addous M, Alkasrawi M (2021) Pyrolysis of solid waste for bio-oil and char production in Refugees' camp: a case study. Energies 14(13):3861

Alfio V, Manzo C, Micillo R (2021) From fish waste to value: an overview of the sustainable recovery of Omega-3 for food supplements. Molecules 26(4):1002

Ángel Siles López J, Li Q, Thompson I (2010) Biorefinery of waste orange peel. Crit Rev Biotechnol 30(1):63–69

Angouria-Tsorochidou E, Teigiserova D, Thomsen M (2021) Environmental and economic assessment of decentralized bioenergy and biorefinery networks treating urban biowaste. Resources Conserv Recycling 176:105898

Arevalo-Gallegos A, Ahmad Z, Asgher M, Parra-Saldivar R, Iqbal H (2017) Lignocellulose: a sustainable material to produce value-added products with a zero waste approach—a review. Int J Biol Macromol 99:308–318

Bittencourt G, Barreto E, Brandão R, Baêta B, Gurgel L (2019) Fractionation of sugarcane bagasse using hydrothermal and advanced oxidative pretreatments for bioethanol and biogas production in lignocellulose biorefineries. Bioresource Technol

Brunklaus B, Riise E (2018) Bio-based materials within the circular economy: opportunities and challenges. In: Benetto E, Gericke K, Guiton M (eds) Designing sustainable technologies, products and policies. Springer, Cham, pp 43–47

Byun J, Han J (2020) Sustainable development of biorefineries: integrated assessment method for co-production pathways. Energy Environ Sci 13(8):2233–2242

Caldeira C, Vlysidis A, Fiore G, De Laurentiis V, Vignali G, Sala S (2020) Sustainability of food waste biorefinery: a review on valorisation pathways, techno-economic constraints, and environmental assessment. Biores Technol 312:123–575

Carus M, Dammer L (2018) The circular bioeconomy—concepts, opportunities, and limitations. Ind Biotechnol 14(2):83–91

Chen Y, Ho S, Nagarajan D, Ren N, Chang J (2018) Waste biorefineries—integrating anaerobic digestion and microalgae cultivation for bioenergy production. Curr Opin Biotechnol 50:101–110

Cherubini F (2010) The biorefinery concept: using biomass instead of oil for producing energy and chemicals. Energy Convers Manage 51(7):1412–1421

Coppola D, Lauritano C, Palma Esposito F, Riccio G, Rizzo C, de Pascale D (2021) Fish waste: from problem to valuable resource. Mar Drugs 19(2):116

Dahiya S, Kumar A, Shanthi Sravan J, Chatterjee S, Sarkar O, Mohan S (2018) Food waste biorefinery: sustainable strategy for circular bioeconomy. Biores Technol 248:2–12

Datatopics.worldbank.org (2021) Trends in solid waste management. [online] Available at: https://datatopics.worldbank.org/what-a-waste/trends_in_solid_waste_management.html

Dávila I, Gullón P, Andrés M, Labidi J (2017) Coproduction of lignin and glucose from vine shoots by eco-friendly strategies: toward the development of an integrated biorefinery. Biores Technol 244:328–337

Dávila I, Remón J, Gullón P, Labidi J, Budarin V (2019) Production and characterization of lignin and cellulose fractions obtained from pretreated vine shoots by microwave assisted alkali treatment. Bioresource Technol 289:121726
De Bhowmick G, Sarmah A, Sen R (2018a) Lignocellulosic biorefinery as a model for sustainable development of biofuels and value added products. Biores Technol 247:1144–1154
De Bhowmick G, Sarmah A, Sen R (2018b) Production and characterization of a value added biochar mix using seaweed, rice husk and pine sawdust: a parametric study. J Clean Prod 200:641–656
De Bhowmick G, Sarmah A, Sen R (2019) Zero-waste algal biorefinery for bioenergy and biochar: a green leap towards achieving energy and environmental sustainability. Sci Total Environ 650:2467–2482
De Jong E, Jungmeier G (2015) Biorefinery concepts in comparison to petrochemical refineries. In: Pandey A, Höfer R, Taherzadeh M, Nampoothiri K, Larroche C (eds) Industrial biorefineries & white biotechnology. Elsevier, pp 3–33
Dutta S, Yu I, Tsang D, Su Z, Hu C, Wu K, Yip A, Ok Y, Poon C (2020) Influence of green solvent on levulinic acid production from lignocellulosic paper waste. Bioresource Technol 298:122544
Ellen MacArthur Foundation (2013) Towards the circular economy. Economic and business rationale for an accelerated transition, pp 1–96
Enriquez J (1998) GENOMICS: genomics and the world's economy. Science 281(5379):925–926
Espinoza Pérez A, Camargo M, Narváez Rincón P, Alfaro Marchant M (2017) Key challenges and requirements for sustainable and industrialized biorefinery supply chain design and management: a bibliographic analysis. Renew Sustain Energy Rev 69:350–359
Fawzy M, Gomaa M (2020) Use of algal biorefinery waste and waste office paper in the development of xerogels: a low cost and eco-friendly biosorbent for the effective removal of congo red and Fe (II) from aqueous solutions. J Environ Manage 262:110380
Fernando S, Adhikari S, Chandrapal C, Murali N (2006) Biorefineries: current status, challenges, and future direction. Energy Fuels 20(4):1727–1737
Filote C, Volf I, Santos S, Botelho C (2019) Bioadsorptive removal of Pb(II) from aqueous solution by the biorefinery waste of Fucus spiralis. Sci Total Environ 648:1201–1209
Galbe M, Wallberg O (2019) Pretreatment for biorefineries: a review of common methods for efficient utilisation of lignocellulosic materials. Biotechnol Biofuels 12(1)
Gao Y, Ozel M, Dugmore T, Sulaeman A, Matharu A (2021) A biorefinery strategy for spent industrial ginger waste. J Hazard Mater 401:123400
Garlapati V, Chandel A, Kumar S, Sharma S, Sevda S, Ingle A, Pant D (2020) Circular economy aspects of lignin: Towards a lignocellulose biorefinery. Renew Sustain Energy Rev 130:109977
Getachew A, Woldesenbet F (2016) Production of biodegradable plastic by polyhydroxybutyrate (PHB) accumulating bacteria using low cost agricultural waste material. BMC Res Notes 9(1)
Gifuni I, Pollio A, Safi C, Marzocchella A, Olivieri G (2019) Current bottlenecks and challenges of the microalgal biorefinery. Trends Biotechnol 37(3):242–252
Gírio F, Marques S, Pinto F, Oliveira A, Costa P, Reis A, Moura P (2017) Biorefineries in the world. Biorefineries. Springer, Cham, pp 227–281
Gnansounou E, Pandey A (2017) Classification of biorefineries taking into account sustainability potentials and flexibility. Life-Cycle Assessment Biorefineries pp 1–39
Gopu C, Gao L, Volpe M, Fiori L, Goldfarb J (2018) Valorizing municipal solid waste: waste to energy and activated carbons for water treatment via pyrolysis. J Anal Appl Pyrol 133:48–58
Hernández D, Solana M, Riaño B, García-González M, Bertucco A (2014) Biofuels from microalgae: lipid extraction and methane production from the residual biomass in a biorefinery approach. Biores Technol 170:370–378
Hingsamer M, Jungmeier G (2019) Biorefineries. In: Lago C, Lechón Y, Caldés N (eds) The role of bioenergy in the emerging bioeconomy. Academic Press, pp 179–222
Ichikawa S, Nishida A, Yasui S, Karita S (2017) Characterization of lignocellulose particles during lignocellulose solubilization by *Clostridium thermocellum*. Biosci Biotechnol Biochem 81(10):2028–2033

Islam M, Wang H, Rehman S, Dong C, Hsu H, Lin C, Leu S (2020) Sustainability metrics of pretreatment processes in a waste derived lignocellulosic biomass biorefinery. Bioresource Technol 298:122558

Jarvis J, Billing J, Corilo Y, Schmidt A, Hallen R, Schaub T (2018) FT-ICR MS analysis of blended pine-microalgae feedstock HTL biocrudes. Fuel 216:341–348

Jin Q, Wang Z, Feng Y, Kim Y, Stewart A, O'Keefe S, Neilson A, He Z, Huang H, (2020) Grape pomace and its secondary waste management: Biochar production for a broad range of lead (Pb) removal from water. Environ Res 186:109442

Joglekar S, Pathak P, Mandavgane S, Kulkarni B (2019) Process of fruit peel waste biorefinery: a case study of citrus waste biorefinery, its environmental impacts and recommendations. Environ Sci Pollut Res 26(34):34713–34722

Judd S, Al Momani F, Znad H, Al Ketife A (2017) The cost benefit of algal technology for combined CO_2 mitigation and nutrient abatement. Renew Sustain Energy Rev 71:379–387

Kamm B, Kamm M (2004) Principles of biorefineries. Appl Microbiol Biotechnol 64(2):137–145

Koutinas A, Vlysidis A, Pleissner D, Kopsahelis N, Lopez Garcia I, Kookos I, Papanikolaou S, Kwan T, Lin C (2014) Valorization of industrial waste and by-product streams via fermentation for the production of chemicals and biopolymers. Chem Soc Rev 43(8):2587

Kumar S, Gupta R, Kumar G, Sahoo D, Kuhad R (2013) Bioethanol production from *Gracilaria verrucosa*, a red alga, in a biorefinery approach. Biores Technol 135:150–156

Lappa I, Papadaki A, Kachrimanidou V, Terpou A, Koulougliotis D, Eriotou E, Kopsahelis N (2019) Cheese whey processing: integrated biorefinery concepts and emerging food applications. Foods 8(8):347

Li Y, Bundeesomchok K, Rakotomanomana N, Fabiano-Tixier A, Bott R, Wang Y, Chemat F (2019) Towards a zero-waste biorefinery using edible oils as solvents for the green extraction of volatile and non-volatile bioactive compounds from rosemary. Antioxidants 8(5):140

Liguori R, Faraco V (2016) Biological processes for advancing lignocellulosic waste biorefinery by advocating circular economy. Biores Technol 215:13–20

Lionetto F, Esposito Corcione C (2021) Recent applications of biopolymers derived from fish industry waste in food packaging. Polymers 13(14):2337

Liu Z, Liao W, Liu Y (2016) A sustainable biorefinery to convert agricultural residues into value-added chemicals. Biotechnol Biofuels 9(1)

Melgosa R, Sanz M, Beltrán S (2021) Supercritical CO_2 processing of omega-3 polyunsaturated fatty acids—towards a biorefinery for fish waste valorization. J Supercrit Fluids 169:105121

Mishra S, Roy M, Mohanty K (2019) Microalgal bioenergy production under zero-waste biorefinery approach: recent advances and future perspectives. Bioresource Technol 292:122008

Mishra S, Mohanty K (2019) Comprehensive characterization of microalgal isolates and lipid-extracted biomass as zero-waste bioenergy feedstock: An integrated bioremediation and biorefinery approach. Biores Technol 273:177–184

Mitra M, Mishra S (2019) Multiproduct biorefinery from *Arthrospira* spp. towards zero waste: current status and future trends. Bioresource Technol 291:121928

Mohan S, Butti S, Amulya K, Dahiya S, Modestra J (2016) Waste biorefinery: a new paradigm for a sustainable bioelectro economy. Trends Biotechnol 34(11):852–855

Monte J, Ribeiro C, Parreira C, Costa L, Brive L, Casal S, Brazinha C, Crespo J (2020) Biorefinery of *Dunaliella salina*: sustainable recovery of carotenoids, polar lipids and glycerol. Bioresource Technol 297:122509

Moretto G, Russo I, Bolzonella D, Pavan P, Majone M, Valentino F (2020) An urban biorefinery for food waste and biological sludge conversion into polyhydroxyalkanoates and biogas. Water Res 170:115371

Nanda S, Berruti F (2021) A technical review of bioenergy and resource recovery from municipal solid waste. J Hazardous Mater 403:123970

Naresh Kumar A, Chatterjee S, Hemalatha M, Althuri A, Min B, Kim S, Venkata Mohan S (2020) Deoiled algal biomass derived renewable sugars for bioethanol and biopolymer production in biorefinery framework. Bioresource Technol 296:122315

Nawaz M, Bilal M, Tariq A, Iqbal H, Alghamdi H, Cheng H (2020) Bio-purification of sugar industry wastewater and production of high-value industrial products with a zero-waste concept. Crit Rev Food Sci Nutrition, pp 1–18

Nizami A, Rehan M, Waqas M, Naqvi M, Ouda O, Shahzad K, Miandad R, Khan M, Syamsiro M, Ismail I, Pant D (2017) Waste biorefineries: enabling circular economies in developing countries. Biores Technol 241:1101–1117

Oliveira A, Navia R (2017) Zero waste bio-refineries: the example of Brazilian pulp and paper sector. Waste Manage Res: J Sustain Circular Econ 35(12):1193–1194

Olsson J, Novy V, Nielsen F, Wallberg O, Galbe M (2019) Sequential fractionation of the lignocellulosic components in hardwood based on steam explosion and hydrotropic extraction. Biotechnol Biofuels 12(1)

Overbeck T, Steele J, Broadbent J (2016) Fermentation of de-oiled algal biomass by *Lactobacillus casei* for production of lactic acid. Bioprocess Biosyst Eng 39(12):1817–1823

Paone E, Fazzino F, Pizzone D, Scurria A, Pagliaro M, Ciriminna R, Calabrò P (2021) Towards the anchovy biorefinery: biogas production from anchovy processing waste after fish oil extraction with biobased limonene. Sustainability 13(5):2428

Pathak P, Mandavgane S, Puranik N, Jambhulkar S, Kulkarni B (2017) Valorization of potato peel: a biorefinery approach. Crit Rev Biotechnol 38(2):218–230

Patsalou M, Samanides C, Protopapa E, Stavrinou S, Vyrides I, Koutinas M (2019) A citrus peel waste biorefinery for ethanol and methane production. Molecules 24(13):2451

Patsalou M, Chrysargyris A, Tzortzakis N, Koutinas M (2020) A biorefinery for conversion of citrus peel waste into essential oils, pectin, fertilizer and succinic acid via different fermentation strategies. Waste Manage 113:469–477

Pimentel D, Pimentel M (2003) World Population, food, natural resources, and survival. World Futures 59(3–4):145–167

Qin S, Shekher Giri B, Kumar Patel A, Sar T, Liu H, Chen H, Juneja A, Kumar D, Zhang Z, Kumar Awasthi M, Taherzadeh M (2021) Resource recovery and biorefinery potential of apple orchard waste in the circular bioeconomy. Bioresource Technol 321:124496

Qiu Y, Yu Y, Lan P, Wang Y, Li Y (2021) An overview on total valorization of *Litsea cubeba* as a new woody oil plant resource toward a zero-waste biorefinery. Molecules 26(13):3948

Rajak R, Jacob S, Kim B (2020) A holistic zero waste biorefinery approach for macroalgal biomass utilization: a review. Sci Total Environ 716:137067

Rajesh Banu J, Kavitha S, Yukesh Kannah R, Bhosale R, Kumar G (2020) Industrial wastewater to biohydrogen: Possibilities towards successful biorefinery route. Bioresource Technol 298;122378

Rodríguez-Valderrama S, Escamilla-Alvarado C, Rivas-García P, Magnin J, Alcalá-Rodríguez M, García-Reyes R (2020) Biorefinery concept comprising acid hydrolysis, dark fermentation, and anaerobic digestion for co-processing of fruit and vegetable wastes and corn stover. Environ Sci Pollut Res 27(23):28585–28596

Romero-Cedillo L, Poggi-Varaldo H, Santoyo-Salazar J, Escamilla-Alvarado C, Matsumoto-Kuwabara Y, Ponce-Noyola M, Bretón-Deval L, García-Rocha M (2020) Biological synthesis of iron nanoparticles using hydrolysates from a waste-based biorefinery. Environ Sci Pollut Res 27(23):28649–28669

Sachdeva N, Mascolo C, Wattiez R, Leroy B (2018) Embedding photosynthetic biorefineries with circular economies: Exploring the waste recycling potential of *Arthrospira* sp. to produce high quality by-products. Biores Technol 268:237–246

Sadhukhan J, Gadkari S, Martinez-Hernandez E, Ng K, Shemfe M, Torres-Garcia E, Lynch J (2019) Novel macroalgae (seaweed) biorefinery systems for integrated chemical, protein, salt, nutrient and mineral extractions and environmental protection by green synthesis and life cycle sustainability assessments. Green Chem 21(10):2635–2655

Salvachúa D, Mohagheghi A, Smith H, Bradfield M, Nicol W, Black B, Biddy M, Dowe N, Beckham G (2016) Succinic acid production on xylose-enriched biorefinery streams by *Actinobacillus succinogenes* in batch fermentation. Biotechnol Biofuels 9(1)

Schwede S, Thorin E, Lindmark J, Klintenberg P, Jääskeläinen A, Suhonen A, Laatikainen R, Hakalehto E (2016) Using slaughterhouse waste in a biochemical-based biorefinery—results from pilot scale tests. Environ Technol 38(10):1275–1284

Shahid A, Ishfaq M, Ahmad M, Malik S, Farooq M, Hui Z, Batawi A, Shafi M, Aloqbi A, Gull M, Mehmood M (2019) Bioenergy potential of the residual microalgal biomass produced in city wastewater assessed through pyrolysis, kinetics and thermodynamics study to design algal biorefinery. Bioresource Technol 289:121701

Shahzad K, Narodoslawsky M, Sagir M, Ali N, Ali S, Rashid M, Ismail I, Koller M (2017) Techno-economic feasibility of waste biorefinery: using slaughtering waste streams as starting material for biopolyester production. Waste Manage 67:73–85

Shukla R, Kumar M, Chakraborty S, Gupta R, Kumar S, Sahoo D, Kuhad R (2016) Process development for the production of bioethanol from waste algal biomass of *Gracilaria verrucosa*. Biores Technol 220:584–589

Squinca P, Badino A, Farinas C (2018) A closed-loop strategy for endoglucanase production using sugarcane bagasse liquefied by a home-made enzymatic cocktail. Biores Technol 249:976–982

Surra E, Bernardo M, Lapa N, Esteves I, Fonseca I, Mota J (2019) Biomethane production through anaerobic co-digestion with Maize Cob Waste based on a biorefinery concept: a review. J Environ Manage 249:109351

Talekar S, Patti A, Vijayraghavan R, Arora A (2018) An integrated green biorefinery approach towards simultaneous recovery of pectin and polyphenols coupled with bioethanol production from waste pomegranate peels. Biores Technol 266:322–334

Tedesco S, Stokes J (2016) Valorisation to biogas of macroalgal waste streams: a circular approach to bioproducts and bioenergy in Ireland. Chem Pap 71(4):721–728

Ubando A, Felix C, Chen W (2020) Biorefineries in circular bioeconomy: a comprehensive review. Bioresource Technol 299:122585

Uellendahl H, Ahring B (2010) Anaerobic digestion as final step of a cellulosic ethanol biorefinery: biogas production from fermentation effluent in a UASB reactor-pilot-scale results. Biotechnol Bioeng 107(1):59–64

Venkata Mohan S, Dahiya S, Amulya K, Katakojwala R, Vanitha T (2019) Can circular bioeconomy be fueled by waste biorefineries—a closer look. Bioresource Technol Reports 7:100277

Venslauskas K, Navickas K, Nappa M, Kangas P, Mozūraitytė R, Šližytė R, Župerka V (2021) Energetic and economic evaluation of zero-waste fish co-stream processing. Int J Environ Res Public Health 18(5):2358

Villacís-Chiriboga J, Elst K, Van Camp J, Vera E, Ruales J (2020) Valorization of byproducts from tropical fruits: Extraction methodologies, applications, environmental, and economic assessment: a review (Part 1: General overview of the byproducts, traditional biorefinery practices, and possible applications). Comprehens Rev Food Sci Food Safety 19(2):405–447

Wahlen B, Wendt L, Murphy A, Thompson V, Hartley D, Dempster T, Gerken H (2020) Preservation of microalgae, lignocellulosic biomass blends by ensiling to enable consistent year-round feedstock supply for thermochemical conversion to biofuels. Front Bioeng Biotechnol 8

Wendt L, Wahlen B, Li C, Kachurin G, Ogden K, Murphy J (2017) Evaluation of a high-moisture stabilization strategy for harvested microalgae blended with herbaceous biomass: part I—Storage performance. Algal Res 25:567–575

Worldometers.info (2021) World population clock: worldometer. Available at: https://www.worldometers.info/world-population/

Wuang S, Khin M, Chua P, Luo Y (2016) Use of Spirulina biomass produced from treatment of aquaculture wastewater as agricultural fertilizers. Algal Res 15:59–64

Yadav G, Dash S, Sen R (2019) A biorefinery for valorization of industrial waste-water and flue gas by microalgae for waste mitigation, carbon-dioxide sequestration and algal biomass production. Sci Total Environ 688:129–135

Yadav B, Pandey A, Kumar L, Tyagi R (2020) Bioconversion of waste (water)/residues to bioplastics—a circular bioeconomy approach. Bioresource Technol 298:122584

Yousuf A, Pirozzi D, Sannino F (2020) Fundamentals of lignocellulosic biomass. In: Yousuf A, Pirozzi D, Sannino F (eds) Lignocellulosic biomass to liquid biofuels. Academic Press, pp 1–15

Zabaniotou A, Kamaterou P (2019) Food waste valorization advocating circular bioeconomy—a critical review of potentialities and perspectives of spent coffee grounds biorefinery. J Clean Prod 211:1553–1566

Zetterholm J, Bryngemark E, Ahlström J, Söderholm P, Harvey S, Wetterlund E (2020) Economic evaluation of large-scale biorefinery deployment: a framework integrating dynamic biomass market and techno-economic models. Sustainability 12(17):7126

Zhai R, Hu J, Chen X, Xu Z, Wen Z, Jin M (2020) Facile synthesis of manganese oxide modified lignin nanocomposites from lignocellulosic biorefinery wastes for dye removal. Bioresource Technol 315:123846

Zheng L, Han X, Han T, Liu G, Bao J (2020) Formulating a fully converged biorefining chain with zero wastewater generation by recycling stillage liquid to dry acid pretreatment operation. Bioresource Technol 318:124077

Zhou L, Santomauro F, Fan J, Macquarrie D, Clark J, Chuck C, Budarin, V (2017) Fast microwave-assisted acidolysis: a new biorefinery approach for the zero-waste utilisation of lignocellulosic biomass to produce high quality lignin and fermentable saccharides. Faraday Discuss 202:351–370.

Zuin V, Budarin V, De bruyn M, Shuttleworth P, Hunt A, Pluciennik C, Borisova A, Dodson J, Parker H, Clark J (2017) Polysaccharide-derived mesoporous materials (Starbon®) for sustainable separation of complex mixtures. Faraday Discuss 202:451–464

Chapter 2
Recent Technologies for Lignocellulose Biomass Conversion to Bioenergy and Biochemicals

Sonika Kag, Neha Kukreti, Rohit Ruhal, Sweeti Mann, Jaigopal Sharma, and Rashmi Kataria

2.1 Introduction

The total world population has reached approximately eight billion by the end of the year 2020, and it will be more than 10 billion by 2057 (Leong et al. 2021). Demand for energy is increasing globally due to the enhancement in population, and it is estimated that by the end of 2040, it will further increase by 28% of its current value. Petrochemical energy resources, including crude oil, coal, and natural gas, are the classical energy sources and need to be replaced by alternatives. There is sharp depleting, high emission of toxic gases, and global warming due to continuous use of traditional energy sources (Mankar et al. 2021). Energy security is not only limited to conventional sources but now renewable sources could be included. The annual production of terrestrial biomass can generate about four times higher energy than total energy demand globally (Meenakshisundaram et al. 2021). This lignocellulosic biomass (LCB) is known as the most prominent and cheapest source for energy and renewable chemicals synthesis (Yiin et al. 2021). LCB, such as energy crops, agriculture residues, and forest leftovers, are the most abundant, non-conventional, renewable, cheap, and sustainable feedstocks for bio-based energy and chemicals generation (Meenakshisundaram et al. 2021). Generally, agricultural leftovers such as crop residues and straw are used as cattle feed and fertilizer applications—environmental pollution results from unutilized biomass dumping and burning in many countries. The LCB is mainly composed of different carbohydrates and lignin polymers, and for its utilization into a biorefinery, high-efficiency pretreatment methods and series of conversion steps are required. Lignin creates a barrier in bioenergy

S. Kag · N. Kukreti · S. Mann · J. Sharma · R. Kataria (✉)
Delhi Technological University (DTU), Shahbad Daulatpur, Bawana Road, Delhi 110042, India
e-mail: rashmikataria@dtu.ac.in

R. Ruhal
Regional Centre of Biotechnology, Faridabad, India

Y. K. Nandabalan et al. (eds.), *Zero Waste Biorefinery*, Energy, Environment, and Sustainability, https://doi.org/10.1007/978-981-16-8682-5_2

production due to its chemical nature, which restricts the interaction between cellulose and enzymes during hydrolysis (Beig et al. 2020). For sustainable transformation of biomass, various pretreatment studies are done alone or with the integration of different methods. In many studies, harsh chemicals are used at high temperatures, leading to the generation of inhibitors and impedes an added cost for the process. These significant challenge-creates hurdles to use them at an industrial scale (Haldar and Purkait 2021). Hence the optimization of pretreatment study for less or non-toxic chemical production has been done and further utilisation of pretreated LCB for hydrolysis and fermentation steps is done for biochemicals production. The overall process of conversion of wastes into valuable products which benefit to environment and development of economy is known as circular economy. Safety of resources, low carbon footprint, green environment concept, zero-waste biorefinery are some advantages of circular bioeconomy (Leong et al. 2021). The present chapter detailed discussion about terrestrial biomass especially lignocellulosic biomass, it's composition, their management strategies, the pivotal role of biorefinery in bioeconomy, different traditional and advanced pretreatment technologies such as irradiation, milling, extrusion, a chemical derived methods (alkali, acidic, organic solvents, ionic liquids, steam mediated and hot water-based treatment), enzymes etc. are discussed with their benefit and limitations. The indispensable role of conversion technologies namely pyrolysis, gasification, transesterification, fermentation, and enzymatic saccharification, are explored for efficient conversion to Bioenergy and value-added chemicals.

2.2 Lignocellulosic Biomass and Composition

Lignocellulosic biomass is mainly composed of cellulose (homopolymer of hexoses), hemicellulose (heteropolymer of pentoses and hexoses), and lignin (Fig. 2.1). It includes the whole plant body, such as roots, leaves, branches, stems, and rhizomes (Houghton 2008). Agriculture, forest, industrial, animal, and municipal solid waste are some commonly known biomass types (Balaman 2019). Approximately 80% of terrestrial plant biomass is forest plant-based biomass which is more attractive among researchers due to its complexity, diversity of habitats, and higher carbon present in dry weight (Houghton and Hole 2008). Since it is a heterogeneous organic matrix and universal availability is the reason for developing biomass-based biorefinery. Other than this, less emission of toxic greenhouse gases, low cost, and easy processing make it as most prominent bioenergy alternative to classical energy resources (Bonechi et al. 2017). The selection of appropriate biomass sources, transportation, handling, storage, efficient pretreatment, and conversion technologies is challenging for biomass-derived bioenergy and biochemicals production (Balaman 2019). Terrestrial plant-based biomass (lignocellulosic biomass) is given more focus in this chapter regarding composition, available pretreatment, conversion technologies, and bioproducts. Chemical composition can differ from plant species to species due to several factors like age, stress condition, and growth stage (Bajpai 2020).

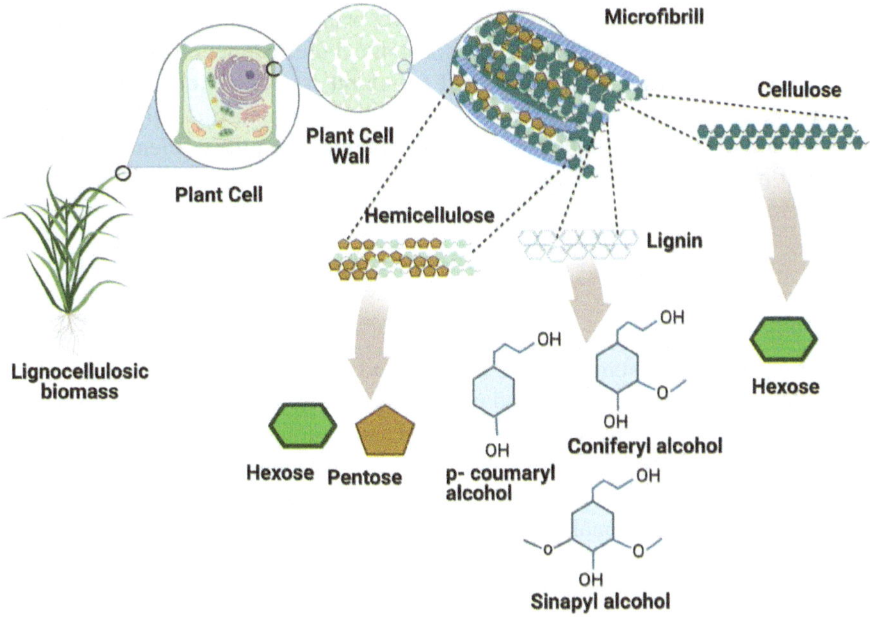

Fig. 2.1 Structural composition of lignocellulosic biomass (Created with www.biorender.com)

2.2.1 *Cellulose*

It is the most abundant natural polymer present on the earth's surface. Polymerization of 10,000 D-glucose units forms a linear, unbranched, and long-chain molecule. A β-1,4-glucoside linkage is present between two glucose monomers. Cellulose accounts for 40–50% of LCB composition (Bonechi et al. 2017). Cellulose is mainly distributed among the cell walls of plants and its crystalline (present in high amount) and amorphous nature (present in less amount). It gives structural support to the plant. Apart from plants, different microbes such as fungi, bacteria, and algae also produce a considerable amount of cellulose. Microfibrils are formed by assembling a few hundred of cellulose chains and covered by lignin and hemicellulose (Bajpai 2020).

2.2.2 *Hemicellulose*

In lignocellulosic biomass, hemicellulose contributes about 20–30% of total biomass. The branched heteropolysaccharide contains approximately 200–500 glucose units. The polymer consists of hexoses such as glucose, mannose, rhamnose galactose, and pentose sugars like arabinose xylose and uronic acids. In the backbone, two types of linkages, β-1,4 and β-1,3-glycosidic linkages are present (Bonechi et al. 2017). It is a are heat-labile polymer, and the monomers recovery from plant biomass enhances the digestibility of cellulose. After thermochemical pretreatment, some types of by-products are also generated, which lead to inhibit microbial fermentation during biomass conversion to energy. Hydroxy-methylfurfural and furfural are known as potent inhibitors for microbial growth and eventually hinder the fermentation process if present in high amounts (Bajpai 2020).

2.2.3 *Lignin*

Aromatic phenolics are cross-linked together and forms lignin polymer. Monolignol molecules, P-coumaryl, sinapyl, and p-coumaryl, undergo oxidative polymerization and produce an end product known as lignin. Lignin provides structural and mechanical support to the plant (Bonechi et al. 2017). Most significant part of lignin produces pollutants like aromatic and polycyclic aromatic hydrocarbons during burning. The presence of multi-functional groups like OH, OCH_3, CO, and COR, seems to be an alternative substrate for producing renewable metabolites lignin is the one of the most abundant polymer in plant biomass, therefore gaining attention from researchers (Nasrullah et al. 2017). Due to its chemical constituents and water-insoluble nature, it acts as a strong barrier to biomass vaporization (Bonechi et al. 2017). Different pretreatment strategies including solubilization of lignin, pulping process, dilute acid hydrolysis are used to separate lignin from rest of the carbohydrate component (Fernández-Rodríguez et al. 2018). Table 2.1 indicates the percentage of cellulose, hemicellulose, and lignin in different lignocellulosic biomass.

Table 2.1 Composition of some lignocellulosic biomass

Name of biomass	Cellulose %	Hemicellulose %	Lignin%	References
Wheat straw	35.97	23.95	19.33	Hammond and Mansell (2018)
Corn cob	45.01	33.12	13.81	Louis and Venkatachalam (2020)
Rice straw	36.1	27.0	13.7	Wu et al. (2020)
Corn stalks	32.9	24.3	18	Liang et al. (2021)
Cotton stalks	39.85	ND	23.92	Wang et al. (2016)
Hemp fibers	75.6	10.05	10.32	Paul and Chakraborty (2019)
Bamboo	37.6	30.4	23.6	Gao et al. (2021)

2.3 Biorefineries for Renewable Products from LCB

Biorefinery is a complex process in which LCB is utilized for sustainable production of bio-chemicals including energy, transport fuel, bioplastics etc. by using comprehensive conversion technologies. This plant biomass is fractionated into sugar, fatty acids, lignin and proteins which are further utilized for value added chemicals Table 2.2. On the basis of feedstock generation, biorefineries are divided into three major categories: energy crops and animal fats fall into first generation; lignocellulosic biomass comes under second generation however, algal feed stock and genetically engineered microorganism feedstocks known to produce third and fourth generation biorefinery respectively (Ng et al. 2017). All types of biorefineries described above have great potential, abundance of lignocellulose biomass as an agricultural waste, it's economic feasibility as well non-competitive to the food crop make second generation biorefinery as an effective approach. Various conversion technologies are used for efficient transformation of lignocellulosic biomass to valuable chemicals (Fig. 2.2). Waste biomass produced from agricultural and agro-industrial sectors is loaded with variety of nutrients including vitamins, proteins, lipids and carbohydrates, serving as potential growth media for microbial enumeration (Sharma et al. 2021). For sustainable development waste biomass attracted by researchers for biotransformation of biomass to zero waste biorefinery. The main aim of a circular economy is to reduce food cost, waste management, and use of renewable resources. "waste-to-wealth" is the goal of a circular economy. Development of novel technologies and Creation of new employment for the safety of the environment (Sharma et al. 2021).

2.4 Pretreatment: An Essential Step for LCB Depolymerization

Pretreatment is an essential process for the complex recalcitrant biomass disintegration from its significant constituents to convert to bioproducts. It is a critical process for the hydrolysis of a cellulosic portion of the biomass. The structure of cellulosic biomass is altered after pretreatment, and there is more availability of the cellulose for enzyme action by removal lignin (Fig. 2.3). Based on several studies, it is observed that removing lignin by suitable pretreatment strategies positively correlated with biomass digestibility (Bajpai 2020). Though effective degradation of biomass should be done in such a way that the monomer constituents should not be affected by the applied pretreatment. Further step of enzymatic hydrolysis is affected by the degree of crystallinity of cellulose (Mankar et al. 2021). Several studies have reported that different pretreatment methods such as chemical, thermochemical, physical, biological, and mechanical have been employed to remove lignin successfully. Also, a combination of these methods are applied to make the hydrolysis process more efficient. Some considerable Factors, including energy requirement and effectiveness of a pretreatment method are crucial criteria for the large-scale feasibility of the approach (Mankar et al. 2021).

Table 2.2 Various conversion technologies and processes

S. No.	Name of pretreatment/conversion technology	Name of biomass used	Process condition	Effect of pretreatment on biomass	References
1	Ball milling	Corn stover	Milling time 10–30 min and temperature 80–100 °C	Reduced particle size of corn stover and disruption of cell wall matrix	Gu et al. (2018a)
2	Microwave	Poplar, pine wheat straw	Temperature 164–192 °C time 5 min Temperature 173–187 °C Time 5–15 min	Yield of product(glucose and furfural) increased up to 90%	Cornejo et al. (2019)
3	Ultrasonic	Wheat straw	Frequency 40 kHz, loading 1/20 (g/ml)	Delignification upto 5.7%	Bussemaker et al. (2013)
4	Mechanical extrusion	Douglas-fir (forest residues)	Moisture content was 30–50% at 0.30 kWh/kg	Enhanced enzymatic hydrolysis for sugar production	Gu et al. (2018b)
5	Acid pretreatment	Wheat straw	79.6% H_3PO_4 + 1.9% H_2O_2 at 40.2 °C for 2.9 h	Cellulose fractionation and lignin production	Liu et al. (2021a)
6	Alkali pretreatment	Wood chip	3% NaOH, $CrCl_3 \cdot 6H_2O$ (catalyst), at 120 °C for two hours	Increased HMF yield (79 mol%)	Nguyen et al. (2016)
7	Organosolv pretreatment	Lignocellulosic biomass	Temperature 180 °C, at 27.2 atm for 40 min. with 50% ethanol and 1.7% H_2SO_4	increase cellulose and hemicellulose, remove lignin	Silva et al. (2017)
8	Ionic liquid		BmimCl 8–9 ml 0.75 g NaOH and H_2O_2 as catalyst for 120 min at 110 °C	High lignin removal without carbohydrate loss	Pang et al. (2016)

(continued)

Table 2.2 (continued)

S. No.	Name of pretreatment/conversion technology	Name of biomass used	Process condition	Effect of pretreatment on biomass	References
9	Ozonolysis	Rice straw	0.006 gO_3/g rice straw, 55 °C, 4d	Reduction of fiber crystallinity of rice straw increased 26% hydrolysis	Patil et al. (2021)
10	Liquid hot water pretreatment	Wheat straw	With hot water at 80–95 °C for 50 min	Release of 41.0–53.0% extractives	Sun et al. (2003)
11	Steam explosion	Palm fibers and sugarcane bagasse	168 °C for 10 min steam treatment plus acetosolv	Improve lignin removal without structural change of biomass	Pereira Marques et al. (2021)
12	Biological pretreatment	Wheat straw	Temperature 22 °C time 35 days, moisture content 70% by *Phanerochaete chrysosporium*	Fibers of wheat straw became loose	Gao et al. (2017)
13	Hydrothermal carbonization	Herbal tea waste	20–300 °C temperature holding time at 5 °C/min	Enhance amorphous structure of biomass	Zhuang et al. (2019)
14	Supercritical fluid	Yellow pine sawdust	Supercritical CO_2 at 3100–4000 psi /112–165 °C for 60 min	Cellulose hydrolysis up to 84%	Kim and Hong (2001)
15	Ammonia fiber explosion	Coastal Bermuda grass	AFEX pretreatment at tem. 100 °C for 30 min	Relocalization of lignin	Lee et al. (2010)

(continued)

Table 2.2 (continued)

S. No.	Name of pretreatment/conversion technology	Name of biomass used	Process condition	Effect of pretreatment on biomass	References
16	Enzymatic hydrolysis	*E. cottonii* seaweed	19 AU cellulase, tem. 40–50 °C for 12 h	Reduction sugar increased up to 7.937 mg/mL for bioethanol fermentation	Puspawati et al. (2015)
17	Fermentation	Wheat straw	After thermochemical pretreatment *Lactobacillus* spp. inoculated on wheat straw containing media ($MgSO_4 \cdot 7H_2O$, $CaCl_2$ and NaCl) at 25 °C for 120 h	Lactic acid production enhanced	Cizeikiene et al. (2018)
18	Pyrolysis	Wood residues	600 °C for 1 h using chemical such as NaOH, KOH	Removal of cadmium from liquid	Aghababaei et al. (2017)
19	Gasification	Corn stover	350 °C for 30 min	Devolatilization of chars and evolution of phenolics	Nsaful et al. (2018)
20	Torrefaction	Sugarcane bagasse	180 °C and 20 min	Fix carbon and heating value of biochar increased upto 18.7 MJ/kg	Zheng et al. (2020)
21	Anaerobic digestion	Forest residue (mixture of spruce, pine, bark)	190 °C temperature with 50% solvent for 60 min	Increased methane yield upto 0.34 m^3 CH_4/kg	Kabir et al. (2015)
22	Transesterification	Algal lipid	Reaction lipase and algal lipid at 40 °C for 12 h	Fatty acid methyl esters yield up to 97%	Wu et al. (2017)

(continued)

Table 2.2 (continued)

S. No.	Name of pretreatment/conversion technology	Name of biomass used	Process condition	Effect of pretreatment on biomass	References
23	Photocatalytic conversion	Organic pollutants	Treatment with $ZnIn_2S_4$ ((5.02 mmol), $NO_3)_3$·(10.00 mmol),thioacetamide (4.01 mmol),at 160 °C for 6 h	Evolution of hydrogen by light irradiation, separation enhanced	Wang et al. (2019)

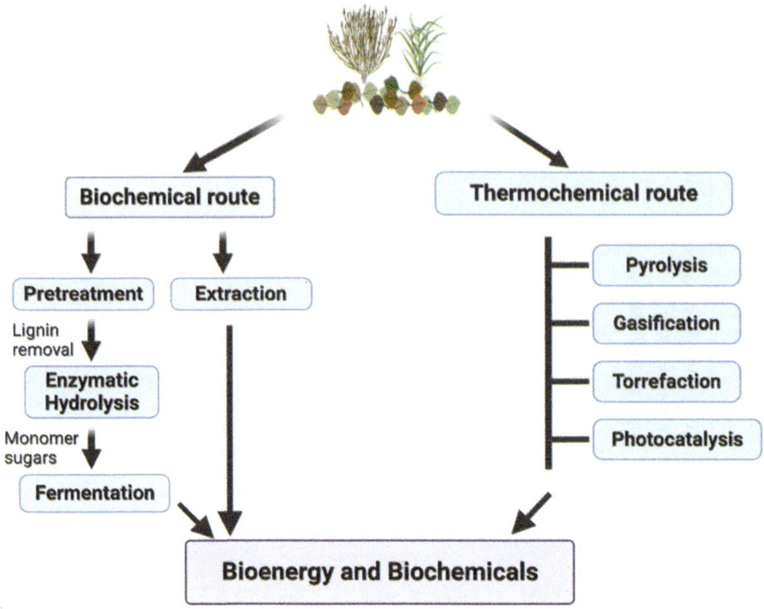

Fig. 2.2 Conversion Technologies for Lignocellulosic biomass (Created with www.biorender.com)

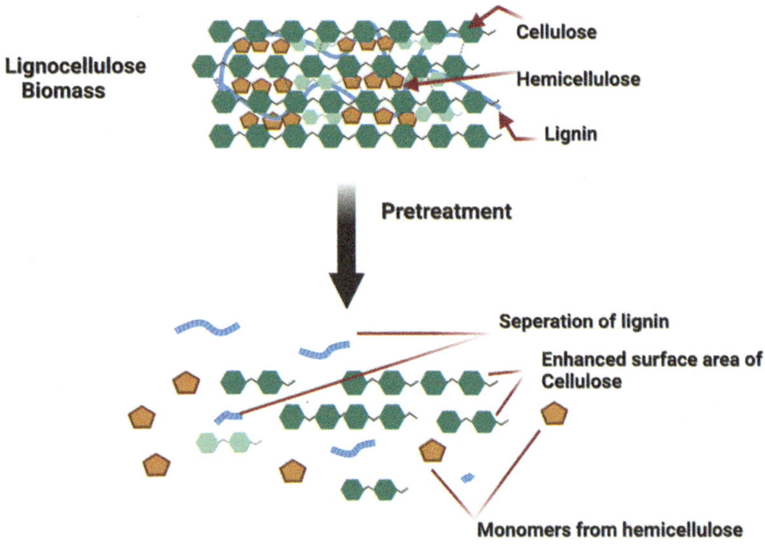

Fig. 2.3 Effect of pretreatment on lignocellulosic biomass (Created with www.biorender.com)

2.4.1 *Physical Pretreatment Method of Lignocellulosic Biomass*

2.4.1.1 Ball Milling

The physical pretreatment technique such as milling are known as effective method as there is no toxic product formation and hence washing and detoxification step is not required after this pretreatment (Gu et al. 2018a). For commercial applications, where lignocellulosic biomass is used as raw material milling is feasible suitable process. Attrition type milling with chemical catalysts is a possible solution to maximize the conversion of feedstock (Gu et al. 2018a). Milling of lignocellulosic feedstock increases the accessibility of cellulose for cellulase enzymes by reducing the particle size and enhancing the part of amorphous cellulose in biomass (Wu et al. 2021). Loud noise during operation, and high energy conception and prolonged reaction time are major limitations of milling (Neikov et al. 2008). It was found that simultaneous physical and biological treatment such as Ball milling and enzymatic saccharification can enhance glucose conversion rate by 30% (Wu et al. 2021). When corn stover was subjected to ball milling for 10–30 min and 80–100 °C, particle size reduction and disruption of the cell wall matrix of biomass was reported (Gu et al. 2018a).

2.4.1.2 Microwave

Microwave irradiation is one of the physical pretreatment methods for lignocellulose pretreatment (Li et al. 2016). This method has advantages like rapid heating, generates heat, uniform heating, and penetrates throughout the volume of the material. In ionic conditions, the sample oscillates towards the forward and backward direction of the dissolved charged particle due to an applied wave (Tsegaye et al. 2019). Microwaves are a substitute for traditional heating. It can rapidly generate intense direct heat and cause the change in the dipole moment of polar molecules (Li et al. 2012). This physical pretreatment has sufficient heating capability and is simple to implement. It has proved that it changes the complex ultrastructure of cellulose molecules, depolymerizes lignin and hemicellulose fraction in biomass, and enhances the enzymatic hydrolysis of LCB (Lu et al. 2011). A high capital cost is required for this process, and moisture parameters should be regulated before pretreatment because it can affect the biomass structure (Puligundla et al. 2016). Rice straw, switchgrass, and wheat straw are subjected to microwave pretreatment to enhance the bioethanol yield (Lu et al. 2011). When switchgrass was treated with microwave radiation and aqueous ammonia, the methane production was improved by 65% compared with untreated one (Li et al. 2012). Wheat straw, pine, and poplar chips were used for the extraction of furfural and glucose with the help of a microwave, and the yield was observed to be 2.5 kg furfural and 12.6 kg of glucose from 50 kg of biomass (Cornejo et al. 2019).

2.4.1.3 Ultrasonic Pretreatment

Ultrasonication pretreatment is based on the cavitation principle through the implementation of ultrasonic radiations. The cavitation generates the force that breaks the complex structure of lignocellulosic biomass and makes the easy extraction of cellulose, hemicellulose, and lignin (Baruah et al. 2018; Farias Silva et al. 2020). The Sonication pretreatment method has the advantage over other pretreatment methods. In this process a lower temperature is required to disrupt the cells, the extraction is faster, suitable for all cell types, no requirement of chemicals or beads, and a low production cost is observed. This method can compete with other procedures like dilute acid treatment as it has excellent performance for enzymatic hydrolysis using microalgal biomass (Farias Silva et al. 2020). A study showed that the combination of different pretreatment methods (electrolysis and ultrasonication) is efficient for methane and organic matter production (Kumar et al. 2017). It is was also reported a considerable delignification in wheat straw biomass when subjected to ultrasonic irradiation at the frequency of 40 kHz (Bussemaker et al. 2013). However High energy requirement for operation is a significant drawback of this process (Savun-Hekimoğlu 2020).

2.4.1.4 Mechanical Extrusion

Extrusion is a thermo-mechanical type of pretreatment. It is a highly versatile and continuous process. It has good mixing and heat transfer capabilities. This process operates at mild temperature and requires lower chemicals and thus does not produce inhibitory compounds which occur in severe conditions. This process works with the help of one or more screws, spins into a tight barrel, and temperature controlled. There are two different types of extrusion machines based on the extruders: single screw extruder and twin screw extruder (Duque et al. 2014). There is an increase in accessibility for the enzyme after extrusion pretreatment. The extrusion processing helps in opening the woody material structures which enhance the access of cellulose to enzymes (Gu et al. 2018b). Further study is required for the co-rotating twin-screw extrusion. Anaerobic digestion improved in five different samples of agricultural biomass with the help of extrusion. After 28 days, methane production increased by 18–70%, and after 90 days, methane production increased by 9–28%. Extrusion plays a role in destroying slowly degrading compounds and otherwise non-degradable compounds (Gu et al. 2018b). Due to extruders, energy yield and methane production increased in biogas plants (Hjorth et al. 2011). The high cost, this technique is still challenging for commercialization (Zheng and Rehmann 2014).

2.4.2 Chemical Pretreatment Methods

Chemical pretreatment effectively depolymerize the complex chemical structure of LCBs. Acidic, alkali ionic liquid and organic solvents, and high-pressure steam are some potent chemicals used in this technique (Jung and Kim 2015). These pretreatment are performed alone or combination with other physical parameters like temperature or pressure for specific duration.

2.4.2.1 Acid Pretreatment

Different pretreatment methods are developed to enhance the bioconversion efficiency for enzymatic hydrolysis (Sheng et al. 2021). Two types of acid pretreatment are used in commercial biorefinery: concentrated acid at low temperature and dilute acid at high temperature. Out of them, the low acid concentration is a more promising method in lignocellulosic material due to its simplicity and cost-effectiveness (Mahmoodi et al. 2018; Sheng et al. 2021). There is generation of various biproducts by like furfural, levinic acid, hydroxymethylfurfural (HMF), and carboxylic acid during acid treatment. They possess an inhibitory effect on desired microbial growth and cell membrane functionality (Mahmoodi et al. 2018). In the chemical pretreatment (acid/alkali) of lignocellulosic biomass, solubilization of generous amounts of fermentable sugars from the hemicellulose and cellulose part into a liquid phase of pretreated feedstock slurry occurs (Jung and Kim 2015). Dilute hydrochloric acid and sulfuric acid are most commonly used for biomass pretreatment. Suitable neutralization methods are essential because pH, state, and chemical composition play a vital role in fermentation after e pretreatment (Gonzales et al. 2017). Wheat straw was subjected to combination of phosphoric acid and hydrogen peroxide with ratio of 79.6% and 1.9% at 40.2 °C for 2.9 h cellulose fractionation and lignin recovery reported (Liu et al. 2021a).

2.4.2.2 Alkali Pretreatment

It is a lignin targeting pretreatment method performed in mild conditions (Nasir et al. 2020). Various types of alkaline chemical, including sodium hydroxide, calcium hydroxide, sodium carbonate, etc., are used as active catalysts for the process of delignification (Zhang et al. 2020a). The delignification process is positively correlated with the severity of treatment, that include alkali concentration and high temperature. High alkali strength leads to high reducing sugar yield after enzymatic hydrolysis. However, higher temperature conditions contribute to inhibitor generation, such as phenolics (Fan et al. 2020). Along with the removal of lignin, it maintains the polymeric structure of cellulose. Sodium hydroxide pretreatment is one of the standard techniques used extensively in the bioconversion of recalcitrant lignocellulosic biomass. It is a very potent chemical that enhances agro-based residues and hardwood

(Xu and Sun 2016). This process is performed at comparatively low pressure and temperature or ambient conditions. When wheat straw was treated with 3% sodium hydroxide with a catalyst $CrCl_3$ at 120 °C for two hours, for delignification and it also yielded HMF significantly (Nguyen et al. 2016). The main limitation of this technique is time, as it can take several hours to days to complete the pretreatment process (Xu and Sun 2016). Still, the use of concentrated alkali and disposal of waste alkaline liquid makes this classical pretreatment method to be challenging (Zhuang et al. 2019).

2.4.2.3 Organosolv Pretreatment

In the organosolv method, removal of hemicellulose and lignin is done by some organic liquids such as alcohols, phenols, ethers, etc. The mixture of organic liquids causes hydrolysis of the internal bonds between hemicellulose and lignin (Hesami et al. 2015). It helps to remove a considerable amount of lignin. Increased surface area and pore size make the enzymatic hydrolysis feasible and lignin, recovered by this process, is pure compared to other methods (Hesami et al. 2015). But high solvents requirement and washing step after pretreatment are the significant challenges of this technique (Zhang et al. 2020a). Different pretreatment technologies were developed based on the type of catalyst used such as acid or alkali (Chen et al. 2015; Silva et al. 2017). It is reported an increased depolymerization of lignocellulose biomass at temperature of 180 °C/ 27.2 atm for 40 min with 50% ethanol and 1.7% sulfuric acid (Silva et al. 2017).

2.4.2.4 Ionic Liquid

Ionic liquids are a group of salts containing organic cations and inorganic anions. They are suitable solvents, non-volatile, chemically inert, and thermally stable, and possess industrial applications such as food, cosmetics, and pharmaceutic (Zhang et al. 2020b). In the solubilization of lignocellulosic biomass towards lignin bridges, the negative ions play a crucial role. Anion makes a nucleophile attack on lignin, and β–O–4 linkage is targeted, which results in release of hydroxyl groups of phenolic compounds at high temperatures (Nakasu et al. 2021). When lignocellulosic biomass is treated with ionic liquid, it produces amorphous cellulose and makes it accessible for enzymatic saccharification without formation of inhibitory products, however ionic liquids are toxic in nature (Vasheghani Farahani et al. 2016). 1-butyl-3-methylimidazolium chloride ([Bmim]Cl) is used with combination of 0.75 g NaOH and H_2O_2 as catalyst for 120 min at 110 °C and reported considerable amount of lignin removal without significant loss of carbohydrate (Pang et al. 2016).

2.4.2.5 Ozonolysis

Ozonation is an effective technique for lignin degradation and it is used to oxidize carbohydrates in biomass. The reaction rate in this process is slower in comparison to other methods (Mulakhudair et al. 2017). Ozone gas is highly reactive. Double bonds containing lignin moieties are prone to oxidize in the process of ozonolysis (Mulakhudair et al. 2017). Variety of biomass such as crop straws, energy grasses, and bagasse are subjected to ozonolysis pretreatment due to considerable delignification, high reducing sugar yield, eco-friendly, and low operational charge (Li et al. 2021). When rice straw was subjected to ozonolysis (0.006 gO_3/g RS) at 55 °C for 4 days, the crystallinity of cellulose reduced and significant amount of delignification reported (Patil et al. 2021).

2.4.2.6 Liquid Hot Water Pretreatment

Above the boiling point, water is known as subcritical water or liquid hot water. At high temperatures and pressure, its physical and chemical properties change; due to this, it shows improved solvation properties (Lamp et al. 2020). Liquid hot water pretreatment, simply known as hydrothermal pretreatment, water is used as a heating medium at a very high temperature (130–240 °C), and high pressure is also maintained without addition of any chemical agent (Zhang et al. 2020c). At high temperatures, hemicellulose containing acetyl is converted into acetic acid, acts as a catalyst, and causes autohydrolysis (Lamp et al. 2020). Depolymerization of complex cell walls occurs, and cellulose gets hydrolyzed to monomers. In the liquid hot water method, liquid state water is responsible for excellent solubilization efficiency compared to steam pretreatment (Zhang et al. 2020c). It is a promising green technology for sustainable biorefineries. It has been used in various feedstocks, including agro-industrial-based products, but high temperature and low yield are drawbacks of this technique (Carvalheiro et al. 2016). Wheat straw treatment with hot water at 80–95 °C for 50 min, leads to release 41.0–53.0% extractives from biomass (Batista et al. 2019).

2.4.2.7 Steam Explosion

It is a physicochemical treatment in which water and crude biomass is used, cellulose and hemicellulose of biomass converted into monomer units that make cellulase more effective for enzymatic hydrolysis (Nasir et al. 2020). For the degradation of recalcitrance biomass, it is an promising alternative technique to less energy consumption and no addition of chemicals (Pereira Marques et al. 2021). In this technique, a reactor is maintained at the optimum time, temperature, and pressure that holds the biomass for a fixed period. Sudden depressurization results in explosive decompression (Pereira Marques et al. 2021). Due to mechanical and chemical effects, the surface area of biomass increases, and depolymerization of cellulose and hemicellulose occurs due to the breaking of carbohydrate-lignin bonds. This technique is used to enhance enzymatic saccharification, second generation biofuels production and bioactive compounds extraction. This technique is eco-friendly and economical but less effective on lignin removal (Nasir et al. 2020). In a study of palm fibers and sugarcane bagasse subjected to steam explosion at 168 °C for 10 min lignin removal without structural change in biomass reported (Nasir et al. 2020).

2.4.3 Biological Pretreatment of Lignocellulosic Biomass

Depolymerization of the complex lignocellulosic biomass (lignin and hemicellulose), with the help of specific microorganisms, known as biological pretreatment. Depolymerase enzymes such as lignin peroxidase, manganese peroxidase, and laccase are some potent lignin-degrading biocatalysts (Ummalyma et al. 2019). Biological pretreatment has attracted researchers as an advanced strategy due to the minimum generation of inhibitory compounds and environmental friendly as no chemical or physical treatment is given (Akyol et al. 2019). Microbes that depolymerize the lignin can be bacteria or fungi. Still, due to less nitrogen requirement and ability to grow in the presence of inhibitory toxic compounds, *basidiomycetes* (white rot) fungi gained interest (Vasco-Correa et al. 2016). Some fungal groups such as soft rot, white rot, and brown rot fungi are potential biological pretreatment agents due to their wood-decaying efficiency of lignocellulosic biomass (Akyol et al. 2019). The presence of peroxidases and laccase, a particular dignifying enzyme in rot white-rot fungi including *Trametes Versicolor*, *Phanerochaete chrysosporium*, *Trichoderma reesei*, etc. make the biological pretreatment process feasible (Akyol et al. 2019). Peroxidases and laccase enzymes are a class of oxidoreductases secreted by microbes that degrade lignin by oxidation either directly or by mediators (Vasco-Correa et al. 2016).

Single enzymes or cocktails of enzymes can be used for the delignification process. PH 3–8 is optimal pH, and 25–80 °C temperature is suitable for the procedure. Saccharification can be combined with physical and chemical methods (Moreno et al. 2019). High enzyme cost and long reaction time is a considerable obstacle to biological pretreatment in industrial scale. To overcome this problem, further research is required to develop an effective biological pretreatment (Moreno et al. 2019). In another study, laccase, β-glucosidase used as efficient conversion of perennial lignocellulosic biomass into bioethanol (Kirupa Sankar et al. 2018). When wheat straw treated with *Phanerochaete chrysosporium* at temperature 22 °C for 35 days with 70% moisture content, it was found that fibers of biomass became loose (Gao et al. 2017).

2.4.4 *Other Special Pretreatment Techniques*

2.4.4.1 Hydrothermal Carbonization (HTC)

It is a powerful thermochemical technology in which high moisture-containing biomass transforms into solid biofuels such as hydrochar at 180–250 °C. This treatment densifies the energy present in biomass (Sztancs et al. 2021). It is the possible alternative to another thermochemical technique, pyrolysis. In pyrolysis, the produced char has high potassium, sodium content and high heating value (Liang et al. 2021). Hydrothermal carbonization overcomes the heating value problem. Apart from this, it enhances the carbon percentage and decreases the ash content in feedstock, moisture-containing samples can be taken, low energy is required, (Liang et al. 2021). In the HTC technique, the breakdown of water occurs into hydronium and hydroxide ions. They are critical factors of hydrolysis of organic substrates. Several acids such as nitric acid/sulfuric acid or hydrochloric acid can be used to lower down reaction time and temperature and help in the modification of hydrochar by adding functional groups to it (Liang et al. 2021). Herbal tea waste is treated at 20–300 °C temperature with holding time at 5 °C/min, a structure change of biomass is reported (Zhuang et al. 2019).

2.4.4.2 Supercritical Fluid (SF)

Any material above the pressure and temperature (critical condition) expresses the property of liquid and gas, for instance, density and compressibility, known as a supercritical fluid. Most often used in SF is carbon dioxide, which is non-toxic, is economical, eco-friendly, with a critical temperature is 31.1 °C, and necessary pressure is 7.36 MPa (Daza Serna et al. 2016). This technique has been reported as an efficient pretreatment method to depolymerize the lignocellulosic composition of vegetal sources. Cellulose containing biomass is maintained in a reactor where carbon dioxide is pressurized at 35 °C, followed by a depressurization mechanism, giving rise to disruption of the cellulose (Putrino et al. 2020). The surface area of lignocellulosic biomass is increased by this process, and biomass can be accessible to the enzymatic saccharification and enhance the sugar yield. This possible alternative technique is feasible as compared to the special techniques such as ammonia and steam explosion due to low operating temperature, and glucose does not degrade by the process (Putrino et al. 2020). When yellow pine sawdust was subjected to cupercritical CO_2 at 3100–4000 psi/112–165 °C for 60 min about 80% of cellulose hydrolysed to monomers (Kim and Hong 2001).

2.4.4.3 Ammonia Fiber Explosion (AFEX)

This is a thermochemical process, causes solubilization of lignin, depolymerization of cellulose into monomer reducing sugar, and hydrolysis of hemicellulose. During this process surface area of LCBs increase and promote the transformation of cellulose and hemicellulose to reducing sugar for fermentation (Teymouri et al. 2005). The process of ammonia fiber explosion works by rapid decompression with a combination of alkaline pretreatment that is similar to steam explosion catalyzed by sulfur dioxide. In this process, the feedstock is treated with liquid ammonia at a temperature (90 to 100 °C) and high pressure for 5 min and then pressure is rapidly released (Lee et al. 2010). The ammonia used in this process can be recycled after the pretreatment is over. Coastal grass pretreatment with AFEX at 100 °C for 30 min cause relocalization of lignin. This process has been applied on corn stover, switchgrass, and rice straw but the high ammonia requirement and its recovery cost make the process inefficient for pilot scale use (Lee et al. 2010).

2.5 Enzymatic Hydrolysis

Hydrolysis of pretreated biomass for the production of fermentable sugars with a cellulase enzyme complex, known as enzymatic hydrolysis. Major steps involved in this process are: (a) transfer of biocatalyst, (b) binding of the enzyme with substrate, (c) cellulose hydrolysis (d) cellobiose hydrolysis in the form of glucose (Fan 2014). Bioconversion of LCBs requires physicochemical pretreatment and enzymatic saccharification to transform polysaccharides into their monomer constituents (Speight 2020). Biological pretreatment or simply enzyme-mediated hydrolysis is carried out with the help of biological agents and complex biomass depolymerize into a range of monomer sugars namely glucose, mannose, galactose, xylose, etc. Carbohydrate-lignin bonds also break due to an effective pretreatment strategy (Speight 2020). In the first step of saccharification, cellulase and hemicellulase (potent enzymes for hydrolysis), depolymerize the cellulose and hemicellulose to sugar monomer. Cellulase, responsible for hydrolyzing cellulose, is of three types (a) exoglucanase (b) endoglucanase, and (c) beta-glucosidase. Xylans, glucomannan, and arabinoxylan are some commonly known hemicellulases (Verardi et al. 2019). In a study seaweed is used as a biomass for enzymatic hydrolysis (19 AU cellulase, temperature. 40–50 °C for 12 h), reducing sugar increased up-to 7.937 mg/mL for bioethanol production (Puspawati et al. 2015).

2.6 Fermentation

In fermentation, substrates get converted into the valuable product by the enzymatic activity of microorganisms including bacteria yeast, and fungi (Vieira et al. 2020).

Bioprocessing is the potent alternative for the conversion of biomass to valuables. LCB generated from various industrial operations attract the formation of a variety of metabolites through the process of fermentation (López-Gómez et al. 2020). Food industry biomass is chemically composed of a considerable amount of carbohydrates, proteins, and other micronutrients, which support the growth of desired microorganisms. The fermentation process is of two types: submerged fermentation and the second is solid-state fermentation (López-Gómez et al. 2020). Industrial production of enzymes can be done by solid-state fermentation of agro-industrial biomass. As improper disposal of agro-industrial biomass causes environmental issues, so the use of this waste as a substrate for microbial fermentation can be a possible alternative to high-cost feedstocks (Meini et al. 2021). It has advantages over submerged due to the minimum requirement of pretreatment technique and less generation of wastewater after the process (Meini et al. 2021). In submerged fermentation, the added nutrients and supplied oxygen can easily dissolve in liquid media and mix evenly all over in the bioreactor, and because of it, biomass mixing and heat transfer takes place evenly (Chakraborty et al. 2019). Wheat straw was used for lactic acid production, after pretreatment biomass was subjected to fermentation with starter culture of *Lactobacillus* spp., at 25 °C for 120 h, lactic acid production reported.

2.7 Pyrolysis

It is a thermochemical conversion method of biomass. When biomass is treated at high temperature (300–600 °C), and with a suitable atmospheric pressure in absence of oxygen, the biomass can be converted into three phases solid (char), liquid (bio-oil), and gaseous (syngas), the process known as pyrolysis (Javed 2020). It has some advantages over other thermochemical methods that is pyrolysis plants can be established in remote locations so that transportation costs can be reduced. The complexity of this process is the biggest challenge of this technique (Rego et al. 2020). Many value-added products such as methane, biochar, paraffin ware produced from lignocellulosic biomass (Javed 2020). In one study wood residues for cadmium removal by pyrolysis performed at operating conditions 600 °C for 1 h using chemical such as NaOH and KOH (Aghababaei et al. 2017).

2.8 Gasification

Gasification is a thermochemical process, where syngas commonly known as producer gas formed by the reaction of gasification of fuel. Few hydrocarbons such as methane, ethane, etc. are the main components of syngas (Zhang et al. 2019). Gasification is a type of combustion. Heat, light, and chemical pollutants are generated as the result of combustion. During the conversion low-grade substrates such as coal, biomass into higher-grade fuel like methane, known as gasification. In this conversion technique, for energy generation, organic waste is processed at a very high temperature. LCB gasification can be achieved with oxygen, steam, carbon dioxide, and supercritical water (Zhang et al. 2019). The gasification takes place inside a closed chamber known as a gasifier. A variety of gasifiers such as fluidized bed, fixed beds and entrained flow gasifiers are often used for this process (Kiang 2018). Different techniques have been applied for lignin-based biorefinery and syngas is most commonly produced, it can be used as for microbial bio alcohol production (Liakakou et al. 2019). Corn stover subjected to gasification at 350 °C for 30 min evolution of phenolic compound reported from biomass (Nsaful et al. 2018).

2.9 Torrefaction

Torrefaction is a thermal approach that is performed at higher temperatures in a range of 200 and 300 °C in an inert environment, which alters the chemical property of lignocellulosic biomass. This process requires the energy supply and leads to the improvement of energy density, improved ignition, lowers the moisture, increases the C/O and C/H ratio, and requires lower grind ability. Hence this increases the combustion for LCB. The terrified material could be used for electricity generation and co-firing with coal. The co-firing lowers the coal utilization which leads to reduced carbon dioxide in the environment. This process gives better storage of biomass and could be used for energy production (Basu 2013). Zheng et al. reported, increased carbon content of biomass and heating value of biochar upto 18.7 MJ/kg, when sugarcane bagasse treated at 180 °C for 20 min (Zheng et al. 2020).

2.10 Anaerobic Digestion

Anaerobic digestion is an effective approach to convert organic components such as lignocellulosic biomass to Bioenergy. It is a process in which microorganisms break down organic material in the absence of oxygen. A large amount of energy can be generated from the wastes by taking into consideration the process of simultaneous anaerobic digestion and co-digestion. Biogas is the product of the bacteria feeding on biodegradable waste and releases methane, carbon dioxide, hydrogen, nitrogen, and hydrogen sulfide. The methane component in the biogas can produce heat and electricity. The agricultural residue which is lignocellulosic material is used to get higher added value (You et al. 2019; Venturin et al. 2018). In a study forest residues such as mixture of spruce, pine, bark treated at at 190 °C temperature with 50% of organic solvent for 60 min and an increased methane yield upto 0.34 m^3 CH4/kg was observed (Kabir et al. 2015).

2.11 Transesterification

Transesterification is a chemical process in which lipids are esterified with acetylating agents and catalysts for biofuel production. Amid different types of biomass, algal-oil biomass is known to be a potential feedstock for biofuel production by transesterification reaction, due to their fast-growing ability and lower nutritional requirement for cultivation (Faraji and Saidi 2021). Algal biomass is composed of 7–22% of lipid, 5–50% protein, and 5–22% of carbohydrate. Chemical composition shows that it can be a perfect feedstock for biochemicals production (Chandra et al. 2019; Knoshaug et al. 2018). Two types of catalysts are used in transesterification, (1) homogenous (acid/alkali) (2) heterogeneous (lipase with solvents). Though base catalyzes processes are faster as compared to acid-based reactions but not suitable for fatty acids, as yield of the acid-based process is high but the reaction rate is very slow (Laurens 2020). In heterogeneous catalysts the separation is easy and high-grade purity is obtained but the process is slow and cost-effective (Makareviciene and Skorupskaite 2019). When subjected to transesterification with lipase at 40 °C for 12 h, fatty acid methyl esters yield up to 97% (Wu et al. 2017).

2.12 Photocatalytic Conversion of Biomass

In biorefinery, the production of platform molecules, from biomass conversion is the major step. Thermochemical and biochemical operations are commonly used techniques but the high cost and generation of toxic compounds are considerable disadvantages of these techniques. Light-driven redox reactions with the use of some homogeneous/heterogeneous photocatalyst, known as photocatalytic conversion. In the process of photocatalytic conversion, the transformation of cellulose using solid catalysts such as TiO_2 and NiS shows considerable cellulose hydrolysis into sugar monomers (Chen et al. 2021). Oxidation of water for generation of OH radicals which strike hexose to form some intermediate. Biochemicals production from glucose oxidation is a promising technique nowadays (Chen et al. 2021). The use of heterogeneous photocatalysts such as TiO_2 is gaining attention by researchers due to less toxicity and economic feasibility In a study organic pollutants were subjected to photocatalytic conversion with $ZnIn_2S_4$ and thioacetamide at 160 °C for 6 h, an evolution of hydrogen by light irradiation enhanced the separation of pollutants (Wang et al. 2019).

2.13 Biochemicals from Lignocellulose Biomass

LCBs are the most abundant biomass and produced more than 150 billion tons yearly, which makes them a potential source to compensate petrochemical resources for the production of a variety of biochemicals. To boost the biorefinery concept, some studies have been done so far for the transformation of LCBs into biomaterials (Shen and Sun 2021). Various types of biochemicals such as biofuels, organic acids, phenolic compounds, sugars, polysaccharides, sugar alcohols, Biosurfactants, and industrially important chemicals (discussed in Table 2.2), can be produced from lignocellulosic biomass (Fig. 2.4). Bioethanol production was observed from the cellulose part of lignocellulose biomass (bamboo) by applying a combination of steam explosion and green solvent pretreatment (Gao et al. 2021). Methane production from wood bark was reported by the anaerobic digestion (Navarro et al. 2020). In a study, co-production of bioplastic and biohydrogen from rice waste under light and dark fermentation with *Bacillus cereus and Rhodopseudomonas*spp. was achieved (Dinesh et al. 2020). Production of 5-hydroxymethylfurfural achieved by cellulose of banana plant waste, where biomass was treated with ball milling and solid catalyst (Flores-Velázquez et al. 2020). Microbial lipid production was studied from corncob by applying chemical pretreatment and *Mortierella isabellina* was used as a potent fungi for the process (Ivančić Šantek et al. 2021). Non detoxified rice straw was used to produce a cellulolytic enzyme (cellulase) by recombinant strain *Trichoderma reesei* RUT C30 (He et al. 2020). Succinic acid production was reported from sugarcane bagasse by the action of *E. coli* (Liu et al. 2013) (Table 2.3).

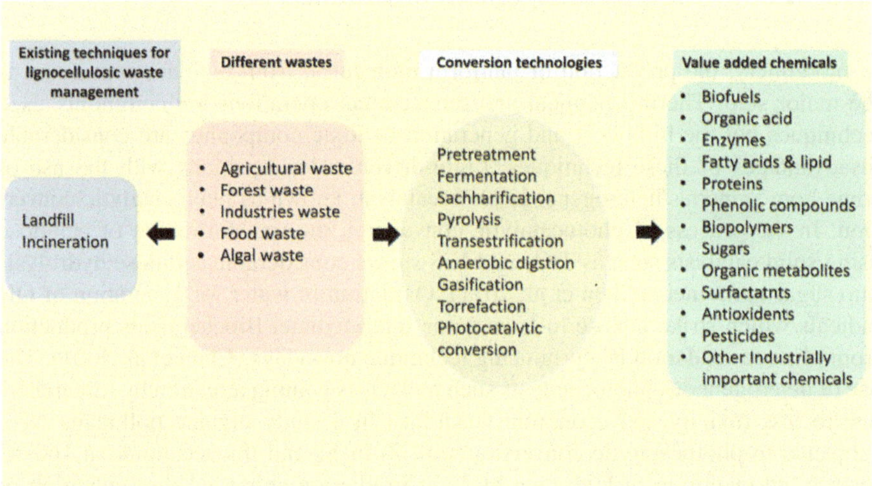

Fig. 2.4 Various lignocellulosic wastes and their conversion technologies and produced biochemicals

Table 2.3 Potential products from lignocellulosic biomass (LCBs)

S. No.	Products	Sources	Microorganisms/treatment	Yield	References
1. Biofuels					
1.1	Biodiesel	Palm	*Aspergillus tubingensis* TSIP9		Intasit et al. (2020)
1.2	Butanol	Cassava bagasse	*Clostridium tyrobutyricum* CtΔack-adhE2	0.30 g/g	Huang et al. (2019)
1.3	Bioethanol	*Sorghum* stover	–	–	Idris et al. (2017)
1.4	Biogas	Forest residue (mixture of spruce, pine, bark)	Anaerobic digestion	0.30 m^3 CH_4/kgVS	Kabir et al. (2015)
2. Organic acids					
2.1	Malic acid	Soybean hull and soy molasses	–	~0.4 g/g and productivity of ~0.5 g/L·h	Cheng et al. (2017)
3. Enzymes					
3.1	Pectin lyase	Pea peel and soybean bagasse	*Penicillium expansum* RSW SEP 1	88.83 U/ml	Atalla et al. (2019)
3.2	Xylanase	Wheat bran, banana peel, millet waste sugarcane bagasse	*Bacillus cereus* TH-050	10545 U/g	Simair et al. (2018)
3.3	Laccase production	Grape seeds, grape stalk and barley bran	*Trametes versicolor*	63.75 113.70 159.75	Lorenzo et al. (2002)
4. Phenolic compounds					
4.1	Levulinic acid	Citrus waste	–	~22% yield	Singh et al. (2019)
4.2	Lignin	Citrus waste	–	7.6 mg, ~8% yield	Singh et al. (2019)

(continued)

Table 2.3 (continued)

S. No.	Products	Sources	Microorganisms/treatment	Yield	References
4.3	Phenolic monomer	Lignin from birch sawdust	Catalytic hydrogenolysis	65% yield	Liu et al. (2021b)
4.4	Polylactide biocomposite	Soda lignin	Grafting	–	Park et al. (2019)
4.5	Adsorbent	Lignin from pre-hydrolysis liquor	Modification	–	Liu et al. (2021c)
4.6	Protocatechuic acid	Depolymerized lignin	–	6.73 mg/L	Nguyen et al. (2021)
4.7	Polyhydroxyalkanoate	Depolymerized lignin	*pseudomonas putida kT2440*	26.75 mg/L	Nguyen et al. (2021)
5. Sugars, polysaccharides sugar alcohols					
5.1	Erythritol	Okara, soybean cake, rape seed cake, peanut press cake and sesame meal	*Y. lipolytica* M53 And *Y. lipolytica* M53S	185.4 mg/gds	Liu et al. (2019)
5.2	Xylose	Citrus waste		~16% yield	Singh et al. (2019)
6. Organic metabolites					
6.1	Isoflavones	Soybeans	Solid state fermentation	2.71 ± 0.27, 0.57 ± 0.1, 0.30 ± 0.05 2.09 ± 0.24 kg/tonn	Carneiro et al. (2020)
6.2	Triterpenic compounds	*Eucalyptus globulus* biomass		–	Domingues et al. (2010)
7. Biosurfactants and Bio-sorbants					
7.1	Cationic Biosurfactant	Rice straw and corn cobs	Achromobacter sp. BP(1)5	0.074 g/L & 0.095 g/L	Ni'matuzahroh et al. (2020)

(continued)

Table 2.3 (continued)

S. No.	Products	Sources	Microorganisms/treatment	Yield	References
8. Industrially important chemicals					
8.1	Fungal pesticide	Rice husk	*Beauveria bassiana* (BB) (CECT 20374) *and Trichoderma harzianum* (CECT 2929) Solid state fermentation	2.0 × 108 to 2.0 × 109 spores g-1dry matter	Sala et al. (2020)
8.2	Hydrogen, Ethanol, 2,3-butanediol	Sugarcane molasses Wheat straw hydrolysate	Dark fermentation	H_2—(73.5 ± 10 cm^3 g^{-1}), Ethanol—(0.24 ± 0.03 g g-1) 2,3-butanediol—(0.42 ± 0.04 g g-1)	Alvarez-Guzmán et al. (2020)

2.14 Conclusions

In biomass-based biorefinery combinations of different pretreatment technologies have been applied till now and each technique has its own pros and cons. Assessment of the number of pretreatment methods indicates the requirement of potent techniques, that can be effective on the complex recalcitrant lignocellulosic biomass, enhance the crystallinity of homopolymer carbohydrates. It could be ensure high sugar yield, reduce the cost of pretreatment and follow the environment safety guidelines, suitable for industrial-scale. Different conversion techniques such as thermochemical and biochemical are based on biomass morphology and chemical structure. Each process has their own advantage and limitations. Hence depending on the biomass type and the expected product the conversion technologies are applied alone or in combination for high biochemicals yield. However, lignin is an obstacle in the way of waste biomass to bioenergy production and a suitable technique for lignin valorization is still the biggest challenge and still an open area of research. The development of novel, eco-friendly, and energy-saving techniques for zero waste biorefinery only can resolve the limitations associated with classical technologies.

Acknowledgements This work is supported by Grant no. BT/RLF/Re-entry/40/2017 from the Department of Biotechnology (DBT), Ministry of Science and technology, and SERB project file no: EEQ/2020/000614 Govt. of India. We acknowledge to CSIR pool scientific scheme Grant no. 9103, Govt. of India. We also acknowledge CSIR and UGC for fellowships support to Ms. Sonika and Ms. Sweeti respectively. Pictures are created by using www.biorender.com.

References

Aghababaei A, Ncibi MC, Sillanpää M (2017) Optimized removal of oxytetracycline and cadmium from contaminated waters using chemically-activated and pyrolyzed biochars from forest and wood-processing residues. Bioresour Technol 239:28–36. https://doi.org/10.1016/j.biortech.2017.04.119

Akyol Ç, Ince O, Bozan M, Ozbayram EG, Ince B (2019) Biological pretreatment with Trametes versicolor to enhance methane production from lignocellulosic biomass: a metagenomic approach. Ind Crops Prod 140. https://doi.org/10.1016/j.indcrop.2019.111659

Alvarez-Guzmán CL, Balderas-Hernández VE, De Leon-Rodriguez A (2020) Coproduction of hydrogen, ethanol and 2,3-butanediol from agro-industrial residues by the Antarctic psychrophilic GA0F bacterium. Int J Hydrogen Energy 45:26179–26187. https://doi.org/10.1016/j.ijhydene.2020.02.105

Atalla SMM, EL Gamal NG, Awad HM, Ali NF (2019) Production of pectin lyase from agricultural wastes by isolated marine Penicillium expansum RSW_SEP1 as dye wool fiber. Heliyon 5:e02302 https://doi.org/10.1016/j.heliyon.2019.e02302

Bajpai P (2020) Biomass composition. Biomass to Energy Convers. Technol 31–40. https://doi.org/10.1016/b978-0-12-818400-4.00004-9

Balaman ŞY (2019) Introduction to biomass—resources, production, harvesting, collection, and storage. Decis Biomass-Based Prod Chain 1–23. https://doi.org/10.1016/b978-0-12-814278-3.00001-7

Baruah J, Nath BK, Sharma R, Kumar S, Deka RC, Baruah DC, Kalita E (2018) Recent trends in the pretreatment of lignocellulosic biomass for value-added products. Front Energy Res 6:1–19. https://doi.org/10.3389/fenrg.2018.00141

Basu P (2013) Chp. 04: Torrefaction

Batista G, Souza RBA, Pratto B, dos Santos-Rocha MSR, Cruz AJG (2019) Effect of severity factor on the hydrothermal pretreatment of sugarcane straw. Bioresour Technol 275:321–327. https://doi.org/10.1016/j.biortech.2018.12.073

Beig B, Riaz M, Raza Naqvi S, Hassan M, Zheng Z, Karimi K, Pugazhendhi A, Atabani AE, Thuy Lan Chi N (2020) Current challenges and innovative developments in pretreatment of lignocellulosic residues for biofuel production: a review. Fuel 287:119670. https://doi.org/10.1016/j.fuel.2020.119670

Bonechi C, Consumi M, Donati A, Leone G, Magnani A, Tamasi G, Rossi C (2017) Biomass: an overview. Elsevier Ltd.

Bussemaker MJ, Xu F, Zhang D (2013) Manipulation of ultrasonic effects on lignocellulose by varying the frequency, particle size, loading and stirring. Bioresour Technol 148:15–23. https://doi.org/10.1016/j.biortech.2013.08.106

Carneiro AM, Moreira EA, Bragagnolo FS, Borges MS, Pilon AC, Rinaldo D, Funari CS (2020) Soya agricultural waste as a rich source of isoflavones. Food Res Int 130:108949. https://doi.org/10.1016/j.foodres.2019.108949

Carvalheiro F, Duarte LC, Gírio F, Moniz P (2016) Hydrothermal/liquid hot water pretreatment (autohydrolysis): a multipurpose process for biomass upgrading

Chakraborty S, Yadav G, Saini JK, Kuhad RC (2019) Comparative study of cellulase production using submerged and solid-state fermentation. Elsevier B.V.

Chandra R, Iqbal HMN, Vishal G, Lee HS, Nagra S (2019) Algal biorefinery: a sustainable approach to valorize algal-based biomass towards multiple product recovery. Bioresour Technol 278:346–359. https://doi.org/10.1016/j.biortech.2019.01.104

Chen H, Zhao J, Hu T, Zhao X, Liu D (2015) A comparison of several organosolv pretreatments for improving the enzymatic hydrolysis of wheat straw: substrate digestibility, fermentability and structural features. Appl Energy 150:224–232. https://doi.org/10.1016/j.apenergy.2015.04.030

Chen H, Wan K, Zheng F, Zhang Z, Zhang H, Zhang Y, Long D (2021) Recent advances in photocatalytic transformation of carbohydrates into valuable platform chemicals. Front Chem Eng 3:1–16. https://doi.org/10.3389/fceng.2021.615309

Cheng C, Zhou Y, Lin M, Wei P, Yang ST (2017) Polymalic acid fermentation by *Aureobasidium pullulans* for malic acid production from soybean hull and soy molasses: fermentation kinetics and economic analysis. Bioresour Technol 223:166–174. https://doi.org/10.1016/j.biortech.2016.10.042

Cizeikiene D, Juodeikiene G, Damasius J (2018) Use of wheat straw biomass in production of L-lactic acid applying biocatalysis and combined lactic acid bacteria strains belonging to the genus Lactobacillus. Biocatal Agric Biotechnol 15:185–191. https://doi.org/10.1016/j.bcab.2018.06.015

Cornejo A, Alegria-Dallo I, García-Yoldi Í, Sarobe Í, Sánchez D, Otazu E, Funcia I, Gil MJ, Martínez-Merino V (2019) Pretreatment and enzymatic hydrolysis for the efficient production of glucose and furfural from wheat straw, pine and poplar chips. Bioresour Technol 288:121583. https://doi.org/10.1016/j.biortech.2019.121583

Da Silva ARG, Errico M, Rong BG (2017) Process alternatives for bioethanol production from organosolv pretreatment using lignocellulosic biomass. Chem Eng Trans 57:1–6. https://doi.org/10.3303/CET1757001

Daza Serna LV, Orrego Alzate CE, Cardona Alzate CA (2016) Supercritical fluids as a green technology for the pretreatment of lignocellulosic biomass. Bioresour Technol 199:113–120. https://doi.org/10.1016/j.biortech.2015.09.078

de Farias Silva CE, Meneghello D, de Souza Abud AK, Bertucco A (2020) Pretreatment of microalgal biomass to improve the enzymatic hydrolysis of carbohydrates by ultrasonication:

yield vs energy consumption. J King Saud Univ - Sci 32:606–613. https://doi.org/10.1016/j.jksus.2018.09.007

Dinesh GH, Nguyen DD, Ravindran B, Chang SW, Vo DVN, Bach QV, Tran HN, Basu MJ, Mohanrasu K, Murugan RS, Swetha TA, Sivapraksh G, Selvaraj A, Arun A (2020) Simultaneous biohydrogen (H2) and bioplastic (poly-β-hydroxybutyrate-PHB) productions under dark, photo, and subsequent dark and photo fermentation utilizing various wastes. Int J Hydrogen Energy 45:5840–5853. https://doi.org/10.1016/j.ijhydene.2019.09.036

Domingues RMA, Sousa GDA, Freire CSR, Silvestre AJD, Neto CP (2010) Eucalyptus globulus biomass residues from pulping industry as a source of high value triterpenic compounds. Ind Crops Prod 31:65–70. https://doi.org/10.1016/j.indcrop.2009.09.002

Duque A, Manzanares P, Ballesteros I, Negro MJ, Oliva JM, González A, Ballesteros M (2014) Sugar production from barley straw biomass pretreated by combined alkali and enzymatic extrusion. Bioresour Technol 158:262–268. https://doi.org/10.1016/j.biortech.2014.02.041

Fan Z (2014) Consolidated Bioprocessing for Ethanol Production. Elsevier B.V.

Fan L, Ruan R, Li J, Ma L, Wang C, Zhou W (2020) Aromatics production from fast co-pyrolysis of lignin and waste cooking oil catalyzed by HZSM-5 zeolite. Appl Energy 263:114629. https://doi.org/10.1016/j.apenergy.2020.114629

Faraji M, Saidi M (2021) Hydrogen-rich syngas production via integrated configuration of pyrolysis and air gasification processes of various algal biomass: process simulation and evaluation using Aspen Plus software. Int J Hydrogen Energy 46:18844–18856. https://doi.org/10.1016/j.ijhydene.2021.03.047

Fernández-Rodríguez J, Erdocia X, Hernández-Ramos F, Alriols MG, Labid J (2018) Lignin separation and fractionation by ultrafiltration

Flores-Velázquez V, Córdova-Pérez GE, Silahua-Pavón AA, Torres-Torres JG, Sierra U, Fernández S, Godavarthi S, Ortiz-Chi F, Espinosa-González CG (2020) Cellulose obtained from banana plant waste for catalytic production of 5-HMF: effect of grinding on the cellulose properties. Fuel 265:116857. https://doi.org/10.1016/j.fuel.2019.116857

Gao W, Tabil LG, Dumonceaux T, Espinel Ríos S, Zhao R (2017) Optimization of biological pretreatment to enhance the quality of wheat straw pellets. Biomass Bioenerg 97:77–89. https://doi.org/10.1016/j.biombioe.2016.12.012

Gao H, Wang Y, Yang Q, Peng H, Li Y, Zhan D, Wei H, Lu H, Bakr MMA, El-Sheekh MM, Qi Z, Peng L, Lin X (2021) Combined steam explosion and optimized green-liquor pretreatments are effective for complete saccharification to maximize bioethanol production by reducing lignocellulose recalcitrance in one-year-old bamboo. Renew Energy 175:1069–1079. https://doi.org/10.1016/j.renene.2021.05.016

Gonzales RR, Kumar G, Sivagurunathan P, Kim SH (2017) Enhancement of hydrogen production by optimization of pH adjustment and separation conditions following dilute acid pretreatment of lignocellulosic biomass. Int J Hydrogen Energy 42:27502–27511. https://doi.org/10.1016/j.ijhydene.2017.05.021

Gu YM, Kim H, Sang B-I, Lee JH (2018a) Effects of water content on ball milling pretreatment and the enzymatic digestibility of corn stover. Water-Energy Nexus. 1:61–65. https://doi.org/10.1016/j.wen.2018.07.002

Gu BJ, Wolcott MP, Ganjyal GM (2018b) Pretreatment with lower feed moisture and lower extrusion temperatures aids in the increase in the fermentable sugar yields from fine-milled Douglas-fir. Bioresour Technol 269:262–268. https://doi.org/10.1016/j.biortech.2018.08.109

Haldar D, Purkait MK (2021) A review on the environment-friendly emerging techniques for pretreatment of lignocellulosic biomass: Mechanistic insight and advancements. Chemosphere 264:128523. https://doi.org/10.1016/j.chemosphere.2020.128523

Hammond GP, Mansell RVM (2018) A comparative thermodynamic evaluation of bioethanol processing from wheat straw. Appl Energy 224:136–146. https://doi.org/10.1016/j.apenergy.2018.04.123

He J, Liu X, Xia J, Xu J, Xiong P, Qiu Z (2020) One-step utilization of non-detoxified pretreated lignocellulose for enhanced cellulolytic enzyme production using recombinant Trichoderma reesei RUT C30 carrying alcohol dehydrogenase and nicotinate phosphoribosyltransferase. Bioresour Technol 310:123458. https://doi.org/10.1016/j.biortech.2020.123458

Hesami SM, Zilouei H, Karimi K, Asadinezhad A (2015) Enhanced biogas production from sunflower stalks using hydrothermal and organosolv pretreatment. Ind Crops Prod 76:449–455. https://doi.org/10.1016/j.indcrop.2015.07.018

Hjorth M, Gränitz K, Adamsen APS, Møller HB (2011) Extrusion as a pretreatment to increase biogas production. Bioresour Technol 102:4989–4994. https://doi.org/10.1016/j.biortech.2010.11.128

Houghton RA, Hole W (2008) Quantities of biomass importance. Small 448–453

Huang J, Du Y, Bao T, Lin M, Wang J, Yang ST (2019) Production of n-butanol from cassava bagasse hydrolysate by engineered Clostridium tyrobutyricum overexpressing adhE2: kinetics and cost analysis. Bioresour Technol 292:121969. https://doi.org/10.1016/j.biortech.2019.121969

Idris ASO, Pandey A, Rao SS, Sukumaran RK (2017) Cellulase production through solid-state tray fermentation, and its use for bioethanol from sorghum stover. Bioresour Technol 242:265–271. https://doi.org/10.1016/j.biortech.2017.03.092

Intasit R, Cheirsilp B, Louhasakul Y, Boonsawang P, Chaiprapat S, Yeesang J (2020) Valorization of palm biomass wastes for biodiesel feedstock and clean solid biofuel through non-sterile repeated solid-state fermentation. Bioresour Technol 298:122551. https://doi.org/10.1016/j.biortech.2019.122551

Ivančić Šantek M, Grubišić M, Galić Perečinec M, Beluhan S, Šantek B (2021) Lipid production by Mortierella isabellina from pretreated corn cobs and effect of lignocellulose derived inhibitors on growth and lipid synthesis. Process Biochem 109:46–58. https://doi.org/10.1016/j.procbio.2021.06.021

Javed MA (2020) Acid treatment effecting the physiochemical structure and thermal degradation of biomass. Renew Energy 159:444–450. https://doi.org/10.1016/j.renene.2020.06.011

Jung YH, Kim KH (2015) Acidic pretreatment. Elsevier B.V.

Kabir MM, Rajendran K, Taherzadeh MJ, Sárvári Horváth I (2015) Experimental and economical evaluation of bioconversion of forest residues to biogas using organosolv pretreatment. Bioresour Technol 178:201–208. https://doi.org/10.1016/j.biortech.2014.07.064

Kiang Y-H (2018) Other and emerging alternative energy technology

Kim KH, Hong J (2001) Supercritical CO_2 pretreatment of lignocellulose enhances enzymatic cellulose hydrolysis. Bioresour Technol 77:139–144. https://doi.org/10.1016/S0960-8524(00)00147-4

Kirupa Sankar M, Ravikumar R, Naresh Kumar M, Sivakumar U (2018) Development of co-immobilized tri-enzyme biocatalytic system for one-pot pretreatment of four different perennial lignocellulosic biomass and evaluation of their bioethanol production potential. Bioresour Technol 269:227–236. https://doi.org/10.1016/j.biortech.2018.08.091

Knoshaug EP, Dong T, Spiller R, Nagle N, Pienkos PT (2018) Pretreatment and fermentation of salt-water grown algal biomass as a feedstock for biofuels and high-value biochemicals. Algal Res 36:239–248. https://doi.org/10.1016/j.algal.2018.10.024

Kumar G, Sivagurunathan P, Zhen G, Kobayashi T, Kim SH, Xu K (2017) Combined pretreatment of electrolysis and ultra-sonication towards enhancing solubilization and methane production from mixed microalgae biomass. Bioresour Technol 245:196–200. https://doi.org/10.1016/j.biortech.2017.08.154

Lamp A, Kaltschmitt M, Lüdtke O (2020) Protein recovery from bioethanol stillage by liquid hot water treatment. J Supercrit Fluids 155. https://doi.org/10.1016/j.supflu.2019.104624

Laurens LML (2020) Whole algal biomass in situ transesterification to fatty acid methyl esters as biofuel feedstocks. Elsevier

Lee JM, Jameel H, Venditti RA (2010) A comparison of the autohydrolysis and ammonia fiber explosion (AFEX) pretreatments on the subsequent enzymatic hydrolysis of coastal Bermuda grass. Bioresour Technol 101:5449–5458. https://doi.org/10.1016/j.biortech.2010.02.055

Leong HY, Chang CK, Khoo KS, Chew KW, Chia SR, Lim JW, Chang JS, Show PL (2021) Waste biorefinery towards a sustainable circular bioeconomy: a solution to global issues. Biotechnol Biofuels 14:1–15. https://doi.org/10.1186/s13068-021-01939-5

Li L, Kong X, Yang F, Li D, Yuan Z, Sun Y (2012) Biogas production potential and kinetics of microwave and conventional thermal pretreatment of grass. Appl Biochem Biotechnol 166:1183–1191. https://doi.org/10.1007/s12010-011-9503-9

Li H, Qu Y, Yang Y, Chang S, Xu J (2016) Microwave irradiation—a green and efficient way to pretreat biomass. Bioresour Technol 199:34–41. https://doi.org/10.1016/j.biortech.2015.08.099

Li C, Wang L, Chen Z, Li Y, Luo X, Zhao F (2021) Ozonolysis of wheat bran in subcritical water for enzymatic saccharification and polysaccharide recovery. J Supercrit Fluids 168:105092. https://doi.org/10.1016/j.supflu.2020.105092

Liakakou ET, Vreugdenhil BJ, Cerone N, Zimbardi F, Pinto F, André R, Marques P, Mata R, Girio F (2019) Gasification of lignin-rich residues for the production of biofuels via syngas fermentation: Comparison of gasification technologies. Fuel 251:580–592. https://doi.org/10.1016/j.fuel.2019.04.081

Liang W, Wang G, Jiao K, Ning X, Zhang J, Guo X, Li J, Wang C (2021) Conversion mechanism and gasification kinetics of biomass char during hydrothermal carbonization. Renew Energy 173:318–328. https://doi.org/10.1016/j.renene.2021.03.123

Liu R, Liang L, Li F, Wu M, Chen K, Ma J, Jiang M, Wei P, Ouyang P (2013) Efficient succinic acid production from lignocellulosic biomass by simultaneous utilization of glucose and xylose in engineered Escherichia coli. Bioresour Technol 149:84–91. https://doi.org/10.1016/j.biortech.2013.09.052

Liu X, Yan Y, Zhao P, Song J, Yu X, Wang Z, Xia J, Wang X (2019) Oil crop wastes as substrate candidates for enhancing erythritol production by modified *Yarrowia lipolytica* via one-step solid state fermentation. Bioresour Technol 294:122194. https://doi.org/10.1016/j.biortech.2019.122194

Liu Z, Wan X, Wang Q, Tian D, Hu J, Huang M, Shen F, Zeng Y (2021a) Performances of a multi-product strategy for bioethanol, lignin, and ultra-high surface area carbon from lignocellulose by PHP (phosphoric acid plus hydrogen peroxide) pretreatment platform. Renew Sustain Energy Rev 150:111503. https://doi.org/10.1016/j.rser.2021.111503

Liu X, Bouxin FP, Fan J, Budarin VL, Hu C, Clark JH (2021b) Microwave-assisted catalytic depolymerization of lignin from birch sawdust to produce phenolic monomers utilizing a hydrogen-free strategy. J Hazard Mater 402:123490. https://doi.org/10.1016/j.jhazmat.2020.123490

Liu Q, Wang F, Zhou H, Li Z, Fu Y, Qin M (2021c) Utilization of lignin separated from pre-hydrolysis liquor via horseradish peroxidase modification as an adsorbent for methylene blue removal from aqueous solution. Ind Crops Prod 167:113535. https://doi.org/10.1016/j.indcrop.2021.113535

López-Gómez JP, Manan MA, Webb C (2020) Solid-state fermentation of food industry wastes. Food Ind Wastes 135–161. https://doi.org/10.1016/b978-0-12-817121-9.00007-3

Lorenzo M, Moldes D, Rodríguez Couto S, Sanromán A (2002) Improving laccase production by employing different lignocellulosic wastes in submerged cultures of Trametes versicolor. Bioresour Technol 82:109–113. https://doi.org/10.1016/S0960-8524(01)00176-6

Louis ACF, Venkatachalam S (2020) Energy efficient process for valorization of corn cob as a source for nanocrystalline cellulose and hemicellulose production. Int J Biol Macromol 163:260–269. https://doi.org/10.1016/j.ijbiomac.2020.06.276

Lu X, Xi B, Zhang Y, Angelidaki I (2011) Microwave pretreatment of rape straw for bioethanol production: focus on energy efficiency. Bioresour Technol 102:7937–7940. https://doi.org/10.1016/j.biortech.2011.06.065

Mahmoodi P, Karimi K, Taherzadeh MJ (2018) Efficient conversion of municipal solid waste to biofuel by simultaneous dilute-acid hydrolysis of starch and pretreatment of lignocelluloses. Energy Convers Manag 166:569–578. https://doi.org/10.1016/j.enconman.2018.04.067

Makareviciene V, Skorupskaite V (2019) Transesterification of microalgae for biodiesel production. Elsevier Inc.

Mankar AR, Pandey A, Modak A, Pant KK (2021) Pretreatment of lignocellulosic biomass: a review on recent advances. Bioresour Technol 334:125235. https://doi.org/10.1016/j.biortech.2021.125235

Meenakshisundaram S, Fayeulle A, Leonard E, Ceballos C, Pauss A (2021) Fiber degradation and carbohydrate production by combined biological and chemical/physicochemical pretreatment methods of lignocellulosic biomass—a review. Bioresour Technol 331:125053. https://doi.org/10.1016/j.biortech.2021.125053

Meini MR, Cabezudo I, Galetto CS, Romanini D (2021) Production of grape pomace extracts with enhanced antioxidant and prebiotic activities through solid-state fermentation by *Aspergillus niger* and *Aspergillus oryzae*. Food Biosci 42:101168. https://doi.org/10.1016/j.fbio.2021.101168

Moreno AD, Tomás-Pejó E, Ballesteros M, Negro MJ (2019) Pretreatment technologies for lignocellulosic biomass deconstruction within a biorefinery perspective. Biomass, Biofuels, Biochem Biofuels Altern Feed Convers Process Prod Liq Gaseous Biofuels 379–399. https://doi.org/10.1016/B978-0-12-816856-1.00016-6

Mulakhudair AR, Hanotu J, Zimmerman W (2017) Exploiting ozonolysis-microbe synergy for biomass processing: application in lignocellulosic biomass pretreatment. Biomass Bioenerg 105:147–154. https://doi.org/10.1016/j.biombioe.2017.06.018

Nakasu PYS, Pin TC, Hallett JP, Rabelo SC, Costa AC (2021) In-depth process parameter investigation into a protic ionic liquid pretreatment for 2G ethanol production. Renew Energy 172:816–828. https://doi.org/10.1016/j.renene.2021.03.004

Nasir A, Chen HZ, Wang L (2020) Novel single-step pretreatment of steam explosion and choline chloride to de-lignify corn stover for enhancing enzymatic edibility. Process Biochem 94:273–281. https://doi.org/10.1016/j.procbio.2020.04.036

Nasrullah A, Bhat AH, Sada Khan A, Ajab H (2017) Comprehensive approach on the structure, production, processing, and application of lignin. Elsevier Ltd.

Navarro RR, Otsuka Y, Matsuo K, Sasaki K, Sasaki K, Hori T, Habe H, Nakamura M, Nakashimada Y, Kimbara K, Kato J (2020) Combined simultaneous enzymatic saccharification and comminution (SESC) and anaerobic digestion for sustainable biomethane generation from wood lignocellulose and the biochemical characterization of residual sludge solid. Bioresour Technol 300:122622. https://doi.org/10.1016/j.biortech.2019.122622

Neikov O, Naboychenko S, Mourachova I (2008) Mechanical crushing and grinding (Chap. 2). Handb Non-Ferrous Met Powders 47–62

Ng DKS, Ng KS, Ng RTL (2017) Integrated biorefineries. Elsevier

Nguyen LT, Tran MH, Lee EY (2021) Co-upgrading of ethanol-assisted depolymerized lignin: a new biological lignin valorization approach for the production of protocatechuic acid and polyhydroxyalkanoic acid. Bioresour Technol 338:125563. https://doi.org/10.1016/j.biortech.2021.125563

Nguyen C Van, Lewis D, Chen WH, Huang HW, ALOthman ZA, Yamauchi Y, Wu KCW (2016) Combined treatments for producing 5-hydroxymethylfurfural (HMF) from lignocellulosic biomass. Catal Today 278:344–349. https://doi.org/10.1016/j.cattod.2016.03.022

Ni'matuzahroh Sari SK, Trikurniadewi N, Ibrahim SNMM, Khiftiyah AM, Abidin AZ, Nurhariyati T (2020) Fatimah: bioconversion of agricultural waste hydrolysate from lignocellulolytic mold into biosurfactant by *Achromobacter* sp. BP(1)5. Biocatal Agric Biotechnol 24:101534. https://doi.org/10.1016/j.bcab.2020.101534

Nsaful F, Collard FX, Görgens JF (2018) Lignocellulose thermal pretreatment and its effect on fuel properties and composition of the condensable products (tar precursors) from char devolatilization for coal substitution in gasification application. Fuel Process Technol 179:334–343. https://doi.org/10.1016/j.fuproc.2018.07.015

Pang Z, Lyu W, Dong C, Li H, Yang G (2016) High selective delignification using oxidative ionic liquid pretreatment at mild conditions for efficient enzymatic hydrolysis of lignocellulose. Bioresour Technol 214:96–101. https://doi.org/10.1016/j.biortech.2016.04.095

Park SY, Kim JY, Youn HJ, Choi JW (2019) Utilization of lignin fractions in UV resistant lignin-PLA biocomposites via lignin-lactide grafting. Int J Biol Macromol 138:1029–1034. https://doi.org/10.1016/j.ijbiomac.2019.07.157

Patil R, Cimon C, Eskicioglu C, Goud V (2021) Effect of ozonolysis and thermal pre-treatment on rice straw hydrolysis for the enhancement of biomethane production. Renew Energy 179:467–474. https://doi.org/10.1016/j.renene.2021.07.048

Paul SK, Chakraborty S (2019) Mixing effects on the kinetics of enzymatic hydrolysis of lignocellulosic Sunn hemp fibres for bioethanol production. Chem Eng J 377. https://doi.org/10.1016/j.cej.2018.10.040

Pereira Marques F, Lima Soares AK, Lomonaco D, Alexandre e Silva LM, Tédde Santaella S, de Freitas Rosa M, Carrhá Leitão R (2021) Steam explosion pretreatment improves acetic acid organosolv delignification of oil palm mesocarp fibers and sugarcane bagasse. Int J Biol Macromol 175:304–312. https://doi.org/10.1016/j.ijbiomac.2021.01.174

Puligundla P, Oh SE, Mok C (2016) Microwave-assisted pretreatment technologies for the conversion of lignocellulosic biomass to sugars and ethanol: a review. Carbon Lett 17:1–10. https://doi.org/10.5714/CL.2016.17.1.001

Puspawati S, Wagiman Ainuri M, Nugraha DA (2015) Haslianti: the production of bioethanol fermentation substrate from *Eucheuma cottonii* seaweed through hydrolysis by cellulose enzyme. Agric Agric Sci Procedia 3:200–205. https://doi.org/10.1016/j.aaspro.2015.01.039

Putrino FM, Tedesco M, Bodini RB, de Oliveira AL (2020) Study of supercritical carbon dioxide pretreatment processes on green coconut fiber to enhance enzymatic hydrolysis of cellulose. Bioresour Technol 309:123387. https://doi.org/10.1016/j.biortech.2020.123387

Rego F, Soares Dias AP, Casquilho M, Rosa FC, Rodrigues A (2020) Pyrolysis kinetics of short rotation coppice poplar biomass. Energy 207:118191. https://doi.org/10.1016/j.energy.2020.118191

Sala A, Artola A, Sánchez A, Barrena R (2020) Rice husk as a source for fungal biopesticide production by solid-state fermentation using *B. bassiana* and *T. harzianum*. Bioresour Technol 296:122322. https://doi.org/10.1016/j.biortech.2019.122322

Savun-Hekimoğlu B (2020) A review on sonochemistry and its environmental applications. Acoustics 2:766–775. https://doi.org/10.3390/acoustics2040042

Sharma P, Gaur VK, Sirohi R, Varjani S, Hyoun Kim S, Wong JWC (2021) Sustainable processing of food waste for production of bio-based products for circular bioeconomy. Bioresour Technol 325:124684. https://doi.org/10.1016/j.biortech.2021.124684

Shen X, Sun R (2021) Recent advances in lignocellulose prior-fractionation for biomaterials, biochemicals, and bioenergy. Carbohydr Polym 261. https://doi.org/10.1016/j.carbpol.2021.117884

Sheng Y, Tan X, Gu Y, Zhou X, Tu M, Xu Y (2021) Effect of ascorbic acid assisted dilute acid pretreatment on lignin removal and enzyme digestibility of agricultural residues. Renew Energy 163:732–739. https://doi.org/10.1016/j.renene.2020.08.135

Simair AA, Qureshi AS, Simair SP, Khushk I, Klykov SP, Ali CH, Lu C (2018) An integrated bioprocess for xylanase production from agriculture waste under open non-sterilized conditions: Biofabrication as fermentation tool. J Clean Prod 193:194–205. https://doi.org/10.1016/j.jclepro.2018.05.020

Singh M, Pandey N, Dwivedi P, Kumar V, Mishra BB (2019) Production of xylose, levulinic acid, and lignin from spent aromatic biomass with a recyclable Br Ønsted acid synthesized from d-limonene as renewable feedstock from citrus waste. Bioresour Technol 293:122105. https://doi.org/10.1016/j.biortech.2019.122105

Speight JG (2020) Production of fuels from nonfossil fuel feedstocks

Sun RC, Salisbury D, Tomkinson J (2003) Chemical composition of lipophilic extractives released during the hot water treatment of wheat straw. Bioresour Technol 88:95–101. https://doi.org/10.1016/S0960-8524(02)00290-0

Sztancs G, Kovacs A, Toth AJ, Mizsey P, Billen P, Fozer D (2021) Catalytic hydrothermal carbonization of microalgae biomass for low-carbon emission power generation: the environmental impacts of hydrochar co-firing. Fuel 300:120927. https://doi.org/10.1016/j.fuel.2021.120927

Teymouri F, Laureano-Perez L, Alizadeh H, Dale BE (2005) Optimization of the ammonia fiber explosion (AFEX) treatment parameters for enzymatic hydrolysis of corn stover. Bioresour Technol 96:2014–2018. https://doi.org/10.1016/j.biortech.2005.01.016

Tsegaye B, Balomajumder C, Roy P (2019) Optimization of microwave and NaOH pretreatments of wheat straw for enhancing biofuel yield. Energy Convers Manag 186:82–92. https://doi.org/10.1016/j.enconman.2019.02.049

Ummalyma SB, Supriya RD, Sindhu R, Binod P, Nair RB, Pandey A, Gnansounou E (2019) Biological pretreatment of lignocellulosic biomass-current trends and future perspectives. Elsevier Inc.

Vasco-Correa, J., Ge, X., Li, Y.: Biological Pretreatment of Lignocellulosic Biomass. Biomass Fractionation Technol. a Lignocellul. Feed. Based Biorefinery. 561–585 (2016). https://doi.org/10.1016/B978-0-12-802323-5.00024-4

Vasheghani Farahani S, Kim YW, Schall CA (2016) A coupled low temperature oxidative and ionic liquid pretreatment of lignocellulosic biomass. Catal Today 269:2–8. https://doi.org/10.1016/j.cattod.2015.12.022

Venturin B, Frumi Camargo A, Scapini T, Mulinari J, Bonatto C, Bazoti S, Pereira Siqueira D, Maria Colla L, Alves SL, Paulo Bender J, Luís Radis Steinmetz R, Kunz A, Fongaro G, Treichel H (2018) Effect of pretreatments on corn stalk chemical properties for biogas production purposes. Bioresour Technol 266:116–124. https://doi.org/10.1016/j.biortech.2018.06.069

Verardi A, Lopresto CG, Blasi A, Chakraborty S, Calabrò V (2019) Bioconversion of lignocellulosic biomass to bioethanol and biobutanol. INC

Vieira P, Ribeiro C, Pinto CA, Saraiva JA, Barba FJ (2020) Application of HPP in food fermentation processes. Elsevier Inc.

Wang M, Zhou D, Wang Y, Wei S, Yang W, Kuang M, Ma L, Fang D, Xu S, Du S (2016) Bioethanol production from cotton stalk: a comparative study of various pretreatments. Fuel 184:527–532. https://doi.org/10.1016/j.fuel.2016.07.061

Wang J, Wang D, Zhang X, Zhao C, Zhang M, Zhang Z, Wang J (2019) An anti-symmetric dual (ASD) Z-scheme photocatalytic system: ($ZnIn_2S_4/Er^{3+}:Y_3Al_5O_{12}@ZnTiO_3/CaIn_2S_4$) for organic pollutants degradation with simultaneous hydrogen evolution. Int J Hydrogen Energy 44:6592–6607. https://doi.org/10.1016/j.ijhydene.2019.01.214

Wu S, Song L, Sommerfeld M, Hu Q, Chen W (2017) Optimization of an effective method for the conversion of crude algal lipids into biodiesel. Fuel 197:467–473. https://doi.org/10.1016/j.fuel.2017.02.040

Wu J, Dong L, Liu B, Xing D, Zhou C, Wang Q, Wu X, Feng L, Cao G (2020) A novel integrated process to convert cellulose and hemicellulose in rice straw to biobutanol. Environ Res 186:109580. https://doi.org/10.1016/j.envres.2020.109580

Wu Y, Ge S, Xia C, Mei C, Kim KH, Cai L, Smith LM, Lee J, Shi SQ (2021) Application of intermittent ball milling to enzymatic hydrolysis for efficient conversion of lignocellulosic biomass into glucose. Renew Sustain Energy Rev 136:110442. https://doi.org/10.1016/j.rser.2020.110442

Xu JK, Sun RC (2016) Recent advances in alkaline pretreatment of lignocellulosic biomass. Biomass Fractionation Technol. a Lignocellul. Feed Based Biorefinery 431–459. https://doi.org/10.1016/B978-0-12-802323-5.00019-0

Yiin CL, Yap KL, Ku AZE, Chin BLF, Lock SSM, Cheah KW, Loy ACM, Chan YH (2021) Recent advances in green solvents for lignocellulosic biomass pretreatment: potential of choline chloride (ChCl) based solvents. Bioresour Technol 333:125195. https://doi.org/10.1016/j.biortech.2021.125195

You Z, Pan SY, Sun N, Kim H, Chiang PC (2019) Enhanced corn-stover fermentation for biogas production by NaOH pretreatment with CaO additive and ultrasound. J Clean Prod 238:117813. https://doi.org/10.1016/j.jclepro.2019.117813

Zhang H, Zhang J, Xie J, Qin Y (2020a) Effects of NaOH-catalyzed organosolv pretreatment and surfactant on the sugar production from sugarcane bagasse. Bioresour Technol 312:123601. https://doi.org/10.1016/j.biortech.2020.123601

Zhang Q, Chen QH, He GQ (2020b) Effect of ultrasonic-ionic liquid pretreatment on the hydrolysis degree and antigenicity of enzymatic hydrolysates from whey protein. Ultrason Sonochem 63:104926. https://doi.org/10.1016/j.ultsonch.2019.104926

Zhang XM, Wang M, Yu Q, Ma ZY, Beauchemin KA, Wang R, Wen JN, Lukuyu BA, Tan ZL (2020c) Liquid hot water treatment of rice straw enhances anaerobic degradation and inhibits methane production during in vitro ruminal fermentation. J Dairy Sci 103:4252–4261. https://doi.org/10.3168/jds.2019-16904

Zhang Y, Cui Y, Chen P, Liu S, Zhou N, Ding K, Fan L, Peng P, Min M, Cheng Y, Wang Y, Wan Y, Liu Y, Li B, Ruan R (2019) Gasification technologies and their energy potentials. Elsevier B.V.

Zheng J, Rehmann L (2014) Extrusion pretreatment of lignocellulosic biomass: a review. Int J Mol Sci 15:18967–18984. https://doi.org/10.3390/ijms151018967

Zheng NY, Lee M, Lin YL (2020) Co-processing textile sludge and lignocellulose biowaste for biofuel production through microwave-assisted wet torrefaction. J Clean Prod 268:122200. https://doi.org/10.1016/j.jclepro.2020.122200

Zhuang X, Zhan H, Song Y, He C, Huang Y, Yin X, Wu C (2019) Insights into the evolution of chemical structures in lignocellulose and non-lignocellulose biowastes during hydrothermal carbonization (HTC). Fuel 236:960–974. https://doi.org/10.1016/j.fuel.2018.09.019

Part II
Lignocellulosic Waste Biorefinery

Chapter 3
Lignocellulosic Waste Treatment in Biorefinery Concept: Challenges and Opportunities

Lukas Kratky

3.1 Introduction

The research and industrial strategies of the twentieth century were associated with the development of coal, petroleum or natural gas-based refineries to produce fossil-based fuels and chemicals. Nevertheless, due to the still-growing population, there is increasing demand for energy and chemical products. Fossil fuels are not regarded as sustainable due to economic and ecological issues (Geissler and Maravelias 2021). This is caused by their potential limitations in supply, declining global reserves, and adverse effects on the climate by GHG emissions especially. Thus, one of the crucial research and industrial challenges strongly connected with the twenty-first century is to cover the growing demand for energy as a primary source to operate industrial processes, transportation, or heating systems. There is, therefore, urgent demand for sustainable and green sources of energy and alternative chemicals (Solarte-Toro et al. 2021).

The lignocellulosic waste biomass with the production of 200 billion tons annually (Mankar et al. 2021) is nowadays viewed as an unused renewable material that has high potential to become a raw material to produce (Roadmap 2021) alternative energy sources (biohydrogen, biomethane, syngas, pyrolytic oil), valuable chemical substances (bio alcohols, fibres, essences, pigments, oils, organic substances), or eco-innovative materials (bio-composites, bioplastics). This waste material can be a traditional part of food and processing waste, agricultural waste, green waste, waste from restaurants and canteens, biodegradable municipal solid waste from households and gardens. However, generally known, up to 80% of biodegradable waste is nowadays landfilled, combusted or composted, resulting in soil, water and air pollutions and adverse effects on human health (Patel and Shah 2021). There are several global

L. Kratky (✉)
Faculty of Mechanical Engineering, Department of Process Engineering, Czech Technical University in Prague, Technicka 4, 16000 Prague 6, Czech Republic
e-mail: Lukas.Kratky@fs.cvut.cz

Y. K. Nandabalan et al. (eds.), *Zero Waste Biorefinery*, Energy, Environment, and Sustainability, https://doi.org/10.1007/978-981-16-8682-5_3

efforts and legislative restrictions to reduce the number of wastes. Regarding EU restrictions 99/31/EC about landfilling of wastes, it has to be landfilled 65% less in 2020 compared to 1995. Therefore, all these restrictions and lignocellulosic material availability make it attractive as raw material to produce biofuels or valuable chemical compounds. Thousands of scientific papers and studies present lignocellulosic waste treatment technologies known as gasification, pyrolysis, enzymatic hydrolysis, or fermentation. However, there is a large gap between laboratory approaches and industrial realizations or implementations of waste treatment technologies due to their un-economic attractiveness. This is because advanced biofuels and bio-products are not cost-competitive with fossil fuels and chemicals in production costs (Pinales-Márquez et al. 2021; Kutsay et al. 2016; Ruiz et al. 2016; Piemonte et al. 2014; Rahimi and Shafiei 2019; Santamaría-Fernández et al. 2018; Padi and Chimphango 2020; Bateni et al. 2014). E.g. Chovau et al. (2013) present production costs of 458 $ m^{-3} for gasoline-based, 651 $ m^{-3} for corn-based, and 800 $ m^{-3} for corn-stover-based ethanol strongly affected by cellulose enzyme price. Cheng et al. (2019) reported production cost 1011 $ m^{-3} and a minimum selling price of 1110 $ m^{-3} for cellulosic ethanol to reach a payback time of 9 years for their proposal. Thus, technologies of waste transformation to advanced biofuels or valuable chemical substances need to be still under intensive development. The main target for cooperating between scientists and industry representatives is to overtake the laboratory results and scale a technology up to a pilot or industrial size.

The biorefinery concept represents a multi-technological processing line, in which biomass is transformed into product portfolio like biofuels, valuable chemicals, and biomaterials, followed by the production of electricity or heat. Such a parallel production of variety in biofuels and bioproducts can improve the waste treatment plant's economy. Thus, waste treatment in the biorefinery concept poses the potential to reduce dependence on fossil resources, reduce the combustion of fossil fuels, reduce carbon dioxide production, or decrease the waste effect on the environment. The chapter presents a technical overview of technological design and operation from a mechanical engineering perspective to identify research needs, public challenges and opportunities for lignocellulosic waste treatment in the biorefinery concept.

3.2 Lignocellulosic Biorefinery

The general block scheme of the lignocellulosic biorefinery was defined by Biorefinery Roadmap (2021), see Fig. 3.1. The lignocellulosic biomass goes through several operating sets, namely pretreatment, primary, secondary processing and tertiary processing, to be fractionated into targeted main and side products. It must also be highlighted that the lignocellulosic waste treatment processing line in the biorefinery concept produces wastes in the form of gases, liquids, or solid residues.

Lignocellulosic waste first undergoes separation and sorting to remove all the potential impurities (sand, glass, metals, plastic) or incompatible waste species following primary or secondary processing treatment steps. Lignocellulosic waste

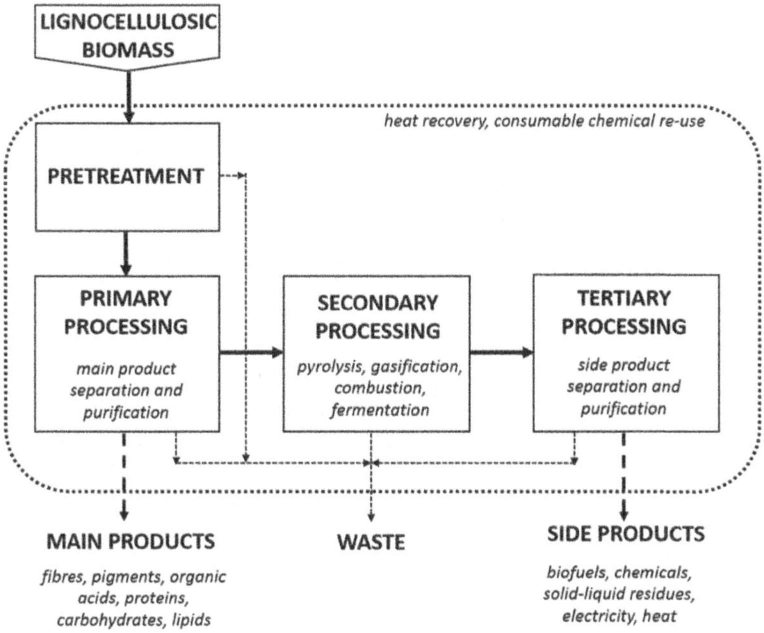

Fig. 3.1 The general concept of lignocellulosic biorefinery

is then handled to the pretreatment section. To achieve a suitable particle size for its subsequent treatment, all the wastes are comminuted in a suitable size reduction machine. Then, physical (mechanical size reduction, sorting, ultrasound, microwave, extrusion), chemical (acidic, alkaline, solvolysis), physiochemical (steam expansion, hydrothermal pretreatment), biological (fungi, moulds, enzymatic hydrolysis) pretreatment follow (Mankar et al. 2021; Rahimi and Shafiei 2019; Santamaría-Fernández et al. 2018; Padi and Chimphango 2020; Bateni et al. 2014; Chovau et al. 2013). The main aim of pretreatment is to destroy a lignocellulosic structure to reach a maximum efficiency to extract a primary product with high added value, i.e. saccharides, proteins, oils, aromas, acids, or natural cellulosic fibres. As the targeted primary product is separated, a residue undergoes a secondary processing step, in which a suitable biochemical (alcoholic fermentation, biogas fermentation, hydrogen production) or thermochemical (combustion, gasification, pyrolysis) follow (Roadmap 2021; Pinales-Márquez et al. 2021) concerning waste moisture, especially. The wetter material, the better to use it for biochemical pathways. Regarding Fig. 3.1, the mutual combination of biochemical and thermochemical are often applied to ensure heat recovery in a complex processing line. Then, secondary treated materials are fed to the tertiary processing step, in which targeted products are separated (Pinales-Márquez et al. 2021), e.g. biofuels (biogas, biomethane, acids, saccharides, pyrolytic oil, syngas). It has to be highlighted that each waste treatment technology also generates wastes. The first one is generated during pre-processing of raw material, i.e. wastewater (washing of trucks, washing of waste), solid impurities

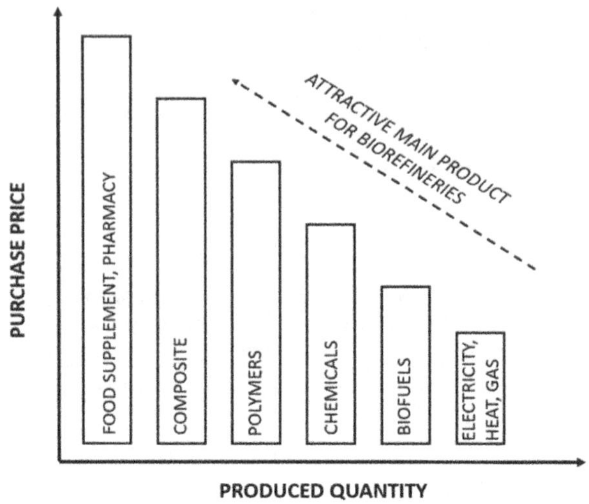

Fig. 3.2 Economic valuable of product related to its quantity

(metals, plastics, glass), or smelly gases from air-conditioning of income buildings and storage systems. Other wastes are generated as solids, liquids or gases through a complex processing line concerning its exact technical and process set-up.

Targeted main products produced from lignocellulosic waste biomass are bio alcohols, biohydrogen, fibres, carbohydrates, oils, organic acid, saccharides, surfactants, resins, dextrins, or solvents. Their market potential can be found in the transportation, energy sector, processing industries (cosmetics, pharmaceutical, chemical, plastics, paper, textile), civil engineering as construction materials, or food processing industry a portfolio of nutrients or various supplements. As lignocellulosic biorefinery is designed and product portfolio is discussed, its production cost related to selling cost and market demand of targeted product must be adequately chosen to reach economically feasible waste processing technology. Only a portfolio composed of main high valued products and side low valued products has the potential to meet the requirement of economic attractiveness, see Fig. 3.2. The primary products should meet market demand in the food industry, process industries (composites, polymers, chemicals, biofuels) followed by secondary ones applicable only as renewable primary energy resources (electricity, heat). The primary products as food supplements, chemicals, or raw material for composites production evince such selling prices, i.e. cellulosic fibres up to 500 EUR t^{-1} as biodegradable reinforcement of bio-composites (Kutsay et al. 2016), 3–30 EUR kg^{-1} of pigment concerning their purity and application as food additives or cosmetics, microalgal lipids serve of 1–2 EUR kg^{-1} of lipids with their application as a food additive or healthcare (Ruiz et al. 2016), glycerol 1.05 Eur kg^{-1} (Piemonte et al. 2014), or salts 0.48 EUR kg^{-1} (Rahimi and Shafiei 2019). The typical fuel selling prices are 729 EUR m^{-3} for biodiesel (Piemonte et al. 2014) or 1.21 EUR Nm^{-3} for biomethane (Rahimi and Shafiei 2019). Nevertheless, it must be noted that green subsidies cover all the "bio" selling prices. Regarding their fossil alternatives, 0.31 EUR L^{-1} of diesel and 0.05

Table 3.1 The "biorefinery" keyword search in the database Web of Science (dated 28.5.2021)

Search	Keyword	Article	Review	Total
#1	"biorefinery"	6771	1196	7967
#2	"lignocellulosic" AND "biorefinery"	1810	398	2208
#3	"green" AND "biorefinery"	740	209	949
#4	"biogas" AND "biorefinery"	432	106	538
#5	"bioethanol" AND "biorefinery"	972	203	1175
#6	"biodiesel" AND "biorefinery"	637	193	830
#7	"gasification" AND "biorefinery"	274	61	483
#8	"pyrolysis" AND "biorefinery"	547	91	608

EUR Nm^{-3} of natural gas are conventional market prices (Rahimi and Shafiei 2019). If only electricity or heat are produced, their selling price is significantly lower than those mentioned above. The selling prices are not higher than 0.13 Eur kWh^{-1} for electricity (Piemonte et al. 2014) and 4.67 EUR GJ^{-1} for heat (Rahimi and Shafiei 2019). Therefore, the production of electricity or heat as a primary biorefinery product cannot provide its economic attractiveness. Only a portfolio of multiple products with different selling prices should reach the prosperous economy of biorefinery.

Scientific teams worldwide offer thousands of reports and studies only scoped to individual technological disciplines, i.e. pretreatment, extraction and separation of valuable products, gasification, pyrolysis, fermentation. Some papers also deal with economic feasibility studies of lignocellulosic biorefineries. Thus, the literature search was done to determine the main scope of biorefineries, see Table 3.1. The keyword "biorefinery" was detected by 6771 times in the honourable database Web of Science. The detailed searches confirmed that the dominant approach of lignocellulosic waste treatment could be found in the concepts of green, biogas, bioethanol, or biodiesel biorefineries, counting up to 76% of search biorefinery varieties. Lignocellulosic waste treatment in biorefinery concept tends to extract targeted primary product with subsequent treatment of residues by fermentation.

A little information about their technological readiness level, no mass and energy balances, economics, environmental impact assessments are presented. Moreover, another technical problem can also be viewed in implementing laboratory results into practice. This is given by the fact that laboratory technologies are based on the treatment of pure materials. The real industrial area serves wastes that can manifest a high level of contamination. Applying an engineering perspective, it is evident that it is necessary to perform fundamental techno-economic studies to apply current knowledge and usability in waste processing in biorefinery concepts, modify existing and develop brand new effective technologies, machines, and equipment for given processing steps. The chapter aims to introduce selected lignocellulosic waste treatment strategies in the biorefinery concept, followed by a discussion about their economic feasibilities, sensitive analysis and investor attractiveness. The chapter's author team made all the above-presented biorefinery approaches. Thus, all the following processes and economic estimations present only the essential findings counting the link to referenced materials. Nevertheless, before this discussion, a general approach to biorefinery's economic feasibility is introduced to the reader.

3.2.1 The Fundamental Techno-Economic Analysis of Biorefinery

As presented in this subchapter, the general approach to perform economic feasibility was adopted from the handbook written by Towler and Sinnot (2013), in which reliable and industrially verified methods to design processes and perform techno-economic studies is defined for chemical, food and waste treatment industries.

Waste treatment in the biorefinery concept become attractive for an investor only if favourable economic feasibility is reached. Economic attractiveness is usually closely associated with the payback period. Its reliable economic evaluation is based on the knowledge of block scheme, process flow diagram, system balances, essential characteristics of all the installed equipment, and fundamental economics (investment cost, operating expenses, cash-flow and simple payback time). Carrying out an economic feasibility study is, therefore, a crucial step to evaluate individual biorefinery proposals. Furthermore, the industrial sector typically demands simple payback time (*SPT*) between 6 and 8 years with a lifetime of up to 30 years. The payback period is given by the ratio of total investments (*CAPEX*) over cash flow. The cash flow is known as the difference between annual income (*AI*) and production cost (*PTD*), see Eq. 3.1.

$$SPT = \frac{CAPEX}{AI - PTD} \tag{3.1}$$

The initial design steps usually need technical knowledge of technological (conversion factor, yield, process conditions) and technical (design, energy demand, service, maintenance, lifetime, local utility streams) factors. Based on this information, the process flow diagram is designed with information about mass and energy flows through biorefinery. Balancing approaches are defined by the law of mass conservation, energy conservation law followed by the assumptions of neglected heat losses. Applying information about stream characteristics and process demand, basic equipment design is proposed, and total investment cost (*CAPEX*) is estimated. Generally known, *CAPEX* summarizes the equipment cost (*ISBL*), offsite costs (*OSBL*), cost for engineering design (EC) and contingencies (*CC*). *ISBL* cost represents the cost of installed equipment in primary process streams. *OSBL* cost is associated with engineering networks, utilities, pipelines, insulations, measurement and control systems, and laboratories. *OSBL* is usually estimated as 0.4·*ISBL* costs for preliminary biorefinery concepts. Engineering cost (*EC*) covers all the engineering services, and 0.3·(*ISBL* + *OSBL*) is usually used as a rough estimation.

Finally, contingency charges *CC* are financial reserve that covers fluctuation in material prices, equipment deliveries, etc. The *CC* value calculated as $0.1 \cdot (ISBL + OSBL)$ is typically used to evaluate the economics of biorefinery. *CAPEX* is dependent only on the price of equipment. The reliable estimation of investment cost can be provided by the Factorial Method of Equipment Cost Estimation (Towler and Sinnott 2013). Its origin is based on the knowledge of the so-called size parameter *S*. Each piece of equipment has a unique size parameter. E.g. volumetric flowrate for pumps, volume for storage tanks, heat transfer surface for heat exchangers, heat transfer surface for dryers, or volume for jacketed reactors equipped with a mechanical mixing system. Knowing size parameter *S*, the equipment cost *CE* (Eur) estimation can be calculated according to such an equation.

$$CE = a + b \cdot S^n \tag{3.2}$$

In which *a*, *b*, *n* represents cost estimating coefficients, as listed in Table 3.2.

The lower the equipment cost, the lower the investment cost, the lower the simple payback time. It means that investment and operating costs can be affected by design optimization in individual equipment. E.g., decrease the needed volume of the reactor because of decrease in needed residence time, lower residence time in the reactor because of intensified mixing or heating system, lower heat transfer surface of heat exchanger because of heat transfer intensification, lower volume in time lower flowrate in the pump, etc. The geometrical configuration of equipment and selection of material is viewed as dominant parameters that affect the investment cost of equipment. Biorefinery equipment can be made of carbon steel, alloy steel, stainless steel, aluminium, nickel or titanium alloys, concerning corrosion, erosion and abrasion. E.g. stainless steel by 1.5 times, or nickel alloys by 1.7 times are more expensive in comparison to carbon steel (Towler and Sinnott 2013). Thus, a choice of

Table 3.2 Characteristic parameters to calculate equipment cost for selected equipment (Towler and Sinnott 2013)

Equipment	Size parameter *S*	*a*	*b*	*n*	Validity ranges of *S*
Reciprocating compressor	Productivity $m^3\ h^{-1}$	260,000	2700	0.75	93–16,800
Pump	Flowrate $l\ s^{-1}$	8000	240	0.9	0.2–126
Hammer mill	Productivity $t\ h^{-1}$	68,400	730	1.0	30–400
Belt dryer	Heat transfer surface m^2	15,000	10,500	0.9	11–180
U-tube heat exchanger	Heat transfer surface m^2	28,000	54	1.2	10–1000
Reactor, jacketed, agitated	Volume m^3	61,500	32,500	0.8	0.5–100
Storage tank, cone roof	Volume m^3	5800	1600	0.7	10–4000

material strongly affects investment cost to equipment. Finally concluded, a balance between corrosion resistance, innovative design and equipment price must be found. From this follows that the proper design of binding machines and equipment can significantly improve total investment cost.

Annual income AI can be easily calculated as the annual product yield multiplied by its purchase price related to local policies or global demand. The higher the annual income, the lower the simple payback time. So only a portfolio composed of main high valued products and side low valued products can positively affect the economic feasibility of biorefinery. *PDT* costs (Towler and Sinnott 2013) summarize direct (energy, personnel costs, supervision, maintenance, consumables, laboratories, reserve), indirect (insurance, corporate directions) and transport costs. The specific perceptual values of 15% of the personal cost for supervision, maintenance of 5% *CAPEX*, consumables of 1% *CAPEX*, laboratories of 15% from maintenance, the reserve of 3% from direct operating cost, insurance of 1% from *CAPEX*, corporates as the sum of personal and supervision cost, are usually used for initial project evaluation.

Such a general approach of techno-economic feasibility study allows the user to define his own scripts (Excel, Octave) without the need for highly professional commercially licenced software like AspenPlus. The chapter's author team applied the above-described method in many works associated with lignocellulosic biorefineries. The introduction and the essential finding of selected techno-economic studies are discussed in subsequent chapters. Nevertheless, detailed information about the studies themselves is provided in referenced literature resource.

3.2.2 Biogas Biorefinery

Multistage lignocellulosic waste treatment in the concept of biogas biorefinery consists of several technological steps, see Fig. 3.3. Biomass initially goes through the section of mechanical pretreatment, in which impurities are separated, biomass is sorted, grinded and milled to a suitable particle size needed in a subsequent processing step. Kutsay et al. (2016) recommend 1–10 mm in biomass particle size to reach effective biodegradability of biochemical treatment. Next, comminuted particles are processed in the extraction section. Primarily targeted products are extracted (green juice, fibres, saccharides, oils, pigments), purified or upgraded (enzymatic hydrolysis, microbial treatment, alcoholic fermentation, transesterification). The solid–liquid residue is mixed with agricultural waste material (straws, manure, corn sillage) to be anaerobically digested to produce biogas. Undigested material is stored and used as solid–liquid fertilizer. Applying solid–liquid separator, solid fermentation residue is removed out of digestate stream and used as solid fertilizer. A liquid residue rich in solubilized organics and minerals, fugate, can undergo membrane separation to generate pure water and concentrated liquid minerals as liquid fertilizer. Generally known, there are two pathways to use biogas. The conventional one deals with its direct combustion in cogeneration unit to parallel produce electricity and heat.

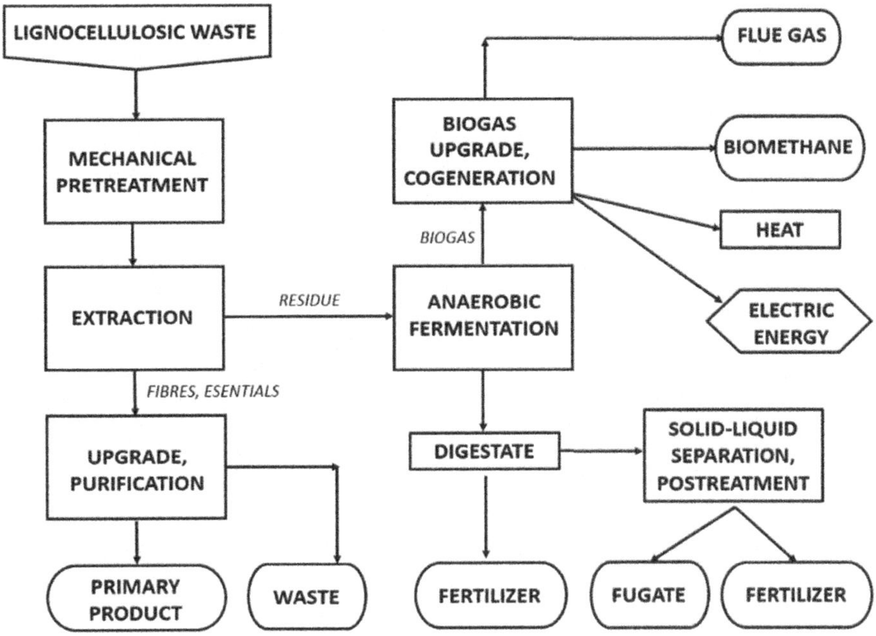

Fig. 3.3 The general block diagram of biogas biorefinery

The second pathway is associated with CO_2 removal to produce biomethane as the alternative to natural gas.

The base literature search was carried out to identify the base directions in biogas biorefineries, see Table 3.3. The reviewed information about waste processing precisely fits the general overview discussed above. Nevertheless, the integration of downstream processing for digestate or biogas upgrades is still missing. The review confirmed that the presented research works are dedicated to experimentally determine process set-up to get the suitable portfolio of available products from a given material, biogas or biomethane production potential. Nevertheless, all the experiments are performed at the lab scale without any respect to the industrial potential of its implementation. Andersen et al. (2018) present the economic study in which protein concentrate and biogas are produced from clover. The technical set-up meets the industrial potential being at TRL 9, but no information about operational cost (OPEX) is provided. Bittencourt et al. (2019) presented total mass and energy balances for sugarcane bagasse treatment to produce bioethanol and biogas in the biorefinery concept. Kaparaju et al. (2009) pointed out only mass balance of wheat straw treatment to bioethanol, biohydrogen and biogas in a biorefinery concept. Thus, the complex techno-economic studies of biogas biorefineries are generally missing.

Kratky et al. (2018) presented the techno-economic study on lignocellulosic waste treatment in biogas biorefinery, see Fig. 3.4. Waste rich in cellulosic fibres is continuously fed to a knife mill, reducing biomass in particle size. Pre-milled biomass is

Table 3.3 The representative characteristics of selected biogas biorefineries

Raw material	Primary processing	Secondary processing	Products	Specific yields to raw material	References
Afalfa	Pressing, fermentation of green juice	Anaerobic fermentation	Protein liquere	0.008 kg kg^{-1}	Santamaría-Fernández et al. (2018)
			Biomethane	0.707 m^3 kg^{-1} VS	
Casawa waste	Extraction, bioethanol production	Anaerobic fermentation	Glucose syrup	0.025 kg kg^{-1}	Padi and Chimphango (2020)
			Bioethanol	0.015 kg kg^{-1}	
			Electricity	643 Wh kg^{-1}	
Cassava waste	Extraction, bioethanol production	Anaerobic fermentation	Succinic acid	0.018 kg kg^{-1}	Padi and Chimphango 2020)
			Bioethanol	0.015 kg kg^{-1}	
			Electric power	626 Wh kg^{-1}	
Castor plant	Oil extraction transesterification	Anaerobic fermentation	Biodiesel	0.061 kg kg^{-1}	Rahimi and Shafiei (2019)
			Biomethane	0.162 m^3 kg^{-1} VS	
			Glycerine	0.016 kg kg^{-1}	
			Heat	6.71 MJ kg^{-1}	
			Salt	0.003 kg kg^{-1}	
Castor plant	Oil extraction transesterification	Anaerobic fermentation	Biodiesel	0.155 kg kg^{-1}	Bateni et al. (2014)
			Biomethane	0.496 m^3 kg^{-1} VS	
Clover grass	Pressing, fermentation of green juice	Anaerobic fermentation	Protein liquere	0.004 kg kg^{-1}	Santamaría-Fernández et al. (2018)
Eruca sativa plant	Oil extraction transesterification	Anaerobic fermentation	Biodiesel	0.061 kg kg^{-1}	Rahimi and Shafiei (2019)
			Biomethane	0.161 kg kg^{-1}	
			Glycerine	0.010 kg kg^{-1}	
			Heat	6.95 MJ kg^{-1}	
			Salt	0.015 kg kg^{-1}	
Marigold	Extraction of valuables	Anaerobic fermentation	Vanillin	5.09 wt % of lignin	Giraldo-Poveda and Cardona Alzate (2021)
			Vanillic acids	1.16 wt % of lignin	
			Biogas	0.162 m^3 kg^{-1} VS	
Olive pomace	Extraction	Anaerobic fermentation	Polyphenols	0.036 kg kg^{-1}	Orive et al. (2021)
			Biomethane	0.149 kg kg^{-1}	

(continued)

Table 3.3 (continued)

Raw material	Primary processing	Secondary processing	Products	Specific yields to raw material	References
Sugarcane bagasse	Hydrothermal pretreatment, bioethanol production	Anaerobic fermentation	Bioethanol	0.150 kg kg^{-1}	Bittencourt et al. (2019)
			Biomethane	0.028 m^3 kg^{-1}	
Wheat straw	Hydrothermal pretreatment, bioethanol production	Anaerobic fermentation	Bioethanol	0.133 kg kg^{-1}	Kaparaju et al. (2009)
			Biogas	0.426 m^3 kg^{-1} VS	

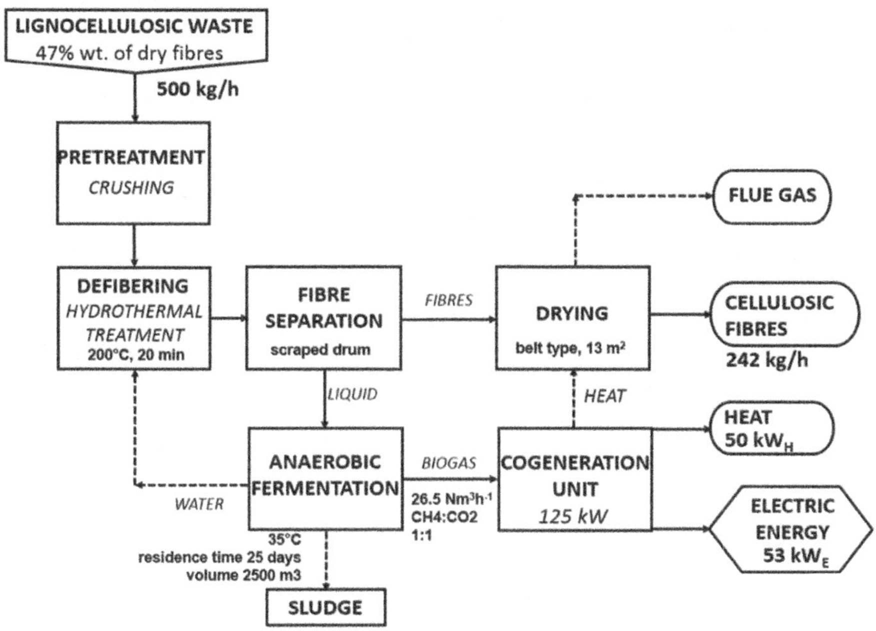

Fig. 3.4 The block scheme of proposed biogas biorefinery (Kratky et al. 2018)

mixed with the hot water in the proportions solid to liquid as 1:10 w/w. The prepared suspension is fed to hydrolyser, which undergoes hydrothermal pretreatment with rapid batch decompression. The batch is indirectly heated to a process temperature of 200 °C, at which the batch is maintained for 20 min. When this residence time elapses, a pressurized batch is rapidly decompressed into an atmospheric storage tank, leading to the solid-liquid separator. The separated cellulosic fibres continue to belt dryer, in which the material becomes dried. The flue gas indirectly preheats drying air from the combustion of biogas in the cogeneration system. The parallel produced electric energy and heat are re-used as primary energy resources in the biorefinery, and they are fully recovered. The main product of the biorefinery is cellulose fibres potentially to be used as insulation materials, electrotechnics, pharmacy, cosmetics. In addition, other fibre applications can meet the demand for biodegradable reinforcement of bio-composites. Side products include residual heat, wastewater sludge, waste vapour from the dryer, and flue gas from the cogeneration unit.

Regarding the technical maturity of the proposed biogas biorefinery, it can be concluded that all the technical steps are available at TRL 9, i.e., industrially viable and verified technologies. Regarding the economic assessment presented by Kratky et al. (2018), the total investment cost of 3.4 million Eur and operating expenses of 3.8 million Eur were identified for such a biorefinery proposal treating referenced 500 kg h^{-1} of waste wood residues. Considering the model selling cost of cellulosic fibres of 2.2 Eur kg^{-1}, the sale income of 4.1 million Eur, taxable profit of 0.3 million Eur $year^{-1}$ and simple payback time of 22 years were reached. Nevertheless, industrial technology usually targets payback time no more than 8–10 years with its minimum lifetime of 20 years. Thus, the sensitivity analysis was done to characterize risk factors crucially influencing the proposed biorefinery cash flow.

The crucial economic risk factors are specific capital investments CAPEX, specific operating expenses OPEX, and cellulosic fibres' selling price. Both specific investment costs are published related to the hourly biomass flow rate. Regarding Fig. 3.5, it is evident that simple payback time sharply increases with specific investment costs. The CAPEX perceptual proportions of the biorefinery proposal were distributed as 35% in pretreatment with fibre separation, 47% in fermentation, 7% in drying, and 11% in accessories. Hydrothermal defibering technology counts several pressure apparatuses, drum screen sieves, heat exchangers, pumps and other handling equipment. Anaerobic fermentation deals with installing large-volume fermentation tanks

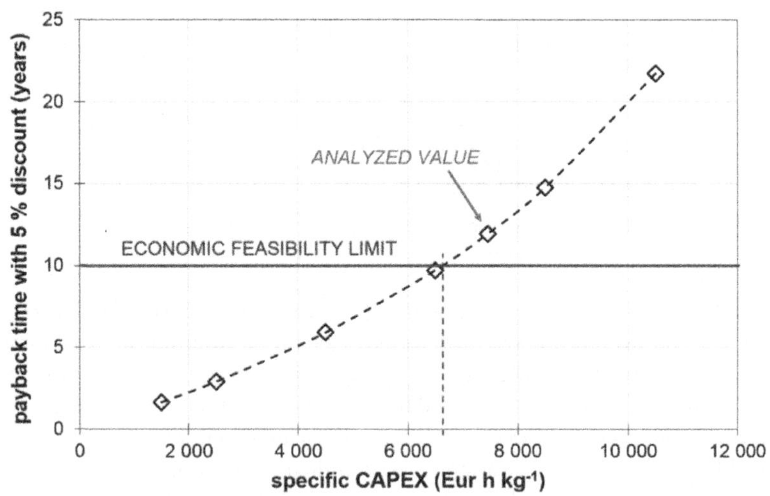

Fig. 3.5 The discounted payback time dependent on the specific capital investment cost (analyzed value represents the cost estimation counting the size parameters of the biorefinery)

due to the need for 25 days in residence time for efficient microbial degrading of feed. Regarding decreasing capital investment cost, the specific value can be decreased by intensive biomass pretreatment in innovative reactors and process optimization regarding the decrease of digestion time during anaerobic fermentation of residues. Figure 3.6 depicts the simple payback time dependent on operating cost. It is evident that payback time sharply increases with increasing specific investment cost, from its value 6,000 Eur h kg^{-1} especially. OPEX values are affected by 41% in energy cost and 42% in distributional costs associated with raw material transportation counting 20 Eur t^{-1} of biomass (Orive et al. 2021). Energy cost covers hydrothermal pretreatment and reaches batch temperature 170–200 °C indirectly heated by a generated steam boiler. A lower process temperature is essential to reduce energy costs. Transporting costs can be reduced by the installation of the biorefinery closed to wood processing industries. The biorefinery proposal considered that wood residues must be bought as raw material. Its model value was set to 0.04 Eur kg^{-1} as the industrial standard for wood chips. Nevertheless, as presented by Fig. 3.7, the purchase price of wooden chips has also not negligible effect on the economic feasibility of the proposal. The lower the purchase price, the decreasing simple payback time. If the purchase cost of wood residues is zero, then simple payback time is reduced from 11.96 to 8.07 years, i.e. by 1.48 times lower. The waste processing cost is always paid for conventional technologies like landfilling or combustion. Suppose the biorefinery proposal is defined as a waste treatment technology. In that case, its price is equal to zero, and a biorefinery operator can reach income from waste handling resulting from improving the economic attractivity of the proposal. The fibre selling price also has a crucial effect on the economic attractiveness of the biorefinery. The general dependence of payback time on the selling price is depicted in Fig. 3.8.

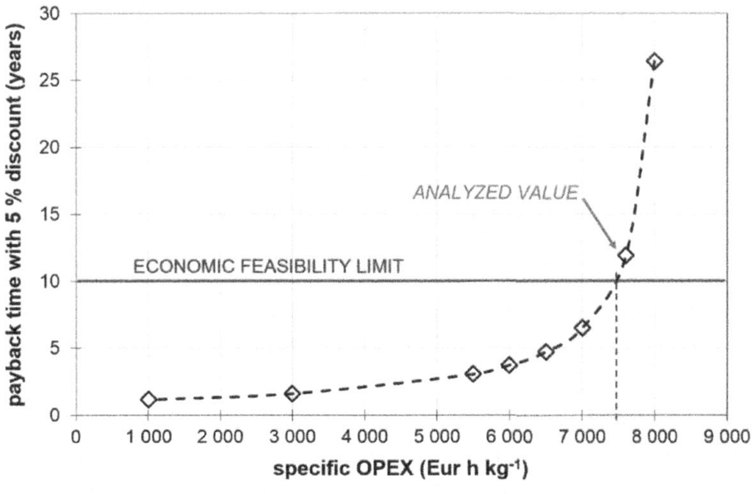

Fig. 3.6 The discounted payback time dependent on the specific operating investment cost (analyzed value represents the cost estimation counting the size parameters of the biorefinery)

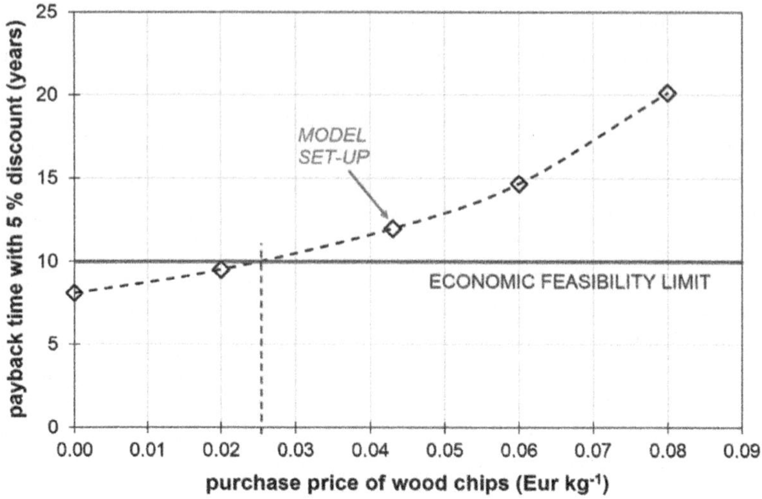

Fig. 3.7 The discounted payback time dependent on the purchase price of wooden chips (analyzed value represents the cost estimation counting the size parameters of the biorefinery)

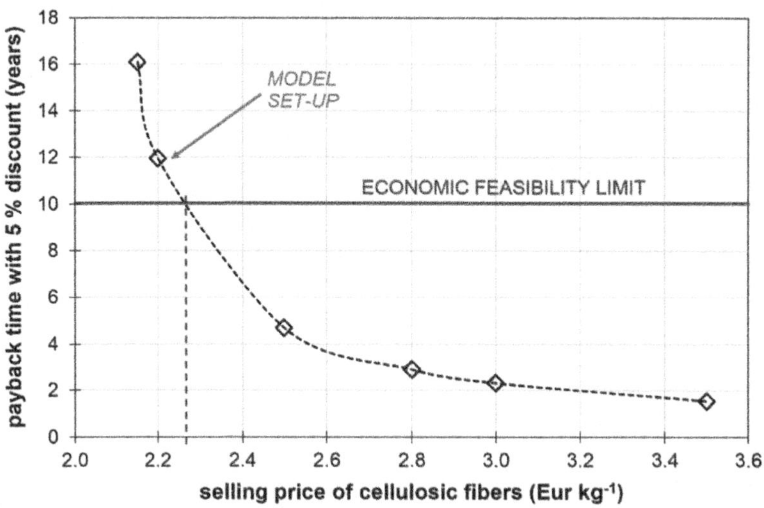

Fig. 3.8 The discounted payback time dependent on the selling price of cellulosic fibres (analyzed value represents the cost estimation counting the size parameters of the biorefinery)

A simple payback time of 22.3 years was identified for this proposal. It means that the technology seems to be at the end of its lifetime because the maximum lifetime of 20–30 years is generally expected for food, chemical or waste processing technologies and biogas plants included. The second note is associated with the selling prices of cellulosic fibres. The maximum selling price of cellulosic fibres

2.2 Eur kg^{-1} ensures a conventional simple payback time of 8 years. Nevertheless, it must be noted that the estimated selling price of cellulosic fibres is an order of magnitude higher than the purchase price of wood residues being around 0.04 Eur kg^{-1}.

Finally stated, Kratky et al. (2018) concluded that critical factors that strongly affect the techno-economic potential of the biorefinery are dominantly associated with the pretreatment and price of cellulosic fibre. Regarding other complex techno-economic studies, Meramo-Hurtado et al. (2021) also found that selling prices of products and raw materials are the crucial factors affecting the economic feasibility of acetone-butanol-ethanol biorefinery. Orive et al. (2021) reported a good profitability and payback time for olive pomace valorization due to the expected selling price of hydroxytyrosol extracts. Nevertheless, its selling price was also found as the crucial risk factor. Padi et al. (2020) present primary product prices (biomethane, bioethanol, glucose syrup, succinic acid), enzyme cost and electricity price as dominant risk factors in their waste biorefinery for casava industries. Rahimi and Shafiei (2019) found seed cost, catalysts, solid residuals and energy for anaerobic fermentation as the crucial risk factors affecting the overall economic balance of biodiesel-biogas-heat based biorefinery. Thus, it is evident that favourable operating costs and product prices could reach biogas biorefinery's economic feasibility and attractivity. Therefore, scientific and research needs should focus on the problems and improve the efficiency of pretreatment technology to reduce investment and operating costs.

3.2.3 Bioethanol Biorefinery

Regarding Table 3.1, bioethanol biorefinery belongs among one of the dominant multistage lignocellulosic waste treatment approaches. Its general block diagram is expressed in Fig. 3.9. The crucial technological steps are mechanical pretreatment, chemical pretreatment, separation and purification of primary products, enzymatic hydrolysis of residue with subsequent alcoholic fermentation, secondary and residual product processing. Lignocellulosic biomass is initially reduced to destroy the lignocellulosic matrix and open cellulosic fibres, increasing the biomass-specific surface to intensify transfer phenomena. The milled raw material is usually mixed with water, and such a batch undergoes pretreatment. Hydrothermal, alkaline, acid, or organosolv pretreatment techniques are applied to allow efficient primary product removal and increase biomass residues' biodegradability. Primary products are extracted, separated and purified after pretreatment. Typical primary products are acetic acid, furfural, xylitol, xylulose, succinic acid, vinasse, phenols, pectin, and silica (Demichelis et al. 2020; Hossain et al. 2019; Morales-Rodriguez et al. 2016; Kuglarz et al. 2016; Gnansounou and Dauriat 2010; Alio et al. 2021; Song et al. 2020; Talekar et al. 2018; Yuan et al. 2018). Biomass residues after primary product extraction are fed into a bioreactor in which undergo enzymatic hydrolysis (converting cellulose to glucose) and alcoholic fermentation (glucose conversion

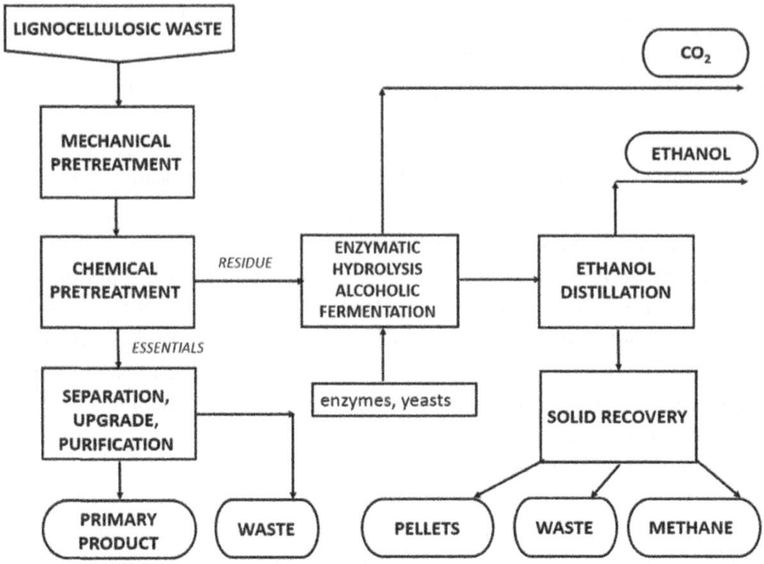

Fig. 3.9 The general block diagram of bioethanol-based biorefinery

to bioethanol and carbon dioxide). Bioethanol is separated of aqueous fermentation batch by sequential distillation followed by adsorption of water residue via molecular sieves to reach commercially demanded bioethanol purity over 99 vol %. The solid-liquid residue after alcoholic fermentation is usually thickened at the set of evaporators and dried to be formed into pellets. Dried distillers solubles are typical by-products that are used as decentralized energy source. Combined heat and power generation from combustion of dried distillers solubles is nowadays viewed as a processing pathway to reach self-sufficient energy biorefinery, profit electricity selling, and decrease waste generation significantly. Dried distillers solubles were also recommended as co-digestion raw material for anaerobic fermentation to biogas. Its subsequent upgrade to biomethane as a targeted renewable energy source could meet biorefinery concepts and improve process economics.

The base literature search was carried out to identify the base directions in bioethanol biorefineries, see Table 3.4. Demichelis et al. (2020) performed deeper techno-economic and LCA analysis for several bioethanol production technologies from sugar cane, manure, rice straw or biodegradable municipal solid waste. Hossain et al. (2019) serve the techno-economic study of bioethanol and furfural simultaneous production by two approaches, i.e. biochemically and thermochemically. The feasibility study of bioethanol and xylitol coproduction from corn stover is presented by Morales-Rodrigues et al. (2016). Alio et al. (2021) simulated operational costs for sawdust-based biorefinery producing ethanol, lignin and electricity. Yuan et al. (2018) present experimental results of pretreatment by an organic solvent to isolated substances of wheat straw in the biorefinery concept. Talekar et al. (2018) experimentally incorporated hydrothermal pretreatment for waste pomegranate peels to recover

Table 3.4 The representative characteristics of selected bioethanol biorefineries

Raw material	Primary processing	Secondary processing	Products	Specific yields to raw material	References
Cattle manure	Enzymatic hydrolysis	Bioethanol fermentation	Acetic acid	0.021 kg kg^{-1}	Demichelis et al. (2020)
			Bioethanol	0.036 kg kg^{-1}	
Corn stover	Acid pretreatment	Bioethanol fermentation	Furfural	0.012 kg kg^{-1}	Hossain et al. (2019)
			Bioethanol	0.034 kg kg^{-1}	
Corn stover	Acid pretreatment	Bioethanol fermentation	Xylitol	0.006 kg kg^{-1}	Morales-Rodriguez et al. (2016)
			Bioethanol	0.124 kg kg^{-1}	
Industrial hemp	Alkaline oxidation	Bioethanol fermentation	Succinic acid	0.123 kg kg^{-1}	Kuglarz et al. (2016)
			Bioethanol	0.159 kg kg^{-1}	
Hardwood chips	Acid pretreatment	Bioethanol fermentation	Furfural	0.091 kg kg^{-1}	Gnansounou and Dauriat (2010)
			Bioethanol	0.146 kg kg^{-1}	
			Electricity	0.298 kWh kg^{-1}	
Rice straw	Enzymatic hydrolysis	Bioethanol fermentation	Bioethanol	0.078 kg kg^{-1}	Demichelis et al. (2020)
			Biomethane	17.940 Nm^3 t^{-1}	
Saw dust	Organosolv	Bioethanol fermentation	Lignin	0.094 kg kg^{-1}	Alio et al. (2021)
			Bioethanol	0.166 kg kg^{-1}	
			Electricity	0.035 kWh kg^{-1}	
Sugarcane	Enzymatic hydrolysis	Bioethanol fermentation	Vinasse	0.079 kg kg^{-1}	Demichelis et al. (2020)
			Bioethanol	0.166 kg kg^{-1}	
			Biomethane	0.635 Nm3 t^{-1}	
Waste bamboo	Acid pretreatment	Bioethanol fermentation	Xylitol	0.089 kg kg^{-1}	Song et al. (2020)
			Xylulose	0.002 kg kg^{-1}	
			Bioethanol	0.121 kg kg^{-1}	
Waste pomegranate peels	Hydrothermal processing	Bioethanol fermentation	Phenolics	0.115 kg kg^{-1}	Talekar et al. (2018)
			Pectin	0.208 kg kg^{-1}	
			Bioethanol	0.080 kg kg^{-1}	
Wheat straw	Organosolv, alkaline	Bioethanol fermentation	Silica	0.049 kg kg^{-1}	Yuan et al. (2018)
			Lignin	0.120 kg kg^{-1}	
			Bioethanol	0.274 kg kg^{-1}	

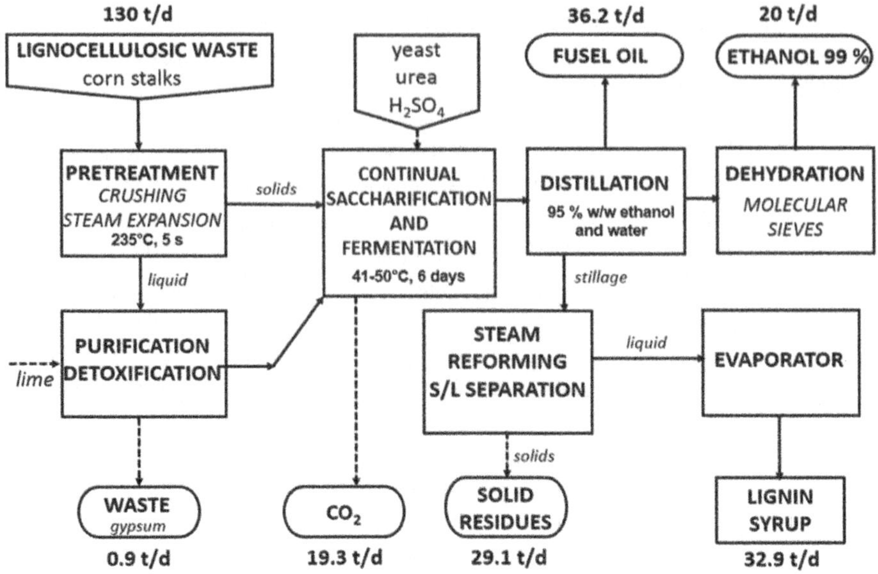

Fig. 3.10 The block scheme of the proposed bioethanol-based biorefinery (Krátký et al. 2018)

phenolics and pectin with bioethanol production. Pratto et al. (2020) present an excellent bioethanol yield for sugar cane straw pretreated using the hydrothermal technique. All the mentioned studies are based on the assumption that a minimum selling price of lignocellulosic ethanol ensures biorefinery payback no higher than 20 years. Sensitivity analysis in terms of capital investments or operating costs concerning market demand, selling prices are superficially analyzed.

The economic feasibility and sensitivity analysis of bioethanol biorefinery was discussed by Seghman and and Krátký (2017) or Krátký et al. 2018), and their results are discussed below. The corn stalks treatment in the biorefinery concept is based on the following ideas, see Fig. 3.10. Biomass is usually rich in impurities like sand, metal, rocks or minerals that must be removed before its treatment by proper washing and sorting technologies to avoid contamination of the whole technology. Then, corn stalks are crushed to the desired particle size for subsequent processing in pretreatment and fermentation sections. Comminuted biomass is handled to the pretreatment section, which undergoes steam expansion pretreatment. The saturated steam of 235 °C is directly fed to biomass. Such an application deals with the solubilization of hemicellulose and lignin portions opens cellulosic fibres. The residence time being several seconds is usually held during such biomass processing. As residence time is over, the batch undergoes rapid batch decompression. Undissolved solid proportion is separated from the liquid. The solid part is handled for its treatment in fermentation tanks. The liquid part is usual acid and toxic. Thus, it is fed into neutralizing reactors for its chemical stabilization by dosing a lime. The stabilised liquid suspensions run through a filter press to hydrolysis tanks mixed with solid parts, yeast, and enzymes. The simultaneous saccharification and fermentation approach, known as the SSF process, is used in the bioethanol biorefinery. The process temperature of 41–65 °C is arranged in fermentation tanks depending on the microbial degrading process phase. As the needed digestion time is over, the fermentation batch runs to a hooper, pumped to the operating set of product separation, cleaning, and purification. The distillation column provides the first stage of bioethanol removal. Vapours, majority rich in ethanol and water, run are condensed in a condenser. The condensed intermediate product is redistilled by the rectifying column that produces ethanol with the azeotropic purity of 95% w/w (ethanol/water) and fusel oil (higher alcohols, organic substances). Molecular sieves provide ethanol purification to 99% w/w. The solid–liquid residue being generated by the first distillation column is thickened by centrifuge. The removed solids are usually used as feed for animals or dried to form pellets as energy sources.

Nowadays, conventional bioethanol plants are old and technically very sophisticated technologies at the highest technological readiness level TRL 9. Starch-based raw materials (corn, cereals, potatoes, sugar beet) evince high biodegradability, close to 90%, without any need for complicated pretreatment of raw materials. The lignocellulosic bioethanol plant differs from the conventional one just in unique pretreatment technologies fitted to treated material. The techno-economic study and sensitivity analysis were performed for the presented lignocellulosic bioethanol biorefinery (Seghman and Krátký 2017; Krátký et al. 2018). The total investment cost of 8,620,000 Eur and the total operating cost of 9,190,000 Eur year^{-1} were identified for the proposal that treats 141 t d^{-1} of corn stalks. The following purchase costs were applied in the biorefinery model—the purchase cost of 25.9 Eur t^{-1} corn silage. In addition, the cost of 148 Eur kg^{-1} yeast and for cellulitic enzymes of 450 Eur t^{-1} were set. Finally, the purchase prices of 1600 Eur t$^{-1\ \mathrm{of}}$ sulphuric acid, 148 Eur t^{-1} of lime, and 240 Eur t^{-1} of urea were used. Considering selling prices for bioethanol of 1150 Eur t^{-1} (99% w/w), 370 Eur t^{-1} of lignin syrup and 110 Eur t^{-1} for solid residues, a taxable income of 2.37 million Eur per year was achieved. A simple payback time of 7.95 was identified for the proposed biorefinery. It was already mentioned that industrial technology usually targets payback time no more than 8–10 years with its minimum lifetime of 20 years. Thus, the modelled payback time makes the biorefinery cost attractive. Nevertheless, selling prices are strongly affected by industrial demand and green subsidies. The sensitivity analysis was performed to identify crucial risk factors that can significantly influence the cash flow of the proposed lignocellulosic bioethanol biorefinery.

Capital investments, operating expenses, energy cost, corn stalk cost, the selling price of bioethanol and the selling price of lignin syrup were identified as the crucial risk factors that dominantly affect the overall economic feasibility of biorefinery proposal. Figure 3.11 depicts discounted payback period depends on the specific capital investments defined as the daily biomass flow rate cost. The capital investments are generally distributed by 23% for pretreatment, 15% for enzymatic hydrolysis and alcoholic fermentation, and 71% for ethanol recovery and residue treatment technologies. The global policy to design innovative reactors and bioreactors that allow intensive treatment of highly concentrated suspensions can meet the

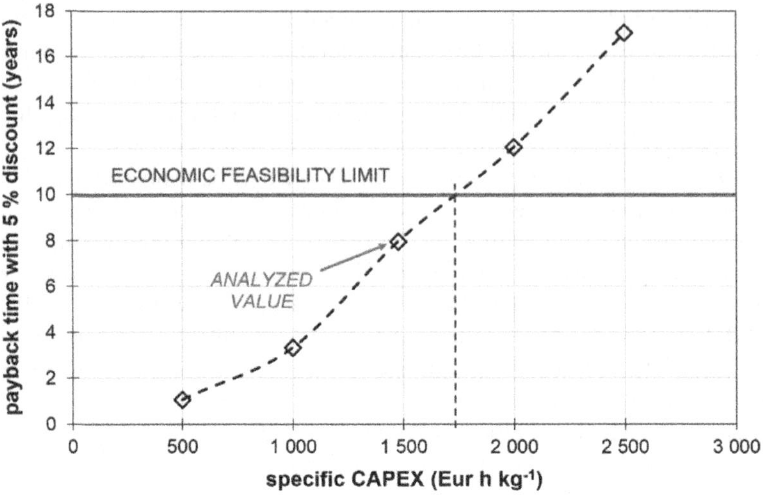

Fig. 3.11 The discounted payback time dependent on the specific capital investment cost

demand to reduce specific investment costs. Figure 3.12 shows that the discounted payback period depends on the specific operating investments related to biomass flow rate. Reducing operating costs, the lower payback period is reached. It was found that energy cost represents 85% of specific operating costs. Energy cost covers the production cost of steam, electricity purchase and cooling power systems.

Energy price policy and decentralized heat recovery can stimulate lower energy costs resulting in a decrease of CAPEX under the feasibility level, as presented by

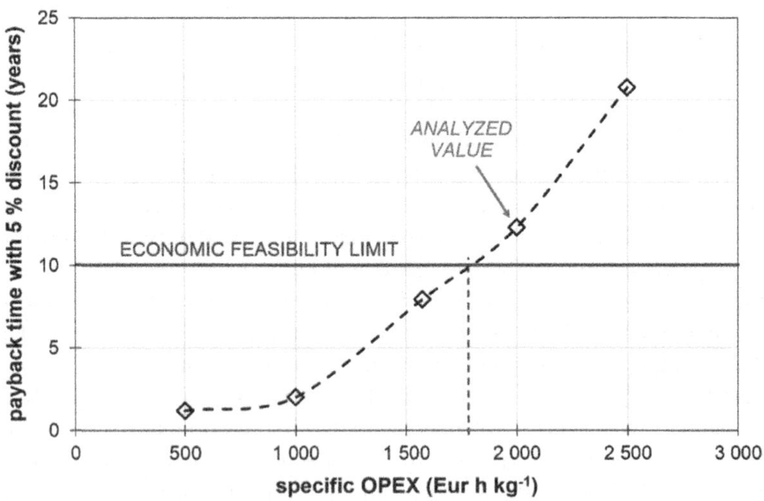

Fig. 3.12 The discounted payback time dependent on the specific operating investment cost

Fig. 3.13. Purchase prices of raw materials and selling prices of products also significantly affect the process economics. Figure 3.14 shows the payback time dependent on the purchase price of corn silage. The biorefinery proposal considered that corn silage must be bought as raw material. The lower the purchase price, the decreasing simple payback time. If the purchase cost of corn silage is zero, then simple payback time is reduced from 7.95 to 3.91 years, i.e. by two times lower. It was mentioned

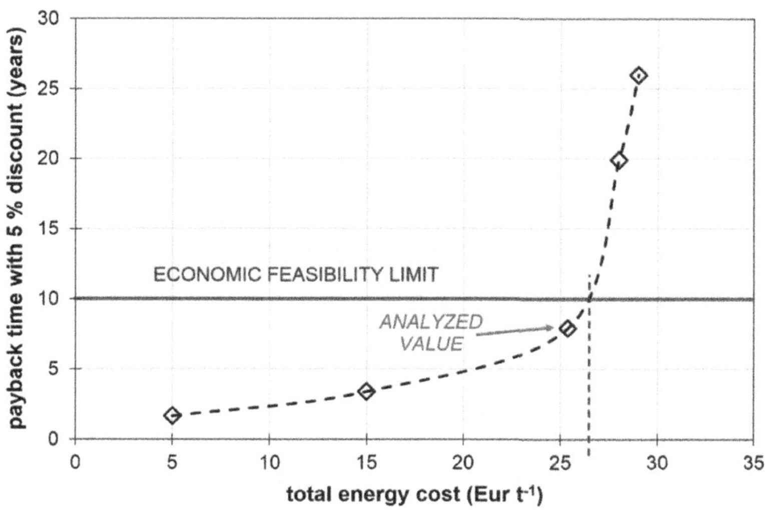

Fig. 3.13 The discounted payback time dependent on the total energy cost

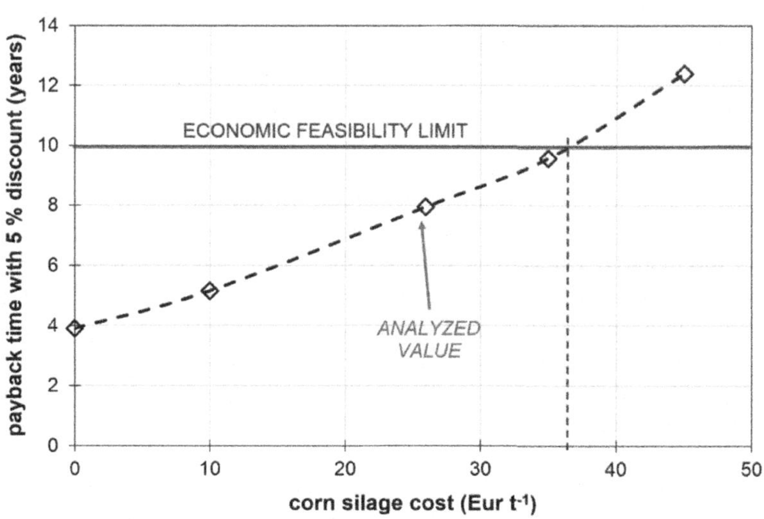

Fig. 3.14 The discounted payback time dependent on the corn silage cost

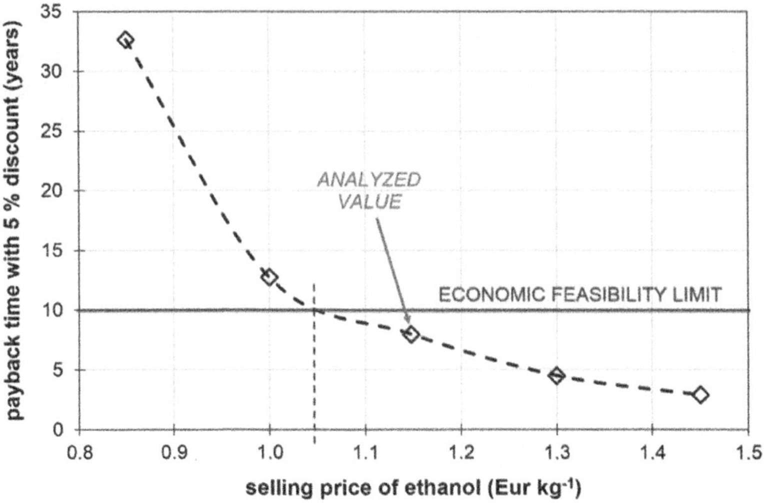

Fig. 3.15 The discounted payback time dependent on the selling price of bioethanol

above that the waste processing cost is always paid for conventional technologies like landfilling or combustion. Supposing the proposal as waste treatment technology, the waste price is equal to zero. Therefore, a biorefinery operator can reach income from waste handling resulting from improving the economic feasibility of the biorefinery.

A sensitivity analysis revealed that selling prices of bioethanol and lignin syrup are the dominant risk parameters affecting the biorefinery's economic attractiveness. Regarding Fig. 3.15, it is evident that the simple payback time sharply increases for the selling price of ethanol being 1000 Eur t^{-1}. Generally known, the ethanol price around 1000 Eur t^{-1} corresponds to the subsidized ethanol price. The standard commercial price of ethanol being 330 Eur t^{-1} can be applied for the economic balance of the proposal. Then it is evident that the proposal losses income, and its simple payback period is closed to its lifetime. Finally, the selling price of ethanol can not be higher than 1100 Eur t^{-1} to keep the biorefinery proposal's simple payback time under nine years. These results precisely fit the finding of Cheng et al. (2019), reporting production cost 1011 \$ m^{-3} and minimum selling price 1110 \$ m^{-3} for cellulosic ethanol to reach a payback time of 9 years for their proposal of lignocellulosic bioethanol biorefinery. Lignin sirup was also identified as the critical parameter that was a significant effect on real economics. Advantageously, its production is roughly 1.5times higher compared to ethanol. Nevertheless, there is the problem of estimating the price of lignin sirup due to its purity. The biorefinery model only shows economic attractivity if selling prices more than 340 Eur t^{-1} are reached, see Fig. 3.16.

Some pilot or industrial lignocellulosic bioethanol biorefinery demonstrators were built as overviewed by Kos and Kratky (2020), see Table 3.5. The lignocellulosic bioethanol biorefinery is limited in technical maturity, pretreatment section, and the SSF process (Seghman and Krátký 2017; Krátký et al. 2018). Nevertheless, the

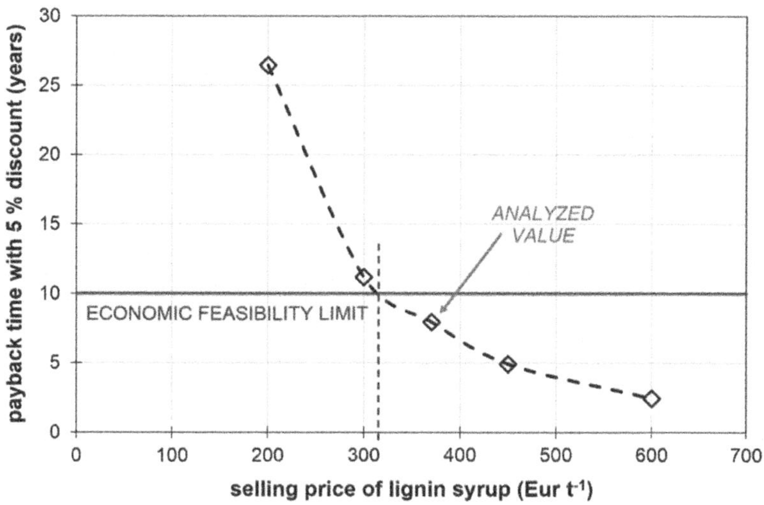

Fig. 3.16 The discounted payback time dependent on the selling price of lignin syrup

presented study showed that the proposed bioethanol-based biorefinery could reach industrial and economic attractivity. Nevertheless, the intensive research and development activities must scope pretreatment techniques, design intensified reactors and bioreactors, and correctly set up and stabilize the SSF process. All these research and development activities tend to improve biomass biodegradability, improve extraction efficiency of the main product, reduce operating expenses by heat recovery in the system, and use base chemical compounds. A similar finding is presented by Zhao et al. (2015). The author's present cellulose-glucose conversion rate, sugar to ethanol production yield, biomass purchase price, and capital investments, affecting the economic feasibility of lignocellulosic ethanol production technologies.

3.3 CO_2 Emission-Free Lignocellulosic Waste Treatment in Biorefinery Concept

Both presented biogas and bioethanol biorefineries produce CO_2 as a secondary product that can be unused and directly emitted to the atmosphere. It is known that CO_2 belongs among greenhouse gases dominantly affecting global warming. Regarding the effort in environmental policies and restrictions, the significant effort presses to reduce CO_2 emissions as proclaimed by the Paris Agreement (2015), Declaration of Katowicze (2018), or The European Green Deal (2019) that aims to reach climate neutrality in 2050. E.g. EU-27 countries reported annual emissions of 3.9×10^9 t CO_2 in 2018 (Chempolis 2021) that resulted in a payment of emission allowances, known as "money to air", equal to 87.5×10^9 Eur annually (respecting

Table 3.5 The overview of lignocellulosic bioethanol biorefineries

Project	Raw material	Bioethanol productivity 10^6 L year^{-1}	Other products	Year of launch×	References
Fuyang Project	Wheat straw, corn stalks	63.0	–	–	Global Database of Biomass Conversion Facilities (2021)
AustroCel Hallein	Wood residues	38.0	–	2020	Global Database of Biomass Conversion Facilities (2021)
ChemCell Ethanol	Wood residues	20.0	Vanilline, cellulose	1938	Global Database of Biomass Conversion Facilities (2021)
Clariant	Agriculture wastes	63.0	–	2020	Clariant Sunliquid Cellulosic Ethanol Technology (2021)
Clariant: Grace Project	Energy crops	1.3	–	2012	Clariant Sunliquid Cellulosic Ethanol Technology (2021)
Domsjö Fabriker AB	Wood residues	17.0	Cellulose, lignin, energy	1940	Domsjö Pulp Mill (2021)
Fiberight LLC: Commercial Plant	BMSW, food wastes	23.0	–	2019	Global Database of Biomass Conversion Facilities (2021)
Fiberight LLC: Integrated Demonstration Plant	BMSW, food wastes	0.004	–	2012	Global Database of Biomass Conversion Facilities (2021)
Chempolis	Wood wastes	6.3	–	2008	Global Database of Biomass Conversion Facilities (2021)
IFP: Futurol	Agriculture waste	0.4	Non specified chemicals	2016	Cellulosic Ethanol (2021)

(continued)

Table 3.5 (continued)

Project	Raw material	Bioethanol productivity 10^6 L year^{-1}	Other products	Year of launch×	References
Iogen Corp	Agriculture waste	2.0	–	2004	Global Database of Biomass Conversion Facilities (2021)
Iogen Corp. & Raízen: Costa Pinto Project	Agriculture waste	40.0	–	2015	Global Database of Biomass Conversion Facilities 2021)
Kanteleen Voima	Wood residues	82.0	Biometane, lignin, fertilizer	2021	Global Database of Biomass Conversion Facilities 2021)
North European Bio Tech	BMSW, food wastes	5.0	–	2015	Cellunolix (2021)
North European Bio Tech	Wood residues	10.0	–	2017	Cellunolix (2021)
Nordfuel	Wood residues	–	Lignin, fertilizer	2021	Nordfuel (2021)
NRL + Fortum	Energy crops	60.0	Furfural, acetic acid, electricity	–	Chempolis (2021)
ORLEN Poludnie: Jedlicze Site	Agriculture waste	32.0	–	–	Global Database of Biomass Conversion Facilities (2021)
POET-DSM Advanced Biofuels: Liberty	Agriculture waste	76.0	–	2014	Global Database of Biomass Conversion Facilities (2021)
Procethol 2G: Futurol	Wood residues	3.4	–	2011	Global Database of Biomass Conversion Facilities (2021)
PTT-RTI	Agriculture waste	0.002	Melasa, lignin	2014	Yuan et al. (2018)
Shandong Zesheng Biotech Co.	Agriculture waste	3.8	–	2006	Yuan et al. (2018)
St1: Bionolix Hameenlinna	BMSW, food wastes	1.0	Biogas	2010	Domsjö Pulp Mill (2021)

(continued)

Table 3.5 (continued)

Project	Raw material	Bioethanol productivity 10^6 L year^{-1}	Other products	Year of launch×	References
Cellunolix Follum	Wood residues	50.0	Terpentyne, vinasse, lignin, furfural, biogas	2021	Domsjö Pulp Mill (2021)
Cellunolix Kajaani 2	Wood residues	50.0	Terpentyne, vinasse, lignin, furfural, biogas	2024	Domsjö Pulp Mill (2021)
Cellunolix Pietarsaari	Wood residues	50.0	Terpentyne, vinasse, lignin, furfural, biogas	2024	Domsjö Pulp Mill (2021)
Etanolix Hamina	BMSW, food wastes	1.3	Liquid feed	2008	Domsjö Pulp Mill (2021)
Etanolix Jokioinen	BMSW, food wastes	8.9	Liquid feed	2011	Domsjö Pulp Mill (2021)
Etanolix Lahti	BMSW, food wastes	1.3	Liquid feed	2009	Domsjö Pulp Mill (2021)
Etanolix Vantaa	BMSW, food wastes	1.3	Liquid feed	2009	Domsjö Pulp Mill (2021)
Versalis: Crescentino restart	Agriculture waste	51.0	Liquid feed	2020	Cellulosic Ethanol (2021)
Woodland Biofuels	Wood residues	0.08	–	2011	Cellunolix (2021)
Woodland Biofuels	Wood residues	0.76	–	2013	Cellunolix (2021)

price 25 Eur t^{-1} CO_2). Naims (ST1 2021) overviewed that fermentation of biomass produced annually 18 × 10^6 t CO_2 in 2014. Although this value represents only 0.11% of global CO_2 emissions, CO_2 recovered from fermentation reactors poses a low-cost feedstock with high purity of 95–99% and the lowest capture cost of 10–30 Eur t^{-1} CO_2.

Carbon Capture and Utilization technologies pose a promising potential of CO_2 emissions treatment to meet the Green Deal plan and overcome CO_2 emission allowances as "money to air". Regarding the laboratory results and studies, chemical or biological technologies can transform CO_2 into various advanced biofuels (biodiesel, bioethanol, methane) or valuable chemicals (formic acid, lipids, carotenoids, pigments, supplementary food products), see Fig. 3.17. These CO_2 based

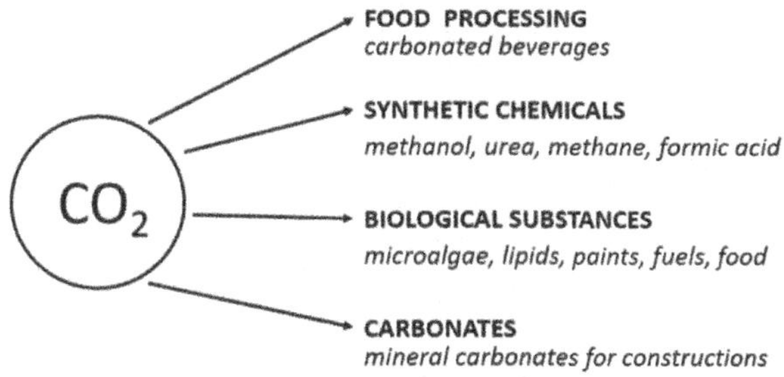

Fig. 3.17 The promising pathways of CO_2 utilization

products are nowadays viewed as an alternative to petrochemical products, and they can become the primary product of CO_2 free technologies, CO_2 free biorefineries. Thus, CO_2 use as a raw material produced by lignocellulosic biorefineries can help to improve product portfolio and improve the economic feasibility of biorefinery proposals.

CO_2 as raw material can be converted by thermochemical or biochemical reaction pathways into a portfolio of CO2based products, as presented by Fig. 3.17. ICEF (2021) overviewed that one ton of CO_2 can thermochemically react with hydrogen and other reactants to ideally form 0.36 t of methane, 0.73 t of methanol, 0.015 t of urea, 0.33 t of octans, or 0.33 t of inorganic carbonates. One ton of CO_2 can also biochemically form 0.546 t of microalgal biomass as an intermediate product to make hydrogen, bioethanol, methane, colouring substances, antioxidants, amino acids, steroids, essential oils, or powder (Aresta and Dibenedetto 2010). Naim (2016) and Styring et al. (Capture and Utilization in the Green Economy 2021) reported selling price 20 Eur kg^{-1} for microalgal feed additive, 50 Eur kg^{-1} fo microalgal food supplement, 625 Eur kg^{-1} of microalgal oil, or 2750 Eur kg^{-1} for microalgal beta-carotene. Therefore, the future of lignocellulosic biorefineries is viewed in mutual material-energy recycling of lignocellulosic wastes that contribute to decreased CO_2 emissions or that can meet the demand of CO2free waste processing technology. These approaches are intensely discussed in the following chapter introducing CO_2 treatment in biogas biorefineries.

3.3.1 CO_2 Utilization in Biogas Biorefinery

The presented biogas biorefinery in Sect. 2.1 can be viewed as CO_2 emitting technology. Kratky et al. (2018) performed the innovative case study, in which CO_2 balance is compared for wood chips combustion, wood chips fermentation to produce biogas with its subsequent combustion in CHP unit, and wood chips treatment in a biorefinery concept. If wooden chips are directly combusted, 1080 kg CO_2 t^{-1} of wood chips is produced. If wooden chips are anaerobically fermented, a biogas yield of 220 Nm^3 t^{-1} is expected. The CO_2 production 110 Nm^3 t^{-1} is reached if the volumetric ratio methane to carbon dioxide 1:1 is supposed as the typical one for polysaccharide substrates. The biogas biorefinery presented that 47% of cellulosic fibres from wooden chips are removed as a primary material. Thus 53% of wooden residues were anaerobically fermented to produce CO_2 specific yield of 58 $Nm^3\ t^{-1}$. CO_2 production by biogas biorefinery is two times lower than a conventional biogas plant and 20 times lower than direct combustion of CO_2. It means that biogas biorefineries can re-use lignocellulosic waste from mutual material and energy points of view and help reduce CO_2 emissions during waste treatment.

Another option is to take CO_2 as feedstock to form advanced biofuels or biochemicals. Several studies were published on CO_2 to X conversion for biogas biorefineries. The base idea of the pathway is to remove CO_2 from biogas to reach methane in quality for its direct injection into the gas grid. Separated carbon dioxide reacts with renewable hydrogen or undergoes a photosynthetic reaction to form a targeted product, see Fig. 3.18. E.g. Baena-Moreno et al. (2020) present a thermochemical pathway to form CH_4 by Power-to-Gas technology. Its origin is based on biogas upgrade by membrane technology to purify CH_4. Separated CO_2 reacts with H_2 in the methanation unit under the Sabatier reaction.

Hydrogen is generated by water electrolysis powered by renewable electricity (solars, wind). The economic attractiveness of their proposal was not reached due to the low natural gas prices. The economically feasible solution can be reached only if green subsidies are achieved. Baena-Moreno et al. (2021) published an option to the previous study where CO_2 reacts with H_2 to form dimethyl ether. The good scenarios were reached only if a high feed-in tariff in green subsidies to dimethyl ether and investment costs were covered. Gutiérrez-Martín et al. (2020) do not use CO_2 separation. Their approach deals with direct methanation of biogas in catalytic fluidized bed reactor. The solution can become cos competitive with the SNG process if biogas and electricity production costs are cheaper. Tozlu (Tozlu in press) present the techno-economic assessment in which CO_2 is captured from biogas. Carbon dioxide reactors with renewable electrolytic hydrogen to generate methanol. The

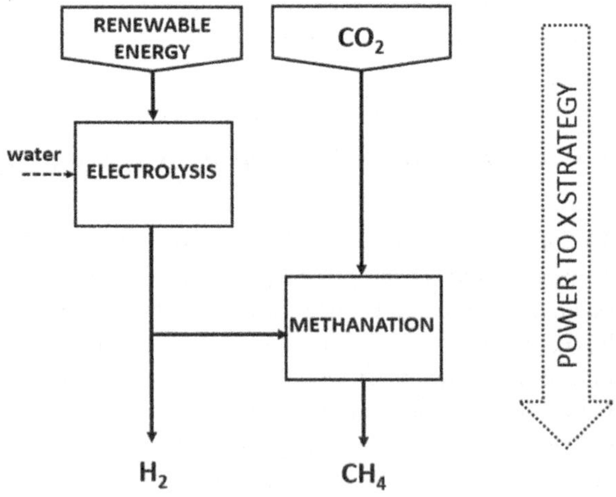

Fig. 3.18 The base of Power-to-X Concept

study serves economically attractive results counting methanol production cost of 0.6935 Eur kg^{-1} under electricity selling price 0.043 Eur kWh^{-1}. Vo et al. (2018) investigated three scenarios of CO_2-to-X strategies regarding biogas upgrade by amine scrubbing and CO_2 thermochemical or biochemical methanation. Their study meets the attractivity only if the surplus in electricity peaks need to be conserved. The crucial factors to meet economic feasibility are an upgrading system with carbon capture, rising methanation efficiency, reduced fixed investment cost for electrolyzer and cost reduction for feedstock, and electricity to produce hydrogen. Hidalgo and Martín-Marroquin (2020) observed that investment, operating cost and process efficiencies are the crucial factors limiting the general implementation of the methanation concept to a low carbon economy.

A little information is provided for biochemical CO_2 conversion pathways for Kutsay et al. (2020) performed the case study for biogas-based CO_2 upgrade to microalgae, see Fig. 3.19. The study supposes that biogas is fired in a cogeneration unit to parallel form electricity and heat. Generated flue gas goes through the technological set of microalgal cultivation that converts emitted CO_2 through microalgae to microalgal powder. The authors concluded that it is impossible to improve the economics even with the diversification of the primary products. The more operating sets, the higher investment cost and operating expenses, especially proposing entirely unfeasible. The study concluded that the production price of microalgae, \$ 0.16 kg^{-1}, must be reached to be competitive in the market—the conventional

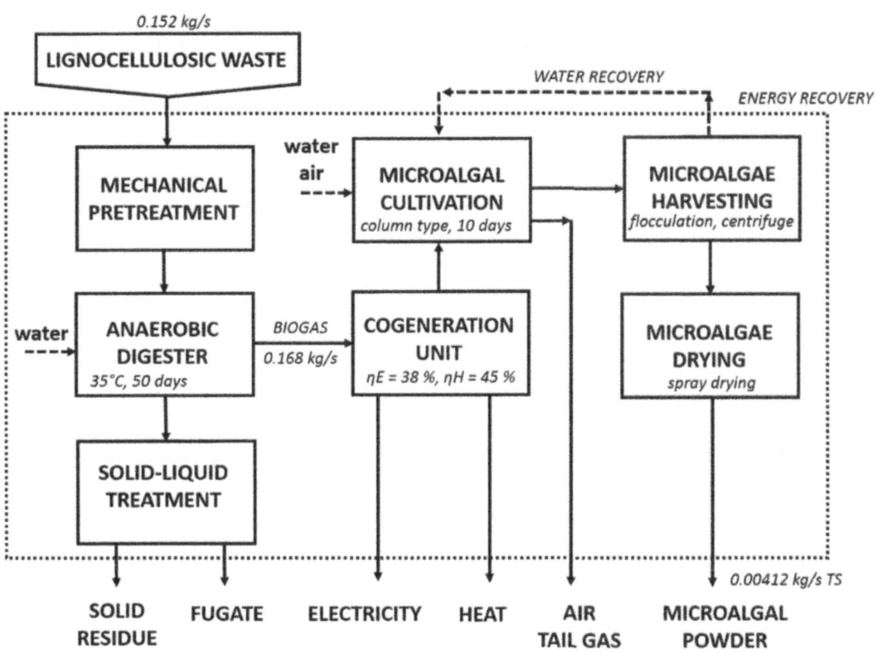

Fig. 3.19 The Block diagram of microalgal biogas biorefinery (Kutsay et al. 2020)

microalgal production price ranges in units or low decimals \$ kg^{-1}. To be realistic, this microalgal biogas biorefinery will never be realized if crude oil prices remain at nowadays levels. Therefore, the intention should be paid preferably on biorefinery strategies with one main product, not an extensive portfolio of main products.

3.4 Challenges and Opportunities for Lignocellulosic Biorefinery

Previous chapters overviewed the approaches of lignocellulosic biomass treatment in the concept of biorefinery. Literature and industrial researches confirmed that there exits still a large gap between laboratory and industrial lignocellulosic-based biorefineries. No complex lignocellulosic waste to X technology was developed concerning energy-efficient and economically feasible on a high-volumed scale. Biobased products are not cost-competitive with the production cost of conventional fuels and petrochemicals. Nevertheless, several biorefinery proposals showed the challenging potential of efficient and economic attractive waste processing in decentralized mode. There are many weaknesses and threats to implement laboratory ideas and results on a commercial scale that define plenty of opportunities for government, industrial and research societies to implement biorefinery concepts on a pilot or industrial scale. SWOT analysis was worked out to define needs and challenges for lignocellulosic biorefinery, see Table 3.6.

Economic sustainability and industrial attractivity of waste treatment in biorefinery concept are strongly dependent on local waste treatment policy and legislative, waste type and stable availability, technological set-up, and efficient equipment design in lignocellulosic biorefineries. Therefore, these challenges, opportunities, recommendations and requirements should be respected when proposing a lignocellulosic waste treatment in biorefinery concept for extensive scale application as follows (Krátký et al. 2018; Kratky et al. 2017).

Demand for waste material

- Availability of waste biomass and sustainability of its production from a local point of view.
- Energy and economic lest demanding solution of waste collection and sorting.
- Waste storage, its time degradation, chemical stability.
- Maximum limitation for production of biofuels from food commodities.

Table 3.6 SWOT analysis of lignocellulosic biorefinery at large scale

Strengths	*Weaknesses*
• Chance to reach a maximum conversion for high-volumed waste biomass treatment to targeted product • Production of single/multiple main products with quality compatible products for different industrial sectors (agricultural, food, chemical, energy) • Biorefinery proposal based on processing pathways of nowadays industrially-operating plants (pulp, ethanol or food plants) • Chance of decentralized waste treatment and direct product use in a given locality respecting local demand and policy	• Quality of final products strongly influenced by the quality of raw material treated waste • Effort to find a marketplace for bio-based products • Competitiveness and compatibility of bioproducts with their conventional alternatives in terms of quality, production cost, and production capacity • Economic feasibility (especially for more sophisticated bioproducts for food, pharmaceutical and chemical industries) • Impossible to design and operate directly industrial scale, the need to develop and test the process in the laboratory, quarter and pilot-scale units
Opportunities	*Threats*
• Minimizing the amount of landfilled waste, combusted waste • Waste treatment in biorefinery concept with direct use of emitted CO_2 (biofuels of third and fourth generation) • Making eco-innovative materials • Ecologic friendly solutions of waste handling • Sustainable development • To meet global policy goals with waste handling and industry decarbonisation (Green Deal, Paris Agreement)	• Demand for such a bioproduct concerning production yield, its grade, capacity and economic behaviour • High investments, high payback period • Not attractive for investors • Strong influence of local government and local policies, uneven changes in decision-making policies

Demand for technology

- Ability to increase the degradability by effective pretreatment of raw material.
- Development of new, more effective enzyme types and their price reduction.
- Ability to process the raw material with a higher load, i.e. at higher raw material concentrations.
- Ability to increase the efficiency of technology in terms of conversion of waste to bioproduct.
- Reduction of investment and operating costs of technology.
- Simultaneous production of biofuels with noble chemical products.
- Demand to ensure compatibility of bioproducts in the current infrastructure.

Demand for equipment

- User-friendly and straightforward device design.
- The corrosion resistance of the apparatus.
- Minimizing the use of precious materials or unique parts.

- Consideration of trade-off—the working volume of the device versus the cost of the device.

Demand for economic policy

- Usage of freely available raw material.
- Use of local ecosystem.
- Strict and not changing rules in environmental protection conditions.
- Fundings to support waste recycling and environmentally friendly waste processing technologies.
- The effective deployment of industrial technologies concerning the availability of raw materials.
- Implementation plan to use advanced biofuels.
- Tax incentives for waste recycling or re-use in industrial sectors and for consumers.

Demand for research and development

- Development of safe and eco-friendly pretreatment and conversion process that generates stable intermediate or final products.
- Take into consideration scale-up limits for industrial equipment during process set-up on a laboratory scale.
- Process intensifications to reach high efficient conversions.
- The use of low-cost consumables, the ability of recovery.
- Less energy-demanding process, the ability for heat recovery.
- Energy requirement reduction during waste processing, e.g., handling, milling, heating, mixing, pumping, and gas compression.
- Developing new active or selective low-cost catalysts, increasing their efficiency and regeneration, reducing thermal and pressure requirements.
- Continuous flow reactor technology must be developed for industrial waste treatment of high volumes.
- Lower CO_2 emission to generate product than it is captured.
- Life cycle assessment with CO_2 balances.
- Investment and operational cost reduction.
- Pilot plant and large-scale testing at a competitive cost.

3.5 Conclusion

Lignocellulosic waste treatment in biorefinery concept is challenging waste processing technology due to its intensive material-energy recycling of biomass. Nowadays, operating biorefineries and case studies show that they do not meet economic attractivity. Lignocellulosic biorefineries cannot compete with conventional fuels and petrochemicals due to high investment cost (green subsidies excluded), operating cost, raw material purchase cost, and low selling prices of products (green subsidies excluded). Nevertheless, it must be highlighted that lignocellulosic biorefineries pose opportunities for waste treatment with a positive impact on the environment reducing biodegradable waste as standardly landfilled or incinerated. The presented study shows that lignocellulosic waste is renewable raw material to produce alternatives, substituents or admixtures for conventional fuels or petrochemicals.

- The overviewed case studies confirmed that bioethanol biorefinery seems to be a promising technology of economic feasible lignocellulosic waste treatment technology.
- Research activities must aim to develop efficient, cost-effective technological steps, emphasizing the need for industrial testing units realizations.
- Innovative technical solutions must be developed that ensure intensive, financially attractive and ecologic waste conversion technology to advanced biofuels and bio-based materials as leading products, heat and electricity as side ones.

However, it is now impossible to make lignocellulosic biorefinery and economically attractive technology without government initiatives and appropriately set legislation, fundings, or ecological boundaries.

Acknowledgements This work was supported by the Ministry of Education, Youth and Sport of the Czech Republic under OP RDE grant number CZ.02.1.01/0.0/0.0/16_019/0000753 "Research centre for low-carbon energy technologies".

References

Alio MA, Marcati A, Pons A, Vial Ch (2021) Modelling and simulation of a sawdust mixture-based integrated biorefinery plant producing ethanol. Bioresource Technol 325:124650–124660

Andersen L, Lamp A, Dieckmann Ch, Baetge S, Schmidt LM (2018) Biogas plants as key units of biorefinery concepts: options and their assessment. J Biotechnol 283:130–139

Aresta M, Dibenedetto A (2010) Industrial utilization of carbon dioxide. Woodhead Publishing Limited, Cambridge

Baena-Moreno FM, Zhang Z, Zhang XP, Reina TR (2020) Profitability analysis of a novel configuration to synergize biogas upgrading and Power-to Gas. Energy Convers Manage 224:113369–113378

Baena-Moreno FM, Gonzalez-Castano M, Arellano-Garcia H, Reina TR (2021) Exploring profitability of bioeconomy paths: dimethyl ether from biogas as a case study. Energy 225:120230–120239

Bateni H, Karimi K, Zamani A, Benakashani F (2014) Castor plant for biodiesel, biogas and ethanol production with a biorefinery processing perspective. Appl Energy 136:14–22

Bittencourt GA, Barreto ES, Brandao RL, Baeta BEL, Gurgel LVA (2019) Fractionation of sugarcane bagasse using hydrothermal and advanced oxidative pretreatments for bioethanol and biogas production in lignocellulose biorefineries. Biores Technol 292:121963–121975

BioRefineries Blog. https://biorrefineria.blogspot.com/2015/04/crescentino-bioethanol-plant.html. Accessed 02 June 2021

Biorefineries Roadmap. https://www.bmbf.de/pub/Roadmap_Biorefineries_eng.pdf. Last accessed 26 May 2021

Carbon Capture and Utilization in the green economy. http://co2chem.co.uk/wp-content/uploads/2012/06/CCU%20in%20the%20green%20economy%20report.pdf. Accessed 03 June 2021

Cellulosic Ethanol. https://www.etipbioenergy.eu/value-chains/products-end-use/products/cellulosic-ethanol. Accessed 02 June 2021
Cellunolix. https://www.st1.eu/cellunolix-ethanol-plant-to-be-built-in-finland. Accessed 02 June 2021
Chempolis. https://chempolis.com/references/. Accessed 02 June 2021
Cheng MH, Wang Z, Dien BS, Slininger PJW, Singh V (2019) Economic analysis of cellulosic ethanol production from sugarcane bagasse using a sequential deacetylation, hot water and disk refining pretreatment. Processes 7:1–15
Chovau S, Degrauwe D, der Bruggen BV (2013) Critical analysis of techno-economic estimates for production cost of lignocellulosic bio-ethanol. Renew Sustain Energy Rev 26:307–321
Clariant sunliquid cellulosic ethanol technology. https://ec.europa.eu/info/sites/info/files/6_markus_rarbach.pdf. Last accessed 02 June 2021
Demichelis F, Laghezza M, Chiappero M, Fiore S (2020) Technical, economic and environmental assessment of bioethanol biorefinery from waste biomass. Biores Technol 277:124111–124127
Domsjö Pulp Mill. https://biorrefineria.blogspot.com/2015/04/domsjo-pulp-mill.html. Last accessed 02 June 2021
Eurostat. https://appsso.eurostat.ec.europa.eu/nui/show.do?dataset=env_air_gge&lang=en. Accessed 03 June 2021
Geissler GH, Maravelias ChT (2021) Economic, energetic and environmental analysis of lignocellulosic biorefineries with carbon capture. Appl Energy 302:117539–117550
Giraldo-Poveda JA, Cardona Alzate CA (2021) A biorefinery for the valorization of marigold (Calendula officinalis) residues to produce biogas and phenolic compounds. Food Bioprod Process 125:91–104
Global database of biomass conversion facilities. https://www.ieabioenergy.com/installations/. Last accessed 02 June 2021
Gnansounou E, Dauriat A (2010) Techno-economic analysis of lignocellulosic ethanol: a review. Biores Technol 101:4980–4991
Gutiérrez-Martín F, Rodríguez-Antón LM, Legrand M (2020) Renewable power to gas by direct catalytic methanation of biogas. Renewable Energy 162:948–959
Hidalgo D, Martín-Marroquin JM (2020) Power-to-methane, coupling CO_2 capture with fuel production. Renew Sustain Energy Rev 132:110057–110070
Hossain MS, Theodoropoulos C, Yousuf A (2019) Techno-economic evaluation of heat integrated second generation and furfural coproduction. Biochem Eng J 144:89–103
ICEF. https://assets.contentful.com/CO2U_ICEF_Roadmap_FINAL_2016_12_07.pdf. Accessed 03 June 2021
Junling N, Yupeng H, Quanguo Z (2011) Effect of different comminution degree on the mesophilic and dry anaerobic fermentation of wheat straw. Acta Energiae Solaris Sinica 32:1683–1686
Kaparaju P, Serrano M, Thomsen AB, Kongjan P, Angelidaki I (2009) Bioethanol, biohydrogen and biogas production from wheat straw in a biorefinery concept. Biores Technol 100:2562–2568
Kos M, Krátký L (2020) Production technology of second-generation ethanol (Bachelor thesis). Czech Technical University in Prague, Faculty of Mechanical Engineering, Department of Process Engineering. https://dspace.cvut.cz/handle/10467/89522. Last accessed 02 June 2021
Kratky L, Jirout T, Kutsay A (2017) Biorefinery: a critical technical review. In: European biomass conference and exhibition, pp 1–6, Stockholm
Kratky L, Azizov S, Jirout T (2018) Lignocellulosic waste treatment in biogas biorefinery with reduced CO_2 production: a techno-economic study. Chem Eng Trans 70:1–6
Krátký L, Azizov S, Seghman P, Formánek R, Jirout T (2018) Economically profitable waste treatment within the concept biorefinery: fiction or reality. Waste Forum, pp 383–394
Kuglarz M, Alvarado-Morales M, Karakashev D, Angelidaki I (2016) Integrated production of cellulosic bioethanol and succinic acid from industrial hemp in a biorefinery concept. Biores Technol 200:639–647
Kutsay A, Krátký L, Jirout T (2016) An economic feasible waste transformation technology. In: 6th International symposium on energy from biomass and waste, pp 1–12, Venice

Kutsay A, Kratky L, Jirout T (2020) Biogas plant upgrade to CO_2 free technology: a techno-economic case study. Chem Eng Technol 43:1981–1993
Mankar AR, Pandey A, Modak A, Pant KK (2021) Pretreatment of lignocellulosic biomass: a review on recent advances. Biores Technol 334:25235–25247
Meramo-Hurtado S, Gonzalez-Delgado A, Rehmann L, Quinones-Bolanos E, Mehvar M (2021) Comparative analysis of biorefinery designs based on acetone-butanol-ethanol fermentation under exergetic, techno-economic, and sensitivity analyses towards a sustainability perspective. J Clean Prod 298:126761–126781
Morales-Rodriguez R, Perez-Cisneros E, Reyes-Heredia JAL, Rodiguez-Gomez D (2016) Evaluation of biorefinery configurations through a dynamic model-based platform: integrated operation for bioethanol and xylitol coproduction from lignocellulose. Renew Energy 89:135–143
Naims H (2016) Economics of carbon dioxide capture and utilization—a supply and demand perspective. Environ Sci Pollut Res 23:22226–22241
NETL. http://www.netl.doe.gov/research/coal/carbon-storage/research-and-development/co2-utilization. Accessed 03 June 2021
Nordfuel. https://nordfuel.fi/en/biorefinery/. Accessed 02 June 2021
Orive M, Cebrián M, Amayra J, Zufía J, Bald C (2021) Techno-economic assessment of a biorefinery plant for extracted olive pomace valorization. Process Saf Environ Prot 147:924–931
Padi RK, Chimphango A (2020) Feasibility of commercial waste biorefineries for cassava starch industries: techno-economic assessment. Biores Technol 29:122461–122473
Patel A, Shah AR (2021) Integrated lignocellulosic biorefinery: gateway for production of second-generation ethanol and value-added products. J Bioresources Bioproducts (in print)
Piemonte V, Paola LD, Gentile A, Masciocchi B, Russo V, Iaquaniello G (2014) Biodiesel production from microalgae: ionic liquid process simulation. Chem Eng Trans 39:379–384
Pinales-Márquez CD, Rodríguez-Jasso RM, Araújo RG, Loredo-Trevino A, Nabarlatz D, Gullón B, Ruiz HA (2021) Circular bioeconomy and integrated biorefinery in the production of xylooligosaccharides from lignocellulosic biomass. Ind Crops Prod 162:113274–113291
Pratto B, Santos-Rocha MSR, Longati AA, Júnior RS, Cruz AJG (2020) Experimental optimization and techno-economic analysis of bioethanol production by simultaneous saccharification and fermentation process using sugarcane straw. Biores Technol 297:122494–122505
Rahimi V, Shafiei M (2019) Techno-economic assessment of a biorefinery based on low-impact energy crops: a step towards commercial production of biodiesel, biogas, and heat. Energy Convers Manage 183:698–707
Ruiz J, Olivieri G, de Vree J, Bosma R, Willems P, Reith JH, Eppink MHM, Kleinegris DMM, Wijffels R, Barbosa MJ (2016) Towards industrial products from microalgae. Energy Environ Sci 9:3036–3044
Santamaría-Fernández M, Molinuevo-Salces B, Lübeck M, Uellendahl H (2018) Biogas potential of green biomass after protein extraction in and organic biorefinery concept for feed, fuel and fertilizer production. Renew Energy 129:769–775
Seghman P, Krátký L (2017) Lignocellulosic bioethanol biorefinery (Master thesis). Czech Technical University in Prague, Faculty of Mechanical Engineering, Department of Process Engineering. https://dspace.cvut.cz/handle/10467/73249. Last accessed 01 June 2021
Solarte-Toro JC, Rueda-Duran CA, Ortiz-Sanchez M, Alzate CAC (2021) A comprehensive review on the economic assessment of biorefineries: the first step towards sustainable biomass conversion. Bioresource Technol Rep 15:100776–100788
Song Y, Lee YG, Cho EJ, Bae HJ (2020) Production of xylose, xylulose, xylitol and bioethanol from waste bamboo using hydrogen peroxide-acetic acid pretreatment. Fuel 278:118247–118256
ST1. www.st1.eu. Accessed 02 June 2021
Talekar S, Patti AF, Vijayranghavan R, Arora A (2018) An integrated green biorefinery approach towards the simultaneous recovery of pectin and polyphenols coupled with bioethanol from waste pomegranate peels. Biores Technol 266:322–334
Towler G, Sinnott R (2013) Chemical engineering design: principles, practise and economics of plant and process design, 2nd ed. Butterworth- Heinemann Elsevier, Waltham

Tozlu A (2021) Techno-economic assessment of a synthetic fuel production facility by hydrogenation of CO_2 captured from biogas. Int J Hydrogen Energy (article in press)

Vo TTQ, Wall DM, Ring D, Rajendran K, Murphy JD (2018) Techno-economic analysis of biogas upgrading via amine scrubber, carbon capture and ex-situ methanation. Appl Energy 212:1191–1202

Woodland Biofuels. https://www.woodlandbiofuels.com/feedstock-opportunities. Accessed 02 June 2021

Yuan Z, Wen Y, Kapu NS, Beatson R (2018) Evaluation of an organosolv-based biorefinery process to fractionate wheat straw into ethanol and co-products. Ind Crops Prod 121:294–302

Zhao L, Zhang X, Xu J, Ou X, Chang S, Wu M (2015) Techno-economic analysis of bioethanol production from lignocellulosic biomass in China: dilute-acid pretreatment and enzymatic hydrolysis of corn stover. Energies 8:4096–4117

Chapter 4 Hydrothermal Processing of Lignocellulosic Biomass to Biofuels

R. Divyabharathi and P. Subramanian

4.1 Introduction

The demand for energy is constantly increasing due to the rapid industrialization and increasing human population. Industrialization increased the consumption of fossil fuels and its associated environmental pollution problems leading to the release of harmful greenhouse gases causing global warming. Global warming is mainly associated with the release of CO_2 in the atmosphere. Unless the released CO_2 is captured, the effects of climate change is going to be inevitable in future. Hence, efforts are needed to find alternate energy solutions for fossil fuel usage and carbon capture like the utilization of biomass as an energy source. Biomass is advantageous since it is renewable i.e., the plants can be regrown within a period of time and it is a clean source of energy (McKendry 2002). The CO_2 released in the atmosphere is taken by plants, utilized as an energy resource, and can be released into the atmosphere, thus making the cycle completely carbon neutral. Lot of research are currently progressing on the utilization of biomass to produce fuels.

Among the various biomass energy conversion technologies, hydrothermal processing has gained recent interest to thermally process biomass directly for energy conversion in the presence of water at high pressure and temperature. Biomass components gets solubilized in water as the dielectric constant of water decreases at elevated reaction conditions, making it a good solvent for non-polar compounds (Elliott 2011). As a result, biomass hydrothermally degrades to produce various solid, liquid and gaseous biofuels via processes called carbonization, liquefaction and gasification respectively. In addition, biomass generally have significant moisture content, which must be dried before use in case of non-hydrothermal processing. This adds further costs for industrial applications. Thus, the energy cost associated

R. Divyabharathi (✉) · P. Subramanian
Department of Renewable Energy Engineering, Tamil Nadu Agricultural University, Coimbatore 641003, India

Y. K. Nandabalan et al. (eds.), *Zero Waste Biorefinery*, Energy, Environment, and Sustainability, https://doi.org/10.1007/978-981-16-8682-5_4

with drying will be significantly reduced in hydrothermal methods, which marks its major importance for research.

This chapter highlights the components of biomass, bioenergy conversion technologies, hydrothermal processing methods, separation and characterization of biofuels from hydrothermal treatment of lignocellulosic biomass.

4.2 Lignocellulosic Biomass as an Energy Source

Biomass are sun-derived organic substances produced by the process of photosynthesis having carbon, hydrogen, nitrogen, oxygen and sulphur as major elements (Vassilev et al. 2013). Among the alternate energy sources, biomass offers a great source of energy potential which constitutes about 50% of world's primary energy demand (Svetlana and Johan 2000). Its major sources include agricultural wastes, forestry residues, industrial wastes, animal-based wastes, algae, sewage etc. Biomass finds its major applications as a raw material for producing chemicals, fuels, thermal and electrical energy. However, it is under-utilized and releases less pollution compared to other sources of energy. Generally, biomass is selected based on its properties like moisture content, calorific value, cellulose, lignin, fixed carbon and volatile content. Based on these properties and type of energy conversion applied, the biofuel product formation varies significantly. The basic components of a lignocellulosic biomass include cellulose, hemi-cellulose, lignin and extractives, whereas algal biomass majorly contain lipids, proteins and carbohydrates. The composition of these organic substances depends on the biomass type, and growing conditions. Inorganic substances which constitute a minor portion in biomass are usually sulphur and nitrogen derived compounds producing ash. The cellulose, hemi-cellulose and lignin compounds help in building mechanical and structural stability of biomass which constitutes for the major chemistry behind biofuel production (Basu and Basu 2010).

The richest natural long chain polymer present in any biomass is the cellulose $(C_6H_{10}O_5)_n$ which is formed by 1,4 β-glycosidic linkage units of D-glucopyranose (Cao et al. 2009). Generally cellulose constitutes about 50% in lignocellulosic biomass and is insoluble in most of the solvents including water at normal temperature of 25 °C, however they become hyperactive and fully soluble at higher sub-critical temperatures. This is due to its tight hydrogen bonding with a high degree of polymerization (Kumar and Gupta 2008; Liu et al. 2010). Hemi-cellulose is a hetero-polymer containing xylan and mannan units with less degree of polymerization (Ren et al. 2010). It constitutes about 35% of lignocellulosic biomass, which varies depending on the type of biomass like woody, herbaceous or agro biomass. Hemi-cellulose binds with cellulose and lignin components in biomass through hydrogen and covalent bonds respectively. When subjected to heat treatment, hemi-cellulose disintegrates easily due to its unstable nature compared to cellulose and lignin. Lignin on the other hand is a natural aromatic polymer composed mainly of phenyl units linked by ether bonds (Savage et al. 2010). It constitutes about 20% of lignocellulosic biomass

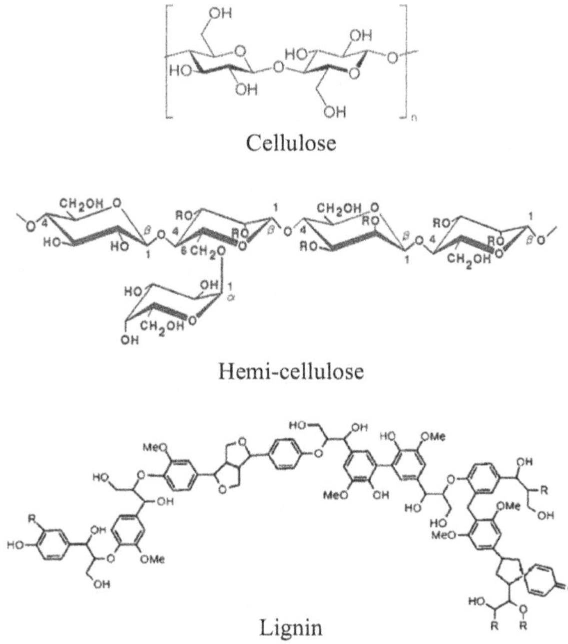

Fig. 4.1 Structure of cellulose, hemi-cellulose and lignin

representing the fibrous nature (Rowell et al. 2005). Lignin helps in strengthening the structure of biomass by storing energy in tightly bound fibers and is higher in woody biomass than herbaceous or agro biomass. Comparatively lignin has higher heating value due to its energy content and is more stable to biological degradation. Followed by cellulose, hemi-cellulose and lignin, there are few heterogeneous substances extracted from biomass composing of proteins, fats, lipids, resins, acids and sugars which vary depending on the type of biomass. Extractives constitutes only about 2–5% of biomass composition (Peng et al. 2010) but they contribute to significant plant characteristics like odor, color, taste etc. It is found more in herbaceous biomass like corns and sugars than in woody biomass. The structures of cellulose, hemi-cellulose and lignin are shown in Fig. 4.1. This lignocellulosic composition has a substantial impact on bioenergy conversion and product yield in terms of both quantity and quality.

4.3 Bioenergy Conversion Technologies

There are two significant pathways of bioenergy conversion to convert biomass to biofuels and valuable chemicals. They are biochemical conversion, which process biomass in the presence of microbes and enzymes and thermochemical conversion which process biomass using heat. Oxygen or air supply during these conversion plays a crucial role determining the end products. Based on the characterization of

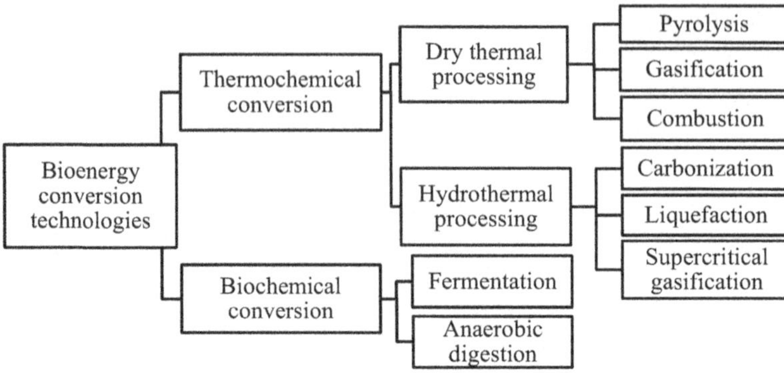

Fig. 4.2 Energy conversion technologies from biomass

biomass, suitable energy conversion technologies can be adopted. A very simple way to select the energy conversion route for a biomass depends on its moisture content. Usually, biomass with moisture content >10% finds it non-economical to process under thermochemical conversion, rather it will be suitable for biochemical conversion. This is because most of the heat supplied will be used for removing the moisture and the chemistry of products formation will be affected. The possible energy conversion routes from biomass are shown in Fig. 4.2.

Biochemical conversion is a natural process carried under aerobic or anaerobic environment. Enzymatic microbial fermentation and anaerobic digestion are the two major biochemical pathways performed by bacteria and enzymes in an anaerobic environment. In these processes, microbes use oxygen integrated in biomass for its chemical conversion producing alcohol, gases and other chemicals. Ethanol is the main product obtained from fermentation of sugars, starchy and lignocellulosic biomass by yeasts and biogas (CH_4 and CO_2) is the main product obtained from anaerobic digestion of wet biomass by hydrolytic, acetogenic and methanogenic bacteria. Usually, pretreatment methods are employed before fermentation for degradation of complex biopolymers like lignocellulose to sugars. This method is used commercially in most of the industries and it adds more to energy costs.

Thermochemical conversion is a heat treatment process carried on biomass with or without the supply of oxygen to produce fuels and chemicals. This conversion is performed from ancient ages where the biomass like wood is burnt for cooking and space heating. They are classified into dry thermal processing where dry biomass with moisture content of 5–10% is used for biofuel production and hydrothermal processing where high moisture content biomass (>10%) are used directly for fuel conversion (Elliott 2011). Dry thermal processing includes pyrolysis, gasification and combustion processes which differ in process and product formation based on the amount of oxygen supplied. Hydrothermal processing includes carbonization, liquefaction and supercritical gasification which differs based on the process conditions of temperature and pressure. In general, dry thermal processing involves high temperature and low pressure, whereas hydrothermal processing involves low temperature and high pressure.

4.4 Dry Thermal Processing

4.4.1 Pyrolysis

Pyrolysis is the thermal disintegration of biomass in the absence of air, mostly in an inert atmosphere like nitrogen or argon. The oxygen/air entrapped between biomass pore spaces is removed by the inert gases ensuring a completely anaerobic environment. At this condition, when the biomass is heated, the cellulose and hemi-cellulose components react vigorously releasing volatiles and organics from biomass which is the major ingredient for bio-oil formation. Lignin on the other hand is responsible for the amount of charred biomass left after volatilization. As a result, pyrolysis yields solid (biochar) as an outcome of release of volatiles, liquid (bio-oil) as an outcome of condensation of volatiles and gas (syngas) which are non-condensable output having low molecular weight gaseous products. Based on the pyrolytic conditions of temperature and reaction time, it is classified into slow pyrolysis, intermediate pyrolysis and fast pyrolysis. Slow pyrolysis occurs at 250–400 °C with residence time between 6 and 24 h at a very slow heating rate of <10 °C/min yielding biochar (35%) as a major product (Frederik et al. 2013). Intermediate pyrolysis occurs at 350–450 °C with 30 min residence time at a heating rate of 1.5 °C/s yielding bio-oil (50%) and syngas (35%) as major products (Deodatus et al. 2021). Fast pyrolysis occurs at 500–900 °C with residence time of <5 s at a rapid heating rate of 1000 °C/s yielding bio-oil (75%) as major product (Czernik and Bridgwater 2004). Biochar has higher carbon content (60–90%) and heating value of 25–30 MJ/kg, hence can be used as a solid fuel or as precursor for activated carbon. It also has significant adsorptive qualities to be used in soil to enhance microbial biomass and soil biology. Bio-oil is a dark brown viscous liquid which can be marketed as liquid smoke, used as a substitute for fuel oil or as a blend with existing fossil fuels, if upgraded. Syngas contains hydrocarbons which can be potentially used for thermal and engine applications. The most commonly used pyrolytic reactors are bubbling fluidized bed pyrolyser, circulating fluidized bed pyrolyser, rotating plate pyrolysis reactor and rotating cone pyrolysis reactor. The yield of pyrolysis products varies depending on the composition of biomass feed, particle size of biomass, residence time, temperature, heating rate and reactor type.

4.4.2 Gasification

Gasification is the thermal decomposition of biomass with limited supply of air/oxygen to yield producer gas or syngas as major product. The process is also referred as partial oxidation, wherein the biomass components like C, H, O, N and S reacts with each other to give gaseous products like CO, CO_2, H_2, CH_4, H_2O and traces of sulphur compounds, which all together referred as producer gas (Antonio et al. 2016). The chemistry of gasification takes place in four zones viz. drying, pyrolysis, oxidation and reduction zones. In drying zone, the moisture content in biomass (10–15%) is evaporated into steam at 100–150 °C. In pyrolysis zone, the organic components are converted at 150–700 °C to charcoal, wood oil, tar and gas. In oxidation zone, the carbon and hydrogen components in biomass reacts with oxygen in the air-steam blast at 1400 °C, to yield CO_2 and H_2O respectively. In reduction zone, the CO_2 and steam coming from the oxidation zone are converted at 800–1000 °C to carbon monoxide and hydrogen, which are the main constituents in producer gas of about 30–60% and 25–30%, respectively (Pratik and Babu 2009). This makes the gas combustible with a calorific value of 950–1200 kcal/m^3. There are two major designs of gasifiers viz. fixed bed gasifiers (updraft, downdraft and cross draft gasifiers) and fluidized bed gasifiers, which find their own applications for direct heating, running engines, chemicals production and power generation. On an average, 1 kg of biomass yields about 1.5–2 m^3 of producer gas, which vary depending on biomass type and gasifier design. The chemistry behind gasification yielding producer gas is given by the following equations.

$$C + O_2 \rightarrow CO_2 + 393,800\ \mathrm{kJ/kg\ mol\ (combustion)} \tag{4.1}$$

$$C + H_2O \rightarrow CO + H_2 - 131,400\ \mathrm{kJ/kg\ mol\ (water\ gas - exothermic)} \tag{4.2}$$

$$CO + H_2O + CO_2 + H_2 + 41,200\ \mathrm{kJ/kg\ mol\ (water\ gas\ shift\ reaction)} \tag{4.3}$$

$$C + CO_2 \rightarrow 2CO - 172,600\ \mathrm{kJ/kg\ mol\ (boudouard\ reaction)} \tag{4.4}$$

$$C + 2H_2O \rightarrow CO_2 + 2H_2 - 78,700\ \mathrm{kJ/kg\ mol} \tag{4.5}$$

$$C + 2H_2 \rightarrow CH_4 + 75,000\ \mathrm{kJ/kg\ mol}\ \mathrm{(methane\ reaction - endothermic)} \tag{4.6}$$

4.4.3 Combustion

Combustion is a thermochemical conversion that takes place in the presence of excess amount of air or oxygen to form CO_2 and H_2O as the ultimate products. The process is exothermic leaving heat as a result of interaction between the biomass components like carbon and hydrogen with oxygen. This process is age old and has been most commonly used for cooking, heating and power generation. The amount of heat released in the process depends on the calorific value of the biomass fuel burnt which in turn depends on the carbon, moisture and ash content (Daniel et al. 2013). Calorific value is directly proportional to the amount of carbon and indirectly proportional to the amount of ash and moisture in the biomass. Combusting the biomass with excessive moisture leads only to the wastage of heat in drawing out the water by evaporation. Hence for an efficient combustion to take place, the factors of temperature, biomass type, time and fuel-air mixing ratio are most influencing. General applications of combustion include fuel combustion in boilers and power plants for steam and power generation. Biomass nowadays is gaining more interest when co-fired with coal in power plants. The process is called co-combustion or co-firing. Co-combustion reduces the dependence on fossil fuels in power plants and produces clean renewable power at limited capital cost, thereby ensuring highest efficiency. The reactors commonly used for co-combustion involve fluidized bed combustor (atmospheric/pressurized), grate combustor and pulverized combustor where the biomass and coal will be directly co-fired, indirectly co-fired or parallel co-fired (Sahu et al. 2014). Using this technology, there will be a net reduction in the emission of NO_x and SO_x due to the distinct interactions between varying chemical composition in biomass and coal.

4.5 Hydrothermal Processing

Hydrothermal processing is a pressurized thermochemical conversion of wet biomass into valuable fuels and chemicals. It occurs at temperatures involving sub-critical and supercritical conditions between 200 and 374 °C at self-generated pressures between 5 and 22 MPa (Peterson et al. 2008). It involves three processes namely hydrothermal carbonization, liquefaction and gasification, which vary depending on the reaction conditions of temperature and pressure. The main advantage of hydrothermal processing of biomass is that it eliminates the need to pretreat the biomass by drying unlike conventional thermochemical processes. Thus, it can directly handle a variety of biomass with moisture content ranging from 20 to 90% like food waste, algae, industrial wastes, sewage sludge etc., which is difficult to process in conventional processes (Yokoyama and Matsumura 2008). Due to this, the energy costs associated with drying is saved and comparatively less energy is required for processing. Since, the processing takes place under pressure, there is an increased yield in terms of products like hydro-char, biocrude and syngas with aqueous phase as byproduct.

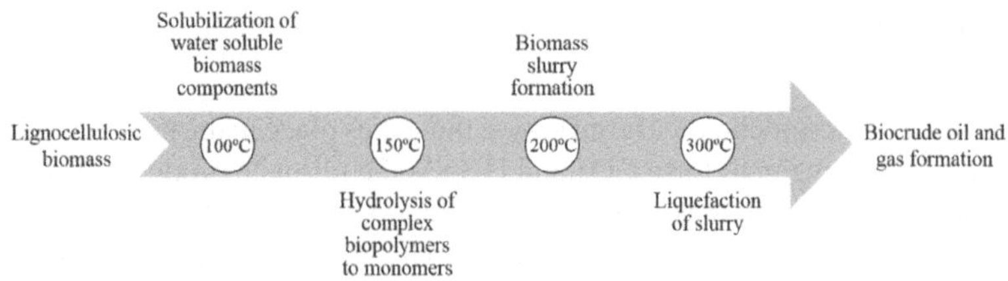

Fig. 4.3 Biomass hydrothermal degradation

Besides these, there are few disadvantages like the need for expensive and high-capacity sturdy reactors, product extraction procedures and up-scaling to industrial level (Zhang et al. 2010a). Hence a lot of research is required on the downstream processing of products from hydrothermal conversion.

The hydrothermal conversion is based on two conditions of sub-critical and super-critical media, which is reliable on water's critical point at 374 °C and 22 MPa. Water behaves differently and has its own distinct characteristics in each condition. For example, water solubilizes the components of cellulose and lignin at high temperatures and pressure, which are insoluble at atmospheric conditions. Also at critical conditions, the complex polymers of biomass are broken easily to their monomers releasing gaseous compounds, which when condensed produce biocrude oil and remaining non-condensable low molecular weight compounds collect as syngas. The biomass decomposition pathway in hydrothermal process is shown in Fig. 4.3. The properties of water can be changed with respect to temperature and pressure. At sub-critical conditions (<374 °C and <2 MPa), the water medium favors acid and base catalyzed reactions due to the increase in ionic product of water and decrease in viscosity (Guo et al. 2010). At supercritical conditions (>374 °C and >22 MPa), water has high diffusion rates (Ogunsola and Berkowitz 1995) making it as a source of hydrogen and involve in hydrolysis reactions. In hydrothermal processing, liquefaction takes place at sub-critical conditions of 200–370 °C with pressures up to 20 MPa, whereas hydrothermal gasification or supercritical gasification takes place at 370–500 °C with pressures up to 30 MPa. Commercially supercritical gasification is used in industries for power generation and removal of toxins from bio-wastes.

4.5.1 Hydrothermal Carbonization

Hydrothermal carbonization (HTC) is an aqueous slow pyrolysis process occurring at a medium temperature and pressure for longer residence time. HTC processes wet biomass at a rapid rate to release volatiles at temperatures from 150 to 250 °C and pressures from 1 to 5 MPa for a longer residence time up to 20 h (Titirici and Antonietti 2010) to yield hydro-char as major product. Biomass with moisture content of 70–90% is ideal for HTC requiring no pretreatment. Hydro-char is formed after complete volatilization of the wet biomass along with aqueous phase as byproduct.

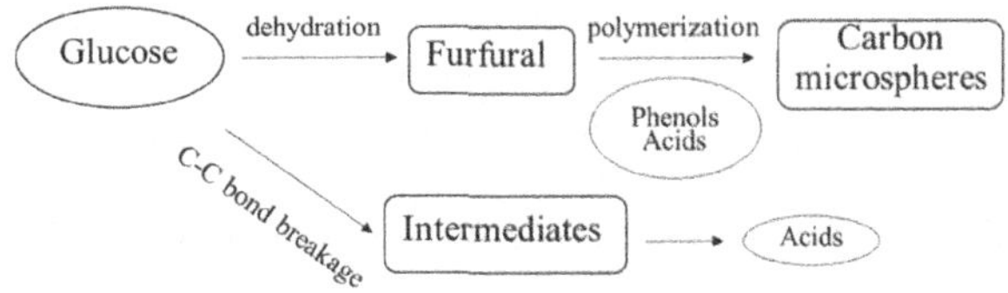

Fig. 4.4 Chemistry of hydrothermal carbonization of glucose

The formation of hydro-char depends on the lignin and fixed carbon content in biomass. For example, if the biomass has high lignin content, it constitutes to the degradation of fibers to carbon monomers thereby increasing the carbon content of material. This results in more char formation after removal of volatiles by decarboxylation and dehydration reactions removing H_2 as H_2O and O_2 as CO_2. Thus hydro-char has high calorific value and carbon content with less sulphur, nitrogen and ash. The carbon-rich hydro-char finds potential applications as fuel alternative for coal, alternative adsorbent or precursor for activated carbon and as an additive in soil for nutrient amendment (Xiao et al. 2012).

HTC process is advantageous over conventional slow pyrolysis by the factors of requirement of medium carbonization temperature, direct wet biomass processing, higher yield of char at less time and comparatively low energy requirement (Sevilla and Fuertes 2009). Several inexpensive and easily available biomass materials like sugars, cellulose and carbohydrate sources have been tested for HTC which has the major chemistry of fractionation to furfurals, polymerization to furans and intermolecular carbonization (Titirici and Antonietti 2010). The end hydrothermal carbon product derived has distinct characteristics of various polar functional groups which makes it to disperse easily in water and other solvents. The chemistry of hydrothermal carbonization of glucose is shown in Fig. 4.4.

4.5.2 *Hydrothermal Liquefaction*

Hydrothermal liquefaction (HTL) also called hydropyrolysis takes place at sub-critical water temperatures <370 °C and at high pressures up to 20 MPa to produce liquid biofuels (Toor et al. 2011). The phenomenon is related to the geological fossil fuel formation happening underground for thousands of years at high temperatures and pressures. Similarly, in HTL, organic liquid fuel is produced at high pressures within a specific period. The biocrude oil formed has comparable properties as that of the conventional crude oil in terms of color, odor, density, heating value etc. HTL technology is welcoming not only for the value of biocrude oil formation but also for a number of reasons like the use of an environment friendly water as solvent, use of wet biomass for thermal treatment without drying, thus increasing the energy efficiency and processing temperatures much lower than the conventional pyrolysis for bio-oil production (Savage et al. 2010).

The chemistry of hydrothermal liquefaction is complex due to the complexity of degradation of biomass components in sub-critical water conditions. The biomass components viz. cellulose, hemi-cellulose, lignin, lipids, fats, proteins etc. undergo hydrolysis, depolymerization, dehydration, decarboxylation and fragmentation reactions to ultimately yield a dark brown viscous liquid called biocrude (Toor et al. 2011). The chemical complexity involved in the hydrothermal liquefaction of lignocellulosic components is given by the following steps.

- Hydrolysis of macro molecules to water soluble oligomers
- Degradation of glycosidic linkages in cellulose to form glucose monomers
- Dissolution of sugar monomers of hemicellulose in water by hydrolysis
- Decomposition of glucose monomers to acids, aldehydes, and furfural derivatives
- Fragmentation of xylose monomers of hemicellulose to furfural
- Dissociation of open chain molecules to ketones, aldehyde, and acids
- Decomposition of lignin to low molecular weight compounds like phenol.

Most of the organic compounds are insoluble in water, however at sub-critical conditions between 200 and 350 °C, they tend to be soluble and water acts both as a reactant and solvent in chemical reactions. The aromatic compounds of ethers, nitriles, amides, and phenyls gain the ability to thermally react at temperatures up to 250 °C, however they dissociate and rapidly hydrolyze to alkenes, phenols and anilines at increased temperatures (Zhang et al. 2010b). This also depends on various factors such as pressure, catalyst, biomass composition, particle size, reaction time and medium. Overall, the chemistry favours an increase in formation of volatiles leading to biocrude oil production at 300–350 °C and gas yield >350 °C. The liquid biocrude formed has high calorific value and less oxygen content compared to the bio-oil obtained from conventional pyrolysis. However, it cannot be directly used as a transportation fuel to its high viscosity, oxygen, water and acid content, which affects its stability on storage. Hence for biocrude to be used as a blend or replacement with existing fossil fuels, it needs to be further upgraded by means of catalytic dehydration or decarboxylation reactions to remove oxygen and water such that the overall calorific value and C/H ratio of biocrude is increased. This improvement of obtaining high quality, stable and high energy biocrude can be made by down-streaming techniques like solvent extraction by using polar solvents namely dichloromethane, diethyl ether, acetone, toluene, hexane, chloroform etc. (Valdez et al. 2011) and by catalytic hydrodeoxygenation at 300–600 °C which combines oxygen in biocrude with hydrogen to form water and saturated hydrocarbons (Huber et al. 2006). Figure 4.5 shows the hydrothermal liquefaction of carbohydrates to valuable chemicals.

Fig. 4.5 HTL pathway of carbohydrates

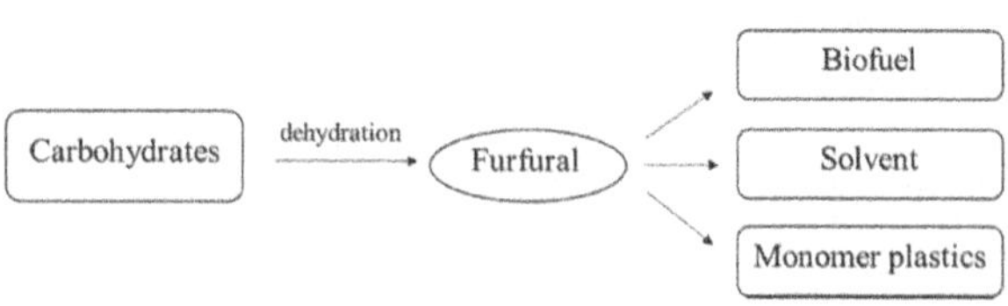

4.5.3 Hydrothermal Gasification

Hydrothermal gasification (HTG) also called as supercritical gasification refers to the thermochemical conversion of biomass to biofuels under hot compressed water. In this environment, water is said to attain supercritical properties such as change in solubility and reactivity, since the temperature and pressure inside the reactor are at elevated levels of >374 °C and >22 MPa respectively. Biomass molecules which are insoluble in water at normal conditions undergo rapid hydrolysis under supercritical conditions, making the fragments of biomass to dissolve, which in turn prevents the formation of hydro-char (Kruse 2009) and paves way for higher gas production. HTG is most commonly used for the production of H_2 and CH_4 rich syngas and it depends on the temperature and catalyst used. The process may produce hydrogen rich gas at temperatures >500 °C under non-catalytic or non-metallic catalytic conditions, or it may produce methane rich gas at temperatures between 374 and 500 °C under metallic catalytic condition or it may produce a combined syngas product using heterogeneous catalysts at near critical condition (Kruse 2008). The chemistry of supercritical biomass degradation is complex such that there will be an occurrence of hydrolysis, fragmentation, methanation, water-gas and water gas shift reactions simultaneously. High temperatures and pressures fragment the bonds of complex biopolymers to produce syngas compounds of H_2, CH_4, CO_2 and CO. To further specialize the process and to enable a sole-rich gas, a catalyst addition is needed. Heterogeneous catalysts such as Ni, Pt, Rh and Rb or homogeneous alkaline catalysts such as Na_2CO_3, K_2CO_3, NaCl and KCl may be used at temperatures between 500 and 800 °C to promote water-gas-shift reactions and other free radical reactions, generating hydrogen and methane rich gas products (Lee et al. 2002). This is also dependable on the biomass composition like carbon and hydrogen content and also on the reaction time of HTG. The Eqs. (4.7) to (4.10) shows the reaction network of hydrothermal gasification of glucose.

$$C_6H_{12}O_6 + 6\,H_2O \rightarrow 6\,CO_2 + 12\,H_2(\text{hydrolysis reaction}) \tag{4.7}$$

$$CH_xO_y + (1 - y)\,H_2O \rightarrow CO + (x/2 + 1 - y)\,H_2(\text{water gas reaction}) \tag{4.8}$$

$$CO + 3H_2 \rightarrow CH_4 + H_2O\ (\text{methanation reaction}) \tag{4.9}$$

$$CO + H_2O \rightarrow CO_2 + H_2(\text{water gas shift reaction}) \tag{4.10}$$

In the HTG study on glucose, the gasification reactions were favorable to H_2 formation and CO formation decreased at increased temperatures >600 °C under non-catalytic conditions, which was due to the rapid rate and dominance of water-gas-shift reaction (Xu et al. 1996). This rate was found to decrease when the concentration of glucose was increased. Hence to alter this, non-metallic catalyst like activated carbon can be added to completely convert glucose to hydrogen rich gas. The remaining aqueous phase may have lignin dissolved which may break down to form phenolic compounds. Phenols on polymerization in hydrothermal media produces formaldehyde which in turn results in cross-linking to form hydro-char compounds (Yoshida and Matsumura 2001). Thus, cellulose and lignin-based biomass should be used with heterogeneous catalysts like Ru in supercritical gasification to completely gasify and form CH_4 and CO_2 (Osada et al. 2004). HTG is therefore advantageous in processing wet biomass to produce H_2 and CH_4 rich gases which can be potentially used in power generation, chemical synthesis and for fuel cell applications.

4.6 Extraction and Analysis of Hydrothermal Products

There are various procedures for extraction of each type of hydrothermal products. In general, after hydrothermal processing, the condensable vapors are collected as liquid products, non-condensable gaseous product is vented or sometimes collected in tedlar bags and the remaining solid product from reactor is filtered. The condensable liquid product has biocrude mixed with water which can be separated by solvent extraction techniques and the solvent can be removed by vacuum filtration (Alba et al. 2012). In cases where homogeneous alkaline catalysts are used, the quality of biocrude extraction can be enhanced by dual solvent extraction process using two typical polar and non-polar solvents (Karagoz et al. 2005). The reaction mixture containing solid product dissolved in water is filtered to get aqueous phase which may contain water soluble organics and the filtrate is obtained as solid hydro-char by drying in hot air oven at 103 ± 2 °C.

The biocrude obtained after solvent extraction is a viscous organic liquid that has high C/H ratio comparable to conventional crude. This can be verified by GC-MS and FT-IR analyses by detecting the functional groups present in the product. The aqueous phase is centrifuged and can be analyzed for total organic carbon (TOC) and for chemical composition by GC-MS. The solid hydro-char can be analyzed by FT-IR and SEM to identify the functional compounds and the formation of pore structures respectively. After the detection of chemical compounds in hydrothermal products by these analyses, the selective extraction of those chemicals is found to be challenging. This is due to the complexity of organics present in the products. However, certain liquid-liquid extraction techniques may selectively recover phenols and certain acids, which may be used for chemical substitution or for industrial applications (Hu et al. 2012). In case of alkaline catalyst usage in hydrothermal processing, it is difficult to separate liquids and solids. This is because catalysts retain in solid product which may in turn retain solids in liquid fraction. This can be rectified by the addition of acids like HCl during extraction (Miller et al. 1999). The separation and extraction of products post hydrothermal processing is shown in Fig. 4.6.

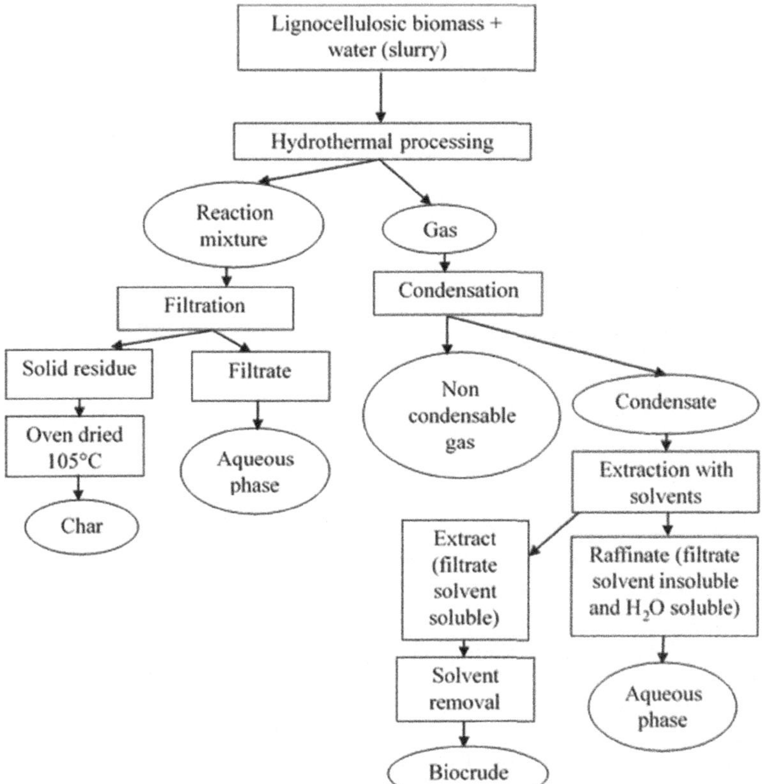

Fig. 4.6 Separation and extraction of hydrothermal products (Divyabharathi and Subramanian 2021a)

4.6.1 Gas Chromatography-Mass Spectroscopy

Gas Chromatography-Mass Spectroscopy (GC-MS) is a significant technique to detect and measure the unknown compounds present in the sample. There are various other chromatography techniques like Liquid chromatography-mass spectrometry (LC-MS), High performance liquid chromatography (HPLC), Tandem mass spectrometry (MS-MS) etc., amongst which the most suitable one for analysis of hydrothermal products is GC-MS. This works by the principle of ionization of gases and measurement of mass:charge (m/z) ratios. The organic samples must be diluted in a proper ratio with suitable solvents before injecting. GC-MS detects and quantifies most of the organics in the sample by comparing with internal standards. However, there will be certain limitations like choking of injector port or non-detection of volatiles in biocrude oils, carbon deposition inside the sampling port and few organics not formed as expected in hydrothermal processing, which solely depends on the biomass processing conditions. The primary compounds in the biocrude as detected

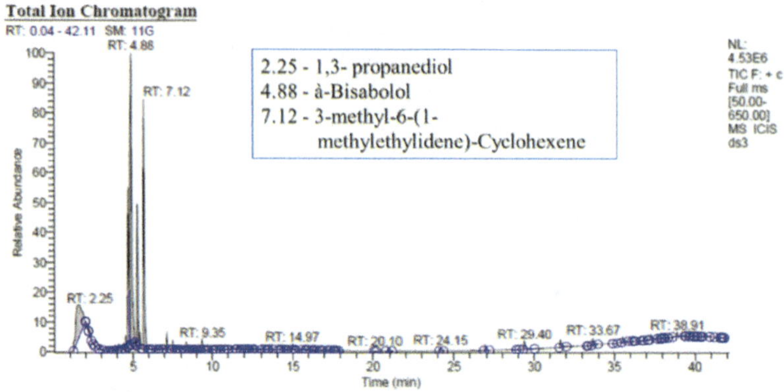

Fig. 4.7 GC-MS chromatogram of paddy straw biocrude at 275 °C (Divyabharathi and Subramanian 2021b)

by GC-MS include phenols, acids, alcohols, alkanes, alkenes, ketones, furfurals etc. which mainly depends on the biomass composition. However, it is challenging to get high quantity of a specific compound when processing a mixed composition biomass. The GC-MS chromatogram obtained for biocrude produced from paddy straw (Divyabharathi and Subramanian 2021b) is shown in Fig. 4.7.

4.6.2 Fourier Transform Infrared Spectroscopy

Fourier Transform Infrared (FTIR) Spectroscopy technique is used to detect the functional groups of compounds present in the sample. It is analyzed using a Fourier Transform Infrared Spectrophotometer which has zinc selenide attenuated total reflection (ATR) crystal mode and KBr beam splitting. Biomass, biocrude, hydro-char and aqueous phase byproducts may be analyzed in the range from 4800 to 400 cm^{-1} at 4 cm^{-1} resolution using origin software. Thin films of samples are applied to dry and the spectra will appear after smearing sample over the surface of crystal. In general, the major spectra shown by GC-MS, indicating typical organic compounds will be confirmed from the results of FTIR. The FTIR spectrum of transmittance plotted against wavelength obtained for biocrude produced from hydrothermal liquefaction of orange peel (Divyabharathi and Subramanian 2021a) is shown in Fig. 4.8.

The spectrum depicts that biocrude is a complex mixture of aliphatic and aromatic derivatives. It presented strong bands at 2875–3000 cm^{-1}, which illustrated the C-H stretching vibrations, which was due to the decomposition of cellulose to levoglucosan, saturated hydrocarbons and acids. The peaks at 1700 cm^{-1} were assigned to the C=O groups which represents the esters and carboxylic acids. Bands at 1300–1490 cm^{-1} were assigned to C=C benzene ring groups especially the phenol derivatives which occur as a result of polymerization. Lower peaks at

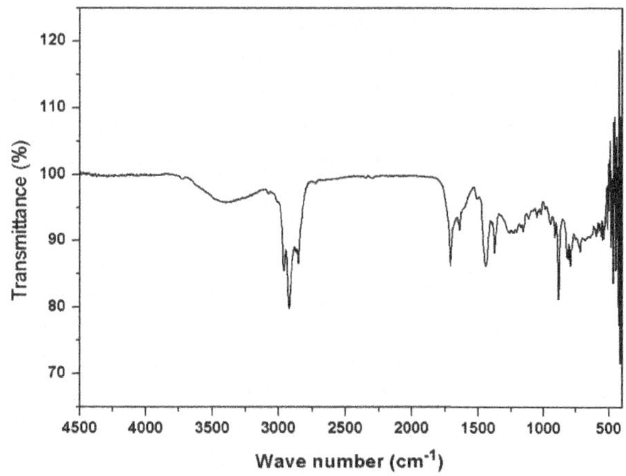

Fig. 4.8 FTIR spectrum of orange peel biocrude at 275 °C (Divyabharathi and Subramanian 2021a)

1100–1270 cm^{-1} were assigned to the C-O groups which represented the carbohydrate derivative compounds due to decarboxylation reactions (Divyabharathi and Subramanian 2021a).

4.6.3 Elemental Analysis

Elemental analysis also called ultimate analysis is used to find the composition of elements like carbon, hydrogen, nitrogen, oxygen and sulphur in the sample. This analysis is significant for finding the heating value of biomass and biofuels which is derived by Dulong's formula. It is also used to determine the C/H and O/H ratios which estimates the chemistry of product formation and the composition of hydrothermal products. In general, C/H ratio and O/H ratio determines the aromatic composition and rate of de-oxygenation reactions taking place during hydrothermal processing (Tekin et al. 2014). Table 4.1. shows the elemental composition and higher heating values of hydro-char obtained by hydrothermal processing of lignocellulosic biomass viz. orange peel, dairy manure and food waste at 250 °C. The obtained

Table 4.1 Elemental composition and heating values of hydro-char obtained from hydrothermal processing of lignocellulosic biomass

Biomass	Hydro-char composition					
	C (%)	H (%)	O (%)*	N (%)	S (%)	HHV (MJ/kg)
Orange peel	70.19	4.25	24.60	0.93	0.02	24.93
Dairy manure	63.41	5.58	30.03	0.92	0.06	24.68
Food waste	70.93	4.08	23.65	1.30	0.03	24.34

* *Calculated by difference*

hydro-char had high oxygen and low carbon content which was due to the incomplete biomass degradation expected due to the increase in the aqueous phase during hydrothermal processing (Caprariis et al. 2017). Hydro-char has higher heating value (25 MJ/kg) which proves that it can be potentially used as solid fuel or as a precursor for activated carbon production (Malins 2017).

4.7 Conclusion

This chapter reviewed the various hydrothermal processing methods viz. carbonization, liquefaction and gasification of biomass into biofuels like hydro-char, biocrude and syngas respectively in comparison with conventional thermochemical conversion processes. Depending upon the target product, suitable hydrothermal processing technology can be adopted. Water in hydrothermal conditions of sub-critical and super-critical media have different physico-chemical properties in terms of solubility and reactivity than at normal conditions, which makes the process and chemistry unique to obtain high-value fuels and chemicals from biomass. Though hydrothermal processing has few shortcomings like the use of high pressure and expensive reactors, difficulty in scaling up of process and upgrading of biofuels, it is a viable technology that can operate with a wide range of wet biomass at low to moderate temperatures and pressures. Separation and extraction of hydrothermal products plays a crucial role and if performed carefully, it is potentially feasible to upgrade and scale up hydrothermal processes in near future for biofuel usage to replace conventional fossil fuels.

References

Alba LG, Torri C, Samorì C, VanderSpek J, Fabbri D, Kersten SRA, Brilman DWF (2012) Hydrothermal treatment (HTT) of microalgae: evaluation of the process as conversion method in an algae biorefinery concept. Energy Fuels 26:642–657

Antonio M, Simeone C, Dino M (2016) Biomass gasification technology: the state of the art overview. J Energy Chem 25(1):10–25

Basu P (2010) Biomass characteristics. In: Basu P (ed) Biomass gasification and pyrolysis. Academic Press, Boston, pp 27–63

Cao Y, Wu J, Zhang J, Li HQ, Zhang Y, He JS (2009) Room temperature ionic liquids (RTILs): a new and versatile platform for cellulose processing and derivatization. Chem Eng J 147:13–21

Caprariis BD, Filippis PD, Petrullo A, Scarsella M (2017) Hydrothermal liquefaction of biomass: influence of temperature and biomass composition on the bio-oil production. Fuel 208:618–625

Czernik S, Bridgwater AV (2004) Overview of applications of biomass fast pyrolysis oil. Energy Fuels 18(2):590–598

Daniel M, Ana S, Javier R, Fernando S (2013) Assessment of CCHP systems based on biomass combustion for small-scale applications through a review of the technology and analysis of energy efficiency parameters. Appl Energy 102:1303–1313

Deodatus K, Justin N, Godlisten K (2021) A review of intermediate pyrolysis as a technology of biomass conversion for coproduction of biooil and adsorption biochar. J Renew Energy 2021:1–10

Divyabharathi R, Subramanian P (2021a) Biocrude production from orange (Citrus reticulata) peel by hydrothermal liquefaction and process optimization. Biomass Convers Biorefinery

Divyabharathi R, Subramanian P (2021b) Hydrothermal liquefaction of paddy straw for biocrude production. Mater Today Proc 45:603–606

Elliott DC (2011) Hydrothermal processing, thermochemical processing of biomass: conversion into fuels. In: Brown RC (eds) Chemicals and power. John Wiley & Sons, Ltd, Chichester, UK, pp 200–231

Frederik R, Sven VH, Dane D, Wolter P (2013) Production and characterization of slow pyrolysis biochar: influence of feedstock type and pyrolysis conditions. GCB Bioenergy 5:104–115

Guo Y, Wang SZ, Xu DH, Gong YM, Ma HH, Tang XY (2010) Review of catalytic supercritical water gasification for hydrogen production from biomass. Renew Sustain Energy Rev 14:334–343

Hu X, Mourant D, Gunawan R, Wu L, Wang Y, Lievensa C, Li CZ (2012) Production of value-added chemicals from bio-oil via acid catalysis coupled with liquid–liquid extraction. RSC Adv 2:9366–9370

Huber GW, Iborra S, Corma A (2006) Synthesis of transportation fuels from biomass: chemistry, catalysts and engineering. Chem Rev 106:4044–4098

Karagoz S, Bhaskar T, Muto A, Sakata Y, Oshiki T, Kishimoto T (2005) Low- temperature catalytic hydrothermal treatment of wood biomass: analysis of liquid products. Chem Eng J 108:127–137

Kruse A (2008) Supercritical water gasification. Biofuels, Bioprod Biorefin 2:415–437

Kruse A (2009) Hydrothermal biomass gasification. J Supercrit Fluids 47:391–399

Kumar S, Gupta RB (2008) Hydrolysis of microcrystalline cellulose in subcritical and supercritical water in a continuous flow reactor. Ind Eng Chem Res 47:9321–9329

Lee I, Kim M, Ihm S (2002) Gasification of glucose in supercritical water. Ind Eng Chem Res 41:1182–1188

Liu CF, Sun RC (2010) Cellulose. In: Sun RC (ed) Cereal straw as a resource for sustainable biomaterials and biofuels. Elsevier, Amsterdam, pp 131–167

Malins K (2017) Production of bio-oil via hydrothermal liquefaction of birch sawdust. Energy Convers Manage 144:243–251

McKendry P (2002) Energy production from biomass (part 1): overview of biomass. Bioresouce Technol 83(1):37–46

Miller JE, Evans L, Littlewolf A, Trudell DE (1999) Batch microreactor studies of lignin and lignin model compound depolymerization by bases in alcohol solvents. Fuel 78:1363–1366

Ogunsola OM, Berkowitz N (1995) Removal of heterocyclic S and N from oil precursors by supercritical water. Fuel 74:1485–1490

Osada M, Sato T, Watanabe M, Adschiri T, Arai K (2004) Low-temperature catalytic gasification of lignin and cellulose with a ruthenium catalyst in supercritical water. Energy Fuels 18:327–333

Peng P, Bian J, Sun RC (2010) Extractives. In: Sun RC (ed) Cereal straw as a resource for sustainable biomaterials and biofuels. Elsevier, Amsterdam, pp 49–72

Peterson AA, Vogel F, Lachance RP, Fröling M, Antal MJ, Tester JW (2008) Thermochemical biofuel production in hydrothermal media: a review of sub- and supercritical water technologies. Energy Environ Sci 1:32–65

Pratik NS, Babu BV (2009) Experimental studies on producer gas generation from wood waste in a downdraft biomass gasifier. Biores Technol 100(12):3127–3133

Ren JL, Sun RC (2010) Hemicelluloses. In: Sun RC (ed) Cereal straw as a resource for sustainable biomaterials and biofuels. Elsevier, Amsterdam, pp 73–130

Rowell RM, Pettersen R, Han JS, Rowell JS, Tshabalala MA (2005) Cell wall chemistry. In: Rowell RM (ed) Hand book of wood chemistry and wood composites. Taylor & Francis Group, Boca Raton, pp 9–40

Sahu SG, Chakraborty N, Sarkar P (2014) Coal–biomass co-combustion: an overview. Renew Sustain Energy Rev 39:575–586

Savage PE, Levine RB, Huelsman CM (2010) Hydrothermal processing of biomass: thermochemical conversion of biomass to liquid. In: Crocker M (ed) Fuels and chemicals. RSC Publishing, Cambridge, pp 192–215

Sevilla M, Fuertes AB (2009) Chemical and structural properties of carbonaceous products obtained by hydrothermal carbonization of saccharides. Chem Eur J 15:4195–4203

Svetlana L, Johan V (2000) Global potential of sustainable biomass for energy. SLU report. Swedish University of Agricultural Sciences, Department of Energy and Technology

Tekin K, Karagöz S, Bektaş S (2014) A review of hydrothermal biomass processing. Renew Sustain Energy Rev 40:673–687

Titirici MM, Antonietti M (2010) Chemistry and materials options of sustainable carbon materials made by hydrothermal carbonization. Chem Soc Rev 39:103–116

Toor SS, Rosendahl L, Rudolf A (2011) Hydrothermal liquefaction of biomass: a review of subcritical water technologies. Energy 36:2328–2342

Valdez PJ, Dickinson JG, Savage PE (2011) Characterization of product fractions from hydrothermal liquefaction of Nanochloropsis sp. and the influence of solvents. Energy Fuels 25:3235–3243

Vassilev SV, Baxter D, Andersen LK, Vassileva CG (2013) An overview of the composition and application of biomass ash. Part1. Phase-mineral and chemical composition and classification. Fuel 105:40–76

Xiao LP, Shi ZJ, Xu F, Sun RC (2012) Hydrothermal carbonization of lignocellulosic biomass. Biores Technol 118:619–623

Xu X, Matsumura Y, Stenberg J, Antal MJ (1996) Carbon-catalyzed gasification of organic feedstocks in supercritical water. Ind Eng Chem Res 35:2522–2530

Yokoyama S, Matsumura Y (2008) The Asian biomass hand book. The Japan Institute of Energy, Tokyo

Yoshida T, Matsumura Y (2001) Gasification of cellulose, xylan, and lignin mixtures in supercritical water. Ind Eng Chem Res 40:5469–5474

Zhang L, Xu C, Champagne P (2010a) Overview of recent advances in thermochemical conversion of biomass. Energy Convers Manage 51:969–982

Zhang Y (2010b) Hydrothermal liquefaction to convert biomass into crude oil. In: Blaschek HP, Ezeji T, Scheffran J (eds) Biofuels from agricultural wastes and byproducts. Wiley-Blackwell, USA, pp 100–276

Chapter 5
De-polymerization/De-fragmentation Aided Extraction of Value-Added Chemicals from Lignin

Parth G. Marakana, Anirban Dey, and Bharti Saini

5.1 Introduction

Value generation from biomass or agro-industry waste to value-added chemicals and materials is one of the most prominent ways to design a sustainable future. Lignocellulosic biomass is the most versatile and widely available renewable resource reservoir in Nature. Lignocellulose is mainly composed of lignin, cellulose, and hemicellulose (Chauhan 2020; Weng et al. 2021). Other biopolymers like pectin, waxes, tannin, etc. are also considered as an inevitable part of lignocelluloses, however, it is found in very small quantities. Among all the biopolymers present in the lignocellulose, cellulose is the most abundantly available biopolymer in the world while lignin is the most abundantly available aromatic biopolymer (which includes phenolic and benzene rings) in the world (Lu et al. 2017; Gellerstedt and Henriksson 2008). The contribution of cellulose, hemicellulose, and lignin in lignocellulosic biomass is around 30–50 wt%, 20–35 wt%, and 15–30 wt% respectively (Lu et al. 2017). Approximately natural production of lignin throughout the year on earth is around $5–36 \times 10^8$ tons/year (Gellerstedt and Henriksson 2008). Although there is a surplus availability of lignin naturally, a major fraction of it remains unutilized. Annually around 50–70 million metric ton of lignin is produced from paper and pulp industry (Norgren and Edlund 2014; Belgacem and Gandini 2008). Indeed it has been calculated that only 1–2% of lignin are getting utilized (Stewart 2008). The rest of the percentage of lignin is either considered as industrial waste or mostly utilized as lower value fuel for boilers in pulp and paper industries. Lignin is a prominent candidate for the production of high-value bio chemicals and fuels due to its desirable traits like high carbon to oxygen ratio and the presence of a strong aromatic matrix. Hence, it is

P. G. Marakana · A. Dey (✉) · B. Saini
Department of Chemical Engineering, Pandit Deendayal Energy University, Gandhinagar, Gujarat 382007, India
e-mail: anirban.dey@sot.pdpu.ac.in

Y. K. Nandabalan et al. (eds.), *Zero Waste Biorefinery*, Energy, Environment, and Sustainability, https://doi.org/10.1007/978-981-16-8682-5_5

highly desirable in the current situation to envisage lignin depolymerization chemistry. This in turn facilitates the development of efficiency as an effective pathway to produce value-generating products from lignin.

Lignin is a three-dimensional amorphous phenolic heteropolymer made up of phenyl propane structural units. The presence of lignin significantly helps the plant cells from severe atmospheric conditions, biological and chemical attacks (Lu et al. 2017; Yang et al. 2021; Vanholme et al. 2010). Lignin bio-synthesis naturally takes place by the phenylpropanoid pathway. Wherein, the multi-enzyme system converts phenylalanine into primary building blocks such as P-courmuryl alcohol, confiryl alcohol, and sinapyl alcohol which are linked by C–C and C–O–C bond to form three dimensional structure (Vanholme et al. 2010; Young et al. 1966). These three primary building blocks are considered monolignols which mainly varies in a methoxyl group composition on phenolic center. Based on these variations, three phenylpropanoids can be derived such as p-hydroxyphenyl (H), guaiacyl (G), and syringyl (S) which are generally linked by C–O or C–C (Vanholme et al. 2012; Nimz et al. 1981). The composition, as well as the concentration of all these monomers, varies depending upon plant species. Hardwood mostly contains coniferyl and sinapyl alcohol. Coumaryl alcohol is generally found in grasses, while softwood mostly contains coniferyl alcohol (Taylor 1993).These monolignol units are interconnected through various linkages such as β–O–4(45–50%), α–O–4(6–8%), β–1(7–10%), β–5(9–12%), 5–5(18–25%) β–β(0–3%) and 4–O–5(4–5%) (Zhao and Dixon 2011) with different ratios which depend on plant species. Among all the linkages, β-O-4 is the commonly occurring linkage which consists of around 50–60% of lignin (Sun et al. 2018; Schuler et al. 2019). Figure 5.1 represents the detailed molecular structure of lignin with common linkages and the existence of three basic monolignols.

Because of the large and complex structure of lignin and variation in lignin structure from source to source, it is very challenging to perceive the knowledge of depolymerization chemistry. Effective depolymerization can deliver very high-value chemicals and fuels. Various studies have indicated that lignin depolymerization could produce phenol and its derivative (phenols, guaiacols, syringols, etc.) based products which can be further processed to produce finer chemicals (Ye et al. 2012; Schutyser et al. 2018). This chapter contains elaborative discussion on lignin depolymerization or defragmentation strategies such as oxidative and reductive depolymerization, acid, and base-catalyzed depolymerization. Along with this, thermal depolymerization operations like hydrothermal liquefaction, microwave-assisted thermal depolymerization, and pyrolysis operation for extracting valuable value-added chemicals and fine chemicals intermediate have been also outlined.

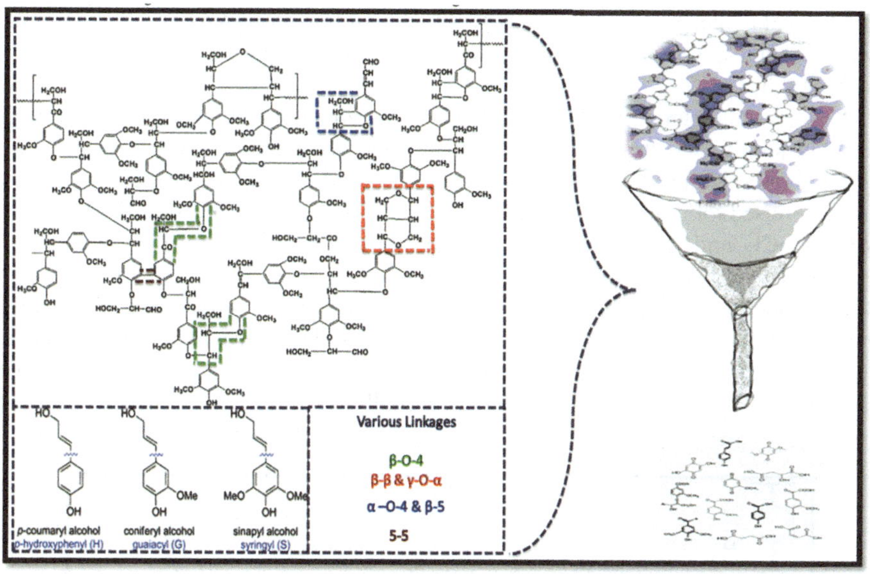

Fig. 5.1 Detailed molecular structure of lignin and its extraction to value-added products

5.2 Various Depolymerization Strategies of Lignin

5.2.1 Oxidative Depolymerization

Oxidative depolymerization is a popular approach for lignin defragmentation. In oxidative depolymerization, oxygen get involves in the reaction medium. Hence this depolymerization conversion takes place in presence of oxidizing agents such as oxygen, hydrogen peroxide, nitrobenzene, cupric oxide enhanced chlorine dioxide ozone, peroxy acid, etc. Oxygen directly is not efficient to oxidize lignin at ambient conditions, it requires an alkaline medium to form phenolate ions from phenolic groups (Heitner et al. 2010; Xu et al. 2014; Ma et al. 2015a). In the oxidative approach, defragmentation of lignin takes place by cleaving the ether side chain, forming phenolic compound, cleaving aromatic ring, and generating aliphatic carboxylic acids. It is delineated from the literature that most studies are inclined towards the production of phenolic monomers while very few studies have focussed on smaller aliphatic and carboxylic products. In the reaction mechanism, oxidation takes place by the electrophilic reaction. Accordingly, electrophilic ions like Cl^+ of chlorine and OH^- of peroxy acids attack high electron-dense locations such as para, ortho position in lignin molecule to deform lignin (Gierer 1986). Other mechanism includes cleavage of C–C bonds, β–O–4 aryl ether bonds, and aromatic rings. The presence of oxygen can lead to further additional functionality which forms many isomers that increase the complexity in lignin structure in terms of its application.

The typical mechanism of oxidative depolymerizations of lignin derived compound which include addition of oxygen functionality to depolymerised products has been presented in Fig. 5.2.

A deep understanding of these mechanisms and influencing reaction parameters can lead to the formation of a wide variety of alcohol, aldehyde, and acids as depolymerized products (Alonso et al. 2012). Various metal-based catalysts have been exercised for lignin depolymerization such as CuO, CeO_2, Fe_2O_3, Al_2O_3, $FeCl_3$, $NiCl_2$, $CoCl_2$, and $CuSO_4$ in various solvents like methanol, ethanol, dioxane, sodium hydroxide, and water as indicated in Fig. 5.3.

In comparison to reductive depolymerization, oxidative depolymerization requires milder conditions around 80–200 °C with varied O_2 pressure (Schutyser et al. 2018; Gale et al. 2020). Table 5.1, summarizes various oxidative depolymerization approaches conducted by the researchers globally.

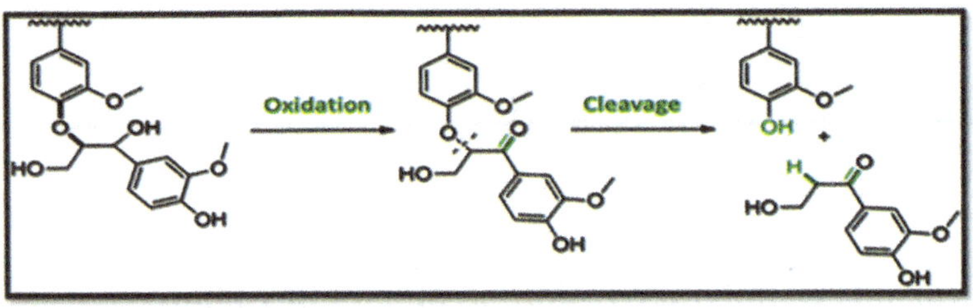

Fig. 5.2 Oxidative depolymerization of lignin-derived compound (Graglia et al. 2015)

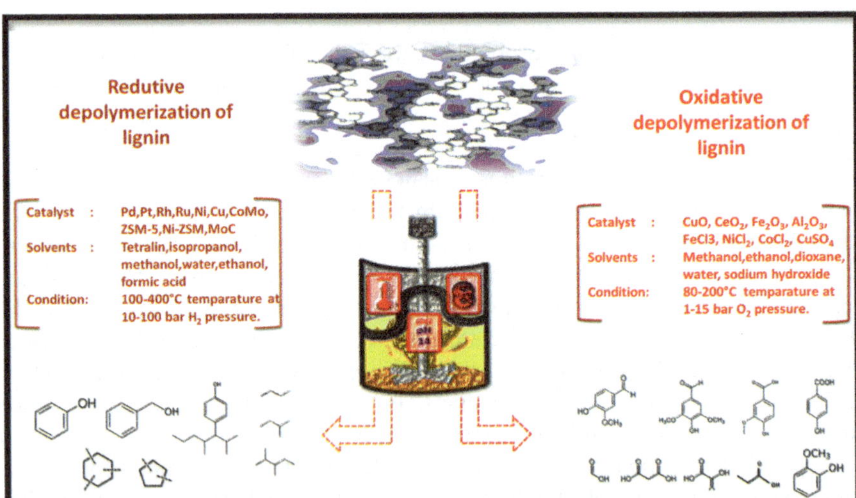

Fig. 5.3 Oxidative and reductive depolymerization approach

Table 5.1 Oxidative epolymerisation of lignin

Lignin source	Reaction conditions	Reactant, solvent, and catalyst	Major product	References
Sodium lignosulphonate lignin	160 °C at 15 bar O_2 for 40 min	H_2O as a solvent, Cu–Mn/δ-Al_2O_3 as Catalyst	Vanillin, *p*-hydroxybenzaldehyde, vanillic acid, *p*-hydroxybenzoic acid	Abdelaziz et al. (2019)
Kraft lignin	44.85 °C for 1–4 h	NaOH-H_2O_2 as a solvent, $Na_2WO_4 \cdot 2H_2O$ as a catalyst	Carboxyl-rich oligomer with lower molecular compounds	Junghans et al. (2020)
Softwood kraft lignin	Ambient temperature (35 °C) for 80 h	H_2O_2 as a solvent	Guaiacol-type, phenol-type, catechol-type carboxylic acids, esters, alcohols	Ahmad et al. (2020)
Lignin	120 °C at 5 bar O_2, for 3 h	NaOH, $LaCoO_3$ as a catalyst	Vanillin, syringaldehyde, p-hydroxybenzaldehyde	Deng et al. (2009)
Lignin	150–200 °C at 10% 1–2 bar O_2/N_2 for 4 h	H_2O as solvent CoFeO mixed oxides as a catalyst	Syringaldehyde, vanillin, acetosyringone, syringone, guaïacol, benzoic acids	Hdidou et al. (2018)
Lignin	30–60 °C for 30–720 min	$C_2H_4O_3$ (paracetic acid) as a solvent, Nb_2O_5 as a catalyst	4-hydroxy-2-methoxy phenol, vanillic acid, 3,4-dihydroxybenzoic acid	Ma et al. (2016)
Wheat alkali Lignin	150 °C for 30 min	$CH_3OH/CH_3CH_2OH/H_2O/C_4H_8O_2$ (dioxane) as a solvent $CuO/Fe_2(SO_4)_3$/NaOH as catalyst	Vanillin, vanillic acid, syringaldehyde, syringic acid	Ouyang et al. (2016)

Abdelaziz et al. (2019) designed Cu, Mn, and Al_2O_3 based catalysts for oxidative depolymerization. The findings of the study suggest that Cu-Mn/δ-Al_2O_3 is catalyzing carbon linkage of the propyl chain in lignin structural units. However, the presence of aldehyde in the product indicates more formation of intermediate products which are considered to be catalyzed by Cu-Mn/δ-Al_2O_3 catalyst (Abdelaziz et al. 2019). The same observation can be inferred from the work of Junghans et al. (2020) where H_2O_2 has been used as a catalyzing agent for the side-chain cleavage, which results in the separation of lower molecular weight products. Alkaline medium is considered as the most favorable condition to form hydroperoxide anion which significantly increases the depolymerization rate of lignin. The combination of sodium hydroxide and hydrogen peroxide results in an overall reduction of 82% in the average molecular weight compared to the starting material (Junghans et al. 2020). Ahmad et al. (2020) conducted the catalyst-free depolymerization study of kraft lignin at room temperature in the presence of hydrogen peroxide for extended time duration of 80 h. The result shows that depolymerization reaction leads to the reduction in the average molecular weight Mw of lignin from 6041 to 1400 Da (Ahmad et al. 2020). This further suggests that hydrogen peroxide plays a key role in the production of phenolic oligomers by breaking ether linkages and demethoxylation of guaiacyl and syringyl nuclei (Shimada et al. 1997). In the reported study of (Deng et al. 2009) work related to the catalytic wet alkali oxidation of lignin, $LaCoO_3$ is more efficient for aldehyde production. It is observed that Co^{3+} and Co^{2+} can be considered as the active redox ions for lignin oxidation. However, due to the stable valence state in the case of lanthanum La, no redox reaction was observed. Nevertheless, $LaCoO_3$ was found to be more efficient than Co^{2+} alone in the same catalytic process. The reaction mechanism involves a reaction between lignin molecule and adsorbed oxygen on the site of catalyst that produces reaction intermediates like quinone methide and hydroperoxide ions which results in lignin depolymerisation (Deng et al. 2009). According to Hdidou et al. (2018) CoFeO mixed metal oxide catalyst can lead to the formation of aromatic acids, alkylbenzene, and ketone around 200 °C temperature. It mainly takes place by extensive demethylation of an intermediate molecule like syringone, syringaldehyde, and guaiacol (Hdidou et al. 2018). Ma et al. (2016) tested oxidative depolymerization of lignin in acidic media of peroxyacetic acid. In the reported work, peroxyacetic acid acted as an oxidant at a very mild condition. The result shows that peracetic acid facilitates in the lignin defragmentation to monophenolic compounds by different mechanism which includes conversion of the propane side chain to carbonyl, ether, or hydroxyl to give better product selectivity. It is also found out that niobium pentoxide as a catalyst significantly improved monophenolic yield up to 47wt% (Ma et al. 2016). Ouyang et al. (2016) reported combination of $CuO/Fe_2(SO_4)_3/NaOH$ as a catalyst in Methanol/ethanol/water/dioxane based solvent and oxidant. Among all the reported solvents, water provides the maximum yield in alkaline conditions compared to the other organic solvents because of higher lignin solubility. While water with methanol triggered hydrothermal liquefaction of lignin, demethoxylation and methoxylation reactions are simultaneously taking place in the reaction medium and depolymerizing lignin. Methanol–water medium resulted in aryl ether linkage cleavage and efficiently inhibiting the Re-polymerization reaction of defragmented products (Ouyang et al. 2016).

5.2.2 *Reductive Depolymerization of Lignin*

The reductive depolymerization focuses more on deoxygenation to form narrow distribution of monomers while oxidative depolymerization, gives functionalized medium-range aromatics. In reductive depolymerization C–O bond replaces with a C–H bond by deoxygenation reaction which cleaves alkyl aryl ether bonds like β–O–4 and α–O–4 along with side-chain functional groups. Around 40–60% of linkage in lignin molecule is β–O–4 linkage. So by targeting it, reductive depolymerization can produce more monomeric yield. The general leading mechanism involves hydrogenolysis of ether bonds, withdrawal of Cγ–OH groups, and elimination of Cα–OH groups (Schutyser et al. 2018; Zaheer and Kempe 2015; Guo et al. 2016). Typical hydrogenation and hydrodeoxygenation of lignin-derived compounds are shown in Fig. 5.4. Hydrogenolysis generally cleaves the C–O bond and establishes hydrogen on cleaved chains wherein the hydrodeoxygenation, the process removes oxygen from oxygen-containing intermediate compounds. Throughout the reductive depolymerization, lignin defragmented or breaks down in the presence of redox catalysts such as noble metal (Pd, Pt, Rh, Ru), base metal (Ni, Cu, CoMo ZSM-5, Ni-ZSM, MoC). Solvents as a reducing agent are also playing a major role such as hydrogen donor; Hydrogen can be imparted in reaction by hydrogen gas or by deriving it from hydrogen donating species (tetralin, isopropanol, formic acid, methanol, and ethanol).

The usage of various catalysts and solvents along with reaction conditions for the reductive depolymerization approach can be retrieved from Fig. 5.3. So in a nutshell, redox catalyst in amalgamation with H_2 or H-donor is essential for the

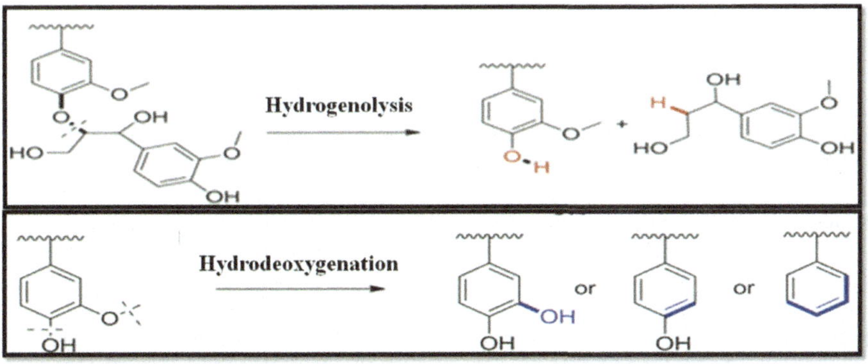

Fig. 5.4 Hydrogenolysis and hydrodeoxygenation reaction in reductive depolymerization approach (Graglia et al. 2015)

reductive depolymerization of lignin (Schutyser et al. 2018). This hydroprocessing of lignin can be further classified into mild, harsh, and bifunctional hydroprocessing. Mild condition operates below 300 °C with the production of p-substituted methoxy phenols, while harsh condition operates above 320 °C and higher pressure of more than 35 bar with the production of phenol, phenols with longer alkyl chains, methylated phenols (Lancefield et al. 2016; Narani et al. 2015). In the case of bifunctional hydroprocessing, acidic catalyst and solvents are disadvantageous due to their corrosive nature. It is also observed by many researchers that the reductive catalytic process inhibits the reactive functional group formation which eventually results in condensation or repolymerization such as carbonyl and alkenyl group. In the reductive depolymerization of lignin, in most cases, cleavages are dependent on hydrogen donating species and solvents. Table 5.2 shows various studies of reductive depolymerization based on different catalysts and solvents. According to Shao et al. (2017)

Table 5.2 Reductive depolymerization of lignin

Lignin source	Reaction conditions	Reactant, solvent, and catalyst	Major product	References
Organosolv lignin	250 °C at 7 bar H_2 for 20 h	H_2O as a solvent, Ru/Nb_2O_5 as catalyst	Arenes, cycloalkanes	Shao et al. (2017)
Alkali Lignin	250 °C, at 10 bar H_2, for 4 h	H_2O, $Ru/\gamma\text{-}Al_2O_3$ and $[Hf(OTf)_4]$ catalytic system	Phenol, benzyl alcohol	Chen et al. (2021)
Lignin	220 °C, at 20 bar H_2 for 5 h	$C_4H_8O_2$ (1,4-dioxane) as a solvent, NiC as a catalyst	Syringylpropanol, guaiacylpropanol, propylsyringol, propylguaiacol	Si et al. (2017)
Organosolv lignin	360 °C at 70 bar H_2 for 3 h	H_2O as a solvent, raney Ni as a catalyst	phenol, 4-propyl phenol	Strüven and Meier (2016)
Lignin	260 °C for 3 h	CH_3OH as a solvent, MoOx/CNT as a catalyst	Syringyls (4-propenylsyrin gol, 4-allylsyringol, and 4-propylsyringol), guaiacol (isoeugenol)	Xiao et al. (2017)
Organosolv lignin	100 °C at 20 bar H_2 for 20 h	$CH_3CH_2OH\text{-}H_2O$ as a solvent, $Pd/NbOPO_4$ as a catalyst	4-ethyl phenol, syringol	Fang et al. (2020)
Lignin	400 °C at 100 bar H_2 for 4 h	NiMoP/AC as a catalyst	Alkylphenols, aromatics, cycloalkanes, noncyclic alkanes	Chowdari et al. (2019)
Kraft lignin	280 °C, 1 bar N_2 bar for 6 h	CH_3CH_2OH as solvent, Mo/Al_2O_3 as a catalyst	Alcohols, esters, monophenols, benzyl alcohols, and arenes	Ma et al. (2015b)

the noble metal catalyst, Ru/Nb_2O_5 provides a complete deduction of oxygen content and produces C_7–C_9 hydrocarbons by depolymerizing lignin. It is also believed that firstly lignin was depolymerized into phenolic monomers by C–O hydrogenolysis of C–O–C ether bonds which result in the formation of arenes and cycloalkanes by further hydrogenolysis on Ru/Nb_2O_5 catalyst (Shao et al. 2017). Chen et al. (2021) reported the application of Hafnium trifluoromethanesulfonate $Hf(OTf)_4$ with Ru/γ-Al_2O_3. It is observed that yield of depolymerized product was around 7.18% at 10 bar while the yield was exponentially increased to 42.98% when the hydrogen pressure was increased to 20 bar.

The high hydrogen pressure promotes higher activity on the surface of metal-based acid catalysts (Chen et al. 2021). It is observed by Si et al. (2017) that β-O-4 linkages in the lignin are linearly transforming into aromatic monomers throughout the catalytic conversion of lignin. It is also highlighted that as temperature increased from 170 to 200 °C monomeric yields were decreased along with the increment in selectivity in the presence of NiC catalyst (Si et al. 2017). In the study of (Strüven and Meier 2016) raney nickel has a worthwhile effect with or without a hydrogen source. Detailed comparison of processes with or without the usage of raney nickel as a catalyst reveals that coke formation was lowered from 48wt% to 20wt% in the absence of hydrogen in the former case (Strüven and Meier 2016). According to Chowdari et al. (2019), depolymerization of lignin at a higher temperature around 400 °C decreases the yield of product and increases the formation of coke or solids which indicates the repolymerization process. Higher temperature and the formation of a high amount of water signify the rapid hydrodeoxygenation reaction. From the GC–MS analysis, sharpness in the peak without any tailing appeared at a higher temperature, which indicates more formation of smaller molecular weight compounds. A similar effect can be achieved by using NiMoP/AC catalyst at a lower temperature (Chowdari et al. 2019). In the study of (Xiao et al. 2017) it is stated that the yield of monomer increased in the case of methanol compared to ethanol, propanol, ethylene glycol, and water. This is due to the higher solubility of H_2 and lignin in the presence of MoOx/CNT catalyst. Product distribution suggests that catalyst particularly results in the breakdown of C-O bond instead of double bonds of phenolic monomers (Xiao et al. 2017).

5.2.3 Acid-Catalyzed Depolymerization of Lignin

The first acid-catalyzed depolymerization of lignin was reported by Hagglund and Bjorkman (1924). Similarly, (Hewson and Hibbert 1943) attempted series of experiments on maple wood by using a varied combination of alcohol (ethanol, ethylene glycol) and acids (formic acid, hydrochloric acid). This further helps to separate water-soluble compounds and insoluble compounds. In these previous studies, it was found that depolymerization using acid gives lower aromatic yields and more re-polymerization of formatted products. However, it was also observed that

Table 5.3 Acid-catalyzed depolymerization of lignin

Lignin source	Reaction conditions	Reactant, solvent, and catalyst	Major product	References
Corn stover Lignin	250 °C for 6 h	CH_3COOH as a solvent Zr-KIT-5 (Si/Zr = 20) as Catalyst	Acetylated phenolic monomers	Nandiwale et al. (2019)
Wheat straw lignin	3600 °C, at 250 bar, for 4 h	CH_3CH_2OH as a solvent, HCOOH as a catalyst	Methoxyphenols, catechols, phenols	Gasson et al. (2012)
Wheat alkali lignin	300 °C, for 2 h	$HCOOH–CH_3CH_2OH–H_2O$ as a solvent, Pd/C as a catalyst	Syringyl and alkyl syringyl, guaiacyl and alkyl guaiacyl	Chen et al. (2018)
Hydrolysis lignin	250 °C at 20 bar N_2 for 1 h	$CH_3CH_2OH–H_2O$ as a solvent, H_2SO_4 as a catalyst	Aliphatic, phenolic monomers	Mahmood et al. (2015)
Cellulolytic enzyme lignin	270 °C at 1 bar N_2 for 30 min	H_2O/CH_3OH as a solvent, acidic zeolites and raney Ni as a catalyst	Phenolic monomers	Jiang et al. (2015)
Lignin	140 °C for 30 min	$(CH_2OH)_2$ and $C_4H_8O_2$ as a solvent, $Fe(OTf)_3$ as a catalyst	Phenolic acetals	Deuss et al. (2017)
Alkaline lignin	100–180 °C, for 10–60 min	$(CH_2OH)_2–C_6H_6O$ (phenol) as a solvent, H_2SO_4 as a catalyst	Guaiacols, syringols	Ouyang et al. (2015)

acidic conditions enhanced the ether bond hydrolysis in the carbohydrate polymers. A summary of different reaction conditions employed in acid-catalyzed depolymerization of lignin is presented in Table 5.3.

In the case of cellulose and hemicelluloses, acidic conditions are prone to solubilize them. Acidic conditions do not promote solubilization of lignin whereas alkaline conditions facilitate lignin solubility, which subsequently influences the balance between depolymerization and re-polymerization of lignin (Pu et al. 2013, 2015). It is considered that α- and β-aryl ether linkages are the weakest linkages followed by aryl-aryl ether bonds. Hydrolytic breakage of α- and β-aryl ether bonds plays a significant role in the acid-catalyzed depolymerization of lignin. Under acidic conditions, lignin is prone to depolymerize by aryl ether bond cleavage (Kim et al. 2011; Awan et al. 2019). In the case of cleavage of β-O-4 ether linkage, primarily β-O-4 acidolysis forms benzylic carbenium ion by the removal of a hydroxyl group from α position. Unstable intermediate carbenium ion gets transferred into enol ethers. This intermediate chemistry depends on used acids (Sturgeon et al. 2014; Imai et al. 2011). For the acid-catalyzed depolymerization, various catalysts such as formic acid, sulphuric acid, Raney Ni, Fe, Cu, Zn, zeolite with different Al/Si ratios are being used in ethanol, methanol, propanol, ethylene glycol, dioxane solvents. The detailed mechanism is presented in Fig. 5.5.

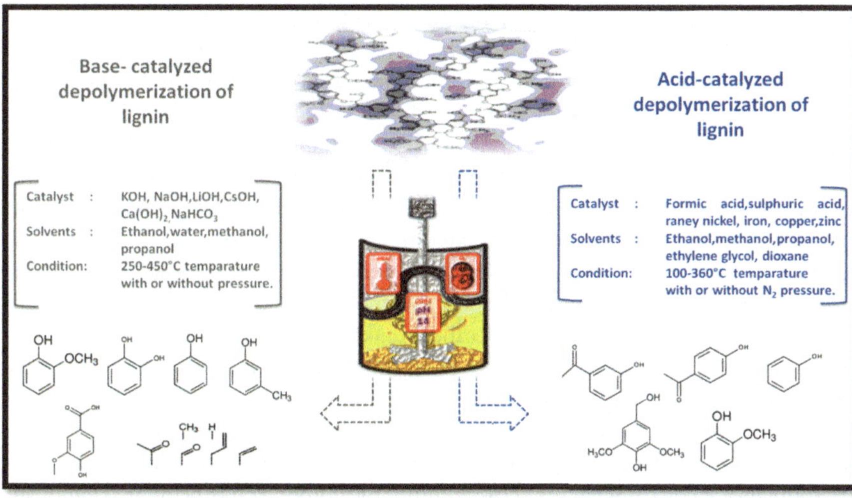

Fig. 5.5 Acid and base-catalyzed depolymerization approach

Various combinations of solvents and catalyst are found to be effective when it is used in proper synchronization. Nandiwale et al. (2019) studied acidic depolymerization of lignin by using solvents such as 1,4-dioxane, glacial acetic acid, and γ-valerolactone with ZrKIT-5 (Si/Zr = 20). In terms of solubility and thermal stability, acetic acid yields the best result among all. This synchronization of optimized solvent-catalyst combination resulted in an increment of yield and selectivity compared to depolymerization without catalyst. With the involvement of catalyst, the yield and selectivity were found to increase from (4 wt % to 11.8 wt %) to (13.5% to 37%) respectively. It has been also observed that Zr-KIT-5 helps in the acetylation of formed phenolic monomers on lewis acid sites which stabilize the product and prevent Re-polymerization (Nandiwale et al. 2019). According to Gasson et al. (2012) the reaction is more temperature-dependent in ethanol and formic acid media. In the starting stage of the reaction, more syringol and guaiacol-type compounds were formed rapidly. These compounds further react via enhanced demethoxylation and deoxygenation step, which results in the formation of more catechol and stable phenols (Gasson et al. 2012). Chen et al. (2018) studied the effect of the initial molecular weight of lignin in an acidic medium. It was found that lignin with higher initial molecular mass, delivered more alkyl compounds such as 1-ethoxy-4-methyl-benzene, 1-methyl-2-propenyl-4-methoxy-benzene as depolymerized products. This indicates that ethylation and methylation take place with the phenolic monomer. In higher molecular weight lignins, the formation of S-type molecules is significantly reduced, which corresponds to the demethoxylation during depolymerization of lignin from alkyl-substituted benzene. When we switch from higher molecular weight lignin to lower molecular weight lignin, higher molecular weight lignin produces guaiacyl and guaiacyl alkyl compounds, while lower molecular weight produces syringyls and alkyl syringyl compounds (Chen et al. 2018). Based on the work of Mahmood et al. (2015) related to solvent and catalyst effects, it has been observed that water alone as the solvent gave a 68% wt yield with a molecular weight distribution of 2030 g/mole. While in the case of water–ethanol, the average yield increased to 70 wt% with a narrow lower molecular weight distribution of 1000 g/mole. This is mainly due to the ability of ethanol to donate hydrogen which greatly enhances the hydrolysis reaction (Mahmood et al. 2015). Comparative study between the catalyst aided and catalyst-free depolymerization reaction shows that upon using sulfuric acid as a catalyst, the efficiency increased to 72 wt%. However, a product distribution of higher molecular weight about 1660 g/mol along with a lower solid residue was obtained. Here, the molecular weight distribution in the presence of a catalyst corresponds to the breakdown of acid-catalyzed alkyl-aryl bond in lignin (Long et al. 2012). Jiang et al. (2015) conducted a study on the effect of acidic zeolites and nickel as catalysts in lignin depolymerization. It was found out that Ni alone produced 12.9 wt% monophenols while zeolite alone produces 5 wt% monophenols. While a combination of Ni and zeolites gives a synergistic effect and produces a 21–27.9 wt% yield of phenolic monomer. It has been also suggested that acidic zeolite promotes nickel's catalytic effect by catalyzing carbonium ions to defragment lignin (Jiang et al. 2015). In the work of Ouyang et al. (2015) the catalytic effect of sulphuric acid and the external addition of phenol in the reaction medium was studied. The results showed that in the presence of sulfuric acid, 13.6 wt% of phenolic monomers were obtained (Ouyang et al. 2015). It was also observed that acidic conditions are more favorable for the cleavage of β-ether bonds. Under acidic conditions, aryl-ether bonds can be easily broken down which results in benzyl carbocation followed by the nucleophilic reaction by alcoholic hydroxyl groups of lignin to detach phenolic monomers. The addition, phenol plays a vital role in hydrogen donation which prevents re-polymerization and condensation of the formed intermediate which led to an increment in the yield of monomeric phenols (Jia et al. 2011; Brebu and Vasile 2010).

5.2.4 *Base Catalyzed Depolymerization of Lignin*

Base catalyzed depolymerization is also a widely studied depolymerization technique to obtain value-added chemicals from lignin. For the alkaline depolymerization technique, various alkali agents are being used such as KOH, NaOH, LiOH, CsOH, $Ca(OH)_2$, $NaHCO_3$, etc. (Schutyser et al. 2018; Chio et al. 2019). Lignin is generally soluble in the alkaline medium which is beneficial here compared to other depolymerization technique which needs additional lignin dissolving expensive organic solvents. Because of the better solubility of lignin in an alkaline medium, most of the lignin isolation work has been carried out in the basic medium. During the process, lignin depolymerization in the alkaline condition allows the breakdown of the carbohydrate bonds of lignin via cleavage of β-O-4 (aryl-glyceryl-β-aryl ether bond) and 4-O-5 (diaryl ether bond) (Yuan et al. 2010; Schmiedl et al. 2012). At lower temperature conditions below 300 °C, methoxy phenols are the anticipated products which are mainly ether substitutes of aromatic aldehyde and acetophenone-based compounds. While at a higher temperature above 300 °C, demethoxylation and demethylation take place which shifts more production of catechols type compounds from initially formed syringol type compounds during depolymerization of lignin (Schmiedl, et al. 2012; Katahira et al. 2016; Kruger et al. 2016). Base-catalyzed depolymerization reaction occurs via carbenium ions. Carbenium ions of syringyl are more stable than guaiacy carbenium ion due to higher concentration of methoxy substituents which results in surface positive charge. Hence, syringyl intermediates are highly stable, still, they can suffer dealkylation, demethylation to form phenolic monomers such as cresol, catechols, guaiacol, and phenols (Miller et al. 2002; Barbier et al. 2012). The schematic representation of the base-catalyzed depolymerization approach is shown in Fig. 5.5.

Base-catalyzed depolymerization is a cost-effective depolymerization technique (Schmiedl, et al. 2012; Katahira et al. 2016). As sodium hydroxide and potassium hydroxide are economical alkaline agents. In addition, due to the high solubility of lignin in alkaline media, an external solvent for lignin dissolution is often not required in most cases (Miller et al. 2002). The different catalysts used along with the reaction conditions in the above process are mentioned in Table 5.4. Chaudhary and Dhepe

Table 5.4 Base-catalyzed depolymerization of lignin

Lignin source	Reaction conditions	Reactant, solvent, and catalyst	Major product	References
Lignin	250–450 °C, for 15 min	KOH	Phenol, catechol, guaiacol, and 4-methylcatechol	Schuler et al. (2017)
Wheat straw lignin	250 °C for 1 h	CH_3CH_2OH–H_2O as a solvent, basic zeolites, NaX (Si/Al = 1.2)/KLTL (Si/Al = 2.7) as catalyst	Guaiacol, acetoguaiacone, 2-methoxy-4-methyl phenol, vanillin, 4-hydroxy benzyl alcohol	Chaudhary and Dhepe (2017)
Steam explosion hemp lignin	300–330 °C at 35 bar	NaOH	Guaiacol, catechol	Lavoie et al. (2011)
Organosolv lignin	300 °C at 90 bar for 40 min	H_2O as solvent, NaOH as a catalyst	Guaiacol, catechol, phenol, cresol	Toledano et al. (2014a)
Organosolv lignin	270–360 °C at 250–300 bar for 20–60 min	NaOH as a catalyst	Syringol, syringyl aldehyde, hydroxyacetophenone, 4-methyl syringol, guaiacol, vanillin	Roberts et al. (2011)
Lignin	300 °C, 40 min	NaOH as a catalyst	Catechol, phenol, m-p-cresol, 4-methyl catechol, guaiacol	Santos et al. (2016)

(2017) studied the base-catalyzed depolymerization in ethanol–water organic solvent using various base catalysts.

The resultant product distribution from their work shows that ethanol, as solvent is not participating in the depolymerization process as a substrate, since most of the compounds contain a methoxy group such as guiacol, 2-methoxy-4-methylphenol, vanillin, acetoguaiacone rather than an ethoxy group. This also indicates that lignin is not depolymerized by alcoholysis, instead is depolymerized using hydrolysis by producing hydroxyl groups on top of aromatic rings such as guaiacol, pyrocatechol, 2,6-dimethoxy phenol etc. When comparing different catalysts, the result showed that basic NaX zeolite (Si/Al = 1.2) to be more effective than others. It has also been observed that an increase in pH leads to a lower product yield. In the case of NaX, a pH value of 9.2 gives better results in terms of catalyst. An increase in pH above 9.2 resulted in a decrease in product yield (Chaudhary and Dhepe 2017).

According to Lavoie et al. (2011) product distribution in NaOH as catalyst and solvent showed substitution of ortho methoxy group in most of the compound. Hydroxyl substitution can be seen for phenol, cresol, guaiacol with an increase in temperature, while methoxy substitution on product compound also led to a decrease in the concentration of 3-methylcathecol, 4,5-dimethylcathecol, 4,5-dimethoxy-2-methyl phenol, and syringaldehyde. After a certain temperature increment, depolymerized products tend to re-polymerize with each other through condensation which is a serious issue. Several studies reported the usage of capping agents to reduce repolymerization (Lavoie et al. 2011). Toledano et al. (2014a) reported the usage of boric acid and phenols as a capping agent to reduce re-polymerization by improving the bio-oil yield. In a comparison between these capping agents, phenol as a capping agent exhibits better performance compared to boric acid. Phenol as a capping agent showed the highest yield of bio-oil by avoiding re-polymerization. It is also observed that excess quantity of phenol does not decrease re-polymerization. In the case of boric acid as a capping agent, the excess dosage is required to achieve an efficient result. Capping agents also had an impact on product distribution as it holds the intermediates for a longer period of time which eventually interfere with the reaction mechanism. In absence of capping agents, the main product was catechol and 4-methyl catechol while in the presence of capping agent, guaiacol yield was decreased (Toledano et al. 2014a). Roberts et al. (2011) also used boric acid as a capping agent. The result showed a 36 wt% increase in product yield with boric acid. Boric acid also served as a catalytic agent for acidic ether hydrolysis along with capping activity. In the experimentation with sodium hydroxide base catalysts, it is observed that β-O-4 ether bond cleavage takes place in which hydroxide as anion and sodium as the cation of sodium hydroxide participated in depolymerization. Sodium-ion catalyzed by polarizing ether linkage to form sodium phenolates which further initiate oligomer formation. On the other side, hydroxide ion leads to the proton abstraction from the carbon atom of β-O-4 ether linkage to form 4–1-hydroxy vinyl-syringol (Roberts et al. 2011). According to (DosSantos et al. 2016) a basic catalyst gives more catechol and guaiacyl compounds compared to syringol compounds due to the more susceptibility of the latter towards base-catalyzed depolymerization (DosSantos et al. 2016). These obtained phenolic monomers, aldehydes, oligomers, arenes, polyols, and other smaller molecular compounds can be used in the production of rubbers and plastic, in pharmaceuticals, paint and pigments, polyurethane and foams, in composite materials.

5.3 Depolymerization Thermal Treatments

5.3.1 Hydrothermal Liquefaction of Lignin

Hydrothermal liquefaction (HTL) is considered a thermo-chemical technique to harvest liquid phase (Bio-oil) products from biomass. This process generally takes place at moderate temperature and high pressurized conditions. Gollakota et al. (2018) and Toor et al. (2011) studied that at varied temperature and pressure conditions, physiochemical properties change which results in distinctive reaction conditions (Gollakota et al. 2018; Toor et al. 2011). Which was also observed by Peterson et al. (2008). HTL can generate low oxygen-containing oil compared to other thermochemical and fast pyrolysis processes. During the process, oxygen-containing smaller organic compounds get hydrolyzed by feedstock which would further separate into an aqueous phase. In this process, the oxygen content of organic feedstock gets reduced from 40 to 10–15% (He et al. 2008). During the process oxygen rapidly mineralizes or oxidizes into water or carbon dioxide, where nitrogen of the heteroatom converts into N_2 or N_2O (Akhtar and Amin 2011). In the HTL depolymerization, feedstock macromolecules are defragmented into fragments of light molecules. Various processing parameters influence the yield of bio-oil during HTL e.g., particle size, solvent density, biomass feedstock, biomass heating rate, pressure, temperature, residence time, etc. (Wang et al. 2016). The mechanism of HTL of lignin is less explored than other treatments such as pyrolysis, gasification, etc.

Several studies have been executed at varied experimental conditions. HTL of lignin mostly gives lignin model compounds from which fine chemicals can be recovered by the selection of specific catalyst, solvent, and reaction conditions. Arturi et al. (2017) conducted HTL of lignin at near supercritical water conditions and studied the influence of phenol. It has been observed that the addition of phenol leads to the production of more single-ringed aromatic compounds. Also, the solubilization of small aromatic compounds increased (Arturi et al. 2017). Jensen et al. (2018) reported HTL of enzymatically hydrolyzed lignin. From the work, it can be outlined that lignin is depolymerizing into methoxy phenols and catechols. At a very higher temperature, due to the rapid demethylation of aromatic methoxyl substances, low or near to zero yield of syringol, methoxyl benzene, guaiacols, and 3-methoxycatechols can be observed (Jensen et al. 2018). Kang et al. (2015) reported the depolymerization of lignin using HTC to get antioxidant products. They have observed the highest yield of the phenolic monomer of more than 85% corresponding to the temperature and reaction time of 320 °C and 30 min respectively. From the GC–MS analysis, the largest spectrum of guaiacol has been identified which can further be narrowed down

Table 5.5 Hydrothermal liquefaction of lignin

Lignin source	Reaction conditions	Solvent and catalyst	Major product	References
Kraft lignin	280–350 °C at 181–240 bar	H_2O, 1.7 wt% K_2CO_3, 9.7wt% phenol	Giuacaol, catechol, alkylphenols, methoxybenzenes	Arturi et al. (2017)
Eucalyptus wood lignin	255–345 °C for 20 min	H_2O, 5.6wt% KOH	Hydroxyl acids, phenols, catechol, phenolic ketones	Jensen et al. (2018)
Alkaline lignin	320 °C for 30 min. At 2 Mpa	H_2O	Guaiacol, catechol, phenols, o-cresol	Kang et al. (2015)
Lignin	350 °C for 2 h	H_2O, HCOOH	Methanol, acetic acid, dimethyl ether, acetone, catechol, phenol	Halleraker and Barth (2020)
Lignin	2000 °C for 30 min	CH_3OH, C_2H_5OH	Phenol substitutes and aromatic ethers	Singh et al. (2014)

to other antioxidant phenols (Kang et al. 2015). Some solvent-based studies have also been reported. Halleraker and Barth (2020) reported the formic acid-assisted HTL of lignin which delivered a higher concentration of methanol in the product mixture (Halleraker and Barth 2020). Singh et al. (2014) carried out phenol substitutes and aromatic ethers from HTL of lignin. It is proposed that aromatic ethers and phenolic substitutes are formed due to the breakage of C–C and C–O–C ether bonds (β-O-4 or/and α-O-4). The presence of ethanol and methanol reacts with remaining OH groups resulting in the formation of methyl and ethyl esters. This signifies that depolymerization can occur for a very short residence time and mild condition in the presence of methanol and ethanol (Singh et al. 2014). Lui et al. (2020) reported the HTL of α-O-4 Aryl-Ether Linkage of lignin. Phenolic and alkyl benzene derivatives were the major products of this study. Methoxy-groups and ether linkages on the 4-positions have a strong response on the rate of decomposition of α-O-4 aryl-ether linkages which resulted in a major production of alkylbenzene and phenolic derivatives (Lui et al. 2020). Table 5.5 represents the summary of various studies conducted on hydrothermal liquefaction of lignin.

5.3.2 *Microwave-Assisted Thermal Depolymerization of Lignin*

Microwave thermal technology has been widely used in lignin liquefaction and manifests to be encouraging technology due to its superiority over other techniques.

Microwave thermal treatment decreases the reaction time and reduces unwanted side reactions compared to other heating treatments. Hence it is widely applied in various synthesis-related applications due to the lower consumption of energy and its high effectiveness (Fan et al. 2018; Morgan et al. 2017). Solvents and catalysts play a key role throughout the Microwave-Assisted liquefaction of lignin. Microwave-assisted thermal degradation of lignin processes includes hydrogenation/hydrodeoxygenation reaction followed by the breakage of C–C, C–O bonds. Phenolic derivatives and liquid fuel are the major products from the microwave-assisted heating of lignin over a variety of catalysts (Zhou et al. 2018a). Microwave heating occurs via ionic conduction and dipolar polarization. Microwave-induced heating at the molecular level during reaction beneficially reduced side reactions or re-polymerizations (Salema and Ani 2011). To form the products from the reaction medium one should focus on the receptivity of microwave radiation in the reaction medium so that reaction can take place. So additional microwave receptors with good absorbance ability must be there in the reaction medium for the effective depolymerization of lignin (Lam et al. 2020). It has been observed that polar solvents are having good receptivity to microwave irradiation. Alcohols possess higher microwave radiation than water (Zhang et al. 2020). So that hydrogen donating polar solvent such as (MeOH, EtOH, iPrOH, BuOH) is very safe and effective to use (Shafaghat et al. 2020). It is also noticed that solvent selection is also influenced by the loss tangent value. The loss tangent, $\tan\delta$, is known as the ratio of dielectric loss factor to dielectric constant, and it analyzes the capability of the solvent to efficiently convert microwave energy to heat energy. Depending on the loss tangent value, solvents are categorized into high microwave absorbing ($\tan\delta > 0.5$), moderate absorbing ($0.1 < \tan\delta < 0.5$) and low absorbing ($\tan\delta < 0.1$) (Gawande et al. 2014). The high absorbency of microwave contributes to more appropriate molecular polarity in the point of view of a free radical reaction which noticeably decreases the reaction time. Microwaves can pierce into the interior of molecules, which results in the formation of dipolar or polar molecules of long-chain oscillation with a very high frequency and hence generate huge energy sources (Yunpu et al. 2016). It is also observed that solvolysis of lignin, catalyst at less than 200 °C gives the best output (Toledano et al. 2014b).

Microwave-assisted thermal treatment of lignin mostly produces lignin model compounds which are medium-chain aromatic compounds like guaiacols, catechol, syringol, vanillin, polyhydroxy phenols, etc. The Proportion of these products and their yields depends on the source and process parameters. Catalyst plays a significant role in obtaining the desired product. Table 5.6 briefly summarizes the studies conducted using microwave-assisted thermal treatment.

Zhou et al. (2018) observed that CuNiAl catalyst promotes a 10–20% increase in yield of liquid products compared to the blank run in methanol solvent (Zhou et al. 2018b). According to Zou et al. (2018) $MnCl_2$ as a catalyst shows a higher catalytic effect among other metal chlorides and results in a higher total yield of monolignols. $MnCl_2$ catalyzed reaction to produce more guaiacly (G-type) (23%) and p-hydroxyphenyl (H-type) (14.8%) type phenolic compounds compared to syringyl (S-type) (11.9%) type (Zou et al. 2018). In the study of Milovanovic et al. (2016) it was stated that the involvement of bifunctional catalyst (NiO/H-ZSM-5) increases the

Table 5.6 Microwave-assisted thermal treatment of lignin

Lignin source	Reaction condition	Solvent and catalyst	Major products	References
Alkaline lignin	100–160 °C at 400 W for 40–80 min	CH_3OH as a solvent, CuNiAl hydrotalcite-based catalysts	Guaiacol, syringaldehyde, p-hydroxy acetophenone, p-hydroxy acetovanillon	Zhou et al. (2018b)
Lignin from black-liquor	100–160 °C at 600 W for 5–60 min	C_3H_8O (isopropanol)	Ethanone	Liu et al. (2017)
Alkaline lignin	120–180 °C at 400 W for 15–45 min	CH_3OH–HCOOH media	2,3-dihydrobenzofuran, *p*-coumaric acid, vanillin (guaiacol, syringol, and phenolic in small fraction)	Shao et al. (2018)
Bamboo lignin	160 °C for 30 min at very low power	HCOOH–CH_3CH_2OH medium in HCl, $MnCl_2$ catalyst	Bio-oil (guaiacol & H-phenolic in high wt% compare to syringol	Zou et al. (2018)
Alkaline lignin	100–140 °C at 600 W for 20–80 min	$(CH_3)_2SO$ (Dimethyl sulfoxide), $(CH_2OH)_2$ media	Syringaldehyde, acetosyringone	Dhar and Vinu (2017)
Bamboo lignin	100–200 °C at 80 W for 20–60 min	HCOOH–CH_3CH_2OH	Bio-oil (guaiacol, vanillin, syringol, etc.)	Duan et al. (2017)
Pine sawdust lignin	180 °C for 15 min	CH_3OH	Phenolics, poly-hydroxy polyols	Xu et al. (2012)
Hardwood lignins	180 °C at 400 W for 60 min	HCOOH, NiO/H-ZSM-5 catalyst	Bio-oil (mesitol, banillin), biochar	Milovanović et al. (2016)
Lignin (sigma aldrich)	130–170 °C at 180 W for 5 min	$HO(CH_2)_4OH$ (1,4-butanediol, glycerol)	Bio polyols	Gosz et al. (2018)
Kraft lignin	626–966 °C at 2.7 kW for 13.33 min	–	Guaiacols, phenols, catechol, p-hydroxy-phenyl	Farag et al. (2014)

bio-oil yield to about 20 folds. During the microwave-assisted thermal degradation process, catalyst composition plays a significant role. It was observed that NiO metal sites were accountable for cleavage of C-O bonding, while H-ZSM-5 acidic sites were responsible for deacylation and dealkylation throughout the process. Along with the catalyst, solvents and their polarity also play remarkable roles (Milovanović et al. 2016). According to Shao et al. (2018) formic acid and lignin ratio can be used in depolymerization and Re-polymerization of lignin. An increase in formic acid content in the reaction mixture increases the production of bio-oil and decreases the residual formation. However, it was also observed that a higher amount of formic acid leads to re-polymerization which increases residual product. So formic acid plays a crucial role in establishing the balance between the depolymerization and re-polymerization reactions. Based on the analysis of the product, it can be inferred that the formic acid was responsible for the cleavage of the C-O bond in lignin (Shao et al. 2018). In the study of Dhar and Vinu (2017) it was stated that ethylene glycol and dimethyl sulfoxide are polar and are characterized by high loss tangent value. In the case of ethylene glycol, at a higher temperature, it becomes transparent to microwave radiation. This results in the requirement of more microwave power to maintain the desired temperature. While the same thermal and depolymerization effect was observed in the production of monomers and oligomers for both the solvent (Dhar and Vinu 2017). Duan et al. (2017) reported the usage of Ethanol-Formic acid media, in which formic acid acted as hydrogen donor solvent while the ethanol worked as a swelling agent. This eventually results in more S-type compounds in the product mixture than G-type compounds (Duan et al. 2017).

5.3.3 Pyrolysis of Lignin

Pyrolysis of any material is a thermo-chemical heating treatment in which material is heated to defragment or decompose in absence of air, vaporize and condense to liquefied products (Hu et al. 2018). The pyrolysis process produces mainly carbon and gaseous products, the condensation of which produces a liquid product known as Bio-oil. For the sake of increasing product yield, a high heating rate around 300–1000 °C/min^{-1} in a very short duration around 1–2 s is advisable (Bridgwater 2012). This thermal defragmentation by pyrolysis includes reactions such as defragmentation, decarboxylation, oxidation, hydrolysis, and dehydration. The depolymerization of lignin occurs in three stages. In the initial stage, pyrolysis occurs around 12–200 °C, followed by the de-fragmentation or rearrangement step. This leads to the formation of free radicals, the release of water molecules, the breakage of bonds, and the formation of carbonyl groups. Then at the middle of the process actual pyrolysis with solid depolymerization occurs which can be usually confirmed by the considerable weight loss from the initial feedstock. Finally, char formation and devolatilization takes place by the breakdown of C-H and C-O linkages. Generally, pyrolysis process conducted in the presence of catalyst for the depolymerisation of lignin. Catalytic fast pyrolysis is gaining attention among the research community, as it is partially giving expected products. Various types of catalyst have been used such as zeolite, metal chloride, metal oxide in the presence of methanol, ethanol, water, acidic and basic agents etc. Pyrolysis of lignin mostly provides phenolic monomers like cathechol, methoxy cathecol, pyrocathechol, cresol, guaiacol, syringol. Various pyrolysis studies using different catalyst and solvents are summarized in Table 5.7.

Various studies have been reported related to the product based on the catalyst used in an acidic and basic condition in fast pyrolysis. Among all the reported catalysts, zeolite and its other forms with different Al/Si ratios are found to be more widely

Table 5.7 Pyrolysis of lignin

Lignin source	Reaction conditions	Solvent and catalyst	Major product	References
Alkali lignin	500–700 °C for 15 min	NaOH and HZSM-5	Benzene, toluene, xylene, phenols	Wu et al. (2020)
Lignin	Ambient to 550 °C at 10 °C/min heating rate	–	PHENOL, pyrocatechol, syrigol, low-molecular-weight alcohols, acids, and esters	Brebu et al. (2011)
Kraft lignin	500 °C for 3 min	Mesoporous Y-zeolite	Toluene, xylene, ethyl benzene, and benzene	Lee et al. (2013)
Lignin	450–600 °C	–	Acetic acid, levoglucosan, guaiacol, syringols, and p-vinylguaiacol	Trinh et al. (2013)
Lignin	400–800 °C at 10–20 °C/min heating rate	Zeolite-X	Phenol, m-cresol, guaiacol, isoeugenol, syringol	Paysepar et al. (2020)
Alkali	600 °C	KCl, $CaCl_2$, $FeCl_3$	Phenol, ketones, ethers, carboxylic acids, aldehydes, alcohols, and 2,3-di-hydrobenzofuran	Wang et al. (2015)

used due to its lignin cracking capability at a very higher temperature. Wu et al. (2020) reported the lignin pyrolysis using HZSM-5 (zeolite) catalyst in presence of alkaline sodium hydroxide medium which resulted in the Bio-oil composition of monomeric aromatic hydrocarbons, polycyclic aromatic hydrocarbon, phenols, methoxy phenols. It is observed that sodium hydroxide in the reaction mixture facilitates the removal of hydrogen and methoxyl groups which results in an increase in monocyclic aromatic hydrogen production from 54.95 to 66.92% (Wu et al. 2020). Brebu et al. (2011) performed fast pyrolysis of lignin without the involvement of any solvent and catalyst which resulted in the production of 3–10 wt% tar, 12.5–16 wt% gases, 27–37 wt% aqueous phase products, and 39–52 wt% char. The aqueous phase product consists mainly of pyrocatechol, syringol, and other sulfur and nitrogen-containing compounds. It is also found that carbohydrate deformation also resulted in the formation of furfural and its derivative which is considered to be better products in catalyst and solvent-free processes (Brebu et al. 2011). Lee et al. (2013) used mesoporous Y-zeolite to catalyze pyrolysis of lignin at very high temperatures for very short residence time. Mesoporous Y-zeolite has more acidic sites which promote multiple reactions on its surface such as oligomerization, cracking, dehydration, isomerization, aromatization, and decarboxylation which resulted in the production of aromatic and polyaromatic hydrocarbons (Lee et al. 2013). Trinh et al. (2013) studied the effect of temperature and residence time on product distribution of lignin pyrolysis. The study shows that a maximum bio-oil yield of 30–31 wt% was observed around 500–550 °C. Further, it has been noticed that there was a gradual decrease in the yield from 31 to 25 wt% when the temperature was further increased from 550 to 600 °C. This decrease in oil yield resulted in an increment of gaseous product yield. Residence time also plays an important role in oil and gas yield. An increase in residence time resulted in, decrease in Bio-oil yield from 31.2 to 17.8 wt % and an increase in gas yield from 11.8 to 24.3 wt%. In processes involving longer residence time, cracking of organic vapors takes place which is a secondary reaction and forms Non-condensable gases (Trinh et al. 2013). According to the study of Paysepar et al. (2020), zeolite-X catalyst shifted the temperature from 600 to 500 °C where maximum product yield was obtained. More water content was also found which suggests the existence of dehydration reaction (Paysepar et al. 2020). Wang et al. (2015) compared the effect of KCl, $CaCl_2$, and $FeCl_3$ in the lignin pyrolysis process. From the study, it can be inferred that $CaCl_2$ and $FeCl_3$ improved bio-oil yield while KCl decreases the bio-oil yield and increases char yield. It is considered that KCl is catalyzing the re-polymerization process of small generated fragments from lignin. A broad range of products was observed which includes 44 compounds of phenol and its derivatives along with carboxylic acids, aldehydes, and ketones. The catechol, p-cresol, and 4-ethylphenol were found in both the catalyst aided as well as catalyst-free processes which suggest that temperature is the only affecting parameter in the formation of these molecules (Wang et al. 2015).

5.4 Challenges and Barriers in Lignin Depolymerization

The selective degradation of lignin is a tedious task. Lignin is formatted to be a robust bio-polymer in lignocellulosic biomass. It acts as the strong adhesive component that gives overall structural stability as well as integrity to plant materials. There is no specific molecular structure for lignin. Lignin is always characterized by its general molecular structure and the presence of monolignol and phenylpropanoid units which again vary from source to source. The separation and purification technology of lignin often leads to varying degrees of structural damage and modification of the lignin molecular matrix. Various pulping processes used in the pulp and paper industry have an impact on the black liquor (major lignin source and by-product of paper and pulp industry) produced and have a great influence on the structure of lignin as well as separation of black liquor. The process of isolation of lignins is classified in two ways, sulfur-containing processes, and sulfur-free processes. The sulfur-containing process is further classified as sulfite and kraft processes, while sulfur-free processes are further classified as soda and organic pulping processes (Mandlekar et al. 2018). These processes left sulfite ions, solvent stains on the surface of lignin, and also results in the structural damage of the lignin surface. This renders difficulties in the lignin defragmentation or depolymerization processes.

Another problem that reduces the efficiency of lignin depolymerization is its condensation and recombination. Multiple cleavages of one large lignin molecule at a time produce more unstable reactive de-fragmented intermediates. This in turn attempts to stabilize themselves by re-polymerizing or attaching irreversibly by forming a stable C–C bond, producing more charcoal or solid residues, and eventually

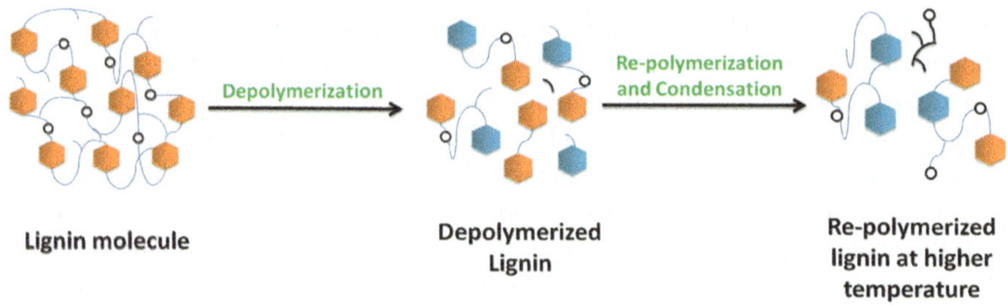

Fig. 5.6 Depolymerization and re-polymerization of lignin

decreases bio-oil yield (Li et al. 2007). Probable condensation and depolymerization of β–O–4 linkage during solvolysis and acidic condition, α–OH of the β–O–4 bond suffer dehydration. This results in the formation of unsaturated C = C linkage that repolymerize with other unstable linkages by oligomerization. While in alkaline conditions, β–O–4 ether can be cleaved which is followed by formaldehyde deduction to form phenylacetaldehyde. This is again an unstable intermediate which in turn is condensed by reaction with nucleophiles. At the ortho and para position of phenol groups, more nucleophilic location forms by deprotonation of phenolic units. Same way quinone methide intermediate can be formed. Whereas Cγ–OH can be removed like formaldehyde to produce an enol ether, that can react with a nucleophile to generate a C–C linkage. This again leads to the formation of large molecules from defragmented small molecules (Shimada et al. 1997; Gierer et al. 1987; Chakar and Ragauskas 2004). The mechanism involving the depolymerization and re-polymerization of formed intermediate products is represented Fig. 5.6.

Several side reactions are taking place throughout the reaction sequence which produces more solid residue, resulting in the net inhibition of Bio-oil yield. This solid residue can be further defragmented into unstable intermediates during the depolymerization of lignin. This in turn again Re-polymerizes and affects the overall depolymerization mechanism to produce bio-oil.

Multiple side reactions and condensation reactions such as polymerization involving formaldehyde, reactive functional groups re-polymerization, radical coupling, and vinyl condensation can be also noticed. A combination of all these reactions can subsequently lead to the formation of char (Bai et al. 2014; Nakamura et al. 2007). Any parameter or factors which yield a higher amount of monomers like high temperature, long reaction time well as high pressure, promotes both depolymerization and re-polymerization (Pińkowska et al. 2012). Variation in these parameters can help in achieving a balance between depolymerization and Re-polymerization.

Other parameters like, presence of varied functional groups such as hydroxyl, methoxyl, phenolic, aldehydes have multiple roles in lignin reactivity and depolymerization mechanism which ultimately create barriers in the selective defragmentation of lignin. One additional difficulty is the solubility of lignin in solvent due to the higher molecular weight structure. Due to the amorphous nature of lignin, it is hard to find a particular solvent that can dissolve lignin at a varying temperature (Wang et al. 2019). So these are the factors that need to be considered during depolymerization. Various new efficient approaches are coming into existence to address these issues.

5.5 Conclusion

In this study, various depolymerization strategies have been discussed such as oxidative and reductive depolymerization and acid & base-catalyzed depolymerization. In the oxidative and reductive depolymerization approach, the addition or removal of oxygen and hydrogen takes place. Oxidative depolymerization generally gives phenolic aldehyde, carboxyl-rich acids, and oligomer products. While in reductive depolymerization approach hydrogenation and hydrodeoxygenation reaction take place, which gives phenolic monomers, cycloalkanes, and arenes. In acid and base-catalyzed depolymerization, the nature of catalyst and temperature plays an important role in depolymerizing lignin. Generally, both techniques give similar kinds of phenolic monomers such as phenol, guaiacol, Catechol, alkylphenols, cresol, syringol, etc. But the range of temperature plays a significant role to distribute products. These smaller molecular products can be used in pharmaceuticals, composite material, plastic, rubber, and in the replacement of petroleum chemicals. To depolymerize lignin, various thermal treatments such as hydrothermal liquefaction, microwave-assisted heating, and pyrolysis have been discussed in the chapter. Hydrothermal liquefaction produces more bio-oil yield compared to pyrolysis. Microwave produces phenolic monomers containing bio-oil. While Pyrolysis gives more aromatic and fine compounds but also results in repolymerization of the product at a higher temperature. Due to the amorphous and complex structure of lignin, the solubility of lignin in a particular solvent is an issue. Also at higher temperature repolymerization and condensation of intermediate products occurs which stand out as a barrier in the process. Due to the uneven structure and source of lignin, uniform distribution of product is challenging. Depolymerization of lignin into the specifically defined compound is a current subject of research. All these points need to be critically considered during the depolymerization of lignin.

References

Abdelaz OY, Meier S, Prothmann J, Turner C, Riisager A, Hulteberg CP (2019) Oxidative depolymerisation of lignosulphonate lignin into low-molecular-weight products with Cu–Mn/Δ-Al_2O_3. Top Catal. https://doi.org/10.1007/s11244-019-01146-5

Ahmad Z, Al Dajani WW, Paleologou M, Xu C (2020) Sustainable process for the depolymerization/oxidation of softwood and hardwood kraft lignins using hydrogen peroxide under ambient conditions. Molecules. https://doi.org/10.3390/molecules25102329

Akhtar J, Amin NAS (2011) A review on process conditions for optimum bio-oil yield in hydrothermal liquefaction of biomass. Renew Sustain Energy Rev. https://doi.org/10.1016/j.rser.2010.11.054

Alonso DM, Wettstein SG, Dumesic JA (2012) Bimetallic catalysts for upgrading of biomass to fuels and chemicals. Chem Soc Rev. https://doi.org/10.1039/c2cs35188a

Arturi KR, Strandgaard M, Nielsen RP, Søgaard EG, Maschietti M (2017) Hydrothermal liquefaction of lignin in near-critical water in a new batch reactor: influence of phenol and temperature. J Supercrit Fluids. https://doi.org/10.1016/j.supflu.2016.12.015

Awan IZ, Tanchoux N, Quignard F, Albonetti S, Cavani F, Di Renzo F (2019) Heterogeneous catalysis as a tool for production of aromatic compounds from lignin. Stud Surf Sci Catal

Bai X et al (2014) Formation of phenolic oligomers during fast pyrolysis of lignin. Fuel. https://doi.org/10.1016/j.fuel.2014.03.013

Barbier J et al (2012) Hydrothermal conversion of lignin compounds. A detailed study of fragmentation and condensation reaction pathways. Biomass Bioenergy. https://doi.org/10.1016/j.biombioe.2012.07.011

Belgacem MN, Gandini A (2008) Monomers, polymers and composites from renewable resources

Brebu M, Vasile C (2010) Thermal degradation of lignin—a review. Cellulose Chem Technol

Brebu M, Cazacu G, Chirila O (2011) Pyrolysis of lignin—a potential method for obtaining chemicals and/or fuels. Cellul Chem Technol

Bridgwater AV (2012) Review of fast pyrolysis of biomass and product upgrading. Biomass Bioenerg. https://doi.org/10.1016/j.biombioe.2011.01.048

Chakar FS, Ragauskas AJ (2004) Review of current and future softwood kraft lignin process chemistry. https://doi.org/10.1016/j.indcrop.2004.04.016

Chaudhary R, Dhepe PL (2017) Solid base catalyzed depolymerization of lignin into low molecular weight products. Green Chem. https://doi.org/10.1039/c6gc02701f

Chauhan PS (2020) Role of various bacterial enzymes in complete depolymerization of lignin: a review. Biocatal Agric Biotechnol. https://doi.org/10.1016/j.bcab.2020.101498

Chen C, Jin D, Ouyang X, Zhao L, Qiu X, Wang F (2018) Effect of structural characteristics on the depolymerization of lignin into phenolic monomers. Fuel. https://doi.org/10.1016/j.fuel.2018.03.041

Chen S, Lu Q, Han W, Yan P, Wang H, Zhu W (2021) Insights into the oxidation–reduction strategy for lignin conversion to high-value aromatics. Fuel. https://doi.org/10.1016/j.fuel.2020.119333

Chio C, Sain M, Qin W (2019) Lignin utilization: a review of lignin depolymerization from various aspects. Renew Sustain Energy Rev. https://doi.org/10.1016/j.rser.2019.03.008

Chowdari RK, Agarwal S, Heeres HJ (2019) Hydrotreatment of kraft lignin to alkylphenolics and aromatics using Ni, Mo, and W phosphides supported on activated carbon. ACS Sustain Chem Eng. https://doi.org/10.1021/acssuschemeng.8b04411

Deng H et al (2009) Activity and stability of perovskite-type oxide LaCoO 3 catalyst in lignin catalytic wet oxidation to aromatic aldehydes process. Energy Fuels. https://doi.org/10.1021/ef8005349

Deuss PJ, Lancefield CS, Narani A, De Vries JG, Westwood NJ, Barta K (2017) Phenolic acetals from lignins of varying compositions: via iron(iii) triflate catalysed depolymerisation. Green Chem. https://doi.org/10.1039/c7gc00195a

Dhar P, Vinu R (2017) Understanding lignin depolymerization to phenols via microwave-assisted solvolysis process. J Environ Chem Eng. https://doi.org/10.1016/j.jece.2017.08.031

Dos Santos PSB, Erdocia X, Gatto DA, Labidi J (2016) Bio-oil from base-catalyzed depolymerization of organosolv lignin as an antifungal agent for wood. Wood Sci Technol. https://doi.org/10.1007/s00226-015-0795-8

Duan D et al (2017) Low-power microwave radiation-assisted depolymerization of ethanol organosolv lignin in ethanol/formic acid mixtures. BioResources. https://doi.org/10.15376/biores.12.3.5308-5320

Fan J et al (2018) The additive free microwave hydrolysis of lignocellulosic biomass for fermentation to high value products. J Clean Prod. https://doi.org/10.1016/j.jclepro.2018.07.088

Fang Q et al (2020) Low temperature catalytic conversion of oligomers derived from lignin in pubescens on Pd/NbOPO$_4$. Appl Catal B Environ. https://doi.org/10.1016/j.apcatb.2019.118325
Farag S, Fu D, Jessop PG, Chaouki J (2014) Detailed compositional analysis and structural investigation of a bio-oil from microwave pyrolysis of kraft lignin. J Anal Appl Pyrolysis. https://doi.org/10.1016/j.jaap.2014.06.005
Gale M, Cai CM, Gilliard-Abdul-Aziz KL (2020) Heterogeneous catalyst design principles for the conversion of lignin into high-value commodity fuels and chemicals. Chemsuschem. https://doi.org/10.1002/cssc.202000002
Gasson JR, Forchheim D, Sutter T, Hornung U, Kruse A, Barth T (2012) Modeling the lignin degradation kinetics in an ethanol/formic acid solvolysis approach. Part 1. Kinetic model development. Ind Eng Chem Res. https://doi.org/10.1021/ie301487v
Gawande MB, Shelke SN, Zboril R, Varma RS (2014) Microwave-assisted chemistry: synthetic applications for rapid assembly of nanomaterials and organics. Acc Chem Res. https://doi.org/10.1021/ar400309b
Gellerstedt GLF, Henriksson EG (2008) Lignins: major sources, structure and properties. In: Monomers, polymers and composites from renewable resources
Gierer J (1986) Chemistry of delignification. Wood Sci Technol. https://doi.org/10.1007/bf00350692
Gierer J, Norén I, Wännström S (1987) Formation of condensation products on treatment of non-phenolic lignin units of the β-aryl ether type with alkali. Model studies on a novel mode of alkaline lignin condensation. Holzforschung. https://doi.org/10.1515/hfsg.1987.41.2.79
Gollakota ARK, Kishore N, Gu S (2018) A review on hydrothermal liquefaction of biomass. Renew Sustain Energy Rev. https://doi.org/10.1016/j.rser.2017.05.178
Gosz K, Kosmela P, Hejna A, Gajowiec G, Piszczyk Ł (2018) Biopolyols obtained via microwave-assisted liquefaction of lignin: structure, rheological, physical and thermal properties. Wood Sci Technol. https://doi.org/10.1007/s00226-018-0991-4
Graglia M, Kanna N, Esposito D (2015) Lignin refinery: towards the preparation of renewable aromatic building blocks. ChemBioEng Reviews. https://doi.org/10.1002/cben.201500019
Guo H et al (2016) Tungsten carbide: a remarkably efficient catalyst for the selective cleavage of lignin C–O bonds. Chemsuschem. https://doi.org/10.1002/cssc.201600901
Hagglund E, Bjorkman C (1924) Lignin hydrochloride. Biochem Z
Halleraker HV, Barth T (2020) Quantitative NMR analysis of the aqueous phase from hydrothermal liquefaction of lignin. J Anal Appl Pyrolysis. https://doi.org/10.1016/j.jaap.2020.104919
Hdidou L, Khallouk K, Solhy A, Manoun B, Oukarroum A, Barakat A (2018) Synthesis of CoFeO mixed oxides via an alginate gelation process as efficient heterogeneous catalysts for lignin depolymerization in water. Catal Sci Technol. https://doi.org/10.1039/c8cy00576a
He W, Li G, Kong L, Wang H, Huang J, Xu J (2008) Application of hydrothermal reaction in resource recovery of organic wastes. Resour Conserv Recycl. https://doi.org/10.1016/j.resconrec.2007.11.003
Heitner C, Dimmel D, Schmidt J (2010) Lignin and Lignans: advances in chemistry
Hewson WB, Hibbert H (1943) Studies on Lignin and Related Compounds. LXV. Re-ethanolysis of Isolated Lignins. J Am Chem Soc. https://doi.org/10.1021/ja01246a044
Hu J, Wu S, Jiang X, Xiao R (2018) Structure-reactivity relationship in fast pyrolysis of lignin into monomeric phenolic compounds. Energy Fuels. https://doi.org/10.1021/acs.energyfuels.7b03593
Imai T, Yokoyama T, Matsumoto Y (2011) Revisiting the mechanism of β-O-4 bond cleavage during acidolysis of lignin IV: dependence of acidolysis reaction on the type of acid. J Wood Sci. https://doi.org/10.1007/s10086-010-1166-6
Jensen MM et al (2018) Hydrothermal liquefaction of enzymatic hydrolysis lignin: biomass pretreatment severity affects lignin valorization. ACS Sustain Chem Eng. https://doi.org/10.1021/acssuschemeng.7b04338

Jia S, Cox BJ, Guo X, Zhang ZC, Ekerdt JG (2011) Hydrolytic cleavage of β-O-4 ether bonds of lignin model compounds in an ionic liquid with metal chlorides. https://doi.org/10.1021/ie101884h

Jiang Y et al (2015) Depolymerization of cellulolytic enzyme lignin for the production of monomeric phenols over Raney Ni and acidic zeolite catalysts. Energy Fuels. https://doi.org/10.1021/ef5022297

Junghans U, Bernhardt JJ, Wollnik R, Triebert D, Unkelbach G, Pufky-Heinrich D (2020) Valorization of lignin via oxidative depolymerization with hydrogen peroxide: towards carboxyl-rich oligomeric lignin fragments. Molecules. https://doi.org/10.3390/molecules25112717

Kang S, Chang J, Fan J (2015) Phenolic antioxidant production by hydrothermal liquefaction of lignin. Energy Sources Part A Recover Util Environ Eff. https://doi.org/10.1080/15567036.2011.585386

Katahira R et al (2016) Base-catalyzed depolymerization of biorefinery lignins. ACS Sustain Chem Eng. https://doi.org/10.1021/acssuschemeng.5b01451

Kim S et al (2011) Computational study of bond dissociation enthalpies for a large range of native and modified lignins. J Phys Chem Lett. https://doi.org/10.1021/jz201182w

Kruger JS et al (2016) Lignin depolymerization with nitrate-intercalated hydrotalcite catalysts. ACS Catal. https://doi.org/10.1021/acscatal.5b02062

Lam SS et al (2020) Engineering pyrolysis biochar via single-step microwave steam activation for hazardous landfill leachate treatment. J Hazard Mater. https://doi.org/10.1016/j.jhazmat.2019.121649

Lancefield CS et al (2016) Investigation of the chemocatalytic and biocatalytic valorization of a range of different lignin preparations: the importance of β-O-4 content. ACS Sustain Chem Eng. https://doi.org/10.1021/acssuschemeng.6b01855

Lavoie JM, Baré W, Bilodeau M (2011) Depolymerization of steam-treated lignin for the production of green chemicals. Bioresour Technol. https://doi.org/10.1016/j.biortech.2011.01.010

Lee HW, Kim TH, Park SH, Jeon JK, Suh DJ, Park YK (2013) Catalytic fast pyrolysis of lignin over mesoporous y zeolite using Py-GC/ms. J Nanosci Nanotechnol. https://doi.org/10.1166/jnn.2013.7421

Li J, Henriksson G, Gellerstedt G (2007) Lignin depolymerization/repolymerization and its critical role for delignification of aspen wood by steam explosion. Bioresour Technol. https://doi.org/10.1016/j.biortech.2006.10.018

Liu Q, Li P, Liu N, Shen D (2017) Lignin depolymerization to aromatic monomers and oligomers in isopropanol assisted by microwave heating. Polym Degrad Stab. https://doi.org/10.1016/j.polymdegradstab.2016.11.016

Long J, Li X, Guo B, Wang F, Yu Y, Wang L (2012) Simultaneous delignification and selective catalytic transformation of agricultural lignocellulose in cooperative ionic liquid pairs. Green Chem. https://doi.org/10.1039/c2gc35105f

Lu Y, Lu YC, Hu HQ, Xie FJ, Wei XY, Fan X (2017) Structural characterization of lignin and its degradation products with spectroscopic methods. J Spectrosc. https://doi.org/10.1155/2017/8951658

Lui MY, Chan B, Yuen AKL, Masters AF, Maschmeyer T (2020) Hydrothermal liquefaction of α-O-4 aryl ether linkages in lignin. Chemsuschem. https://doi.org/10.1002/cssc.201903263

Ma R, Xu Y, Zhang X (2015a) Catalytic oxidation of biorefinery lignin to value-added chemicals to support sustainable biofuel production. Chemsuschem. https://doi.org/10.1002/cssc.201402503

Ma X, Cui K, Hao W, Ma R, Tian Y, Li Y (2015b) Alumina supported molybdenum catalyst for lignin valorization: effect of reduction temperature. Bioresour Technol. https://doi.org/10.1016/j.biortech.2015.05.032

Ma R et al (2016) Peracetic acid depolymerization of biorefinery lignin for production of selective monomeric phenolic compounds. Chem A Eur J. https://doi.org/10.1002/chem.201600546

Mahmood N, Yuan Z, Schmidt J, Xu CC (2015) Hydrolytic depolymerization of hydrolysis lignin: effects of catalysts and solvents. Bioresour Technol. https://doi.org/10.1016/j.biortech.2015.04.074

Mandlekar N et al (2018) An overview on the use of lignin and its derivatives in fire retardant polymer systems. In: Lignin—trends and applications

Miller J, Evans L, Mudd JE, Brown KA (2002) Batch microreactor studies of lignin depolymerization by bases. 2. Aqueous Solvents. Sandia Natl Rep

Milovanović J et al (2016) Insights into the microwave-assisted mild deconstruction of lignin feedstocks using NiO-containing ZSM-5 zeolites. ACS Sustain Chem Eng. https://doi.org/10.1021/acssuschemeng.6b00825

Morgan HM et al (2017) A review of catalytic microwave pyrolysis of lignocellulosic biomass for value-added fuel and chemicals. Bioresour Technol. https://doi.org/10.1016/j.biortech.2017.01.059

Nakamura T, Kawamoto H, Saka S (2007) Condensation reactions of some lignin related compounds at relatively low pyrolysis temperature. J Wood Chem Technol. https://doi.org/10.1080/02773810701515143

Nandiwale KY, Danby AM, Ramanathan A, Chaudhari RV, Subramaniam B (2019) Dual function Lewis acid catalyzed depolymerization of industrial corn Stover lignin into stable monomeric phenols. ACS Sustain Chem Eng. https://doi.org/10.1021/acssuschemeng.8b05077

Narani A et al (2015) Efficient catalytic hydrotreatment of Kraft lignin to alkylphenolics using supported NiW and NiMo catalysts in supercritical methanol. Green Chem. https://doi.org/10.1039/c5gc01643f

Nimz HH, Nemr M, Robert D, Faix O (1981) Carbon-13 NMR spectra of lignins, 8). Structural differences between lignins of hardwoods, softwoods, grasses and compression wood. Holzforschung. https://doi.org/10.1515/hfsg.1981.35.1.16.

Norgren M, Edlund H (2014) Lignin: recent advances and emerging applications. Curr Opin Colloid Interface Sci. https://doi.org/10.1016/j.cocis.2014.08.004

Ouyang X, Zhu G, Huang X, Qiu X (2015) Microwave assisted liquefaction of wheat straw alkali lignin for the production of monophenolic compounds. J Energy Chem. https://doi.org/10.1016/S2095-4956(15)60286-8

Ouyang X, Ruan T, Qiu X (2016) Effect of solvent on hydrothermal oxidation depolymerization of lignin for the production of monophenolic compounds. Fuel Process Technol. https://doi.org/10.1016/j.fuproc.2015.12.019

Paysepar H, Venkateswara Rao KT, Yuan Z, Shui H, Xu C (2020) Production of phenolic chemicals from hydrolysis lignin via catalytic fast pyrolysis. J Anal Appl Pyrolysis. https://doi.org/10.1016/j.jaap.2020.104842

Peterson AA, Vogel F, Lachance RP, Fröling M, Antal MJ, Tester JW (2008) Thermochemical biofuel production in hydrothermal media: a review of sub- and supercritical water technologies. Energy Environ Sci. https://doi.org/10.1039/b810100k

Pińkowska H, Wolak P, Złocińska A (2012) Hydrothermal decomposition of alkali lignin in sub- and supercritical water. Chem Eng J. https://doi.org/10.1016/j.cej.2012.01.092

Pu Y, Hu F, Huang F, Davison BH, Ragauskas AJ (2013) Assessing the molecular structure basis for biomass recalcitrance during dilute acid and hydrothermal pretreatments. Biotechnol Biofuels. https://doi.org/10.1186/1754-6834-6-15

Pu Y, Hu F, Huang F, Ragauskas AJ (2015) Lignin structural alterations in thermochemical pretreatments with limited delignification. Bioenergy Res. https://doi.org/10.1007/s12155-015-9655-5

Roberts VM, Stein V, Reiner T, Lemonidou A, Li X, Lercher JA (2011) Towards quantitative catalytic lignin depolymerization. Chem A Eur J. https://doi.org/10.1002/chem.201002438

Salema AA, Ani FN (2011) Microwave induced pyrolysis of oil palm biomass. Bioresour Technol. https://doi.org/10.1016/j.biortech.2010.09.115

Schmiedl D et al (2012) Base catalyzed degradation of lignin for the generation of oxy-aromatic compounds—possibilities and challenges. Erdol Erdgas Kohle incl Oil Gas Eur Mag

Schuler J, Hornung U, Kruse A, Dahmen N, Sauer J (2017) Hydrothermal liquefaction of lignin. J Biomater Nanobiotechnol. https://doi.org/10.4236/jbnb.2017.81007

Schuler J, Hornung U, Dahmen N, Sauer J (2019) Lignin from bark as a resource for aromatics production by hydrothermal liquefaction. GCB Bioenergy. https://doi.org/10.1111/gcbb.12562

Schutyser W, Renders T, Van Den Bosch S, Koelewijn SF, Beckham GT, Sels BF (2018) Chemicals from lignin: an interplay of lignocellulose fractionation, depolymerisation, and upgrading. Chem Soc Rev. https://doi.org/10.1039/c7cs00566k

Shafaghat H et al (2020) In-situ hydrogenation of bio-oil/bio-oil phenolic compounds with secondary alcohols over a synthesized mesoporous Ni/CeO_2 catalyst. Chem Eng J. https://doi.org/10.1016/j.cej.2019.122912

Shao Y et al (2017) Selective production of arenes via direct lignin upgrading over a niobium-based catalyst. Nat Commun. https://doi.org/10.1038/ncomms16104

Shao L, Zhang Q, You T, Zhang X, Xu F (2018) Microwave-assisted efficient depolymerization of alkaline lignin in methanol/formic acid media. Bioresour Technol. https://doi.org/10.1016/j.biortech.2018.05.083

Shimada K, Hosoya S, Ikeda T (1997) Condensation reactions of softwood and hardwood lignin model compounds under organic acid cooking conditions. J Wood Chem Technol. https://doi.org/10.1080/02773819708003118

Si X et al (2017) A strategy for generating high-quality cellulose and lignin simultaneously from woody biomass. Green Chem. https://doi.org/10.1039/c7gc02492d

Singh R et al (2014) Hydrothermal conversion of lignin to substituted phenols and aromatic ethers. Bioresour Technol. https://doi.org/10.1016/j.biortech.2014.02.076

Stewart D (2008) Lignin as a base material for materials applications: chemistry, application and economics. Ind Crops Prod. https://doi.org/10.1016/j.indcrop.2007.07.008

Strüven JO, Meier D (2016) Hydrocracking of organosolv lignin in subcritical water to useful phenols employing various Raney nickel catalysts. ACS Sustain Chem Eng. https://doi.org/10.1021/acssuschemeng.6b00342

Sturgeon MR et al (2014) A mechanistic investigation of acid-catalyzed cleavage of aryl-ether linkages: implications for lignin depolymerization in acidic environments. ACS Sustain Chem Eng. https://doi.org/10.1021/sc400384w

Sun Z, Fridrich B, De Santi A, Elangovan S, Barta K (2018) Bright side of lignin depolymerization: toward new platform chemicals. Chem Rev. https://doi.org/10.1021/acs.chemrev.7b00588

Taylor K (1994) The encyclopaedia of food science, food technology and nutrition. In: Macrae R, Robinson RK, Sadler MJ (eds) Academic Press Limited, London. ISBN 0-12-226850-4. Price £1100, Food Chem. https://doi.org/10.1016/0308-8146(94)90266-6

Toledano A, Serrano L, Labidi J (2014a) Improving base catalyzed lignin depolymerization by avoiding lignin repolymerization. Fuel. https://doi.org/10.1016/j.fuel.2013.08.071

Toledano A, Serrano L, Pineda A, Romero AA, Luque R, Labidi J (2014b) Microwave-assisted depolymerisation of organosolv lignin via mild hydrogen-free hydrogenolysis: catalyst screening. Appl Catal B Environ. https://doi.org/10.1016/j.apcatb.2012.10.015

Toor SS, Rosendahl L, Rudolf A (2011) Hydrothermal liquefaction of biomass: a review of subcritical water technologies. Energy. https://doi.org/10.1016/j.energy.2011.03.013

Trinh TN et al (2013) Fast pyrolysis of lignin using a pyrolysis centrifuge reactor. Energy Fuels. https://doi.org/10.1021/ef400527k

Vanholme R, Demedts B, Morreel K, Ralph J, Boerjan W (2010) Lignin biosynthesis and structure. Plant Physiol. https://doi.org/10.1104/pp.110.155119

Vanholme R et al (2012) Metabolic engineering of novel lignin in biomass crops. New Phytol. https://doi.org/10.1111/j.1469-8137.2012.04337.x

Wang WL, Ren XY, Chang JM, Cai LP, Shi SQ (2015) Characterization of bio-oils and bio-chars obtained from the catalytic pyrolysis of alkali lignin with metal chlorides. Fuel Process Technol. https://doi.org/10.1016/j.fuproc.2015.06.048

Wang H et al (2016) ZnCl2 induced catalytic conversion of softwood lignin to aromatics and hydrocarbons. Green Chem. https://doi.org/10.1039/c5gc02967h

Wang H, Pu Y, Ragauskas A, Yang B (2019) From lignin to valuable products–strategies, challenges, and prospects. Biores Technol. https://doi.org/10.1016/j.biortech.2018.09.072

Weng C, Peng X, Han Y (2021) Depolymerization and conversion of lignin to value-added bioproducts by microbial and enzymatic catalysis. Biotechnol Biofuels. https://doi.org/10.1186/s13068-021-01934-w
Wu J, Chang G, Li X, Li J, Guo Q (2020) Effects of NaOH on the catalytic pyrolysis of lignin/HZSM-5 to prepare aromatic hydrocarbons. J Anal Appl Pyrolysis. https://doi.org/10.1016/j.jaap.2020.104775
Xiao LP et al (2017) Catalytic hydrogenolysis of lignins into phenolic compounds over carbon nanotube supported molybdenum oxide. ACS Catal. https://doi.org/10.1021/acscatal.7b02563
Xu J, Jiang J, Hse C, Shupe TF (2012) Renewable chemical feedstocks from integrated liquefaction processing of lignocellulosic materials using microwave energy. Green Chem. https://doi.org/10.1039/c2gc35805k
Xu C, Arancon RAD, Labidi J, Luque R (2014) Lignin depolymerisation strategies: towards valuable chemicals and fuels. Chem Soc Rev. https://doi.org/10.1039/c4cs00235k
Yang Z, Feng J, Cheng H, Liu Y, Jiang J (2021) Directional depolymerization of lignin into high added-value chemical with synergistic effect of binary solvents. Bioresour Technol. https://doi.org/10.1016/j.biortech.2020.124440
Ye Y, Zhang Y, Fan J, Chang J (2012) Novel method for production of phenolics by combining lignin extraction with lignin depolymerization in aqueous ethanol. Ind Eng Chem Res. https://doi.org/10.1021/ie202118d
Young MR, Towers GHN, Neish AC (1966) Taxonomic distribution of ammonia-lyases for l-phenylalanine and L-tyrosine in relation to lignification. Can J Bot. https://doi.org/10.1139/b66-040
Yuan Z, Cheng S, Leitch M, Xu CC (2010) Hydrolytic degradation of alkaline lignin in hot-compressed water and ethanol. Bioresour Technol. https://doi.org/10.1016/j.biortech.2010.06.140
Yunpu W, Leilei D, Liangliang F, Shaoqi S, Yuhuan L, Roger R (2016) Review of microwave-assisted lignin conversion for renewable fuels and chemicals. J Anal Appl Pyrol. https://doi.org/10.1016/j.jaap.2016.03.011
Zaheer M, Kempe R (2015) Catalytic hydrogenolysis of aryl ethers: a key step in lignin valorization to valuable chemicals. ACS Catal. https://doi.org/10.1021/cs501498f
Zhang Y, Wang Z, Feng J, Pan H (2020) Maximizing utilization of poplar wood by microwave-assisted pretreatment with methanol/dioxane binary solvent. Bioresour Technol. https://doi.org/10.1016/j.biortech.2019.122657
Zhao Q, Dixon RA (2011) Transcriptional networks for lignin biosynthesis: more complex than we thought? Trends Plant Sci. https://doi.org/10.1016/j.tplants.2010.12.005
Zhou M, Xu J, Jiang J, Sharma BK (2018a) A review of microwave assisted liquefaction of lignin in hydrogen donor solvents: effect of solvents and catalysts. Energies. https://doi.org/10.3390/en11112877
Zhou M, Sharma BK, Liu P, Xia H, Xu J, Jiang JC (2018b) Microwave assisted depolymerization of alkaline lignin over hydrotalcite-based CuNiAl mixed oxides. ACS Sustain Chem Eng. https://doi.org/10.1021/acssuschemeng.8b01697
Zou R et al (2018) Microwave-assisted depolymerization of lignin with metal chloride in a hydrochloric acid and formic acid system. BioResources. https://doi.org/10.15376/biores.13.2.3704-3719

Chapter 6
Recent Advances in Packed-Bed Gasification of Lignocellulosic Biomass

V. M. Jaganathan and S. Varunkumar

6.1 Introduction

Biomass is any organic substance derived from plant and animal matter. Biomass like cow dung, night soil, kitchen and food waste, which can undergo biodegradation at a faster rate compared to ligno-cellulosic biomass, are better suited for bio-chemical conversion techniques like anaerobic digestion (which can handle high moisture i.e. >30% and cannot generally handle lignin). Biomass derived from plant matter or in other words lignocellulosic biomass is the other category which is abundantly available (around 230 million metric tonne per year in India) (Mnre 2021). Ligno-cellulosic biomass typically contains less than 30% moisture and also contains more than 10% lignin. Therefore, it is more suitable for thermo-chemical conversion than other routes like bio-chemical and physio-chemical conversion techniques.

Gasification is one of the three major thermo-chemical conversion routes for ligno-cellulosic biomass (combustion and pyrolysis being the other two). The process of gasification is a combination of partial oxidation and char-reduction, leading to formation of products rich in carbon-monoxide and hydrogen (with traces of methane and other higher hydrocarbons); the reactive gases just mentioned will be present along with inert components like nitrogen, carbon-dioxide and water vapor; the fraction of the inerts will depend on the choice of the gasification medium (air vs mixtures of oxygen-carbondioxide-steam). In India in the 1980s, biomass air gasification was used as a technological solution for rural electrification—the product gases are burned in internal combustion engines to produce electric power. Now with

V. M. Jaganathan (✉)
Department of Energy and Environment, National Institute of Technology, Tiruchirappalli, Tamil Nadu, India
e-mail: vmjagan@nitt.edu

S. Varunkumar
Department of Mechanical Engineering, Indian Institute of Technology Madras, Chennai, Tamil Nadu, India

Y. K. Nandabalan et al. (eds.), *Zero Waste Biorefinery*, Energy, Environment, and Sustainability, https://doi.org/10.1007/978-981-16-8682-5_6

the emergence of solar PV and wind energy technologies, the role of gasification in power generation is perhaps negligible. But large quantities of surplus biomass is generated annually (about 500 MMT in India alone) and therefore it is crucial to identify avenues for value addition—potential candidates include the use of products of gasification for synthesis of pure hydrogen, liquid fuels and other platform chemicals. Gasification with mediums other than air, like with mixtures of O_2–CO_2–steam (the term oxy-fuel gasification is generally used to refer to the use of these mixtures), are needed for tailoring product gas stoichiometry for downstream applications and process intensification; use of these mixtures also have significance from the point of view of pre-combustion CO_2 capture and storage. Gasification can also be used for synthesis of high quality activated carbon in an environmentally friendly and economical way (IndiaToday 2019). With this very brief background, lets move on to a discussion of packed bed gasification process, the focus of the current chapter.

6.1.1 Packed Bed Biomass Gasifier Configuration

Packed bed is the simplest configuration in which the process of gasification can be arranged—by supporting a bed of biomass particles (for instance, wood chips, densified agro-residue pellets, cocount shells etc.) on a grate and allowing the gasification medium to flow through the bed, a self-sustained reaction zone can be established and controlled (flow rate of the gasification medium being the predominat control variable). Depending on the relative motion of the gasification medium and the reaction zone from a frame of reference attached to the fresh biomass, two variants are possible—counter-current (flow and reaction zone move in opposite directions) and co-current (flow and reaction zone move in the same direction). Over the years, counter-current mode has become the preferred choice owing to its superior performance, mainly in terms of tar levels in the exit gas—that the products of oxidation pass through a hot bed of char in this configuration leads to significantly lower tar levels compared to co-current mode (where the products of oxidation pass through the fresh biomass). The current chapter will focus only on counter-current mode propagation. This mode of gasification can be realized in practice using either bottom-lit (see Fig. 6.1a) or top-lit (see Fig. 6.1c) arrangement. Bottom-lit configuration is preferred for continuous operation as the left over char and ash can be easily extracted from below the grate. Top-lit configurations, while easier to construct compared to the bottom-lit one, can be used only for batch operations—though predominatly used for laboratory studies, this configuration found an interesting application as high efficiency low emissions gasfier stove known by the name Oorja (Mukunda et al. 1994; Mukunda 2011; Varunkumar et al. 2012).

It is important to mention here that, the major constituents of any biomass are, by mass, cellulose (28–45%), hemi-cellulose (10–29%), lignin (0–40%), extractives (crude proteins and oil, less than 5%) and ash (less than 5%). Most of the cellulose and hemi-cellulose form 'volatile' matter (about 60–80% by weight) which will evaporate and burn in gas phase when temperatures go beyond 200 °C and lignin

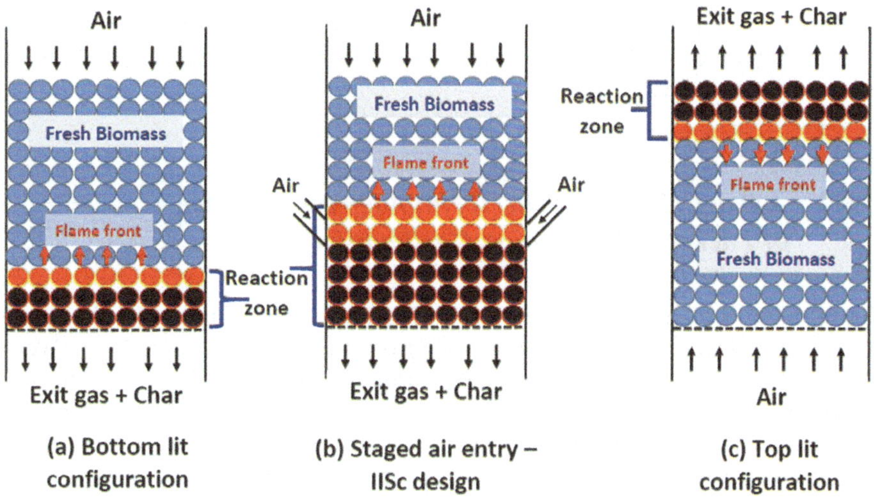

Fig. 6.1 Schematic of flame propagation in packed-bed configurations

forms as fixed carbon (about 20% by weight). The remaining other constituents are moisture (varies from 10 to 20% by weight) and ash (less than 5% by weight). The fraction of these constituents is more or less invariant for most of the biomass except in some special class of fuels like coconut shells, corn cob and bagasse (with extractives more than 10%) and rice husk (with ash more than 20%). In other words, on moisture and ash free basis, all biomass are similar from the point of view of proximate and ultimate analysis (Mukunda 2011).

A few important conclusions can be drawn from this insight. The first is the following—the elemental constitution of ligno-cellulosic biomass, on dry and ash free (DAF) basis, can be taken as $CH_{1.6}O_{0.9}$; this gives the overall stoichiometric air requirement as 4.5 kg/kg of biomass. This fact combined with the inputs from proximate analysis implies that the amount of oxidizer required for complete combustion of 'volatiles' is about 60% of total stoichiometry requirement; this is termed as 'volatile stoichiometry' (corresponding to volatiles equivalence ratio, $\phi_v = 1$). If the amount of oxidizer supplied is such that $\phi_v \geq 1$, the 'volatiles' present in the biomass can be either fully or partially oxidized. Further, these 'volatiles' oxidation products can undergo reduction (aka gasification) with char present to form syngas and this process is termed as gasification. If $\phi_v < 1$, char also participates in oxidation reaction and hence it is no longer available for gasification reactions. Operation under such conditions is termed as combustion or simultaneous volatile and char oxidation. If $\phi_v > 5$, which means that the oxidizer presence is very less, biomass just decomposes into volatile gases and char and this type of thermo-chemical conversion is close to pyrolysis.

Before moving on to a discussion of recent results in packed bed biomass gasification, a brief history of the technology and science of gasification relevant to the current chapter is presented.

6.1.2 A Brief History of the Technology and Science of Packed Bed Biomass Gasification

Gasification of biomass emerged as a technological solution to handle the energy crisis faced by Germany due to its isolation from the rest of the world during World War II. The most commonly known design from this era is the 'Imbert gasifier'—a top closed design with radial side air nozzles and a restriction in the mid-section to aid material flow. It was operated under suction, a natural choice for use in automotive propulsion, where the suction stroke of the reciprocating engine was used to draw air through the side nozzles. Several variants of it was manufactured in Europe and later in other parts of the world mainly for automotive propulsion and power generation. Though not much of quantitative data is available to assess the performance of this design, a good overview of the technical and historical aspects of this design can be found in Keith and Saenz (2021). Many designs of gasifiers deployed around the world are based on the Imbert design.

The next noteworthy design is the 'Chinese rice hull gasifiers' (Mukunda 2011; Mukunda et al. 1994; Varunkumar 2014). These were developed in the 1980s and featured a top open design (as opposed to the top closed design of Imbert type) with no side air-nozzles. Air is drawn in through the top by suction and the bed of biomass is ignited close to the grate at the bottom (refer (a) of Fig. 6.1). The reaction zone thus formed will consist of a 'volatiles oxidation zone' adjoining the fresh biomass; the 'volatiles oxidation zone' will be followed by the hot char zone, where reduction and other cracking reactions take place. Generally, in the Imbert type and the Chinese design, it is known now that the temperature and the residence time in the hot char zone is not adequate to bring the levels of higher hydrocarbons (aka tars) to acceptable levels for use in stationary applications (power generation principally). With the aim of overcoming this limitation, the IISc gasifier (refer (b) of Fig. 6.1) design introduced the idea of simultaneous air intake from the top and the side; referred to as the re-burn process, the side air helps in increasing the temperature in the char zone to aid tar cracking. This feature makes the IISc design one of the lowest tar level packed bed gasifier system and therefore also the one which requires minimal downstream gas cleaning (Mukunda 2011).

It is important to recognize that almost all these designs were developed based on an intuitive understanding of the reaction zone structure and the observed variation in its propagation rate with air flow rate. The general behavior of flame propagation in all the three configurations discussed earlier is as follows—as the air flow rate is increased, the reaction zone moves upwards and the rate of movement increases initially, a maximum and then starts decreasing; it will decrease with further increase

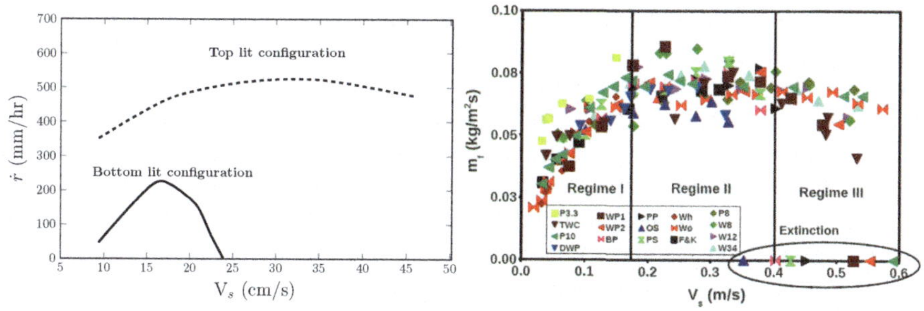

(a) bottom vs top lit (b) operating regimes

Fig. 6.2 Variation of flame propagation rate with superficial velocity in bottom-lit and top-lit configurations and operating regimes; data taken from Rönnbäck et al. (2000), Mukunda (2011), Varunkumar et al. (2013)

in air flow rate, reach 0 (stationary reaction front) and then becomes negative (indicating downward propagation). Figure 6.2 shows the variation of propagation rate ($\dot{r}$) versus superficial velocity for a top lit and bottom lit configurations [data taken from Rönnbäck et al. (2000), Mukunda (2011)].

The ideal operating condition is the one with the stationary reaction zone; but in practice it is not always possible to maintain a stationary reaction front; therefore, an air flow slightly lower than what is needed for a stationary reaction front is used. This will create conditions in which there is a slowly propagating upward flame and the reaction front is reset to the required level by periodic char extraction. It should be noted that in the case of the re-burn gasifier (IISc design), it is significantly easier to maintain a stationary reaction zone by adjusting the area ratio between the top and the side air entries. The same strategy is used to tailor the reburn-gasification system to either maximize gas yield (with minimum tar) or charcoal yield. In the configurations described above, the reaction zone location is tracked using equally spaced thermocouples along the length of the reactor. Therefore, the measured propagation rate variation with air flow is with respect to the laboratory flame of reference. A quantity of fundamental importance is the propagation rate with respect to fresh biomass (like flame speed); from now onward we will refer this quantity as propagation rate. This can be obtained by adding the rate of downward movement of biomass to the measured propagation rate. Another simpler way to obtain this is to measure the propagation rate in a top-lit configuration (refer (c) in Fig. 6.1) in which, the biomass bed is fixed with respect to the laboratory frame and hence the propagation rate measured by the thermocouples is directly the propagation rate w.r.t fresh biomass. Also, the top-lit configuration can be easily mounted on a weighing balance and the fuel consumption rate (and fuel flux) can be measured from the mass loss with time. The fuel flux is also equal to the product of flame propagation rate and bed packing density.

In terms of fuel flux variation with air flux, the top-lit and the bottom-lit configurations are equivalent—this was first explicitly brought out in Rönnbäck et al. (2000).

Several experimental studies have reported the fuel flux variation with air flow in the top-lit configuration with different types of biomass—see for instance, Fatehi and Kaviany (1994), Gort (1995), Porteiro et al. (2010), Ronnback et al. (2008), Yang et al. (2004), Ryu et al. (2007), Collazo et al. (2012). There are also a number of modeling studies which have attempted to predict the observed fuel flux variation with air flux and through this have proposed controlling mechanisms for different ranges of air flux. A review of these studies show that not all studies agree on the controlling mechanisms and a detailed discussion of these conflicting elucidations and other aspects that can be found in Varunkumar (2014). In an attempt to resolve the conflicts in literature on the regimes of propagation and the corresponding controlling mechanisms, Varunkumar et al. (2013) compiled the fuel flux data for over a dozen biomass types from literature along with their own data for four types as shown in Fig. 6.2b.

After correcting for variations in the moisture and ash content, it became clear that the flue flux variation with air flux follows a *universal curve* (aka universal flame propagation behaviour, UFP in short)—steady propagation starts at an air flux (a function of V_s) of 30 g/m^2-s (corresponding fuel flux is about 20 g/m^2-s); the fuel flux increases more or less linearly with air flux ($\dot{m}''_f$ reaches about 70 g/m^2-s at $\dot{m}''_a$ of 180 g/m^2-s) and then remains constant till sudden extinction (caused by convective cooling). The extinction air flux varies between 350 and 500 g/m^2-s (the large variation in extinction air flux is due to the sensitivity of the extinction process to minute changes in heat loss, which in turn is dependent on several factors including biomass shape and size, level of insulation etc.). The bed operation in the linear part of the curve is referred to as the 'gasification' regime and the flat part is referred to as the 'char oxidation' regime. The classification is based on the fact that the onset of *char oxidation* beyond about 180 g/m^2-s of air flux makes the hot char inaccessible for reduction reactions which are essential to the gasification process.

Recognition of the universality of flame propagation process leads to two fundamental questions—(1) which processes control the flame propagation phenomenon in the *gasification* and *char oxidation* regimes? and (2) could a mathematical model be constructed to explain and predict the observed variation in magnitude of fuel flux with air flux? A definitive answer to the first question is given in Varunkumar et al. (2013)—a plot of temperature versus oxygen concentration, measured simultaneously as the flame propagates through a bed of fresh biomass, shows a near instantaneous drop in oxygen concentration from 23 to 0% with a simultaneous increase in temperature from 300 K to about 1000 K for the entire range of operation; this is a clear demonstration of the fact that the flame propagation to a fresh layer of biomass is diffusion controlled in both regimes [see Fig. 6.8 in Varunkumar et al. (2013)].

Varunkumar et al. (2013) also provided answer for the second question—a model was constructed to track the evolution of volume averaged temperature of a fresh biomass particle upstream of the propagating flame front. The particle is heated by radiation from the hot char particles in the de-volatilization zone and cooled by the air flowing through the bed. The net heating rate determines the rate of increase of particle temperature with time—once the particle reaches ignition temperature (which depends on whether the propagation is in the gasification or char regime) the flame is considered to have moved to the fresh layer. Therefore, the propagation rate is equal to the ratio of particle size to ignition time. The char emissivity is taken to be 0.9 in the gasification regime; transition to char oxidation regime leads to a sudden drop in the emissivity to about 0.2 due to the formation of ash layer over the particles; the emissivity further decreases with increase in air flux (and hence average bed temperature). Incorporation of this emissivity variation is a crucial ingredient of the model—this explains the leveling off of the fuel flux beyond an air flux of 180 g/m^2-s. The question, why the gasification to char oxidation transition occurs at around 180 g/m^2-s is not fully addressed in Varunkumar et al. (2013). All that is mentioned on this question is the observation that the transition point coincides with the point of *volatiles stoichiometry*—the air-flux at which the equivalence ratio of oxidation of volatiles alone is around unity.

It is clear from the above discussion that, with air as oxidizer, flame propagation through biomass packed beds exhibits universality; that is, on a dry and ash free basis, the fuel flux is not dependent on biomass type, but only on air flux. Fuel flux ($\dot{m}''_{ox}$) increases with increase in superficial velocity in the 'gasification' regime and saturates in 'char oxidation' regime and reduces to zero at extinction. That the propagation behavior is universal clearly established the top-lit packed bed system as the canonical configuration for addressing a wide range of questions, both fundamental and applied. Insights from studies on top-lit packed bed configuration has proved to be crucial to understanding the thermo-chemical conversion of biomass particles. The questions, the approach developed to address those and the insights obtained from these studies form the rest of this chapter.

6.1.3 Open Issues in Packed Bed Biomass Gasification

The open issues in packed bed biomass gasification are best brought out from the point of view of the Universal Flame Propagation model (UFP model). Recall that in the UFP model, the flame propagation rate is calculated as the ratio of the particle size to its ignition time. Therefore, the following assumption is implicit in the model—that the ignition time (t_{ig}) is rate controlling. It is to be noted that there is another time scale, the de-volatilization time (t_v), that is relevant and its magnitude relative to the ignition time must be estimated to check this assumption. It will be shown subsequently that the ratio of igntion time to de-volatilization time determine the regimes of propagation—as will be shown later, there are 5 distinct ones. Of these 5, the UFP model as outlined in Varunkumar et al. (2013), describes only 2; predicting the propagation behavior in the other 3 regimes require additional considerations and are brought out and discussed later. Two common situations in which the UFP model assumption is not valid are as follows—(1) starting from conditions of steady propagation, if the oxygen fracftion in the oxidizer is increased (for instance, while operating under oxy-fuel conditions), t_{ig} can become much lower than t_v leading to the so called 'flame jump' and (2) for a fixed oxidizer composition and flux corresponding to steady propagation, if the particle size exceeds a critical value, t_{ig} can become lower than t_v; this critical value is a function of O_2 fraction and mass flux of the oxidizer. Hence, these cases need further exploration and are addressed here.

It is important to recognize that the operation of packed beds under conditions, that might be close to the limits of *steady flame propagation* are relevant from the perspective of oxy-fuel gasification (that is gasification with mixtures of O_2–CO_2–steam). Other pertinent question concerns the intrinsic levels of net conversion of CO_2 and steam with char; this question is one of the important ones that will be addressed here. Results from our recent lab studies (described herein) provide answers for all of the above questions. Before discussing the results from these studies, a brief discussion of relevant literature is presented.

6.1.4 Earlier Studies on Oxy-Fuel Biomass Systems

It is pertinent to point out that the questions posed above cannot be answered using analytical methods like TGA, DTA etc. In fact, we can go one step further and state that these important questions itself could be posed only because of the canonical platform provided by counter-current packed bed configuration. The reasons for this is explained in some detail in the following paragraphs.

A number of studies have been carried out to explore the role of O_2–CO_2–steam–N_2 (in varying proportions) in biomass/coal gasification/combustion reactors. Lab scale TGA/DTA (Dai et al. 2017; Huo et al. 2014; Sircar et al. 2014; Butterman and Castaldi 2009; Lin and Strand 2013; Bouraoui et al. 2015) and Drop Tube Reactor (DTR) studies (Kyotani et al.1993); Riaza et al. 2014; Billaud et al. 2016; Ranzi et al. 2013) are the predominant ones among these. While TGA/DTA studies are conducted in the 1–100 K/min heating rate range and 700–1400 °C peak temperature range, DTR studies use ultra high heating rates of $O(10^4)$ K/s. On the other hand, heating rates expereinced by particled in practical configurations are in the range of a few 1000 K/min. This makes results of the TGA/DTA and DTR studies less relevant in actual applications.

Studies with practical configurations like fixed bed (updraft and downdraft configuration) and fluidized bed systems are also been reported in literature. See for instance, Nilsson et al. (2012), Kramb et al. (2017), Bu et al. (2017), Sandeep et al. (2011), Gao et al. (2008), Lucas et al. (2004), Umeki et al. (2010), Alex et al. (2011), Yang et al. (2014, Balu et al. (2015), Nipattummakul et al. (2010), Sandeep and Dasappa (2013, 2014), Cerone et al. (2016, 2017), Lahijani et al. (2014). The following general observations apply to all the studies listed earlier—(1) wide range of oxygen fractions and oxidizer flux covering gasification to combustion regime and limits of propagation are not studied, (2) conditions for steady propagation are not clearly brought out, (3) net CO_2 and steam conversion are not quantified and has discrepancies in the reported CO and H_2 yield, (4) fuel rich operation was always chosen in order to get high fractions of CO and H_2 in syngas in spite of high tar fractions in the syngas, (5) in a few studies steam at very high temperatures (more than 800 °C) were used to overcome the issue of excessive tar in the exit gas and (6) some systems use external heating to sustain the flame propagation limiting their applicability.

Another aspect that has received little attention in literature is flame propagation behavior in biomass packed beds with oxidizers other than air. Superficial velocity (V_s) was found to be the most influential parameter to affect fuel flux ($\dot{m}''_f$), bed temperature (T_{pb}) and syngas composition. Other fuel linked properties like size, density, moisture and ash content are shown to be have only secondary/no effects. Extensive discussion on literature on other flame propagation models from Fatehi and Kaviany (1994), Gort (1995), Porteiro et al. (2010, Ronnback et al. (2008), Yang et al. (2004), Ryu et al. (2007), Collazo et al. (2012) is given in Varunkumar (2014). The conclusion from Varunkumar (2014) is that, biomass conversion in packed beds is governed by heat transfer and hence, models with detailed kinetics to predict the burn rate is of less relevance. Models to predict the propagation rate is important to design practical gasification/combustion systems; one example is the design of traveling grate boilers, where the 'universal flame propagation model' was used to set out gasification/combustion parameters like fuel layer height, grate velocity, quantity of air to be supplied and distribution strategies (Varunkumar et al. 2015). As mentioned earlier, the important conclusion from (Varunkumar 2014) is that the propagation rate is controlled by radiation heat transfer from flame front, which in turn determines the ignition time of fresh biomass upstream of the flame front. This behavior is shown to be due to ignition time (t_{ig}) being the rate limiting step in flame propagation and the ignition time to particle density ratio being independent of biomass type. But when the oxygen fraction in the oxidizer is increased well beyond 21%, which might be required for enhancing char conversion with O_2–CO_2 and O_2–steam mixtures, there is the possibility of devolatilization time (t_v) becoming rate limiting—this causes the so called 'flame jump', a type of unsteady propagation. Another condition under which this can happen is as follows—starting from conditions of steady propagation, if the particle size is increased beyond a critical size (which is a strong function of oxygen fraction in the oxidizer), the ignition time can become smaller than devolatalization time. These aspects, which go beyond the scope of 'universal flame propagation' model, need further investigation for identifying optimal conditions for steady flame propagation.

The next important aspect of theoretical modeling is to predict the syngas composition at gasification conditions and this is mostly approached through kinetic rate models and thermodynamic equilibrium models, in earlier literature. Kinetic models predict the rate of thermal pyrolysis and gasification using kinetic rate expressions obtained from experiments at controlled conditions like TGA (Wang and Kinoshita 1993; Di Blasi 2000; Fiaschi and Michelini 2001; Babu and Sheth 2006; Gobel et al. 2007; Sharma 2008). Kinetic parameters obtained from such studies show sensitivity to biomass type, catalyst present in the form of inorganic matter and temperature. When the particle size is increased (say to a few mm), heat and mass transfer processes become rate limiting rather than kinetics and the conversion is truly an 'aero-thermal-chemical' process (Mukunda 2011). As mentioned earlier, the oxidation process in a packed bed flame propagation phenomenon is clearly diffusion controlled. Hence, kinetics of oxidation of volatiles is not relevant in these configurations. Gasification process is predominantly a reaction of char with CO_2 and H_2O. Single particle experiments were done by Dasappa (1999) with char of larger sizes (i.e. equivalent spherical diameter $d_p > 4$ mm) with different oxidizers and at different temperatures. Mass loss with time and temperature were recorded and the conversion times with air, O_2, H_2O and CO_2 were determined. The conversion time is proportional to d^2 for air and O_2 which implies that char oxidation is diffusion limited and the exponent is lower for CO_2 and H_2O which shows combined diffusion-reaction control for reduction reactions of char (the reduction reactions will transition from reaction to diffusion control at higher temperatures).

Equilibrium modeling to predict the exit gas composition is the other area that has received a lot of attention in gasification literature (Gomez-Barea and Leckner 2010; Li et al. 2001; Melgar et al. 2007; Jarungthammachote and Dutta 2007, 2008); Yoshida et al. 2008; Karamarkovic and Karamarkovic 2010; Huang and Ramaswamy 2009). Volatile oxidation is diffusion controlled and hence, equilibrium approach to predict volatile oxidation products is appropriate. Char oxidation is also diffusion controlled and hence, equilibrium modeling is still applicable here too. But, in gasification regime, char kinetics plays a key role and equilibrium calculations to predict the overall exit gas composition in gasification regime is not the right choice. Also, as shown by Varunkumar (2014), char conversion is complete for cases near 'gasification' regime. This is due to limitation of char availability in biomass to around 20% by mass. In brief, biomass contains 80% 'volatiles' and releases 70% of the total heat due to 'volatiles' oxidation, and drives the flame propagation which is governed by heat transfer. Hence, models based on detailed kinetics to predict the composition is of less relevance. Hence, a new methodology was developed by combining the ideas of 'volatiles' equilibration with standard elemental and energy balances to address these issues. Relevant details of the experimental setup, methodology, results and analysis are presented in subsequent sections.

6.2 An Overview of the Experimental Methods and Materials

In the studies reported in Jaganathan and Varunkumar (2019), Jaganathan et al. (2019, 2020), Jaganathan (2019) four types of biomass are used—two types of pellets (sawdust and groundnut shells), coconut shells and one type of briquette (groundnut shells). Other types of biomass are used in the studies of Cerone et al. (2016, 2017), Cerone and Zimbardi 2018)—noteworthy among the biomass used is the biorefinery hydrolytic residues. But given the *universality* of the behavior of various types of biomass in packed beds, insights obtained for a few types of biomass can be extended in a straight forward way to other kinds. An extensive discussions of the results in

Table 6.1 Experimental nomenclature

S. No.	Oxidizer (% mass)	Experiments	S. No.	Oxidizer (% mass)	Experiments
1	Air	PA, CA and WPA	7	33% O_2–67% CO_2	P33N
2	19% O_2–81% CO_2	P19C and C19C	8	43% O_2–57% CO_2	P43N
3	23% O_2–77% CO_2	P23C, C23C and WP23C	9	23% O_2–77% Steam	WP23
4	32% O_2–68% CO_2	P32C and C32C	10	30% O_2–70% Steam	WP30
5	42% O_2–58% CO_2	P42C	11	40% O_2–60% Steam	WP40 and P40
6	28% O_2–72% CO_2	P28N			

P—Agro-residue pellet; C—Coconut shell; WP—Wood pellet

Cerone et al. (2016, 2017), Cerone and Zimbardi (2018), including their validity for downdraft systems and their limitations, are presented in Jaganathan et al. (2019). Further discussions are limited to the results in Jaganathan (2019).

6.2.1 *Packed Bed Experiments*

All data used here were obtained from standardized experimental setup reported in Jaganathan and Varunkumar (2019, 2020), Jaganathan et al. (2019, 2020). One of the important outcomes of the studies reported in Jaganathan and Varunkumar (2019, 2020), Jaganathan et al. (2019, 2020) is the standardization of the experimental procedure—the method of ignition, mass and temperature acquisition and its interpreration, exit gas analysis and method of reactor quenching can be directly adopted by other researchers. List of oxidizer combinations and the nomenclature used in discussions is reproduced in the Table 6.1 for easy reference.

6.2.2 *Oxidizers—Estimation of 'Volatiles Stoichiometry'*

Three different oxidizers are used for the study—(1) O_2–CO_2, (2) O_2–N_2 (air falls under this category) and (3) O_2–steam. As indicated earlier, experiments covering the entire range of oxidizer flux, that is, from the lowest possible steady propagation rate to the other limit of *flame jump* was explored. The parameter that emerged as the unifying factor in interpretation of experimental results is the so called '*volatiles equivalence ratio' as defined below.

$$\phi_v = \frac{(F_v/O)_{act}}{(F_v/O)_{stoi}}$$

Extensive analysis of the entire set of experimental data revealed that, in general, the point of *volatiles stoichiometry* (that is $\phi_v \sim 1$) conincides with the point of transition from gasification to char oxidation. But some exceptions to this general rule were identified as well which perhaps are operating conditions of very little relevance to practical applications rendering the rule a useful guideline for practioners.

6.2.3 Single Particle Experiments

As discussed earlier, to validate the particle size effects on t_{ig} and t_v and their influence on UFP behaviour, single particle experiments becomes essential. Also, to understand the transition mechanism (details later) of 'gasification' to 'char oxidation' regime near 'volatiles' stoichiometry single particle experiments are carried out. Principal aim of the single particle experiments is to determine the variation of ignition (t_{ig}) and devolatalization time (t_v) with particle sizes and to extend the understanding to packed beds. Single particle burner consists of 100 mm steel reactor with LPG and air inlets and the reactants are premixed in the burner chamber and a fine mesh of 0.1 mm is placed at 50 mm from the top to ensure a flat flame. LPG flame is established at the mesh surface and the amount of air and LPG are regulated with mass flow controllers and it is ensured that the mixture is lean so that the O_2 fraction in the exhaust stream is around 10–12 % (by vol). The O_2 fraction in the exhaust stream is crosschecked with NDIR based gas analyzer and found to be consistent with the estimates from equilibrium calculations. The particle is suspended by fine needle and mounted on a weighing balance. The mass loss is recorded through RS 232 data logger. The hot gas temperature is measured with a K type thermocouple. Single particles are hung by a needle/necessary support and ignited with quiescent flame/LPG burner for small/large particles. Ignition source in general, is removed after ignition but, in some cases not removed away from the burning particle. Both conditions are studied to elucidate the effect of ambient stream. For more details, refer Jaganathan et al. (2017). Mass loss with time was tracked with a weighing balance (1 mg accuracy). As is very well known from earlier studies, the mass loss with time plot shows three distinct regimes–(1) ignition, (2) de-volatilization and (3) char oxidation. Ignition is characterized by an accelerating mass loss rate and typically occurs up to 15% loss in the initial mass. This is followed by devolatilization phase characterized by constant mass loss rate (linear mass loss with time). During this phase a diffusion flame envelopes the biomass particle. Transition from devolatilization to char oxidation can be identified as a distinct change in slope in the mass loss curve accompanied by the disappearance of the diffusion flame surrounding the particle. Ignition time (t_{ig}) is taken to be the time for 15% loss in initial mass and de-volatilization time (t_{ig}) is taken as the time from start of ignition to the point on the mass loss curve with a distinct change in the slope (indicating transition to char mode). More details on this are presented latter in the chapter.

6.2.4 Analysis Framework and Algorithm

Development of a standard analysis framework for packed beds is another important outcome of the studies reported in Jaganathan (2019). The algorithm used for calculating various quantities was developed with the definition of Net CO_2 conversion (NCC) as the starting point. The formula for NCC is given by Eq. 6.1.

$$NCC \quad = \frac{(m_{CO_2,v} + m_{CO_2,inlet} - m_{CO_2,outlet})}{m_b} \tag{6.1}$$

where, m_b is mass of biomass (kg), $m_{CO_{2,v}}$ is the mass of CO_2 (g) released through 'volatiles' oxidation, $m_{CO_{2,inlet}}$ is the mass of CO_2 in the O_2–CO_2 mixture supplied (g) and $m_{CO_{2,outlet}}$ is the mass of CO_2 in exit gas (g). That the CO_2 released from oxidation of 'volatiles' will react with char to generate CO and the associated NCC is substantial even with air as oxidizer is important to recognize and quantify. Following similar approach, net steam and char conversion are defined; corresponding gas yields are calculated from the analysis of experimental data for specific cases.

In air gasification systems, the possible sources of CO are through (1) partial oxidation of 'volatiles' carbon and (2) reduction of CO_2 and other gases formed during 'volatiles' oxidation by char. When N_2 in air is replaced with CO_2 like in the experiments with mixtures of O_2 and CO_2, further augmentation of the char reduction reactions is expected; there will be an associated increase in NCC compared to air cases. The general principle that char becomes unavailable for reduction when undergoing oxidation (which generally occurs when $\phi_v < 1$) applies to all the oxidizer combinations employed in packed bed experiments. It has emerged from laboratory studies that the fuel conversion follows the pathway of release of volatiles, followed by its oxidation (partial or complete depending on the ϕ_v), then reduction or simulataneous oxidation of char and volatiles (again, depending on the ϕ_v)); ash is taken to be inert in these calculations.

General algorithm used for calculating the net CO_2/steam conversion, CO yield, H_2 yield etc., is reproduced in Fig. 6.3. Note that the algorithm requires inputs from proximate and ultimate analysis, results of equilibrium composition and temperature of products of oxidation of volatiles and the amount of left over char. The convergence criteria for the iterative procedure is set based on the imbalance in atom count and energy (limit set as 5%). Readers interested in using the algorithm are referred to Jaganathan and Varunkumar (2019) for details; in particular the discussion and data given in Table 3 of Jaganathan and Varunkumar (2019) for the P30C case with oxidizer mass flux of 31 g/m^2-s. In the discussions that follow, only the general characteristics of NCC and other performance measures of packed bed gasification, as inferred from extensive analysis of data for a wide range of conditions are discussed. For details of specific cases, readers can refer to our earlier publications (Jaganathan and Varunkumar 2019; Jaganathan et al. 2019; Jaganathan 2019).

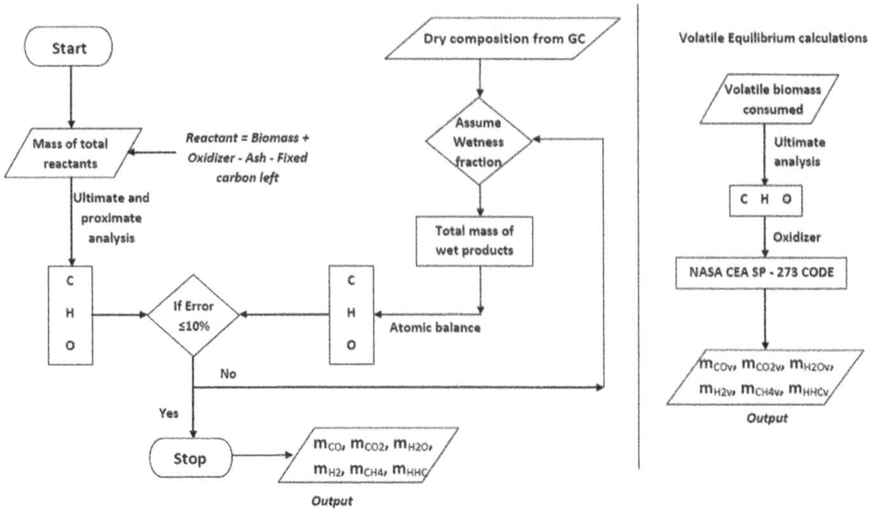

Fig. 6.3 Algorithm used for calculation

6.3 Universal Characteristics of Performance Measures of Packed ...

As mentioned earlier, the focus of this section will be on the universal characteristics of performance measures of packed bed gasification systems; for specific data points corresponding to a particular biomass and/or operating conditions, interested readers are referred to Appendix E of Jaganathan (2019). With 'volatiles equivalence ratio' emerging as the unifying parameter in interpretation of data, variation of quantities like NCC, NSC, H_2 and CO yield, char conversion and higher hydrocarbon fraction in the exit gas will be presented as a function of ϕ_v; to aid interpretation a separate plot showing the variation of ϕ_v with oxidizer mass flux is shown in Fig. 6.4. It can be clearly seen that the general trend is followed for all oxidizer combinations and all experimental data is sandwiched between the 23 and 42% O_2 limits.

Net CO_2 Conversion (NCC) peaks around 'volatiles stoichiometry' ($\phi_v \approx 1$) i.e. the transition point of gasification to combustion and falls off to near zero on either side of ϕ_v 1. This trend is shown in Fig. 6.5a. The magnitude of NCC is bounded on the lower side by the air case and on the upper side by 23% O_2–77% CO_2 case. The magnitude of net steam conversion (NSC) also shows similar trends and is shown in Fig. 6.5b.

The general behavior of the NCC and NSC variation with ϕ_v is due to the interplay of bed temperature, gas residence time through the char bed and the available amount of char. The reduction of CO_2 by the carbon in the char is the principal contributor to NCC and NSC. The rate of this endothermic reaction is negligible up to about 700 K and then it starts increasing. Beyond about 1450 K this reaction

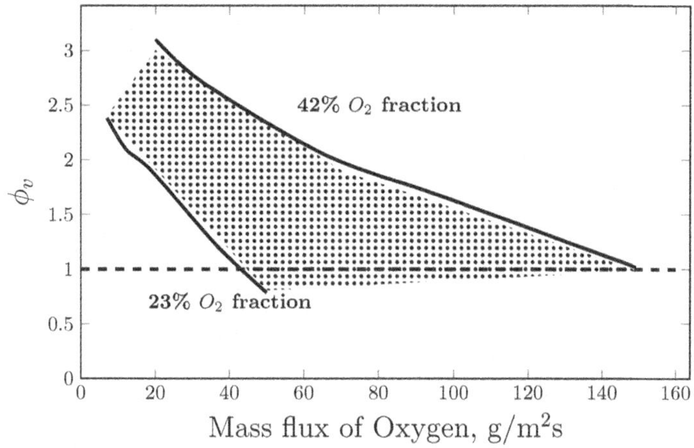

Fig. 6.4 General trend of 'Volatiles equivalence ratio' versus oxidizer mass flux

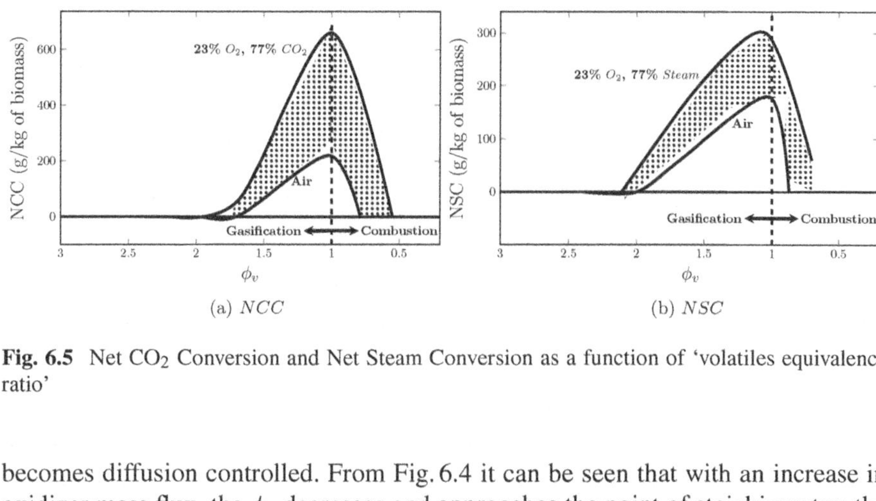

(a) *NCC* (b) *NSC*

Fig. 6.5 Net CO_2 Conversion and Net Steam Conversion as a function of 'volatiles equivalence ratio'

becomes diffusion controlled. From Fig. 6.4 it can be seen that with an increase in oxidizer mass flux, the ϕ_v decreases and approaches the point of stoichiometry; the average bed temperature will increase as ϕ_v approaches the point of stoichiometry (the magnitude of temperature is dependent on the oxidizer composition) and the reaction rate increases correspondingly. This leads to an increase in NCC as ϕ_v approaches stoichiometry from the rich side as can be seen in Fig. 6.5a. Once the point of stoichiometry is reached, further increase in oxidizer flux initiates char oxidation which in turn blocks the char availability for reduction reactions. This phenomenon causes the NCC to sharply decrease to 0 as the oxidizer mass flux is increased beyond the point of 'volatiles stoichiometry'. The peak value is about 250 g/kg with air as oxidizer and is about 600 g/kg with O_2–CO_2 as oxidizer.

A pertinent question is how far the maximum values observed are away from the theoretical maximum? For this, the percentage char conversion values must be considered—the general behavior of % char conversion is shown in Fig. 6.6a. That for

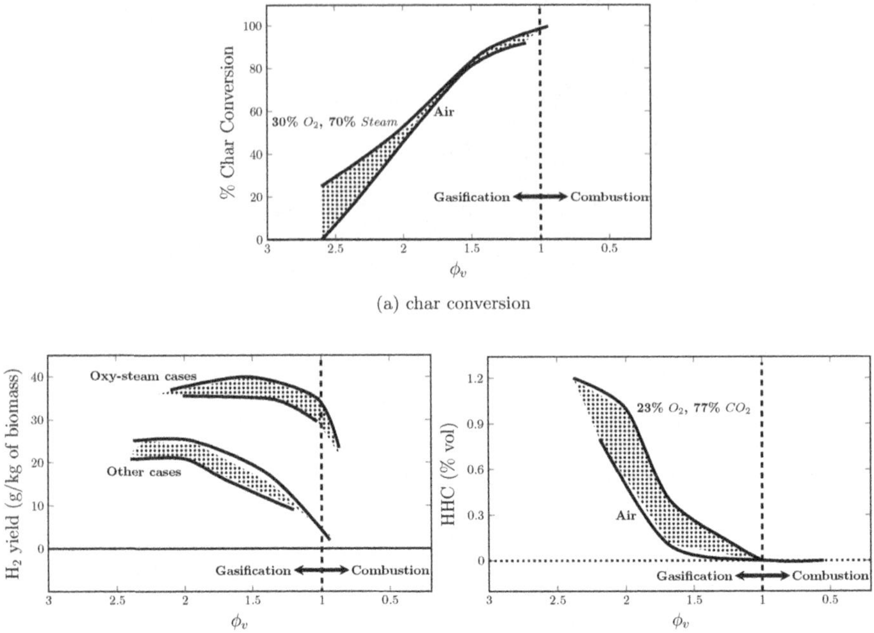

(a) char conversion

(b) H_2 yield

(c) HHC

Fig. 6.6 Yield of hydrogen and higher hydrocarbons as a function of 'volatiles equivalence ratio'

all cases, the char conversion approaches the maximum value of 100% is indicative of the fact that the maximum observed NCC is close to the theoretical maximum. This combined with the fact that the residence time decreases with increase in oxidizer flux (from as high as 80 ms it decreases to 6 ms as $\phi_v \rightarrow 1$), that is, as the point of 'volatiles stoichiometry' is approached, clearly indicates that the NCC and other performance characteristics are limited by the char availability (which is typically 10–20% of the biomass weight). Therefore, near ϕ_v 1 on the rich side, the conditions in the reaction zone are conducive for both C +H_2O and C + CO_2 reactions. An important point to note is the role of diffusion on the rate of this reaction which is significantly higher at $\phi_v \sim 1$ ($T_{pb} \sim 1500$ K) as compared to lower temperatures at $\phi_v > 2$ ($T_{pb} \sim 1000$ K). That the NCC and other performance parameters appear constrained by the char avaialbility indicates that there is significant diffusion control on reduction reactions close to $\phi_v = 1$.

The next quantity of interest is the hydrogen yield. Here focus is on the yield of hydrogen only in oxy-steam cases; while H_2 yield from O_2–CO_2 experiments is also quantified, it is not discussed here. The H_2 yield is in the range of 30–40 g/kg of biomass for all the O_2–steam cases over the entire gasification range (i.e. $2.1 \geq \phi_v \geq 1$) as shown in Fig. 6.6b.

This observation may seem to be in contradiction with the points brought out earlier, through which it was established that operation close to ϕ_v 1 should be preferred. If the yiels of H_2 is more or less invariant over the entire gasification regime, why operate close to ϕ_v 1? Also, it is important to note that all the cases in literature spscitepch10sandeepsteam,ch10lv2007hydrogen were restricted to fuel rich regime (ϕ_v> 2), which seems to bring the usefulness of operating close to ϕ_v 1 into question. Before clarifying this apparent contradiction, the invariance of the yield of H_2 in the gasification regime is first explained. For $\phi_v > 3$, the source of almost all hydrogen in the exit gas is the volatiles. As $\phi_v \rightarrow 1$, almost all the hydrogen in the volatiles is oxidized leaving no hydrogen; hence, the direct contribution of volatiles to exit hydrogen close to ϕ_v 1 is negligible. Almost all the exit gas hydrogen obtained when operating close to ϕ_v 1 is from the $C + H_2O$ reaction. It was brought out earlier that the rate of this reaction increases with increase in bed temperature (which occurs as $\phi_v \rightarrow$ 1) and is limited only by the availability of char. Therefore as $\phi_v \rightarrow 1$, the reduction in hydrogen due to oxidation of volatiles is more or less exactly compensated by the hydrogen from $C + H_2O$ reaction. One may still wonder why not operate at $\phi_v > 3$? This is not preferable as in addition to hydrogen from volatiles, there will also be higher hydrocarbons (HHCs) in the exit gas (tars) under rich conditions (which increases the complexity of downstream gas scrubbing process). On the other hand, while operating close to $\phi_v \rightarrow 1$, almost all the volatiles are oxidized without leaving behind any HHCs and hence eliminating the need for extensive gas scrubbing systems. This expected behavior of HHC concentration with ϕ_v is borne out by experimental data as can be seen from the trends shown in Fig. 6.6c. All the hydrogen in the exit gas is from the $C + H_2O$ reaction and hence without any HHCs. The role of volatiles under this conditions is to generate high temperature steam in-situ to sustain the endothermic $C + H_2O$ reaction only.

At this point it is important to mention that the cold gas efficiency, a measure of the overall performance of a gasification system, for the optimal operating conditions is in the range of 60–70% (Jaganathan 2019). This is considered good for a top-lit packed bed system; the same conditions when adopted for a continuous system is expected to perform with a cold gas efficiency in excess of 75% due to additional residence time for char reactions and better thermal environment of larger throughput systems.

Going back to the relationship between 'volatiles stoichiometry' and *gasification to char oxidation transition*, it is very clear that the transition between 'gasification' to 'char oxidation' occurs close to 'volatiles stoichiometry' for all biomass-oxidizer combinations of practical interest. This is a very useful rule of thumb in understanding the dynamics of biomass thermo-chemical systems. The underlying mechanism of this behavior is brought out using a combination of single particle experiments and combustion-theory inspired analysis in Mani Kalyani et al. (2021). A summary of the same is given here.

6.4 Mechanism of Gasification to Char Oxidation Regime Transition in Packed Beds

Following up on an earlier hypothesis that the gasification to char oxidation transition could be linked to the transition of envelope flame to wake flame for particles in a packed bed and that this transition is driven by extinction of the *volatiles flame in the forward stagnation point*, Mani Kalyani et al. 2021) have reported estimates for strain rate experienced by particles in packed beds and the behavior of single particles in corresponding conditions. But in addition to the two states, that is envelope and wake flame, a third flame-state, *simultaneous volatiles-char oxidation state*, was identified in the experiments of Mani Kalyani et al. (2021). Figure 6.7, shows all the three possible states of the flame around a particle in a stream of oxidizer.

It is shown in Mani Kalyani et al. (2021) that the bed operates in gasification regime if the particle flame state is *envelope flame*; this is consistent with the fact that only in the existence of an envelope flame, the char will be shielded from oxygen and hence will be available for reduction reactions (which is essential for gasification). The operating regime transitions from gasification to char oxidation when the envelope flame transitions to either the wake flame (Fig. 6.7b) or the simultaneous volatiles-char oxidation flame state (Fig. 6.7c). Results reported in Mani Kalyani et al. (2021) show that for oxidizers with O_2 fraction (v/v) up to about 25%, the wake flame transition is observed; for O_2 fraction >30%, the simultaneous volatiles-char oxidation flame state transition is observed. The implications of these insights are as follows—if one wishes to know apriori the oxidizer mass flux at which a packed bed will transition from gasification to char oxidation, all that one needs to do is to determine the flame transition velocities for single particles.

With this the discussion of insights from experiments and analysis is complete. What follows is a review of the advances in the theoretical aspects of flame propagation in packed beds.

6.5 A Theoretical Framework for the Phenomenon of Flame Jump

The recognition of the *universal flame propagation* behavior, by Varunkumar et al. (2013), in packed beds of biomass with air as oxidizer and *the corresponding transport controlled model* has remained the principal theoretical result for over a decade on this topic. Therefore any attempts at extending the theoretical framework for this problem should start by identifying the shortcomings of the *universal flame propagation model*. One obvious shortcoming is the use of data with only air as oxidizer; this was because of the lack of systematic data with other oxidizers. Now that data for other oxidizers have become available, the first task would be to determine if the '*universality*' could be extended to the entire data set. The answer is yes and the approach used to establish this fact is described below.

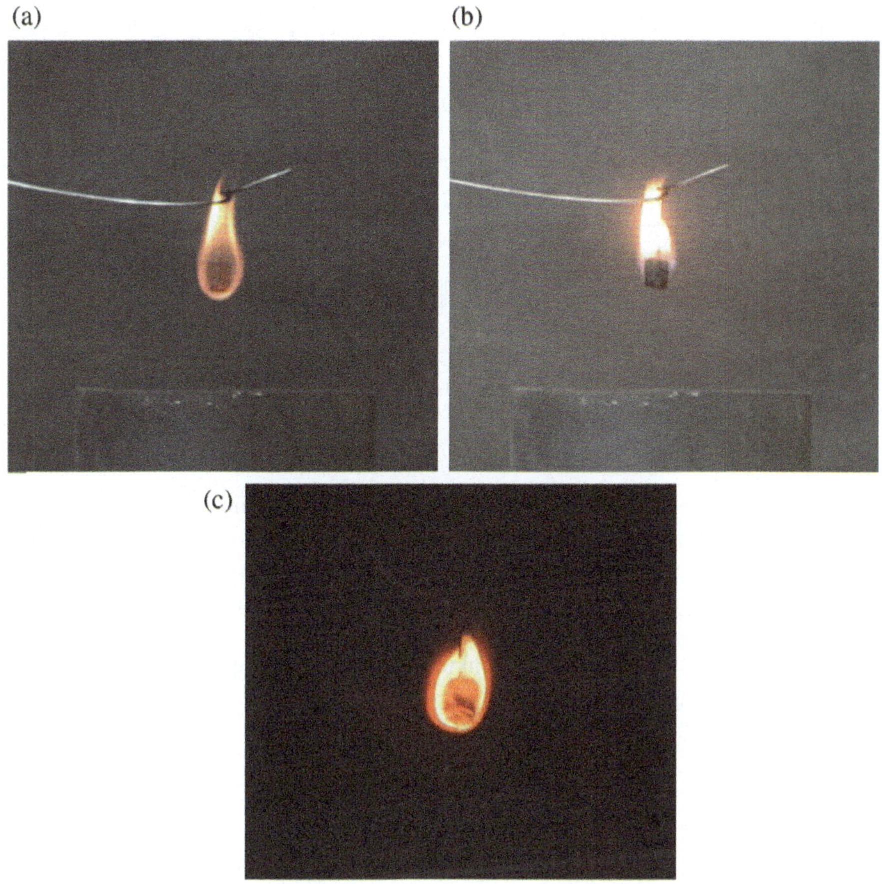

Fig. 6.7 Three possible flame-states around a biomass particle in a stream of oxidizer—**a** envelope flame, **b** wake flame and **c** simultaneous volatiles-char oxidation. *Source* Mani Kalyani et al. (2021)

6.5.1 Normalized Fuel Flux (NFF)

Universality under steady propagation conditions is expected even when the oxidizer is changed from air to mixtures of O_2–CO_2 and O_2–N_2. Raw data (not corrected for ash and moisture) showing the variation of $\dot{m}''_f$ with $\dot{m}''_{ox}$ for the cases of O_2–CO_2, O_2–N_2 and air experiments is shown in Fig. 6.8a.

Similar to air gasification, the *gasification* and *char oxidation* flame propagation regimes are observed in O_2–CO_2 and O_2–N_2 cases too. In other words, the fuel flux increases with increase in oxidizer mass flux in the *gasification* regime and saturates in *char oxidation* regime. As brought out earlier, the transition can be identified using 'volatiles' stoichiometry. The fuel flux saturates in the char oxidation regime due to the following—as the oxygen mass flux increases, char layer gets oxidized and

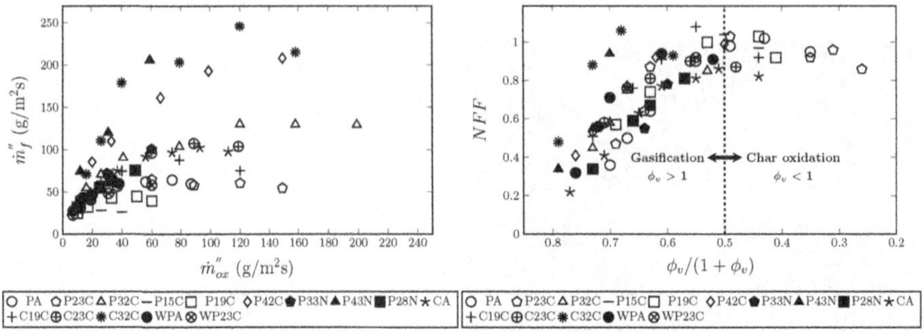

(a) Raw fuel flux data (b) NFF

Fig. 6.8 Raw data of fuel mass flux variation with $\dot{m}''_{ox}$; uncertainty $\leq \pm 10\%$; taken from Jaganathan (2019); normalized fuel flux with scaled equivalence ratio is shown

the increasing temperature is compensated by decreasing emissivity of the char-ash layer formed at the char surface which reduces the radiative heat transfer to fresh layer of biomass; hence, the $\dot{m}''_f$ reaches saturation Varunkumar et al. (2013). Higher oxygen fraction (more than 40% of O_2, by mass) shows enhanced fuel flux (as high as 208 g/m^{2}s at a oxygen flux of 149 g/m^{2}s of P42C case) due to increase of 'volatiles' heat release rate. Jaganathan and Varunkumar (2019) proposed a corrected fuel flux (CFF) given by Eq. 6.2.

$$CFF = \left(\frac{\dot{m}''_{fa}}{\dot{m}''_{f_{ref}}}\right)\left(\frac{(HY_{ox}/s)_{ref}}{(HY_{ox}/s)_a}\right)\left(\frac{\rho_a}{\rho_{ref}}\right) \tag{6.2}$$

where, H, Y_{ox}, s and ρ represent heating value, oxidizer fraction, stoichiometric coefficient and bed density respectively. Suffixes (ref) and (a) represent reference and actual cases. Here agro residue based pellets with air as reactant was taken as the reference case. In Eq. 6.2, the s_a was calculated by using actual oxidizer to fuel ratio of particular experiment and ρ was taken to be bed density. Fuel like wood chips have calorific value comparable to pellets (around 16 MJ/kg) but, the bed density of such fuels is lesser like coconut shells. The fuel flux ($\dot{m}''_f$) variation with $\dot{m}''_{ox}$ of wood chips is similar to that of pellets (Varunkumar 2014). CFF calculated from Eq. 6.2 for such fuels will reduce further as compared to pellets. Also, the non-linear dependence of $\dot{m}''_f$ with bed temperature is not captured in Eq. 6.2. Hence the Eq. 6.2 requires re-assessment. In the light of this, a modified equation was sought. The 'normalized fuel flux' as defined in Eq. 6.3 was found to be appropriate.

$$NFF = \left(\frac{\dot{m}''_{fa}}{\dot{m}''_{f_{ref}}}\right)\left(\frac{(HY_{ox}/s)_{ref}}{(HY_{ox}/s)_a}\right)^2\left(\frac{\epsilon_{b,act}}{\epsilon_{b,ref}}\right) \tag{6.3}$$

where, ϵ_b represent the bed voidage i.e. $1-(\rho_b/\rho_p)$. In NFF, s is the stoichiometric coefficient which is invariant for a particular fuel with experimental conditions unlike

s_a used in CFF. Bed voidage is taken as the reference in NFF unlike bed density used in CFF which can correct for the fuel shape and density effects effectively than density ratio as used in CFF. Also, exponent 2 for the second R.H.S term of Eq. 6.3 captures the non-linear dependence of fuel flux enhancement due to bed temperature rise at higher O_2 fractions. A non-dimensional plot of NFF versus $\phi_v/(1+\phi_v)$ is shown in Fig. 6.8b which takes into account all the required parameters of normalization. The term $\phi_v/(1+\phi_v)$ is used in the plot to have a symmetric axis. The NFF plot clearly shows that, biomass do exhibit 'universal behavior' of fuel flux increase till *gasification* regime and saturation of fuel flux in *combustion* regime. Note that the transition from *gasification* to *combustion* occurs at 'volatiles stoichiometry' ($\phi_v \sim 1$).

6.5.2 The Phenomenon of flame Jump

As discussed earlier, conditions of steady flame propagation were established over a range of O_2 mass fractions (from 19 to 42%) with O_2–CO_2, O_2–N_2 and O_2–steam mixtures. For each oxidizer, experiments were conducted to cover a range of O_2 mass flux—usually covering the overall equivalence ratio range of relevance to gasification (5–1.4). However, experiments with O_2–CO_2 and O_2–steam mixtures, especially because of the elevated oxygen levels and temperatures used (120–150 °C in case of O_2–steam mixtures) showed that the propagation is not always steady [see Jaganathan and Varunkumar (2019), Jaganathan et al. (2019)]. Such 'unsteady' regimes were identified and avoided as operating points in our earlier studies as these points are not suitable for stable gasifier operation. In particular, when the oxygen fraction in the oxidizer is increased well beyond 21%, which is required for enhancing char conversion with O_2–CO_2 and O_2–steam mixtures, there is the possibility of devolatilization time (t_v) becoming rate limiting instead of ignition time. This leads to incomplete conversion of fuel before the ignition front jumps to the next layer of fresh biomass. This is referred to as *flame jump*.

Pictorial representation of *flame jump* is shown in Fig. 6.9a. To elucidate the characteristic features of *flame jump* vis-a-vis steady propagation, spatial temperature profiles across the reaction front at different times for two cases from our work are shown in Fig. 6.9b, c. The data was obtained by placing six R-type thermocouples along the reactor at equal distance interval of 75 mm. Once the bed is lit at the top, the flame front travels and reaches the first thermocouple (T1) and gradually moves downwards to successive thermocouples i.e. from T2 to T6. Figure 6.9b shows the data for steady propagation of agro-residue based pellets at O_2 mass fraction of 23% (rest CO_2) and oxygen mass flux ($\dot{m}''_{ox}$) of 120 g/m^{2}s. Time when the flame front reaches T2 is shown as zero here. Similar temperature profiles and uniform time interval across the flame front are the characteristics of steady flame propagation as seen from Fig. 6.9b. Figure 6.9c shows *flame jump*, observed with coconut shells at 42% O_2 mass fraction (rest CO_2) and oxidizer flux of 20 g/m^{2}s—that is, flame front jumps to the next layer in chaotic manner without complete devolatilization of the

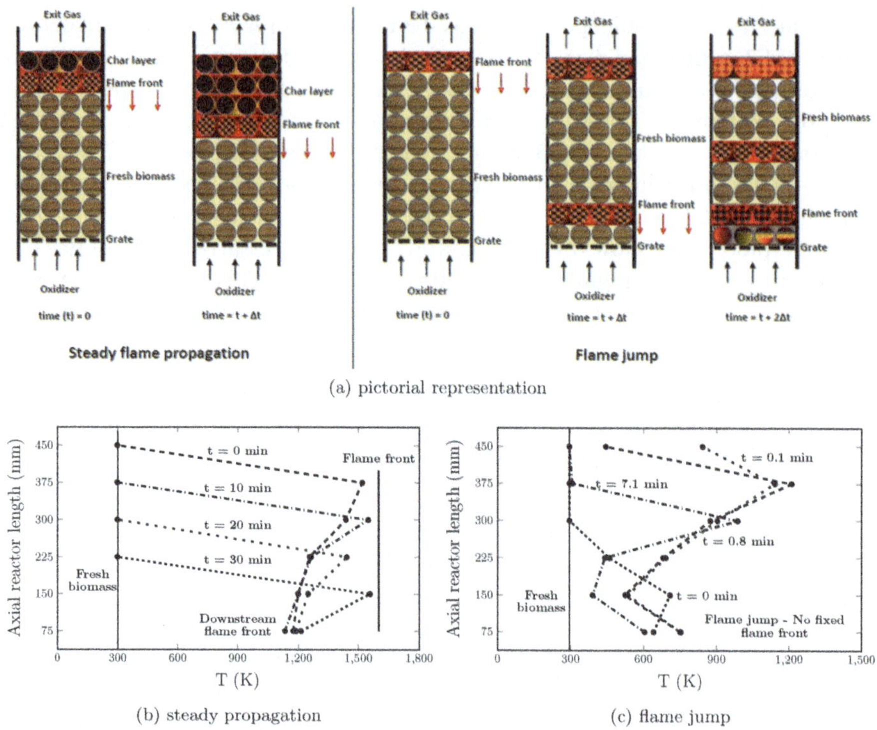

Fig. 6.9 The phenomenon of flame jump in packed beds

current layer. This type of propagation generally occurs when the oxygen fraction is increased beyond 42%. It can be inferred from the Fig. 6.9c that, the temperature profiles are not similar and time interval is not a constant or in other words, the flame front jumps from one layer to the other in a random fashion (refer Fig. 6.9a). Some important characteristic features of *flame jump* are—(1) intermittent propagation of reaction front, (2) incomplete fuel conversion and (3) reactor melting and ash fusion issues. Such operating points were excluded from analysis in our earlier studies for the obvious reason that a practical gasification system must operate under steady propagation conditions.

Another condition under which a mild form of *flame jump* can occur is as follows—starting from conditions of steady propagation, if the particle size is increased beyond a critical size (which is a strong function of oxygen fraction in the oxidizer), the ignition time can become smaller than devolatilization time. Effect of particle size on *flame jump* is not as severe as high O_2 fractions except that the fuel mass flux with larger particles; for instance, with air as the oxidizer, fuel flux remains independent of particle size till 20 mm and starts decreasing beyond 20 mm (Jaganathan et al. 2017). This is due to the overlap of multiple devolatilization layers as compared to single layer conversion as observed in lower particle sizes. Particle size effects with air as oxidizer was investigated as a part of a study undertaken to optimize briquette fired grate furnaces used for steam generation in process industries. Single particle and packed bed experiments (in 500 mm reactor) were conducted to determine the optimum particle size and superficial velocity required for maximizing fuel flux and minimizing particulate carry over. Only parts relevant to the role of t_{ig} and t_v in flame propagation will be presented here. Other details can be found in Varunkumar et al. (2015).

After identifying the parameter ranges for steady operation in terms of oxidizer mass flux and O_2 fraction in the oxidizer, attempts were made to determine the thermo-chemical mechanism controlling the propagation dynamics. The ratio of ignition to de-volatalization times for single particles subject to thermal and flow conditions in the packed bed is shown to emerge as the controlling parameter in determining the nature of propagation in these systems (Jaganathan et al. 2020). In their work, to further elucidate the role of relative magnitudes of t_{ig} and t_v, the single particle model of Mukunda et al. (1984) is extended to predict t_{ig}. The extended version is called the 'unified ignition-devolatilization model' (details can be found in Jaganathan et al. (2017). This model is further extended to study the unsteady flame propagation characteristics like 'flame jump' in a packed bed in Jaganathan et al. (2020). The main results and the conclusions of the theoretical analysis are summarized here.

6.5.3 *Flame Propagation Regimes—A Review*

6.5.3.1 Steady Propagation

Effect of oxygen mass fraction in the oxidizer on t_{ig} and t_v and its role in steady propagation versus *flame jump* will be discussed in this section. Steady flame propagation is essential for any thermo-chemical packed bed biomass conversion systems. However, in experiments with O_2–CO_2 and O_2–N_2 it is observed that when the O_2 fraction is less than 15%, propagation ceases. This is also in agreement with (Dasappa 1999; Dasappa et al. 1994), which states, 20% O_2 by volume (corresponds to 15% oxygen mass fraction for O_2–CO_2 case) is the minimum volume fraction for combustion to sustain and this is fixed as the lower limit. On the other end, if the O_2 fraction is higher than 42%, *flame jump* occurs. Steady propagating flame front exhibits characteristics as shown in Fig. 6.9b. When the coordinate frame is fixed to the flame front (as in the case of practical downdraft configuration, where the flame front is fixed and the biomass bed moves toward the flame front zone), the temperature profiles will be same at any instant of time. In other words, the heat flux received by the fresh layer of biomass from the flame front is constant in steady propagation. During *flame jump*, the time interval is no more a constant (refer Fig. 6.9c). For a steady flame, from 'Universal flame propagation, UFP' model (refer Varunkumar et al. 2013), the total time required for the temperature of the fresh biomass particle to increase from ambient temperature (T_∞) to the ignition temperature (T_{ig}) is computed from Eq. 6.4 and propagation rate $\dot{r}$ is calculated as d_p/t_{ig}. The surface energy balance of a fresh biomass particle upstream of a steady flame front is given as,

$$\frac{mC_p}{A_c}\frac{dT_p}{dt} = f\epsilon\sigma(T_s^4 - T_p^4) - h(T_p - T_0) \tag{6.4}$$

where, m, C_p, A_c, h, f, T_p, T_s and T_0 represent mass, specific heat, cross–sectional area, heat transfer coefficient, view factor, particle, surface and ambient temperature respectively (Varunkumar et al. 2013). An order of magnitude estimate of the L.H.S of Eq. 6.4 gives

$$\frac{mC_p}{A_c}\frac{dT_p}{dt} \sim \frac{mC_p(T_{ig} - T_\infty)}{A_c t_{ig}} \tag{6.5}$$

as inferred from UFP model, the time taken for the fresh biomass particle of diameter d_p to reach the pyrolysis temperature governs propagation rate $\dot{r}$. Hence, substituting t_{ig} as $d_p/\dot{r}$ and using $m = \rho_b V$

$$\frac{mC_p}{A_c}\frac{dT_p}{dt} \sim \frac{\rho_b V \dot{r} C_p(T_{ig} - T_\infty)}{d_p A_c} \tag{6.6}$$

volume $V = \delta_{critical} A_c$ where, $\delta_{critical}$ is the critical thickness of the fresh biomass layer to be raised to the ignition temperature (T_{ig}) for flame propagation.

$$\frac{mC_p}{A_c}\frac{dT_p}{dt} \sim \frac{\rho_b \dot{r} C_p(T_{ig} - T_\infty)}{\frac{d_p}{\delta_{critical}}} \tag{6.7}$$

Substituting $\delta_{critical}/d_p$ as K, Eq. 6.4 finally reduces into

$$K\frac{mC_p}{A_c}\frac{dT_p}{dt} \sim \rho_b \dot{r} C_p(T_{ig} - T_\infty) \tag{6.8}$$

where, factor K is to account for the critical mass of biomass to be raised to the ignition temperature and $\rho_b \dot{r}$ is the fuel flux ($\dot{m}''_f$) and the R.H.S of Eq. 6.8 becomes $\dot{m}''_f C_p(T_{ig} - T_0)$ which denotes the flux received by fresh biomass. The ignition time variation is dependent on incident flux which is a function of $\dot{m}''_f$ of the flux term $\dot{m}''_f C_p(T_{ig} - T_0)$.

Steady propagation data for all the experiments with air, O_2–CO_2, O_2–N_2 and O_2–steam for different fuels were available from Jaganathan and Varunkumar (2019), Jaganathan et al. (2019). Taking T_{ig} as pyrolysis temperature i.e. 473 K, the ignition flux term $\dot{m}''_f C_p(T_{ig} - T_0)$ is calculated. Using UID model, replacing the ignition source term with $\dot{m}''_f C_p(T_{ig} - T_0)$, t_{ig} for all the cases is calculated.

6.5.3.2 Validation of the Ignition Criterion

Ignition time (t_{ig}) is defined as the time taken for a fresh biomass particle to have a steady sustained flame. The ignition criteria of 15% mass loss as chosen in the model can be validated by two methods, (1) using $t_{ig} = G/\dot{r}$ (where $\dot{r}$ is flame propagation rate and G is the geometric parameter of the fuel particle) from packed bed experiments and (2) single particle studies. Out of these methods, calculation of ignition time from packed bed experiments is most appropriate due to the following reasons, (1) under the conditions of steady propagation, the measurement of flame velocity or propagation rate can be accurately calculated from the temperature-time data, (2) the fresh biomass mass particle enjoys better thermal environment in a bed with minimal heat loss which gives a good estimate of heat flux unlike single particle studies and (3) effect of oxygen fraction in the oxidizer on ignition time can be validated with available experimental data from our earlier studies.

Ignition time from bed is given as $t_{ig} = G/\dot{r}$ where, $\dot{r}$ is calculated from temperature-time data. As mentioned in Mukunda (2011), equivalent spherical diameter (d_{eq}) of pellets is calculated and G value of $d_{eq}/6$ is used for pellets and for coconut shell it is taken as its thickness. Values of $\dot{r}$ is derived from temperature-time data for all the cases reported here (refer Jaganathan and Varunkumar (2019), Jaganathan (2019), Jaganathan et al. (2019) for more details). Ignition time from packed bed experiments is calculated using this procedure. Corresponding value from the model is calculated by imposing the heat flux ($\dot{q}''_w=\dot{m}''_f C_p(T_{ig} - T_0)$) obtained from the experimental data as described in previous section. Figure 6.10 shows the variation of density scaled ignition time calculated from packed bed experiments and the model. The agreement between the model and the experiments is very good (error $\leq \pm$ 15%). Also, single particle studies were carried out with different fuels

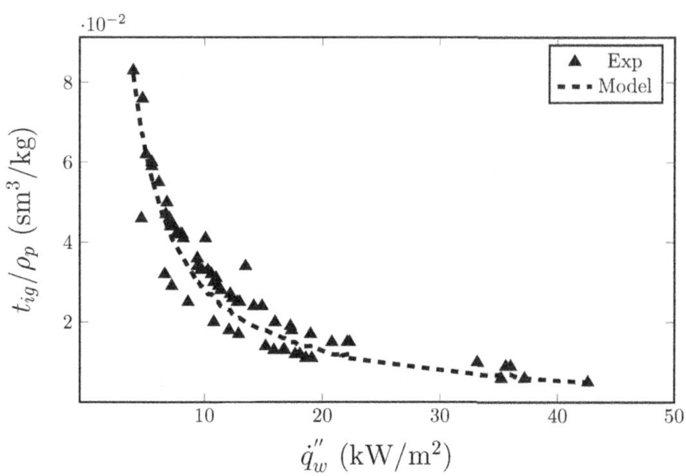

Fig. 6.10 Ignition time as a function of heat flux from experiments and UID model

namely, (1) agro-residue based pellets (d_p = 10 mm), (2) groundnut shell based briquettes (10 mm < d_p < 60 mm) and (2) bagasse briquettes (20 mm < d_p < 70 mm). All these fuels when subjected to ignition heat flux of around 10–15 kW/m^2 gets ignited (self sustained burning) when critical mass loss is about 15%. This is also consistent with critical mass for ignition of m_0/6 (i.e. around 16.6 % weight loss) where, m_0 is the initial mass of fresh layer of biomass to be ignited in packed bed systems as reported by Varunkumar (2014), Mukunda (2011). From the above discussion it is clear that, the choice of 15% mass loss as ignition criterion is valid for all possible fuel-oxidizer combinations. More importantly, the increase of O_2 fraction in the oxidizer increases the volatile heat release rate ($H_v Y_{ox,\infty}/s$) which is reflected as heat flux increase is captured by the UID model.

6.5.3.3 O_2 Fraction and *Flame Jump*

A typical example of O_2–N_2 experiments with agro-residue pellets is taken to study the role of O_2 fraction in propagation. Fuel mass flux ($\dot{m}''_f$) obtained with same oxidizer (in this case O_2–N_2 mixture) mass flux of around 75 g/m^{2}s corresponding to O_2 fractions of 23, 28, 33 and 43 % (by mass) were taken. This is used to calculate the incident heat flux for a fresh biomass layer to be ignited (i.e. $\dot{q}'' = \dot{m}''_f C_p(T_{ig} - T_0)$). The variation of $\dot{q}''$ with O_2 fraction and results of t_{ig} and t_v calculated using UID model are shown in Fig. 6.11.

It can be seen from Fig. 6.11 that, t_{ig} is higher than t_v for air case (i.e. 23.2% O_2) and as the O_2 fraction is increased, the heat influx ($\dot{q}''$) increases, the ignition time becomes smaller than t_v and the ratio of t_v/t_{ig} increases from unity to around 2.3 at 43% till which steady propagation is observed. As seen from Fig. 6.11, with increase in O_2 fraction, t_v decrease is insignificant unlike t_{ig}. Increase in heat flux with increasing O_2 fraction enhances the volatile generation rate that prevent heat ingress

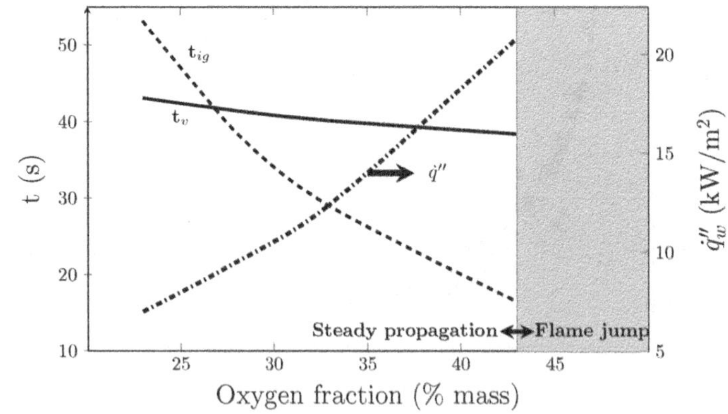

Fig. 6.11 Variation of t_{ig}, t_v and $\dot{q}''_w$ with increasing O_2 fraction

due to what is termed 'blocking effect' (Mukunda 2011). Due to this, the gas phase zone thickness is enhanced due to enhancement in flow from surface of the particle and hence, reduce the thermal gradients near the surface and convective heat transfer to the surface (Mukunda 2011; Spalding 1979). When O_2 fraction exceeds 43% the t_v/t_{ig} ratio goes beyond 2.3 and *flame jump* is observed. This is consistent with visual observation of *flame jump* in a transparent reactor by Varunkumar (2014) where flaming to ignition time ratio around 2 for dense biomass like pellets is reported. In *flame jump* zone UFP model is not applicable as the phenomenon becomes unsteady.

6.5.3.4 Operational Regimes

It is shown from experiments that 'volatiles' stoichiometry (i.e. $\phi_v \sim 1$) is the transition point from gasification to combustion (Jaganathan and Varunkumar 2019, Jaganathan et al. 2019). A plot of the ratio of ignition to devolatilization time versus 'volatiles' equivalence ratio is shown in Fig. 6.12 for all cases listed in Table 6.1.

Based on this plot, the propagation regimes in a packed bed of biomass can be classified in to five zones. Some important flame characteristics of these zones are listed in the Table 6.2. In Zone I, oxidizer flux is low and conditions are highly fuel rich ($\phi_v > 2$). Hence, bed temperatures are less in this zone owing to low propagation rates (< 25 g/m^{2}s). Also, here the time of ignition is high as compared to devolatilization time ($t_v/t_{ig} < 0.75$). Zone II represents the optimum zone of operation with $0.75 \leq t_v/t_{ig} \leq 2$ and bed operation near 'volatiles' stoichiometry which enhances the bed temperature and favors reduction reactions of CO_2 and steam with char. Most of the cases which have shown steady propagation fall under this category. It is important ro note here that, as mentioned in the Table 6.2, slow propagation refers to fuel flux lesser that 25 g/m^{2}s and steady propagation refers to fuel flux more than

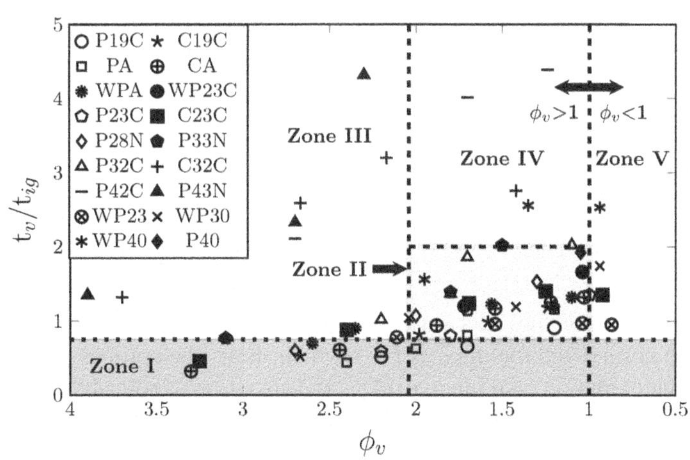

Fig. 6.12 Zones of operation of a packed bed of biomass

Table 6.2 Characteristics of flame propagation zones of packed bed systems

Zone characteristics

Zones	t_v/t_{ig}	ϕ_v	Flame propagation	T_{pb} (K)	% Char Conversion	CO/H 2 yield
Zone I	≤0.75	–	Slow/Propagation ceases	≤ 1000	0–18 (Char reduction)	Low
Zone II	0.75–2	2–1	Steady	1000–1500	20–100 (Char reduction)	High
Zone III	0.75–5	≥2	Steady/Unsteady	≥ 1000	0–50	Low
Zone IV	2–5	2–1	Unsteady	1500–1800	21–100	Low
Zone V	≥0.75	≤1	Steady/Unsteady	≥ 1500	100 (Char oxidation)	Low

25 g/m^{2}s with ratio of t_v/t_{ig} not exceeding 2. Zone III represents fuel rich zone where incomplete combustion of 'volatiles' is observed and the ratio t_v/t_{ig} is comparatively higher than Zone I and Zone II which shows tendency towards *flame jump*. Zone IV represents the *flame jump* zone (i.e. $t_v/t_{ig} > 2$). Except for coconut shell cases (which shows *flame jump* around 32% O_2 fraction for which possible reasons are discussed in Jaganathan (2019), P42C case fall in this zone where the O_2 fraction is more than 40%. It is also important to note that in this zone, the peak bed temperatures are high ($1500 \leq T_{pb} \leq 1800$ K) which leads to melting of reactors and ash fusion problems. Zone V represents the char oxidation zone ($\phi_v < 1$).

Particle size is another variable that brings in effects that need to be considered in addition to those discussed so far. The key result from Jaganathan et al. (2017) is that, with air as oxidizer, the cut-off size for the condition $t_{ig} > t_v$ to be satisfied (that is, for steady propagation) is 20 mm. Beyond this size, a mild case of flame jump is observed. Corresponding particle sizes for other oxidizer combinations can be estimated using the procedure outlined in Jaganathan et al. (2017), Jaganathan (2019).

6.6 Implications for Practical Gasification Systems

Some important implications of the results discussed so far for practical systems are listed below.

1. Normalization of fuel flux variations from different O_2 fractions is addressed through normalized fuel flux (NFF), which shows the regimes of 'gasification' and 'char oxidation' as a function of ϕ_v and also shows applicability of 'universal behavior' to all fuel–oxidizer combinations. This is an important result to design and operate biomass packed bed systems. By fixing a ϕ_v value for operation,

designers can quickly determine the power throughput, exit gas composition, net CO_2 and/or steam conversion possible and CO and H_2 yield for such systems.

2. Role and relative magnitudes of t_{ig} and t_v are clearly brought out to explain the phenomenon of *flame jump* which occurs when t_v/t_{ig} exceeds 2. From UID model and experimental results, a non dimensional plot of t_v/t_{ig} versus ϕ_v showing different operational zones of biomass gasification systems is presented. Five zones are identified from this map, out of which Zone II ($0.75 < t_v/t_{ig} < 2$) is identified as the optimum operating zone for steady flame propagation, moderate bed temperatures and optimal conversion of CO_2 and steam (since, the operation is close to 'volatiles stoichiometry'). Also, syngas from this regime of operation is tar free which minimizes the requirement of downstream gas cleaning equipments and operating cost involved.
3. It seen from experiments that when the oxidizer mixture temperature exceeds 200 °C which is equivalent to the pyrolysis temperature (i.e. 473 K) of the biomass assumed in UID model, the total bed gets de-volatilized and flame jump ensues. Hence, with systems involving O_2–steam as oxidizer, for steady propagation, one should maintain the oxidizer mixture temperature between 120 and 150 °C.
4. The effect of particle size on steady flame propagation is brought out. To ensure complete conversion of fuel and better efficiency of the plant, the biomass size shall not exceed 30 mm. This is an important result for designing the fuel preparation unit to pelletize or chop the fuel to required size for 100% fuel conversion.

6.7 Summary

Studies presented in this chapter have contributed to the fundamental understanding of counter-current flame propagation characteristics in a biomass packed bed system. Effects of different oxidizer combinations, namely, O_2–CO_2, O_2–N_2 and O_2–steam on propagation rate, CO and H_2 yield and HHC fraction in the exit gas are brought out. 'Volatiles equivalence ratio' is shown to emerge as an unifying parameter for analyzing thermo-chemical conversion of biomass in packed bed reactors. Variables including, fuel mass flux, CO and H_2 yield, char conversion and HHC fraction in exit gases share similar characteristics when expressed as functions of ϕ_v, irrespective of the oxidizer. For all fuel and oxidizer combinations, in general, $\phi_v > 1$ corresponds to *gasification* regime and $\phi_v < 1$ corresponds to *char oxidation* regime, that is, transition from *gasification* to *char oxidation* occurs at $\phi_v = 1$. *Normalized fuel flux* exhibits universality irrespective of the oxidizers and fuel type as a function of ϕ_v. Syngas yield (CO and H_2) is found to be maximum around $\phi_v \rightarrow 1$ due to enhanced conversion of CO_2 and steam with char. Complete char conversion is achieved at a slightly rich ϕ_v (usually around 1.5), indicating participation of char in reduction reactions. In general, the HHC fractions are near zero around $\phi_v \sim 1$ which implies that the syngas is tar free. Maximum net CO_2 conversion of 400–600 g/kg of biomass and CO yield around 600–800 g/kg of biomass are observed for O_2–CO_2 cases, which is 2–2.5 times higher than the corresponding air cases. Intrinsic H_2 yield of 30–40 g/kg of biomass is observed in entire 'gasification' regime and this is 2-2.5 times higher than the corresponding air cases. The gasification efficiency of all O_2–CO_2 and O_2–steam cases are higher than corresponding air cases and this is due to the reactive contribution of CO_2 and steam unlike inert N_2 in air and O_2–N_2 cases. Though under conditions of steady propagation, fuel flux exhibits universality with ϕ_v, explanation of unsteady propagation phenomenon ('flame jump') needed exploration beyond the scope of *universal flame propagation model*. Also, single particle experiments have shown that the t_v becomes greater than t_{ig} beyond 20 mm with air as oxidizer and hence, t_v can limit the fuel flux. In the light of this, a *unified ignition-devolatalization* model is developed to predict t_v and t_{ig}. With air as oxidizer, beyond d_p of 20 mm $t_v > t_{ig}$ and t_v influences the flame propagation. This is consistent with results of large reactor experiments using GSB. 'Flame jump' is shown to occur when t_v/t_{ig} exceeds 2. From UID model and experimental results, a non dimensional plot of t_v/t_{ig} versus ϕ_v showing different operational zones of biomass gasification systems is presented. Five zones are identified from this map, out of which Zone II is identified as the optimum operating zone for gasification. Results presented here can be readily used by practitioners to build commercially scalable systems and choose operating conditions for a particular application.

References

Babu BV, Sheth PN (2006) Modeling and simulation of reduction zone of downdraft biomass gasifier: Effect of char reactivity factor. Energy Convers Manage 47(15):2602–2611. ISSN 0196-8904. https://doi.org/10.1016/j.enconman.2005.10.032. http://www.sciencedirect.com/science/article/pii/S0196890405002840

Balu E, Lee U, Chung JN (2015) High temperature steam gasification of woody biomass—a combined experimental and mathematical modeling approach. Int J Hydrogen Energy 40(41):14104–14115. ISSN 0360-3199. https://doi.org/10.1016/j.ijhydene.2015.08.085. http://www.sciencedirect.com/science/article/pii/S0360319915022326

Billaud J, Valin S, Peyrot M, Salvador S (2016) Influence of H_2O, CO_2 and O_2 addition on biomass gasification in entrained flow reactor conditions: Experiments and modelling. Fuel 166:166–178. ISSN 0016-2361. https://doi.org/10.1016/j.fuel.2015.10.046. http://www.sciencedirect.com/science/article/pii/S0016236115010510

Bouraoui Z, Jeguirim M, Guizani C, Limousy L, Dupont C, Gadiou R (2015) Thermogravimetric study on the influence of structural, textural and chemical properties of biomass chars on CO_2 gasification reactivity. Energy 88:703–710. ISSN 0360-5442. https://doi.org/10.1016/j.energy.2015.05.100. http://www.sciencedirect.com/science/article/pii/S0360544215007021

Bu C, Gomez-Barea A, Leckner B, Chen X, Pallares D, Liu D, Lu P (2017) Oxy-fuel conversion of sub-bituminous coal particles in fluidized bed and pulverized combustors. Proc Combust Inst 36(3):3331–3339. ISSN 1540-7489. https://doi.org/10.1016/j.proci.2016.08.075. http://www.sciencedirect.com/science/article/pii/S1540748916304643

Butterman HC, Castaldi MJ (2009) CO_2 as a carbon neutral fuel source via enhanced biomass gasification. Environ Sci Technol 43(23):9030–9037

Cerone N, Zimbardi F, Villone A, Strjiugas N, Kiyikci EG (2016) Gasification of wood and torrefied wood with air, oxygen, and steam in a fixed-bed pilot plant. Energy Fuels 30(5):4034–4043. https://doi.org/10.1021/acs.energyfuels.6b00126

Cerone N, Zimbardi F (2018) Gasification of agroresidues for syngas production. Energies 11(5). ISSN 1996-1073. http://www.mdpi.com/1996-1073/11/5/1280

Cerone N, Zimbardi F, Contuzzi L, Prestipino M, Carnevale MO, Valerio V (2017) Air-steam and oxy-steam gasification of hydrolytic residues from biorefinery. Fuel Process Technol 167:451–461. ISSN 0378-3820. https://doi.org/10.1016/j.fuproc.2017.07.027. http://www.sciencedirect.com/science/article/pii/S0378382017308913

Chang ACC, Chang H-F, Lin F-J, Lin K-H, Chen C-H (2011) Biomass gasification for hydrogen production. Int J Hydrogen Energy 36(21):14252–14260. ISSN 0360-3199. https://doi.org/10.1016/j.ijhydene.2011.05.105. http://www.sciencedirect.com/science/article/pii/S0360319911013176. 2010 Asian/APEC BioH2

Collazo J, Porteiro J, Patino D, Granada E (2012) Numerical modeling of the combustion of densified wood under fixed-bed conditions. Fuel 93:149–159. ISSN 0016-2361. https://doi.org/10.1016/j.fuel.2011.09.044. http://www.sciencedirect.com/science/article/pii/S0016236111006041

Dai B, Hoadley A, Zhang L (2017) Characteristics of high temperature C-CO_2 gasification reactivity of victorian brown coal char and its blends with high ash fusion temperature bituminous coal. Fuel 202:352–365. ISSN 0016-2361. https://doi.org/10.1016/j.fuel.2017.04.044. http://www.sciencedirect.com/science/article/pii/S0016236117304507

Dasappa S (1999) Experiments and modelling studies on gasification of wood–char. Ph.D. thesis, Ph D thesis, Indian Institute of Science, Bangalore

Dasappa S, Paul PJ, Mukunda HS, Shrinivasa U (1994) The gasification of wood-char spheres in CO_2 - N_2 mixtures?: analysis and experiments. Chem Eng Sci 49:223–232

Di Blasi C (2000) Dynamic behaviour of stratified downdraft gasifiers. Chem Eng Sci. 55(15):2931–2944. ISSN 0009-2509. https://doi.org/10.1016/S0009-2509(99)00562-X. http://www.sciencedirect.com/science/article/pii/S000925099900562X

Fatehi M, Kaviany M (1994) Adiabatic reverse combustion in a packed bed. Combust Flame 99(1):1–17. ISSN 0010-2180. https://doi.org/10.1016/0010-2180(94)90078-7. http://www.sciencedirect.com/science/article/pii/0010218094900787

Fiaschi D, Michelini M (2001) A two-phase one-dimensional biomass gasification kinetics model. Biomass Bioenergy 21(2):121–132. ISSN 0961-9534. https://doi.org/10.1016/S0961-9534(01)00018-6. http://www.sciencedirect.com/science/article/pii/S0961953401000186

Gao N, Li A, Quan C, Gao F (2008) Hydrogen-rich gas production from biomass steam gasification in an updraft fixed-bed gasifier combined with a porous ceramic reformer. Int J Hydrogen Energy 33(20):5430–5438. ISSN 0360-3199. https://doi.org/10.1016/j.ijhydene.2008.07.033. http://www.sciencedirect.com/science/article/pii/S0360319908008197

Gobel B, Henriksen U, Jensen TV, Qvale B, Houbak N (2007) The development of a computer model for a fixed bed gasifier and its use for optimization and control. Bioresour Technol 98(10):2043–2052. ISSN 0960-8524. https://doi.org/10.1016/j.biortech.2006.08.019. http://www.sciencedirect.com/science/article/pii/S0960852406004214

Gomez-Barea A, Leckner B (2010) Modeling of biomass gasification in fluidized bed. Progr Energy Combust Sci 36(4):444–509. ISSN 0360-1285. https://doi.org/10.1016/j.pecs.2009.12.002. http://www.sciencedirect.com/science/article/pii/S0360128509000707

Gort R (1995) On the propagation of a reaction front in a packed bed. Thermal conversion of municipal waste and biomass. Ph.D. thesis, University of Twente

Huang H-J, Ramaswamy S (2009) Modeling biomass gasification using thermodynamic equilibrium approach. Appl Biochem Biotechnol 154(1):14–25

Huo W, Zhou Z, Wang F, Wang Y, Yu G (2014) Experimental study of pore diffusion effect on char gasification with CO_2 and steam. Fuel 131:59–65. ISSN 0016-2361. https://doi.org/10.1016/j.fuel.2014.04.058. http://www.sciencedirect.com/science/article/pii/S0016236114003937

IndiaToday (2019) Carbon zero challenge. https://www.indiatoday.in/education-today/news/story/iit-madras-hosts-2nd-edition-of-carbon-zero-challenge-to-tackle-environmental-problems-in-india-1544445-2019-06-07

Jaganathan VM, Ambatipudi MK, Varunkumar S (2020) The phenomenon of flame jump in counter-current flame propagation in biomass packed beds experiments and theory. Combust

Sci Technol 0(0), 1–14. doi: https://doi.org/10.1080/00102202.2020.1804886.https://doi.org/10.1080/00102202.2020.1804886

Jaganathan VM (2019) Syngas synthesis using gasification of biomass with O_2-CO_2 and O_2-steam mixtures. Ph.D. thesis, Indian Institute of Technology Madras

Jaganathan VM, Mohan O, Varunkumar S (2019) Intrinsic hydrogen yield from gasification of biomass with oxy-steam mixtures. Int J Hydrogen Energy 44(33):17781–17791. ISSN 0360-3199. https://doi.org/10.1016/j.ijhydene.2019.05.095. https://www.sciencedirect.com/science/article/pii/S0360319919319512

Jaganathan VM, Varunkumar S (2019) Net carbon-di-oxide conversion and other novel features of packed bed biomass gasification with O_2/CO_2mixtures. Fuel 244:545–558. ISSN 0016-2361. https://doi.org/10.1016/j.fuel.2019.01.171. http://www.sciencedirect.com/science/article/pii/S0016236119301723

Jaganathan VM, Varunkumar S (2020) A novel self-sustained single step process for synthesizing activated char from ligno-cellulosic biomass. Fuel Process Technol 208:106516. ISSN 0378-3820. https://doi.org/10.1016/j.fuproc.2020.106516. https://www.sciencedirect.com/science/article/pii/S0378382020308079

Jaganathan VM, Mani Kalyani A, Varunkumar S (2017) Unified ignition-devolatilization model for fixed bed biomass gasification/combustion. Energy Procedia 120:643–648

Jarungthammachote S, Dutta A (2007) Thermodynamic equilibrium model and second law analysis of a downdraft waste gasifier. Energy 32(9):1660–1669. ISSN 0360-5442. https://doi.org/10.1016/j.energy.2007.01.010. http://www.sciencedirect.com/science/article/pii/S0360544207000199

Jarungthammachote S, Dutta A (2008) Equilibrium modeling of gasification: Gibbs free energy minimization approach and its application to spouted bed and spout-fluid bed gasifiers. Energy Convers Manage 49(6):1345–1356. ISSN 0196-8904. https://doi.org/10.1016/j.enconman.2008.01.006. http://www.sciencedirect.com/science/article/pii/S019689040800023X

Karamarkovic R, Karamarkovic V (2010) Energy and exergy analysis of biomass gasification at different temperatures. Energy 35(2):537–549. ISSN 0360-5442. https://doi.org/10.1016/j.energy.2009.10.022. http://www.sciencedirect.com/science/article/pii/S0360544209004551. ECOS 2008

Keith W, Saenz C (2021) Drive on wood. https://www.driveonwood.com/library/imbert-gasifier/

Kramb J, Gomez-Barea A, DeMartini N, Romar H, Doddapaneni TR, Konttinen J (2017) The effects of calcium and potassium on CO_2 gasification of birch wood in a fluidized bed. Fuel 196:398–407. ISSN 0016-2361. https://doi.org/10.1016/j.fuel.2017.01.101. http://www.sciencedirect.com/science/article/pii/S0016236117301199

Kyotani T, Kubota K, Cao J, Yamashita H, Tomita A (1993) Combustion and CO_2 gasification of coals in a wide temperature range. Fuel Process Technol 36(1):209–217. ISSN 0378-3820. https://doi.org/10.1016/0378-3820(93)90029-4. http://www.sciencedirect.com/science/article/pii/0378382093900294. In: Third international Rolduc symposium on coal science and technology and related processes

Lahijani P, Zainal ZA, Mohamed AR, Mohammadi M (2014) Microwave-enhanced CO_2 gasification of oil palm shell char. Bioresour Technol 158:193–200. ISSN 0960-8524. https://doi.org/10.1016/j.biortech.2014.02.015. http://www.sciencedirect.com/science/article/pii/S0960852414001862

Li X, Grace JR, Watkinson AP, Lim CJ, Ergudenler A (2001) Equilibrium modeling of gasification: a free energy minimization approach and its application to a circulating fluidized bed coal gasifier. Fuel 80(2):195–207. ISSN 0016-2361. https://doi.org/10.1016/S0016-2361(00)00074-0. http://www.sciencedirect.com/science/article/pii/S0016236100000740

Lin L, Strand M (2013) Investigation of the intrinsic CO_2 gasification kinetics of biomass char at medium to high temperatures. Appl Energy 109:220–228. ISSN 0306-2619. https://doi.org/10.1016/j.apenergy.2013.04.027. http://www.sciencedirect.com/science/article/pii/S0306261913003218

Lucas C, Szewczyk D, Blasiak W, Mochida S (2004) High-temperature air and steam gasification of densified biofuels. Biomass Bioenergy 27(6):563–575. ISSN 0961-9534.

https://doi.org/10.1016/j.biombioe.2003.08.015. http://www.sciencedirect.com/science/article/pii/S0961953404000984. Pellets 2002. The first world conference on pellets

Lv P, Yuan Z, Ma L, Wu C, Chen Y, Zhu J (2007) Hydrogen-rich gas production from biomass air and oxygen/steam gasification in a downdraft gasifier. Renew Energy 32(13):2173–2185

Mani Kalyani A, Jaganathan VM, Varunkumar S (2021) Mechanisms governing the gasification to char oxidation transition in counter-current flame propagation in packed beds: insights from single-particle experiments. Particulate Sci Technol 39(4):490–494. https://doi.org/10.1080/02726351.2020.1767245

Melgar A, Pérez JF, Laget H, Horillo A (2007) Thermochemical equilibrium modelling of a gasifying process. Energy Convers Manage 48(1):59–67. ISSN 0196-8904. https://doi.org/10.1016/j.enconman.2006.05.004. http://www.sciencedirect.com/science/article/pii/S0196890406001646

Ministry of New and Renewable Energy (2021) Bio energy. https://mnre.gov.in/bio-energy/current-status

Mukunda HS, Dasappa S, Paul PJ, Shrinivasa U, Rajan NKS (1994) Gasifiers and combustors for biomass – technology and field studies. Energy Sustain Dev 1(3):27–38. ISSN 0973-0826. https://doi.org/10.1016/S0973-0826(08)60047-2. http://www.sciencedirect.com/science/article/pii/S0973082608600472

Mukunda HS (2011) Understanding clean energy and fuels from biomass. Wiley India Pvt, Limited

Mukunda HS, Paul PJ, Shrinivasa U, Rajan NKS (1984) Combustion of wooden spheres. Proc Combust Instit 20:1619–1628

Nilsson S, Gomez-Barea A, Cano DF (2012) Gasification reactivity of char from dried sewage sludge in a fluidized bed. Fuel 92(1):346–353. ISSN 0016-2361. https://doi.org/10.1016/j.fuel.2011.07.031. http://www.sciencedirect.com/science/article/pii/S0016236111004285

Nipattummakul N, Ahmed II, Kerdsuwan S, Gupta AK (2010) Hydrogen and syngas production from sewage sludge via steam gasification. Int J Hydrogen Energy 35(21):11738–11745. ISSN 0360-3199. https://doi.org/10.1016/j.ijhydene.2010.08.032. http://www.sciencedirect.com/science/article/pii/S036031991001671X. VIII symposium of the Mexican Hydrogen Society

Porteiro J, Patino D, Collazo J, Granada E, Moran J, Miguez JL (2010) Experimental analysis of the ignition front propagation of several biomass fuels in a fixed-bed combustor. Fuel 89(1):26–35. ISSN 0016-2361. https://doi.org/10.1016/j.fuel.2009.01.024. http://www.sciencedirect.com/science/article/pii/S0016236109000477

Rönnbäck M, Axell M, Gustavsson L, Thunman H, Leckner B (2008) Combustion processes in a biomass fuel bed-experimental results. Wiley Sons Ltd, pp 743–757. ISBN 9780470694954. https://doi.org/10.1002/9780470694954.ch59. https://onlinelibrary.wiley.com/doi/abs/10.1002/9780470694954.ch59

Ranzi E, Maffei T, Khatami R, Pierucci S, Faravelli T, Levendis YA (2013) Experimental and modeling study of single coal particle combustion in O_2/N_2 and oxy-fuel O_2/CO_2 atmospheres. Combust Flame 160(11):2559–2572. ISSN 0010-2180. https://doi.org/10.1016/j.combustflame.2013.06.002. http://www.sciencedirect.com/science/article/pii/S0010218013002125

Riaza J, Khatami R, Levendis YA, Álvarez L, Gil MV, Pevida C, Rubiera F, Pis JJ (2014) Combustion of single biomass particles in air and in oxy-fuel conditions. Biomass Bioenergy 64:162–174. ISSN 0961-9534. https://doi.org/10.1016/j.biombioe.2014.03.018. http://www.sciencedirect.com/science/article/pii/S0961953414001366

Rönnbäck M, Axell M, Gustavsson L, Thunman H, Leckner B (2000) Combustion processes in a biomass fuel bed-experimental results. Progr Thermochem Biomass Convers, pp 17–22

Ryu C, Phan AN, Yang YB, Sharifi VN, Swithenbank J (2007) Ignition and burning rates of segregated waste combustion in packed beds. Waste Manage 27(6):802–810. ISSN 0956-053X. https://doi.org/10.1016/j.wasman.2006.04.013. http://www.sciencedirect.com/science/article/pii/S0956053X06001371

Sandeep K, Dasappa S (2013) Oxy-steam gasification of biomass for hydrogen rich syngas production using downdraft reactor configuration. Int J Energy Res 38(2):174–188

Sandeep K, Dasappa S (2014) First and second law thermodynamic analysis of air and oxy-steam biomass gasification. Int J Hydrogen Energy 39(34):19474–19484. ISSN 0360-3199. https://doi.org/10.1016/j.ijhydene.2014.09.134. http://www.sciencedirect.com/science/article/pii/S0360319914027268

Sandeep K, Snehesh S, Dasappa S (2011) Carbon dioxide capture through biomass gasification. In: Proceedings of the 19th European biomass conference and exhibition, pp 1127–1133

Sharma AK (2008) Equilibrium and kinetic modeling of char reduction reactions in a downdraft biomass gasifier: a comparison. Solar Energy 82(10):918–928. ISSN 0038-092X. https://doi.org/10.1016/j.solener.2008.03.004. http://www.sciencedirect.com/science/article/pii/S0038092X08000649

Sircar I, Sane A, Wang W, Gore J (2014) Experimental and modeling study of pinewood char gasification with CO_2. Fuel 119:38–46. ISSN 0016-2361. https://doi.org/10.1016/j.fuel.2013.11.026. http://www.sciencedirect.com/science/article/pii/S0016236113010727

Spalding DB (1979) Combustion and mass transfer. Pergamon Press. ISBN 0-08-022105-X

Umeki K, Yamamoto K, Namioka T, Yoshikawa K (2010) High temperature steam-only gasification of woody biomass. Appl Energy 87(3):791–798. ISSN 0306-2619. https://doi.org/10.1016/j.apenergy.2009.09.035. http://www.sciencedirect.com/science/article/pii/S0306261909004279

Varunkumar S (2014) Packed bed gasification-combustion in biomass based domestic stoves and combustion systems. Ph.D. thesis, G25323

Varunkumar S, Rajan NKS, Mukunda HS (2013) Universal flame propagation behavior in packed bed of biomass. Combust Sci Technol 185(8):1241–1260

Varunkumar S, Jaganathan VM, Ajey N (2015) Application of universal flame propagation behaviour in biomass packed bed to design of grate furnaces. In: Proceedings of the 24th national conference on I.C. engines and combustion

Varunkumar S, Rajan NKS, Mukunda HS (2012) Experimental and computational studies on a gasifier based stove. Energy Convers Manage 53(1):135–141. ISSN 0196-8904. https://doi.org/10.1016/j.enconman.2011.08.022. http://www.sciencedirect.com/science/article/pii/S0196890411002366

Wang Y, Kinoshita CM (1993) Kinetic model of biomass gasification. Solar Energy 51(1):19–25. ISSN 0038-092X. https://doi.org/10.1016/0038-092X(93)90037-O. http://www.sciencedirect.com/science/article/pii/0038092X9390037O

Yang YB, Sharifi VN, Swithenbank J (2004) Effect of air flow rate and fuel moisture on the burning behaviours of biomass and simulated municipal solid wastes in packed beds. Fuel 83(11):1553–1562. ISSN 0016-2361. https://doi.org/10.1016/j.fuel.2004.01.016. http://www.sciencedirect.com/science/article/pii/S0016236104000286

Yoshida H, Kiyono F, Tajima H, Yamasaki A, Ogasawara K, Masuyama T (2008) Two-stage equilibrium model for a coal gasifier to predict the accurate carbon conversion in hydrogen production. Fuel 87(10):2186–2193. ISSN 0016-2361. https://doi.org/10.1016/j.fuel.2008.01.009. http://www.sciencedirect.com/science/article/pii/S0016236108000148

Zhang Y, Gong X, Zhang B, Liu W, Xu M (2014) Potassium catalytic hydrogen production in sorption enhanced gasification of biomass with steam. Int J Hydrogen Energy 39(9):4234–4243. ISSN 0360-3199. https://doi.org/10.1016/j.ijhydene.2014.01.015. http://www.sciencedirect.com/science/article/pii/S0360319914000640

Chapter 7
Sustainable Production of Biochar, Bio-Gas and Bio-Oil from Lignocellulosic Biomass and Biomass Waste

Rohit Dalal, Roshan Wathore, and Nitin Labhasetwar

7.1 Introduction

Bioenergy is the world's largest renewable energy resource, accounting for 70% of the total renewable supply. In 2017, 55.6 EJ of total primary energy supply was generated from biomass, a significant drop compared to previous years. This downtrend still persists due to competition from other renewables. However, most developing nations in Asia and Africa are still dependent on biomass for heating and cooking purposes, contributing majorly to total biomass supply, as indicated in Fig. 7.1 (Batchelor et al. 2019; Sen and Ganguly 2017). The shift towards use in modern biomass solutions like liquid biofuel, pellets, and biogas may significantly contribute to future renewable energy (Global Bioenergy Statistics 2019; Stančin et al. 2020).

7.1.1 Biomass Supply

According to the International Energy Agency (IEA) in 2017, an estimated amount of 5.6 EJ biomass was utilized for energy applications, out of which 86% was used for primary solid biofuels (in particular wood pellets, wood chips, fuelwood for cooking and heating), and 7% biomass was used as liquid biofuels (Global Bioenergy Statistics 2019).

Solid biofuels are in lower demand than liquid biofuels and biogas as a result of the shift to cleaner cooking (Ravindra et al. 2019); this became the driving force

R. Dalal · R. Wathore (✉) · N. Labhasetwar (✉)
Energy and Resource Management Division, CSIR-National Environmental Engineering Research Institute, CSIR-NEERI, Nagpur, Maharashtra 440020, India
e-mail: r.wathore@neeri.res.in

N. Labhasetwar
e-mail: nk_labhsetwar@neeri.res.in

Y. K. Nandabalan et al. (eds.), *Zero Waste Biorefinery*, Energy, Environment, and Sustainability, https://doi.org/10.1007/978-981-16-8682-5_7

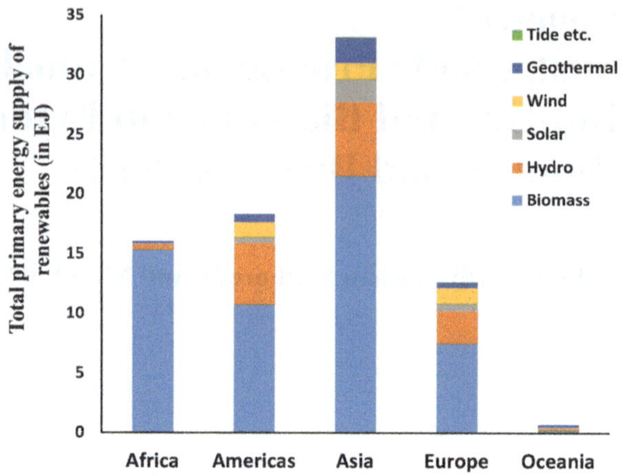

Fig. 7.1 Bar graph representing the continent wise total primary energy supply of renewables (EJ) in 2017 (Global Bioenergy Statistics 2019). Copyright granted

for the reduction of biomass supply for the first time in 2017 (Global Bioenergy Statistics 2019).

Asia and Africa are the major contributors to domestic biomass supply, as both these countries use biomass for their heating and cooking needs (Mehetre et al. 2017). The USA and Brazil have the highest share in biofuels, which is majorly attributed to bioethanol production from corn and sugarcane (Jacobus et al. 2021; Mekonnen et al. 2018). On the other hand, Europe has a maximum contribution of about 18 billion m^3 methane accounting for approximately 50% of the global biogas supply because of their extensive use of Waste-to-Energy (WtE) technologies like incineration and gasification (Scarlat et al. 2018). This trend can be observed in Fig. 7.2 (Global Bioenergy Statistics 2019).

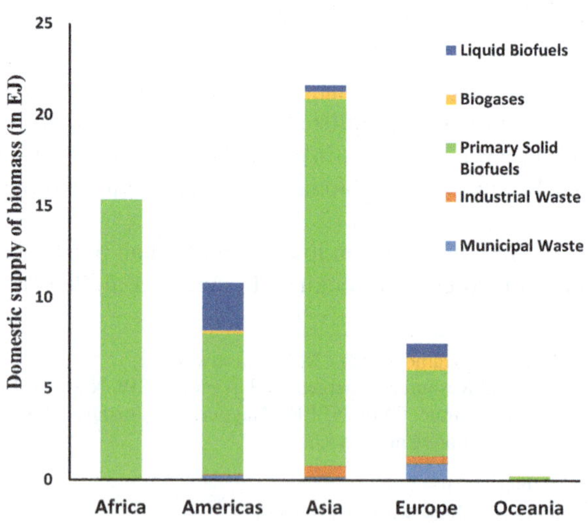

Fig. 7.2 Bar graph representing the continent wise domestic supply of biomass (EJ) in 2017 (Global Bioenergy Statistics 2019). Copyright granted

7.1.2 Biomass Potential in India

Historically, biomass has always been a vital and indispensable energy source for India; currently, it accounts for about 32% of the total primary energy use (Sharma et al. 2020). Most biomass is used for cooking purposes because of lack of access to clean cooking, i.e., LPG or improved cookstoves. This shift can potentially displace up to 25% of traditional biomass used for cooking with other purposes, which is well understood from Fig. 7.3. It is estimated that 660 million in 2019 and 580 million Indians in 2030 use or will be using solid fuels. The Indian Government has initiated several programmes for efficient utilization of biomass and biomass-derived waste across various sections, such as bagasse based co-generation of sugar and power etc. (Ibrahim et al. 2021).

India is one of the major contributors to global bioenergy, has a yearly biomass availability of around 500 million metric tonnes; 30% of which is surplus and majorly generated from agricultural and forestry residues and account for an estimated potential of 18,000 MW. In addition, other biomass sources, primarily generated as waste from various agricultural and industrial sectors, could also be potentially utilized as an energy source for the generation of useful products and byproducts. Plant-based waste materials classified as lignocellulosic biomass have a high potential to replace fossil fuels such as crude oil for the production of biofuels and biochemicals, which may significantly reduce greenhouse gases (GHG) (Daioglou et al. 2015).

Figure 7.4 summarises the various biomass sources and thermochemical conversion options (combustion, gasification and pyrolysis) to derive useful value-added products. Major products formed are heat, syngas, bio-oil and biochar, with the pyrolysis process resulting in the majority of the products (Tripathi et al. 2016).

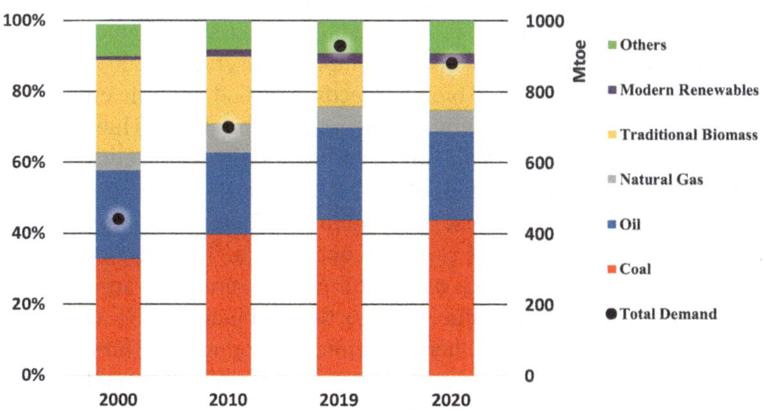

Fig. 7.3 Total primary energy demand distribution (Bar graph) and total primary energy demand (Mtoe) are represented in circular markers in India (Ibrahim et al. 2021). Copyright granted

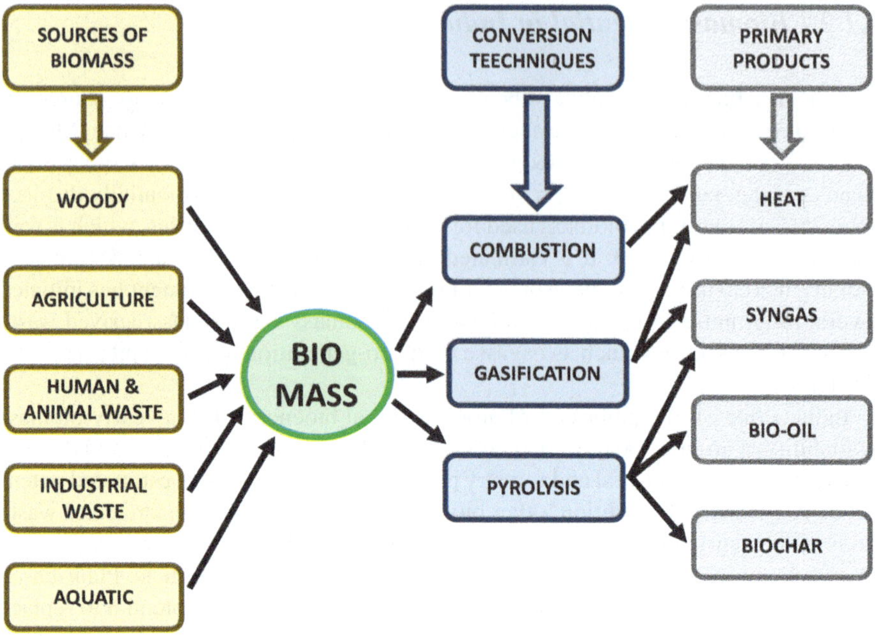

Fig. 7.4 Sources of biomass, their conversion techniques and potential products obtained from them (Tripathi et al. 2016). Copyright granted

7.1.3 Concept of a Zero-Waste Biorefinery

In view of the above significant generation of biomass and thermal conversion techniques, an integrated biorefinery for processing biomass to produce biofuels and biochemicals is an attractive alternative to compete with crude-based refineries. Such biorefineries could use a range of biomass feedstocks and multiple efficient conversion processes, including solid, liquid, and gaseous products (Huber et al. 2006; Ragauskas 2006; Aghbashlo et al. 2021; Awasthi 2020; Bhowmick et al. 2018). Lignocellulosic biomass such as agricultural waste and woodchips would be an inexpensive resource for such a large-scale biorefinery. This kind of biomass is mainly available and can be a source of carbon sequestration. The targeted products are biochar, bio-oil and syngas, which can be further converted to various compounds that can potentially be used in biofuel and biochemical industries (Kang et al. 2021). The conceptualized scheme for a thermochemical biorefinery is shown in Fig. 7.5.

This chapter primarily discusses the various options for sustainable and optimal processes to produce biochar, bio-oil and syngas from thermochemical conversion processes of lignocellulosic, woody biomass and residue. Further, the feasibility for a close to zero-waste biorefinery producing a variety of biochemical with global and India's perspective is also discussed.

The proposed biorefinery can be classified into four main features as follows;

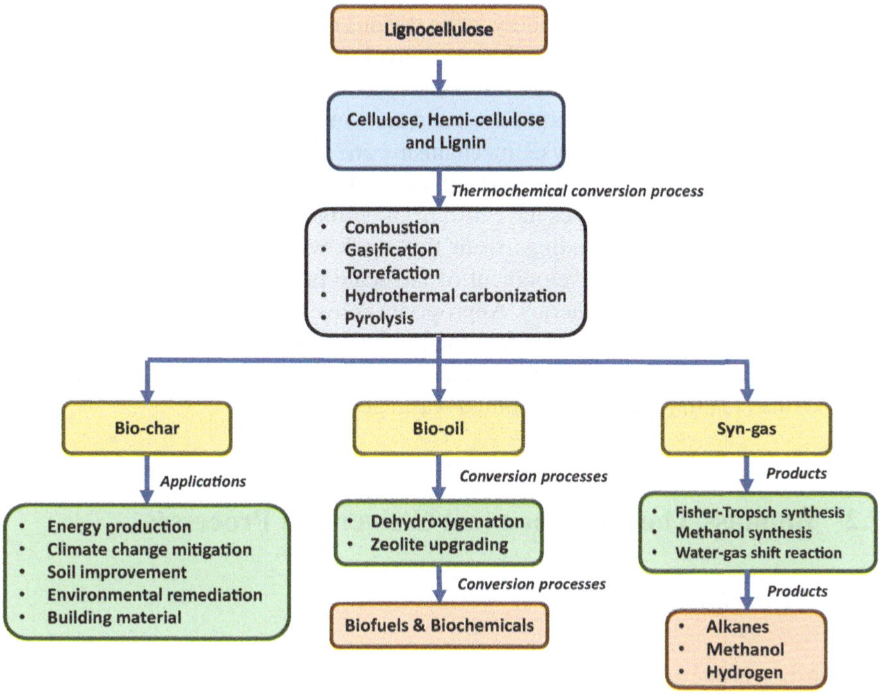

Fig. 7.5 Proposed scheme for a thermochemical biorefinery for the generation of useful products from biochar, bio-oil and biogas

(1) **Platforms**: Platforms are the intermediates that link feedstock and the final product. The proposed concept is a three-platform biorefinery as it deals with solid char, oils and syngas to produce various biofuels and biochemicals (Cherubini et al. 2009).

(2) **Products**: The products obtained from a biorefinery can be divided into two main categories; energy-driven and material-driven. The proposed biorefinery would produce products from both classes in the form of biofuels and biochemicals (Cherubini et al. 2009).

(3) **Feedstock**: In a biorefinery, the feedstock is a renewable raw resource (biomass) that has been transformed into marketable goods. Lignocellulose crops (such as wood, short-rotation poplar, switchgrass etc.) or lignocellulosic residue (such as agricultural crop residues, sawmill residues etc.) can be used in the proposed case (Cherubini et al. 2009).

(4) **Processes**: Thermochemical conversion processes can be used to obtain desired products in the proposed scheme. These processes include combustion, gasification, pyrolysis, torrefaction, hydrothermal carbonization etc., when the feedstock is subjected to extreme operating parameters (high temperature and/or pressure, with or without catalytic means) (Cherubini et al. 2009).

The vital aspect of this biorefinery is the various thermochemical process options involved, which can affect the quality and quantity of the desired products. In this chapter, various thermochemical conversion processes—combustion, gasification, torrefaction, hydrothermal carbonization and most importantly, pyrolysis is explored and discussed. Multiple pyrolysis mechanisms are also explored. These processes are then summarized, where the pros and cons of each are identified. In addition, various pre-treatment processes have also been outlined and summarized,

It is expected that the findings from this work will help pave the way to aid in the conceptual design, development of lab-scale processes, their validation and eventual upscaling of biorefineries. Appropriate conversion processes and operating parameters can be implemented as per the product requirement. The idea is that all the products obtained (solid, liquid and gas) are sustainably produced and effectively utilized, thus leading to zero or minimal waste generation.

7.2 Biomass Thermo-Chemical Conversion Processes

7.2.1 Combustion

Combustion is one of the oldest conversion processes to produce energy by converting chemical energy stored in biomass into heat by direct heating or burning in the presence of air or oxygen. Common examples of devices where combustion takes place are stoves, boilers, furnaces and steam turbines. In the combustion process, an oxidizing agent (air) is used, where complete oxidation of biomass gives heat and flue gas having CO_2 and H_2O with a significant quantity of NO_2 in it and a small amount of unconverted energy into ash. Combustion takes place in a temperature range of 800–1000 °C. Any biomass with less than 50% moisture content can be treated via combustion (Goyal et al. 2008). Biomass with high moisture content can be converted by biological conversion processes (McKendry 2002a).

In many cases, direct combustion of biomass is not suitable, and pre-treatment becomes necessary. This includes drying, chipping or sizing; these pre-treatment processes increase the efficiency of the process and increase operating costs (McKendry 2002a).

7.2.2 Gasification

Gasification, unlike combustion, is a process for biomass conversion into combustible gas mixture or fuel by partial oxidation of biomass at high temperatures. It is usually done at temperatures between 700 and 900 °C in a gaseous medium such as air, O_2, CO_2, N_2 or their mixtures (Neves et al. 2011). Partial oxidation extracts energy out of biomass and stores it in gaseous chemical mixtures. In gasification, the chemical energy of carbon in produced flue gases can be used more easily and efficiently than raw biomass (McKendry 2002b). Gasification yields syngas (hydrogen + carbon monoxide), carbon dioxide and water. This process is generally used for the production of electricity and heat. The residue in this process is char, ash, tar and some oil. Also, at low temperatures, gasification may yield hydrocarbons (Morrin et al. 2012; Hagemann et al. 2018).

Due to the biomass composition, some challenges may be faced in producing high energy containing biomass. The O/C ratio plays an important role in determining gasification efficiency. To obtain higher efficiency in gasification, low O/C ratio is preferred. In general, the O/C ratio can be reduced by torrefaction. So, torrefaction can be regarded as a pre-treatment process for gasification for improving product quality (Tripathi et al. 2016).

Dutta et al. (2014) studied equilibrium modelling of gasification process of various lignocellulosic biomass to investigate the influence of moisture on gaseous product composition. This biomass includes Bamboo (*Banbusea Tulda*), Gulmohar (*Delonix regia*), Neem (*Melia Azedarach L*), Dimaru (*Ficus lepidosa wall*), and Shisham (*Delbergia sissoo*). The ultimate and proximate evaluation showed that Bamboo gives maximum gas yield contributing to the calorific value of 18.40 MJ/kg with fixed carbon of 48.69%, giving the best producer gas quality. At the same time, Shisham has the highest carbon percentage and calorific value among the four woody biomass. In comparison, Gulmolhar yielded maximum hydrogen production (24.5%).

7.2.3 Torrefaction

Torrefaction is a non-oxidizing process operated at atmospheric pressure at a temperature of 200–300 °C. Removing volatiles and decreasing the O/C ratio from biomass is the fundamental step in this process by different decomposition reactions. This decomposition reaction may include (a) devolatilization and carbonization of hemicelluloses, (b) depolymerization and devolatilization of cellulose and lignin (Sri et al. 2020).

Torrefaction increases the calorific value and hydrophobicity of biomass and enriches ultimate/proximate composition characteristics of biomass. Torrified biomass also displays enhanced particle size and shape distributions, mechanical strength, palatability, and grindability properties. Also, biomass porosity is increased by torrefaction than conventional gasification (Tripathi et al. 2016; Shankar Tumuluru et al. 2011). This technology can be used to obtain flue gases from biomass and biomass waste more cost-effectively and efficiently. After torrefaction, the product has approximately 30% higher energy density than traditional biomass (Khoo et al. 2013; Weber and Quicker 2018).

7.2.4 Hydrothermal Carbonisation (HTC)

Hydrothermal carbonization can be regarded as a pre-treatment process of biomass operated at a lower temperature of around 200–300 °C, which converts biomass feedstock into a solid product with high carbon content. The solid carbonaceous from HTC can be called Hydrochar. HTC produces better efficiency of fuel, homogenization of biomass and biomass waste and is more energy-efficient than pyrolysis. If biomass contains a smaller particle size, HTC is favoured over pyrolysis due to higher char yield in HTC (Reza et al. 2014). The HTC process starts with hydrolysis, followed by dehydration and decarboxylation, thereby undergoing polymerization, condensation and aromatization. HTC results in producing Hydrochar, water, water-soluble compounds and a gas (mainly CO_2) (Correa and Kruse 2018). Hydrochar is a solid substance that is stable, hydrophobic, and brittle, with a fuel value similar to lignite coal. In conventional coal-fired power plants, Hydrochar has the potential to replace coal. Hydrochar also has a lot of stable carbon and other nutrients in it, which are important for soil improvement. Furthermore, the HTC system liquid contains potentially harmful chemicals such as phenols, furfurals, and their derivatives, that provide possibilities for anaerobic digestion to produce biogas, especially if a short retention period is used. Hydrochar has comparable thermal properties to lignite coal but with reduced slagging and fouling indices, making it more suitable for combustion (Reza et al. 2014). Hydrochar in powdered form can be an excellent precursor to produce activated carbon (Correa and Kruse 2018).

7.2.5 Pyrolysis

Pyrolysis is the thermochemical breakdown or cracking of biomass or comparable feedstock into a variety of usable products also without any oxidizing agents or in the presence of a restricted supply that prevents gasification (Basu 2018). The application of pyrolysis for the production of tar and embalming agents can be dated back to ancient Egypt. Studies done in the 1980's show increased liquid yield and faster vapour condensation of the biomass feed when heated at a rapid rate (Vamvuka 2011; Hosoya et al. 2007).

Pyrolysis is a forerunner to both combustion and gasification. However, unlike the latter, it does not necessitate external factors like oxygen and steam. Cracking, carbonization, torrefaction, devolatilization etc., all have parallels or overlaps, but gasification, which involves chemical reactions with an external agent known as the gasification medium, does not (Basu 2018). The pyrolysis process is generally carried out in a range of 400–1200 °C (Tripathi et al. 2016).

Pyrolysis is heart of all thermochemical processes and can be used to make petroleum-like compounds from biomass. Pyrolysis is significant in reaction kinetics and, as a result, in reactor design and evaluating product distribution, composition, and characteristics in all thermochemical processes. For obtaining solid biochar, pyrolysis can be regarded as one of the most cost-effective and efficient methods while also offering less pollution than combustion, making it more adaptable than other thermochemical conversion processes (Tripathi et al. 2016). This has been achieved because of the following benefits of pyrolysis;

- Pyrolysis is an effective method for reducing waste and converting it into solid biochar, liquid bio-oil and syngas as the gaseous product.
- This process is easy to optimize (i.e., process parameters) according to the desired results, e.g., to get a high yield of bio-oil, fast pyrolysis can be used, while slow pyrolysis results in a high yield of solid biochar.
- Pyrolysis can treat a wide variety of biomass and waste directly without any major difficulty subject to pre-treatment processes in some cases. It provides versatility in terms of feedstock type and operation circumstances, resulting in the desired product quality.
- Large scale pyrolysis units can make the process economical.
- Lower content of sulfur and NO_x gases in pyrolysis makes it more environmentally friendly, thereby reducing GHG emissions. Therefore, it helps in reducing global warming (Tripathi et al. 2016).

Principle and Mechanism

Pyrolysis is a complex process having various reactions in the reacting system (Tripathi et al. 2016). In pyrolysis, the biomass is heated so that it does not get exposed to oxygen. Lignocellulosic biomass mainly consists of cellulose, hemicellulose and lignin, which generally are long-chain polymeric compounds. Also, biomass has moisture content to some extent. The products obtained from pyrolysis are governed by the decomposition of these constituents (Collard 2014; Yaashikaa et al. 2019). Figure 7.6 explains the decomposition behaviour of lignocellulosic biomass constituents at different temperature and their potential product forming tendency.

Pyrolysis consists of various reactions—dehydration, depolymerization, isomerization, aromatization, decarboxylation and charring, categorized into primary and secondary reactions (Vamvuka 2011; Collard 2014; Lange 2007). The pyrolysis reaction can be represented as;

$$\underset{\text{Biomass}}{(C_6H_6O_6)_n} \rightarrow \underset{\text{gases}}{(H_2+CO+CH_4+\ldots+C_5H_{12})}+\underset{\text{liquid}}{(H_2O+CH_3OH+CH_3COOH+\ldots)}+\underset{\text{char}}{C} \tag{7.1}$$

- The cleaving of chemical bonds within polymers occurs when different chemical linkages within the polymers are broken during biomass heating, followed by the release of volatile chemicals and recombination reactions within the residue structure (Fig. 7.7). These reactions are considered primary mechanisms. Primary reactions include. (a) Charring reaction: This reaction occurs by forming polycyclic aromatic hydrocarbons (PAHs) during pyrolysis producing char. (b) Depolymerisation reaction: This reaction is an important step as it results in volatiles and gases

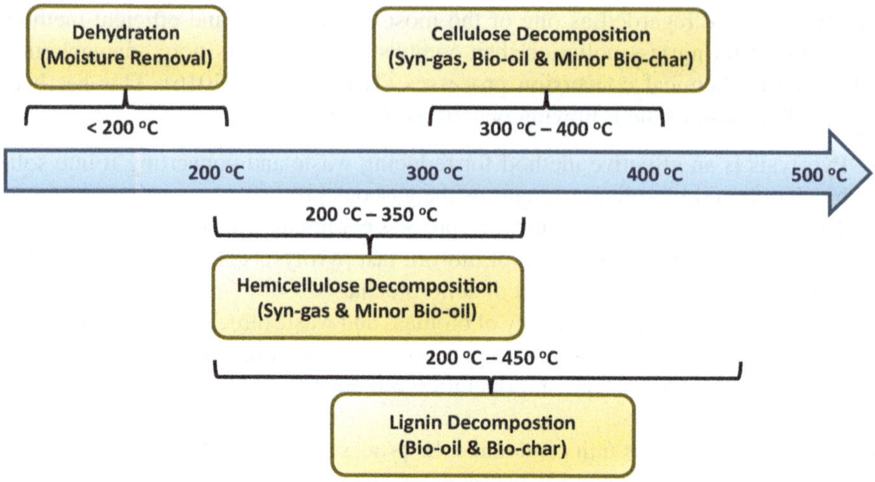

Fig. 7.6 Decomposition behavior of different biomass constituents at different temperatures ranges with their respective product formation tendency (Collard 2014)

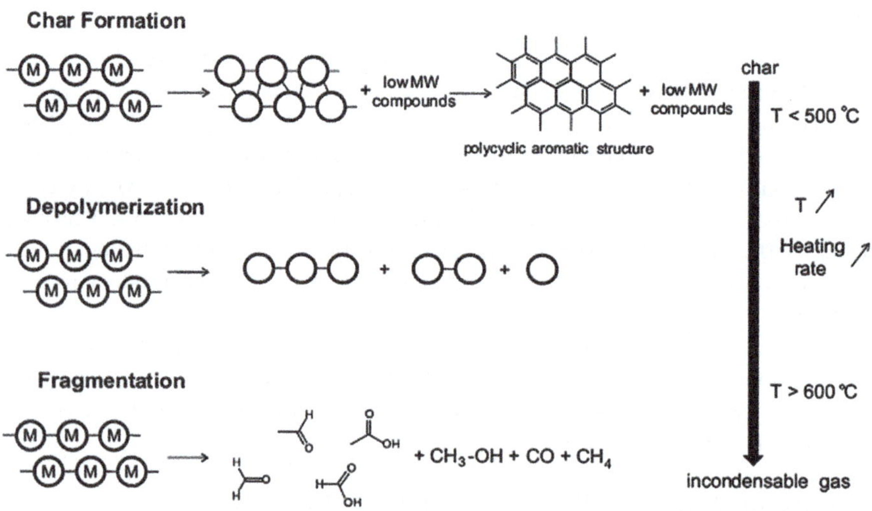

Fig. 7.7 Primary reaction mechanisms for char formation, depolymerization and fragmentation in pyrolysis process (M: monomer; MW: molecular weight) (Collard 2014). Copyright granted

production, and it involves breaking of linkages in monomers. (c) Fragmentation: In this reaction unit, monomer or polymer undergoes fragmentation. Short-chain compounds and non-condensable gases get formed during this (Hosoya et al. 2007; Collard 2014; Velden et al. 2010; Hu and Gholizadeh 2019; Garcia-Perez et al. 2007; Mullen and Boateng 2011; Morf et al. 2002).

- Some of the primary compounds formed during primary reactions are not stable, resulting in secondary reactions. These reactions comprise cracking as well as recombination reactions and cracking results in lighter fractions. At the same time, recombination results in the deposition of heavier compounds on the surface of char (Morf et al. 2002; Zhang et al. 2018). Hence, Secondary reactions can be called controlling reactions as they decide specific properties of the product (Hu and Gholizadeh 2019).
- Pyrolysis reactions can substantially get affected by different process conditions. Temperature, heating rate and residence time are the most important parameters in deciding product profile in pyrolysis. For instance, lower temperature favours char formation, while high temperature results in volatiles and condensable bio-oil. High heating rate and little residence time inhibit char formation resulting in bio-oil and non-condensable gases (Garcia-Perez et al. 2007; Shen et al. 2009).

Different types of pyrolysis are detailed below:

Fast Pyrolysis

Fast pyrolysis is a thermochemical conversion process carried out at a moderately high temperature where biomass is rapidly heated at a higher rate in the absence of an oxidizing agent. The temperature range for fast pyrolysis is between 400 and 600 °C; the heating rate is about 1000 °C/s and the residence time of 2 s or less (Pattiya 2018). When biomass is exposed to these circumstances, it decomposes into pyrolysis vapours and solid residue, often known as char or biochar. The pyrolysis vapours are then condensed into a dark brown coloured fuel oil or bio-oil with nearly half the heating value of conventional fuel oil, with the non-condensable gaseous product remaining (Bridgwater 2000). A well-homogenized feed with moisture <10% and size around 5 mm is ideal for a high pyrolysis yield (Sri et al. 2020; Novotny et al. 2015).

Fast pyrolysis can be achieved using a fluidized bed, entrained flow, ablative, auger reactor, vacuum, rotating cone, moving bed reactor (Bridgwater 2000). The major result is bio-oil from fast pyrolysis; solid biochar and non-condensable gaseous products are minor but valuable (Pattiya 2018).

To achieve maximum yield of liquid bio-oil, the responsible factors are;

(i) To overcome mass and heat transfer limitations, a freshly shredded biomass feed is essential to facilitate high heating and heat transfer rates
(ii) A highly controlled temperature of roughly 500 °C and 400–450 °C for vapours,
(iii) Short vapour residence time of typically <2 s, and
(iv) Rapid cooling of the pyrolysis vapours (Bridgwater 2003).

Because of the very low pH value, the liquid product, i.e., bio-oil obtained from fast pyrolysis can be highly corrosive. Also, up-gradation of bio-oil becomes necessary as it has half the high heating value (HHV) compared to crude oil (Xu and Etcheverry 2008).

If biochar is the primary product desired from fast pyrolysis, the biomass must be thermally cracked at a higher temperature with a shorter vapour exposure time. On the other hand, fast pyrolysis transforms biomass into liquid bio-oil before reacting to generate char products due to the high heating rate (Mohan et al. 2006).

Xiao et al. (2007) performed fast pyrolysis on an unknown bamboo species to explore the effect of the products within a range of 400–700 °C. Results showed that the solid products formed at 700 °C had increased high heating value (19%), fixed carbon (14%) and ash content (49%) than products formed at 400 °C; alternatively, there were reductions observed in volatile (52%), hydrogen (51%) and oxygen (64%) contents. On the other hand, liquid product (tar) formed at higher temperature (700 °C) had higher moisture content (7% difference) and 24% lower high heating value compared to 400 °C. Kato et al. (2014) experimented on fast pyrolysis of Moso Bamboo to study the products of bio-oil production in greater detail and the influence of operating temperature on the composition of liquid products. It was found that an increase in temperature resulted in higher yields of compounds having furan structures and lower yields of phenolic compounds. The chemical characteristics of bio-oil were influenced by biomass source, pyrolysis condition and type of pyrolyzer.

Fast pyrolysis for bio-oil production also has advantages such as process easiness, atmospheric pressure execution, and the primary product being a liquid that can be easy to store and transported, projected low cost, thermally efficient, as well as limited fossil fuel inputs, likely to result in a lower carbon fuel (Oasmaa et al. 2003). Fast pyrolysis has a wide variety of applications in food flavours and speciality chemicals (Tripathi et al. 2016). When the expected output is primarily liquid and gaseous products, fast pyrolysis is recommended. This process requires woody biomass to be chipped for uniformity. The entire energy balance of fast biomass pyrolysis can yield 60–75 wt% liquid bio-oil, 15–25 wt% solid char, and 10–20 wt% non-condensable gases, based on the feedstock employed, with low environmental emissions resulting in low sulphur and nitrogen proportion of the source biomass (Bridgwater 2003; Oasmaa et al. 2005).

Slow Pyrolysis

Slow pyrolysis is considered a conventional pyrolysis process and is mainly done to produce solid products; char or biochar. It is one of the oldest processes for producing biochar, used for thousands of years (Basu 2018). This process is also called carbonization (Tripathi et al. 2016; Demirbas and Arin 2002). Slow pyrolysis is characterized by a slow heating rate of 0.1 to 1 °C/s and a longer residence time of 5–30 min; the biomass is heated up to 400–500 °C (Tripathi et al. 2016). Slow pyrolysis is preferred for more yield of solid biochar and retention of soil nutrients, although bio-oil and syngas are also formed in small fractions (Sri et al. 2020; Lehmann and Joseph 2010). In slow pyrolysis, a longer residence period and lower heating rate can favour completion of secondary reactions to form solid biochar (Tripathi et al. 2016). High functional biochar can be produced at lower temperatures and low residence time (Sri et al. 2020).

The operating conditions of the pyrolysis process, such as particle size, pressure, temperature, purging gas and heating rate, have a strong influence on the product yield and characteristics and can be adjusted to meet the product requirements (Antal and Grønli 2003; Enders 2012; Lee et al. 2013). The efficiency of this process can be improved by increasing the operating pressure and/or utilization of the CO_2 from the flue gas as an inert carrier gas, the latter of which, upon implementation, significantly increased the yield of CO (Pilon and Lavoie 2013; Manyà et al. 2018).

Hernández-Mena et al. (2014) performed experimental research on slow pyrolysis of Bamboo biomass (species Dendrocalamus giganteus Munro) to determine biochar yield as a function of final reaction temperature. Figure 7.8 summarises their findings; the maximum yield of 80% is observed for biochar at 300 °C. Bio-oil yield increased from 300 to 500 °C, but not beyond. Higher temperature led to an increase in secondary reactions of volatile compounds resulting in a higher gas yield. A high porosity product was obtained, which could be further used as activated carbon.

Flash Pyrolysis

Flash pyrolysis is a modified and improved version of fast pyrolysis. Special reactor configurations are required where operating include a temperature range of 900–1200 °C, heat pulse of 0.1–1 s, a heating rate of around 1000 °C or more and a reaction time lasting several seconds (or less). Factors such as the reaction kinetics, phase transition behaviour of the feed, heat and mass transfer process can play a significant part in the final product distribution. Due to the very high hearing rates, low residence times and high temperatures, the liquid yield is high, but the char yield is lower (Tripathi et al. 2016). Two suitable reactors would be a fluidized-bed reactor and an entrained flow reactor (Canabarro et al. 2013).

The configuration of a reactor to be used in flash pyrolysis, which can provide a high heating rate and little residence time on a large industrial scale, can be challenging due to less thermal stability and corrosiveness of the bio-oil due to the char or ash present. Also, the char present can catalyze the polymerization reaction in bio-oil, increasing the viscosity of oil (Tripathi et al. 2016; Canabarro et al. 2013; Jahirul et al. 2012).

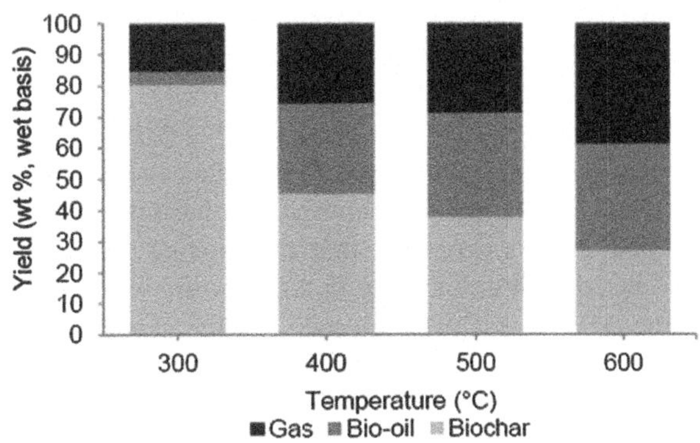

Fig. 7.8 Variation of the product yield with respect to reactor operating temperature for slow pyrolysis (Hernandez-Mena et al. 2014)

Vacuum pyrolysis

Vacuum pyrolysis includes a combination of operating parameters of slow and fast pyrolysis operated at very low pressure in the absence of oxygen. Vacuum pyrolysis is performed between 450 and 600 °C and 5–20 kPa, so operating under vacuum reduces vapour residence time in the reactor, thereby limiting secondary reactions and increasing bio-oil yield. Reduction in heat transfer requirement enables using larger particle size in vacuum pyrolysis than fast pyrolysis. Approximately 35–50% of dry feed is obtained as bio-oil with a higher char yield than fast pyrolysis without a need for carrier gas. But, large pressure vessels and pipings are required because of high vacuum, making this process more complex and costly (Bridgwater 2012; Carrier et al. 2011; Benallal et al. 1995).

Hydro-pyrolysis

The Hydro-pyrolysis process combines pyrolysis and hydrocracking integrated into a single system by adding hydrogen at high pressure (ranging 5–20 MPa) to the pyrolysis reactor. This process reduces oxygen content by adding a reducing agent, hydrogen, at high pressure, thereby improving bio-oil quality. High quality liquid products can be obtained from this process, from which a variety of biofuels can be produced. The only disadvantage of this process is in increased consumption of hydrogen, increasing processing cost. If hydrogen production becomes cheaper, this process can become commercially successful for lignocellulose biomass and carbon conversion, leading to zero waste (Bhaskar et al. 2011; Thangalazhy-Gopakumar et al. 2011; Melligan et al. 2012; Marker et al. 2012).

Microwave-assisted Pyrolysis

Microwave-assisted pyrolysis or MAP is an effective heating process of biomass on the principle of microwave dielectric effect of heating. This process can be done in the presence or absence of a catalyst. The two important mechanisms in MAP are ionic conduction and dipolar polarisation (Yin 2012). The solid product obtained in the MAP process provides heat, while gaseous products such as gaseous hydrocarbon and hydrocarbon oils are used to dry feedstock. Different types of MAP technologies for conversion of various types of biomass, such as agricultural residue sawdust, forestry residue, wood, algal biomass and sewage into biofuels, has been explored in literature (Sri et al. 2020; Yin 2012; Puligundla et al. 2016; Wang et al. 2016; Zaker et al. 2019).

Zhang et al. (2017) assessed MAP of Moso Bamboo and Rice Husk at different pyrolysis temperatures, fixed heating rate and fixed carrier gas flow rate. The average yield and properties of bamboo biochar (BC) and rice husk biochar (RHC) was studied. It was observed that the biochar yield for both bamboo and rice husk decreased with an increase in temperature while fixed carbon content, high heating value (HHV) and pH increased. At the same temperature, BC had higher carbon content, HHV and pH than RHC making it more suitable as fuel and carbon sequester. However, ash content was higher in RHC, which can potentially be used in extracting pure amorphous silica. Also, both BC and RHC had a high bet surface area of 259.89

m^2/g and 331.23 m^2/g, respectively, which could potentially be used in activated carbon production, and fuel cell material.

MAP process is fast, easy to control, have lower operating cost, is energy-efficient and gives better quality of products. However, capital costs for setting up and upscaling the equipment, including provision for pre-treatment of the feed, are the main limiting factors for MAP (Yin 2012).

A systematic comparison of all pyrolysis types discussed earlier, including process parameters, advantages, disadvantages and other remarks is summarized in Table 7.1.

Pre-treatment processes in pyrolysis

By altering the lignocellulosic biomass structure, pre-treatment methods can effectively enhance pyrolysis efficiency. Such pre-treatment processes are discussed below.

Physical pre-treatment: Biomass is not an excellent heat conductor. As a result, the pyrolysis mechanism has a temperature gradient. More char is formed when the particle size is larger. However, if the particle size is decreased too much, the operation cost increases. As a result, selecting the ideal particle is critical (Sri et al. 2020).

(1) Particle Size: The impact of particle size on the pyrolysis process and cost shows that more char and gases are created when particle size rises, but bio-oil production drops. Furthermore, particle size impacts product yield; as particle size rises, so does water yield. (Shen et al. 2009; Bennadji et al. 2014; Luo et al. 2010). Larger particles produce more char and fumes, while bio-oil production decreases.

(2) Density: Density of biomass particle also affect product yield and composition. Pelletisation of cotton straw and high densification of woody biomass produces an economical feedstock for pyrolysis. Bio-oil yield can be increased by increasing the density of biomass (Li and Liu 2000; Mani et al. 2006; Ndiema et al. 2002). Pellet mill, Briquette press, screw extruder are some densification systems for pyrolysis.

(3) Dry torrefaction: It is a thermal biomass pre-treatment process. Depending upon the temperature applied, it can be done in the light, mild and severe mode, degrading hemicellulose, cellulose, and lignin. This improves biomass structure and produces a better quality of bio-oil (Mamvura 2020; Zadeh et al. 2020).

Chemical pre-treatment: The presence of inorganic minerals in biomass feed can affect pyrolysis. Chemical pre-treatment can be used for removing such materials and improving product quality. This can be done in the following ways:

(4) Acid and alkali treatment: Minerals such as carbonates, phosphates, sulfates and chlorides are found in biomass, affect the pyrolysis process such as causing corrosion of the reactor (Miller and Bellan 1997; Ranzi et al. 2008). Moreover, the presence of cations can positively affect char formation, thus reducing bio-oil yield. It also acts as an ageing agent for bio-oil (Yu et al. 2014). For

Table 7.1 Comparison of different types of pyrolysis processes parameters, their advantages, disadvantages and other remarks

Type of pyrolysis	Process parameters				Remarks	References
	Temperature (°C)	Pressure (MPa)	Heating rate (°C/s)	Residence time (s)		
Fast	800–1250	0.1	10–200	0.5–10	• Feasible to carry out in a wide range of reactor types • Atmospheric pressure process condition making it simple • Low operating cost • High thermal efficiency • Main product being liquid, is easy to store and transport	Tripathi et al. (2016), Bridgwater (2000), Oasmaa et al. (2003), Demirbas and Arin (2002), Bahng et al. (2009),
Slow	400–500	0.1	0.1–1	300–550	• Favourable for producing a solid product • High quality of biochar • Low operating process parameters • High pressure can be applied to improve efficiency	Tripathi et al. (2016), Sri et al. (2020), Demirbas and Arin (2002), Lehmann and Joseph (2010), Pilon and Lavoie (2013), Bahng et al. (2009)

(continued)

Table 7.1 (continued)

Type of pyrolysis	Process parameters				Remarks	References
	Temperature (°C)	Pressure (MPa)	Heating rate (°C/s)	Residence time (s)		
Flash	900–1200	0.1	>1000	<1	• Require special type of reactor configuration • High heating rate & very less residence time • Process parameters play an important role in determining products • Lower stability and quality of bio-oil	Tripathi et al. (2016), Demirbas and Arin (2002) Demirbas and Arin (2002), Canabarro et al. (2013), Bahng et al. (2009)

(continued)

Table 7.1 (continued)

Type of pyrolysis	Process parameters				Remarks	References
	Temperature (°C)	Pressure (MPa)	Heating rate (°C/s)	Residence time (s)		
Vacuum	450–600	0.01–0.02	1–10	2–30	• High bio-oil and biochar production • Larger biomass particles can be used • No requirement of carrier gas • High processing cost	Bridgwater (2012), Carrier et al. (2011), Benallal et al. (2015), Zaman et al. (2017)
Hydro-pyrolysis	800–1000	5–20	10–100	<10	• High quality of bio-oil • Bio-oil with lower oxygen content • A variety of biofuels can be produced • Complete utilization of carbon content • High processing cost	Bhaskar et al. (2011), Thangalazhy-Gopakumar et al. (2011), Melligan et al. (2012), Marker et al. (2012)

(continued)

Table 7.1 (continued)

Type of pyrolysis	Process parameters				Remarks	References
	Temperature (°C)	Pressure (MPa)	Heating rate (°C/s)	Residence time (s)		
Microwave-assisted Pyrolysis	–	–	–	–	• Liquid & gases are the main products • Energy efficient • Fast and easy to control • Less operating cost • High-quality products	Sri et al. (2020), Yin (2012), Puligundla et al. (2016), Wang et al. (2016), Zaker et al. (2019)

eliminating these minerals, acid treatment is often utilized. For acid treatment, dilute sulfuric acid, also known as is utilized (Jiang et al. 2013). Acid treatment can increase bio-oil output by up to 30% (Bozell 2001). Acid treatments can be divided into two types: the high temperature for a short time (above 180 °C., 1–5 min) and low temperature for a long time (below 120 °C temperature, 30–90 min) (Hu and Gholizadeh 2019). A low-temperature alkaline solution such as concentrated NaOH can be quite beneficial for partially removing hemicellulose and lignin from biomass. Acids and alkalis can react to form salts that catalyze char formation (Carrillo et al. 2005; Lee et al. 2005; Mosier 2005).

(5) Hydrothermal pre-treatment: Aso known as wet torrefaction. In this process, biomass is treated with hot compressed water or subcritical water, where the temperature is in the range of 180–260 °C and pressure of around 47 bar. This high-pressure water can enter the cellulose structure and hydrate it, solubilizing and removing hemicellulose and a small amount of lignin. Hemicellulose generally produces water, so by doing this process, water production can be reduced, improving bio-oil yield (Kumar et al. 2020).

(6) Steam explosion: In woody biomass, steam explosion may provide feedstock for the secondary conversion process. In this process, woody chips are treated in a vessel at 285 °C and 35 bar for 2 min. In the next 5 s, pressure is increased up to 70 bar (Zadeh et al. 2020). The rigid structure of the biomass fibres is ruptured, and the biomass is transformed into a dispersed fibrous solid. One of the most environmentally friendly and energy-efficient pre-treatment techniques for lignocellulose is steam explosion (Shrotri et al. 2017).

(7) Ammonia fibre expansion: This is one of the most effective techniques for improving biomass structure, density and compressibility. It is a combination of physical and chemical pre-treatment. This biomass is exposed to ammonia at a high temperature of around 80–150 °C and pressure of 200–400 psi (14-28 bar). This process combines physical (temperature and pressure) and chemical (ammonia) processes, generally carried out in a specific reactor with a temperature/pressure controller and ammonia flow. Ammonia and biomass are mixed in a ratio of 1:1 or 1:2 for 10–60 min and then treated at required parameters (Sundaram et al. 2015).

Biological pre-treatment: Biological pre-treatment is one of the most efficient methods for no temperature/pressure or chemicals. In this method, biomass is treated with micro-organisms and bacteria, which degrades the linkages between cellulose or lignin and hemicellulose (Yu et al. 2013).

7.3 Discussion

Biorefining is indeed the sustainable processing of biomass into a range of commercial products and energy, as stated by the International Energy Agency's (IEA)

Bioenergy Task 42. This biorefinery system can be classified at different levels of the process, platform, feedstock and potential product (Cherubini et al. 2009; Sonnenberg et al. 2007). A combination of different thermochemical processing of lignocellulosic biomass with the potential of producing a range of biofuels and biochemicals has been discussed in this chapter. Appropriate combination of processes should be selected to develop such biorefinery. Table 7.2 summarizes the processes the technological challenges involved in different thermos-chemical processes before selection.

Among other discussed processes, pyrolysis is a flexible process explored in this chapter wherein the distribution and quality of the desired products can be altered by adjusting operating parameters. Biochar yield shows a strong dependence on the biomass composition, temperature and residence time. The presence of high lignin content helps in the formation of char, while cellulose in biomass favours tar formation. The high moisture content of the biomass inhibits the formation of char. High temperature tends to liquid and gaseous yield, while lower temperature, lower retention time and larger particle size result in char. Char yield can be increased by increasing pressure. If carrier gas flow rate is kept high, vapours get little residence time, resulting in lower char yield. Also, a high heating rate in the process results in bio-oil and non-condensable gases production. Biomass with a high value of lignin, larger particle size, lower moisture content, and longer residence time should be employed in pyrolysis for increased biochar production from an extensive assessment of the conditions.

Biochar is gaining importance because of its renewable nature, and this opens ways to explore the properties of biochar, techno-economic details, emission control and its applications and effect on soil characteristics. Bio-oil obtained from the primary process can be chemically upgraded to various biochemicals having their individual application. At the same time, the gaseous products have the potential to produce various hydrocarbons, including hydrogen.

Because of the technological and commercial development of the pyrolysis process, it could be regarded as the most feasible process for setting up a biorefinery to sustainably and effectively produce products. Proper upgradation and conversion techniques like Fisher-tropsch synthesis, dehydroxylation, hydrogenation, esterification, fermentation, catalytic (zeolite) upgradation can be incorporated on the obtained solid, liquid and gaseous products to produce useful chemicals such as fertilizers, hydrocarbons, alcohols, biofuels like biodiesel, aromatics and various others. Complete utilization of products/byproducts, producing marketable commercial products, would lead to zero or minimal waste production.

While not discussed in this work, catalyst use in pyrolysis is also an effective way for achieving higher char yield. The mixing of zinc-based acidic catalysis has been shown to enhance biochar formation and improve its characteristics (Lyu et al. 2020; Nanda et al. 2016).

Table 7.2 Selection of thermo-chemical process

Type of process	Products	Remarks	References
Combustion	Heat	• Require excess air supply and no need for pre-treatment • Low operating cost • Higher emissions • No or fewer value products obtained • Not feasible for all types of biomasses as uniformity is required	Tripathi et al. (2016), Goyal et al. (2008), McKendry (2002a, b), Demirbas and Arin (2002)
Gasification	Heat, Electricity, Biochar, Bio-oil and Gases	• Marginal air supply pre-treatment required • High operating cost • Lower emissions • Soot or aerosol formation may happen • Formation of tar may promote corrosion • Mainly used for heat and electricity production • Higher O/C required for higher efficiency	Tripathi et al. (2016), Neves et al. (2011), McKendry (2002b), Morrin et al. (2012), Hagemann et al. (2018), Kreuzeder et al. (2007), Lateh et al. (2019), Hussain et al.)2011)
Torrefaction	Biochar	• No air supply or pre-treatment required • Lower operating cost • Lower emissions • May be considered as a pre-treatment process for gasification • Enhance physical properties of biomass as well as product	Sri et al. (2020; Shankar Tumuluru et al. (2011), Khoo et al. (2013), Weber and Quicker (2018), Deng et al. (2009), Prins and Ptasinski (2006)
Hydrothermal Carbonisation	Biochar, water-soluble gases and gases	• No air supply as well as pre-treatment required • Lower operating cost • Lower emissions • Require high pressure so feeding raw material may get difficult • Heat recovery of process water and post-treatment may be required for char	Sri et al. (2020), Funke and Ziegler (2010), Libra et al. (2011), Liu and Balasubramanian (2012), Stemann et al. (2013)

(continued)

Table 7.2 (continued)

Type of process	Products	Remarks	References
Pyrolysis	Biochar, Bio-oil and gases	• No air supply but pre-treatment required • High operating cost • Lower emissions • Challenges may occur in maintaining high operating parameters • Rapid and effective removal of products may be difficult in case of fast pyrolysis	Tripathi et al. (2016), Sri et al. (2020), Demirbas and Arin (2002), Jensen et al. (2000), Bridgwater et al. (2002)

7.4 Conclusion

The concept of waste to energy (WtE) and close to zero-waste biorefineries to produce value-added products such as biochar, bio-oil syngas and major platform chemicals from woody or lignocellulosic biomass and biomass waste which can be integrated into commercial, industrial supply chains is presented in this chapter. Special attention was focused on the production of chemicals obtained from the thermochemical conversion of biomass. These products can be further converted to biochemicals by after-treatment and upgradation processes.

Owning great biomass potential in India, such biorefineries can be set up in both rural and urban areas where a variety of products can be produced as per the local and regional demands. Ideally, such biorefineries should be "modular" in design, meaning that it has the provisions for varying the operating parameters—reactor temperature/pressure, heating rate and residence time, availability of oxygen in the reactor, carrier gas selection and flow and input power. Additionally, the provision to incorporate different sources and types of biomasses, including their blending, particle size selection, and moisture control and other pre-treatment methods, can add further flexibility to the physio-chemical properties of the desired products. Biofuel and biochemical production can be optimized as required when such feedstock is used with a biorefinery approach because of the production of a wide range of products and byproducts, making the whole process sustainable with minimal waste generation.

However, it is essential to acknowledge the impacts of the all the processes involved in the conversion, for which extensive cost–benefit analysis, life cycle assessments and estimations of environmental damages averted is required. The variety of products obtained would likely require further processing before they are commercialized. The high energy demands of the thermochemical processes can significantly contribute to the global warming potential, especially if the system is powered by coal or coal-based electricity which is dominant in India; although utilization of renewable energy sources can close the gap. Furthermore, government interventions in the form of policy supports, promotion of technology adoption and spreading a positive sentiment would further catalyze the growth of these sustainable near-zero-waste biorefineries.

Acknowledgements Authors thankfully acknowledge Council of Scientific and Industrial Research (CSIR), India and its constituent laboratory National Environmental Engineering Research Institute (NEERI) for supporting the research under the project CNP-2-2469. The manuscript is checked for plagiarism using licensed iThenticate software through CSIR-NEERI Knowledge Resource Centre [KRC], KRC No. CSIR-NEERI/KRC/2021/JULY/ERMD/1.

References

Aghbashlo M, Almasi F, Jafari A, Nadian MH, Soltanian S, Lam SS, Tabatabaei M (2021) Describing biomass pyrolysis kinetics using a generic hybrid intelligent model: a critical stage in sustainable waste-oriented biorefineries. Renew Energy 170:81–91. https://doi.org/10.1016/j.renene.2021.01.111

Antal MJ, Grønli M (2003) The art, science, and technology of charcoal production †. Ind Eng Chem Res 42:1619–1640. https://doi.org/10.1021/ie0207919

Awasthi MK (2020) Refining biomass residues for sustainable energy and bio-products: an assessment of technology, its importance, and strategic applications in circular bio-economy. Renew Sustain Energy Rev 18

Bahng M-K, Mukarakate C, Robichaud DJ, Nimlos MR (2009) Current technologies for analysis of biomass thermochemical processing: a review. Anal Chim Acta 651:117–138. https://doi.org/10.1016/j.aca.2009.08.016

Basu P (2018.) Biomass gasification, pyrolysis and torrefaction: practical design and theory. Academic press, p 33

Batchelor S, Brown E, Scott N, Leary J (2019) Two Birds, one stone—reframing cooking energy policies in Africa and Asia. Energies 12:1591. https://doi.org/10.3390/en12091591

Benallal B, Roy C, Pakdel H, Chabot S, Poirier MA (1995) Characterization of pyrolytic light naphtha from vacuum pyrolysis of used tyres comparison with petroleum naphtha, p 6

Bennadji H, Smith K, Serapiglia MJ, Fisher EM (2014) Effect of particle size on low-temperature pyrolysis of woody biomass. Energy Fuels 28:7527–7537. https://doi.org/10.1021/ef501869e

Bhaskar T, Bhavya B, Singh R, Naik DV, Kumar A, Goyal HB (2011) Thermochemical conversion of biomass to biofuels. In: Biofuels. Elsevier, pp 51–77

Bozell JJ (ed) (2001) Chemicals and materials from renewable resources. American Chemical Society, Washington, DC

Bridgwater A (2000) Fast pyrolysis processes for biomass. Renew Sustain Energy Rev 4:1–73. https://doi.org/10.1016/S1364-0321(99)00007-6

Bridgwater AV (2012) Review of fast pyrolysis of biomass and product upgrading. Biomass Bioenerg 38:68–94. https://doi.org/10.1016/j.biombioe.2011.01.048

Bridgwater AV, Toft AJ, Brammer JG (2002) A techno-economic comparison of power production by biomass fast pyrolysis with gasification and combustion. Renew Sustain Energy Rev 6:181–246. https://doi.org/10.1016/S1364-0321(01)00010-7

Bridgwater AV (200) Renewable fuels and chemicals by thermal processing of biomass. Chem Eng J 16

Canabarro N, Soares JF, Anchieta CG, Kelling CS, Mazutti MA (2013) Thermochemical processes for biofuels production from biomass. Sustain Chem Processes 1:22. https://doi.org/10.1186/2043-7129-1-22

Carrier M, Hugo T, Gorgens J, Knoetze H (2011) Comparison of slow and vacuum pyrolysis of sugar cane bagasse. J Anal Appl Pyrol 90:18–26. https://doi.org/10.1016/j.jaap.2010.10.001

Carrillo F, Lis MJ, Colom X, López-Mesas M, Valldeperas J (2005) Effect of alkali pretreatment on cellulase hydrolysis of wheat straw: kinetic study. Process Biochem 40:3360–3364. https://doi.org/10.1016/j.procbio.2005.03.003

Cherubini F, Jungmeier G, Wellisch M, Willke T, Skiadas I, Van Ree R, de Jong E (2009) Toward a common classification approach for biorefinery systems. Biofuels Bioprod Biorefin 3:534–546. https://doi.org/10.1002/bbb.172

Collard F-X (2014) A review on pyrolysis of biomass constituents_ Mechanisms and composition of the products obtained from the conversion of cellulose, hemicelluloses and lignin. Renew Sustain Energy Rev 15

Correa C, Kruse A (2018) Biobased functional carbon materials: production, characterization, and applications—a review. Materials 11:1568. https://doi.org/10.3390/ma11091568

Daioglou V, Wicke B, Faaij APC, van Vuuren DP (2015) Competing uses of biomass for energy and chemicals: implications for long-term global CO_2 mitigation potential. GCB Bioenergy 7:1321–1334. https://doi.org/10.1111/gcbb.12228

De Bhowmick G, Sarmah AK, Sen R (2018) Lignocellulosic biorefinery as a model for sustainable development of biofuels and value added products. Biores Technol 247:1144–1154. https://doi.org/10.1016/j.biortech.2017.09.163

Demirbas A, Arin G (2002) An overview of biomass pyrolysis. Energy Sour 24:471–482. https://doi.org/10.1080/00908310252889979

Deng J, Wang G, Kuang J, Zhang Y, Luo Y (2009) Pretreatment of agricultural residues for co-gasification via torrefaction. J Anal Appl Pyrol 86:331–337. https://doi.org/10.1016/j.jaap.2009.08.006

Dutta PP, Pandey V, Das AR, Sen S, Baruah DC (2014) Down draft gasification modelling and experimentation of some indigenous biomass for thermal applications. Energy Procedia 54:21–34. https://doi.org/10.1016/j.egypro.2014.07.246

Enders A (2012) Characterization of biochars to evaluate recalcitrance and agronomic performance. Bioresour Technol 10

Funke A, Ziegler F (2010) Hydrothermal carbonization of biomass: a summary and discussion of chemical mechanisms for process engineering. Biofuels Bioprod Biorefin 4:160–177. https://doi.org/10.1002/bbb.198

Garcia-Perez M, Chaala A, Pakdel H, Kretschmer D, Roy C (2007) Characterization of bio-oils in chemical families. Biomass Bioenerg 31:222–242. https://doi.org/10.1016/j.biombioe.2006.02.006

Global Bioenergy Statistics (2019) World Bioenergy Association: Stockholm, Sweden. Available online: https://worldbioenergy.org/uploads/191129%20WBA%20GBS%202019_LQ.pdf. Accessed on 17 Aug 2021

Goyal HB, Seal D, Saxena RC (2008) Bio-fuels from thermochemical conversion of renewable resources: a review. Renew Sustain Energy Rev 12:504–517. https://doi.org/10.1016/j.rser.2006.07.014

Hagemann N, Spokas K, Schmidt H-P, Kägi R, Böhler M, Bucheli T (2018) Activated carbon, biochar and charcoal: linkages and synergies across pyrogenic carbon's ABCs. Water 10:182. https://doi.org/10.3390/w10020182

Hernandez-Mena LE, Pécoraa AA, Beraldob AL (2014) Slow pyrolysis of bamboo biomass: analysis of biochar properties. Chem Eng Trans 37:115–120. https://doi.org/10.3303/CET1437020

Hosoya T, Kawamoto H, Saka S (2007) Pyrolysis behaviors of wood and its constituent polymers at gasification temperature. J Anal Appl Pyrol 78:328–336. https://doi.org/10.1016/j.jaap.2006.08.008

Hu X, Gholizadeh M (2019) Biomass pyrolysis: a review of the process development and challenges from initial researches up to the commercialisation stage. J Energy Chem 39:109–143. https://doi.org/10.1016/j.jechem.2019.01.024

Huber GW, Iborra S, Corma A (2006) Synthesis of transportation fuels from biomass: chemistry, catalysts, and engineering. Chem Rev 106:4044–4098. https://doi.org/10.1021/cr068360d
Hussain A, Guiot SR, Mehta P, Raghavan V, Tartakovsky B (2011) Electricity generation from carbon monoxide and syngas in a microbial fuel cell. Appl Microbiol Biotechnol 90:827–836. https://doi.org/10.1007/s00253-011-3188-4
Ibrahim IA, Ötvös T, Gilmanova A, Rocca E, Ghanem C, Wanat M (2021) International energy agency. Kluwer Law International BV. OECD
Jacobus AP, Gross J, Evans JH, Ceccato-Antonini SR, Gombert AK (2021) Saccharomyces cerevisiae strains used industrially for bioethanol production. Essays Biochem 65:147–161. https://doi.org/10.1042/EBC20200160
Jahirul M, Rasul M, Chowdhury A, Ashwath N (2012) Biofuels production through biomass pyrolysis—a technological review. Energies 5:4952–5001. https://doi.org/10.3390/en5124952
Jensen PA, Frandsen FJ, Dam-Johansen K, Sander B (2000) Experimental investigation of the transformation and release to gas phase of potassium and chlorine during straw pyrolysis. Energy Fuels 14:1280–1285. https://doi.org/10.1021/ef000104v
Jiang L, Hu S, Sun L, Su S, Xu K, He L, Xiang J (2013) Influence of different demineralization treatments on physicochemical structure and thermal degradation of biomass. Biores Technol 146:254–260. https://doi.org/10.1016/j.biortech.2013.07.063
Kang K, Klinghoffer NB, ElGhamrawy I, Berruti F (2021) Thermochemical conversion of agroforestry biomass and solid waste using decentralized and mobile systems for renewable energy and products. Renew Sustain Energy Rev 149:111372. https://doi.org/10.1016/j.rser.2021.111372
Kato Y, Kohnosu T, Enomoto R, Akazawa M, Yoon S-L, Kojima Y (2014) Chemical properties of bio-oils produced by fast pyrolysis of bamboo. Trans Mater Res Soc Japan 39:491–498. https://doi.org/10.14723/tmrsj.39.491
Khoo HH, Koh CY, Shaik MS, Sharratt PN (2013) Bioenergy co-products derived from microalgae biomass via thermochemical conversion—Life cycle energy balances and CO_2 emissions. Biores Technol 143:298–307. https://doi.org/10.1016/j.biortech.2013.06.004
Kreuzeder A, Pfeifer C, Hofbauer H (2007) Fluid-dynamic investigations in a scaled cold model for a dual fluidized bed biomass steam gasification process: solid flux measurements and optimization of the cyclone. Int J Chem Reactor Eng 5. https://doi.org/10.2202/1542-6580.1394
Kumar R, Strezov V, Weldekidan H, He J, Singh S, Kan T, Dastjerdi B (2020) Lignocellulose biomass pyrolysis for bio-oil production: a review of biomass pre-treatment methods for production of drop-in fuels. Renew Sustain Energy Rev 123:109763. https://doi.org/10.1016/j.rser.2020.109763
Lange J-P (2007) Lignocellulose conversion: an introduction to chemistry, process and economics. Biofuels Bioprod Biorefin 1:39–48. https://doi.org/10.1002/bbb.7
Lateh H, Taweekun J, Maliwan K, Zainal ZA, Rattanawilai S, Thongpat W (2019) Thermo-catalytic decomposition of biomass tar models through microwave irradiation. IOP Conf Ser Mater Sci Eng 505;012129. https://doi.org/10.1088/1757-899X/505/1/012129
Lee K-H, Kang B-S, Park Y-K, Kim J-S (2005) Influence of reaction temperature, pretreatment, and a char removal system on the production of bio-oil from rice straw by fast pyrolysis, using a fluidized bed. Energy Fuels 19:2179–2184. https://doi.org/10.1021/ef050015o
Lee Y, Park J, Ryu C, Gang KS, Yang W, Park Y-K, Jung J, Hyun S (2013) Comparison of biochar properties from biomass residues produced by slow pyrolysis at 500 °C. Biores Technol 148:196–201. https://doi.org/10.1016/j.biortech.2013.08.135
Lehmann J, Joseph S (eds) (2010) Biochar for environmental management: science and technology. Earthscan, London
Li Y, Liu H (2000) High-pressure densiÿcation of wood residues to form an upgraded fuel. Biomass Bioenergy, 10
Libra JA, Ro KS, Kammann C, Funke A, Berge ND, Neubauer Y, Titirici M-M, Fühner C, Bens O, Kern J, Emmerich K-H (2011) Hydrothermal carbonization of biomass residuals: a comparative review of the chemistry, processes and applications of wet and dry pyrolysis. Biofuels 2:71–106. https://doi.org/10.4155/bfs.10.81

Liu Z, Balasubramanian R (2012) Hydrothermal carbonization of waste biomass for energy generation. Procedia Environ Sci 16:159–166. https://doi.org/10.1016/j.proenv.2012.10.022

Luo S, Xiao B, Hu Z, Liu S, Guan Y, Cai L (2010) Influence of particle size on pyrolysis and gasification performance of municipal solid waste in a fixed bed reactor. Biores Technol 101:6517–6520. https://doi.org/10.1016/j.biortech.2010.03.060

Lyu H, Zhang Q, Shen B (2020) Application of biochar and its composites in catalysis. Chemosphere 240:124842. https://doi.org/10.1016/j.chemosphere.2019.124842

Mamvura TA (2020) Biomass torrefaction as an emerging technology to aid in energy production, 17

Mani S, Tabil LG, Sokhansanj S (2006) Specific energy requirement for compacting corn stover. Bioresour Technol 7

Manyà JJ, Azuara M, Manso JA (2018) Biochar production through slow pyrolysis of different biomass materials: Seeking the best operating conditions. Biomass Bioenerg 117:115–123. https://doi.org/10.1016/j.biombioe.2018.07.019

Marker TL, Felix LG, Linck MB, Roberts MJ (2012) Integrated hydropyrolysis and hydroconversion (IH2) for the direct production of gasoline and diesel fuels or blending components from biomass, part 1: Proof of principle testing. Environ Prog Sustain Energy 31:191–199. https://doi.org/10.1002/ep.10629

McKendry P (2002a) Energy production from biomass (part 2): conversion technologies. Biores Technol 83:47–54. https://doi.org/10.1016/S0960-8524(01)00119-5

McKendry P (2002b) Energy production from biomass (part 3): gasification technologies. Biores Technol 83:55–63. https://doi.org/10.1016/S0960-8524(01)00120-1

Mehetre SA, Panwar NL, Sharma D, Kumar H (2017) Improved biomass cookstoves for sustainable development: a review. Renew Sustain Energy Rev 73:672–687. https://doi.org/10.1016/j.rser.2017.01.150

Mekonnen MM, Romanelli TL, Ray C, Hoekstra AY, Liska AJ, Neale CMU (2018) Water, energy, and carbon footprints of bioethanol from the U.S. and Brazil. Environ Sci Technol 52:14508–14518. https://doi.org/10.1021/acs.est.8b03359

Melligan F, Hayes MHB, Kwapinski W, Leahy JJ (2012) Hydro-pyrolysis of biomass and online catalytic vapor upgrading with Ni-ZSM-5 and Ni-MCM-41. Energy Fuels 26:6080–6090. https://doi.org/10.1021/ef301244h

Miller RS, Bellan J (1997) A Generalized biomass pyrolysis model based on superimposed cellulose, hemicelluloseand liqnin kinetics. Combust Sci Technol 126:97–137. https://doi.org/10.1080/00102209708935670

Mohan D, Pittman CU, Steele PH (2006) Pyrolysis of wood/biomass for bio-oil: a critical review. Energy Fuels 20:848–889. https://doi.org/10.1021/ef0502397

Morf P, Hasler P, Nussbaumer T (2002) Mechanisms and kinetics of homogeneous secondary reactions of tar from continuous pyrolysis of wood chips, p 11

Morrin S, Lettieri P, Chapman C, Mazzei L (2012) Two stage fluid bed-plasma gasification process for solid waste valorisation: Technical review and preliminary thermodynamic modelling of sulphur emissions. Waste Manage 32:676–684. https://doi.org/10.1016/j.wasman.2011.08.020

Mosier N (2005) Features of promising technologies for pretreatment of lignocellulosic biomass. Biores Technol 96:673–686. https://doi.org/10.1016/j.biortech.2004.06.025

Mullen CA, Boateng AA (2011) Characterization of water insoluble solids isolated from various biomass fast pyrolysis oils. J Anal Appl Pyrol 90:197–203. https://doi.org/10.1016/j.jaap.2010.12.004

Nanda S, Dalai AK, Berruti F, Kozinski JA (2016) Biochar as an exceptional bioresource for energy, agronomy, carbon sequestration, activated carbon and specialty materials. Waste Biomass Valorization 7:201–235. https://doi.org/10.1007/s12649-015-9459-z

Ndiema CKW, Manga PN, Ruttoh CR (2002) Influence of die pressure on relaxation characteristics of briquetted biomass. Energy Convers Manage 43:2157–2161. https://doi.org/10.1016/S0196-8904(01)00165-0

Neves D, Thunman H, Matos A, Tarelho L, Gómez-Barea A (2011) Characterization and prediction of biomass pyrolysis products. Prog Energy Combust Sci 37:611–630. https://doi.org/10.1016/j.pecs.2011.01.001

Novotny EH, Carvalho M, Madari BE (2015) Biochar: pyrogenic carbon for agricultural use—a critical review, p 24

Oasmaa A, Peacocke C, Gust S, Meier D, McLellan R (2005) Norms and standards for pyrolysis liquids. End-user requirements and specifications. Energy Fuels 19:2155–2163. https://doi.org/10.1021/ef040094o

Oasmaa A, Kuoppala E, Gust S, Solantausta Y (2003) Fast pyrolysis of forestry residue. 1. Effect of extractives on phase separation of pyrolysis liquids. Energy Fuels 17:1–12. https://doi.org/10.1021/ef020088x

Pattiya A (2018) Fast pyrolysis. In: Direct thermochemical liquefaction for energy applications. Elsevier, pp. 3–28

Pilon G, Lavoie J-M (2013) Pyrolysis of switchgrass (Panicum virgatum L.) at low temperatures within N_2 and CO_2 environments: product yield study. ACS Sustain Chem Eng 1:198–204. https://doi.org/10.1021/sc300098e

Prins MJ, Ptasinski KJ (2006) Torrefaction of wood Part 2. Analysis of products. J Anal Appl Pyrolysis 6

Puligundla P, Oh S-E, Mok C (2016) Microwave-assisted pretreatment technologies for the conversion of lignocellulosic biomass to sugars and ethanol: a review. Carbon Lett 17:10

Ragauskas AJ (2006) The path forward for biofuels and biomaterials. Science 311:484–489. https://doi.org/10.1126/science.1114736

Ranzi E, Cuoci A, Faravelli T, Frassoldati A, Migliavacca G, Pierucci S, Sommariva S (2008) Chemical kinetics of biomass pyrolysis, 9

Ravindra K, Kaur-Sidhu M, Mor S, John S (2019) Trend in household energy consumption pattern in India: a case study on the influence of socio-cultural factors for the choice of clean fuel use. J Clean Prod 213:1024–1034. https://doi.org/10.1016/j.jclepro.2018.12.092

Reza MT, Andert J, Wirth B, Busch D, Pielert J, Lynam JG, Mumme J (2014) Hydrothermal carbonization of biomass for energy and crop production. Appl Bioenergy 1. https://doi.org/10.2478/apbi-2014-0001

Scarlat N, Dallemand J-F, Fahl F (2018) Biogas: developments and perspectives in Europe. Renew Energy 129:457–472. https://doi.org/10.1016/j.renene.2018.03.006

Sen S, Ganguly S (2017) Opportunities, barriers and issues with renewable energy development—a discussion. Renew Sustain Energy Rev 69:1170–1181. https://doi.org/10.1016/j.rser.2016.09.137

Shankar Tumuluru J, Sokhansanj S, Hess JR, Wright CT, Boardman RD (2011) Review: a review on biomass torrefaction process and product properties for energy applications. Ind Biotechnol 7:384–401. https://doi.org/10.1089/ind.2011.7.384

Sharma S, Basu S, Shetti NP, Aminabhavi TM (2020) Waste-to-energy nexus for circular economy and environmental protection: recent trends in hydrogen energy. Sci Total Environ 713:136633. https://doi.org/10.1016/j.scitotenv.2020.136633

Shen J, Wang X-S, Garcia-Perez M, Mourant D, Rhodes MJ, Li C-Z (2009) Effects of particle size on the fast pyrolysis of oil mallee woody biomass. Fuel 88:1810–1817. https://doi.org/10.1016/j.fuel.2009.05.001

Shrotri A, Kobayashi H, Fukuoka A (2017) Catalytic conversion of structural carbohydrates and lignin to chemicals. In: Advances in catalysis. Elsevier, pp. 59–123

Sonnenberg A, Baars J, Hendrickx P (2007) IEA bioenergy task 42 biorefinery. Avantium, Biomass Research and Wageningen University and Research Centre, Wageningen, The Netherlands

Sri SS, Palanivelu K, Ramachandran A, Raghavan V (2020) Biochar from biomass waste as a renewable carbon material for climate change mitigation in reducing greenhouse gas emissions—a review. Biomass Convers Biorefin. https://doi.org/10.1007/s13399-020-00604-5

Stančin H, Mikulčić H, Wang X, Duić N (2020) A review on alternative fuels in future energy system. Renew Sustain Energy Rev 128:109927. https://doi.org/10.1016/j.rser.2020.109927

Stemann J, Erlach B, Ziegler F (2013) Hydrothermal carbonisation of empty palm oil fruit bunches: laboratory trials, plant simulation, carbon avoidance, and economic feasibility. Waste Biomass Valorization 4:441–454. https://doi.org/10.1007/s12649-012-9190-y

Sundaram V, Muthukumarappan K, Kamireddy SR (2015) Effect of ammonia fiber expansion (AFEX^{TM}) pretreatment on compression behavior of corn stover, prairie cord grass and switchgrass. Ind Crops Prod 74:45–54. https://doi.org/10.1016/j.indcrop.2015.04.027

Thangalazhy-Gopakumar S, Adhikari S, Gupta RB, Tu M, Taylor S (2011) Production of hydrocarbon fuels from biomass using catalytic pyrolysis under helium and hydrogen environments. Biores Technol 102:6742–6749. https://doi.org/10.1016/j.biortech.2011.03.104

Tripathi M, Sahu JN, Ganesan P (2016) Effect of process parameters on production of biochar from biomass waste through pyrolysis: a review. Renew Sustain Energy Rev 55:467–481. https://doi.org/10.1016/j.rser.2015.10.122

Vamvuka D (2011) Bio-oil, solid and gaseous biofuels from biomass pyrolysis processes—an overview. Int J Energy Res 35:835–862. https://doi.org/10.1002/er.1804

Van de Velden M, Baeyens J, Brems A, Janssens B, Dewil R (2010) Fundamentals, kinetics and endothermicity of the biomass pyrolysis reaction. Renew Energy 35:232–242. https://doi.org/10.1016/j.renene.2009.04.019

Wang X, Qin G, Chen M, Wang J (2016) Microwave-assisted pyrolysis of cotton stalk with additives. BioResources 11;6125–6136. https://doi.org/10.15376/biores.11.3.6125-6136

Weber K, Quicker P (2018) Properties of biochar. Fuel 217:240–261. https://doi.org/10.1016/j.fuel.2017.12.054

Xiao G, Ni M, Huang H, Chi Y, Xiao R, Zhong Z, Cen K (2007) Fluidized-bed pyrolysis of waste bamboo. J Zhejiang Univ Sci A 8:1495–1499. https://doi.org/10.1631/jzus.2007.A1495

Xu C, Etcheverry T (2008) Hydro-liquefaction of woody biomass in sub- and super-critical ethanol with iron-based catalysts. Fuel 87:335–345. https://doi.org/10.1016/j.fuel.2007.05.013

Yaashikaa PR, Senthil Kumar P, Varjani SJ, Saravanan A (2019) Advances in production and application of biochar from lignocellulosic feedstocks for remediation of environmental pollutants. Biores Technol 292:122030. https://doi.org/10.1016/j.biortech.2019.122030

Yin C (2012) Microwave-assisted pyrolysis of biomass for liquid biofuels production. Biores Technol 120:273–284. https://doi.org/10.1016/j.biortech.2012.06.016

Yu Y, Zeng Y, Zuo J, Ma F, Yang X, Zhang X, Wang Y (2013) Improving the conversion of biomass in catalytic fast pyrolysis via white-rot fungal pretreatment. Biores Technol 134:198–203. https://doi.org/10.1016/j.biortech.2013.01.167

Yu H, Zhang Z, Li Z, Chen D (2014) Characteristics of tar formation during cellulose, hemicellulose and lignin gasification. Fuel 118:250–256. https://doi.org/10.1016/j.fuel.2013.10.080

Zadeh ZE, Abdulkhani A, Aboelazayem O, Saha B (2020) Recent insights into lignocellulosic biomass pyrolysis: a critical review on pretreatment, characterization, and products upgrading. Processes 8:799. https://doi.org/10.3390/pr8070799

Zaker A, Chen Z, Wang X, Zhang Q (2019) Microwave-assisted pyrolysis of sewage sludge: a review. Fuel Process Technol 187:84–104. https://doi.org/10.1016/j.fuproc.2018.12.011

Zaman CZ, Pal K, Yehye WA, Sagadevan S, Shah ST, Adebisi GA, Marliana E, Rafique RF, Johan RB (2017) Pyrolysis: a sustainable way to generate energy from waste, p 34

Zhang L, Hu X, Hu K, Hu C, Zhang Z, Liu Q, Hu S, Xiang J, Wang Y, Zhang S (2018) Progress in the reforming of bio-oil derived carboxylic acids for hydrogen generation. J Power Sources 403:137–156. https://doi.org/10.1016/j.jpowsour.2018.09.097

Zhang Y, Ma Z, Zhang Q, Wang J, Ma Q, Yang Y, Luo X, Zhang W (2017) Comparison of the physicochemical characteristics of bio-char pyrolyzed from Moso Bamboo and Rice Husk with different pyrolysis temperatures. BioResources 12:4652–4669. https://doi.org/10.15376/biores.12.3.4652-4669

Chapter 8
Perspectives of Agro-Waste Biorefineries for Sustainable Biofuels

M. S. Dhanya

8.1 Introduction

The global production of crop residues throughout the year is in surplus quantity, but the major portion has been remained unutilized (Prasad et al. 2020; Kosre et al. 2021; Singh et al. 2020). The judicious management of the agro-residues is very much needed to address associated environmental and health concerns (Chen et al. 2020; Sharma et al. 2020a; Klai et al. 2021). The immense potential of biomass residues for various purposes are well-documented (Prasad et al. 2020; Daioglou et al. 2016; Honorato-Salazar and Sadhukhan 2020; Xu et al. 2020; Venkatramanan et al. 2021a). The biomass wastes generated from agriculture sector is mainly from post harvesting of crops and post processing (Sadh et al. 2018). The abundance, renewability, biodegradability and inexpensiveness make the agricultural crop wastes and agro-processing wastes as a viable feedstock for eco-friendly and sustainable products (Nguyen 2017; Duque-Acevedo et al. 2020; Ibitoye et al. 2021). The biorefineries based on agro-residues were reported by many researchers (Serna-Loaiza et al. 2018; Fito et al. 2019; El-Ramady et al. 2020; Redondo-Gómez et al. 2020).

The major lignocellulosic crop residues in Asia are generated from rice, wheat and maize, in Europe and North America are from wheat and maize respectively (Kim and Dale 2004; Sharma et al. 2020b). The major residues generated from the crop harvesting are husk, straw, leaves and stalks (Deshwal et al. 2021). The agricultural processing has also added to agro-residue production by means of husk, hull, shell, bagasse and fibres (Go et al. 2019). The small intervals between harvesting of one crop and sowing of next crop, lack of land for post harvesting and post processing crop residue disposal, long time required for biodegradation and less suitability as fodder exert pressure on farmers to choose more easier method of open burning of

M. S. Dhanya (✉)
Department of Environmental Science and Technology, Central University of Punjab, Bathinda, Punjab 151401, India
e-mail: ms.dhanya@cup.edu.in

Y. K. Nandabalan et al. (eds.), *Zero Waste Biorefinery*, Energy, Environment, and Sustainability, https://doi.org/10.1007/978-981-16-8682-5_8

crop residues (Pandiyan et al. 2019; Dai et al. 2020). The effective and sustainable utilization of agro-residues is urgently required to reduce the impacts of open burning and emissions of Greenhouse gases from open dumping (Bhardwaj et al. 2019; Prasad et al. 2020). The developing countries are more benefitted from the use of agro-residues with improved air quality and ensured energy security (Ullah et al. 2015; Naqvi et al. 2020). These secondary feedstock at the harvesting or processing site saves transportation and handling cost compared to first generation raw materials making it a more potential candidate for production of bioenergy and biochemicals (Hassan et al. 2019; Raud et al. 2019; Kumar et al. 2020a). The development of effective biorefinery approach with technologies enable the agrarian countries rich in agricultural residues to achieve food and fuel sustainability (Nanda et al. 2015; Solarte-Toro and Alzate 2021). This chapter aimed at the perspectives and challenges of agro-residues for biorefinery of biofuels and biochemicals with main focus on India with different production techniques.

8.2 Second Generation Lignocellulosic Biorefinery

The conflict on food-fuel made the focus stronger on the second generation biorefineries over the biorefineries depend on first generation feedstock (Sadhukhan et al. 2018; Olguin-Maciel et al. 2020; Patel and Shah 2021). India being an agrarian country huge scope lies in the utilization of lignocellulosic agro-wastes for biorefinery (Banu et al. 2021; Govil et al. 2020). The estimated crop residue production in our country as per Ministry of New and Renewable Energy is around 500 million tonnes with more than 60% from rice, wheat, sugarcane and cotton (MNRE 2014). The annual surplus of biomass available for bioenergy production was 34% of the total crop residue generation with of 234 MT (Hiloidhari et al. 2014). The crop residue production varies with the states in India with Uttar Pradesh at topmost position (NPMCR 2014).

The lignocellulosic residues produced from crops after harvesting and conversion to various products by processing in India is given in Fig. 8.1. The major wastes produced after the post harvesting of crops are straw or stalks, for the cereals, pulses

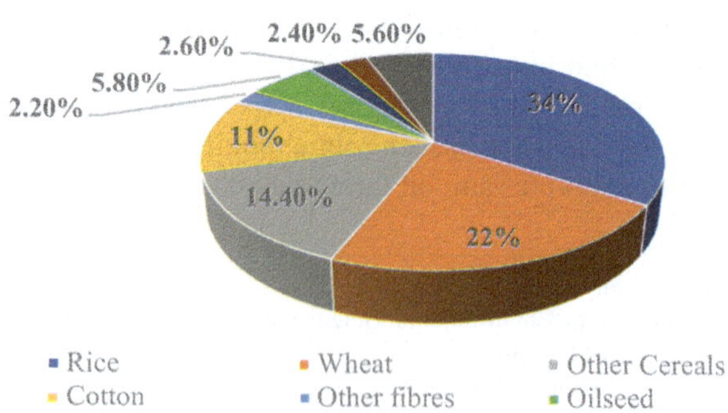

Fig. 8.1 Percent contribution of major crops to total residue generation in India. Calculated from MNRE (2009)

and oilseed crops. The post-processing of crops had generated residues such as husk, bagasse and pressed cakes. Hiloidhari et al. (2014) estimated the bioenergy potential of 26 crops cultivated in India from its 39 residues. Chowdhury et al. (2018) evaluated the scope of straw and husk of rice, straw of wheat, press cakes of oilseeds, bagasse from sugarcane, coconut shells, pseudo-stem of banana and banana peels for second generation biorefineries. The bioethanol production potential of 5.42 billion litres were reported from the straw of rice, wheat and maize and bagasse of sugarcane with its 50% conversion efficiency and estimated as surplus to meet 10% bioethanol blending (Sukumaran et al. 2010).

The Fig 8.2 shows common lignocellulosic residues generated from major crops in India after harvesting and processing. The leaves, stalk, husk, fruit bunches and vegetable peels have low economic value but has immense potential for production of various value-added products (Sadh et al. 2018; Kumar and Yaashikaa 2020; Molina-Guerrero et al. 2020; Jusakulvijit et al. 2021).

The integrated approach based biorefineries for the utilization of agro-wastes for biofuels, bioenergy, and other high-value chemicals have been needed for sustainable development (López-Molina et al. 2020; Philippini et al. 2020; Leong et al. 2021).

8.3 Composition of Agro-Residues

The constituents of agro-residues play an important role in biorefinery for biofuel and biochemical production. The major components of these lignocellulosic wastes are cellulose, hemicellulose (xyloglucan, rhamnogalactan or glucurono-arabinoxylan), lignin, proteins, sugars soluble in water, resins and pigments soluble in ethers and alcohols (Kumar et al. 2020b; Rishikesh et al. 2021; Šelo et al. 2021). The structural components of agro-residues differ depending on genetic and environmental factors (Kosre et al. 2021; Paschos et al. 2020). These lignocellulosic materials (Fig. 8.1) had major fraction of cellulose followed by hemicellulose and lignin (Fig. 8.3). The cellulosic micro-fibrils in bundles provide cell wall strength and flexibility which was covered by resistant lignin with hemicellulose in between (Berglund et al. 2020).

The agro-residues had 25–50% of cellulose, with its monomer glucose having β-1, 4 linkage (Rishikesh et al. 2021; Bhatnagar et al. 2020; Lojkova et al. 2020); 15–35% of hemicellulose, a heteropolysaccharide contains D-glucose, D-xylose, D-galactose, D-mannose and L-arabinose (Kumla et al. 2020; Huang et al. 2021; Liu et al. 2021) and 12–20% of lignin consists of phenolic biopolymer with 4-hydroxyphenyl, guaiacyl, and syringyl subunits (De et al. 2020; Domínguez-Robles et al. 2021). The structural composition of residues from major crops and after processing are compiled from Gupte and Madamwar (1997); Prasad et al. (2007); Furlan et al. (2012); Kumar et al. (2018a); Nasir et al. (2019); Ibarra-Díaz et al. (2020); Kumar et al. (2020b) and summarized in Table 8.1.

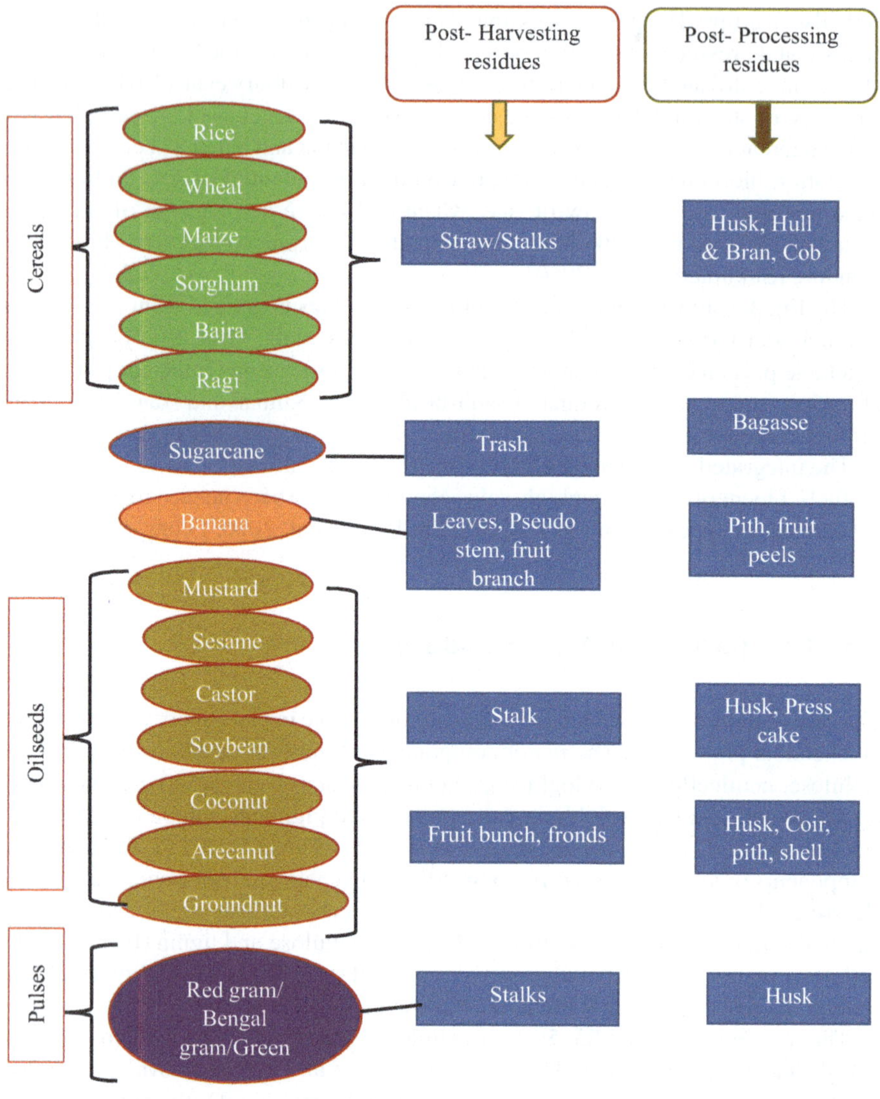

Fig. 8.2 Lignocellulosic residues from major crops in India after harvesting and processing (Singh and Gu 2010; Kumar et al. 2015; López-Molina et al. 2020)

8.4 Environmental Benefits of Second Generation Biorefinery from Agro-Residues

The fossil fuel burning and biomass burning are major anthropogenic contributors to air pollution, climate change and global warming (IPCC 2018; Bhattacharyya and Barman 2018; Fagodiya et al. 2019; Kaushal and Prashar 2021). The emissions

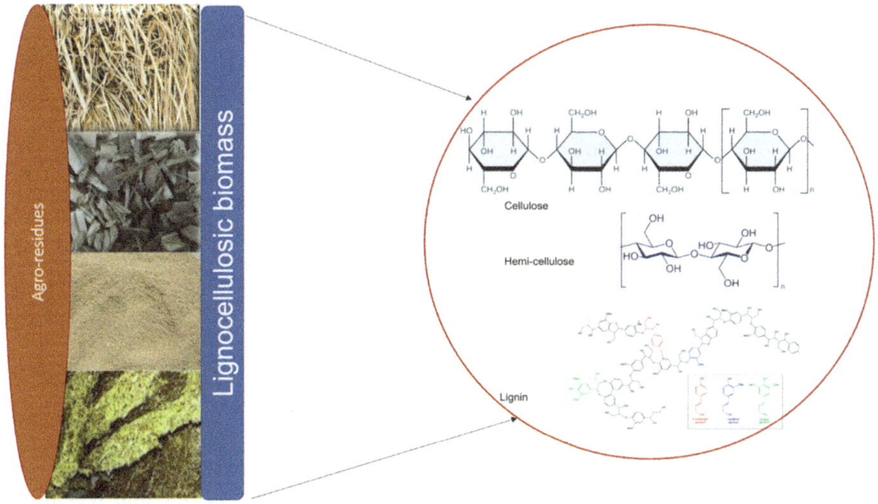

Fig. 8.3 Major constituents of agro-residues

Table 8.1 Composition of common agricultural residues

Agro-residues	Cellulose (%)	Hemicellulose (%)	Lignin (%)
Rice straw	26–40	18–28	12–14
Wheat straw	33–41	20–32	13–20
Maize stover	38–40	28	7–21
Barley straw	31–45	27–38	14–19
Nut shells	25–30	25–30	30–40
Sorghum straw	32	24	13
Bagasse from sugarcane	39	37	21
Bagasse from Sweet sorghum	34–45	18–28	14–22
Rice husk	35–40	15–20	20–25
Rice bran	34	28.2	24.8
Corn cobs	45	35	15
Wheat husk	36	18	16
Rye husk	26	16	13

of methane and nitrous oxide from crop residue burning in India during 2017–18 were 176.1 Tg and 313.9 Gg of CO_2 and CH_4 respectively (Venkatramanan et al. 2021b). The open biomass burning contributes to around 40% of total emissions of carbon dioxide and 38% of tropospheric ozone along with ammonia, volatile organic compounds and particulate matter (Levine 1991; Bhattacharyya and Barman 2018;

Table 8.2 Emission pattern (g/km) of biofuel blends in automotive fuel

Pollutant	Ethanol		Biodiesel		Biogas	Diesel
	22% ethanol blend	100% Ethanol	10% biodiesel	15% biodiesel		
Carbon monoxide	0.76	0.65	0.65	0.62	0.08	0.77
NO_x	0.45	0.34	0.83	0.89	–	0.79
Particulate matter	0.08	0.02	0.093	0.08	0.015	0.129
Unburned Hydrocarbons	0.004	0.02	0.22	0.16	0.35	0.37

Compiled from Prasad et al. (2007), Demirbas and Karslioglu 2007)

Fagodiya et al. 2019; Kumari et al. 2018; Bali et al. 2021). The 80% of total crop residue burned in 2017–18 in India was generated from straw of rice and wheat and sugarcane trash (Venkatramanan et al. 2021b). The burning of residue from rice was reported highest from Punjab, Haryana and Western Uttar Pradesh (NPMCR 2014; Porichha et al. 2021).

The energy consumption in India was 20.5 million tonnes in Dec, 2020 making it one of the largest importers of crude oil in the world (Trading.com. 2021). The total energy demand is going to increase further over the years and is projected India will be in top three importing countries by 2025 (Kumar et al. 2020a; Lin and Raza 2020). The use of alternative sources of transportation like biofuel reduce the emissions to the atmosphere and mitigate various environmental concerns (Bhuvaneshwari et al. 2019; Sahoo et al. 2020). The power generation in India from biomass, bagasse and waste to energy accounts 1.6, 0.46 and 0.23% of total potential of renewable energy sources (MoSPI 2020). The judicious management of the agro-wastes as potential feedstock for various biorefineries added value to it as resources (Sadh et al. 2018; Philippini et al. 2020). The possibility of energy recovery from agro-wastes were studied by Vaish et al. (2016) and Borja and Fernández-Rodríguez (2021). Table 8.2 compares the emission pattern of biofuel blends with diesel as automotive fuel. The emission reduction was observed from use of bioethanol, biodiesel and biogas as compared to diesel (Sharma et al. 2020a; Bušić et al. 2018; Turkcan 2020).

The high value -added products from agro-residues are produced in cost effective way (Ibitoye et al. 2021; Kumar et al. 2020a; Bhatnagar et al. 2020). These products found applications in pharmaceuticals, medical and cosmetic industries, chemical, food/feed, and pharmaceuticals (Philippini et al. 2020; Alexandri et al. 2021; Brandão et al. 2021).

8.5 Conversion Technologies in Agro-Waste Biorefinery

Biorefineries are alternative to petro-based refineries with the conversion of biomass to thermo-chemical energy and biobased products such as biopolymer and high-value chemicals (Ubando et al. 2019; Ong and Wu 2020). The conversion routes and technoeconomic feasibility of different biorefinery products mainly focused on biofuels and biochemicals from agro-wastes have been well discussed (Kumar et al. 2020; Kover et al. 2021; Ginni et al. 2021).

The various biorefinery technologies for straw from rice, wheat and maize and agro- processing wastes such as rice husks, vegetable and fruit peels, sugarcane bagasse and oil cakes include fermentation, anaerobic digestion, pyrolysis, gasification and transesterification (Awasthi et al. 2020; Hiloidhari et al. 2020). Further the products are recovered by stabilization, dewatering and refining (Liu et al. 2021; Kumar et al. 2018b; Paul and Dutta 2018). The conversion of underutilized agro-wastes to clean biofuels (Fig. 8.4) by thermal, chemical and biological routes help to meet energy demands and also the climate change mitigation (Prasad et al. 2020; Kumar et al. 2020a; Koytsoumpa et al. 2021).

8.5.1 Bioethanol Biorefinery from Agro-wastes

The global leaders in bioethanol production from agro-wastes are Asia, Europe, and North America (Najafi et al. 2009; Das et al. 2020). The bioethanol production from second generation feedstock is more eco-friendly without compromising on food (Niphadkar et al. 2018). The globally produced agro wastes had high estimated ethanol production potential which was sixteen times more than ethanol produced (Kim and Dale 2004). Holmatov et al. (2021) also estimated potential of residues produced from 123 crops in the world and its replacement power for fossil-based fuel. The ethanol blends with gasoline saves the fossil fuel consumption and improves the environmental quality (Ali et al. 2019; Barbosa et al. 2021). The residue after the fermentation which are lignin rich able to produce electricity and steam (Mathew et al. 2018; Zhao et al. 2021).

The pre-treatment is needed for lignocellulosic ethanol production in comparison to sugary feedstock (Jansen et al. 2017; Mathew et al. 2018). The stages in ethanol production from agro-residues are pre-treatment, saccharification, fermentation of ethanol and finally the distillation (Carpio and de Souza 2017; Nogueira et al. 2020). Bechara et al. (2018) reviewed the sugarcane biorefinery for ethanol and electricity production.

The metabolism of hexoses undergoes phosphorylation and two moles of ethanol is produced (Eq. 8.1) along with carbon dioxide in fermentation process (Ingram et al. 1998). The three molecules of pentose generated (Eq. 8.2) five molecules each of ethanol and CO_2 are produced by utilizing the carbohydrates from the agro-wastes by fermentation.

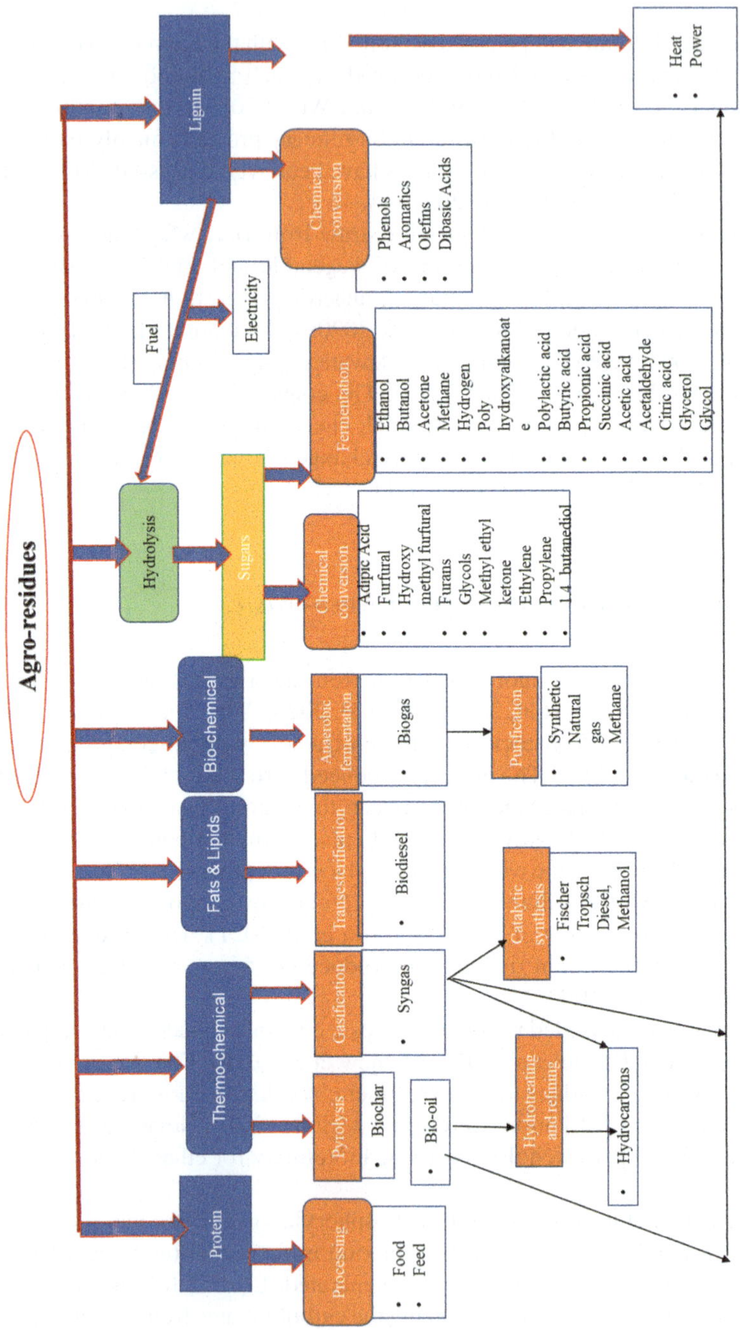

Fig. 8.4 Sustainable biorefinery for agro-residues (Wyman 1990)

$$C_6H_{12}O_6 \rightarrow 2C_2H_5OH + 2CO_2 \quad (8.1)$$

$$3C_5H_{10}O_5 \rightarrow 5\ C_2H_5OH + 5CO_2 + \text{energy (as ATP)} \quad (8.2)$$

Saccharomyces cerevisiae widely ferment and convert C6 carbon sugars to ethanol (Zhang et al. 2016). The yeasts such as *Pichia stipitis, Candida shehatae* and *Pachysolan tannophilus* are reported with ethanol producing capability from pentoses and hexoses (Rastogi and Shrivastava 2017; Ntaikou et al. 2018). The immobilization of *Saccharomyces cerevisiae* and *Pichia stipitis* improved the ethanol production from rice straw pre-treated with microwave assisted ferric chloride by 1.27 times (Lü and Zhou 2015). Watanabe et al. (2012) confirmed ethanol production of 38 g/L from alkali pre-treated rice straw by *Saccharomyces cerevisiae* immobilized on photocrosslinkable resin beads. The ethanol production form sugarcane molasses by *S. cerevisiae* was 85–90% from batch fermentation and found increased to 94.5% in continuous fermentation (Sánchez and Cardona 2008). The schematic representation of process integration (Kumar et al. 2020; Duque et al. 2021) in bioethanol production from agro-wastes are given in Fig. 8.5.

8.5.2 Biobutanol Biorefinery from Agro-Wastes

The butanol based biorefinery from various agricultural residues was studied by Nanda et al. (2015), Pereira et al. (2015) and Pinto et al. (2021).

Qureshi et al. (2007) reported the butanol production of 12 g/L from wheat straw hydrolysate by *Clostridium beijerinckii* P260. The potential of rice bran, fibres of maize, stover from maize, and bagasse from cassava for butanol production were already reported (Baral et al. 2018; Huang et al. 2019). Molina-Guerrero et al. (2020) estimated the sustainable butanol production potential of major residues of maize, wheat, barley and sorghum as well as waste generated from beans, coffee and sugarcane produced in Mexico. The electricity generation from these residues were also reported in high amounts.

Sanchez et al. (2017) demonstrated biorefinery with value added products namely butanol, acetone, ethanol, hydrogen and biogas with the help of mixed culture by consolidated bioprocessing technology. The butanol biorefinery based on sweet sorghum bagasse following fed- batch fermentation was reported by Su et al. (2020). Dutta et al. (2021) demonstrated biobutanol biorefinery along with heat and power from the rice straw. The spent coffee grounds were used for producing butanol and also gallic acid and catechin (López-Linares et al. 2021).

Many researchers studied the enhancement of ABE fermentation by immobilized cell technology (Gupta et al. 2020; Meramo-Hurtado et al. 2021). The immobilization improved butanol or bioethanol tolerance with high viable cell densities, increasing fermentation rate and productivity (Jiang et al. 2009).

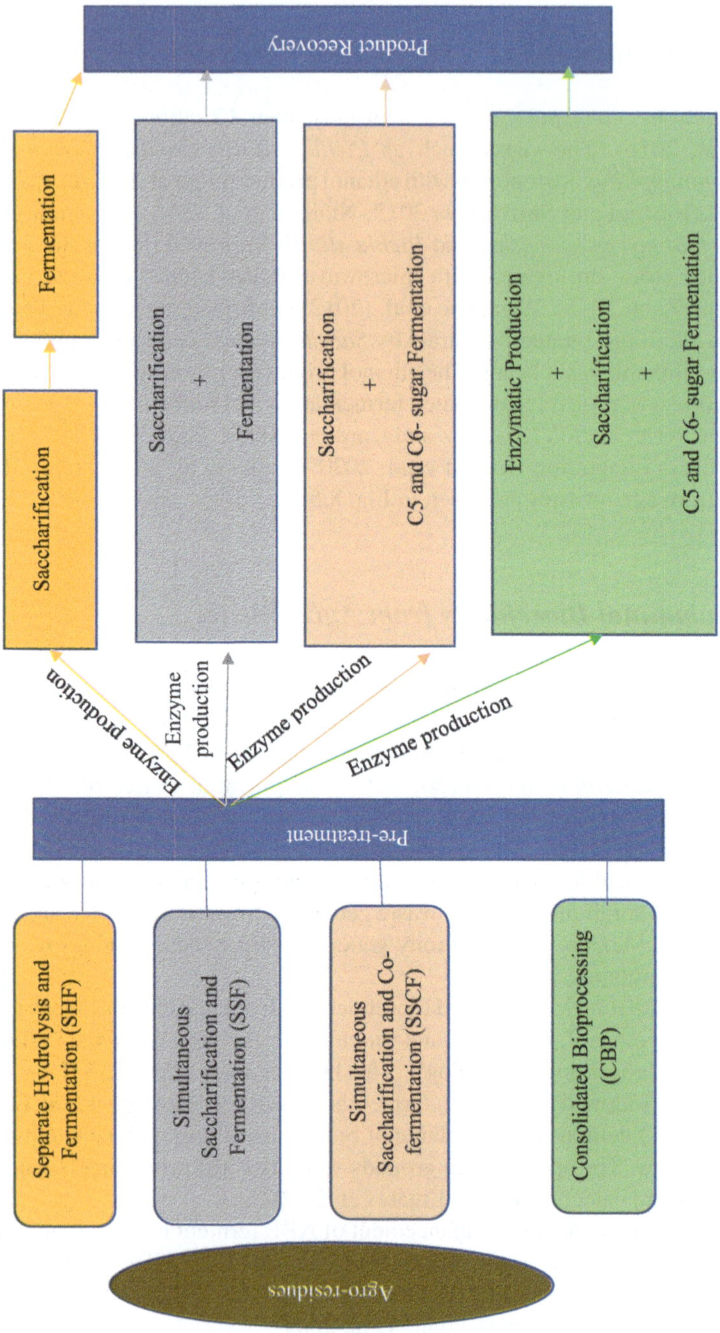

Fig. 8.5 Different steps in conversion processes for bio-alcohol production from agro-residues

Qureshi et al. (2010) demonstrated the butanol production of 14.5 g/L from corn stover with *Clostridium beijerinckii* P260. The corn cob produced 12.3 g/L butanol by *C. saccharobutylicum* DSM 13864 (Dong et al. 2016). The butanol production from wheat straw and barley straw by *C. beijerinckii* P260 was studied by Qureshi et al. (2007) and Qureshi et al. (2014) respectively. The *C. saccharoperbutylacetonicum* N1-4 able to produce 7.72 g/L butanol using rice bran (Al-Shorgani et al. 2012). The alkali pre-treated rice straw improved butanol fermentation to 45.2 g, acetone and ethanol production of 17.7 g and 1.2 g respectively (Moradi et al. 2013). The organosolv pre-treatment of rice straw for butanol production was reported by Amiri et al. (2014) and Tsegaye et al. (2020). Razi and Sasmal (2020) had also used organic solvents as pre-treatment agent for butanol production form groundnut shells.

8.5.3 Biogas from Crop Residues Biorefinery

Various agro-wastes were studied by different researchers for anaerobic digestion and biogas production (Kumar et al. 2018a; Bušić et al. 2018). The lack of direct competition with other conventional renewable feedstocks increased its importance as feedstock for biogas production.

The biogas production from wheat straw was studied by Somayaji and Khanna (1994) and Labatut et al. (2011). Mancini et al. (2018) reported enhancement in biogas generation by chemical pre-treatments from wheat straw. The biogas production of 65 L/kg VS in 23 days (Isci and Demirer 2007) and 149 L/kg in 30 days (Sahito et al. 2013) from cotton stalk were reported. The cotton gin waste was studied for biogas production by Labatut et al. (2011). The cotton stalk was also studied for methane production by Zhang et al. (2018).

Liew et al. (2012) demonstrated biogas production of 81.2 L kg^{-1} and 66.9 L kg^{-1} from corn stover and wheat straw respectively. The whole stillage (85%) codigested with cattle manure (15%) produced biogas of 0.31 L CH_4 g^{-1} VS_{added} (Westerholm et al. 2012) in 45 days. The co-digestion of goat manure with corn stalks, rice straw and wheat straw enhanced biogas production (Zhang et al. 2013). The methane production by co-digestion of wastes from banana, canola, cotton, rice, sugarcane and wheat with buffalo dung was also demonstrated by Sahito et al. (2013). The residues of maize, sorghum and bedding straw from cattle farm were studied for biogas production and found methane yields of 267 L CH_4 kg^{-1} VS (Kalamaras and Kotsopoulos 2014). Afif et al. (2020) demonstrated the improved biogas generation from cotton stalks pre-treated with supercritical carbon dioxide.

The other agro-wastes used for methane production by anaerobic fermentation were food and vegetable wastes, oat straw, olive husk and maize stalks (Paul and Dutta 2018; Nwokolo et al. 2020).

8.5.4 Biohydrogen Biorefinery from Agro-Wastes

The biohydrogen is a renewable, alternative fuel combined with electricity production from fuel cell technology makes it a cleanest and efficient fuel source (Bundhoo 2019). The biochemical route of biohydrogen production from agro-residues was by bio-photolysis and anaerobic fermentative processes, either dark or photo fermentation (Bušić et al. 2018; Anwar et al. 2019). The major microbes associated with biohydrogen production are micro algae, cyanobacteria, photosynthetic and fermentative bacteria (Goswami et al. 2021; Rather and Srivastav 2021). The biohydrogen yield from acetate and butyrate fermentation was 4 mol and 2 mol of hydrogen respectively per molecule of glucose (Vijayaraghavan and Soom 2006). Kannah et al. (2018) demonstrated the efficient biohydrogen production from rice straw by combinative dispersion thermochemical disintegration. Lopez-Hidalgo et al. (2017) demonstrated the biohydrogen production from mixture of cheese whey and wheat straw hydrolysate. Mohanto et al. (2020) reported the fruit waste can be used for hydrogen production by *Clostridium* strain BOH3. Tinpranee et al. (2018) screened blue green algae collected from coastal waters of Thailand for the production of hydrogen.

8.5.5 Biodiesel Biorefinery from Agro-Wastes

The agro-residues with lipid content mainly oilseed crops are suitable feedstock for biodiesel production. The biodiesel is commonly produced by transesterification in the presence of a catalyst and alkali with production of mono-alkyl esters (Bušić et al. 2018; Demirbas and Karslioglu 2007). The stearic acid, oleic acid, linoleic acid and linolenic acid are common fatty acids found in oilseeds with 18 carbon atoms and no, mono, di and tri- unsaturated bonds respectively, and a few oilseeds also have palmitic acid (16:0), lauric acid (12:0) type of fatty acids. The chemical reaction representing the trans-esterification for biodiesel production is shown in Fig. 8.6 as follows:

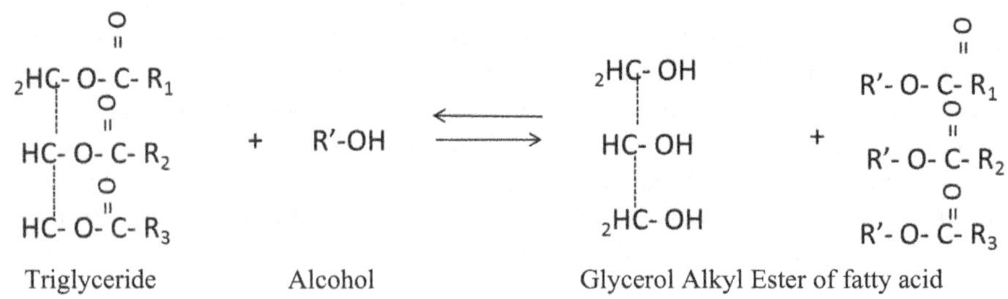

Fig. 8.6 Biodiesel production by transesterification of agro-wastes with oil content (Bušić et al. 2018; Demirbas et al. 2016)

The 'R_1, R_2 and R_3' presented in the reaction are carbon chains of fatty acid and 'R' is the alkyl group of alcohol used in transesterification.

The oilcake produced after oil expulsion also contains oil and can be a suitable raw material for biodiesel production (Jain and Naik 2018; Budžaki et al. 2019). The rice bran, a by-product of rice mill had oil content with biodiesel production potential (Mazaheri et al. 2018; Hoang et al. 2021). Quayson et al. (2020) demonstrated the palm kernel shells and palm oil mill effluent to biodiesel with immobilized lipase from recombinant *Fusarium heterosporum*. Khounani et al. (2021) produced bioethanol, biodiesel, biogas and other bioproducts from safflower biorefinery.

8.5.6 Syngas and Fischer–Tropsch Derivatives from Agro-Waste Biorefinery

The gasification for syngas production from agro-wastes is a thermo-chemical conversion technology and it contains carbon monoxide, hydrogen and other hydrocarbons. The syngas has its direct application in internal combustion engines and for combined heat and power generation. The syngas had been converted to liquid fuel such as methanol, dimethyl ether and hydrogen and hydro-cabons produced by the Fischer–Tropsch process (Cerone and Zimbardi 2018; Dieterich et al. 2020). Fischer–Tropsch process is a polymerization process with catalyst (iron, cobalt or ruthenium) with high temperature Fischer–Tropsch Synthesis (FTS) resulting in short-chain hydrocarbons and light olefins and waxes and long-chain hydrocarbons from low temperature FTS (Yahyazadeh et al. 2021). The gaseous fuels produced from agro-residues are more economical and sustainable (Mallikarjuna et al. 2021).

8.5.7 Value-Added Chemicals from Agro-Wastes

The biorefinery based on biofuels and biochemicals from agro-residues are well studied by various researchers (Bilal et al. 2017; Bhatnagar et al. 2020). The organic acids are an important group of bio-chemicals produced from agro-wastes (Porro and Branduardi 2017; Sheldon 2018). The major organic acids generated by fermentation of lignocellulosics were succinic acid, adipic acids, glutamic acid, aspartic acids, itaconic acid and 3-hydroxypropionic acids (Corato et al. 2018; Misra et al. 2021). The value-added chemicals like sorbitol, p-xylene, hydroxymethyl furfural, furans, levulinic acids were derived from glucose units of cellulosic materials (Pattnaik et al. 2021). The glycerol is produced as a byproduct in transesterification of oil containing agro-residues (Chilakamarry et al. 2021). The biochemicals which can be produced as derivatives of glycerol are propylene glycol, 1,3 propane-di-ol, acrolein and epichlorohydrin (Lima et al. 2021). The lignin derivatives such as syngas, hydrocarbons, phenols, vanillin, vanillic acids and carbon fibre fillers, composites, adhesives, etc. are also value-added products from agro-wastes (Adhikari et al. 2018; Bhatnagar et al. 2020).

8.6 Challenges in Agro-Waste Based Biorefinery

The economic viability is major limitation of biomass based biorefinery (Paul and Dutta 2018; Hassan et al. 2019; Raud et al. 2019; Bhatnagar et al. 2020; Kumar et al. 2020a). The major challenges are

- High capital cost and operational costs
- Biomass supply chain
- Less-developed conversion techniques
- Problems associated with scale up processes
- Constraints involved in down-stream processing.

One another bottleneck in viability of agro-waste based biorefineries is the pre-treatment for fermentation processes (Philippini et al. 2020). The pre-treatment of agro-wastes is essential to reduce recalcitrance of lignocellulosics for enzymatic hydrolysis (Baral et al. 2018; Mancini et al. 2018; Mahmood et al. 2019). The technical hurdles in transportation of agro-wastes to the biorefinery, processing and product recovery were assessed by Chandel et al. (2018) and Kumar et al. (2020a).

The cellulosic ethanol has been commercialized by multi-national companies like Raizen (Piracicaba, Brazil), Gran-bio (S. João dos Milagres, Brazil) and POET-DSM (Emmetsburg, USA). The POET-DSM has also been marketing bio-based succinic acid. The chemicals from cellulosic materials are also produced and commercialized by Dow-DuPont (1,3-Propanediol, 1,4-Butanediol), Clariant-Global bioenergies-INEOS (bio-isobutene), Braskem (Ethylene, polypropylene). The torrefaction, steam explosion and hydrothermal carbonization improved the fuel efficiency of pellets from agro-residues. The advanced techniques like solid-liquid extraction (SLE), and pulsed electric fields (PEF) and membrane and resin techniques were applied for extraction and production of bioactive compounds from agro residues (Nguyen 2017).

The National Policy on biofuels (2018) aimed at a target of achieving 20% of biofuels in our country by 2030. The policy focussed on agro-residues to reach the target mainly by use of damaged food grains of rice and wheat unsuitable for human consumption to be diverted to fermentation process for producing ethanol. This will meet the requirement of ethanol in blending with gasoline and helps in mitigating environmental pollution as well as attaining energy security.

The biorefineries are seemed to be more sustainable for biochemical production than the biofuels.

8.7 Conclusion

The agro-residues from the crop harvesting and processing mainly composed of cellulose, hemicellulose and lignin polymers. These compounds by means of thermo-chemical and biochemical conversion routes are able to convert to biofuels and biochemicals. The fermentation technology is the highly recommended technique involved in the production of bioethanol, biobutanol, biohydrogen and biogas. The gasification process had syngas as products that can further transformed to bio-chemicals. The oil containing agro wastes are good source for biodiesel and glycerol production. The biorefineries based on agro residues are second generation biorefineries and reduce the negative impacts on environment and help to move towards sustainable development. The advanced and effective technologies with an integrated and zero-waste approach overcome the technical and economic challenges presently faced by the agro-waste based biorefineries and enable towards commercialization.

References

Adhikari S, Nam H, Chakraborty JP (2018) Conversion of solid wastes to fuels and chemicals through pyrolysis. In: Bhaskar T, Pandey A, Venkata Mohan S, Lee DJ, Khanal SK (eds) Waste biorefinery. Elsevier, pp.239–263. https://doi.org/10.1016/B978-0-444-63992-9.00008-2

Afif RA, Wendland M, Amon T, Pfeifer C (2020) Supercritical carbon dioxide enhanced pre-treatment of cotton stalks for methane production. Energy 194:116903. https://doi.org/10.1016/j.energy.2020.116903

Alexandri M, Maina S, Tsouko E, Papapostolou H, Koutinas A, Kourmentza K (2021) Valorization of fruit processing by-product streams into integrated biorefinery concepts: extraction of value-added compounds and bioconversion to chemicals. In: Bhat R (ed) Valorization of agri-food wastes and by-products. Academic Press, pp 927–945. https://doi.org/10.1016/B978-0-12-824044-1.00014-3

Ali M, Saleem M, Khan Z, Watson IA (2019) The use of crop residues for biofuel production. In: Verma D, Fortunati E, Jain S, Zhang X (eds) In Woodhead Publishing Series in Composites science and engineering, biomass, biopolymer-based materials, and bioenergy. Woodhead Publishing, pp 369–395. https://doi.org/10.1016/B978-0-08-102426-3.00016-3

Al-Shorgani NK, Kalil MS, Yusoff WM (2012) Biobutanol production from rice bran and de-oiled rice bran by Clostridium saccharoperbutylacetonicum N1–4. Bioprocess Biosyst Eng 35(5):817–826. https://doi.org/10.1007/s00449-011-0664-2

Amiri H, Karimi K, Zilouei H (2014) Organosolv pre-treatment of rice straw for efficient acetone, butanol, and ethanol production. Bioresour Technol 152:450–456. https://doi.org/10.1016/j.biortech.2013.11.038

Anwar M, Lou S, Chen L, Li H, Hu Z (2019) Recent advancement and strategy on bio-hydrogen production from photosynthetic microalgae. Bioresour Technol 292:121972. https://doi.org/10.1016/j.biortech.2019.121972

Awasthi MK, Sarsaiya S, Patel A, Juneja A, Singh RP, Yan B, Awasthi SK, Jain A, Liu T, Duan Y, Pandey A, Zhang Z, Taherzadeh MJ (2020) Refining biomass residues for sustainable energy and bio-products: An assessment of technology, its importance, and strategic applications in circular bio-economy. Renew Sustain Energy Rev 127:109876. https://doi.org/10.1016/j.rser.2020.109876

Bali K, Kumar A, Chourasiya S (2021) Emission estimates of trace gases (VOCs and NOx) and their reactivity during biomass burning period (2003–2017) over Northeast India. J Atmos Chem 78:17–34. https://doi.org/10.1007/s10874-020-09413-6

Banu JR, Preethi, Kavitha S, Tyagi VK, Gunasekaran M, Karthikeyan OP, Kumar G (2021) Lignocellulosic biomass based biorefinery: A successful platform towards circular bioeconomy. Fuel 302:121086. https://doi.org/10.1016/j.fuel.2021.121086.

Baral NR, Quiroz-Arita C, Bradley TH (2018) Probabilistic lifecycle assessment of butanol production from corn stover using different pre-treatment methods. Environ Sci Technol 52(24):14528–14537. https://doi.org/10.1021/acs.est.8b05176

Barbosa TP, Eckert JJ, Roso VR, Pujatti FJP, da Silva LAR, Horta Gutiérrez JC (2021) Fuel saving and lower pollutants emissions using an ethanol-fueled engine in a hydraulic hybrid passengers vehicle. Energy 235:121361. https://doi.org/10.1016/j.energy.2021.121361

Bechara R, Gomez A, Saint-Antonin V, Schweitzer J, Maréchal F, Ensinas A (2018) Review of design works for the conversion of sugarcane to first and second-generation ethanol and electricity. Renew Sustain Energy Rev 91:152–164. https://doi.org/10.1016/j.rser.2018.02.020

Berglund J, Mikkelsen D, Flanagan BM, Dhital S, Gaunitz S, Henriksson G, Lindström ME, Yakubov GE, Gidley MJ, Vilaplana F (2020) Wood hemicelluloses exert distinct biomechanical contributions to cellulose fibrillar networks. Nat Commun 11:4692. https://doi.org/10.1038/s41467-020-18390-z

Bhardwaj S, Bhardwaj NK, Negi YS (2019) Cleaner approach for improving the papermaking from agro and hardwood blended pulps using biopolymers. J Clean Prod 213:134–142. https://doi.org/10.1016/j.jclepro.2018.12.143

Bhatnagar A, Tolvanen H, Konttinen J (2020) Potential of stepwise pyrolysis for on-site treatment of agro-residues and enrichment of value-added chemicals. Waste Manage 118:667–676. https://doi.org/10.1016/j.wasman.2020.09.022

Bhattacharyya P, Barman D (2018) Crop residue management and greenhouse gases emissions in tropical rice lands. In: Ángeles Muñoz M, Zornoza R (eds) Soil management and climate change. Academic Press, pp. 323–335. https://doi.org/10.1016/B978-0-12-812128-3.00021-5

Bhuvaneshwari S, Hettiarachchi H, Meegoda JN (2019) Crop residue burning in India: policy challenges and potential solutions. Int J Environ Res Public Health 16(5):832. https://doi.org/10.3390/ijerph16050832

Bilal M, Asgher M, Iqbal HMN, Hu H, Zhang X (2017) Biotransformation of lignocellulosic materials into value-added products—a review. Int J Biol Macromol 98:447–458. https://doi.org/10.1016/j.ijbiomac.2017.01.133

Borja R, Fernández-Rodríguez MJ (2021) Energy recovery as added value from food and agricultural solid wastes. In: Kumar S, Kumar R, Pandey A (eds) Current developments in biotechnology and bioengineering. Elsevier, pp. 151–174. https://doi.org/10.1016/B978-0-12-821009-3.00007-5

Brandão AS, Gonçalves A, Santos JMRCA (2021) Circular bioeconomy strategies: From scientific research to commercially viable products. J Clean Prod 295:126407. https://doi.org/10.1016/j.jclepro.2021.126407

Budžaki S, Sundaram S, Tišma M, Hessel V (2019) Cost analysis of oil cake-to-biodiesel production in packed-bed micro-flow reactors with immobilized lipases. J Biosci Bioeng 128(1):98–102. https://doi.org/10.1016/j.jbiosc.2019.01.004

Bundhoo ZMA (2019) Potential of bio-hydrogen production from dark fermentation of crop residues: a review. Int J Hydrog Energy 44(32):17346–17362. https://doi.org/10.1016/j.ijhydene.2018.11.098

Bušić A, Kundas S, Morzak G, Belskaya H, Marđetko N, Ivančić Šantek M, Komes D, Novak S, Šantek B (2018) Recent trends in biodiesel and biogas production. Food Technol Biotechnol 56(2):152–173. https://doi.org/10.17113/ftb.56.02.18.5547

Carpio LGT, de Souza FS (2017) Optimal allocation of sugarcane bagasse for producing bioelectricity and second generation ethanol in Brazil: Scenarios of cost reductions. Renew Energy 111:771–780. https://doi.org/10.1016/j.renene.2017.05.015

Cerone N, Zimbardi F (2018) Gasification of agro-residues for syngas production. Energies 11(5):1280. https://doi.org/10.3390/en11051280

Chandel AK, Garlapati VK, Singh AK, Antunes FAF, da Silva SS (2018) The path forward for lignocellulose biorefineries: bottlenecks, solutions, and perspective on commercialization. Bioresour Technol 264:370–381. https://doi.org/10.1016/j.biortech.2018.06.004

Chen P, Kang S, Tripathee L, Panday AK, Rupakheti M, Rupakheti D, Zhang Q, Guo J, Li C, Pu T (2020) Severe air pollution and characteristics of light-absorbing particles in a typical rural area of the Indo-Gangetic Plain. Environ Sci Pollut Res 27:10617–10628. https://doi.org/10.1007/s11356-020-07618-6

Chilakamarry CR, Mimi Sakinah AM, Zularisam AW, Pandey A, Vo DVN (2021) Technological perspectives for utilisation of waste glycerol for the production of biofuels: a review. Environ Technol Innov 24:101902. https://doi.org/10.1016/j.eti.2021.101902

Chowdhury R, Ghosh S, Debnath B, Manna D (2018) Indian agro-wastes for 2G biorefineries: strategic decision on conversion processes. In: Sustainable energy technology and policies, pp 353–373. https://doi.org/10.1007/978-981-10-7188-1_16

Corato UD, Bari ID, Viola E, Pugliese M (2018) Assessing the main opportunities of integrated biorefining from agro-bioenergy co/by-products and agroindustrial residues into high-value added products associated to some emerging markets: a review. Renew Sustain Energy Rev 88:326–346. https://doi.org/10.1016/j.rser.2018.02.041

Dai Y, Zheng H, Jiang Z, Xing B (2020) Comparison of different crop residue-based technologies for their energy production and air pollutant emission. Sci Total Environ 707:136122. https://doi.org/10.1016/j.scitotenv.2019.136122

Daioglou V, Stehfest E, Wicke B, Faaij APC, Vuuren D (2016) Projections of the availability and cost of residues from agriculture and forestry. GCB Bioenerg. 8:456–470. https://doi.org/10.1111/gcbb.12285

Das PK, Das BP, Dash P (2020) Potentials of postharvest rice crop residues as a source of biofuel. In: Praveen Kumar R, Gnansounou E, Raman JK, Baskar G (eds) Refining biomass residues for sustainable energy and bioproducts. Academic Press, pp 275–301. https://doi.org/10.1016/B978-0-12-818996-2.00013-2

De S, Mishra S, Poonguzhali E, Rajesh M, Tamilarasan K (2020) Fractionation and characterization of lignin from waste rice straw: Biomass surface chemical composition analysis. Int J Biol Macromol 145:795–803. https://doi.org/10.1016/j.ijbiomac.2019.10.068

Demirbas A, Bafail A, Ahmad W, Sheikh M (2016) Biodiesel production from non-edible plant oils. Energy Explor Exploit 34(2):290–318. https://doi.org/10.1177/0144598716630166

Demirbas A, Karslioglu S (2007) Biodiesel production facilities from vegetable oils and animal fats. Energy Sour Part A Recov Util Environ Eff 29(2):133–141. https://doi.org/10.1080/009083190951320

Deshwal GK, Alam T, Panjagari NR, Bharadwaj A (2021) Utilization of cereal crop residues, cereal milling, sugarcane and dairy processing by-products for sustainable packaging solutions. J Polym Environ 29:2046–2061. https://doi.org/10.1007/s10924-020-02034-w

Dieterich V, Buttler A, Hanel A, Spliethoff H, Fendt S (2020) Power-to-liquid via synthesis of methanol, DME or Fischer–Tropsch-fuels: a review. Energy Environ Sci 13:3207–3252. https://doi.org/10.1039/D0EE01187H

Domínguez-Robles J, Cárcamo-Martínez Á, Stewart SA, Donnelly RF, Larrañeta E (2021) Fabrication of lignin-based hydrogels and their applications. In: Santos H, Figueiredo P (eds) Lignin-based materials for biomedical applications. Elsevier, pp 371–394. https://doi.org/10.1016/B978-0-12-820303-3.00008-4

Dong JJ, Ding JC, Zhang Y, Ma L, Xu GC, Han RZ, Ni Y (2016) Simultaneous saccharification and fermentation of dilute alkaline-pretreated corn stover for enhanced butanol production by *Clostridium saccharobutylicum* DSM 13864. FEMS Microbiol Lett 363(4):fnw003. https://doi.org/10.1093/femsle/fnw003

Duque A, Álvarez C, Doménech P, Manzanares P, Moreno AD (2021) Advanced bioethanol production: from novel raw materials to integrated biorefineries. Processes 9(2):206. https://doi.org/10.3390/pr9020206
Duque-Acevedo M, Belmonte-Ureña LJ, Cortés-García FJ, Camacho-Ferre F (2020) Agricultural waste: review of the evolution, approaches and perspectives on alternative uses. Global Ecol Conserv 22:e00902. https://doi.org/10.1016/j.gecco.2020.e00902
Dutta S, Ghosh S, Manna D, Chowdhury R (2021) Energy and environmental performance of a near-zero-effluent rice straw to butanol production plant. Indian Chem Eng 63(2):139–151. https://doi.org/10.1080/00194506.2020.1831406
El-Ramady H, El-Henawy A, Amer M, Omara A, Elsakhawy T, Elbasiouny H, Elbehiry F, Abou Elyazid D, El-Mahrouk M (2020) Agricultural waste and its nano-management: Mini review. Egyptian J Soil Sci 60(4):349–364. https://doi.org/10.21608/ejss.2020.46807.1397
Fagodiya RK, Pathak H, Bhatia A, Jain N, Gupta DK, Kumar A, Malyan SK, Dubey R, Radhakrishanan S, Tomer R (2019) Nitrous oxide emission and mitigation from maize–wheat rotation in the upper Indo-Gangetic Plains. Carbon Manage 10(5):489–499. https://doi.org/10.1080/17583004.2019.1650579
Fito J, Tefera N, Van Hulle SWH (2019) Sugarcane biorefineries wastewater: bioremediation technologies for environmental sustainability. Chem Biol Technol Agric 6:6. https://doi.org/10.1186/s40538-019-0144-5
Furlan FF, Costa CBB, Cruz AJG, Secchi AR, Soares RP, Giordano RC (2012) Integrated tool for simulation and optimization of a first and second generation ethanol-from-sugarcane production plant. In: Bogle IDL, Fairweather M (eds) Computer aided chemical engineering, vol 30, pp 81–85. https://doi.org/10.1016/B978-0-444-59519-5.50017-4
Ginni G, Kavitha S, Kannah RY, Bhatia SK, Kumar SA, Rajkumar M, Kumar G, Pugazhendhi A, Chi NTL, Banu JR (2021) Valorization of agricultural residues: Different biorefinery routes. J Environ Chem Eng 9(4):105435. https://doi.org/10.1016/j.jece.2021.105435
Go AW, Conag AT, Igdon RMB, Toledo AS, Malila JS (2019) Potentials of agricultural and agro-industrial crop residues for the displacement of fossil fuels: a Philippine context. Energy Strateg Rev 23:100–113. https://doi.org/10.1016/j.esr.2018.12.010
Goswami RK, Mehariya S, Obulisamy PK, Verma P (2021) Advanced microalgae-based renewable biohydrogen production systems: a review. Bioresour Technol 320(Part A):24301. https://doi.org/10.1016/j.biortech.2020.124301
Govil T, Wang J, Samanta D, David A, Tripathi A, Rauniyar S, Salem DR, Sani RK (2020) Lignocellulosic feedstock: a review of a sustainable platform for cleaner production of nature's plastics. J Clean Prod 270:122521. https://doi.org/10.1016/j.jclepro.2020.122521
Gupta O, Roy S, Mitra S (2020) Low temperature recovery of acetone–butanol–ethanol (ABE) fermentation products via microwave induced membrane distillation on carbon nanotube immobilized membranes. Sustain Energ Fuels 4:3487–3499. https://doi.org/10.1039/D0SE00461H
Gupte A, Madamwar D (1997) Solid state fermentation of lignocellulosic waste for cellulase and glucosidase production by cocultivation of Aspergillus elipticus and Aspergillus fumigatus. Biotechnol Prog 13:166–169. https://doi.org/10.1021/bp970004g
Hassan SS, Williams GA, Jaiswal AK (2019) Moving towards the second generation of lignocellulosic biorefineries in the EU: drivers, challenges, and opportunities. Renew Sustain Energy Rev 101:590–599. https://doi.org/10.1016/j.rser.2018.11.041
Hiloidhari M, Das D, Baruah D (2014) Bioenergy potential from crop residue biomass in India. Renew Sustain Energy Rev 32:504–512. https://doi.org/10.1016/j.rser.2014.01.025
Hiloidhari M, Bhuyan N, Gogoi N, Seth D, Garg A, Singh A, Prasad S, Kataki R (2020) Agroindustry wastes: biofuels and biomaterials feedstocks for sustainable rural development. In: Praveen Kumar R, Gnansounou E, Raman JK, Baskar G (eds) Refining biomass residues for sustainable energy and bioproducts. Academic Press, pp 357–388. https://doi.org/10.1016/B978-0-12-818996-2.00016-8

Hoang AT, Tabatabaei M, Aghbashlo M, Carlucci AP, Ölçer AI, Le AT, Ghassemi A (2021) Rice bran oil-based biodiesel as a promising renewable fuel alternative to petrodiesel: a review Renew. Sustain Energy Rev 135:110204. https://doi.org/10.1016/j.rser.2020.110204

Holmatov B, Schyns JF, Krol MS, Gerbens-Leenes PW, Hoekstra AY (2021) Can crop residues provide fuel for future transport? Limited global residue bioethanol potentials and large associated land, water and carbon footprints. Renew Sustain Energy Rev 149:111417. https://doi.org/10.1016/j.rser.2021.111417

Honorato-Salazar JA, Sadhukhan J (2020) Annual biomass variation of agriculture crops and forestry residues, and seasonality of crop residues for energy production in Mexico. Food Bioprod Process 119:1–19. https://doi.org/10.1016/j.fbp.2019.10.005

Huang J, Du Y, Bao T, Lin M, Wang J, Yang ST (2019) Production of n-butanol from cassava bagasse hydrolysate by engineered Clostridium tyrobutyricum overexpressing adhE2: kinetics and cost analysis. Bioresour Technol 292:121969. https://doi.org/10.1016/j.biortech.2019.121969

Huang L, Ma M, Ji X, Choi S, Si C (2021) Recent developments and applications of hemicellulose from wheat straw: a review. Front Bioeng Biotechnol 9:440. https://doi.org/10.3389/fbioe.2021.690773

Ibarra-Díaz N, Castañón-Rodríguez JF, Gómez-Rodríguez J, Aguilar-Uscanga MG (2020) Optimization of peroxide-alkaline pre-treatment and enzymatic hydrolysis of barley straw (Hordeum vulgare L.) to produce fermentable sugars using a Box–Behnken design. Biomass Conv Bioref. https://doi.org/10.1007/s13399-020-00853-4

Ibitoye SE, Jen TC, Mahamood RM, Akinlabi ET (2021) Densification of agro-residues for sustainable energy generation: an overview. Bioresour Bioprocess 8:75. https://doi.org/10.1186/s40643-021-00427-w

Ingram LO, Gomez PF, Lai X, Moniruzzaman M, Wood BE, Yomano LP, York SW (1998) Metabolic engineering of bacteria for ethanol production. Biotechnol Bioeng 58(2–3):204–214. https://doi.org/10.1002/(SICI)1097-0290(19980420)58:2/3%3c204::AID-BIT13%3e3.0.CO;2-C

IPCC (2018) Global Warming of 1.5 °C. An IPCC Special Report on the impacts of global warming of 1.5 °C above pre-industrial levels and related global greenhouse gas emission pathways, in the context of strengthening the global response to the threat of climate change, sustainable development, and efforts to eradicate poverty. In: Masson-Delmotte V, Zhai P, Pörtner HO, Roberts D, Skea J, Shukla PR, Pirani A, Moufouma-Okia W, Péan C, Pidcock R, Connors S, Matthews JBR, Chen Y, Zhou X, Gomis MI, Lonnoy E, Maycock T, Tignor M, Waterfield T (eds)

Isci A, Demirer GN (2007) Biogas production potential from cotton wastes. Renew Energy 32(5):750–757. https://doi.org/10.1016/j.renene.2006.03.018

Jain R, Naik SN (2018) Adding value to the oil cake as a waste from oil processing industry: production of lipase in solid state fermentation. Biocatal Agric Biotechnol 15:181–184. https://doi.org/10.1016/j.bcab.2018.06.010

Jansen MLA, Bracher JM, Papapetridis I, Verhoeven MD, de Bruijn H, de Waal PP, van Maris AJA, Klaassen P, Pronk JT (2017) Saccharomyces cerevisiae strains for second-generation ethanol production: from academic exploration to industrial implementation. FEMS Yeast Res 17(5):fox044. https://doi.org/10.1093/femsyr/fox044

Jiang L, Wang J, Liang S, Wang X, Cen P, Xu Z (2009) Butyric acid fermentation in a fibrous bed bioreactor with immobilized Clostridium tyrobutyricum from cane molasses. Bioresour Technol 100:3403–3409. https://doi.org/10.1016/j.biortech.2009.02.032

Jusakulvijit P, Bezama A, Thrän D (2021) The availability and assessment of potential agricultural residues for the regional development of second-generation bioethanol in Thailand. Waste Biomass Valor. https://doi.org/10.1007/s12649-021-01424-y

Kalamaras SD, Kotsopoulos TA (2014) Anaerobic co-digestion of cattle manure and alternative crops for the substitution of maize in South Europe. Bioresour Technol 172:68–75. https://doi.org/10.1016/j.biortech.2014.09.005

Kannah RY, Kavitha S, Sivashanmugam P, Kumar G, Nguyen DD, Chang SW, Banu R (2018) Biohydrogen production from rice straw: Effect of combinative pre-treatment, modelling assessment and energy balance consideration. Int J Hydrog Energy 44:2203–2215. https://doi.org/10.1016/j.ijhydene.2018.07.201

Kaushal LA, Prashar A (2021) Agricultural crop residue burning and its environmental impacts and potential causes—case of northwest India. J Environ Plan Manage 64(3):464–484. https://doi.org/10.1080/09640568.2020.1767044

Khounani Z, Hosseinzadeh-Bandbafha H, Nazemi F, Shaeifi M, Karimi K, Tabatabaei M, Aghbashlo M, Lam SS (2021) Exergy analysis of a whole-crop safflower biorefinery: a step towards reducing agricultural wastes in a sustainable manner. J Environ Manage 279:111822. https://doi.org/10.1016/j.jenvman.2020.111822

Kim S, Dale B (2004) Global potential bioethanol production from wasted crops and crop residues. Biomass Bioenerg 26:361–375. https://doi.org/10.1016/j.biombioe.2003.08.002

Klai N, Yadav B, Hachimi OE, Pandey A, Sellamuthu B, Tyagi RD (2021) Agro-industrial waste valorization for biopolymer production and life-cycle assessment toward circular bioeconomy. In: Pandey A, Tyagi RD, Varjani S (eds) Biomass, biofuels, biochemicals. Elsevier, pp. 515–555. https://doi.org/10.1016/B978-0-12-821878-5.00007-6

Kosre A, Koreti D, Mahish PK, Chandrawanshi NK (2021) Current perspective of sustainable utilization of agro waste and biotransformation of energy in mushroom. In: Singh P, Singh S, Kumar K, Baweja P (eds) Energy: crises, challenges and solutions. Wiley & Sons Ltd. https://doi.org/10.1002/9781119741503.ch15

Kover A, Kraljić D, Marinaro R, Rene ER (2021) Processes for the valorization of food and agricultural wastes to value-added products: recent practices and perspectives. Syst Microbiol Biomanuf. https://doi.org/10.1007/s43393-021-00042-y

Koytsoumpa EI, Magiri–Skouloudi D, Karellas S, Kakaras E (2021) Bioenergy with carbon capture and utilization: A review on the potential deployment towards a European circular bioeconomy. Renew Sustain Energy Rev 152:111641. https://doi.org/10.1016/j.rser.2021.111641

Kumar A, Kumar N, Baredar P, Shukla A (2015) A review on biomass energy resources, potential, conversion and policy in India. Renew Sustain Energy Rev 45:530–539. https://doi.org/10.1016/j.rser.2015.02.007

Kumar PS, Yaashikaa PR (2020b) Sources and operations of waste biorefineries. In: Kumar RP, Gnansounou E, Raman JK, Baskar G (eds) Refining biomass residues for sustainable energy and bioproducts: technology, advances, life cycle assessment and economics. Academic Press, Massachusetts, pp 111–133

Kumar S, Paritosh K, Pareek N, Chawade A, Vivekanand V (2018a) De-construction of major Indian cereal crop residues through chemical pre-treatment for improved biogas production: an overview. Renew Sustain Energy Rev 90:160–170. https://doi.org/10.1016/j.rser.2018.03.049

Kumar M, Oyedun AO, Kumar A (2018b) A review on the current status of various hydrothermal technologies on biomass feedstock. Renew Sustain Energy Rev 81(2):1742–1770. https://doi.org/10.1016/j.rser.2017.05.270

Kumar B, Bhardwaj N, Agrawal K, Chaturvedi V, Verma P (2020a) Current perspective on pretreatment technologies using lignocellulosic biomass: an emerging biorefinery concept. Fuel Process Technol 199:106244. https://doi.org/10.1016/j.fuproc.2019.106244

Kumar JSP, Kumar SNS, Chintagunta AD (2020b) Bioethanol production from cereal crops and lignocelluloses rich agro-residues: prospects and challenges. SN Appl Sci 2:1673. https://doi.org/10.1007/s42452-020-03471-x

Kumari S, Verma N, Lakhani A, Tiwari S, Kandikonda MK (2018) Tropospheric ozone enhancement during post-harvest crop-residue fires at two downwind sites of the Indo-Gangetic Plain. Environ Sci Pollut Res 25:18879–18893. https://doi.org/10.1007/s11356-018-2034-y

Kumla J, Suwannarach N, Sujarit K, Penkhrue W, Kakumyan P, Jatuwong K, Vadthanarat S, Lumyong S (2020) Cultivation of mushrooms and their lignocellulolytic enzyme production through the utilization of agro-industrial waste. Molecules 25(12):2811. https://doi.org/10.3390/molecules25122811

Labatut RA, Angenent LT, Scott NR (2011) Biochemical methane potential and biodegradability of complex organic substrates. Bioresour Technol 102:2255–2264. https://doi.org/10.1016/j.biortech.2010.10.035

Leong HY, Chang CK, Khoo KS, Chew KW, Chia SR, Lim JW, Chang J, Show PL (2021) Waste biorefinery towards a sustainable circular bioeconomy: a solution to global issues. Biotechnol Biofuels 14:87. https://doi.org/10.1186/s13068-021-01939-5

Levine JS (1991) Global biomass burning: atmospheric, climatic, and biospheric implications. The MIT Press, Cambridge, MA

Liew LN, Shi J, Li Y (2012) Methane production from solid-state anaerobic digestion of lignocellulosic biomass. Biomass Bioenerg 46:125–132. https://doi.org/10.1016/j.biombioe.2012.09.014

Lima PJM, da Silva RM, Neto CACG, e Silva NCG, da Silva Souza JE, Nunes YL, dos Santos JCS (2021) An overview on the conversion of glycerol to value-added industrial products via chemical and biochemical routes. Biotechnol Appl Biochem. https://doi.org/10.1002/bab.2098

Lin B, Raza MY (2020) Analysis of energy security indicators and CO_2 emissions. A case from a developing economy. Energy 200:117575. https://doi.org/10.1016/j.energy.2020.117575

Liu K, Du H, Zheng T, Liu H, Zhang M, Zhang R, Li H, Xie H, Zhang X, Ma M, Si C (2021) Recent advances in cellulose and its derivatives for oilfield applications. Carbohydr Polym 117740. https://doi.org/10.1016/j.carbpol.2021.117740

Lojkova L, Vranová V, Formánek P, Drápelová I, Brtnicky M, Datta R (2020) Enantiomers of carbohydrates and their role in ecosystem interactions: a review. Symmetry 12(3):470. https://doi.org/10.3390/sym12030470

Lopez-Hidalgo A, Sánchez A, León-Rodríguez A (2017) Simultaneous production of bioethanol and biohydrogen by Escherichia coli WDHL using wheat straw hydrolysate as substrate. Fuel 188. https://doi.org/10.1016/j.fuel.2016.10.022

López-Linares JC, García-Cubero MT, Coca M, Lucas S (2021) A biorefinery approach for the valorization of spent coffee grounds to produce antioxidant compounds and biobutanol. Biomass Bioenerg 147:106026. https://doi.org/10.1016/j.biombioe.2021.106026

López-Molina A, Sengupta D, Shi C, Aldamigh E, Alandejani M, El-Halwagi MM (2020) An integrated approach to the design of centralized and decentralized biorefineries with environmental, safety, and economic objectives. Processes 8(12):1682. https://doi.org/10.3390/pr8121682

Lü JL, Zhou PJ (2015) Ethanol production from microwave-assisted FeCl3 pretreated rice straw. Energy Sources Part A Recov Util Environ Eff 37(21):2367–2374. https://doi.org/10.1080/15567036.2011.649335

Mahato RK, Kumar D, Rajagopalan G (2020) Biohydrogen production from fruit waste by Clostridium strain BOH3. Renew Energy 153(C):1368–1377. https://doi.org/10.1016/j.renene.2020.02.092

Mahmood H, Moniruzzaman M, Iqbal T, Khan MJ (2019) Recent advances in the pre-treatment of lignocellulosic biomass for biofuels and value-added products. Curr Opin Green Sustain Chem 20:18–24. https://doi.org/10.1016/j.cogsc.2019.08.001

Mallikarjuna C, Dash RR, Bhunia P (2021) Sustainability of gaseous biofuels: Potential uses, technological constraints and environmental concerns. In: Pandey A, Tyagi AD, Varjani S (eds) Biomass, biofuels, biochemicals. Elsevier, pp 247–268. https://doi.org/10.1016/B978-0-12-821878-5.00025-8.

Mancini G, Papirio S, Lens PNL, Esposito G (2018) Increased biogas production from wheat straw by chemical pre-treatments. Renew Energy 119:608–614. https://doi.org/10.1016/j.renene.2017.12.045

Mathew AK, Abraham A, Mallapureddy KK, Sukumaran RK (2018) Lignocellulosic biorefinery wastes, or resources? In: Bhaskar T, Pandey A, Venkata Mohan S, Lee D, Khanal SK (eds) Waste biorefinery. Elsevier, pp 267–297. https://doi.org/10.1016/B978-0-444-63992-9.00009-4

Mazaheri H, Ong HC, Masjuki HH, Amini Z, Harrison MD, Wang CT, Kusumo F, Alwi A (2018) Rice bran oil based biodiesel production using calcium oxide catalyst derived from Chicoreus brunneus shell. Energy 144:10–19. https://doi.org/10.1016/j.energy.2017.11.073

Meramo-Hurtado SI, González-Delgado A, Rehmann L, Quinones-Bolanos E, Mehvar M (2021) Comparative analysis of biorefinery designs based on acetone-butanol-ethanol fermentation under exergetic, techno-economic, and sensitivity analyses towards a sustainability perspective. J Clean Prod 298:126761. https://doi.org/10.1016/j.jclepro.2021.126761

Misra SK, Pathak D, Pathak K (2021) Bioprivileged Molecules. In: Inamuddin BR, Ahamed MI, Khan A (eds) Advances in green synthesis. Advances in science, technology & innovation (IEREK interdisciplinary series for sustainable development). Springer, Cham. https://doi.org/10.1007/978-3-030-67884-5_6

MNRE (Ministry of New and Renewable Energy Resources) (2009) Govt. of India, New Delhi. www.mnre.gov.in/biomassrsources

MNRE (2014) MNRE annual report. Retrieved from http://mnre.gov.in/filemanager/annual-report/2014-2015/EN/Chapter%201/chapter_1.htm. On 5 Sept 2021

Molina-Guerrero CE, Sanchez A, Vázquez-Núñez E (2020) Energy potential of agricultural residues generated in Mexico and their use for butanol and electricity production under a biorefinery configuration. Environ Sci Pollut Res 27:28607–28622. https://doi.org/10.1007/s11356-020-08430-y

Moradi F, Amiri H, Soleimanian-Zad S, Ehsani MR, Karimi K (2013) Improvement of acetone, butanol and ethanol production from rice straw by acid and alkaline pre-treatments. Fuel 112:8–13. https://doi.org/10.1016/j.fuel.2013.05.011

MoSPI (2020) Energy statistics 2020. Ministry of Statistics and Programme Implementation. http://mospi.nic.in/publication/energy-statistics-india-2021. Retrieved on 6 June 2021

Najafi G, Ghobadian B, Tavakoli T, Buttsworth DR, Yusaf TF, Faizollahnejad M (2009) Performance and exhaust emissions of a gasoline engine with ethanol blended gasoline fuels using artificial neural network. Appl Energy 86(5):630–639. https://doi.org/10.1016/j.apenergy.2008.09.017

Nanda S, Azargohar R, Dalai AK, Kozinski JA (2015) An assessment on the sustainability of lignocellulosic biomass for biorefining. Renew Sustain Energy Rev 50:925–941. https://doi.org/10.1016/j.rser.2015.05.058

Naqvi SR, Ali I, Nasir S, Taqvi SAA, Atabani AE, Chen W (2020) Assessment of agro-industrial residues for bioenergy potential by investigating thermo-kinetic behavior in a slow pyrolysis process. Fuel 278:118259. https://doi.org/10.1016/j.fuel.2020.118259

Nasir M, Khali DP, Jawaid M, Tahir PM, Siakeng R, Asim M, Khan TA (2019) Recent development in binderless fiber-board fabrication from agricultural residues: a review. Constr Build Mater 211:502–516. https://doi.org/10.1016/j.conbuildmat.2019.03.279

National Policy on biofuels (2018) http://petroleum.nic.in/sites/default/files/biofuels.pdf

Nguyen VT (2017) Potential, uses and future perspectives of agricultural wastes. In: Nguyen VT (ed) Recovering bioactive compounds from agricultural wastes. Wiley & Sons Ltd. https://doi.org/10.1002/9781119168850.ch1

Niphadkar S, Bagade P, Ahmed S (2018) Bioethanol production: insight into past, present and future perspectives. Biofuels 9(2):229–238. https://doi.org/10.1080/17597269.2017.1334338

Nogueira KMV, Mendes V, Carraro CB, Taveira IC, Oshiquiri LH, Gupta VK, Silva RN (2020) Sugar transporters from industrial fungi: Key to improving second-generation ethanol production. Renew Sustain Energy Rev 131:109991. https://doi.org/10.1016/j.rser.2020.109991

NPMCR (2014) Retrieved from https://agricoop.nic.in/sites/default/files/NPMCR_1.pdf. On 10 Sept 2021

Ntaikou I, Menis N, Alexandropoulou M, Antonopoulou G, Lyberatos G (2018) Valorization of kitchen biowaste for ethanol production via simultaneous saccharification and fermentation using co-cultures of the yeasts Saccharomyces cerevisiae and Pichia stipites. Bioresour Technol 263:75–83. https://doi.org/10.1016/j.biortech.2018.04.109

Nwokolo N, Mukumba P, Obileke K, Enebe M (2020) Waste to energy: a focus on the impact of substrate type in biogas production. Processes 8(10):1224. https://doi.org/10.3390/pr8101224

Olguin-Maciel E, Singh A, Chable-Villacis R, Tapia-Tussell R, Ruiz HA (2020) Consolidated bioprocessing, an innovative strategy towards sustainability for biofuels production from crop residues: an overview. Agronomy 10(11):1834. https://doi.org/10.3390/agronomy10111834

Ong VZ, Wu TY (2020) An application of ultrasonication in lignocellulosic biomass valorisation into bio-energy and bio-based products. Renew Sustain Energy Rev 132:109924. https://doi.org/10.1016/j.rser.2020.109924

Pandiyan K, Singh A, Singh S, Saxena AK, Nain L (2019) Technological interventions for utilization of crop residues and weedy biomass for second generation bio-ethanol production. Renew Energy 132:723–741. https://doi.org/10.1016/j.renene.2018.08.049

Paschos T, Louloudi A, Papayannakos N, Kekos D, Mamma D (2020) Potential of barley straw for high titer bioethanol production applying pre-hydrolysis and simultaneous saccharification and fermentation at high solid loading. Biofuels. https://doi.org/10.1080/17597269.2020.1760688

Patel A, Shah AR (2021) Integrated lignocellulosic biorefinery: Gateway for production of second generation ethanol and value added products. J Bioresour Bioprod 6(2):108–128. https://doi.org/10.1016/j.jobab.2021.02.001

Pattnaik F, Tripathi S, Patra BR, Nanda S, Kumar V, Dalai AK, Naik SN (2021) Catalytic conversion of lignocellulosic polysaccharides to commodity biochemicals: a review. Environ Chem Lett. https://doi.org/10.1007/s10311-021-01284-x

Paul S, Dutta A (2018) Challenges and opportunities of lignocellulosic biomass for anaerobic digestion. Resour Conserv Recycl 130:164–174. https://doi.org/10.1016/j.resconrec.2017.12.005

Pereira LG, Dias MOS, Mariano AP, Maciel Filho R, Bonomi A (2015) Economic and environmental assessment of n-butanol production in an integrated first and second generation sugarcane biorefinery: fermentative versus catalytic routes. Appl Energy 160:120–131. https://doi.org/10.1016/j.apenergy.2015.09.063

Philippini RR, Martiniano SE, Ingle AP, Franco Marcelino PR, Silva GM, Gonçalves BF, dos Santos JC, da Silva SS (2020) Agroindustrial byproducts for the generation of biobased products: alternatives for sustainable biorefineries. Front Energy Res 152. https://doi.org/10.3389/fenrg.2020.00152

Pinto T, Flores-Alsina X, Gernaey KV, Junicke H (2021) Alone or together? A review on pure and mixed microbial cultures for butanol production. Renew Sustain Energy Rev 147:111244. https://doi.org/10.1016/j.rser.2021.111244

Porichha GK, Hu Y, Rao KTV, Xu CC (2021) Crop residue management in India: Stubble burning vs. other utilizations including bioenergy. Energies 4281. https://doi.org/10.3390/en14144281

Porro D, Branduardi P (2017) Production of organic acids by yeasts and filamentous fungi. In: Sibirny A (ed) Biotechnology of yeasts and filamentous fungi. Springer, Cham. https://doi.org/10.1007/978-3-319-58829-2_7

Prasad S, Singh A, Joshi HC (2007) Ethanol as an alternative fuel from agricultural, industrial and urban residues. Resour Conserv Recycl 1854:1–39. https://doi.org/10.1016/j.resconrec.2006.05.007

Prasad S, Singh A, Korres N, Rathore D, Sevda S, Pant D (2020) Sustainable utilization of crop residues for energy generation: a life cycle assessment (LCA) perspective. Bioresour Technol 303:122964. https://doi.org/10.1016/j.biortech.2020.122964

Quayson E, Amoah J, Rachmadona N, Morita K, Darkwah L, Hama S, Yoshida A, Kondo A, Ogino C (2020) Valorization of palm biomass waste into carbon matrices for the immobilization of recombinant fusarium heterosporum lipase towards palm. Biodiesel Synth Biomass Bioenerg 142:105768. https://doi.org/10.1016/j.biombioe.2020.105768

Qureshi N, Saha BC, Cotta MA (2007) Butanol production from wheat straw hydrolysate using Clostridium beijerinckii. Bioprocess Biosyst Eng 30(6):419–427. https://doi.org/10.1007/s00449-007-0137-9

Qureshi N, Cotta MA, Saha BC (2014) Bioconversion of barley straw and corn stover to butanol (a biofuel) in integrated fermentation and simultaneous product recovery bioreactors. Food Bioprod Process 92:298–308. https://doi.org/10.1016/j.fbp.2013.11.005

Qureshi N, Saha BC, Hector RE, Dien B, Hughes S, Liu S, Iten L, Bowman MJ, Sarath G, Cotta MA (2010) Production of butanol (a biofuel) from agricultural residues: Part II—Use of corn stover and switchgrass hydrolysates. Biomass Bioenerg 34(4):566–571. https://doi.org/10.1016/j.biombioe.2009.12.023

Rastogi M, Shrivastava S (2017) Recent advances in second generation bioethanol production: An insight to pre-treatment, saccharification and fermentation processes. Renew Sustain Energy Rev 80:330–340. https://doi.org/10.1016/j.rser.2017.05.225
Rather MA, Srivastav AK (2021) A study on biohydrogen production based on biophotolysis from cyanobacteria. Ann Romanian Soc Cell Biol 25(6):12500–12509
Raud M, Kikas T, Sippula O, Shurpali NJ (2019) Potentials and challenges in lignocellulosic biofuel production technology. Renew Sustain Energy Rev 111:44–56. https://doi.org/10.1016/j.rser.2019.05.020
Razi MO, Sasmal S (2020) Organosolv pre-treatment of groundnut (Arachis hypogaea) shell and its upshot. Biomass Conv Bioref. https://doi.org/10.1007/s13399-020-00920-w
Redondo-Gómez C, Rodríguez Quesada M, Vallejo Astúa S, Murillo Zamora JP, Lopretti M, Vega-Baudrit JR (2020) Biorefinery of biomass of agro-industrial banana waste to obtain high-value biopolymers. Molecules 25(17):3829. https://doi.org/10.3390/molecules25173829
Rishikesh MS, Harish S, Mahendran Prasanth S, Gnana Prakash D (2021) A comprehensive review on lignin obtained from agro-residues: potential source of useful chemicals. Biomass Conv Bioref. https://doi.org/10.1007/s13399-021-01637-0
Sadh PK, Duhan S, Duhan JS (2018) Agro-industrial wastes and their utilization using solid state fermentation: a review. Bioresour. Bioprocess. 5:1. https://doi.org/10.1186/s40643-017-0187-z
Sadhukhan J, Martinez-Hernandez E, Murphy RJ, Ng DKS, Hassim MH, Ng KS, Kin WY, Jaye IFM, Hang MYLP, Andiappan V (2018) Role of bioenergy, biorefinery and bioeconomy in sustainable development: strategic pathways for Malaysia. Renew Sustain Energy Rev 81(2):1966–1987. https://doi.org/10.1016/j.rser.2017.06.007
Sahito AR, Mahar R, Khan B (2013) Anaerobic biodegradability and methane potential of crop residue co-digested with buffalo dung. Mehran Uni Res J Eng Technol 32:509–518
Sahoo A, Kumar S, Mohanty KA (2020) Comprehensive characterization of non-edible lignocellulosic biomass to elucidate their biofuel production potential. Biomass Conv Bioref. https://doi.org/10.1007/s13399-020-00924-6
Sanchez A, Valdez-Vazquez I, Soto A, Sánchez S, Tavarez D (2017) Lignocellulosic n-butanol co-production in an advanced biorefinery using mixed cultures. Biomass Bioenerg 102:1–12. https://doi.org/10.1016/j.biombioe.2017.03.023
Sánchez OJ, Cardona CA (2008) Trends in biotechnological production of fuel ethanol from different feedstocks. Bioresour Technol 99(13):5270–5295. https://doi.org/10.1016/j.biortech.2007.11.013
Šelo G, Planinić M, Tišma M, Tomas S, Koceva Komlenić D, Bucić-Kojić A (2021) A comprehensive review on valorization of agro-food industrial residues by solid-state fermentation. Foods 10(5):927. https://doi.org/10.3390/foods10050927
Serna-Loaiza S, Carmona-Garcia E, Cardona CA (2018) Potential raw materials for biorefineries to ensure food security: the Cocoyam case. Ind Crops Prod 126:92–102. https://doi.org/10.1016/j.indcrop.2018.10.005
Sharma A, Singh Y, Ansari NA, Pal A, Lalhriatpuia S (2020a) Experimental investigation of the behaviour of a DI diesel engine fuelled with biodiesel/diesel blends having effect of raw biogas at different operating responses. Fuel 279:118460. https://doi.org/10.1016/j.fuel.2020.118460
Sharma A, Sharma P, Singh J, Singh S, Nain L (2020b) Prospecting the potential of agro-residues as substrate for microbial flavor production. Front Sustain Food Syst 4. https://doi.org/10.3389/fsufs.2020.00018
Sheldon RA (2018) Chemicals from renewable biomass: a renaissance in carbohydrate chemistry. Curr Opin Green Sustain Chem 14:89–95. https://doi.org/10.1016/j.cogsc.2018.08.003
Singh J, Gu S (2010) Biomass conversion to energy in India—a critique. Renew Sustain Energy Rev 14(5):1367–1378. https://doi.org/10.1016/j.rser.2010.01.013
Singh S, Biswal A, Mor S, Ravindra K, Singh V, Mor S (2020) A high-resolution emission inventory of air pollutants from primary crop residue burning over Northern India based on VIIRS thermal anomalies. Environ Pollut 266(1):115132. https://doi.org/10.1016/j.envpol.2020.115132

Solarte-Toro JC, Alzate CAC (2021) Biorefineries as the base for accomplishing the sustainable development goals (SDGs) and the transition to bioeconomy: Technical aspects, challenges and perspectives. Bioresour Technol 340:125626. https://doi.org/10.1016/j.biortech.2021.125626

Somayaji D, Khanna S (1994) Biomethanation of rice and wheat straw. World J Microbiol Biotechnol 10(5):521–523. https://doi.org/10.1007/BF00367657

Su C, Qi L, Cai D, Chen B, Chen H, Zhang C, Si Z, Wang Z, Li G, Qin P (2020) Integrated ethanol fermentation and acetone-butanol-ethanol fermentation using sweet sorghum bagasse. Renew Energy 162:1125–1131. https://doi.org/10.1016/j.renene.2020.07.119

Sukumaran RK, Vikram JS, Raveendran S, Binod P, Usha KJ, Kuttavan VS, Kuni PR, Pandey A (2010) Lignocellulosic ethanol in India: prospects, challenges and feedstock availability. Bioresour Technol 101(13):4826–4833. https://doi.org/10.1016/j.biortech.2009.11.049

Tinpranee N, Incharoensakdi A, Phunpruch S (2018) Screening cyanobacteria from marine coastal waters of Thailand for biohydrogen production. J Appl Phycol. https://doi.org/10.1007/s10811-018-1490-6

Trading.com. Ministry of Commerce and Industry, GoI. https://tradingeconomics.com/india/imports-of-crude-oil. Retrieved on June 2021

Tsegaye B, Balomajumder C, Roy P (2020) Organosolv pre-treatments of rice straw followed by microbial hydrolysis for efficient biofuel production. Renew Energy 148:923–934. https://doi.org/10.1016/j.renene.2019.10.176

Turkcan A (2020) The effects of different types of biodiesels and biodiesel-bioethanol-diesel blends on the cyclic variations and correlation coefficient. Fuel 261:116453. https://doi.org/10.1016/j.fuel.2019.116453

Ubando AT, Felix CB, Chen WH (2019) Biorefineries in circular bioeconomy: a comprehensive review. Bioresour Technol 299:122585. https://doi.org/10.1016/j.biortech.2019.122585

Ullah K, Sharma VK, Dhingra S, Braccio G, Ahmad M, Sofia S (2015) Assessing the lignocellulosic biomass resources potential in developing countries: a critical review. Renew Sustain Energy Rev 51:682–698. https://doi.org/10.1016/j.rser.2015.06.044

Vaish B, Srivastava V, Singh P, Singh A, Singh PK, Singh RP (2016) Exploring untapped energy potential of urban solid waste. Energ Ecol Environ 1:323–342. https://doi.org/10.1007/s40974-016-0023-x

Venkatramanan V, Shah S, Prasad S, Singh A, Prasad R (2021a) Assessment of bioenergy generation potential of agricultural crop residues in India. Circ Econ Sustain. https://doi.org/10.1007/s43615-021-00072-7

Venkatramanan V, Shachi S, Kumar RA, Ram P (2021b) Nexus between crop residue burning, bioeconomy and sustainable development goals over North-Western India. Front Energy Res 8:392. https://doi.org/10.3389/fenrg.2020.614212

Vijayaraghavan K, Soom MAM (2006) Trends in bio-hydrogen generation—a review. Environ Sci 3(4):255–271. https://doi.org/10.1080/15693430601049660

Watanabe I, Miyata N, Ando A, Shiroma R, Tokuyasu K, Nakamura T (2012) Ethanol production by repeated-batch simultaneous saccharification and fermentation (SSF) of alkali-treated rice straw using immobilized Saccharomyces cerevisiae cells. Bioresour Technol 123:695–698. https://doi.org/10.1016/j.biortech.2012.07.052

Westerholm M, Hansson M, Schnürer A (2012) Improved biogas production from whole stillage by co-digestion with cattle manure. Bioresour Technol 114:314–319. https://doi.org/10.1016/j.biortech.2012.03.005

Wyman CE (1990) Biological production of chemicals from renewable feedstocks, National Meeting, American Chemical Society, Washington, D.C.

Xu XL, Chen HH, Li Y (2020) Exploring the influencing factors of continuous crop residue supply: from the perspective of a sustainable and bioenergy-oriented crop cultivation. Energ Sustain Soc 10:35. https://doi.org/10.1186/s13705-020-00267-0

Yahyazadeh A, Dalai AK, Ma W, Zhang L (2021) Fischer-Tropsch synthesis for light olefins from syngas: a review of catalyst development. Reactions 2:227–257. https://doi.org/10.3390/reactions203001

Zhang T, Liu L, Song Z, Ren G, Feng Y, Han X, Yang G (2013) Biogas production by co-digestion of goat manure with three crop residues. PLoS ONE 8(6):e66845. https://doi.org/10.1371/journal.pone.0066845

Zhang Q, Weng C, Huang H, Achal V, Wang D (2016) Optimization of bioethanol production using whole plant of water hyacinth as substrate in simultaneous saccharification and fermentation process. Front Microbiol 6:1411. https://doi.org/10.3389/fmicb.2015.01411

Zhang H, Ning Z, Khalid H, Zhang, Liu G, Chen C (2018) Enhancement of methane production from cotton stalk using different pre-treatment techniques. Sci Rep 8:3463. https://doi.org/10.1038/s41598-018-21413-x

Zhao F, Yi Y, Lü X (2021) Essential process and key barriers for converting plant biomass into biofuels. In: Lü X (ed) Woodhead Publishing Series in Energy, Advances in 2nd generation of bioethanol production. Woodhead Publishing, pp 53–70. https://doi.org/10.1016/B978-0-12-818862-0.00009-1

Chapter 9
Bioconversion of Agricultural Residue into Biofuel and High-Value Biochemicals: Recent Advancement

Pawan Kumar Rose

9.1 Introduction

The excessive burning of fossil fuel has posed detrimental environmental impacts on the different ecosystems which demand the development of environment-friendly alternatives such as bioethanol, biobutanol, etc. The selection of appropriate feedstock and associated bioconversion technologies are the bottleneck in the sustainable production of biofuel. The utilization of agricultural residues as feedstock attributes sustainable waste management and production of biofuel and high-value biochemicals (Qureshi et al. 2020). Lignocellulosic components (cellulose, hemicellulose, and lignin) share a maximum portion of the agricultural residue and their use in cost-effective bioconversion to various value-added chemicals is highly challenging due to their recalcitrance nature (Patel and Shah 2021; Banu et al. 2021). Pre-treatment of lignocellulosic biomass is a primary, essential, and most critical stage which breaks the recalcitrance of feedstock for effective and economical utilization prior to conversion. Numerous conventional pre-treatment techniques require the application of harsh chemicals, use of high energy, and generation of undesirable compounds that interfere in the downstream process (Hassan et al. 2018). These limitations shifting the movement in the development of more energy-efficient and eco-friendly techniques such as microwave irradiation, ultrasound, high hydrostatic pressure, etc. (Hassan et al. 2018). After pre-treatment, the subsequent step in the bioconversion of lignocellulosic biomass includes saccharification and fermentation. Saccharification transforms cellulose and hemicellulose into their simple sugar monomers via chemical and/or biological routes. Fermentation transforms sugar monomers into biofuel and other platform chemicals. However, both have their limitation and associated benefits. One of the strategies to overcome various limitations associated with the saccharification and fermentation process is their integration (Patel and Shah

P. K. Rose (✉)
Department of Energy and Environmental Sciences, Chaudhary Devi Lal University, Sirsa, Haryana, India

Y. K. Nandabalan et al. (eds.), *Zero Waste Biorefinery*, Energy, Environment, and Sustainability, https://doi.org/10.1007/978-981-16-8682-5_9

2021). The pre-treatment and the subsequent process of lignocellulosic biomass conversion to biofuel co-produce various compounds known as inhibitors. Initially, these co-products such as 5-HMF, furfural, xylitol, levulinic acid, etc., are categorized as undesired products. However, they have proven their potential as platform chemicals in the synthesis of high-value biochemicals (Takkellapati et al. 2018). Lignocellulosic biorefinery generates about 200 biochemicals of diverse applications (Isikgor and Becer 2015). The overarching objective of this chapter is to highlight the challenges and recent research for the possibility of integration of biofuel production with high-value biochemicals for zero-waste biorefinery. This chapter intends to provide recent progress in biofuel production (bioethanol and biobutanol) and platform chemical (5-HMF, Furfural, Xylitol, Levulinic acid) based synthesis of high-value biochemicals.

9.2 Agricultural Residues

9.2.1 Source and Availability

According to Food and Agriculture Organization (FAO), the globally produced top ten agricultural commodities in the year 2019 were cereals (2.98 billion tonnes), sugar crops primary (2.23 billion tonnes), sugarcane (1.94 billion tonnes), maize (1.14 billion tonnes), vegetable primary (1.13 billion tonnes), oil crops (1.10 billion tonnes), fruits primary (0.88 billion tonnes), roots and tubers (0.86 billion tonnes), wheat (0.76 billion tonnes) and rice (0.75 billion tonnes) (see Fig. 9.1) (FAOSTAT 2021). Agricultural practices such as harvesting and post-harvesting to obtain major crops and processed products, respectively produce a significant quantity of solid residues (Prasad et al. 2020). The agricultural residues are crop-based (primary biomass residue) non-edible parts left after harvesting of the main crop such as root, stem, stalk, straw, leaves, etc., and processed-based residues (secondary biomass residue) generated after the processing of the main crop such as husk, bran, bagasse, etc. (Mande et al. 2005). The residue generates after the use of processed materials is known as tertiary biomass residue (Santana-Méridas et al. 2012). The poor post-harvesting facilities cause approximately 20% loss of agriculture-based products every year (Lakshmi et al. 2017; Prasad et al. 2020). Further, orthodox approaches in the management of agricultural residues such as cattle feeding (e.g. stalk, husk, peel, pulp, leaves, etc.) (Avadhanam et al. 2020), animal bedding (e.g. cereal straw, sawdust, corn stover, sugarcane bagasse, etc.) (Sanchis-Sebastiá et al. 2020), soil mulching (e.g. straw, husk, grasses, bagasse, cover crops, sawdust, etc.) (Iqbal et al. 2020), thatching for rural homes (e.g. cereal straw, grasses, etc.) (Prasad et al. 2020), fuel for domestic needs (e.g. rice husk, cotton stalk, corn stover, bagasse, sawdust, etc.) (Vitali et al. 2013) are wasting the potential residue. No doubt, these practices are associated with numerous regional economic and environmental benefits. Even after use in orthodox practices, about a 30% surplus of agricultural residue remains

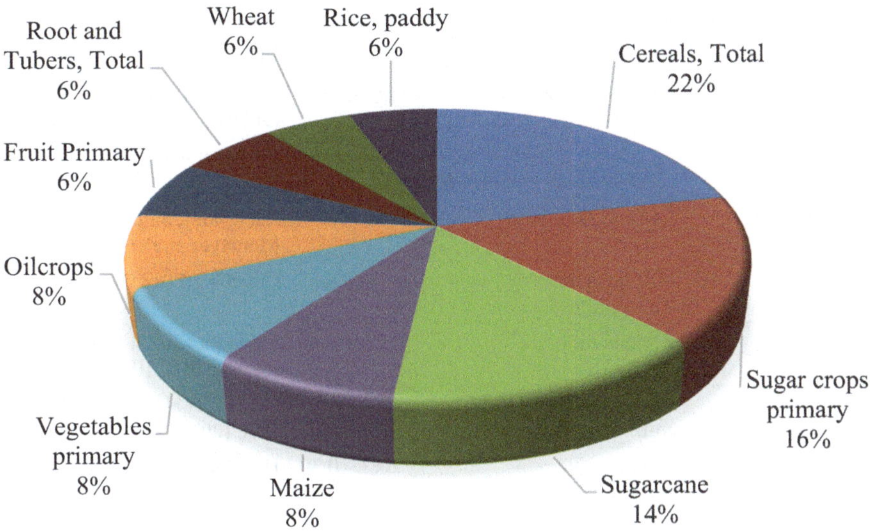

Fig. 9.1 Worldwide produced top ten commodities in 2019 (FAOSTAT 2021)

unutilized (Devi et al. 2017). In the rural context of the developing world, agricultural residues are often either burned onsite to clear the lands or dispose openly (Vitali et al. 2013). Onsite burning is a common practice for cereals crops residue which generates numerous pollutants of potential human health risk and global warming potential while wasting a potential resource with numerous valuable applications (Vitali et al. 2013; Kapoor et al. 2016).

Precise information on the availability of agricultural residues is vital for the formulation and adaptation of sustainable waste management strategies. Despite this, no comprehensive information is available on official or non-official statistical sources about the global annual production of agricultural residues (Santana-Méridas et al. 2012). However, yield production and crop to residue ratio are worldwide adopted common strategies for the estimation of agriculture-based residue availability (Table 9.1). Considering this aspect, an approximate value of agricultural residue has been calculated using the data of agriculture crop production from FAO for the year 2019 (FAOSTAT 2021) and an average value of residue ratio from Table 9.1 for major agriculture crops which are frequently used in the production of biofuels and high-value biochemicals i.e. rice straw (0.755 billion tonnes), rice husk (0.211 billion tonnes), wheat straw (1.07 billion tonnes), wheat bran (0.12 billion tonnes), corn stover (1.15 billion tonnes), corn cob (0.252 billion tonnes), sugarcane bagasse (0.536 billion tonnes). The bioconversion route of lignocellulosic biomass (agricultural residues) into biofuel and high-value biochemicals involves various stages start from pre-treatment followed by saccharification, and fermentation (see Fig. 9.2).

Table 9.1 Some common agricultural residues and their associated residue ratio

Agricultural residue	Criterion	Ratio	References
Rice straw	Straw to grain ratio	0.7–1.5	Lal (2005), Kapoor et al. (2016)
Rice husk	Processing of 1 kg white rice generates 0.28 kg of the husk	0.28	Millati et al. (2016), Knowledge and Bank (2021)
Wheat straw	Straw to grain ratio	1.3–1.4	Lal (2005), Saha et al. (2005), Momayez et al. (2019)
Wheat bran	Milling process of 1 kg wheat grain (w/w)	14–19%	Cui et al. (2013)
Corn stover	Residue ratio	1.01	Mazurkiewicz et al. (2019), Ruan et al. (2019)
Corn cob	Processing of 1.0 kg of ear corn	0.22	Basalan et al. (1995), Millati et al. (2016)
Sugarcane bagasse	Residue ratio	0.25–0.30	Hernández et al. (2019)

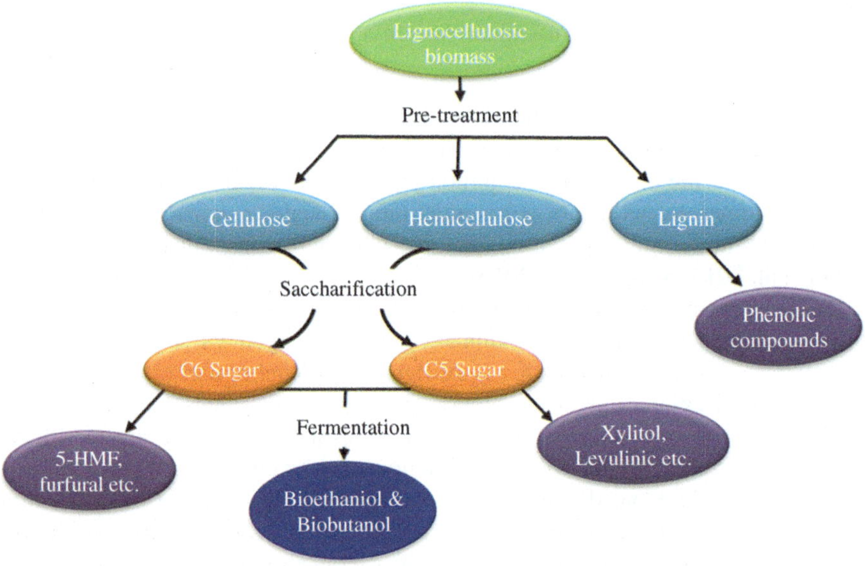

Fig. 9.2 Bioconversion routes of lignocellulosic biomass in biorefinery

9.2.2 *Composition*

Generally, agricultural residue consists of water-insoluble components jointly known as "lignocellulosic materials". These include cellulose, hemicellulose, and lignin and are commonly called lignocellulosic components. The relative proportion and structural complexity of lignocellulosic components depend on the nature of the lignocellulosic materials such as residue source, harvesting, and processing practices (Pasangulapati et al. 2012; Singhvi and Gokhale 2019). The lignocellulosic materials, the most common natural polymer accounted for half of the world's total available biomass (Tsoutsos 2010; Haldar and Purkait 2020). On a dry weight basis, typically lignocellulosic components share 85–90% portion of biomass and the rest are balanced by proteins, ash, pectin (Baruah et al. 2018). Generally, herbaceous crops consist of 25–95% of cellulose, 20–50% of hemicellulose, and 0–40% of lignin (Tarasov et al. 2018).

Cellulose, a prominent universal organic polymer comprises several hundred to ten thousand anhydrous D-glucose units in linear chain unified via β-1,4-glycosidic linkages, and cellobiose acts as repeating units with several degrees of polymerization. This configuration provides high strength to cellulose by developing a strong microfibrils structure (Bai et al. 2019; Sharma et al. 2019; Havstad 2020). Cellulose structure has intertwined crystalline and amorphous regions and the latter is more susceptible to chemical and enzymatic attacks (Singhvi and Gokhale 2019). Hemicellulose is a complex carbohydrate, branched amorphous biopolymer composed chiefly of xylans (pentose sugar) and others hexoses sugar (glucose, mannose, galactose), and uronic acid connected by β-1,4-glycosidic and/or β-1,3-glycosidic and further bonded with cellulose and lignin via hydrogen bond that's why commonly known as "cross-linking glucans" (Pasangulapati et al. 2012; Baruah et al. 2018; Sharma et al. 2019). Hardwood hemicellulose mainly consists of xylans and softwood consists of glucomannans (Singhvi and Gokhale 2019; Sharma et al. 2019). Lignin is the most complex structure and second most abundant organic polymer on the earth, after cellulose. Lignin is a water-insoluble amorphous hetero-three-dimensional biopolymer chiefly comprises a long chain of phenylpropanoid units known as *p*-hydroxyphenyl, guaiacyl, and syringyl (Sharma et al. 2019; Kumla et al. 2020). These phenylpropanoid units provide structural strength and resistance to lignin against physical and biological intrusions. Lignin maintains the integrity of the biomass and causes hindrance in the economic bioconversion of lignocellulosic biomass (Singhvi and Gokhale 2019).

9.3 Necessity and Available Strategies for Pre-treatment of Agricultural Residue

The recalcitrant nature is the major hindrance in the worthy utilization of the agricultural residue in biorefinery. The recalcitrance is a function of the cellulose crystallinity, degree of lignification, structural multiformity, and intricacy of cell-wall ingredients which provide resistance to chemical and biological breakdown (Guerriero et al. 2016; Yogalakshmi et al. 2021). The cellulose resistance comes from the degree of polymerization and crystallinity of cellulose. The number of repeating units in the polymer molecule defines the degree of polymerization (Shrivastava

2018) and in cellulose polymer the repeating unit is cellobiose. The nature of crystallinity is defined by the number of inter-and intra-molecular hydrogen bonds in cellulose and currently, four different crystalline allomorphs have been identified i.e. celluloses I, II, III, and IV. The most abundant crystalline form in nature is cellulose I (Park et al. 2010). The high degree of polymerization and/or high crystallinity limits cellulose degradation by inhibiting the diffusion of the liquefying agent (Jasiukaitytė-Grojzdek et al. 2012). The low solubility of cellulose in water or even in acid under an ambient environment advocates the existence of hydrophobic interactions which makes cellulose a hygroscopic material (Budtova and Navard 2016; Jędrzejczyk et al. 2019). However, high temperature increases the cellulose solubility in both water and acid (Roy et al. 2009; Jędrzejczyk et al. 2019). The alkaline reagents such as sodium hydroxide, phosphoric acid are more efficient in cellulose digestion and sodium hydroxide-water is also known as cellulose solvent (Budtova and Navard 2016).

Hemicellulose, act as a cementing material in the cell wall and create a matrix structure with cellulose microfibrils (Nayak et al. 2021). Three main features that differentiate hemicellulose from cellulose are (1) various dissimilar sugar units, (2) significant chain branching, and (3) 10–100 times less degree of polymerization (Patel and Parsania 2018; Nayak et al. 2021). Hemicellulose has a substantial degree of chain branching due to the presence of several diverse sugar monomers which allow the formation of an open structure to attract water molecules (Benaimeche et al. 2020). This condition makes hemicellulose more hygroscopic and less crystalline than cellulose and less inherent physical strength and chemical resistance (Li et al. 2013). Hemicellulose is extremely hydrophilic (little soluble in water at ambient temperature), soluble in alkali, and easily hydrolyzed by acids (Patel and Parsania 2018). A dilute alkali and water can easily dissolve around three-quarters of hemicellulose at a temperature of more than 180 °C (Li et al. 2013). Further, significant degradation of hemicellulose has been observed at temperatures range from 150–180 °C than cellulose (200–230 °C) (Nayak et al. 2021).

Lignin is a complex, hierarchical, and a highly recalcitrant component of lignocellulosic biomass (Baruah et al. 2018), and the reason is in the polymerization mechanism which involves both chain branching and inter/intra-chain coupling results in a bulky heterogeneous and complex three-dimensional structure (Ruiz-Dueñas and Martínez 2009). The lignin adjoint the cellulose and hemicellulose by covalent binding and forms a protected boundary which provides a limited surface area for biological and chemical intrusion (Baruah et al. 2018; Sharma et al. 2019). Further, lignin exhibits two major mechanisms to resist biological hydrolysis i.e. (1) restrict polysaccharides accessibility which acts like a physical barrier, and (2) non-productive binding with enzymes as an act of inhibition (Li et al. 2016). Lignin is an amorphous polymer structure formed by the monolignols deposition on a protein template in a random pattern. This assembly of hydrocarbon provides hydrophobicity and nonspecific structure to lignin that limits the enzymatic action (Horwath 2007). Lignin is a hydrophobic polymer which makes it water-insoluble at neutral conditions however, high temperature (more than 180 °C) enhance solubility can be

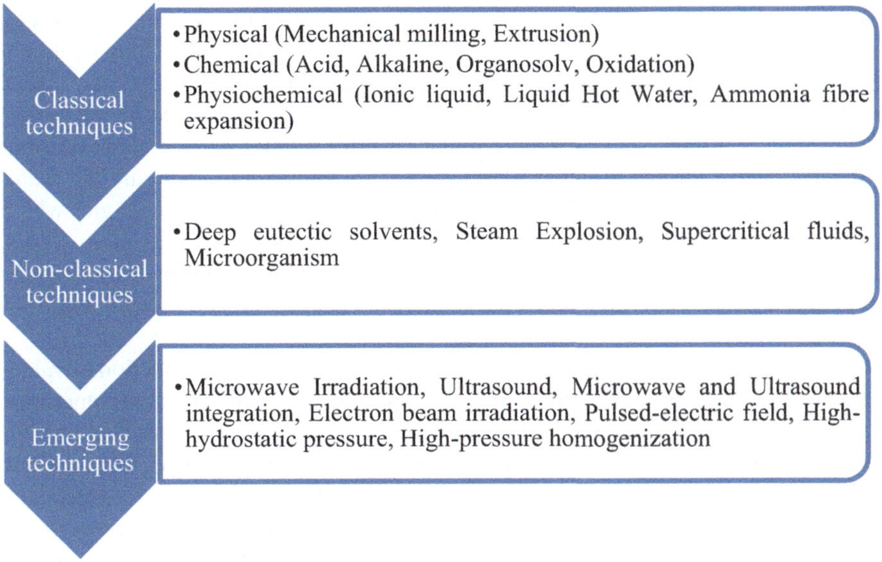

Fig. 9.3 Pre-treatment technologies for lignocellulosic biomass

further improved in the presence of acids (Jędrzejczyk et al. 2019). Therefore, the pre-treatment of agricultural residue is an obligatory necessity to disturb the recalcitrant structure of lignocellulose. The intrusion results in the cracking of lignin encapsulate and allows hemicellulose degradation and reduction in cellulose crystallinity and degree of polymerization (Chen et al. 2017; Baruah et al. 2018). Moreover, these conditions increase the porosity and surface area for better downstream processing of lignocellulosic biomass (Jędrzejczyk et al. 2019).

Numerous pre-treatment techniques have been formulated and tested to improve accessibility for lignocellulosic biomass (see Fig. 9.3) (Hassan et al. 2018). Various studies enlisted the key factors that should be considered to reduce the operational cost, enhance the effectiveness of the entire pre-treatment process, and intensifies the possible recovery of specific high-value biochemicals platform chemicals (Jędrzejczyk et al. 2019; Bhatia et al. 2020). The acid and alkali pre-treatment and saccharification generate various co-products commonly known as inhibitors are of two types (1) process-derived inhibitors (e.g. furan derivatives) and (2) feedstock-inherited inhibitors (e.g. acetic acid and phenolic compounds) (Verardi et al. 2020; Patel and Shah 2021). These inhibitors significantly affect microbial performance during the fermentation process (Moreno et al. 2019; Bhatia et al. 2020). Pre-treatment of lignocellulosic biomass has been explored for nearly two centuries. The non-sustainable classical techniques such as physical, chemical, and physiochemical work differently to separate lignocellulosic components (Hassan et al. 2018). However, these techniques demand huge operational energy, generate toxic chemicals which interfere in the microbial functionality during the fermentation process, and are not even environment friendly (Ravindran and Jaiswal 2016; Hassan et al. 2018;

Kavitha et al. 2020). The biological treatment, known as sustainable green technology is environment friendly but not suitable for commercialization due to long treatment duration, loss of carbohydrates, and low yield. Limitations associated with classical approaches can be eliminated by the use of green technologies based on the "green chemistry". Although, they are environment-friendly but high production costs, poor efficiency, and shortage in the availability of commercial equipment are the major challenges for industrial-level processing. Nowadays, emerging technologies which are the chemical approach combined with recent techniques such as microwaves, ultrasound, gamma-ray, electron beam, pulsed-electric field, high hydrostatic pressure, and high-pressure homogenization can lead to possible commercial solutions (Hassan et al. 2018).

Currently, the microwave irradiation technique is used in the production of value-added chemicals from lignocellulosic biomass via depolymerization at a temperature greater than 200 °C. The microwave irradiation technique provides fast heat transfer in a short time with volumetric uniformity. This technique is energy-efficient and generates low co-product than conventional heating (Hassan et al. 2018). Nuchdang et al. (2020) observed 137.3% increment in reducing sugars with microwave irradiation at 120 °C for 30 min. The lignocellulosic biomass demands strong activation energy to lower the crystallinity of cellulose, for example, about 8 kJ/mol of energy is required to release glucan from crystalline cellulose (Cho et al. 2011). Ultrasound pre-treatment provides high energy as well as maintains a mild reaction environment. Moreover, short residence time and easy integration with other classical methods (chemical pre-treatment) enable commercial upscaling (Haldar and Purkait 2020). The integration of ultrasound technique with alkali treatment increased the delignification of lignocellulosic biomass from 80 to 100% over conventional alkali treatment (Subhedar et al. 2018). The coupling of microwaves and ultrasound techniques increased the surface area of lignocellulosic biomass and enhanced the degradation of hemicellulose (North 2016). The electronic beam irradiation technique has been recently employed on lignocellulosic biomass and has shown the ability to depolymerize cellulose only. Further, hemicellulose and lignin can be hydrolyzed via the integration of the electronic bean technique with conventional methods such as steam explosion or alkali (Xiang et al. 2017). High-pressure homogenization generates a hollow structure in lignocellulosic biomass and preserves the hemicellulose for efficient conversion to sugar in comparison to alkali heat pre-treatment (Jin et al. 2015). High-hydrostatic pressure reduces the processing time and enhanced the activities of the enzymes in comparison to available classical techniques (Eisenmenger and Reyes-De-Corcuera 2009; Hassan et al. 2018).

9.4 Saccharification and Fermentation

9.4.1 Saccharification

Saccharification (also known as hydrolysis) is the conversion of holocellulose (cellulose and hemicellulose) into simple sugar monomers such as glucose, xylose, etc., via two different routes (1) chemical, and (2) biological. The chemical methods that use acid or alkali solutions cause toxicity to enzymes during the fermentation process, however, biological methods utilize enzymes for hydrolysis. Biological methods offer various advantages over chemical methods such as low energy requirement, low toxicity to microorganisms, generates high yields, require mild operation conditions (temperature of 40–50 °C and pH of 4–5), and lower maintenance costs (Saini et al. 2015; Wang and Lü 2021). However, biological saccharification requires substrate-specific enzymes such as cellulase and hemicellulase which elevate the cost of the bioconversion process (Toor et al. 2020). Therefore, the selection of adequate enzymatic saccharification procedure is a major bottleneck for economically viable utilization of agricultural residues in biorefinery. Recent findings suggest that auxiliary enzymes such as lytic polysaccharide monooxygenases and cellobiose de-hydrogenases can work synergistically with cellulase and improve saccharification efficiency (Barbosa et al. 2020; Patel and Shah 2021). Nowadays, enzymatic saccharification shifts the focus on ligninolytic enzymes instead of holocellulolytic enzymes, because the removal of lignin facilitates the easy degradation of holocelluosic components via cellulase and hemicellulase enzymes. The laccase-cellulase mixture allows concurrent pre-treatment and saccharification of lignocellulosic biomass which further reduces the entire cost of the bioconversion process (Masran et al. 2020; Patel and Shah 2021).

9.4.2 Fermentation

Solid-state fermentation and submerged fermentation are the leading techniques for industrial-level production of biofuel and high-value biochemicals and their efficiency depends on the nature of substrates, environmental conditions, and microorganisms. Solid-state fermentation is a recent technique, whereas submerged fermentation is a traditional method that has been in practice for a long time.

Submerged fermentation. Submerged fermentation is performed in a liquid medium having substrate either in the dissolved or suspended form. This technique is generally used for enzyme production at diverse operation modes namely batch, fed-batch, and continuous (Confortin et al. 2019). The process parameters such as pH, temperature, moisture content, and aeration can easily be controlled in submerged fermentation (Martínez-Medina et al. 2019). Further, genetically modified organisms can thrive better in the liquid medium. The submerged fermentation technique allows easy purification and recovery of the platform chemicals for the synthesis of high-value biochemicals from the liquid medium (Suriya et al. 2016). However, this technique requires a constant supply of nutrients in form of substrate because growing microorganisms rapidly consume the substrate and high moisture content can also cause the possibilities of contamination of liquid medium (Subramaniyam and Vimala 2012). Submerged fermentation is a well-established technique for bacterial fermentation of substrate. Submerged fermentation requires well-defined culture media and inoculum for better control on the production of desired compounds in adequate quantity and quality. However, the high liquid to solid ratio in the fermentation reactor enables the downstream processing more difficult and expensive. This condition generates a large quantity of liquid waste which causes dumping problems and further puts a burden on waste management practices. The submerged fermentation technique is commonly employed in the western world to produce enzymes, antibiotics, and other high-value biochemicals (Martínez-Medina et al. 2019).

Solid-state fermentation. Solid-state fermentation requires insoluble solid substrate which provides controlled release of the nutrient for a longer period. The microorganism can be natural microflora especially filamentous fungi as they demand less moisture can lower the contamination possibilities of the substrate. The product of interest is generated in concentration form which enables downstream processing easier, less costly, and less time-consuming. Further, the solid nature of waste generated after the processing offers better waste management. The most important advantage of solid-state fermentation to its counterpart is the nature of the feedstock. The residual biomass from agricultural practices such as straw, stalk, bran, bagasse, etc., can be a potential feedstock for biorefinery. The agricultural residues are less expensive, available in an abundant quality, rich in nutrients, and recycled in solid-state fermentation offers sustainable production of biofuels and high-value biochemicals especially in developing countries (Suriya et al. 2016; Subramaniyam and Vimala 2012; Martínez-Medina et al. 2019).

9.4.3 Integration of Saccharification and Fermentation

There are various fermentation processes integrated with saccharification such as Simultaneous Saccharification and Fermentation (SSF), Co-Fermentation (CF), Simultaneous Saccharification, and Co-Fermentation (SSCF), and Consolidated Bioprocessing (CBP) (see Fig. 9.4) (Patel and Shah 2021). The two most common techniques applied on agricultural residue are SSF and SHF (Separate Hydrolysis and Fermentation) (Tomás-Pejó et al. 2008). Both techniques are highly efficient and commonly utilized on pre-treated agricultural residues but the enzymatic system is the key factor in the productivity evaluation of the entire process (Cannella and Jørgensen 2014; Nitsos et al. 2018).

SHF technique provides better control on enzymatic saccharification and fermentation but hindrance by inhibitors on cellulase activities is a major drawback that can be overcome by separation of inhibitors prior to fermentation process or application of SSF process. The SSF technique provides more economic viability by offering high

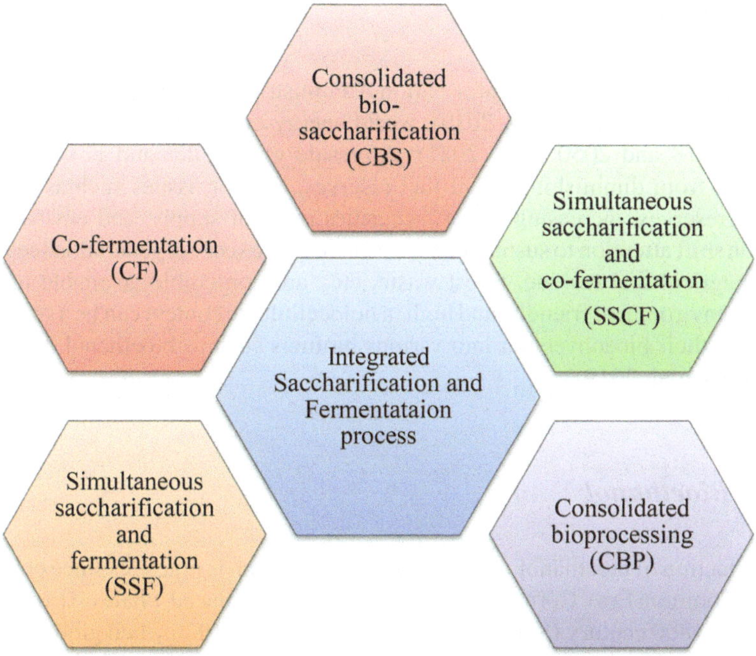

Fig. 9.4 Various integration of saccharification and fermentation process

feedstock loading (Hoyer et al. 2013) but difficulties attributed by high sold loading such as the increase in viscosity, production of more inhibitors, mixing problems, and maintaining optimal conditions for efficient working of enzymes are the limiting factors (Patel et al. 2019). These problems can be resolve with the application of pre-hydrolysis and SSCF. SSCF technique is analogous to SSF and provides better separation possibilities of inhibitors after pre-hydrolysis which subsequently facilitates high biofuel production such as bioethanol by promoting microorganisms to consume both hexose and pentose sugar simultaneously (Chen and Fu 2016; Toor et al. 2020). CF technique is attributed to the simultaneous utilization of hexose and pentose sugar during the fermentation process. CBP technique integrates the enzyme generation, lignocellulosic biomass hydrolysis, and fermentation process in one reactor facilitated with a microbial consortium. CBP has shown superiority over SHF, SSF, and SSCF by offering low energy requirements and exogenous enzymes costs (Agbor et al. 2014). However, the selection of suitable microorganisms is the limiting factor (Li et al. 2020; Patel and Shah 2021).

9.5 Biofuel

According to the U.S. Energy Information Administration report (EIA) i.e. IEO2019 (International Energy outlook 2019), world energy consumption will rise by 50% between 2018 and 2050 (U.S. 2021). Aforesaid energy demand is very difficult to achieve from diminishing fossil fuel reserves. Further, issues such as depletion of fossil reserves, increasing prices, securities issue of supply, and environmental pollution shift attention to sustainable alternatives. The second-generation feedstocks such as agricultural residue, forest waste, etc., are non-edible, available in ample quantity, environment friendly, and high in holocellulosic content can be a sustainable choice for their bioconversion into various biofuels such as bioethanol, biobutanol (Kushwaha et al. 2019).

9.5.1 Bioethanol

The production of bioethanol via sugarcane fermentation is practices since 6000 BCE (Before Common Era). Corn feedstock-based ethanol replaced whale oil (as lamp oil) around the 1800 century (Sticklen 2007). It was the year 1970 when global oil crises shift the focus on the development of alternatives of petroleum products and that upturned the ethanol market (Demirbas 2010). In 2005, the United States incorporate the "*Renewable Fuels Standard*" in their policy aim to touch the production of ethanol from 4 billion gallons in 2006 to 36 billion gallons yearly by 2022 (Bagheri et al. 2016; Halder et al. 2019). In 2016, the United States generated about 57,700 million liters of ethanol from corn (first generation feedstock) and Brazil generates about 27,600 million liters of ethanol from sugarcane (Mohanty and Swain 2019). The worldwide bioethanol production in 2018 was 110 billion liters and anticipate to touch the 140 billion liters mark in 2022 with a compound annual growth rate (CAGR) of 7.6% (Sharma et al. 2020; Edeh 2020). The corn stover-based advanced bioethanol plants cost up to US$ 1.82 to 2.5/litre/year of production capacity (Humbird et al. 2011) and capital cost breakdown is shown in Fig. 9.6. Bioethanol has the largest market in the fuel sector due to its potential as a primary fuel or gasoline additive (Dutt 2016) (see Fig. 9.5).

The bioconversion routes of bioethanol production primarily depend on the source of feedstock, and the efficiency of bioconversion steps i.e. pre-treatment, saccharification, fermentation, and recovery process. The pre-treatment ensures holocellulose availability for the saccharification process by shattering the recalcitrant lignocellulosic structure and share 30% cost per gallon of ethanol (Edeh 2020). This makes pre-treatment the costliest step in lignocellulose biomass-based bioethanol production. The utilization of acid and alkali reagents has been commercially accepted in pre-treatment techniques. However, these classical techniques generate varieties of

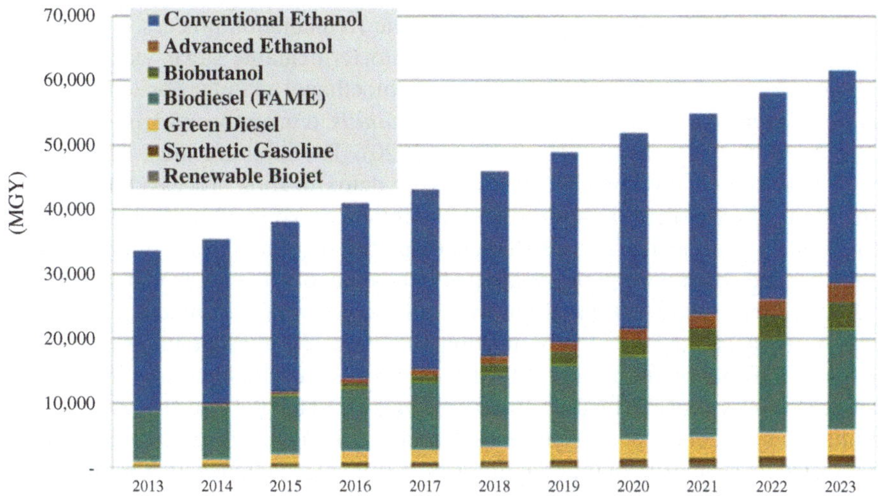

Fig. 9.5 The total biofuels production by fuel type, world markets: 2013–2023 (Bharathiraja et al. 2017)

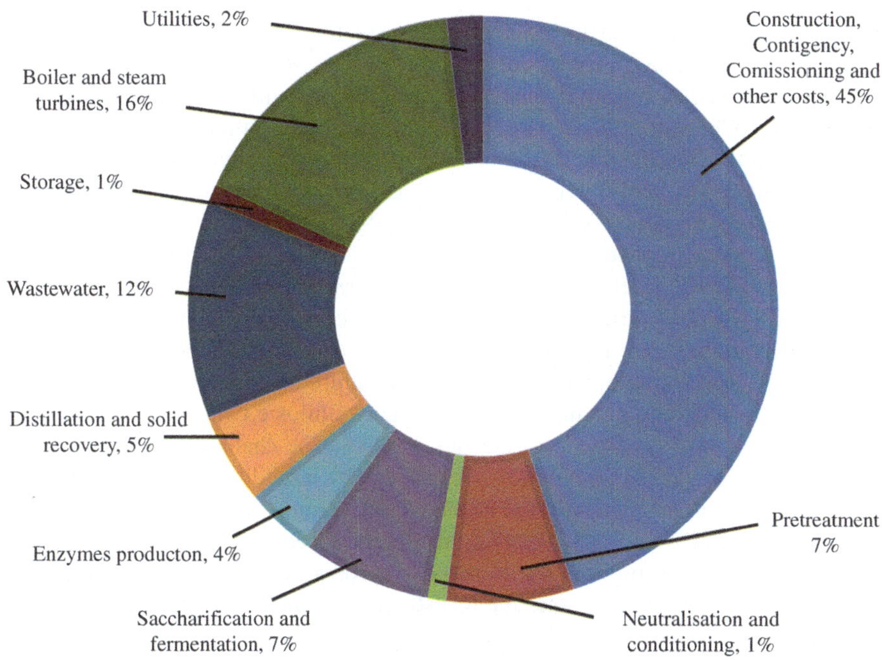

Fig. 9.6 Capital cost breakdown for bioethanol production from corn stover (Humbird et al. 2011)

inhibitors that affect microbial efficiency in the fermentation process. The saccharification process transforms holocellulose into fermentable sugar via lignocellulolytic enzymes subsequently transform into bioethanol via microbial strains such as *Saccharomyces cerevisiae*, *Zymomonas mobilis* (currently used in industries) (Nikolić et al. 2009; Dutt 2016; Su et al. 2020). The inhibitor must be separated before the fermentation process (known as the detoxification process) for the better functioning of the microbial strains. The adsorption, pressure swin adsorption, pervaporation, distillation, reactive distillation, extractive distillation, azeotrophic distillation, etc., are common techniques applied in bioethanol recovery from fermentation reactor (Karimi et al. 2021). High processing costs which include pre-treatment, enzymatic saccharification, detoxification, recovery, and huge capital investment in terms of transportation, pilling of feedstock, process reactors, product distribution, and narrowness between feedstock and bioethanol cost are the three major barriers in profitable lignocellulosic biomass-based biorefineries (Zhang 2008; Patel and Shah 2021). However, the use of the emerging pre-treatment techniques such as microwave irradiation, ultrasound, integrated microwave and ultrasound, electron beam irradiation, etc., can overcome the various issues associated with the classical methods in the development of more efficient and economically viable biorefinery (Hassan et al. 2018).

The inhibitors generated during classical pre-treatment techniques such as organic acids, furan derivatives, phenolic compounds, etc., significantly affect the microbial cell functionality that results in the reduction of cell viability, bioethanol yield, and productivity (Chandel et al. 2011). The most commonly employed detoxification process include (1) physical methods such as evaporation, membrane separation, ion-exchange resins, and active charcoal, (2) chemical methods such as neutralization, extraction via organic solvent, sugaring-out extraction, and salting-out extraction, (3) biological methods such as microbial pre-treatment, microbial acclimatization, in-situ microbial detoxification via *Trichoderma reesei* and enzymatic clarification, and (4) technological integration (Chandel et al. 2011; Kumar et al. 2020a). The genetic engineering approaches and directed evolution methods are the latest approaches to develop competitive strains having inhibitors tolerance (Fu et al. 2011). Also, plant genetic engineering to alter the lignocellulosic composition and lignin syringyl/guaiacyl (S/G) ratio are a curious practice. The change in lignin S/G ratio can reduce the syringyl-rich based recalcitrance which facilitates easy saccharification and low inhibitor generation (Fu et al. 2011; Kumar et al. 2020a). Deshavath et al. (2018) reported 11.6–17.7% of lignin content in genetically modified sorghum.

The conventional process of bioethanol production includes pre-treatment followed by SHF which increases the investment cost and production of inhibitors. Nowadays, saccharification and fermentation processes can be integrated based on the target product and with associated benefits and limitations (Su et al. 2020; Karimi et al. 2021; Alkasrawi et al. 2021). Distillation and dehydration are the two main essential steps in the recovery of anhydrous bioethanol. Distillation is a conventional method that can recover about 92.4 wt% of bioethanol from fermentation beer and the leftover mixture is further processed by dehydration methods such as extractive distillation, azeotropic distillation, pervaporation, adsorption, or complex hybrid separation methods to obtain anhydrous bioethanol (Miranda et al. 2020). Conventional distillation and adsorption were the most reported methods in different literature for bioethanol recovery in the year 2010. In 2020, distillation maintained its position but adsorption was almost replaced by reactive distillation followed by extractive distillation methods for the dehydration according to reported literature (Karimi et al. 2021). The distillation is a commercially approved technique shown dominancy over other available recovery techniques because of proven principles, easy functioning, and realizable expertise (Zhang 2008; Khalid et al. 2019).

9.5.2 *Biobutanol*

In the year 2014, the global bio-butanol market size was valued US$ 7.86 billion which is expected to reach 17.78 billion by 2022 (Grand view research 2021) because of the increasing demand for butanol isomers in various applications. Butanol has four naturally occurring isomers i.e. 1-butanol, 2-butanol, tert-butanol, and iso-butanol and each has its specific application (see Fig. 9.7) (Jin et al. 2011; Rocha et al. 2015; Nawab et al. 2020). In 1861, Louis Pasteur obtained biobutanol as a by-product of starch fermentation via *Vibrion butyrique.* 1-Butanol as complete fuel has many pros over bioethanol such as high energy content (27 MJ/L than 19.6 MJ/L), compatibility with the existing engine, low flammability, hydrophobicity, no or low corrosiveness, good miscibility with gasoline, safety, lesser emission of NO_x, and

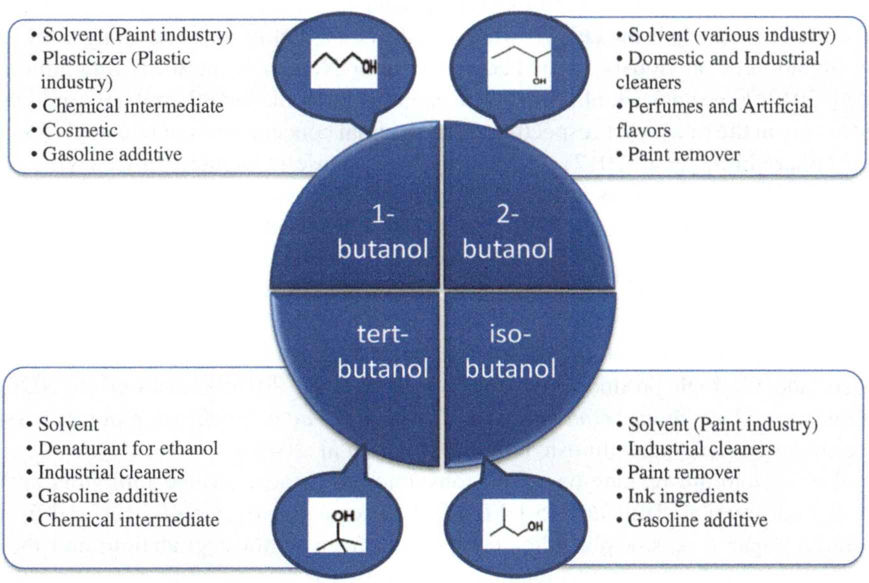

Fig. 9.7 Molecular structure and application of butanol isomers

shoot particles (Nawab et al. 2020; Tsai et al. 2020). Commonly, biobutanol is a by-product of ABE (acetone, butanol, ethanol) process where sugar monomers derived from different agriculture residues are fermented with the diverse genus of solventogenic *Clostridium* species such as *Clostridium acetobutylicum*, *Clostridium beijerinckii*, *Clostridium saccharoperbutylacetonicum*, *Clostridium saccharoperbutylicum*, *Clostridium sporogenes*, *Clostridium perfrigens*, *Clostridium pasteurianum*, *Clostridium carboxidivorus*, *Clostridium tetanomorphum*, *Clostridium aurantibutyricum*, *Clostridium cadaveris*, *Clostridium butylicum*, etc. (Dutt 2016; Verardi et al. 2020). However, only four species of *Clostridium* family have shown the ability to produce high biobutanol yield at optimum conditions i.e. *Clostridium acetobutylicum*, *Clostridium beijerinckii*, *Clostridium saccaroperbutylacetonicum*, and *Clostridium saccharoacetobutylicum*. Among these, *Clostridium acetobutylicum* is the epitome of first-generation feedstock, while *Clostridium beijerinckii* is an ideal candidate for second-generation feedstock (Li et al. 2020). *Clostridium acetobutylicum* potential has been broadly explored with first-generation feedstock such as molasses, sugar beets, corn, cassava, millet, and potato for biobutanol production (Dutt 2016; Kolesinska et al. 2019; Verardi et al. 2020).

Biobutanol production using agricultural residue (second-generation feedstock) involves pre-treatment, saccharification, detoxification, fermentation, and recovery stages. Qureshi et al. (2020) reported a total capital investments cost of US$ 198.16 $\times 10^6$ with SSFR plant (simultaneous saccharification, fermentation, and recovery) of a total plant capacity of 170,000–171,000 metric tons/year with sweet sorghum bagasse ($50/dry ton) as feedstock. The fermentation step is a two-phase process namely (1) acidogenesis, where acetic and butyric acids are produced along with optimum cell growth that decreases pH to around 4.5, and (2) solventogenesis, where low pH of media halt the cell growth and allows solvent production along with re-assimilation of previously formed acids (Amador-Noguez et al. 2011; Bharathiraja et al. 2017; Kolesinska et al. 2019). The acetone, butanol, and ethanol are produced generally in the ratio 3:6:1 respectively and the final concentration of butanol is about 3% (Bharathiraja et al. 2017). The acidogenesis phase is facilitated by enzymes of *Clostridia* species such as amylase, saccharase, glucosidase, glucoamylase, pullulanase, and amylopullulanase (Kolesinska et al. 2019) and solventogenesis phase by aldehyde dehyrogenase, butanol dehydrogenase, alcohol dehydrogenase (Bharathiraja et al. 2017). There are three major challenges in the commercial production of biobutanol from agricultural residues i.e. (1) high cost of pre-treatment process and generation of inhibitors, (2) low butanol titer because of low bacterial tolerance, and (3) high product recovery cost (Xin et al. 2016; Nawab et al. 2020). Moreover, *Clostridium beijerinckii* and *Clostridium acetobutylicum* requires harsh pre-treatment on lignocellulosic biomass (Jiang et al. 2019).

The agriculture residue-based bioconversion generates diverse inhibitors such as 5-(hydroxymethyl)furfural (5-HMF) from hexose sugar (glucose), furfural from pentose sugar (xylose), phenolics compounds from lignin degradation, and their separation based on current technologies are very expensive (see Fig. 9.2) (Bharathiraja et al. 2017). The conventional biobutanol production process (ABE fermentation) with native *Clostridail* species generates low butanol yields i.e. 16–17 g/L (Kolesinska et al. 2019). The major field of research for sustainable production of biobutanol from agricultural residues must focus on (1) selection of low-cost carbon source, (2) improvement in the process design such as integration of saccharification and fermentation step, (3) selection of a potential native microbial strain or development of genetically improved strain, and (4) cost-effective separation of inhibitors and recovery of biobutanol. Agricultural residues such as soya straw, sugarcane leaves, corn husk, and straw are high in holocellulose and low in lignin content and can consider as a suitable feedstock for biobutanol production. Further, alkali pretreatment of agricultural residues significantly reduces the lignin and improves the availability of holocellulose for ABE fermentation (Dong et al. 2019; Raita et al. 2021). The alkali treatment increased the cellulose content from 36.8% to 71.20% for rice straw (Nguyen et al. 2010; Cheng et al. 2012), 44.2% to 74% for soyabean straw (Kim 2018), 37.7% to 71.4% for sorghum straw (Dong et al. 2019), 36.6% to 64% for corn stover (Saha et al. 2016; Yoav et al. 2017).

The common strategies used to improve *Clostridium* sp. efficiency are genetic modification, metabolic engineering, and mutation (Gao et al. 2020). Zhang et al. (2018) reported the highest biobutanol titer i.e. 26.2 g/L from metabolic engineered *Clostridium tyrobutyricum* with glucose as substrate. Similarly, metabolic engineered *Clostridium thermocellum* tolerate butanol toxicity up to 15 g/L i.e. 300% higher than wild strain (Tian et al. 2019). Glucose and xylose are the major monosaccharides of lignocellulosic biomass hydrolysates (Tsai et al. 2020). The *Clostridium* sp. preferred glucose as a carbon source with rapid and efficient utilization (Tsai et al. 2020) and left another simple, second abundant sugar, xylose because of carbon catabolite repression in presence of high glucose concentration which results in the reduction of biobutanol yield (Raita et al. 2021). The co-culturing of *Clostridium* sp. with xylose utilizing species such as *Thermoanaerobacterium* sp. can lower carbon catabolite repression and allows efficient utilization of xylose results in high total biobutanol yield. Jiang et al. (2018) co-cultured *Clostridium acetobutylicum* NJ4 with *Thermoanaerobacterium* sp. M5 and reported improvement in xylose consumption and an increase in biobutanol production to 8.34 g/L via CBP technique under thermophilic conditions (55 °C). Nowadays, some nonsolvent-producing species are engineered to synthesize biobutanol such as *Escherichia coli, Saccharomyces cerevisiae, Pseudomonas putida*, and *Bacillus subtilis* (Li et al. 2020). These microorganisms exhibit well-characterized genetic backgrounds and well-established genetic manipulation systems for biobutanol production at an industrial scale (Li et al. 2020). Distillation, liquid–liquid separation, adsorption, gas-stripping, pervaporation, pertraction and reverse osmosis as individual and/or combinations are the major biobutanol recovery techniques and each has its advantages and limitations. Distillation, liquid–liquid extraction, and adsorption are the traditional techniques and due to economic viability, distillation has been terminated. Membrane separation techniques (pervaporation, pertraction, and reverse osmosis) and gas stripping are emerging novel techniques that have shown superiority over traditional techniques. However, heavy inter-competition in terms of associated advantages over each other in different ways should be explored in this excerpt (Bharathiraja et al. 2017).

9.6 High-Value Biochemicals

The biofuel production from agricultural residues generates various additional compounds (co-products) along with targeted products during different stages of bioconversion. These additional compounds exhibited the potential to hinder the proper functioning of microorganisms during the fermentation process and are designated as inhibitors. However, can be assigned a resource because acts as platform chemicals for the synthesis of numerous high-value biochemicals or itself acts as a product. The high-value biochemicals precursor (platform chemicals) can be obtained in two different ways (1) recovery of co-products during bioconversion process and transform into high-value biochemicals and (2) agricultural residue directly used as feedstock for the production of specific high-value biochemicals. In 2004, the Department of Energy (DOE) acknowledged 12 chemical building blocks of biomass origin, as potential platform chemicals (Werpy and Petersen 2004) further revised the list in 2008 (Bozell and Petersen 2010). The revised list of platform chemical list includes ethanol, furfural, hydroxymethylfurfural, 2,5-furandicarboxylic acid, glycerol, isoprene, succinic acid, 3-hydroxypropionic acid/aldehyde, levulinic acid, lactic acid, sorbitol, and xylitol (Takkellapati et al. 2018). In the following section, the potential of some major platform chemicals obtained from holocellulose hydrolysates for the synthesis of high-value biochemicals are discussed and their application in the various commercial sector has been highlighted.

9.6.1 Cellulose

Cellulose is a highly stable homogeneous crystalized polymer that produces glucose (a hexose sugar) after saccharification. Along with biofuel (target product of biorefinery), a major co-product after the organic acids produced from the glucose is hydroxymethylfurfural, also known as 5-(hydroxymethyl)furfural (5-HMF) (Singhvi and Kim 2020). In recent years, 5-HMF production gains comprehensive consideration because directly produced from cellulose as well as glucose (Rosatella et al. 2011). 5-HMF is a valuable lignocellulosic biomass-derivative chemical synthesized through the dehydration process of fructose (traditional method) and glucose (modern method) via acid catalytic reaction (Ab Rasid et al. 2020). However, the direct transformation of glucose to 5-HMF has various limitations (Shrotri et al. 2017). Therefore, glucose first isomerized to fructose and later converted to 5-HMF (Ab Rasid et al. 2020). Chemically, 5-HMF consists of a furan ring with a hydroxyl group, and a formyl group as functional groups that allow various reactions such as oxidation–reduction, esterification, etc., for the production of numerous high-value biochemicals (Wang et al. 2018).

Some of the common derivatives of 5-HMF synthesized via hydrogenation reaction are 2,5-dimethylfuran, 2-5 dimethyltetrahydrofuran, 2-hydroxymethylfuran, 2-methylfuran, 2,5-bis(hydroxymethyl)furan, and via rehydration reaction are levulinic acid, formic acid, and via oxidation reaction is 2-5 furandicarboxylic acid (Kohli et al. 2019). 5-HMF is the most popular platform chemical obtained from renewable lignocellulosic feedstock such as agricultural residue for the synthesis of next-generation plastic and biofuel (Kohli et al. 2019). In 2004, the United States Department of Energy released a historic report titled "Top Value-Added Chemicals from Biomass," based on potential as a platform chemical for the production of products that could serve as a building block of biorefinery (Bozell and Petersen 2010). The list consists 5-HMF, together with furfural, levulinic acid, 2,5-furandicarboxylic acid as furan-derivatives compounds (Wang et al. 2018). Both monomeric and polymeric carbohydrates can be efficiently utilized for the 5-HMF synthesis (Teong et al. 2014). 5-HMF is a potential and versatile platform chemical for the production of green polymers, resins, solvents, fungicides, fuels, etc. (Putten et al. 2013; Wang et al. 2018). These chemicals have various applications in industries such as packaging, construction, fabric, cosmetics, food, and health (as nutraceuticals agent), herbicides, bioplastics, etc. (see Fig. 9.8) (Zhou et al. 2016a). The most promising high-value biochemicals derived from 5-HMF is 2,5-dimethylfuran (2,5-DMF). 2,5-DMF is the most valuable biofuel synthesized via catalytic hydrogenolysis and exhibits potential as a drop-in replacement for both gasoline and diesel due to high octane number, high energy density, higher boiling point, lower water miscibility, and lower volatility (Wang et al. 2018). Similarly, 5-ethoxymethylfurfural (5-EMF) exhibits potential as a biofuel substitute compound synthesized from 5-HMF (Li et al. 2019). 2,5-bis(hydroxymethyl)furan and 2-5 furandicarboxylic acid act as monomers for the

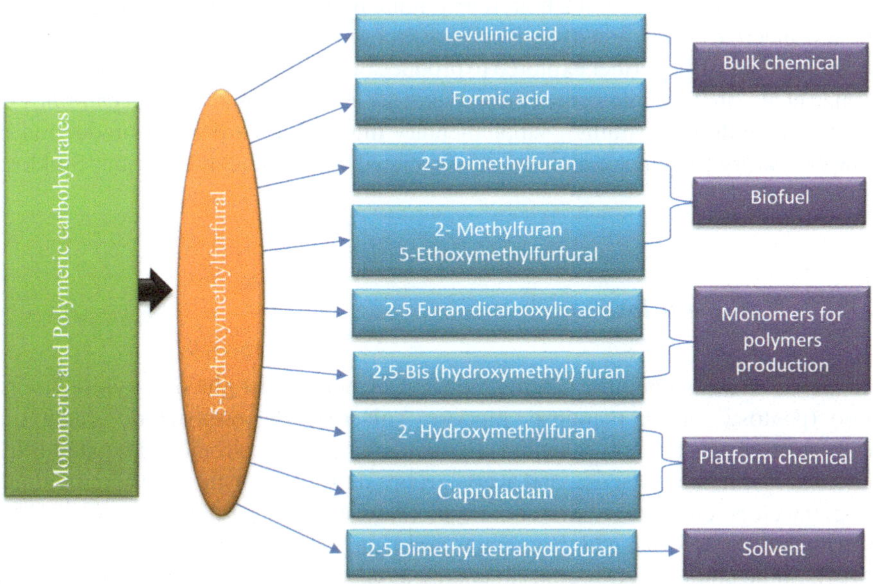

Fig. 9.8 Lignocellulosic biomass-derived 5-hydroxymethylfurfural as a platform chemical for the synthesis of high-value biochemicals

fabrication of polyesters, polyamides, and polyurethane polymers which exhibits superior quality and are also biodegradable (Teong et al. 2014; Kohli et al. 2019). 2-5 furandicarboxylic acid (FDCA) exhibit structural and chemical similarities with petrol-based terephthalic acid offer sustainable replacement of terephthalic acid in the synthesis of viable poly(ethylene 2,5-furandicarboxylate) (PEF), a potential alternative for poly(ethylene terephthalate) (PET) (Li et al. 2019; Yuan et al. 2020a). PEF can be used in the synthesis of biodegradable fibers, films, potable bottles, and food packaging materials (Yuan et al. 2020a).

5-HMF also acts as a building block for the manufacturing of caprolactam, a precursor to Nylon 6,6, (Zhou et al. 2016a). Therefore, 5-HMF is considered a "sleeping giant" in the platform chemicals sector of renewable feedstocks origin (Kohli et al. 2019). However, 5-HMF is a highly unstable compound under the acidic condition and easily rehydrated and transforms into levulinic acid and formic acid (also highly useful matter) or humin (a completely useless matter) (Muranaka et al. 2020). The selectivity and production yield of 5-HMF ameliorate with more effective pre-treatment techniques such as ionic liquids however, the high cost and difficulty in the separation reduce the economic viability and upscaling to industries (Ito et al. 2016; Zhou et al. 2016a; Muranaka et al. 2020). The selection of an appropriate combination of anions and cations for ionic liquid offers a more efficient and cost-effective solution for lignocellulose biomass dissolution (Hou et al. 2017). Khan et al. (2020) reported that the ionic liquid performs as both catalyst and solvent for the synthesis of 5-HMF and levulinic acid from lignocellulosic biomass. Ionic liquids proved their efficiency for wheat straw and rice husks focusing on cellulose (Diaz et al. 2018). In the context of a biorefinery, ionic liquids demonstrated unique potential for selective dissolution of lignocellulosic components or bring physicochemical variation, that can be further explored to produce biofuel and high-value biochemicals (Rajan et al. 2020). While genetic improvements have been made on *Saccharomyces cerevisiae* (industrial strains), supplementary high-value co-product production is now obligatory for generating revenue associated with biofuel production (Stoklosa et al. 2019).

9.6.2 Hemicellulose

Hemicellulose is a complex heterogeneous polymer consist glucose, xylose, arabinose (pentose sugar), mannose, glucose, galactose (hexose sugar), glucuronic acid, galacturonic acid, 4-*O*-methyl-d-glucuronic acid (hexuronic acids), and trace amounts of rhamnose, fucose, and acetyl groups depending on the type of agriculture crop, developmental stage, and tissue type (Anwar et al. 2014; Zhou et al. 2016b). The agricultural residue mainly contains arabinan, galactan, and xylan (Anwar et al. 2014). Some common co-products generated during hemicellulose saccharification are acetic acid (hydrolysis of an acetyl group), levulinic acids (degradation of sugar), furfural (dehydration of pentose sugar under acidic and thermal conditions) (Rosales-Calderon and Arantes 2019). Hemicellulose degradation generates pentose sugar via acid hydrolysis, steam explosion, and organosolv treatment, and is least utilized during the bioconversion of lignocellulosic biomass to biofuel. Hemicellulose is more amenable to hydrolysis due to the absence of crystalline region and produces furfural depending on the pre-treatment methods (Rosatella et al. 2011). The furfural, acetic acid, 5-HMF are generated during dilute acid pre-treatment and found missing in ammonia fibre expansion pre-treatment (Rosales-Calderon and Arantes 2019).

Furfural. Furfural, an important derivative of furan obtained via dehydration of xylose. Furfural is also known as 2 furancarboxyaldehyde, furaldehyde, 2-furanaldehyde, fural, and furfuraldehyde (Mathew et al. 2018). Chemically, furfural consists of a furan ring with a reactive aldehyde functional group at the second carbon position (Mathew et al. 2018; Patel and Shah 2021). Furan-ring system and side aldehyde functional group permit varieties of reaction which make furfural a promising platform chemical for diverse high-value biochemicals synthesis (see Fig. 9.9) (Mathew et al. 2018). Furfural derivatives includes furfuryl alcohol (via hydrogenation reaction), tetrahydrofuran (via hydrogenation reaction), levulinic acid, 2-methylfuran (via decarbonylation reaction), furan (via decarboxylation reaction), furoic acid (via oxidation reaction), furfurylamine (via reductive amination reaction) (Mathew et al. 2018; Patel and Shah 2021). The Quaker Oats is the traditional method for industrial production of furfural (Romero-García et al. 2019). Furfural applications include inks, plastics, antacids, adhesives, nematicides, fungicides, fertilizers, and flavoring compounds (Raman and Gnansounou 2015). Furfural acts as a solvent to extract dienes in the petrochemical refining process (Mathew et al. 2018). Furfuryl alcohol, a derivative of furfural act as a monomer for the synthesis of resins either itself or with others chemicals such as phenol, acetone, or urea (Mathew et al. 2018). Furfuryl-based resins are very hard in structure which permits its use in making fiberglass, and parts in the automotive and aviation sector. Furfuryl alcohol is also used for the synthesis of levulinic acid.

Furfural acts as an intermediate chemical for furan and tetrahydrofuran (THF) synthesis. THF serves as a solvent in the production of photosensitive resins and

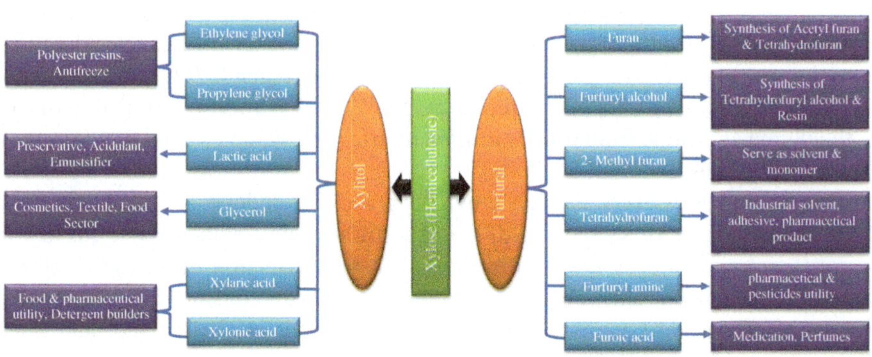

Fig. 9.9 Lignocellulosic biomass-derived furfural and xylitol as a platform chemical for the synthesis of high-value biochemicals

many polymeric materials. THF is an intermediate chemical for the synthesis of nylon 6, nylon-6,6, and synthetic pesticides, e.g. fenbutatin (Kamm et al. 2013). THF has a high reputation in the paint industry due to its application as surface coatings and anti-corrosion coatings. Carbetapentane, rifamycin, progesterone, and some hormone drugs are also synthesized from THF (Ram et al. 2019). Recently, pyrrole and D-proline (high-value N-containing chemicals) are reported from lignocellulose biomass can boost the economic viability of biorefinery. Pyrrole and D-proline have shown structure similarities with furans and furfural can be transformed into pyrrole and subsequently into high-value D-proline. Moreover, L/D-proline can be used either directly, or as a precursor for pharmaceuticals application (Song et al. 2020). The expected market of L/D-proline would reach US$ 310 million in 2025 (Song et al. 2020) and could raise the economic viability of lignocellulosic biomass-based biorefinery and zero waste biorefinery.

Xylitol. Xylitol is a naturally occurring alcoholic compound in plant biomass in low quantities but produced in significant quantities from its precursor xylose (Benahmed et al. 2020). The enzymatic saccharification (an important step in biorefinery) of the agricultural residue that produces xylose offers economic bioconversion to xylitol via yeast or bacteria strains. Therefore, hemicellulosic hydrolysates serve as primary raw materials, and recombinant *Escherichia coli* is a potential strain for the economical production of xylitol (Benahmed et al. 2020; Yuan et al. 2020b). Xylitol is a polyol that includes a five-carbon skeleton with a hydroxyl group at each carbon atom. Xylitol-based toothpaste and chewing gum exhibit antiplaque properties which prevent dental caries because the microorganisms attributed to dental plaque are incapable to ferment the xylitol. Henceforth, xylitol can be an active ingredient in chewing gums, toothpaste, and antimicrobial mouthwashes, especially in candidiasis conditions (Talattof et al. 2018; Benahmed et al. 2020; Lugani et al. 2021). Xylitol can suppress inflammatory reactions and also exhibit anti-carcinogenic properties (Lugani et al. 2021). Xylitol uses in the production of polyesters such as poly(xylitol sebacate) PXS having adjustable mechanical and degradation properties (Mohd Sani et al. 2021). Xylitol is a low-calorie sugar and can be a potential ingredient in the food industry (Irmak et al. 2017). Due to the promising aspect of xylitol in the food and beverages sector as a low-calorie sweetener for health and weight conscious consumers, the global market of xylitol can reach 0.2665 million metric tons valued at around US$ 1000 million by 2022, at 5.7% robust CAGR in terms of both volume and value from 0.1909 million metric tons, valued at US$ 725.9 million in 2016 (Research and markets 2021).

The xylitol derivatives include ethylene glycol, propylene glycol, lactic acid, gylcerol, xylaric acid, xylonic acid, hydroxy furans, etc., with diverse utilities in various commercial sectors (see Fig. 9.9). The bioconversion of xylose-to-xylitol is an energy-efficient and eco-friendly approach and can consider as a sustainable route for large-scale commercial production (Cortez et al. 2016; Patel and Shah 2021). The application of biotechnologically engineered strains of yeasts, fungi, and bacteria can further boost the viability of the zero-waste biorefinery. Recently, *Zymomonas mobilis* ATCC ZW658, a biotechnologically engineered strain exhibits a higher utilization rate for xylose, about 1.65-fold more than the parent ones, and further shown the ability in co-utilization of glucose and xylose (Sarkar et al. 2020). A recombinant yeast *Saccharomyces cerevisiae* efficiently consumes wheat stalk hydrolysate as a source of xylose for xylitol production (Reshamwala and Lali 2020). Recombinant *Escherichia coli* and engineered strain, *Candida tropicalis* XZX-B4ZG also shown promising results for economical production of xylitol (Yuan et al. 2020b; Zhang et al. 2021a). The xylose reductase and xylitol dehydrogenase are the two key enzymes in the transformation of xylose into xylitol (Mohamad et al. 2015). Xylose reductase improved the ability of xylose extraction (21.5–116.7 g/L) from various agricultural residues such as rice straw, cane bagasse, corncob, and corn stover (Azizah 2019). Recently, xylose reductase from *Barnettozyma populi* Y-12728 strain generates arabitol-free xylitol from undetoxified corn stover hemicellulosic hydrolysate (Saha and Kennedy 2020).

Levulinic acid. Levulinic acid (also known as 4-oxo-pentanoic acid) is produced from two different precursors, namely 5-HMF (via rehydration of 5-HMF) and furfural alcohol (via hydrogenation of furfural produced furfuryl alcohol which further converted to levulinic acid) (Khromova et al. 2016; Audemar et al. 2020). The chemical structure of levulinic acid comprises an acetone group and a carboxylic acid group that favors diverse reactions. Levulinic acid offers high versatility as a potential building block chemical for the production of diverse organic chemicals with numerous potential applications in diverse sectors such as agricultural (herbicides), petroleum (biofuel and fuel additives), cosmetics, food, and beverages (flavors, plant growth simulator), pharmaceutical (capsules, pill, injection), tobacco (nicotine), rubber and plastic (commodity chemical and also known as the forefather of these industries), etc. (Pileidis and Titirici 2016; Kumar et al. 2020b). Levulinic acid is a keto-acid and acts as a precursor for the synthesis of resins, plasticizers, textile, coatings, and fuel additives (Nzediegwu et al. 2021). Levulinic acid derivatives include γ-valerolactone (GVL) (via hydrogenation reaction), 2-methyltetranhydrofuran (MTHF) (via hydrogenation reaction), ethyl levulinate (via esterification reaction), succinic acid, 2-pentanol, 2-butanol, etc. (Kumar et al. 2020b). Holocellulose degradation generates levulinic acid that acts as a potential precursor for biofuel and high-value biochemicals (see Fig. 9.10).

Levulinic acid derivatives such as γ-valerolactone (GLV), 2-methyltetrahydrofuran (2-MTHF), 2-methylene GVL, esters of levulinic acid, hydroxyvaleric acid (HA), angelicalactone, etc., can used as a fuel additive. Among these, 2-methyl-tetrahydrofuran and γ-valerolactone (GLV) can gladly blend with petroleum products such as gasoline and diesel (Rackemann and Doherty 2011; Escobedo et al. 2019; Al-Lami et al. 2020). The 5-aminolevulinic acid is an innocuous and eco-friendly agrochemical that exhibits herbicides properties. 5-aminolevulinic acid and its derivatives act as natural precursors for the synthesis of diverse compounds having wide application in photodynamic therapy (PDT), especially in the treatment of various cancers (Li et al. 2019; Zhang et al. 2021b). The levulinic acid market is expected to reach a valuation of US$ 41.2 Mn in 2027 demonstrates a CAGR of 5% during the forecast period (2019–2027) (Transparency market research 2021). Lignocellulosic biomass is a tremendously low-cost feedstock that is less than 5% of the cost of maleic anhydride, currently the main

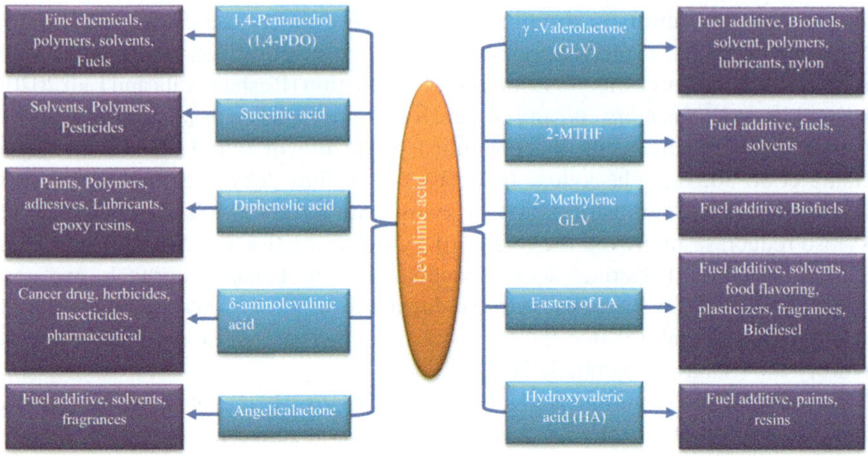

Fig. 9.10 Lignocellulosic biomass-derived levulinic acid as a platform chemical for the synthesis of high-value biochemicals

feedstock for the production of levulinic acid of high purity (analytical grade) (Rackemann and Doherty 2011). The efficient production of levulinic acid is highly dependent on the nature of the feedstock and various agricultural residues proved their ability for levulinic acid production such as bagasse, corn straw, corn stalks, corn strove, sorghum grains, cotton stems, wheat straw, rice straw, and rice husk (Rackemann and Doherty 2011; Kumar et al. 2020b; Thakkar et al. 2021; Junior et al. 2021).

9.7 Challenges and Future Prospective

Presently lignocellulosic biomass-based biorefinery is target product-oriented where sugar monomers of holocellulose are transformed into biofuel and residue of holocellulose and lignin are utilized for the generation of value-added chemicals and burned to generate steam electricity. This approach demands huge capital investment and high processing costs that reduce the economical margin. Other crucial factors are cost, availability of single feedstock throughout the year, and transportation of feedstock from field to industries in the commercialization of lignocellulose biorefineries (Patel and Shah 2021). Junqueira et al. (2016) reported that the 50% cost in cellulose-based ethanol production is shared by feedstock price. Further, the profitable biorefinery has to overcome the diverse technical challenges associated with steps in the route of bioconversion. Preliminary and utmost challenges associated with pre-treatment of lignocellulosic biomass due to low economical viability of conventional (classical) techniques. However, the application of emerging pre-treatment techniques such as microwave irradiation, ultrasound, etc., can overcome the limitation of classical techniques as well as improved the lignocellulosic biomass digestibility (Hassan et al. 2018). Recently, CBP and SSCF gained recognition in resolving the problems associated with the most commonly employed saccharification and fermentation techniques in industries i.e. SHF and SSF (Diaz et al. 2018). However, in industrial perspectives, identification of natural high tolerance microbial species to their end metabolic product or designing of genetically improved species are the limiting factor in the commercialization of CBP techniques (Agbor et al. 2014; Patel and Shah 2021). The results from numerous researches have shown promising and encouraging aspects regarding the economical viability of lignocellulosic biorefineries. Lignol Energy Corporation, British Columbia; Verenium Corporation, San Diego, California, and Mascoma Corporation, Lebanon, New Hampshire are some of the promising companies which commercialized agricultural residue-based production of advanced biofuel and high-value biochemicals (Isikgor and Becer 2015). The major drivers in the commercialization of lignocellulosic biorefinery and zero-waste biorefinery are the suitable feedstock, low pre-treatment cost, low utility cost, sustainable waste management, adequate yield, tolerant microbial strains, and synthesis of high value biochemical from co-product of bioconversion. The government policies need further reforms to promote the production and use of biofuel and other high-value biochemicals. For the successful commercialization of agriculture residue-based biorefinery, the triangle among producer (farmers), improviser (research institute), and the recipient (industries) solely depends on the interaction among them and is supported by the government policies (Ginni et al. 2021).

9.8 Conclusion

Biofuel production using 1st generation feedstock has a long history. However, the global oil crisis in 1970 shifted the focus to develop a sustainable alternative of petroleum products that upturned the biofuel market such as bioethanol, biobutanol, etc. The agricultural residue (2nd generation feedstock) is renewable and potential biomass for biofuel production due to its lignocellulosic components that are generally wasted by various orthodox practices. Lignocellulosic biomass consists of holocellulose that can be converted into biofuel via fermentation process. However, the recalcitrance nature of lignocellulosic biomass is the major hurdle in their sustainable utilization in biorefinery. Various techniques are employed to break the recalcitrance known as pre-treatment. However, this increases the cost of the bioconversion process and generates toxic chemicals with various environmental issues. Numerous novel techniques are developed for a cost-effective pre-treatment and to improve the viability of the biorefinery. The pre-treatment and the subsequent process of biofuel production using agricultural residue generates various co-product (known as inhibitors) initially considered as waste products. These co-products such as 5-HMF, furfural, xylitol, levulinic acid, etc., have been universally acknowledged as a promising platform chemical for the synthesis of various high-value biochemicals. These co-products can be extracted during the bioconversion process which makes agricultural biomass a complete utility for bioconversion to biofuel and high-value biochemicals. The commercial utility of these platform chemicals raised the economic viability of agricultural biomass-based biorefinery by providing additional income from waste resources and also offer simultaneously waste management which promotes the concept of zero-waste biorefinery.

References

Ab Rasid NS, Zainol MM, Amin NAS (2020) Pretreatment of agroindustry waste by ozonolysis for synthesis of biorefinery products. In: Kumar P, Gnansounou E, Raman JK, Baskar G (eds) Refining biomass residues for sustainable energy and bioproducts. Academic Press, pp 303–336. https://doi.org/10.1016/B978-0-12-818996-2.00014-4

Agbor V, Carere C, Cicek N, Sparling R, Levin D (2014) Biomass pretreatment for consolidated bio processing (CBP). In: Waldron K (ed) Advances in biorefineries. Elsevier inc, New York, pp 234–258.https://doi.org/10.1533/9780857097385.1.234

Alkasrawi M, Al-Othman A, Tawalbeh M, Doncan S, Gurram R, Singsaas E, Almomani F, Al-Asheh S (2021) A novel technique of paper mill sludge conversion to bioethanol toward sustainable energy production: effect of fiber recovery on the saccharification hydrolysis and fermentation. Energy 223:120018. https://doi.org/10.1016/j.energy.2021.120018

Al-Lami M, Havasi D, Batha B, Pusztai E, Mika LT (2020) Isobaric vapor-liquid equilibria for binary mixtures of biomass-derived γ-valerolactone+ tetrahydrofuran and 2-methyltetrahydrofuran. J Chem Eng Data 65(6):3063–3071. https://doi.org/10.1021/acs.jced.0c00084

Amador-Noguez D, Brasg IA, Feng XJ, Roquet N, Rabinowitz JD (2011) Metabolome remodeling during the acidogenic-solventogenic transition in Clostridium acetobutylicum. Appl Environ Microbiol 77(22):7984–7997. https://doi.org/10.1128/AEM.05374-11

Anwar Z, Gulfraz M, Irshad M (2014) Agro-industrial lignocellulosic biomass a key to unlock the future bio-energy: a brief review. J Radiat Res Appl Sci 7(2):163–173. https://doi.org/10.1016/j.jrras.2014.02.003

Audemar M, Wang Y, Zhao D, Royer S, Jérôme F, Len C, De Oliveira Vigier K (2020) Synthesis of furfuryl alcohol from furfural: a comparison between batch and continuous flow reactors. Energies 13(4):1002. https://doi.org/10.3390/en13041002

Avadhanam S, Polumat A, Ellamraju S (2020) Usage of crop residue for making livestock feed. ILRI project report. Nairobi, Kenya: International Livestock Research Institute (ILRI)

Azizah N (2019) Biotransformation of xylitol production from xylose of lignocellulose biomass using xylose reductase enzyme. J Food Life Sci 3(2):103–112. https://doi.org/10.21776/ub.jfls.2019.003.02.06

Bagheri H, Espi E, Cacho MTL (2016) Fuel ethanol at a key economic and technological junction in the USA. SSRN, pp 1–11. https://doi.org/10.2139/ssrn.2789612

Bai FW, Yang S, Ho NWY (2019) Fuel ethanol production from lignocellulosic biomass. In: Moo-Young M (ed) Comprehensive biotechnology, vol 3. Elsevier Inc, New York, pp 49–65. https://doi.org/10.1016/B978-0-444-64046-8.00150-6

Banu JR, Kavitha S, Tyagi VK, Gunasekaran M, Karthikeyan OP, Kumar G (2021) Lignocellulosic biomass based biorefinery: a successful platform towards circular bioeconomy. Fuel 302:121086. https://doi.org/10.1016/j.fuel.2021.121086

Barbosa FC, Silvello MA, Goldbeck R (2020) Cellulase and oxidative enzymes: new approaches, challenges and perspectives on cellulose degradation for bioethanol production. Biotech Lett 42(6):875–884. https://doi.org/10.1007/s10529-020-02875-4

Baruah J, Nath BK, Sharma R, Kumar S, Deka RC, Baruah DC, Kalita E (2018) Recent trends in the pretreatment of lignocellulosic biomass for value-added products. Front Energy Res 6:141. https://doi.org/10.3389/fenrg.2018.00141

Basalan M, Bayhan R, Secrist DS, Hill J, Owens FN, Witt M, Kreikemeier K (1995) Corn maturation: changes in the grain and cob. Animal Science Research Report, pp 92–98

Benahmed AG, Gasmi A, Arshad M, Shanaida M, Lysiuk R, Peana M, Pshyk-Titko I, Adamiv S, Shanaida Y, Bjørklund G (2020) Health benefits of xylitol. Appl Microbiol Biotechnol 104:7225–7237. https://doi.org/10.1007/s00253-020-10708-7

Benaimeche O, Seghir NT, Sadowski T, Mellas M (2020) The utilization of vegetable fibers in cementitious materials. In: Hashmi S, Choudhury IA (eds) Encyclopedia of renewable and sustainable materials. Elsevier Inc, New York, pp 649–662. https://doi.org/10.1016/B978-0-12-803581-8.11596-6

Bharathiraja B, Jayamuthunagai J, Sudharsanaa T, Bharghavi A, Praveenkumar R, Chakravarthy M, Yuvaraj D (2017) Biobutanol-an impending biofuel for future: a review on upstream and downstream processing tecniques. Renew Sustain Energy Rev 68:788–807. https://doi.org/10.1016/j.rser.2016.10.017

Bhatia SK, Jagtap SS, Bedekar AA, Bhatia RK, Patel AK, Pant D, Banu JR, Rao CV, Kim YG, Yang YH (2020) Recent developments in pretreatment technologies on lignocellulosic biomass: effect of key parameters, technological improvements, and challenges. Biores Technol 300:122724. https://doi.org/10.1016/j.biortech.2019.122724

Bozell JJ, Petersen GR (2010) Technology development for the production of biobased products from biorefinery carbohydratesthe US department of energy's "Top 10" revisited. Green Chem 12:539–554. https://doi.org/10.1039/B922014C

Budtova T, Navard P (2016) Cellulose in NaOH-water based solvents: a review. Cellulose 23(1):5–55. https://doi.org/10.1007/s10570-015-0779-8

Cannella D, Jørgensen H (2014) Do new cellulolytic enzyme preparations affect the industrial strategies for high solids lignocellulosic ethanol production? Biotechnol Bioeng 111(1):59–68. https://doi.org/10.1002/bit.25098

Chandel AK, da Silva SS, Singh OV (2011) Detoxification of lignocellulosic hydrolysates for improved bioethanol production. In: dos Santos Bernardes MA (ed) Biofuel production—recent developments and prospects. IntechOpen, p 225. https://doi.org/10.5772/16454

Chen H, Fu X (2016) Industrial technologies for bioethanol production from lignocellulosic biomass. Renew Sustain Energy Rev 57:468–478. https://doi.org/10.1016/j.rser.2015.12.069

Chen H, Liu J, Chang X, Chen D, Xue Y, Liu P, Lin H, Han S (2017) A review on the pretreatment of lignocellulose for high-value chemicals. Fuel Process Technol 160:196–206. https://doi.org/10.1016/j.fuproc.2016.12.007

Cheng CL, Che PY, Chen BY, Lee WJ, Lin CY, Chang JS (2012) Biobutanol production from agricultural waste by an acclimated mixed bacterial microflora. Appl Energy 100:3–9. https://doi.org/10.1016/j.apenergy.2012.05.042

Cho HM, Gross AS, Chu JW (2011) Dissecting force interactions in cellulose deconstruction reveals the required solvent versatility for overcoming biomass recalcitrance. J Am Chem Soc 133(35):14033–14041. https://doi.org/10.1021/ja2046155

Confortin TC, Spannemberg SS, Todero I, Luft L, Brun T, Alves EA, Kuhn RC, Mazutti MA (2019) Microbial enzymes as control agents of diseases and pests in organic agriculture. In: Gupta VK, Pandey A (eds) New and future developments in microbial biotechnology and bioengineering. Elsevier Inc, New York, pp 321–332. https://doi.org/10.1016/B978-0-444-63504-4.00021-9

Cortez DV, Mussatto SI, Roberto IC (2016) Improvement on D-xylose to xylitol biotransformation by Candida guilliermondii using cells permeabilized with Triton X-100 and selected process conditions. Appl Biochem Biotechnol 180:969–979. https://doi.org/10.1007/s12010-016-2146-0

Cui SW, Wu Y, Ding H (2013) The range of dietary fibre ingredients and a comparison of their technical functionality. In: Delcour JA, Poutanen K (eds) Fibre-rich and wholegrain foods: improving quality. Elsevier Inc, New York, pp 96–119. https://doi.org/10.1533/9780857095787.1.96

da Rocha RC, Alviso DA, Darabiha N, dos Santos RG (2015) Comparison between N-Butanol and ethanol combustion kinetic models through one-dimensional premixed flame simulations. In: Proceedings of the 23rd ABCM international congress of mechanical engineering. Rio de Janiero, Brazil, pp 6–11

Demirbas A (2010) Biorefineries: for biomass upgrading facilities. Springer, London. https://doi.org/10.1007/978-1-84882-721-9

Deshavath NN, Mahanta S, Goud VV, Dasu VV (2018) Chemical composition analysis of various genetically modified sorghum traits: pretreatment process optimization and bioethanol production from hemicellulosic hydrolyzates without detoxification. J Environ Chem Eng 6(4):5625–5634. https://doi.org/10.1016/j.jece.2018.08.002

Devi S, Gupta C, Jat SL, Parmar MS (2017) Crop residue recycling for economic and environmental sustainability: the case of India. Open Agric 2(1):486–494. https://doi.org/10.1515/opag-2017-0053

Diaz AB, Blandino A, Caro I (2018) Value added products from fermentation of sugars derived from agro-food residues. Trends Food Sci Technol 71:52–64. https://doi.org/10.1016/j.tifs.2017.10.016

Dong M, Wang S, Xu F, Wang J, Yang N, Li Q, Chen J, Li W (2019) Pretreatment of sweet sorghum straw and its enzymatic digestion: insight into the structural changes and visualization of hydrolysis process. Biotechnol Biofuels 12(276):1–11. https://doi.org/10.1186/s13068-019-1613-6

Dutt D (2016) Biotechnological transformation of lignocellulosic biomass in to industrial products: an overview. Adv Biosci Biotechnol 7(3):149–168. https://doi.org/10.4236/abb.2016.73014

Edeh I (2020) Bioethanol production: an overview. Intechopen. https://doi.org/10.5772/intechopen.94895

Eisenmenger MJ, Reyes-De-Corcuera JI (2009) High pressure enhancement of enzymes: a review. Enzyme Microb Technol 45(5):331–347. https://doi.org/10.1016/j.enzmictec.2009.08.001

Escobedo JLG, Mäkelä E, Braunschweiler A, Lehtonen J, Lindblad M, Puurunen RL, Karinen R (2019) Solvent-free hydrodeoxygenation of γ-nonalactone on noble metal catalysts supported on zirconia. Top Catal 62(7):724–737. https://doi.org/10.1007/s11244-019-01161-6

FAOSTAT—Food and Agriculture Organization Corporate Statistical Database. Food and Agriculture Organization of the United Nations (FAO). World Production/Crops. http://www.fao.org/faostat/en/?#data/QC, last accessed 08 July 2021.

Fu C, Mielenz JR, Xiao X, Ge Y, Hamilton CY, Rodriguez M, Chen F, Foston M, Ragauskas A, Bouton J, Dixon RA (2011) Genetic manipulation of lignin reduces recalcitrance and improves ethanol production from switchgrass. Proc Natl Acad Sci 108(9):3803–3808. https://doi.org/10.1073/pnas.1100310108

Gao Y, Zhang M, Zhou X, Guo X, Lei C, Li W, Lu D (2020) Effects of carbon ion beam irradiation on butanol tolerance and production of Clostridium acetobutylicum. Front Microbiol 11(12):1–13. https://doi.org/10.3389/fmicb.2020.602774

Ginni G, Kavitha S, Kannah Y, Bhatia SK, Kumar A, Rajkumar M, Kumar G, Pugazhendhi A, Chi NTL (2021) Valorization of agricultural residues: different bio refinery routes. J Environ Chem Eng 9:105435. https://doi.org/10.1016/j.jece.2021.105435

Grand view research: https://www.grandviewresearch.com/industry-analysis/bio-butanol-industry, last accessed on 16 July 2021

Guerriero G, Hausman JF, Strauss J, Ertan H, Siddiqui KS (2016) Lignocellulosic biomass: biosynthesis, degradation, and industrial utilization. Eng Life Sci 16:1–16. https://doi.org/10.1002/elsc.201400196

Haldar D, Purkait MK (2020) A review on the environment-friendly emerging techniques for pretreatment of lignocellulosic biomass: mechanistic insight and advancements. Chemosphere 128523. https://doi.org/10.1016/j.chemosphere.2020.128523

Halder P, Azad K, Shah S, Sarker E (2019) Prospects and technological advancement of cellulosic bioethanol ecofuel production. In: Azad K (ed), Advances in eco-fuels for a sustainable environment. Elsevier inc, New York, pp 211–236. https://doi.org/10.1016/B978-0-08-102728-8.00008-5

Hassan SS, Williams GA, Jaiswal AK (2018) Emerging technologies for the pretreatment of lignocellulosic biomass. Biores Technol 262:310–318. https://doi.org/10.1016/j.biortech.2018.04.099

Havstad MR (2020) Biodegradable plastics. In: Letcher TM (ed) Plastic waste and recycling. Elsevier Inc, New York, pp 97–129. https://doi.org/10.1016/B978-0-12-817880-5.00005-0

Hernández C, Escamilla-Alvarado C, Sánchez A, Alarcón E, Ziarelli F, Musule R, Valdez-Vazquez I (2019) Wheat straw, corn stover, sugarcane, and agave biomasses: chemical properties, availability, and cellulosic-bioethanol production potential in Mexico. Biofuels, Bioprod Biorefin 13(5):1143–1159. https://doi.org/10.1002/bbb.2017

Horwath W (2007) Carbon cycling and formation of soil organic matter. In: Paul EA (ed) Soil microbiology, ecology and biochemistry, 3rd edn. Academic Press, pp 303–339. https://doi.org/10.1016/B978-0-08-047514-1.50016-0

Hou Q, Ju M, Li W, Liu L, Chen Y, Yang Q (2017) Pretreatment of lignocellulosic biomass with ionic liquids and ionic liquid-based solvent systems. Molecules 22(3):490. https://doi.org/10.3390/molecules22030490

Hoyer K, Galbe M, Zacchi G (2013) The effect of prehydrolysis and improved mixing on high-solids batch simultaneous saccharification and fermentation of spruce to ethanol. Process Biochem 48(2):289–293. https://doi.org/10.1016/j.procbio.2012.12.020

Humbird D, Davis R, Tao L, Kinchin C, Hsu D, Aden A, Schoen P, Lukas J, Olthof B, Worley M, Sexton D (2011) Process design and economics for biochemical conversion of lignocellulosic biomass to ethanol: dilute-acid pretreatment and enzymatic hydrolysis of corn stover (No. NREL/TP-5100-47764). National Renewable Energy Lab. (NREL), Golden, CO (United States). https://doi.org/10.2172/1013269

Iqbal R, Raza MAS, Valipour M, Saleem MF, Zaheer MS, Ahmad S, Toleikiene M, Haider I, Aslam MU, Nazar MA (2020) Potential agricultural and environmental benefits of mulches-a review. Bull Nat Res Centre 44:1–16. https://doi.org/10.1186/s42269-020-00290-3

Irmak S, Canisag H, Vokoun C, Meryemoglu B (2017) Xylitol production from lignocellulosics: are corn biomass residues good candidates? Biocatal Agric Biotechnol 11:220–223. https://doi.org/10.1016/j.bcab.2017.07.010

Isikgor FH, Becer CR (2015) Lignocellulosic biomass: a sustainable platform for the production of bio-based chemicals and polymers. Polym Chem 6(25):4497–4559. https://doi.org/10.1039/C5PY00263J

Ito R, Miyafuji H, Miyazaki Y, Kawai T (2016) Production of 5-hydroxymethylfurfural from wood by ionic liquid treatment. J Wood Sci 62(4):349–355. https://doi.org/10.1007/s10086-016-1554-7

Jasiukaitytė-Grojzdek E, Kunaver M, Poljanšek I (2012) Influence of cellulose polymerization degree and crystallinity on kinetics of cellulose degradation. BioResources 7(3):3008–3027

Jędrzejczyk M, Soszka E, Czapnik M, Ruppert AM, Grams J (2019) Physical and chemical pretreatment of lignocellulosic biomass. In: Basile A, Dalena F (eds) Second and third generation of feedstocks. Elsevier Inc, New York, pp 143–196. https://doi.org/10.1016/B978-0-12-815162-4.00006-9

Jiang Y, Lv Y, Wu R, Sui Y, Chen C, Xin F, Zhou J, Dong W, Jiang M (2019) Current status and perspectives on biobutanol production using lignocellulosic feedstocks. Bioresour Technol Rep 7:100245. https://doi.org/10.1016/j.biteb.2019.100245

Jiang Y, Guo D, Lu J, Dürre P, Dong W, Yan W, Zhang W, Ma J, Jiang M, Xin F (2018) Consolidated bioprocessing of butanol production from xylan by a thermophilic and butanologenic Thermoanaerobacterium sp. M5. Biotechnol Biofuels 11(89):1–14. https://doi.org/10.1186/s13068-018-1092-1

Jin C, Yao M, Liu H, Chia-fon FL, Ji J (2011) Progress in the production and application of n-butanol as a biofuel. Renew Sustain Energy Rev 15(8):4080–4106. https://doi.org/10.1016/j.rser.2011.06.001

Jin S, Zhang G, Zhang P, Fan S, Li F (2015) High-pressure homogenization pretreatment of four different lignocellulosic biomass for enhancing enzymatic digestibility. Biores Technol 181:270–274. https://doi.org/10.1016/j.biortech.2015.01.069

Junior CCS, Rambo MCD, Teófilo RF, Cardoso WJ, Bertuol DA, Rambo MKD (2021) Production of levulinic acid from coconut residues (cocos nucifera) using differents approaches. Waste Biomass Valorization 1–12. https://doi.org/10.1007/s12649-021-01484-0

Junqueira TL, Cavalett O, Bonomi A (2016) The virtual sugarcane biorefinery-a simulation tool to support public policies formulation in bioenergy. Ind Biotechnol 12(1):62–67. https://doi.org/10.1089/ind.2015.0015

Kamm B, Gerhardt M, Dautzenberg G (2013) Catalytic processes of lignocellulosic feedstock conversion for production of furfural, levulinic acid, and formic acid-based fuel components. In: Suib SL (ed) New and future developments in catalysis. Elsevier Inc, New York, pp 91–113. https://doi.org/10.1016/B978-0-444-53878-9.00005-9

Kapoor M, Panwar D, Kaira GS (2016) Bioprocesses for enzyme production using agro-industrial wastes: technical challenges and commercialization potential. In: Dhillon GS, Kaur S (eds) Agro-industrial wastes as feedstock for enzyme production. Elsevier Inc, New York, pp 61–93. https://doi.org/10.1016/B978-0-12-802392-1.00003-4

Karimi S, Karri RR, Yaraki MT, Koduru JR (2021) Processes and separation technologies for the production of fuel-grade bioethanol: a review. Environ Chem Lett 2873–2890. https://doi.org/10.1007/s10311-021-01208-9

Kavitha S, Kannah RY, Kasthuri S, Gunasekaran M, Pugazhendi A, Rene ER, Pant D, Kumar G, Banu JR (2020) Profitable biomethane production from delignified rice straw biomass: the effect of lignin, energy and economic analysis. Green Chem 22(22):8024–8035. https://doi.org/10.1039/D0GC02738C

Khalid A, Aslam M, Qyyum MA, Faisal A, Khan AL, Ahmed F, Lee M, Kim J, Jang N, Chang IS, Bazmi AA, Yasin M (2019) Membrane separation processes for dehydration of bioethanol from fermentation broths: recent developments, challenges, and prospects. Renew Sustain Energy Rev 105:427–443. https://doi.org/10.1016/j.rser.2019.02.002

Khan AS, Man Z, Nasrullah A, Ullah Z, Muhammad N, Rahim A, Bustam A, Idris A (2020) Conversion of biomass to chemicals using ionic liquids. In: Inamuddin AM, Asiri A, Kanchi S (eds) Green sustainable process for chemical and environmental engineering and science. Elsevier inc, New York, pp 1–30. https://doi.org/10.1016/B978-0-12-817386-2.00001-9

Khromova SA, Bykova MV, Bulavchenko OA, Ermakov DY, Saraev AA, Kaichev VV, Venderbosch RH, Yakovlev VA (2016) Furfural hydrogenation to furfuryl alcohol over bimetallic Ni-Cu sol-gel catalyst: a model reaction for conversion of oxygenates in pyrolysis liquids. Top Catal 59(15):1413–1423. https://doi.org/10.1007/s11244-016-0649-0

Kim S (2018) Evaluation of alkali-pretreated soybean straw for lignocellulosic bioethanol production. Int J Polymer Sci 1–7. https://doi.org/10.1155/2018/5241748

Kohli K, Prajapati R, Sharma BK (2019) Bio-based chemicals from renewable biomass for integrated biorefineries. Energies 12(2):233. https://doi.org/10.3390/en12020233

Kolesinska B, Fraczyk J, Binczarski M, Modelska M, Berlowska J, Dziugan P, Antolak H, Kaminski ZJ, Witonska IA, Kregiel D (2019) Butanol synthesis routes for biofuel production: trends and perspectives. Materials 12(3):350. https://doi.org/10.3390/ma12030350

Kumar V, Yadav SK, Kumar J, Ahluwalia V (2020a) A critical review on current strategies and trends employed for removal of inhibitors and toxic materials generated during biomass pretreatment. Biores Technol 299:122633. https://doi.org/10.1016/j.biortech.2019.122633

Kumar A, Shende DZ, Wasewar KL (2020b) Production of levulinic acid: a promising building block material for pharmaceutical and food industry. Materials Today: Proc 29:790–793. https://doi.org/10.1016/j.matpr.2020.04.749

Kumla J, Suwannarach N, Sujarit K, Penkhrue W, Kakumyan P, Jatuwong K, Vadthanarat S, Lumyong S (2020) Cultivation of mushrooms and their lignocellulolytic enzyme production through the utilization of agro-industrial waste. Molecules 25(12):2811. https://doi.org/10.3390/molecules25122811

Kushwaha D, Srivastava N, Mishra I, Upadhyay SN, Mishra PK (2019) Recent trends in biobutanol production. Rev Chem Eng 35(4):475–504. https://doi.org/10.1515/revce-2017-0041

Lakshmi MV, Goutami N, Kumari AH (2017) Agricultural waste concept, generation, utilization and management. Int J Multi Advan Res Trends 4(3):1–4

Lal R (2005) World crop residues production and implications of its use as a biofuel. Environ Int 31(4):575–584. https://doi.org/10.1016/j.envint.2004.09.005

Li M, Pu Y, Ragauskas AJ (2016) Current understanding of the correlation of lignin structure with biomass recalcitrance. Front Chem 4:45. https://doi.org/10.3389/fchem.2016.00045

Li X, Xu R, Yang J, Nie S, Liu D, Liu Y, Si C (2019) Production of 5-hydroxymethylfurfural and levulinic acid from lignocellulosic biomass and catalytic upgradation. Ind Crops Prod 130:184–197. https://doi.org/10.1016/j.indcrop.2018.12.082

Li S, Huang L, Ke C, Pang Z, Liu L (2020) Pathway dissection, regulation, engineering and application: lessons learned from biobutanol production by solventogenic clostridia. Biotechnol Biofuels 13(39):1–25. https://doi.org/10.1186/s13068-020-01674-3

Li L, Rowbotham JS, Greenwell HC, Dyer PW (2013) An introduction to pyrolysis and catalytic pyrolysis: versatile techniques for biomass conversion. In: Suib SL (ed) New and future developments in catalysis. Elsevier Inc, New York, pp 173–208. https://doi.org/10.1016/B978-0-444-53878-9.00009-6

Lugani Y, Puri M, Sooch BS (2021) Recent insights, applications and prospects of xylose reductase: a futuristic enzyme for xylitol production. Eur Food Res Technol 247(4):921–946. https://doi.org/10.1007/s00217-020-03674-x

Mande S (2005) Biomass gasifier-based power plants: potential, problems, and research needs for decentralized rural electrification. In: Lal B, Reddy MRVP (eds) Wealth from waste: trends and technologies. TERI press, New Delhi, pp 1–28

Martínez-Medina GA, Barragán AP, Ruiz HA, Ilyina A, Hernández JLM, Rodríguez-Jasso RM, Hoyos-Concha JL, Aguilar-González CN (2019) Fungal proteases and production of bioactive peptides for the food industry. In: Kuddus M (ed) Enzymes in food biotechnology. Academic Press, pp 221–246. https://doi.org/10.1016/B978-0-12-813280-7.00014-1

Masran R, Bahrin EK, Ibrahim MF, Phang LY, Abd-Aziz S (2020) Simultaneous pretreatment and saccharification of oil palm empty fruit bunch using laccase-cellulase cocktail. Biocatal Agric Biotechnol 29:101824. https://doi.org/10.1016/j.bcab.2020.101824

Mathew AK, Abraham A, Mallapureddy KK, Sukumaran RK (2018) Lignocellulosic biorefinery wastes, or resources?. In: Thallada Bhaskar T, Pandey A, Mohan SV, Lee DJ, Khanal SK (eds) Waste biorefinery. Elsevier Inc, New York, pp 267–297. https://doi.org/10.1016/B978-0-444-63992-9.00009-4

Mazurkiewicz J, Marczuk A, Pochwatka P, Kujawa S (2019) Maize straw as a valuable energetic material for biogas plant feeding. Materials 12(23):3848. https://doi.org/10.3390/ma12233848

Millati R, Cahyono RB, Ariyanto T, Azzahrani IN, Putri RU, Taherzadeh MJ (2016) Agricultural, industrial, municipal, and forest wastes: An Overview. In: Taherzadeh MJ, Bolton K, Wong J, Pandey A (eds) Sustainable resource recovery and zero waste approaches. Elsevier Inc, New York, pp 1–22. https://doi.org/10.1016/B978-0-444-64200-4.00001-3

Miranda NT, Maciel Filho R, Maciel MRW (2020) Comparison of complete extractive and azeotropic distillation processes for anhydrous ethanol production using aspen plus TM simulator. Chem Eng Trans 80:43–48. https://doi.org/10.3303/CET2080008

Mohamad NL, Mustapa Kamal SM, Mokhtar MN (2015) Xylitol biological production: a review of recent studies. Food Rev Intl 31(1):74–89. https://doi.org/10.1080/87559129.2014.961077

Mohanty SK, Swain MR (2019) Bioethanol production from corn and wheat. In: Ramesh CR, Ramachandran S (eds) Bioethanol production from food crops. Academic Press, pp 45–59. https://doi.org/10.1016/B978-0-12-813766-6.00003-5

Mohd Sani NF, Jafri NA, Majid NA, Rosaidi NA, Onn M (2021) Effect of glutamic acid as additional monomer in biodegradable poly (xylitol sebacate glutamate) polymer. Egypt J Chem 64(6):2783–2787. https://doi.org/10.21608/EJCHEM.2021.54273.3128

Momayez F, Karimi K, Horváth IS (2019) Sustainable and efficient sugar production from wheat straw by pretreatment with biogas digestate. RSC Adv 9(47):27692–27701. https://doi.org/10.1039/C9RA05285B

Moreno AD, Carbone A, Pavone R, Olsson L, Geijer C (2019) Evolutionary engineered Candida intermedia exhibits improved xylose utilization and robustness to lignocellulose-derived inhibitors and ethanol. Appl Microbiol Biotechnol 103(3):1405–1416. https://doi.org/10.1007/s00253-018-9528-x

Muranaka Y, Matsubara K, Maki T, Asano S, Nakagawa H, Mae K (2020) 5-Hydroxymethylfurfural synthesis from monosaccharides by a biphasic reaction-extraction system using a microreactor and extractor. ACS Omega 5(16):9384–9390. https://doi.org/10.1021/acsomega.0c00399

Nawab S, Wang N, Ma X, Huo YX (2020) Genetic engineering of non-native hosts for 1-butanol production and its challenges: a review. Microb Cell Fact 19(79):1–16. https://doi.org/10.1186/s12934-020-01337-w

Nayak S, Barik S, Jena PK (2021) Eco-friendly, bio-degradable and compostable plates from areca leaf. In: Saba N, Jawaid M, Thariq M (eds) Biopolymers and biocomposites from agro-waste for packaging applications. Elsevier inc, New York, pp 127–139. https://doi.org/10.1016/B978-0-12-819953-4.00002-1

Nguyen TAD, Kim KR, Han SJ, Cho HY, Kim JW, Park SM, Park JC, Sim SJ (2010) Pretreatment of rice straw with ammonia and ionic liquid for lignocellulose conversion to fermentable sugars. Biores Technol 101(19):7432–7438. https://doi.org/10.1016/j.biortech.2010.04.053

Nikolić S, Mojović L, Rakin M, Pejin D (2009) Bioethanol production from corn meal by simultaneous enzymatic saccharification and fermentation with immobilized cells of Saccharomyces cerevisiae var. ellipsoideus. Fuel 88(9):1602–1607. https://doi.org/10.1016/j.fuel.2008.12.019

Nitsos C, Matsakas L, Triantafyllidis K, Rova U, Christakopoulos P (2018) Investigation of different pretreatment methods of Mediterranean-type ecosystem agricultural residues: characterisation of pretreatment products, high-solids enzymatic hydrolysis and bioethanol production. Biofuels 9(5):545–558. https://doi.org/10.1080/17597269.2017.1378988

North PH (2016) Method and system for fractionation of lignocellulosic biomass. US20160068920 A1

Nuchdang S, Thongtus V, Khemkhao M, Kirdponpattara S, Moore EJ, Setiabudi HDB, Phalakornkule C (2020) Enhanced production of reducing sugars from paragrass using microwave-assisted alkaline pretreatment. Biomass Convers Biorefinery 1–13. https://doi.org/10.1007/s13399-020-00624-1

Nzediegwu E, Portillo-Perez G, Dumont MJ (2021) Valorization of decationized newsprint to levulinic acid. Cellulose 1–19. https://doi.org/10.1007/s10570-021-04061-9(0123456789

Park S, Baker JO, Himmel ME, Parilla PA, Johnson DK (2010) Cellulose crystallinity index: measurement techniques and their impact on interpreting cellulase performance. Biotechnol Biofuels 3(1):1–10. https://doi.org/10.1186/1754-6834-3-10

Pasangulapati V, Ramachandriya KD, Kumar A, Wilkins MR, Jones CL, Huhnke RL (2012) Effects of cellulose, hemicellulose and lignin on thermochemical conversion characteristics of the selected biomass. Biores Technol 114:663–669. https://doi.org/10.1016/j.biortech.2012.03.036

Patel JP, Parsania PH (2018) Characterization, testing, and reinforcing materials of biodegradable composites. In: Shimpi NG (ed) Biodegradable and biocompatible polymer composites. Elsevier inc, New York, pp 55–79. https://doi.org/10.1016/B978-0-08-100970-3.00003-1

Patel A, Patel H, Divecha J, Shah AR (2019) Enhanced production of ethanol from enzymatic hydrolysate of microwave-treated wheat straw by statistical optimization and mass balance analysis of bioconversion process. Biofuels 1–8. https://doi.org/10.1080/17597269.2019.1608037

Patel A, Shah A (2021) Integrated lignocellulosic biorefinery: Gateway for production of second generation ethanol and value added products. J Bioresour Bioprod 6(2):108–128. https://doi.org/10.1016/j.jobab.2021.02.001

Pileidis FD, Titirici MM (2016) Levulinic acid biorefineries: new challenges for efficient utilization of biomass. Chemsuschem 9:1–22. https://doi.org/10.1002/cssc.201501405

Prasad M, Ranjan R, Ali A, Goyal D, Yadav A, Singh TB, Shrivastav P, Dantu PK (2020) Efficient transformation of agricultural waste in India. In: Naeem M, Ansari A, Gill S (eds) Contaminants in agriculture. Springer, Cham, pp 271–287. https://doi.org/10.1007/978-3-030-41552-5_13

Qureshi N, Lin X, Liu S, Saha BC, Mariano AP, Polaina J, Ezeji TC, Friedl A, Maddox IS, Klasson KT, Dien BS (2020) Global view of biofuel butanol and economics of its production by fermentation from sweet sorghum bagasse, food waste, and yellow top presscake: application of novel technologies. Fermentation 6(2):58. https://doi.org/10.3390/fermentation6020058

Rackemann DW, Doherty WO (2011) The conversion of lignocellulosics to levulinic acid. Biofuels, Bioprod Biorefin 5(2):198–214. https://doi.org/10.1002/bbb.267

Raita S, Spalvins K, Blumberga D (2021) Prospect on agro-industrial residues usage for biobutanol production. Agron Res 19(S1):877–895. https://doi.org/10.15159/AR.21.084

Rajan K, Elder T, Abdoulmoumine N, Carrier DJ, Labbé N (2020) Understanding the in situ state of lignocellulosic biomass during ionic liquids-based engineering of renewable materials and chemicals. Green Chem 22(20):6748–6766. https://doi.org/10.1039/D0GC02582H

Ram VJ, Sethi A, Nath M, Pratap R (2019) Five-membered heterocycles. In: Ram VJ, Sethi A, Nath M, Pratap R (eds) The chemistry of heterocycles. Elsevier Inc, New York, pp 149–478. https://doi.org/10.1016/B978-0-08-101033-4.00005-X

Raman JK, Gnansounou E (2015) Furfural production from empty fruit bunch-A biorefinery approach. Ind Crops Prod 69:371–377. https://doi.org/10.1016/j.indcrop.2015.02.063

Ravindran R, Jaiswal AK (2016) A comprehensive review on pre-treatment strategy for lignocellulosic food industry waste: challenges and opportunities. Biores Technol 199:92–102. https://doi.org/10.1016/j.biortech.2015.07.106

Research and markets: https://www.researchandmarkets.com/reports/4033114/xylitol-a-global-market-overview, last accessed on 20 July 2021

Reshamwala SMS, Lali AM (2020) Exploiting the NADPH pool for xylitol production using recombinant Saccharomyces cerevisiae. Biotechnol Prog 36(3):e2972. https://doi.org/10.1002/btpr.2972

Rice Knowledge Bank: http://www.knowledgebank.irri.org/step-by-step-production/postharvest/rice-by-products/rice-husk. last accessed 13 July 2021.

Romero-García AG, Prado-Rúbio OA, Contreras-Zarazúa G, Ramírez-Márquez C, Segovia-Hernández JG (2019) Simultaneous design and controllability optimization for the reaction zone for furfural bioproduction. In: Kiss AA, Zondervan E, Lakerveld R, Özkan L (eds) Computer aided chemical engineering, vol 46. Elsevier Inc, New York, pp 133–138. https://doi.org/10.1016/B978-0-12-818634-3.50023-0

Rosales-Calderon O, Arantes V (2019) A review on commercial-scale high-value products that can be produced alongside cellulosic ethanol. Biotechnol Biofuels 12(1):1–58. https://doi.org/10.1186/s13068-019-1529-1

Rosatella AA, Simeonov SP, Frade RF, Afonso CA (2011) 5-Hydroxymethylfurfural (HMF) as a building block platform: biological properties, synthesis and synthetic applications. Green Chem 13(4):754–793. https://doi.org/10.1039/C0GC00401D

Roy D, Semsarilar M, Guthrie JT, Perrier S (2009) Cellulose modification by polymer grafting: a review. Chem Soc Rev 38(7):2046–2064. https://doi.org/10.1039/b808639g

Ruan Z, Wang X, Liu Y, Liao W (2019) Corn. In: Pan Z, Zhang R, Zicari S (eds) Integrated processing technologies for food and agricultural by-products. Elsevier Inc, New York, pp 59–72. https://doi.org/10.1016/B978-0-12-814138-0.00003-4

Ruiz-Dueñas FJ, Martínez ÁT (2009) Microbial degradation of lignin: how a bulky recalcitrant polymer is efficiently recycled in nature and how we can take advantage of this. Microb Biotechnol 2(2):164–177. https://doi.org/10.1111/j.1751-7915.2008.00078.x

Saha BC, Kennedy GJ (2020) Production of xylitol from mixed sugars of xylose and arabinose without co-producing arabitol. Biocatal Agric Biotechnol 29:101786. https://doi.org/10.1016/j.bcab.2020.101786

Saha BC, Iten LB, Cotta MA, Wu YV (2005) Dilute acid pretreatment, enzymatic saccharification and fermentation of wheat straw to ethanol. Process Biochem 40(12):3693–3700. https://doi.org/10.1016/j.procbio.2005.04.006

Saha BC, Qureshi N, Kennedy GJ, Cotta MA (2016) Biological pretreatment of corn stover with white-rot fungus for improved enzymatic hydrolysis. Int Biodeterior Biodegradation 109:29–35. https://doi.org/10.1016/j.ibiod.2015.12.020

Saini JK, Saini R, Tewari L (2015) Lignocellulosic agriculture wastes as biomass feedstocks for second-generation bioethanol production: concepts and recent developments. 3 Biotech 5(4):337–353. https://doi.org/10.1007/s13205-014-0246-5

Sanchis-Sebastiá M, Erdei B, Kovacs K, Galbe M, Wallberg O (2020) Analysis of animal bedding heterogeneity for potential use in biorefineries based on farmyard manure. Waste Biomass Valorization 11(6):2387–2395. https://doi.org/10.1007/s12649-018-00578-6

Santana-Méridas O, González-Coloma A, Sánchez-Vioque R (2012) Agricultural residues as a source of bioactive natural products. Phytochem Rev 11(4):447–466. https://doi.org/10.1007/s11101-012-9266-0

Sarkar P, Mukherjee M, Goswami G, Das D (2020) Adaptive laboratory evolution induced novel mutations in Zymomonas mobilis ATCC ZW658: a potential platform for co-utilization of glucose and xylose. J Ind Microbiol Biotechnol 47(3):329–341. https://doi.org/10.1007/s10295-020-02270-y

Sharma HK, Xu C, Qin W (2019) Biological pretreatment of lignocellulosic biomass for biofuels and bioproducts: an overview. Waste Biomass Valorization 10(2):235–251. https://doi.org/10.1007/s12649-017-0059-y

Sharma B, Larroche C, Dussap CG (2020) Comprehensive assessment of 2G bioethanol production. Biores Technol 313:123630. https://doi.org/10.1016/j.biortech.2020.123630

Shrivastava A (2018) Polymerization. In: Shrivastava A (ed) In: Plastics design library, introduction to plastics engineering. William Andrew Publishing, pp 17–48. https://doi.org/10.1016/B978-0-323-39500-7.00002-2

Shrotri A, Kobayashi H, Fukuoka A (2017) Catalytic conversion of structural carbohydrates and lignin to chemicals. In: Song C (ed) Advances in Catalysis, vol 60. Academic Press, pp 59–123. https://doi.org/10.1016/bs.acat.2017.09.002

Singhvi MS, Gokhale DV (2019) Lignocellulosic biomass: hurdles and challenges in its valorization. Appl Microbiol Biotechnol 103(23):9305–9320. https://doi.org/10.1007/s00253-019-10212-7

Singhvi M, Kim BS (2020) Current developments in lignocellulosic biomass conversion into biofuels using nanobiotechology approach. Energies 13(20):5300. https://doi.org/10.3390/en13205300

Song S, Fung Kin Yuen V, Di L, Sun Q, Zhou K, Yan N (2020) Integrating biomass into the organonitrogen chemical supply chain: production of pyrrole and d-proline from furfural. Angew Chem 132(45):20018–20022. https://doi.org/10.1002/ange.202006315

Sticklen MB (2007) Feedstock crop genetic engineering for alcohol fuels. Crop Sci 47:2238–2248. https://doi.org/10.2135/cropsci2007.04.0212

Stoklosa RJ, Nghiem NP, Latona RJ (2019) Xylose-enriched ethanol fermentation stillage from sweet sorghum for xylitol and astaxanthin production. Fermentation 5(4):84. https://doi.org/10.3390/fermentation5040084

Su T, Zhao D, Khodadadi M, Len C (2020) Lignocellulosic biomass for bioethanol: recent advances, technology trends, and barriers to industrial development. Curr Opin Green Sustain Chem 24:56–60. https://doi.org/10.1016/j.cogsc.2020.04.005

Subhedar PB, Ray P, Gogate PR (2018) Intensification of delignification and subsequent hydrolysis for the fermentable sugar production from lignocellulosic biomass using ultrasonic irradiation. Ultrason Sonochem 40:140–150. https://doi.org/10.1016/j.ultsonch.2017.01.030

Subramaniyam R, Vimala R (2012) Solid state and submerged fermentation for the production of bioactive substances: a comparative study. Int J Sci Nat 3(3):480–486

Suriya J, Bharathiraja S, Krishnan M, Manivasagan P, Kim SK (2016) Marine microbial amylases: properties and applications. Adv Food Nutr Res 79:161–177

Takkellapati S, Li T, Gonzalez MA (2018) An overview of biorefinery-derived platform chemicals from a cellulose and hemicellulose biorefinery. Clean Technol Environ Policy 20(7):1615–1630. https://doi.org/10.1007/s10098-018-1568-5

Talattof Z, Azad A, Zahed M, Shahradnia N (2018) Antifungal activity of xylitol against Candida albicans: an in vitro study. J Contemp Dent Pract 19(2):125–129. https://doi.org/10.5005/jp-journals-10024-2225

Tarasov D, Leitch M, Fatehi P (2018) Lignin–carbohydrate complexes: properties, applications, analyses, and methods of extraction: a review. Biotechnol Biofuels 11(1):1–28. https://doi.org/10.1186/s13068-018-1262-1

Teong SP, Yi G, Zhang Y (2014) Hydroxymethylfurfural production from bioresources: past, present and future. Green Chem 16(4):2015–2026. https://doi.org/10.1039/C3GC42018C

Thakkar A, Shell KM, Bertosin M, Rodene DD, Amar V, Bertucco A, Gupta RB, Shende R, Kumar S (2021) Production of levulinic acid and biocarbon electrode material from corn stover through an integrated biorefinery process. Fuel Process Technol 213:106644. https://doi.org/10.1016/j.fuproc.2020.106644

Tian L, Cervenka ND, Low AM, Olsin DG, Lynd LR (2019) A mutation in the Adhe alcohol dehydrogenase of Clostridium thermocellum increases tolerance to several primary alcohols, including isobutanol, n-butanol and ethanol. Sci Rep 9(1):1–7. https://doi.org/10.1038/s41598-018-37979-5

Tomás-Pejó E, Oliva JM, Ballesteros M, Olsson L (2008) Comparison of SHF and SSF processes from steam-exploded wheat straw for ethanol production by xylose-fermenting and robust glucose-fermenting Saccharomyces cerevisiae strains. Biotechnol Bioeng 100(6):1122–1131. https://doi.org/10.1002/bit.21849

Toor M, Kumar SS, Malyan SK, Bishnoi NR, Mathimani T, Rajendran K, Pugazhendhi A (2020) An overview on bioethanol production from lignocellulosic feedstocks. Chemosphere 242:125080. https://doi.org/10.1016/j.chemosphere.2019.125080

Transparency market research. https://www.transparencymarketresearch.com/levulinic-acid-market.html, last accessed on 21 July 2021

Tsai TY, Lo YC, Dong CD, Nagarajan D, Chang JS, Lee DJ (2020) Biobutanol production from lignocellulosic biomass using immobilized Clostridium acetobutylicum. Appl Energy 277:115531. https://doi.org/10.1016/j.apenergy.2020.115531

Tsoutsos T (2010) Modelling hydrolysis and fermentation processes in lignocelluloses-to-bioalcohol production. In: Waldron K (ed) Bioalcohol production. Elsevier Inc, New York, pp 340–362. https://doi.org/10.1533/9781845699611.4.340

U.S. Energy Information Administration, Independent statistic and analysis https://www.eia.gov/todayinenergy/detail.php?id=41433, last accessed on 17 July 2021

van Putten RJ, Van Der Waal JC, De Jong ED, Rasrendra CB, Heeres HJ, de Vries JG (2013) Hydroxymethylfurfural, a versatile platform chemical made from renewable resources. Chem Rev 113(3):1499–1597. https://doi.org/10.1021/cr300182k

Verardi A, Lopresto CG, Blasi A, Chakraborty S, Calabrò V (2020) Bioconversion of lignocellulosic biomass to bioethanol and biobutanol. In: Yousuf A, Pirozzi D, Sannino F (eds) Lignocellulosic biomass to liquid biofuels. Academic Press, pp 67–125. https://doi.org/10.1016/B978-0-12-815936-1.00003-4

Vitali F, Parmigiani S, Vaccari M, Collivignarelli C (2013) Agricultural waste as household fuel: techno-economic assessment of a new rice-husk cookstove for developing countries. Waste Manage 33(12):2762–2770. https://doi.org/10.1016/j.wasman.2013.08.026

Wang Y, Brown CA, Chen R (2018) Industrial production, application, microbial biosynthesis and degradation of furanic compound, hydroxymethylfurfural (HMF). AIMS Microbiol 4(2):261–273. https://doi.org/10.3934/microbiol.2018.2.261

Wang T, Lü X (2021) Overcome saccharification barrier: advances in hydrolysis technology. In: Lü X (ed) Advances in 2nd generation of bioethanol production. Elsevier inc, New York, pp 137–159. https://doi.org/10.1016/B978-0-12-818862-0.00005-4

Werpy T, Petersen G (2004) Top value added chemicals from biomass: volume I—results of screening for potential candidates from sugars and synthesis gas. United States https://doi.org/10.2172/15008859

Xiang Y, Xiang Y, Wang L (2017) Electron beam irradiation to enhance enzymatic saccharification of alkali soaked Artemisia ordosica used for production of biofuels. J Environ Chem Eng 5(4):4093–4100. https://doi.org/10.1016/j.jece.2017.07.058

Xin F, Wang C, Dong W, Zhang W, Wu H, Ma J, Jiang M (2016) Comprehensive investigations of biobutanol production by a non-acetone and 1, 3-propanediol generating Clostridium strain from glycerol and polysaccharides. Biotechnol Biofuels 9(220):1–12. https://doi.org/10.1186/s13068-016-0641-8

Yoav S, Barak Y, Shamshoum M, Borovok I, Lamed R, Dassa B, Hadar Y, Morag E, Bayer EA (2017) How does cellulosome composition influence deconstruction of lignocellulosic substrates in Clostridium (Ruminiclostridium) Thermocellum DSM 1313? Biotechnol Biofuels 10(1):1–16. https://doi.org/10.1186/s13068-017-0909-7

Yogalakshmi KN, Devi TP, Sivashanmugam P, Kavitha S, Kannah RY, Varjani S, Kumar SA, Kumar G, Banu JR (2021) Lignocellulosic biomass-based pyrolysis: a comprehensive review. Chemosphere 286(2):131824. https://doi.org/10.1016/j.chemosphere.2021.131824

Yuan H, Liu H, Du J, Liu K, Wang T, Liu L (2020a) Biocatalytic production of 2, 5-furandicarboxylic acid: recent advances and future perspectives. Appl Microbiol Biotechnol 104(2):527–543. https://doi.org/10.1007/s00253-019-10272-9

Yuan X, Tu S, Lin J, Yang L, Shen H, Wu M (2020b) Combination of the CRP mutation and ptsG deletion in Escherichia coli to efficiently synthesize xylitol from corncob hydrolysates. Appl Microbiol Biotechnol 104(5):2039–2050. https://doi.org/10.1007/s00253-019-10324-0

Zhang YP (2008) Reviving the carbohydrate economy via multi-product lignocellulose biorefineries. J Ind Microbiol Biotechnol 35(5):367–375. https://doi.org/10.1007/s10295-007-0293-6

Zhang J, Zong W, Hong W, Zhang ZT, Wang Y (2018) Exploiting endogenous CRISPR-Cas system for multiplex genome editing in Clostridium tyrobutyricum and engineer the strain for high-level butanol production. Metab Eng 47(3):49–59. https://doi.org/10.1016/j.ymben.2018.03.007

Zhang L, Chen Z, Wang J, Shen W, Li Q, Chen X (2021a) Stepwise metabolic engineering of Candida tropicalis for efficient xylitol production from xylose mother liquor. Microb Cell Fact 20(105):1–12. https://doi.org/10.1186/s12934-021-01596-1

Zhang M, Wang N, Liu J, Wang C, Xu Y, Ma L (2021) A review on biomass-derived levulinic acid for application in drug synthesis. Crit Rev Biotechnol 1–34. https://doi.org/10.1080/07388551.2021.1939261

Zhou J, Tang Z, Jiang X, Jiang R, Shao J, Han F, Xu Q (2016a) Catalytic conversion of glucose into 5-hydroxymethyl-furfural over chromium-exchanged bentonite in ionic liquid-dimethyl sulfoxide mixtures. Waste Biomass Valorization 7(6):1357–1368. https://doi.org/10.1007/s12649-016-9525-1

Zhou X, Broadbelt LJ, Vinu R (2016) Mechanistic understanding of thermochemical conversion of polymers and lignocellulosic biomass. In: Kevin M, Geem V (eds) Advances in chemical engineering, vol 49. Academic Press, pp 95–198. https://doi.org/10.1016/bs.ache.2016.09.002

Chapter 10
A Sustainable Biorefinery Approach to Valorize Corn Waste to Valuable Chemicals

Neha Kukreti, Sonika Kag, Rohit Ruhal, and Rashmi Kataria

10.1 Corn Biomass-As an Agricultural Waste

Agricultural wastes are produced in vast amounts of two billion tons (Millati et al. 2019) every year in the world. This waste has a diverse chemical composition, and after cultivation and manufacturing processes, a considerable amount of biomass is produced from major crops. In the agricultural sector, crop waste is generated in the form of corn-stalk, fruit peel, vegetable peel, sugarcane-bagasse, sunflower–stalk, and rice-stalk, rice husk, wheat straw and sugarcane etc. (Millati et al. 2019; Sud et al. 2008). Among these, corn waste, also known as corn stover, consists of maize (corn) stalks, leaves, cobs and these lignocellulosic wastes constitute lignin, cellulose and hemicellulose. Cellulose, a homopolysaccharide made of a linear chain of ß-1,4-d-glucose units, is present in a significant amount in plant cell walls (Tathod and Dhepe 2015; Baruah et al. 2018). Hemicellulose is a heteropolysaccharide of xylose, arabinose, mannose, galactose, glucose, fructose, glucuronic acid and galacturonic acid and present in variable amounts in plant cells (Baruah et al. 2018; Machmudah et al. 2017). Lignin is a complex polymer composed of aromatic monomers like synapyl alcohol, coniferyl alcohol and p-coumaryl alcohol (Tathod and Dhepe 2015).

Corn cob is cut apart from corn kernels either manually or using machines during manufacturing processes (Millati et al. 2019). The composition of corn stalks is 35.0 cellulose (wt%), 14.4 hemicelluloses (wt%) and 21.5 lignin (wt%). The corn cobs are composed of 33.7 cellulose (wt%), 31.7 hemicellulose (wt%) and 6.1 lignin (wt%) (Ifeanyi et al. 2016) (Table 10.1). Corn is also rich in silica and carbon compounds (Millati et al. 2019). Corn stover is a sustainable resource for second-generation

N. Kukreti · S. Kag · R. Kataria (✉)
Delhi Technological University (DTU), Shahbad Daulatpur, Bawana Road, Delhi 110042, India
e-mail: rashmikataria@dtu.ac.in

R. Ruhal
Regional Centre of Biotechnology, Faridabad, India

Y. K. Nandabalan et al. (eds.), *Zero Waste Biorefinery*, Energy, Environment, and Sustainability, https://doi.org/10.1007/978-981-16-8682-5_10

Table 10.1 Composition of different types of corn wastes

S. No.	Corn biomass type	Cellulose (%)	Hemicellulose (%)	Lignin (%)
1.	Corn stover	8–40	24–26	7–19 Tayyab et al. (2017)
2.	Corn cob	42–45	35–39	14–15 Tayyab et al. (2017)
3.	Corn stalk	35	14.4	21.5 Xu et al. (2018)
4.	Corn leaves	26.93	13.27	15.18 Tayyab et al. (2017)

biofuel production (Table 10.1). It has massive potential for bioenergy and biochemical production through biorefinery rather than being left in the field as a discard. The leftovers like corn stalks and husks are burnt on the field in developing nations which causes severe air pollution. Hence, there is a need for proper utilization and disposal of agricultural waste. This waste could be utilized for biofuel and biochemical production (Shafiei Alavijeh et al. 2020). Different steps involved in corn waste utilizations are shown in (Figs. 10.1 and 10.2). According to a report by the United States Agriculture Department, in 2017 the global mass production of corn was 1061 million tons, and the approximate corn cob was around 230 million tons. The top three countries that are the significant producers of corn globally are the United States, China, and Brazil. Thus, there is an abundance of corn waste, and there is a need for technological innovation to efficiently utilize corn waste, improve deep processing and overcome generic methods (Millati et al. 2019). The main challenge with the utilization of corn waste, as it is highly recalcitrant in characteristics due to the presence of lignin. It is highly resistant; therefore is difficult enzymatic breakdown of the carbohydrate portion.

A pretreatment step is essential to enhance the interaction of biomass with enzymes such as cellulases, hemicellulases, and ligninases for effective hydrolysis. Different chemicals, including alkali solutions, are used for the pretreatment of corn biomass which leads to solubilization of the lignin component in the alkali solution. During acidic pretreatment, the disruption of glucosidic bonds between hemicellulose and cellulose occurs (Baruah et al. 2018). The main objective of all the pretreatment methods is to optimize the yield of the fermentable sugars by removing lignin from biomass. Microorganisms such as *Saccharomyces cerevisiae*, *Pichia stiptis*, *Clostridium* are used to ferment sugars into acids or alcohols. The sugar yield depends not only on the characteristics of the biomass but also on the interaction with the pretreatment conditions and formulations of enzymes. The different physio-chemical pretreatment for efficient sugar recovery includes grind milling, auto/hot water hydrolysis, steam explosion, acid and alkali treatment, hydrothermal hydrolysis (Baruah et al. 2018; Moodley and Gueguim Kana 2019; Timung et al. 2015).

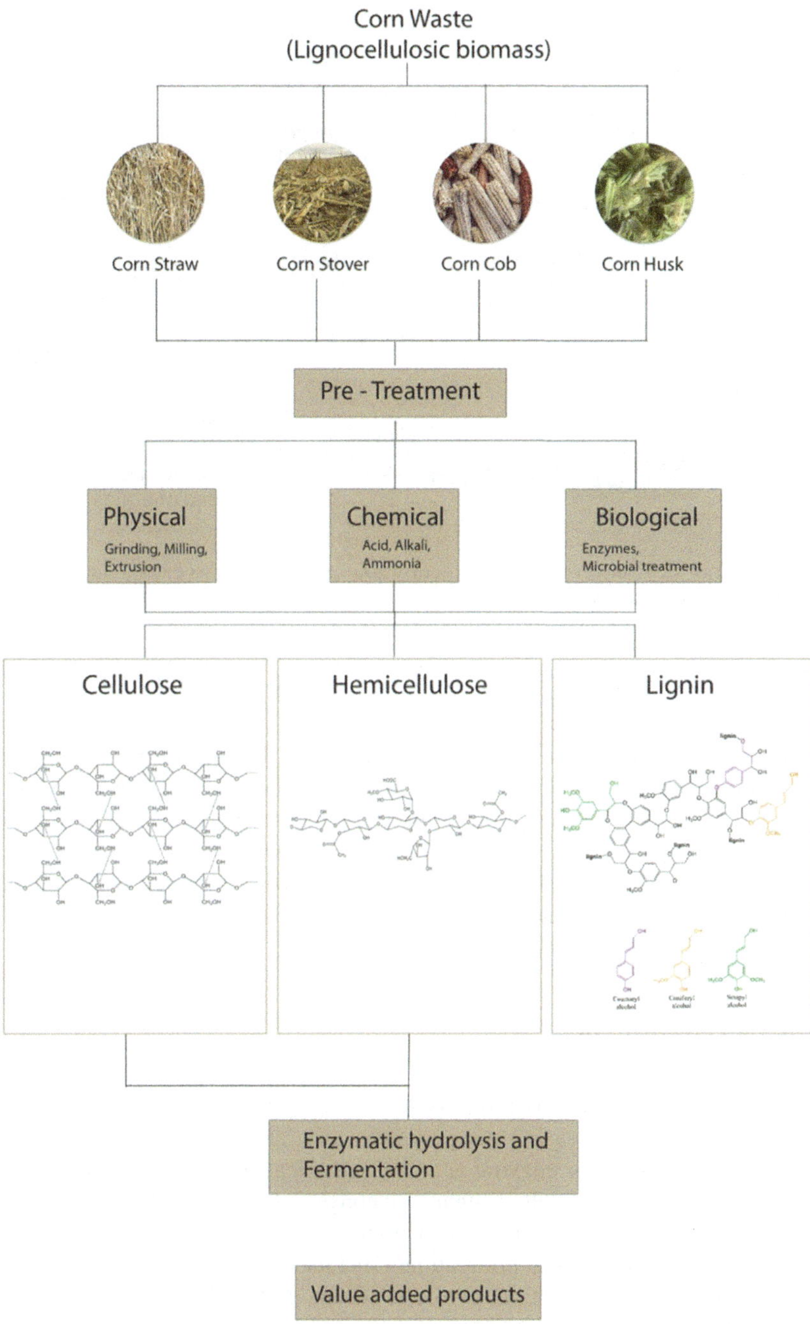

Fig. 10.1 Pretreatment of corn waste (corn straw, stover, con and husk) to yield carbohydrate fraction and lignin for value-based chemicals

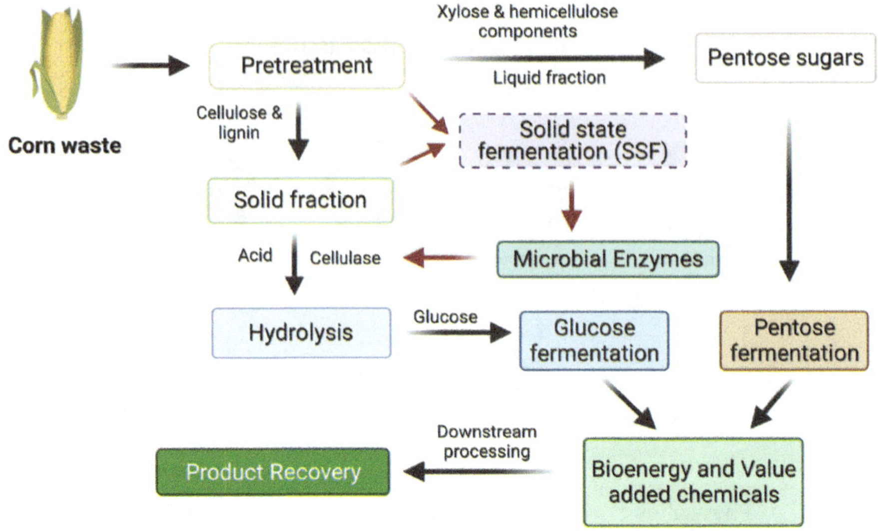

Fig. 10.2 Overall steps involved in the of corn waste carbohydrates to energy and metabolites production (Created with www.biorender.com)

10.2 Processes Involved in the Conversion of Corn Waste

The biochemical steps for utilization of corn waste involve are pretreatment, saccharification of pretreated corn biomass and using enzymes like cellulases, xylanases and the last step is the fermentation of monosaccharide sugars. Different microbes are used for the production of desirable products (Khare et al. 2015). The corn waste biomass is also utilized for solid-state fermentation for microbial production of enzymes. The overall scheme is shown in (Fig. 10.3).

10.2.1 Pretreatment

Pretreatment of lignocellulosic biomass such as corn biomass is an important step for removing lignin, decreasing the polymerization and crystallinity of cellulose, increasing surface area for enzyme activity, and releasing several reducing sugars (Behera et al. 2014). The resulted sugars are converted to value-added products and metabolites. There are different types of pretreatments, like physical, chemical or biological (Fig. 10.1). Physical pretreatment includes milling, grinding, mechanical extrusion, microwave, ultrasound, pyrolysis (Chen et al. 2019). Chemical pretreatment uses alkali, acid, oxidizing agents to degrade recalcitrant lignocellulose in corn waste with or without combination to temperature (Behera et al. 2014). There are

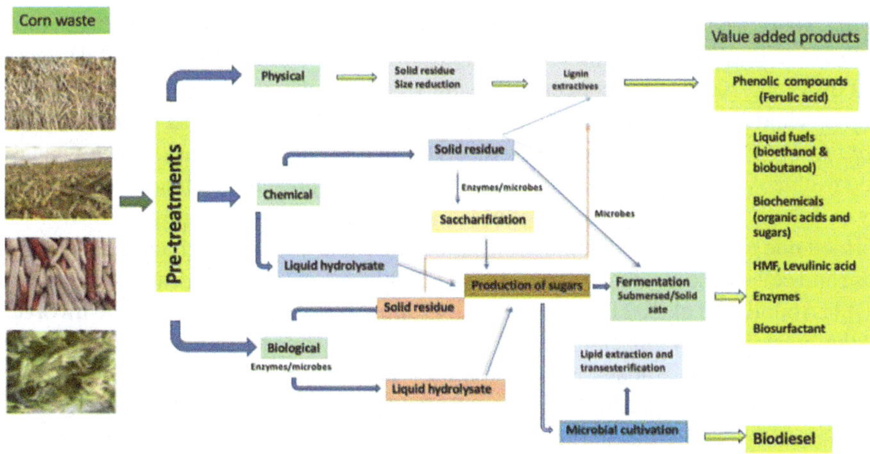

Fig. 10.3 A roadmap of the trends and technologies involved in the corn waste valorization (created with MS Powerpoint)

structural changes in corn waste after the pretreatment step, which affects the enzymatic hydrolysis and microbial fermentation. The degradation of cellulose leads to the formation of glucose sugars, and hemicellulose degradation forms glucose, xylose, arabinose, mannose, galactose, and rhamnose sugars. Dilute acid pretreatment with HCl is the efficient industrially proven method for lignocellulosic biomass conversion and production of low-cost fermentable sugars (Zviely 2013). Acidic pretreatments lead to disruption of van der Waals forces, hydrogen bonds and covalent bonds that hold the corn biomass components, which leads to the disruption and solubilization of hemicellulose fractions. Hemicellulose xylan is hydrolyzed during the acidic chemical pretreatment method. In the alkali chemical pretreatment, acetate group from the hemicellulose fraction is removed. Thus an easy action of hydrolytic enzymes on carbohydrates polymer can take place; therefore, this process is effective for the delignification of corn waste (Bhutto et al. 2017). Biological pretreatment of lignocellulosic biomass is done with either microorganisms or enzymes and requires less energy. It is eco-friendly as there is no release of toxic compounds, effluents and fermentation inhibitors (Sindhu et al. 2016). Several enzymes involved in the pretreatment of corn waste are laccase, lignin peroxidase, manganese peroxidase and other microorganisms.

10.2.2 Hydrolysis

The production of value-added products from corn waste involves saccharification, leading to the bioconversion of complex sugars into simple sugars (Nagoor Gunny et al. 2019). The hydrolysis can be performed by enzymes such as cellulase, hemicellulases, ligninase or acids such as H_2SO_4, HNO_3 or alkali such as NaOH, KOH (Abdu et al. 2020). The concept of using essential enzymatic cocktails to have a higher yield during lignocellulosic biomass saccharification. Three major enzymes are involved in cellulosic biomass hydrolysis that is endo-glucanase, exo-glycanase and β-glucosidase. The repeating units of glucose are present in a linear chain linked by β-1,4- glycosidic bonds in cellulose structure. Different oligosaccharides are formed due to the action of endo -1,4-β glucanase and exo-1,4-β-glucanase on cellulose and cellobiose. β-glucosidase acts on cellobiose and forms glucose (Dhillon et al. 2011; Abdullah et al. 2021). Hemi-cellulases are a group of enzymes that are involved in the breakdown of galactans by D- galactanases, xylans by D-xylanases, mannans by D-manases and arabans by L-arabinose. The enzymes endo-xylanase and 1,4-β-D- xylan xylanohydolase hydrolyze the β-D- xylano pyranosyl linkages in xylan and forms xylo-oilgosaccharides. β-D- xylosidase hydrolyzes xylobiose to form D-xylose (Meena et al. 2017). The nature of the substrates and source of enzymes play a role in biomass hydrolysis (Maitan-Alfenas et al. 2015). The lignin part of lignocellulosic biomass is degraded by lignin peroxidase, manganese peroxidase and laccase enzymes (Khare et al. 2015).

10.2.3 Fermentation

Fermentation is a process having a series of chemical reactions that convert sugars into alcohols or acids with the help of yeast and bacteria. Corn waste, after physical and chemical pretreatment, forms liquid hydrolysate, which is composed of monomeric sugars, it is further fermented by microorganisms into valuable products (Khan et al. 2013). There are different types of fermentation methods such as solid-state fermentation, submerged fermentation, dark fermentation and photo-state fermentation. The solid-state fermentation (SSF) process uses solid support to produce various microbial products like antibiotics, single-cell protein, PUFA's, enzymes, organic acids, biopesticides, biofuels (Sukumaran et al. 2009; Bhargav et al. 2008). Submerged fermentation (SmF) uses free liquids such as molasses and broth where the product formed after fermentation is present in the broth (Suriya et al. 2016). The dark fermentation process occurs in the absence of light and uses anaerobic bacteria to degrade the organic content (Borole and Greig 2019). In the Photo-fermentation process, sunlight is the energy source and green algae provide electrons for photosynthesis by breaking down the endogenous substrate (Sağır and Hallenbeck 2019).

10.2.4 Transesterification

Transesterification is a conversion process in which triglycerides of oils are converted into fatty acid esters (Elkady et al. 2015). Alkali and acids are the most commonly used catalyst for transesterification and lipase is also used as a catalyst which is cost-effective. Heterogeneous catalysts have great potential due to their reusability and are easy to handle in nature. Corn stover is used for biodiesel production using *Mucor indicus CCUG22424* following transesterification reaction (Mondala et al. 2009) (Table 10.2). Nowadays, nanoparticles such as ZnO, CaO, Fe_2O_3 also seem like an excellent alternative to acid or base because of surface area and strong interaction between catalyst and reactant (Soliman et al. 2019).

10.3 Value-Added Products from Corn Waste

Different value-added products are produced from corn waste (Table 10.2). This includes Biofuels, organic acids, bio enzymes, phenolic compounds, biopolymers, and industrially essential chemicals (Fig. 10.4).

10.3.1 Biofuels

Biodiesel

Corn stover is a sustainable feedstock for second-generation biofuel and biochemicals production. Biodiesel is mono-alkyl esters of long-chain fatty acids that are derived from lipid feedstocks which are produced by a transesterification reaction with the help of a catalyst to speed up the reaction (Mondala et al. 2009). It is the process to transform the triglycerides with alcohol and catalyst to yield biodiesel. The by-product obtained is glycerol (Elkady et al. 2015; Mondala et al. 2009; Demirbas 2007)

$$\text{Triglycerides} + \text{Monohydric alcohol} \rightarrow \text{Glycerin} + \text{mono alkyl esters (biodiesel)}$$

The importance of biodiesel is that it has a flash point above 200 °F making it safer, easier to use, and store for a long time. Thus, agricultural waste can be valorized into useful and eco-efficient products (Mondala et al. 2009). Dilute acid pretreated corn stover hydrolysate was utilized by the fungal cells *Mucor indicus* for lipid accumulation and 2.2 g biodiesel was obtained from 4 g of accumulated lipid from 100 g dry corn stover (Shafiei Alavijeh et al. 2020) (Mondala et al. 2009).

Table 10.2 Corn waste to different value added chemicals

S. No.	Bioproduct	Biowaste	Microorganism	Parameters/conditions	Yield	References
1. Biofuels						
1.A	Biodiesel	Corn stover	*Mucor indicus* CCUG22424	Grinding, Dilute sulfuric acid pretreatment (0–2%), time 20–120 min, Aerobic cultivation (32 °C,120 rpm,5 days), Transesterification (48 ml of 5%v/v HCl in methanol,1 h, 85 °C)	2.2 g/L	Mondala et al. (2009)
1.B	Biobutanol	Corn stover	*Clostridium saccharobutylicum* DSM 13,864	Sulfuric acid pretreatment (0.89% w/w, 160 °C, 5 min and 10% solid loading)	9.02 g/L	Hijosa-Valsero et al. (2020)
1.C	Bioethanol	Corn stalk	*Saccharomyces cerevisiae*	Compressed hot water pretreatment (100,120,140,160,180and 200 °C, 10 min), Bioreactor at 30 °C	31.06 g/L	Adekunle et al. (2020)
1.D	Bioethanol	Corn stover	*Mucor indicus* CCUG22424	Dilute acid pretreatment (1.8%v/v H_2SO_4, 22 min and 121 °C), Enzymatic hydrolysis (cellulase 3.9 g),anaerobic cultivation of *M. indicus* (glucose 24.75 g)	21.4 g/L	Shafiei Alavijeh et al. (2020)

(continued)

Table 10.2 (continued)

S. No.	Bioproduct	Biowaste	Microorganism	Parameters/conditions	Yield	References
2. Organic acids						
2.A	Succinic acid	Corn stover	*Actinobacillus succinogenes* CMCC1593	Pretreatment (0.75%H_2SO_4 v/v, 1%NaOH at 121°C, 1 h), alkaline peroxide (2% v/v H_2O_2, 30 °C, 16 h), aqueous ammonia (10%ammonia v/v, 30C, 24 h), Enzymatic hydrolysis (50 °C, 80 rpm, 36 h), Fermentation (substrate conc. 40–90 g/l, cellulase 10–35 FPU/g substrate, temperature 38–42 °C)	47.4 g/L	Zheng et al. (2010)
2.B	Lactic acid	Corn cobs	*Lactobacillus delbrueckii sp. bulgaricus*	Pretreatment (1% sulfuric acid, 2% HCl, 1.5% solid acid solutions, 120 °C, 80 min), Enzymatic hydrolysis (cellulase 10–50 FPU/g cellulose, 50 °C, 48 h)	18 g/L	Wang et al. (2017)
2.C	Citric acid	Corn stover °C	*Aspergillus niger* SIIM M288	Pretreatment (5% w/w H_2SO_4), Enzymatic hydrolysis (15FPUcellulase/gram dry corn stover, 48 h, 50 °C), Fermentation (250–400 rpm, 33 °C,8 days)	100.04 g/L	Zhou et al. (2017b)
2.D	Propionic acid	Corn waste	*Propionibacterium acidipropionici* ATCC 4875	Pretreatment (0.8% w/w dilute sulfuric acid, 160 °C, 10 min), Fermentation (300 rpm, 30 °C, pH 6	35 g/L	Mohan and Sivaprakasam (2016)

(continued)

Table 10.2 (continued)

S. No.	Bioproduct	Biowaste	Microorganism	Parameters/conditions	Yield	References
3. **Bio enzymes**						
3.A	Xylanase	Corn cob	*Bacillus cereus* TH-050	SmF (Xylan, 48 h, 50 °C, pH 8.0, sterilized) SSF (Wheat bran, 4 days, pH 8.0, open nonsterilized fermentation)	5991 U/mL 11,400 U/g	Simair et al. (2018)
3.B	Laccase	Corn steep liquor	*Trametes versicolor* CICC 14001	SmF (liquid medium with CSL per liter KH_2PO_4, $CaCl_2$ $2H_2O$, $MgSO_4$ $7H_2O$, $FeSO_4 \cdot 7H_2O$, glucose, $CuSO_4 \cdot 5H_2O$, pH 4, rpm 150r/min, 26 °C)	633.3 U/L	Wang et al. (2014)
3.C	Beta-glucosidase	Corn cob	*Aspergillus niger*	SSF (6 days, 30 °C, pH 7, inoculum solution 3% v/v)	95.02 U/ml	Aliyah et al. (2017)
3.D	Alpha-amylase	Corn cob	*Aspergillus niger*	SSF (6 days, 30 °C, pH 7, inoculum solution 3% v/v)	81.86 U/ml	Aliyah et al. (2017)
4. Sugars						
4.A	Xylitol	Corn cob	*Kluyveromyces marxianus* CICC 1727–5	Pretreatment (0.5% (w/w) H_2SO_4 and 1.5% (w/w) H_3PO_4, 12.5% solid loading), Fermentation (40 °C, pH 5, 200 rpm, 0.2–2 vvm, 48 h)	24.2 g/L	Du et al. (2020)
4.B	Xylitol	Corn cob	*Saccharomyces cerevisiae* PE-2	Acid hydrolysis (0.5% w/w sulfuric acid, 165 min, 125 °C), SSF (5% solid, 6 and 12 FPU/g)	0.93 g/g	Baptista et al. (2018)

(continued)

Table 10.2 (continued)

S. No.	Bioproduct	Biowaste	Microorganism	Parameters/conditions	Yield	References
5. Phenolic compounds						
5.A	Ferulic acid	Corn bran	–	Alkaline hydrolysis (NaOH concentrations (0.2, 0.5, 0.75 and 1.0 mol/L), ethanol concentrations (30%, 50%, 70% and 90% v/v) solid-to-liquid ratios (1:5, 1:10, 1:15 and 1:20), extraction temperatures (65, 75, 85 and 95 °C) and extraction times (1.5, 2.0, 2.5 and 3.0 h)	7.31 mg/g	Wang et al. (2014)
6. Biosurfactant						
6.A	Cationic Biosurfactant	corn steep liquor	*Bacillus subtilis*	Culture medium—5%,10%,15% v/v corn steep liquor, \Fermentation (37 °C and 200 rpm)	1.3 g/L	Gudiña et al. (2015)
7. Industrially important chemical						
7.A	Sodium gluconate	Corn cob	*Gluconobacter oxydans* DSM 2003	Dilute acid pretreatment(5% w/w), Enzymatic hydrolysis (50C, pH 4.8, 48 h, cellulase15 FPU/g DM), Flask fermentation (30 °C, 220 rpm for 24–72 h)	132.46 g/L	Han et al. (2018)

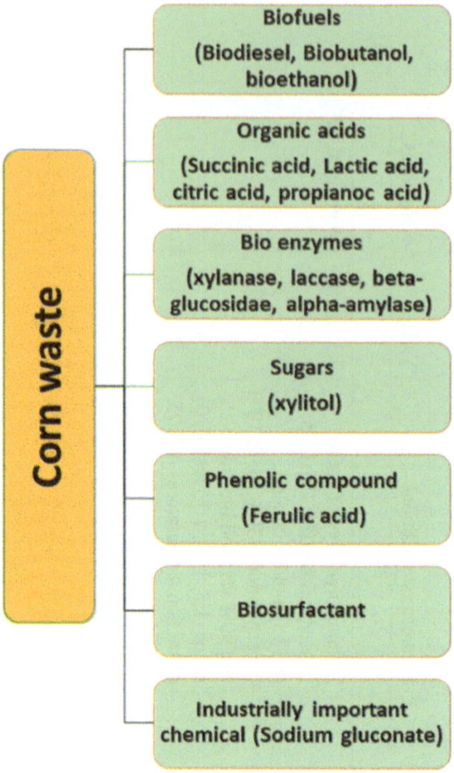

Fig. 10.4 Corn waste to value added chemicals. (Created with MS Powerpoint)

Biobutanol

Corn stover is a potential alternate substrate due to high carbohydrate content and easy availability and therefore, these characteristics have attracted the attention of researchers for utilizing this feedstock to produce biobutanol. Biobutanol is four-carbon alcohol and produced by the fermentation of biomass. It is considered for transportation fuel due to its properties like octane number, energy density, air–fuel ratio and heat of vaporization similar to gasoline (Huang et al. 2019). Corn stover before utilization first undergoes physiochemical pretreatment with sulfuric acid with 0.89% w/w, at 160 °C for 5 min with 10% solid loading followed by enzymatic hydrolysis (Table 10.2, 1.B) to release monomer sugars, *Clostridium saccharobutylicum* utilizes sugars for fermentation and produces 9.02 g/L biobutanol. Corn stovers produce second generation biobutanol by scaling up to industrial scale (Hijosa-Valsero et al. 2020). It has an industrial interest as it is used in solvents for different compound production and the energy content of bio-butanol is higher than that of ethanol. It has better compatibility with unmodified current car engines. It is less corrosive and less hygroscopic and has lower volatility (Qin et al. 2018). Hence, it is considered an advanced transportation fuel (Huang et al. 2019).

Bioethanol

Cornstalk, an agricultural waste, is utilized in large quantities for bioethanol production and it poses no harm to the food chain. Corn biomass is a lignocellulosic biomass for bioenergy production which is predicted as an excellent resource. cornstalk is utilized following compressed hot water pretreatment at 100, 120, 140, 160, 180 and 200 °C for 10 min and 31.06 g/L bioethanol production by *Saccharomyces cerevisiae* was observed (Adekunle et al. 2020). In a study *Mucor indicus* was to produce bioethanol production followed by dilute acid pretreatment (with 1.8% v/v sulfuric acid at 121 °C for 22 min) with high bioethanol yield (Shafiei Alavijeh et al. 2020). Bioethanol is a renewable biofuels considered better than gasoline due to a clean-burn engine, high compression ratio, and less burning time. It also has a higher heat of vaporization, broader flammability and more flame speed. It is composed of an oxygen content of around 35%, due to which ethanol after combustion emits less nitric oxide, particulate matter and hydrocarbons (Adekunle et al. 2020).

10.3.2 Organic Acids

Succinic acid

Succinic acid is an intermediate of the tricarboxylic acid (TCA) cycle (Olajuyin et al. 2019) and is a key precursor for the production of 1,4-butanediol and polybutylene. It is an essential organic acid with various industrial applications including pharmaceutical, polymer, cosmetic, and pesticide industries. *Actinobacillus succinogenes* is used for succinic acid production with cover stover as a substrate following solid-state fermentation (Zheng et al. 2010) (Table 10.2, 2.A). Corn stover is pedretreatment with 0.75% v/v sulfuric acid and 1% NaOH at 121 °C for 1 h, alkaline peroxide pretreatment at with 2% v/v H_2O_2 at 30 °C for 16 h and aqueous ammonia pretreatment with 10% ammonia v/v at 30 °C for 24 h and enzymatic hydrolysis at 50 °C, 80 rpm for 36 h is required to increase the effective production of succinic acid by *A. succinogenes*. Solid-state fermentation is the best method to use cheap biomass like corn for potential succinic acid production (Zheng et al. 2010). The biobased succinic acid production depends on different factors like cheaper sustainable feedstock such as corn husk, corn stover, strain improvement (Babaei et al. 2019).

Lactic Acid

Lactic acid is an organic acid and important chemical widely used in foods, cosmetics, drugs and chemical industries. It is extensively used to synthesize polymers like polylactic acid, a bioplastic forming polymer (Bahry et al. 2019). Its chemical synthesis requires by-products from petrochemical industries which is costly and not an eco-friendly process. Calcium carbonate is the best catalyst to convert corn cobs into lactic acid using hydrothermal reaction and the method of fermentation is complex and sensitive, especially the recovery process and different parameters need to be monitored such as pH and temperature (Sánchez et al. 2012). Researchers have shifted the focus on microbial fermentation of D-lactic acid using glucose as substrate, which is obtained from lignocellulosic biomass such as corn stover (Wang et al. 2017). According to Rivas et al. (2004), corn cob substrate is treated with water at high temperature, known autohydrolysis pretreatment (Table 10.2, 2.B) to remove hemicellulose and lignin. The pretreated fraction have cellulose-rich residue for lactic acid production. The steps involved in acid production are corn cob collection, biomass analysis for cellulose, hemicellulose and lignin, depolymerization of hemicellulose and cellulose, simultaneous saccharification and selection of microbes for fermentation. The optimum temperature for maximum yield of lactic acid was 45 °C, yeast concentration 5–20 g /L, enzyme to substrate ratio is 35–55 FPU/g, reaction time 122 h (Rivas et al. 2004).

Citric Acid

Citric acid is an organic acid that is naturally produced in the Krebs cycle as an intermediate and widely distributed among citrous fruits. It has great value in various industries including food and pharmaceutical and is widely used as a flavoring, acidulant agent (Kumar et al. 2003), construction and chemical industries. Chemically, it is synthesized by submerged fermentation utilizing sucrose or starch as a substrate (Kumar et al. 2003). In a study by Zhou et al. (2017), corn stover feedstock was used for citric acid production using a fungus strain *Aspergillus niger* following dilute acid pretreatment with 5% w/w sulfuric acid and enzymatic hydrolysis with 15 FPU cellulase/gm dry corn stover at 50 °C for 48 h and the addition of extra nutrients did not show any increase in citric acid formation. The yield of citric acid from corn stover was 100.04 g/L and 94.11% conversion yield. Other feedstocks used for citric acid production are corn cobs, coffee husk, orange waste and apple pomace (Zhou et al. 2017a).

Propanoic Acid

Propionic acid is an organic molecule with economic importance and is commonly used as an antimicrobial agent, supplied as poultry feed to decrease acidogenesis in the metabolism of ruminants. In a study, corn stover hydrolysate was used to produce propanoic acid, a probable building block for C_3 based chemicals using *Proionibacterium acidipropionici* (Mohan and Sivaprakasam 2016). Different culture conditions and parameters were studied and experimentally optimized to have a maximum yield of propanoic acid. Corn stover was collected and milled. Fibers were treated with dilute sodium hydroxide solution 0.4% w/w at 80 °C for 2 h. The solid part obtained after the pretreatment was further mixed with dilute sulfuric acid 0.8% w/w. The microorganism used was a native propanoic acid-producing bacteria, *P. acidipropionici* ATCC 4875. Fermentation conditions were performed to analyze the propanoic acid titer and productivity in batch and fed-batch fermentation. PA titer and productivity were 32.1 g/L, 0.31 g/L h and 64.7 g/L, 0.77 g/L h in batch and fed-batch fermentation. Commercial production of propanoic acid is mainly achieved through chemical synthesis from petroleum feedstocks. Glycerol, a petroleum industry byproduct used as a carbon source for the production of propionic acid and chemical methods has several drawbacks, including non-environmentally friendly, high cost and accumulation of toxic byproducts. It has various industrial applications such as food preservation, pharmaceutical industries, antibacterial agents etc. The use of cost-effective calcium derivatives like calcium alginate, polygalacturonate makes this process non-feasible for large-scale production (Mohan and Sivaprakasam 2016).

Levulinic Acid

Corn stover is a potential agricultural waste with applications in the growing economy and levulinic acid is the top value-added product generated from biomass like corn stover. It is produced through acid-catalyzed dehydration and hydrolysis of hexose sugars. In a study, an integrated biorefinery process of using corn stover for levulinic acid production was performed. The first step for levulinic acid production from corn biomass includes acidic or alkali hydrothermal pretreatment for removal of hemicellulose fraction from the corn biomass. The liquid hydrolysate formed from pretreatment of lignocellulosic biomass mainly contains sugars from hemicelluloses and celluloses, different inhibitor products like furfural lignin-derived phenolic compounds. In a study, the first step involved the hydrothermal pretreatment of corn stover with potassium carbonate solution in a flow-through reactor following the second step by batch acid hydrolysis of pretreated corn biomass using sulfuric acid to form levulinic acid (Thakkar et al. 2021). Levulinic acid is a keto acid, also known as 4-Oxopentanoic acid with two functional groups carboxylic acid and keto group (Malu et al. 2019). It can be produced by acid hydrolysis of sugars including glucose, sucrose, fructose, galactose and starch. It has various pharmaceutical and food applications (Ghorpade and Hanna 1997). Various waste biomass is used as a feedstock for levulinic acid production including corn stover (Dutta et al. 2020).

10.3.3 Enzymes

Xylanase

Xylan is the second most abundant polysaccharide and is a major component of the plant cell wall. The binding properties are mediated by covalent and noncovalent interactions with lignin, cellulose and other polymers (Subramaniyan and Prema 2002). The hydrolysis of xylan yields xylose. Plant feedstock including corn cob is composed of C_6 sugars from cellulose and C_5 sugars from hemicellulose. The production of corn cob from the processing of maize crop are nearly 180 kg of corn cob from each tonne of maize. The microorganisms for xylanase production are *Bacillus* sp., *Bacillus megaterium, Streptomyces viridochromoge, Aspergillus terricola, Aspergillus ochraceus, Humicola brevis, Penicillium* sp., *Aspergillus fumigatus* (Simair et al. 2018). Corn stover media is used for cellulase and xylanase production with white-rot fungi (Tirado-González et al. 2016). In a study, corn cob was used for xylanase production using wild type and UV mutated *Aspergillus niger*. The bio enzyme production from cheaper substrates such as corn cob is beneficial and reduces the negative environmental effects of these wastes. Xylanase can be produced from agricultural waste with new thermophilic *Bacillus cereus* strain TH-050 by solid substrate fermentation and non-sterilized solid substrate fermentation (Ire et al. 2021). The application of xylanase includes in paper and pulp industry, clarification of fruit juices and food industries. This enzyme is also used in agribusiness in animal feed to break down, such as arabinoxylans that reduce the viscosity of raw material (Simair et al. 2018).

Laccase

Laccases are multicopper proteins found in higher plants, fungi, insects and bacteria. Cornstalk is used as the substrate for laccase production using solid-state fermentation. It's easy available and low cost play a major role in extracellular enzyme production from microorganisms. Pretreatment of the substrate is necessary before the fermentation process. The maximum laccase production from corn stalk waste observed as 6.88 U/m L using steam explosion pretreatment and solid-state fermentation (Perdani et al. 2020). Corn steep liquor also demonstrated as a nutrient for improving laccase production by *Trametes versicolor* (Wang et al. 2014). Laccase has application in delignification of pulp, oxidation of organic pollutants, decolorization of and detoxification of industrial effluents. The application of laccase is in higher plants for lignification process and degradation of humic acids. The reactive radicals lead to the formation of monomers by the cleavage of covalent bonds. The ring cleavage of aromatic compounds is catalyzed by laccase and used to degrade xenobiotics like synthetic dyes and nitroaromatics (Kuddus et al. 2013).

Beta-glucosidase

The steps involved in the production of beta-glucosidase from corn cob are pretreatment, enzymatic hydrolysis and fermentation. There are mild conditions for enzymatic hydrolysis such as temperature 30–50 °C and pH 5. In a study corn cobs cut into small pieces, dried and pretreatment in autoclave conditions that were 121 °C for 15 min to obtain carbohydrate fraction from the corncob biomass to remove lignin compounds. Fermentation was done using *Aspergillus niger* following solid-state fermentation. The nutrients and substrate were in the ratio 1:15. The crude enzyme fraction was extracted using 0.1 M phosphate buffer and pH maintained at 7. The samples for the analysis of beta-glucosidase during fermentation analyzed for enzyme activity. Fermentation using corn cob produced beta-glucosidase with enzyme activity 95.01 U/ml (Aliyah et al. 2017). Beta-glucosidase is a component of the cellulase enzyme complex and is responsible for the complete hydrolysis of cellulose into glucose (Bai et al. 2013). It has industrial applications such as improved the conversion rate, easier separation of product from the reactant broth, low cost of commercial production (Hati et al. 2020). This enzyme cleaves the beta-glycosidic linkage in disaccharides. It has other applications in medical, biotechnological, agricultural, industrial (Jeng et al. 2011).

Alpha-amylase

Alpha- amylases are the enzymes that hydrolyze the internal alpha 1,4-glycosidic bond which converts starch into glucose, maltose and maltotriose units and is low molecular weight products. It is endo-acting and hydrolyzes alpha-1,4 bonds and alpha-1,6 linkages in amylopectin and glycogen. Corn cob is used for alpha-amylase production using *Aspergillus niger* by solid-state fermentation. The alpha-amylase production involves the steps such as pretreatment, enzymatic hydrolysis and fermentation. The various other parameters that affect the enzyme production using solid-state fermentation are substrate used by microorganisms, pretreatment process of biomass, substrate moisture content, particle size, type and size of inoculum (Aliyah et al. 2017). The industrial sector mainly uses enzymes derived from fungi, bacteria and yeast. It has industrial importance and is used in 25% of the world enzyme market. The fungal strain *Aspergillus* is thoroughly studied to produce alpha-amylase for industrial use (Mikawlrawng 2016).

10.3.4 Phenolic Compound

Ferulic Acid

Ferulic acid is a hydroxycinnamic acid present in cell walls of plants cross-linked with cell wall and lignin through ether or ester linkage (Ferri et al. 2020). It is a potent antioxidant and has great industrial value in pharmaceuticals, food, and cosmetics. Its derivatives are used as a flavouring agent and as a bioplastic forming material. It can be extracted from plant-based biomass by acid or alkali pretreatments which leads to breakage of lignin/phenolic-carbohydrate complexes in a corn cob and corn stover used for ferulic acid production. The other cheaper raw material for ferulic acid production includes agricultural residues such as wheat bran and maize bran. Enzymatic hydrolysis is an alternate method to produce ferulic acid (Wang et al. 2020a). The steps involved in ferulic acid production from corn bran include corn bran hydrolysis, which forms solid residue and liquid hydrolysate. The alkaline hydrolysis with various concentrations of sodium hydroxide mixed with ethanol. The different NaOH concentrations used were 0.2, 0.5, 0.75 and 1.0 mol/L, ethanol concentrations (30%, 50%, 70% and 90% v/v), solid-to-liquid ratios (1:5, 1:10, 1:15 and 1:20), extraction temperatures (65, 75, 85 and 95 °C) and extraction times (1.5, 2.0, 2.5 and 3.0 h) was used to obtain optimal extraction of acid. Further, the liquid hydrolysate obtained after the pretreatment undergoes evaporation, ultrafiltration followed by nanofiltration to separate the phenolic compounds. The extraction steps for ferulic acid include the addition of 0.25 mol/L NaOH aqueous solution and 50% ethanol (v/v) to corn bran at 75 °C for 2 h released 81% of the bound ferulic acid (Zhou et al. 2017a). The enzymes and microbes seem to be a promising approach for extraction (Ferri et al. 2020) (Zhou et al. 2017a).

10.3.5 *Sugar Alcohol*

Xylitol

Xylitol is a type of sugar alcohol with great importance in the food and pharmaceutical-based industries as an alternative to sugar due to its natural sweetness. It can be used in chemical industries for polymer synthesis (Baptista et al. 2018). It is commercially prepared by chemical or enzymatic hydrolysis of xylan, a cost-effective and non-eco-friendly method. Corn cob was used for ethanol and xylitol production using thermotolerant yeast *Kluyveromyces marxianus* CICC 1727–5. The steps and parameters include corn cob collection and physical pretreatment, dilute acid pretreatment, seed preparation, two-stage fermentation and sugars and other products analysis by High-Performance Liquid Chromatography (HPLC). Corn cob was treated with 0.5% w/w sulfuric acid and 1.5% w/w/ phosphoric acid and kept at 128 °C for 1 h. The yeast strain was cultivated at YPD media and cultured at 30 °C for 24 h at 150 rpm. The acid pretreated corn cob slurry was used for fermentation. Simultaneous saccharification and co-fermentation approach used for ethanol was produced to valorize corn cob and xylose (Du et al. 2020).

10.3.6 *Biosurfactants*

Biosurfactants are amphipathic compounds having the property of both hydrophobic and hydrophilic moieties (Almeida et al. 2016). Biosurfactants are classified into six types—lipoproteins, glycolipids, phospholipids, fatty acids, particulate and polymeric surfactants (Sharma and Melkania 2017). In Glycolipids, the carbohydrate moiety is attached to a fatty acid moiety. Biosurfactants have advantages over chemically synthesized biosurfactants as they are non-toxic, biodegradable, mild, environment friendly. They continue to have activity at a higher temperature, pH and salinity (Almeida et al. 2016; Sharma and Melkania 2017; Wang et al. 2020b). The production cost of biosurfactants is high so agricultural and other waste can be used for biosurfactant production and microbial cultures and microorganisms including *Pseudomonas app., Bacillus spp., Acinetobacter spp. And Candida spp.* There are various applications of biosurfactants like increasing the transportation of crude oil through pipelines, cleaning of oil storage tanks, demulsifying agents, anti-corrosive agents, treatment of oil waste, and control of sulfate-reducing bacteria. In recent times, food waste is being used to form value-added products such as lactic acid, bio-colourants, succinic acid, reactive lignin, microbial surfactants (Wang et al. 2020b).

10.3.7 *Industrially Important Chemicals*

Sodium gluconate

Sodium gluconate is synthesized using corn stover as substrate after dilute acid pretreatment by *Aspergillus niger.* To maintain the enzyme activity, a suitable pH environment was maintained by adding sodium hydroxide slowly in the solution leading to sodium gluconate formation. It showed that corn cob residue after dilute acid pretreatment (5% w/w) and enzymatic hydrolysis yield a145.80 g/L glucose, from which 166.87 g/L sodium gluconate produced using *Gluconobacter oxydans* DSM 2003. Thus, corn waste can produce sodium gluconate, which has several industrial applications as, it is used as surface cleaning agent of glass or steel, super-plasticizer in the cement industry, chelating agent, a high-efficiency retarder (Han et al. 2018).

10.4 Benefits and Challenges in Converting Corn Waste to Value-Added Products

The advantages of using corn waste are: high biomass productivity and availability of economic, effective, renewable, easily available and carbon–neutral biomass (Bhutto et al. 2017). There is no additional use of land to produce corn stover, corn cob and corn straw and no direct competition of corn stover with food as this is not useful as food for human beings. There is an additional benefit to the farmers as the straw or husk is used for high-value products. Valorizing corn waste is important rather than disposing of it as waste; it serves as an alternate substrate in place of depleting fossil fuels (Loong et al. 2021). The limitation of using corn stover biomass is due to the recalcitrant and resistant nature of lignocellulosic biomass cellulose, hemicellulose and lignin (Bhutto et al. 2017). This structure of corn stover biomass restricts the interaction between the enzymes or microbial strain with the cellulose and thus is not easily converted to various metabolites. The main important step before hydrolysis, fermentation and purification steps is the pretreatment method which disrupts the crystalline and polymeric structure of corn stover. Extensive research is required for the cost-effective utilization of corn stover and corn cob for the viable production of chemicals and bioenergy. The saccharification process is still a critical bottleneck and an ideal method should be generated for stoichiometric amounts of fermentable monomeric sugars out of the lignocellulosic complex. The other problem which needs to be solved is regarding the search for kinetically more efficient cellulase (Khare et al. 2015).

10.5 Conclusion

Agricultural production has increased over the last 50 years and corn waste is an abundant, renewable carbon resource that may be utilized which otherwise causes resource waste or environmental deterioration. The burning of crop waste is the major cause of worsening air quality in different countries worldwide. Biorefinery is a major direction for crop straw use reasonably and efficiently. Corn waste is un-fermentable due to the primary cellulosic, hemicellulosic and lignin components and most microbes cannot degrade it and thus, pretreatment plays a vital role in overcoming the recalcitrant and resistant nature of the biomass. Wide conversion technologies are used for making value-added products from corn waste. Biorefineries enable sustainable processing of biowaste into a wide range of marketable products, bioenergy and biobased products. Nowadays, research focus on the high yield of products with less energy consumption. Different types of products just as biofuels, organic acids, bio enzymes, phenolic compounds, bio-based polymers, sugar alcohols, and other industrially important chemicals are obtained from corn waste. Optimizing, scaling, implementing and integrating parameters and techniques will lead to enhanced and efficient bioproducts for corn waste.

Acknowledgements This work is supported by Grant no. BT/RLF/Re-entry/40/2017 from the Department of Biotechnology (DBT), Ministry of Science and technology and SERB project file no: EEQ/2020/000614 Govt. of India. We acknowledge to CSIR pool scientific scheme Grant no. 9103, Govt. of India. We also acknowledge CSIR for fellowships support to Ms. Sonika. The figures are created by www.biorender.com.

References

Abdu HI, Eid K, Abdullah AM, Han Z, Ibrahim MH, Shan D, Chen J, Elzatahry AA, Lu X (2020) Unveiling one-pot scalable fabrication of reusable carboxylated heterogeneous carbon-based catalysts from eucalyptus plant with the assistance of dry ice for selective hydrolysis of eucalyptus biomass. Renew Energy 153:998–1004. https://doi.org/10.1016/j.renene.2020.02.034

Abdullah R, Tahseen M, Nisar K, Kaleem A, Iqtedar M, Saleem F, Aftab M (2021) Statistical optimization of cellulases by Talaromyces thermophilus utilizing Saccharum spontaneum, a novel substrate. Electron J Biotechnol 51:79–87. https://doi.org/10.1016/j.ejbt.2021.03.007

Adekunle AE, Rabeya T, Jehadin F, Asad MA, Ayodele OO, Islam MS (2020) Compressed hot water pretreatment enhanced bioethanol production from corn stalk. Bioresour Technol Reports 12:100595. https://doi.org/10.1016/j.biteb.2020.100595

Aliyah A, Alamsyah G, Ramadhani R, Hermansyah H (2017) Production of α-amylase and β-glucosidase from Aspergillus Niger by solid state fermentation method on biomass waste substrates from rice husk, bagasse and corn cob. Energy Procedia 136:418–423. https://doi.org/10.1016/j.egypro.2017.10.269

Babaei M, Tsapekos P, Alvarado-Morales M, Hosseini M, Ebrahimi S, Niaei A, Angelidaki I (2019) Valorization of organic waste with simultaneous biogas upgrading for the production of succinic acid. Biochem Eng J 147:136–145. https://doi.org/10.1016/j.bej.2019.04.012

Bahry H, Abdalla R, Pons A, Taha S, Vial C (2019) Optimization of lactic acid production using immobilized Lactobacillus Rhamnosus and carob pod waste from the Lebanese food industry. J Biotechnol 306:81–88. https://doi.org/10.1016/j.jbiotec.2019.09.017

Bai A, Zhao X, Jin Y, Yang G, Feng Y (2013) A novel thermophilic β-glucosidase from Caldicellulosiruptor bescii: characterization and its synergistic catalysis with other cellulases. J Mol Catal B Enzym 85–86:248–256. https://doi.org/10.1016/j.molcatb.2012.09.016

Baptista SL, Cunha JT, Romaní A, Domingues L (2018) Xylitol production from lignocellulosic whole slurry corn cob by engineered industrial Saccharomyces cerevisiae PE-2. Bioresour Technol 267:481–491. https://doi.org/10.1016/j.biortech.2018.07.068

Baruah J, Nath BK, Sharma R, Kumar S, Deka RC, Baruah DC, Kalita E (2018) Recent trends in the pretreatment of lignocellulosic biomass for value-added products. Front Energy Res 6:1–19. https://doi.org/10.3389/fenrg.2018.00141

Behera S, Arora R, Nandhagopal N, Kumar S (2014) Importance of chemical pretreatment for bioconversion of lignocellulosic biomass. Renew Sustain Energy Rev 36:91–106. https://doi.org/10.1016/j.rser.2014.04.047

Bhargav S, Sanjrani MA, Javed S (2008) Solid-state fermentation: an overview solid-state fermentation: an overview

Bhutto AW, Qureshi K, Harijan K, Abro R, Abbas T, Bazmi AA, Karim S, Yu G (2017) Insight into progress in pretreatment of lignocellulosic biomass. Energy 122:724–745. https://doi.org/10.1016/j.energy.2017.01.005

Borole AP, Greig AL (2019) Life-cycle assessment and systems analysis of hydrogen production. Biohydrogen. 485–512. https://doi.org/10.1016/b978-0-444-64203-5.00020-4

Chen Y, Dong L, Miao J, Wang J, Zhu C, Xu Y, Chen G, Liu J (2019) Bioresource technology hydrothermal liquefaction of corn straw with mixed catalysts for the production of bio-oil and aromatic compounds. Bioresour Technol 294:122148. https://doi.org/10.1016/j.biortech.2019.122148

De Almeida DG, de Soares Da Silva RCF, Luna JM, Rufino RD, Santos VA, Banat IM, Sarubbo LA (2016) Biosurfactants: promising molecules for petroleum biotechnology advances. Front Microbiol 7:1–14. https://doi.org/10.3389/fmicb.2016.01718

Demirbas A (2007) Importance of biodiesel as transportation fuel. Energy Policy 35:4661–4670. https://doi.org/10.1016/j.enpol.2007.04.003

Dhillon GS, Oberoi HS, Kaur S, Bansal S, Brar SK (2011) Value-addition of agricultural wastes for augmented cellulase and xylanase production through solid-state tray fermentation employing mixed-culture of fungi. Ind Crops Prod 34:1160–1167. https://doi.org/10.1016/j.indcrop.2011.04.001

Du C, Li Y, Zong H, Yuan T, Yuan W, Jiang Y (2020) Production of bioethanol and xylitol from non-detoxified corn cob via a two-stage fermentation strategy. Bioresour Technol 310:123427. https://doi.org/10.1016/j.biortech.2020.123427

Dutta S, Yu IKM, Tsang DCW, Su Z, Hu C, Wu KCW, Yip ACK, Ok YS, Poon CS (2020) Influence of green solvent on levulinic acid production from lignocellulosic paper waste. Bioresour Technol 298:122544. https://doi.org/10.1016/j.biortech.2019.122544

Elkady MF, Zaatout A, Balbaa O (2015) Production of biodiesel from waste vegetable oil via KM micromixer. J Chem 2015. https://doi.org/10.1155/2015/630168

Ferri M, Happel A, Zanaroli G, Bertolini M, Chiesa S, Commisso M, Guzzo F, Tassoni A (2020) Advances in combined enzymatic extraction of ferulic acid from wheat bran. N Biotechnol 56:38–45. https://doi.org/10.1016/j.nbt.2019.10.010

Ghorpade V, Hanna M (1997) Industrial applications for levulinic acid. Cereals. 49–55. https://doi.org/10.1007/978-1-4757-2675-6_7

Gudiña EJ, Fernandes EC, Rodrigues AI, Teixeira JA, Rodrigues LR (2015) Biosurfactant production by Bacillus subtilis using corn steep liquor as culture medium. 6:1–8. https://doi.org/10.3389/fmicb.2015.00059

Han X, Liu G, Song W, Qu Y (2018) Production of sodium gluconate from delignified corn cob residue by on-site produced cellulase and co-immobilized glucose oxidase and catalase. Bioresour Technol 248:248–257. https://doi.org/10.1016/j.biortech.2017.06.109

Hati S, Ningtyas DW, Khanuja JK, Prakash S (2020) β-Glucosidase from almonds and yoghurt cultures in the biotransformation of isoflavones in soy milk. Food Biosci 34:100542. https://doi.org/10.1016/j.fbio.2020.100542

Hijosa-Valsero M, Garita-Cambronero J, Paniagua-García AI, Díez-Antolínez R (2020) A global approach to obtain biobutanol from corn stover. Renew Energy 148:223–233. https://doi.org/10.1016/j.renene.2019.12.026

Huang Q, Niu CH, Dalai AK (2019) Production of anhydrous biobutanol using a biosorbent developed from oat hulls. Chem Eng J 356:830–838. https://doi.org/10.1016/j.cej.2018.09.067

Ifeanyi N, Uchechi NE, Ngozi GA, Chinyere RO (2016) Bioconversion of agricultural and food processing wastes to value-added products using solid state fermentation technology. A review. IOSR J Pharm Biol Sci 11:68–71. https://doi.org/10.9790/3008-1105036871

Ire FS, Chima IJ, Ezebuiro V (2021) Enhanced xylanase production from UV-mutated Aspergillus niger grown on corn cob and sawdust. Biocatal Agric Biotechnol 31. https://doi.org/10.1016/j.bcab.2020.101869

Jeng WY, Wang NC, Lin MH, Lin CT, Liaw YC, Chang WJ, Liu CI, Liang PH, Wang AHJ (2011) Structural and functional analysis of three β-glucosidases from bacterium Clostridium cellulovorans, fungus Trichoderma reesei and termite Neotermes koshunensis. J Struct Biol 173:46–56. https://doi.org/10.1016/j.jsb.2010.07.008

Khan Z, Dwivedi AK, Engineering C, College UE (2013) Fermentation of biomass for production of ethanol: a review abstract: 2. Potential of biomass. Univers J Environ Res Technol 3:1–13

Khare SK, Pandey A, Larroche C (2015) Current perspectives in enzymatic saccharification of lignocellulosic biomass. Biochem Eng J 102:38–44. https://doi.org/10.1016/j.bej.2015.02.033

Kuddus M, Joseph B, Wasudev P (2013) Production of laccase from newly isolated pseudomonas putida and its application in bioremediation of synthetic dyes and industrial effluents biocatalysis and agricultural biotechnology production of laccase from newly isolated pseudomonas putida and its application in bioremediation of synthetic dyes and industrial effluents. Biocatal Agric Biotechnol 2:333–338. https://doi.org/10.1016/j.bcab.2013.06.002

Kumar D, Jain VK, Shanker G, Srivastava A (2003) Utilization of fruits waste for citric acid production by solid state fermentation. Process Biochem 38:1725–1729. https://doi.org/10.1016/S0032-9592(02)00253-4

Loong C, Liang K, Zi A, Ku E, Lai B, Chin F, Sow S, Lock M, Wai K, Chun A, Loy M, Herng Y (2021) Bioresource technology recent advances in green solvents for lignocellulosic biomass pretreatment: potential of choline chloride (ChCl) based solvents. Bioresour Technol 333:125195. https://doi.org/10.1016/j.biortech.2021.125195

Machmudah S, Wahyudiono, Kanda H, Goto M (2017) Hydrolysis of biopolymers in near-critical and subcritical water. Elsevier Inc.

Maitan-Alfenas GP, Visser EM, Guimarães V, Ria M (2015) Enzymatic hydrolysis of lignocellulosic biomass: converting food waste in valuable products. Curr Opin Food Sci 1:44–49. https://doi.org/10.1016/j.cofs.2014.10.001

Malu TJ, Manikandan K, Cheralathan KK (2019) Levulinic acid-a potential keto acid for producing biofuels and chemicals. Elsevier B.V

Meena M, Zehra A, Dubey MK, Aamir M, Upadhyay RS (2017) Penicillium enzymes for the food industries. Elsevier B.V

Mikawlrawng K (2016) Aspergillus in biomedical research. Elsevier B.V

Millati R, Cahyono RB, Ariyanto T, Azzahrani IN, Putri RU, Taherzadeh MJ (2019) Agricultural, industrial, municipal, and forest wastes. Elsevier B.V

Mohan N, Sivaprakasam S (2016) Process design and optimization for platform chemical biorefinery. Elsevier Inc

Mondala A, Liang K, Toghiani H, Hernandez R, French T (2009) Biodiesel production by in situ transesterification of municipal primary and secondary sludges. Bioresour Technol 100:1203–1210. https://doi.org/10.1016/j.biortech.2008.08.020

Moodley P, Gueguim Kana EB (2019) Bioethanol production from sugarcane leaf waste: effect of various optimized pretreatments and fermentation conditions on process kinetics. Biotechnol. Reports. 22:e00329. https://doi.org/10.1016/j.btre.2019.e00329

Nagoor Gunny AA, Arbain D, Javed M, Baghaei-Yazdi N, Gopinath SCB, Jamal P (2019) Deep eutectic solvents-halophilic cellulase system: an efficient route for in situ saccharification of lignocellulose. Process Biochem. 81:99–103. https://doi.org/10.1016/j.procbio.2019.03.003

Olajuyin AM, Yang M, Thygesen A, Tian J, Mu T, Xing J (2019) Effective production of succinic acid from coconut water (Cocos nucifera) by metabolically engineered Escherichia coli with overexpression of Bacillus subtilis pyruvate carboxylase. Biotechnol Rep 24:e00378. https://doi.org/10.1016/j.btre.2019.e00378

Perdani MS, Margaretha G, Sahlan M, Hermansyah H (2020) Science direct solid state fermentation method for production of laccase enzyme with bagasse, cornstalk and rice husk as substrates for adrenaline biosensor. Energy Rep 6:336–340. https://doi.org/10.1016/j.egyr.2019.08.065

Qin Z, Duns GJ, Pan T, Xin F (2018) Consolidated processing of biobutanol production from food wastes by solventogenic Clostridium sp. strain HN4. Bioresour Technol 264:148–153. https://doi.org/10.1016/j.biortech.2018.05.076

Rivas B, Moldes AB, Dom JM, Parajó JC (2004) Lactic acid production from corn cobs by simultaneous saccharification and fermentation: a mathematical interpretation. 34:627–634. https://doi.org/10.1016/j.enzmictec.2004.01.011

Sağır E, Hallenbeck PC (2019) Photofermentative hydrogen production. Biohydrogen. 141–157. https://doi.org/10.1016/b978-0-444-64203-5.00006-x

Sánchez C, Egüés I, García A, Llano-Ponte R, Labidi J (2012) Lactic acid production by alkaline hydrothermal treatment of corn cobs. Chem Eng J 181–182:655–660. https://doi.org/10.1016/j.cej.2011.12.033

Shafiei Alavijeh R, Karimi K, van den Berg C (2020) An integrated and optimized process for cleaner production of ethanol and biodiesel from corn stover by Mucor indicus. Elsevier Ltd

Sharma P, Melkania U (2017) Biosurfactant-enhanced hydrogen production from organic fraction of municipal solid waste using co-culture of E. coli and Enterobacter aerogenes. Bioresour. Technol. 243:566–572. https://doi.org/10.1016/j.biortech.2017.06.182

Simair AA, Qureshi AS, Simair SP, Khushk I, Klykov SP, Ali CH, Lu C (2018) An integrated bioprocess for xylanase production from agriculture waste under open non-sterilized conditions: biofabrication as fermentation tool. J Clean Prod 193:194–205. https://doi.org/10.1016/j.jclepro.2018.05.020

Sindhu R, Binod P, Pandey A (2016) Biological pretreatment of lignocellulosic biomass—an overview. Bioresour Technol 199:76–82. https://doi.org/10.1016/j.biortech.2015.08.030

Soliman MMA, Karmakar A, Alegria ECBA, Ribeir APC, Rúbio GMDM, Saraiva MS, Guedes MFC, Pombeiro AJL (2019) ZnO nanoparticles: an efficient catalyst for transesterification reaction of α—keto carboxylic esters. https://doi.org/10.1016/j.cattod.2019.08.053

Subramaniyan S, Prema P (2002) Biotechnology of microbial xylanases: enzymology, molecular biology, and application. Crit Rev Biotechnol 22:33–64. https://doi.org/10.1080/07388550290789450

Sud D, Mahajan G, Kaur MP (2008) Agricultural waste material as potential adsorbent for sequestering heavy metal ions from aqueous solutions—a review. Bioresour Technol 99:6017–6027. https://doi.org/10.1016/j.biortech.2007.11.064

Sukumaran RK, Singhania RR, Mathew GM, Pandey A (2009) Cellulase production using biomass feed stock and its application in lignocellulose saccharification for bio-ethanol production. Renew Energy 34:421–424. https://doi.org/10.1016/j.renene.2008.05.008

Suriya J, Bharathiraja S, Krishnan M, Manivasagan P, Kim SK (2016) Marine microbial amylases: properties and applications. Elsevier Inc

Tathod AP, Dhepe PL (2015) Efficient method for the conversion of agricultural waste into sugar alcohols over supported bimetallic catalysts. Bioresour Technol 178:36–44. https://doi.org/10.1016/j.biortech.2014.10.036

Tayyab M, Noman A, Islam W, Waheed S, Arafat Y, Ali F, Zaynab M, Lin S, Zhang H, Lin W (2017) Bioethanol production from lignocellulosic biomass by environment friendly pretreatment methods: a review 16:225–249. https://doi.org/10.15666/aeer/1601_225249

Thakkar A, Shell KM, Bertosin M, Rodene DD, Amar V, Bertucco A, Gupta RB, Shende R, Kumar S (2021) Production of levulinic acid and biocarbon electrode material from corn stover through an integrated biorefinery process. Fuel Process Technol 213:106644. https://doi.org/10.1016/j.fuproc.2020.106644

Timung R, Mohan M, Chilukoti B, Sasmal S, Banerjee T, Goud VV (2015) Optimization of dilute acid and hot water pretreatment of different lignocellulosic biomass: a comparative study. Biomass Bioenerg 81:9–18. https://doi.org/10.1016/j.biombioe.2015.05.006

Tirado-González DN, Jáuregui-Rincón J, Tirado-Estrada GG, Martínez-Hernández PA, Guevara-Lara F, Miranda-Romero LA (2016) Production of cellulases and xylanases by white-rot fungi cultured in corn stover media for ruminant feed applications. Anim Feed Sci Technol 221:147–156. https://doi.org/10.1016/j.anifeedsci.2016.09.001

Wang F, Hu JH, Guo C, Liu CZ (2014) Enhanced laccase production by Trametes versicolor using corn steep liquor as both nitrogen source and inducer. Bioresour Technol 166:602–605. https://doi.org/10.1016/j.biortech.2014.05.068

Wang X, Wang G, Yu X, Chen H, Sun Y, Chen G (2017) Pretreatment of corn stover by solid acid for D-lactic acid fermentation. Bioresour Technol 239:490–495. https://doi.org/10.1016/j.biortech.2017.04.089

Wang H, Tsang CW, To MH, Kaur G, Roelants SLKW, Stevens CV, Soetaert W, Lin CSK (2020b) Techno-economic evaluation of a biorefinery applying food waste for sophorolipid production—a case study for Hong Kong. Bioresour Technol 303:122852. https://doi.org/10.1016/j.biortech.2020.122852

Wang R, Yang J, Jang JM, Liu J, Zhang Y, Liu L, Yuan H (2020) Efficient ferulic acid and xylo-oligosaccharides production by a novel multi-modular bifunctional xylanase/feruloyl esterase using agricultural residues as substrates. Elsevier Ltd

Xu X, Tu R, Sun Y, Wu Y, Jiang E, Zhen J, Jiang E (2018) The Influence of combined pretreatment with surfactant/ultrasonic and hydrothermal carbonization on fuel properties, pyrolysis and combustion behavior of corn stalk Bioresour. Technol. https://doi.org/10.1016/j.biortech.2018.09.066

Zheng P, Fang L, Xu Y, Dong JJ, Ni Y, Sun ZH (2010) Succinic acid production from corn stover by simultaneous saccharification and fermentation using Actinobacillus succinogenes. Bioresour Technol 101:7889–7894. https://doi.org/10.1016/j.biortech.2010.05.016

Zhou PP, Meng J, Bao J (2017b) Fermentative production of high titer citric acid from corn stover feedstock after dry dilute acid pretreatment and biodetoxification. Bioresour Technol 224:563–572. https://doi.org/10.1016/j.biortech.2016.11.046

Zhou M, Yan B, Wong JWC, Zhang Y (2017a) Sino-forest applied research centre for pearl river delta environment Bioresour Technol https://doi.org/10.1016/j.biortech.2017.06.121

Zviely M (2013) Converting lignocellulosic biomass to low-cost fermentable sugars. https://doi.org/10.1007/978-3-642-32735-3

Part III
Algal Biorefinery

Chapter 11
Algal Biorefinery: A Paradigm to Sustainable Circular Bioeconomy

Rimjhim Sangtani, Smrity Sonbhadra, Regina Nogueira, and Bala Kiran

11.1 Status-Quo of Algal Biorefinery

The preponderance appertaining to plummeting fossil fuel reservoirs and burgeoning greenhouse gas emissions has expedited microbial cell factories scrutinization as a carbon–neutral origin of bioenergy and biochemical products (Bhatia et al. 2017; Kumar et al. 2020; Xu et al. 2018). While taking into consideration the fringe benefits of photosynthetically proficient and carbon-driven algae, issues including energy crunch and pollution predicaments can be unraveled promisingly. Algal cells have been ascertained as a renowned carbon–neutral and renewable repertoire of multifarious biotechnologically essential bioenergy products, including biofuels, bioplastic, pigments, animal feed, vitamins, therapeutic and nutraceutical metabolites, etc. (Sangtani et al. 2020; Jacob-Lopes et al. 2015; Anand et al. 2019; Kiran et al. 2014; Rahman et al. 2015; Kiran and Venkata Mohan 2021).

However, the utilization of algal biomass can further be ameliorated, and emissions and waste churning out owing to the manufacturing of bioenergy products can be reduced with the stratagem of biorefinery, which eventually proves to be beneficial in the context of environmental sustainability and economic feasibility (Hemalatha et al. 2019; Sarkar et al. 2015; Bastiaens et al. 2017; Venkata Mohan et al. 2016). More importantly, in recent years, the research in the field of algae has deviated from the

R. Sangtani · S. Sonbhadra · B. Kiran (✉)
Department of Biosciences and Biomedical Engineering, Indian Institute of Technology Indore, Indore, Madhya Pradesh 453552, India
e-mail: kiranb@iiti.ac.in

R. Nogueira
Institute for Sanitary Engineering and Waste Management, Leibniz Universitat Hannover, Hannover, Germany

Y. K. Nandabalan et al. (eds.), *Zero Waste Biorefinery*, Energy, Environment, and Sustainability, https://doi.org/10.1007/978-981-16-8682-5_11

single product to multi-product integrated manufacturing by focusing on the biorefinery model superintending on the techniques and tools maneuvering both down and upstream processing (Hemalatha et al. 2019; Ghosh and Das 2016; Bhowmick et al. 2018).

A concatenated methodology encompassing a torrent of procedures, targeting the remodeling of whole algal biomass into imperative bio-products without compromising any bit of the raw materials, envisages the fundamentals of biorefinery. In accordance with the Task 42 of the International Energy Agency (IEA), bioenergy biorefinery comprehends to be a sustainable, cost-effective and environment-friendly maneuvering of the biomass into a gamut of market worthy bio-based (bioplastic, feed, nutraceuticals, etc.) and bioenergy (power, fuels, heat) products. Biorefinery not only focuses on the complete deployment of algae but also contemplates obligatory environmental and socio-economic parameters accompanying several algae bioprocessing technologies (International Energy Agency Bioenergy 2021; Subhadra 2010; Cherubini 2010). Unlike petroleum oil refinery, in which numerous fossil fuel-derived products are produced, the biorefinery concept encompasses the usage of biomass as raw material for the bioproducts production. Production of petroleum-based products as compared to bio-based products has led to severe environmental challenges such as greenhouse gas emissions, global warming, etc., along with the depletion of non-renewable feedstocks, thus paving the way for adapting the emerging biorefinery approach. Biorefinery not only includes the utilization of renewable biomass for the production of various classes of products but also keeps a check on the sustainability of the process opted for the production of multiple products (Cherubini 2010).

Several researchers have investigated various pathways and models to achieve algal biorefinery considering cyanobacteria, microalgae and macroalgae for the sustainable production of numerous high-valued bio-based products (Jiang et al. 2016; Venkata Mohan et al. 2019; Meixner et al. 2018). Concurrently, diverse bioprocessing techniques such as bioelectrochemical system, photosynthetic and carbon fixation approach, nutrient invigorating wastewater treatment and acidogenesis have been explored by the algal community in order to produce diverse bio-based products. The global benefits of algae as well that of various bioprocessing techniques have been depicted in Fig. 11.1. However, more environmentally sound and energy-efficient upstream and downstream processing strategies are still in their infancy in nexus with the integrated biorefinery approach.

Henceforth, this chapter sheds light on the pertinence of algal biopolymers in the biorefinery concept, followed by comprehending the extravagant bio-based algal metabolites. It also recapitulates the know-how of different bioprocessing strategies in relation to biorefinery as well the quintessential aspect of the algal circular bioeconomy.

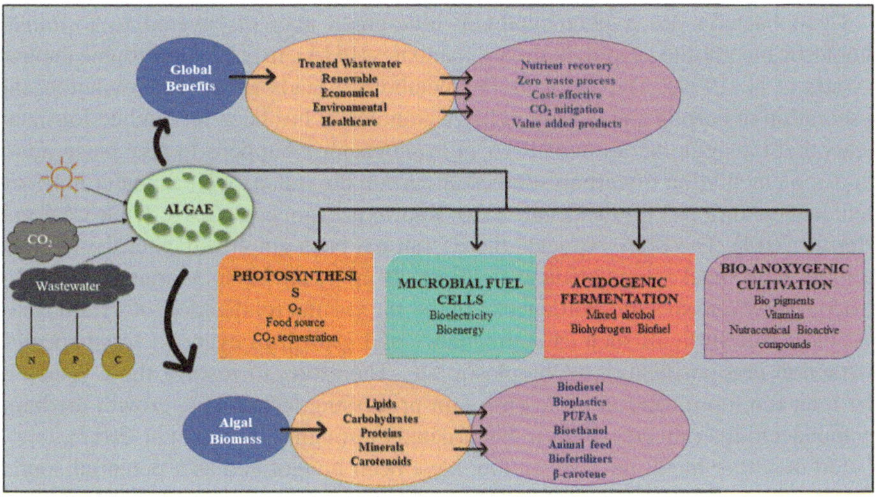

Fig. 11.1 Comprehending different branches of algal biorefinery integrated to achieve circular bioeconomy

11.2 Biopolymers as a Baseline for Algal Biorefinery

An exuberant and amplified consumption of plastic has been a global menace to human health as well as the ecosystem; having said that, the outbreak of the COVID-19 pandemic has further contributed enormously to the environmental threat. A sudden hike in demand for plastic healthcare commodities for the treatment and prevention of coronavirus has inevitably led to its manufacturing and the generation of an immense amount of non-biodegradable plastic waste (Patrício Silva et al. 2021; Klemeš et al. 2020). Furthermore, taking into account the current scenario of plastic pollution, 53 million metric tons per annum of plastic emissions have been projected to occur globally by the year 2030 (Borrelle et al. 2020) and in order to cope up with the environmental impact of plastic, in addition to the plastic's end-of-life management, the hunt for potential and efficient source or feedstock of biodegradable bio-based plastic becomes the pertinent matter of concern.

Henceforth, bioplastics derived from algal biomass come into the picture. Algae have proved to be an excellent source of numerous bioproducts, and their efficiency to produce bioplastic metabolites has been enquired substantially by the researchers. Owing to several advantages algae beholds certain noteworthy characteristics such as exceptional photosynthetic and CO_2 sequestration efficiency, the prerequisite of non-arable land and efficacious growth in residual waste streams that makes algae an exceptional feedstock for biopolymer formulation (Karan et al. 2019). Further, the life cycle impact analysis and techno-economic analysis of the process included in the conversion of microalgae into protein based bioplastic feedstock as well as other fuel products demonstrates that the algal biorefinery coming through it proves to be economically feasible and environmentally sustainable (Beckstrom et al. 2020).

Cyanobacteria are a photosynthetic blue-green alga discovered to formulate bioplastic metabolite i.e. polyhydroxyalkanoates (PHA), in photoautotrophic manner (Asada et al. 1999). The intracellular metabolism leads to the accumulation and reservation of carbon and energy in PHA granules. It has been researched for many years that the agglomeration of PHA in cyanobacteria happens to take place under the supplementation of carbon sources or carbon dioxide as well as under nutrient-deficient conditions (Tarawat et al. 2020; Kaewbai-ngam et al. 2016; Jendrossek and Pfeiffer 2014). The PHA extracted there from has been employed extensively in the field of food, packaging, medical field etc. (Bini et al. 2016; Morais et al. 2016, 2015). However, specific bottlenecks make the commercialization of cyanobacterial bioplastic difficult such as maintenance and growth of cyanobacterial species, extraction and purification of bioplastic etc. Therefore, to resolve these issues in the long run, optimization of downstream processing techniques, growth medium, cyanobacterial strain etc. needs to be targeted thoroughly (Singh et al. 2017).

On the other hand, microalgae too, have been invigorated as a potential source of biopolymers. For instance, the PHA producing ability of microalgae *Microcystis aeruginosa* was detected by the scientists and they illustrated that microalgal biomass tend to accumulate PHA and more research need to be performed to understand the metabolism of microalgae in order to attain economical feasible large-scale production of bioplastic (Abdo and Ali 2019). *Scenedesmus* sp. individually as well as with *Desmodesmus* sp. in consortia were also explored in the nutrient-deficient environment and wastewater, respectively for the production of PHA. It was discovered that the species holds a tremendous amount of potential to accumulate PHA under crucial and compromised circumstances such as salinity, nitrate, phosphate, iron and glucose (López Rocha et al. 2020; García et al. 2021).

The bioplastic producing potency of seaweeds or macroalgae too have been unraveled by scientists from the past few decades (Rajendran et al. 2012). Polysaccharides derived from seaweeds such as carrageenan, fucoidan, alginate, sulfated galactan, xylans etc. have been known to be a type of biopolymer efficient to be commercialized for manufacturing bioplastic commodities (Otari and Jadhav 2021). Seaweeds including *Ulva* sp., *Sargassum* sp., *Ecklonia* sp. etc. have been recognized as a proficient source of metabolites which eventually formulates into environmentally sound, biodegradable, and biocompatible biopolymer (Jahan and Strezov 2019; Flórez-Fernández et al. 2019; Steinbruch et al. 2020; Jumaidin et al. 2017).

The TEA and LCA performed for the formulation of bioplastic from algal biomass along with the fuel co-products has been discovered to be economically and environmentally sound biorefinery approaches (Beckstrom et al. 2020). Howbeit, more research and optimization still need to be performed for the commercialization of bio-based plastic with an integrated biorefinery approach and to curb our reliance on conventional plastic completely.

11.3 High-Value Products Derived from Algal Biorefinery

11.3.1 Vitamins

Algae have always contributed substantially to the diet of human beings owing to their enormous protein composition and expedient amino acid characteristics. Additionally, algae have also been projected as a possible source of vitamins, PUFAs and pigments etc. (Buono et al. 2014; Wells et al. 2017). Vitamins, though they are consumed in trace amounts, form an indispensable part of our diet. They have been known to play a imperative role in energy metabolism in the human body either by acting as a coenzyme or by intervening as an electron or proton carrier; however, they are not directly synthesized by the body. Indeed, their deficiency in the body grounds several diseases like rickets, anemia, night-blindness, scurvy etc. (Koyande et al. 2019). Based on solubility, vitamins have been classified broadly into two types viz. fat-soluble (vitamin A, D, E, K) (Schubert et al. 2018) and water-soluble (vitamin B complex and C) (Chawla and Kvarnberg 2014). Because of their rapid growth and cheaply accessible nature, several algal species have been explored for decades to find out the possibilities of extracting essential vitamins from them.

Microalgae, a non-conventional source of health food products is marketed in the forms of tablets, powders, liquids etc. (Kovač et al. 2013). Researchers have scrutinized several commercial microalgae powders in order to ascertain their vitamin proportion and they have epitomized the presence of active and pseudo vitamin B12, riboflavin, niacin, vitamin K, E, B6, provitamin A etc. (Edelmann et al. 2019; Belay et al. 1993). It has also been researched that in comparison to the vitamin content (vitamins C, B1, B2, B3, and B9) of traditional-vitamin-source, higher plants, microalgae stand out in terms of the vitamin composition (Kay and Barton 1991; Fabregas and Herrero 1990).

Few scientists investigated the vitamin content of 5 different aquaculture microalgae as well as one macroalga and concluded that *Tetraselmis suecica,* a green marine microalga, comprises provitamin A and vitamin E in immense quantity; *Isochrysis galbana* exhibited a high amount of vitamin PP (composed of niacin and niacinamide) and vitamin B6 (pyridoxine); *Pavlova lutheri* had vitamin B12 and vitamin C and *Skeletonema costatum* contained vitamin B1 (thiamine); whereas macroalgae, *Sargassum muticum* was observed to accumulate vitamin K and vitamin C in a reasonable quantity (Roeck-Holtzhauer et al. 1991). Subsequently, acknowledging the fact that culture conditions, growth stage, diverse abiotic factors, harvesting stage and the algal species predominantly determine the vitamin composition of the algae (Fabregas and Herrero 1990; Brown et al. 1999; Galasso et al. 2019).

The presence of vitamin B12 in microalgae, as well as macroalgae has also been illustrated in the literature, thus enriching the diet of vegan people (Watanabe et al. 2013). *Chlorella* and *Dunaliella* also unravel as rich sources of group B and lipid-soluble vitamins; *Chlorella stigmatophora* encompasses higher concentrations of vitamin C, E and some B-complexes like B5 (pantothenic acid), B7 (biotin) and B3 (nicotinic acid). In contrast, *Dunaliella tertiolecta* proves to be quite rich in vitamin B2, B12, B9 (folic acid), C, B3, and E (Fabregas and Herrero 1990). Chlorella strains (9–18%) were found to be a enriched origin of vitamin B12 in a study performed by few coworkers (Shim et al. 2008). Besides, Chlorella is a significant contributor to vitamin K (required for blood coagulation); it is also proclaimed to cause abdominal cramping, flatulence, diarrhea, green stools, and also specific allergic reactions like asthma, anaphylaxis in many cases (Bethesda 2006).

A red microalga, *Porphyridium cruentum* has also been determined to consist of a good quantity of vitamin E (tocopherols), C (ascorbic acid), and provitamin A (β carotene) (Santhosh et al. 2016). Seaweeds are also ascertained to be opulent sources of vitamin B2, B12, C and E (Borowitzka 1998).

To commercialize algal vitamins, selecting proper algal strain and time of culling of vitamins produced is equally significant and imperative. Since most of the vitamins are anti-oxidants, it is estimated that one can achieve their higher yield in certain stress conditions like nutrient (nitrogen) deficiency, osmotic and oxidative stress (Chandra et al. 2019).

11.3.2 Therapeutic Compounds

Algae are recognized to accumulate wide array of therapeutic compounds for human welfare. These include antibiotics, antivirals, antifungals, algicides, anti-inflammatory, antidiabetic, anticarcinogenic, plant growth regulators (PGRs) and many more. Several kinds of research have been carried out to date to discern and explore algal species that could potentially elicit these essential pharmaceuticals. Algal antibiotic entities generally comprise carbohydrates, fatty acids, organic acids and solvents, polysaccharides, terpenoids etc. (Parsaeimehr and Lutzu 2016).

To mention a few, the green algal class Chlorophyta stands in good support of producing Chlorellin, a fatty acid mixture derived from *Chlorella* species and that has been used as an antibacterial agent (Pratt et al. 1944). *Spirogyra* species, accumulating gallotannin, ultimately acting as an antiviral (Misra et al. 2013). *Chlamydomonas reinhardii*, produces fatty acids that act as an algicidal compound (McCracken et al. 1980). A few diatoms like *Chaetoceros lauderi* have a polysaccharide compound working as an antifungal; *Fragilaria pinnata* gives a type of peptide functioning that acts as an antifungal (Pesando and Caram 1984). Some Chrysophyta (golden-brown) members, including *Ochromonas danica* produces fatty acid used as fungicidal (Aaronson et al. 1967); *Phaeocystis pouchetii* producing acrylic acid has been used as an antibacterial agent (Sieburth 1960). Furthermore, cyanobacteria, also have been extensively worked upon to come up with such examples as *Calothrix brevissima* accumulating bromophenols has been known as a potential antibiotic (Pedersén and DaSilva 1973). *Anabaena flosaquae* is known to produce an anti-protozoan agent (Snell et al. 1983) and Aponin extracted out from *Gomphosphaeria aponica* has also been utilized as an algicidal (Trevan and Mak 1988). Antiviral activity of the microalgal and cyanobacterial extract in response to the influenza virus was investigated and it was summarized that the extract thus obtained from various algal species has an immense anti-influenza therapeutic potential (Silva et al. 2018).

Dunaliella salina, a green microalga, exhibits ant carcinogenic effects due to the presence of β-carotenoid, unveiling its anti-oxidant effects (El-baky et al. 2004). The mightiest of all other anti-oxidants produced by algae include vitamins, phycobiliproteins and polyphenols, (Plaza et al. 2008). Astaxanthin, a reddish pigment belonging to carotenoids (also found in certain microalgal species), was discovered as a viable chemopreventive agent in an experiment performed by few researchers (Tanaka et al. 1994), attributing to its anti-oxidant effects, thereby suppressing carcinogenesis and cell proliferation. Carotenoids, which can be efficiently derived from algal species, have also been displayed to boost up the immune response and safeguard the cell membranes and cellular DNA from mutation (Bendich 1989).

A few brown and red marine algal species have been speculated to express antidiabetic roles (Unnikrishnan and Jayasri 2018). The seaweed *Undaria pinnatifida*, rich in compounds fucoxanthin and neoxanthin, was found to promote cell apoptosis, hence declining the risk of human prostate cancer when ingested in experiments carried by certain scientists (Wang et al. 2014). There is still a lot of work to be done to identify, seclude and characterize the unexplored macroalgae and microalgae species, as some reports have proclaimed their use as an anti-inflammatory in neuroinflammation (Barbalace et al. 2019) and as antidiabetic agents. Apparently, the antiviral capability of metabolites derived from algae has been discussed in detail, explicitly comprehending the plausibility of utilizing algal metabolites against coronavirus, human immunodeficiency virus, dengue virus etc. (Sangtani et al. 2020). Hence, more chemical and pharmacological studies should be done in order to attain the sustainable extraction, purification and management of bioactive therapeutic compounds derived from algae while considering the benefits of algal biorefinery.

11.3.3 Feed and Fertilizer

The application of algal biomass in animal feed is still in its infancy; thus, more knowledge needs to be gathered to incorporate this branch into the algal biorefinery. As per literature, almost 30% of the algae produced worldwide has been believed to be deployed as animal feed. Many of the nutritional experiments vividly support the asset of using these algal species with the high nutritional value among the animal population of cows, sheep, pigs, poultry farms, other domestic animals, and also several aquatic organisms (Yamaguchi 1996; Richmond 2008). Algal biomass feeding to the animals has many divergent benefits like high immune response, improved lipid metabolism, development of stress resistance, an improved gut function that subsequently increased weight in animals, increased number of eggs production in poultry, better reproductive performance and decline in cholesterol levels (Madeira et al. 2017; Shields and Lupatsch 2013).

A study performed on chicken population fed with microalga *Porphyridium* species biomass showed declined cholesterol levels by 10%, essential fats, arachidonic acid and linolenic acid were enhanced by 24% and 29%, respectively, and darkened egg yolk color due to enhanced carotenoid levels (Ginzberg et al. 2000). Microalgal biomass also plays a chief role in being utilized as a feed for aquatic organisms. For instance, they are predominantly being used to feed for the larval and juvenile stages of shell and finfish and raise the zooplankton population, which is used to feed other aquatic animals (Benemann 1992).

It has also been revealed that ruminants, as they can digest even unprocessed algal cell walls, tend to be the most suitable animal species for having algae as their feed (McCauley et al. 2020). Furthermore, algae have also been worked upon for analyzing their significance as a fish feed and it has been established that a fish diet supplemented with microalgal biomass indeed contributes to the growth and development of fish and also enhances its nutritional value (Souza et al. 2020; Sarker et al. 2020). The detailed discussion regarding the role of macroalgae in the form of good quality feed for livestock as well as fish has been done by many scientists in their work (Wan et al. 2019; Makkar et al. 2016). Further scrutinization of algal biomass as animal feed could confer with many other options, lowering the price for animal feed and solving the issue of saving plants that hold the first place to be used as food to a much larger extent.

11.3.4 Nutraceuticals

Nutraceuticals refer to the nutritionally functional foods that aid in gaining good health. These valuable products include PUFAs (polyunsaturated fatty acids) that contain more than 2 double bonds along with acyl chains, vitamins, minerals, etc. Fishes are the conventional source of PUFAs, but as they hold a limited capacity to produce them, there's an absolute need to switch for an alternative source. For the human population, because of their extensive rise in numbers and strain for availing nutritious food, algal biomass serves as a promising source and a worthier alternative, and hence, they possess a striking source for such compounds (Wells et al. 2017). Species like *Porphyra* sp., *Chondrus crispus, Himanthalia elongate, Undaria pinnatifida, Spirulina* and *Chlorella* genera have proved to be propitious sources in this regard; carrying low calories and enriching the diet with high protein, vitamins, minerals, PUFAs and dietary fiber content (Plaza et al. 2008). *Spirulina* is known to be the prime origin of nutritional components and various bioactive compounds. Biochemical products derived from microalgal biomass have been utilized in different forms in the nutraceutical industries and could also be made available by entailing it with different food products. Lately, a few microalgal species *Isochrysis galbana* and *Diacronema vlkianum* have been studied to accumulate PUFAs (chiefly EPA and DHA) as oil droplets (Gouveia et al. 2008). Other familiar examples include; *Schizochytrium limacinum* culminating DHA (22:6); *Nannochloropsis* and *Phaeodactylum* sp. producing (EPA, 20:5 n-6); *Parietochloris incisa,* a freshwater green microalga yielding out arachidonic acid (ARA, 20:4 n-6), etc. (Tababa et al. 2012; Cheng-Wu et al. 2002; Remmers et al. 2017; Pyle et al. 2008; Hoffmann et al. 2010). Also, *Spirulina* and *Chlorella* are seen as promising sources of nutraceutical compounds for commercial purposes. Additionally, like cyanobacteria, the blue-green algae lack the polysaccharide (cellulosic) cell wall; unlike higher eukaryotic algae, it becomes more convenient for the human population to feed upon them (Richmond and Preiss 1980).

Chlorella is known to be rich in the biologically active compound β-1,3-glucan, which acts as an active immune-stimulator and a blood lipid-lowering agent (Benemann 1992; Lavens and Sorgeloos 1996). Forbye, marine microalgal species that are often cultivated to provide EPA (eicosapentaenoic acid) and DHA (docosahexaenoic acid) enter the food web for aquaculture nutrition purposes (Rocha et al. 2021).

Thenceforth, the approach was further substantiated by a group of researchers while determining the nutritional value of seaweeds/macroalgae as an enriched source of fatty acids for promoting the good health of human beings. They postulated that the addition of seaweeds to the human diet give positive results in this direction, thus making new opportunities for the development of algal biorefinery; where not only products like bioplastic are derived from the algae but simultaneously, certain high-valued metabolites proving beneficial as nutraceuticals can also be obtained (Du et al. 2007).

11.4 Biorefinery in Conjunction with the Bioprocess Based Systematic Strategies

11.4.1 Bioelectrochemical Technique

The concept of "Bio-Electrochemical cell (BES)" alludes to a mechanism where microorganism stimulates the biochemical reaction, leading to the generation of electricity from redox chemical reactions; microorganism here plays the role of a catalyst by either accepting or donating the electron from the electrodes. Microbial Fuel Cells (MFC) have been recognized as a type of BES where energy stored in the chemical bonds of substrates is oxidized and harnessed into electric energy, and the reaction is instigated by the involvement of microbes (Du et al. 2007).

The detailed recapitulation of the mechanism and the microbes endured for the functioning of BES has been reviewed by researchers elsewhere (Zheng et al. 2020; Harnisch and Schröder 2010; Logan et al. 2019). Briefly, the general set-up and working of an MFC can be explicated as the existence of 2 electrodes, viz. cathode and anode. Generally, microorganisms in the form of biocatalyst are employed in the anode chamber of MFC, where organic and inorganic substrates are oxidized due to the microbial intracellular metabolism. The electrons there from liberated move to the anode electrode and get entrapped in the cathode electrode through the medium of the external circuit, which acts as an electron acceptor or results in the reduction of oxygen (Logan et al. 2006). Supplementarily, a unique ion exchange membrane between anode and cathode selectively permits the protons produced at the anodic chamber to reach out the cathode. At the cathode, oxygen is supplied pumping air, or the cathode is directly made into contact with external air. Subsequently, oxygen is being reduced to water by the interaction of protons with electrons at the cathodic chamber, ultimately leading to electricity production (Logan et al. 2006; Kakarla et al. 2015). The role of selectively permeable ion exchange membrane can also be very well inferred as it does not allow oxygen of cathodic chamber to reach out the anode where electrons are produced, to prevent the ready uptake of electrons. Excess CO_2 stockpile and oxygen delivery often becomes the chief limiting element for the efficient working of MFCs (Gajda et al. 2013).

MFC can be further categorized into various types depending upon its constructional design, nature of work, number of compartments, and other factors, including single-chambered, stacked, up-flow mode, double-chambered MFC, etc. (Du et al. 2007). Amongst all, single-chambered MFC has been considered as the simplest model, wherein the cathode directly interacts with external air to trap oxygen whereas, in double-chambered MFC, the air has been pumped to the cathode. Howbeit, the Up-flow MFC system has been known to work in a continuous flow mode and is considered as the most appropriate method for the treatment of wastewater; whereas in stacked MFC, various MFCs are attached in series and in parallel to enhance the current output (Venkata Mohan et al. 2014).

According to literature, the microbial fuel cells have been utilized from the early 1990s and 2000 for the production of numerous bioproducts, generate energy and treat wastewater due to their enhanced substrate degradation rate (Habermann and Pommer 1991; Liu and Logan 2004). The nutrients and organic compounds existing in the waste serve as the oxidizing substrate for the microbial population. Henceforth, there is no need to supply biomass from an external source, making the overall system economically more favorable with minimum energy input demand and streets ahead environment-friendly (Rashid et al. 2021; Wilberforce et al. 2021; Munoz-Cupa et al. 2021; Venkata Mohan et al. 2008).

A novel and promising scheme of integrating algae with MFC has been endeavored, favoring the treatment process. The algal population is entrenched in the cathodic chamber, where they play the part of efficient electron acceptors and uptake the CO_2 coming from the anodic chamber and reduce it to algal biomass. The basic idea behind this underlies in the fact of algae being able to effectuate oxygen via the process of photosynthesis by utilizing solar energy, so there is no need for provision of oxygen from an outside source and collaborating with a bacterial population that provides carbonates to algae, enriching more feasibility to the system. They are capable of converting CO_2 to produce varieties of organic compounds and biomass; meanwhile, in the dark phase of their photosynthesis, they can consume some amount of oxygen produced to yield energy by directly oxidizing the stored organic materials within them (Wang et al. 2010; Liu et al. 2015; Wu et al. 2013).

Some of the different parameters affecting the gross productivity of microalgae-MFCs include light intensity, illuminating period, nature of electrode, oxygen availability, and concentration of CO_2 (Gouveia et al. 2014; Bazdar et al. 2018; Gonzalez

Olias et al. 2019; Varanasi et al. 2020). Furthermore, work done by many scientists has demonstrated the use of photosynthetic cyanobacteria in the anodic chamber as biocatalysts instead of anaerobic bacteria, considering that the BGA (blue-green algae) could effectively oxidize substrates without giving out oxygen and yielding better electrogenic function (Parlevliet and Moheimani 2014; Rosenbaum et al. 2010). To give an instance, some workers showed the use of *Spirulina platensis* as an anode biocatalyst. Additionally, they also do not require electron mediators to pass on the electrons generated by them to make them finally available at the cathode (Fu and Holtzapple 2010).

Evidently, the microalgae- MFC or microalgal fuel cells perform explicitly in terms of bioelectricity production, wastewater treatment, bioremediation, CO_2 sequestration and can even be employed for the production of biosensors; however, there exist specific environment and cost-related predicaments including high energy requirements for the growth as well as harvesting coupled which eventually makes the process cost-ineffective. The technical issues related to pH gradient along the membrane also exist; thus, the solution to all these problems needs to be practiced so as to make the process sustainable (Liu et al. 2015; Jaiswal et al. 2020). Microalgal cultivation coupled with microbial fuel cells has also been explored and reviewed by some scientists recently (Hou et al. 2016). They concluded that this phenomenal integration of two techniques paves the way for the integrated algal biorefinery in a sustainable and economically sound manner (Chiranjeevi and Patil 2020). However, several process optimization and design research need to be performed yet along with the determination of appropriate algal strain in order to make the biorefinery process feasible on a commercial scale in both the upstream and downstream processing aspects.

11.4.2 Photosynthetic Approach

"Photosynthesis" serves as such an opportune mechanism that footholds the entire living world. Plants, algae including macroalgae (seaweeds) and microalgae (diatoms, some red algae, yellow-green algae etc.), along with some exclusive group of bacteria known as cyanobacteria, are proficient enough to carry out such a process. The necessities for carrying out photosynthesis include solar energy, water, and carbon dioxide. Photosynthesis comprises of light followed by dark phase reactions wherein CO_2 (inorganic substrates) along with other energy intermediates (ATP, NADPH) are converted into complex organic compounds enduring CBB (Calvin Benson-Bassham) cycle, using the key enzyme Ribulose-1,5-bisphosphate carboxylase oxygenase (RuBisCO) that catalyzes the carboxylation of RUBP (a 5-C compound) into 2 molecules of 3 PGA (phosphoglyceric acid) via reduction step in the C3 cycle (occurring in all photosynthetically active organisms), of which one molecule is guided in the chief metabolic pathway, and the other is engaged in the continuation of cycle via the process of regeneration (Venkata Mohan et al. 2015). There has been a constant elevation in the levels of CO_2 in the atmosphere, comprising 20% of the total greenhouse impact (Lacis et al. 2010). The cause for this includes some of the natural means comprising of volcanic eruptions, autotrophic as well as heterotrophic respiration (Sharma et al. 2012) and mainly due to unrestrained population growth resulting in increasing dependency on fossil fuels and their combustion, puffed up levels of industrial pollutions, deforestations, and various other anthropogenic actions, thereby disturbing the global carbon cycle (Grace 2004).

Though numerous biological and physical strategies for CO_2 sequestration have been concocted like the physical methods, comprise scrubbing, mineral carbonation, oceanic injection etc. (Mirjafari et al. 2007). However, these methods have turned out to carry several drawbacks as they are expensive and deal with the challenges of accumulation, transportation, storing of CO_2 gas, as well as intimidation of its accidental leakage (Bachu et al. 2000). Consequently, biological methods came into the picture, including CO_2 sequestration via the process of photosynthesis performed by aquatic and terrestrial plants and algae, which subsist as a promising approach to global warfare warming (Guldhe et al. 2015). It has been reported that 1 kg of dry algal biomass can utilize upto 1.83 kg of CO_2, reflecting their great potential of sequestering CO_2 gas (Chisti 2007). Furthermore, *Chlorella* and *Scenedesmus* sp. have significantly shown high carbon fixation capacity (Toledo-Cervantes et al. 2013; Fulke et al. 2010).

Microalgae tend to cultivate in divergent nutritional models such as; heterotrophic, autotrophic, and mixotrophic systems (Prathima Devi et al. 2013). To further facilitate the process of photosynthesis, these organisms are explored to perform a biorefinery process to utilize every product produced, promoting cogency of the process (Craigie et al. 1966). To maximize the full capability of CO_2 sequestration by microalgal species, the process has been integrated with wastewater treatment with parallel microalgal biomass production, including biofuel and several other value-added products (pharmaceuticals, nutraceuticals, vitamins, anti-oxidants, antibiotics, PHA/PHB, animal feed, lipids, PUFA, Omega 3 fatty acids, antibiotics, etc.) bestowing a very efficient scheme altogether of industrial, ecological and economical significance (Zhao and Su 2014; Singh and Ahluwalia 2013; Kumar and Singh 2019).

It has been known that over 20–30% of primary photosynthetic activities occurring around the earth have been carried out by a prokaryotic micro-organism, blue-green algae, also known as cyanobacteria. They, unlike higher plants and algae, contain the photosynthetic pigment in their cytoplasm. Besides, algae carry pyrenoids containing proteins in the chloroplast that isolates the RuBisCO enzyme (the primary CO_2 acceptor) from stroma, whereas in cyanobacteria, the RuBisCO is confined inside carboxysomes that are the proteinaceous microcompartments having polyhedral protein shells (Borkhsenious et al. 1998; Pisciotta et al. 2010). Though the RuBisCO of cyanobacteria displays a lower affinity for CO_2, these were still these were reported to show high growth at lower CO_2 concentrations because of carbonic anhydrase action. Hence, due to their simple structure and metabolism, cyanobacteria are considered more effective for capturing CO_2, carrying photosynthesis and yielding higher biomass (including biofuel and other value-added products) (Oliver et al. 2014; Rabinovitch-Deere et al. 2013).

Many researchers have explored the CO_2 fixation capability of algae integrated with the scrutinization of biorefinery potential of the same. For instance, the CO_2 biofixation ability of two microalgal species *Chlorella* and *Tetraselmis* sp. have been investigated and afterward, It was suggested that the algal species could be utilized efficiently for sequestering carbon dioxide along with the production of essential biofuels, including bioethanol; organic solvents like acetone and butanol; and can be utilized as raw material for synthesis of organic acids such as butyric acid and acetic acid (Kassim and Meng 2017). The photosynthetic consumption of greenhouse gases, specifically CO_2 by cyanobacteria for its conversion into biodegradable plastic, has been ascertained and has been a field of research to understand the polyhydroxybutyrate production efficiency of cyanobacteria (Panda and Mallick 2007; Carpine et al. 2018).

In light of this, seaweed's potential has also been worked upon so as to evaluate the possibility of seaweed being utilized for carbon capture as well as for bioenergy production (Hughes et al. 2012). Various factors have been reported that can affect CO_2 sequestration and, ultimately, the photosynthetic efficiency of algae, including light intensity, temperature, appropriate algal species selection, nutrient availability, optimum growth condition, pH, etc. (Cheng et al. 2013). Even though there is the substantial growth of algae accounting for the synthesis of high-value algal biomass and biofuels generation, there still lies notable limitations in the downstream processes like harvesting of biomass, dewatering etc. which are cost and energy derived processes that need to be further investigated for more profitable outcomes.

11.4.3 Wastewater Based Nutrient Recovery Strategy

There are plethora of causes, mainly involving human activities like massive rise in urbanization, industrialization, growing factories at sky-high rates, deforestation, chemicals like herbicides, insecticides, fertilizers etc., that are widely being used, thereby adversely affecting the aquatic ecosystem. The major chemical contributors due to these anthropogenic actions are nitrogen, phosphorus, and carbon compounds (that can even serve as crucial elements for sustaining different life forms). Besides, nitrogen and phosphorus play a significant role in driving food chains and contribute vitally to plant growth and metabolism (Sengupta et al. 2015). Nevertheless, these chemicals lead to an accelerated rate of eutrophication of water bodies to a substantial level (Mennaa et al. 2015). Additionally, the types of pollutants dwelling in the water body also depend upon its source. For instance, municipal wastewater is found to be rich in heavy metals concentrations like; copper, zinc, and lead in comparison to wastewaters from other sources, on the other hand, animal wastewater is viewed to be predominantly rich in compounds of nitrogen and phosphorus; agricultural runoff could bring dissolved chemicals, pathogens etc. and depend upon the source of an operation, the industrial effluents could also be a rich source of heavy metals, carcinogens, chemicals, etc., henceforth degrading the water quality to its worst (Ahluwalia and Goyal 2007; Aida et al. 2016).

Compared to the non-point sources, point sources of water pollution are easy to monitor and could be better controlled by putting sincere and effective technological treatment schemes (Cai et al. 2013). Therefore, to overcome this challenge, scientists eventually elicit three nutrient recovery treatment strategies: physical, chemical (conventional methods), and biological (the growing area). Physical processes enroll techniques like; ion-exchange, membrane-based recovery, adsorption of contaminants, air-stripping etc., whereas the chemical methods account for chemical extraction via precipitate formation, crystallization etc. of nitrate and phosphate compounds (Ye et al. 2020). Nonetheless, these conventional practices remain to be in limited applications because of their expensive nature. Then comes into practice the biological methods like electrochemical systems catalyzed by microbes that work on the principle of converting complex organic substances (pollutants) into simpler inorganic forms, chiefly by microbial degradation method (Venkata Mohan et al. 2014).

Another very promising approach in this field of wastewater treatment that has lately emerged is of using microalgae for the recovery of nutrients via wastewater treatment. The scheme aims at the extraction of organic and inorganic nutrients, which are being consumed by algae for their essential metabolic pathways and growth, combined with the accumulation of various other valuable products (Craggs et al. 1995; Abdel-Raouf et al. 2012). This novel method invites various other perks along with, including, decreased cost of tertiary water treatment process, as microalgae require minimum nutrients for their survival (which could be derived from the wastewater itself), lesser sludge production, easy to grow etc. (Mennaa et al. 2015). Moreover, microalgae also bear the capacity of CO_2 sequestration, thereby balancing its level in the atmosphere and controlling the greenhouse effect to an appreciable level. Also, the lipids extracted from microalgal biomass cultivation and their by-products have shown to serve multiple roles of yielding highly demanding biofuels, protein-rich animal feed, bioplastics, vitamins, and other high-value compounds (Spolaore et al. 2006). Hence, this approach favors environmental, economical, ecological as well as healthier means of achieving multiple goals at a time.

As a result, bioremediation of inorganic compounds like nitrogen, phosphorus and toxic metals present in wastewater by the cultivation of microalgae have been verified to be effective (Boelee et al. 2012; Zhou et al. 2012; Khan and Yoshida 2008). The review by Cai and team (2013) articulates the different wastewater compositions derived from various water pollution sources, nutrient uptake mechanisms, and its further utilization for significant algal biomass production, nutrient disposal efficiency by different microalgal species other wastewater treatment strategies. This scientific approach has brought a turnaround in looking up to the waste generated (nutrients extracted) as not just a scrap, but as a treasure to make the best use of it by extracting several valuable products, following up the biorefinery practice (Aslan and Kapdan 2006).

Further research work is required to be performed to have escalated microalgal biomass production, economically feasible downstream processing of that algal biomass (biofuel, high-value products like; single-cell protein, vitamins, antioxidants, PUFAs, bioplastics like PHA/PHB, etc.), and along with that an integrated wastewater treatment schemes with maximum efficiency.

11.4.4 Acidogenic Technique

As fermentation is a biologically mediated process wherein organic compounds are anaerobically digested to yield energy. Acidogenic fermentation (also known as dark fermentation) involves the production of compounds like hydrogen (H_2) which can be used as an energy carrier, CO_2, volatile fatty acids (VFAs), and short-chain fatty acids (SCFA), along with a few other by-products in a bit (Sarkar and Mohan 2016; Ramos-Suarez et al. 2021). The VFAs and SCFAs consist of carbon atoms C_2-C_5 in their fatty acid chains. VFAs form a building block to many vital industrial products, including medicines/pharmaceuticals, chemicals, functional food items, agricultural entities, etc. (Baumann 2016). VFAs can also be converted into hydrogen via microbial electrolysis (Uyar et al. 2009).

There could be multiple ways of synthesizing VFAs, such as; the petrochemical processes, which is known to cause perilous effects on the environment and also involves conditions like high temperatures, high pressures as well as catalysts for the production of VFAs, making it a cost-intensive process (Dionisi and Silva 2016; Huang et al. 2002). Then comes into option another environment-friendly and economically efficient process as a potential alternative for VFAs production via the employment of wastes like (food wastes, animal manure, agricultural wastes, biodegradable organic wastes, sludge, etc.) through the process of acidogenic fermentation (Fu and Holtzapple 2010). In contrast to renowned petrochemical based or electrolyte hydrogen synthesis, this approach has been verified to be economically viable (Azwar et al. 2014). There are some well-known acidogenic bacteria, which upon fermentation, produce short-chain VFAs like valeric, acetic, propionic, and butyric acid consisting of carbon numbers like 2,3,4 and 5 respectively (Dahiya et al. 2015).

Interestingly, in recent times microalgae have also been researched for their VFA accumulation potential, and it has been postulated that acidogenic fermentation of microalgae leads to the production of short-chain fatty acids. The studies have determined that different abiotic factors such as temperature, pH etc. influence the quality and quantity of the VFAs being formulated by the acidogenesis of microalgae (Li et al. 2013). The microalgae cultivated in wastewater could be employed to produce VFA, and the residual further can be utilized to produce other bio-based energy products. Various experiments have till now been performed and reported to find the best suitable conditions for microalgal biomass to carry out fermentation of waste fed to the feedstock and obtain a maximum concentration of VFAs and their ultimate bio-conversion into valuable products. Magdalena et al. reviewed the potential of microalgae as a valuable source of volatile fatty acids along with the knowledge of various operational parameters affecting the accumulation of VFA. Further, it has also been determined that the VFA thus produced proves to be an essential alternative for creating biodegradable polyhydroxyalkanoate-based plastic and other bioproducts (Magdalena and González-Fernández 2019).

The operational parameters affecting the acidogenic fermentation process include variables like; pH, temperature, substrate concentration, retention time, feeding strategy etc.asnd their cumulative effect greatly impacts the microbial activity, substrate degradation, VFA yields, and product allocation. Also, the selection of pertinent inoculum, using mixed consortia as biocatalyst, significantly influences the acidogenic process efficiency, declining its cost further and thus adding to its value (Nikhil et al. 2014). For instance, in results shown by Jose Antonio Magdalena (2018), the best conditions for VFAs production by *Chlorella* sp. biomass as substrate was found to lie in the mesophilic temperature ranges at neutral pH values, and psychrophilic temperature ranges at low pH respectively that yielded in the change of the initial COD (from waste provided) into 48% VFAs. Out of the total VFAs produced, about 73% acetic acid and propionic acid (the most abundant products) were found. Furthermore, a study carried by Gruhn et al. (2016) was done to compare the acidogenic fermentation of *Scenedesmus* sp. concerning the production of VFA from algal biomass, concerning the inoculum, pH, and nutrients under mesophilic (35 °C) and thermophilic (55 °C) conditions. The result showed a higher value of VFA production on using bovine manure as inoculum, operating at mesophilic conditions and the pH of 4.5, affecting the overall process positively (Gruhn et al. 2016; Magdalena et al. 2018)].

Consequently, few microalgal species have been discovered to be potent in utilizing these VFAs like carboxylic acids as their carbon source backbone and elongating them into polyunsaturated fatty acids (PUFAs) (Perez-Garcia et al. 2011). Thus, focusing on the deployment of VFAs as a phenomenal growth stimulator and carbon substrate, microalgae such as *Chlorella vulgaris*, *Chlorella sorokiniana* and *Auxenochlorella protothecoides* have been studied and executed successfully for manufacturing of diverse valuable products (Patel et al. 2021; Liu et al. 2013). In a study performed by You et al. (2021), an approach converging acidogenic fermentation (AF) and algal growth was maneuvered for the treatment of marine culture wastewater (MW) with the aid of *Chlorella vulgaris*, where via AF, organic compounds present in MW were primarily converted into VFAs followed by the utilization of them by the microalgae for the production of highly enriched algal biomass. It was also determined that simpler carboxylic acids or volatile fatty acids tend to be utilized faster by the algae, and later the complex acids are assimilated for the growth of algae.

Hence, microalgae have turned out to be a novel and an efficient source of producing biofuels, bioplastics (PHA/PHB), alcohol, microalgal lipids, biohydrogen, bioelectricity, and other high-value products like omega-3 fatty acids and exopolysaccharides etc. by subsequently using VFAs and their bioconversion, extracted from the fermentation of waste products and other renewable resources, thereby bridging the gap between the recovery of product and bioremediation of waste. to attain sustainability in the future (Chalima et al. 2017).

11.5 Felicitous Aspects of Holistic Algal Circular Bioeconomy

A circular bioeconomy has been anticipated to attain sustainability in terms of feed-stock consumption by following the principle of 3R's that is Resource, Recovery, Recycle. It basically encompasses the economical feasibility of the process involved in the conversion of resources to products until their disposal and back to their reuse (Carus and Dammer 2018).

As aforementioned, algal biomass has proven to be a potent source of several biochemical metabolites, including bioplastic precursors, biofuels, high-valued and bioenergy products. However, when biorefinery is accomplished linearly, then it leads to economical and environmental imbalance along with the accumulation of waste; thus, to overcome this circular bioeconomy has been practiced (Choi et al. 2021). The pictorial description of the integrated approach comprising of wastewater treatment by cultivation of microalgae along with the production of bio-products have been depicted in Fig. 11.2. Indenting to attain sustainable circularity in the economy of products obtained from algae, a bioplastic derived from algae amounts to be of utmost importance. When bioplastic commodities are utilized thoroughly and processed for decomposition or recycled, they are converted back into the natural components, which can be consumed later as a feed or substrate for microbes to survive.

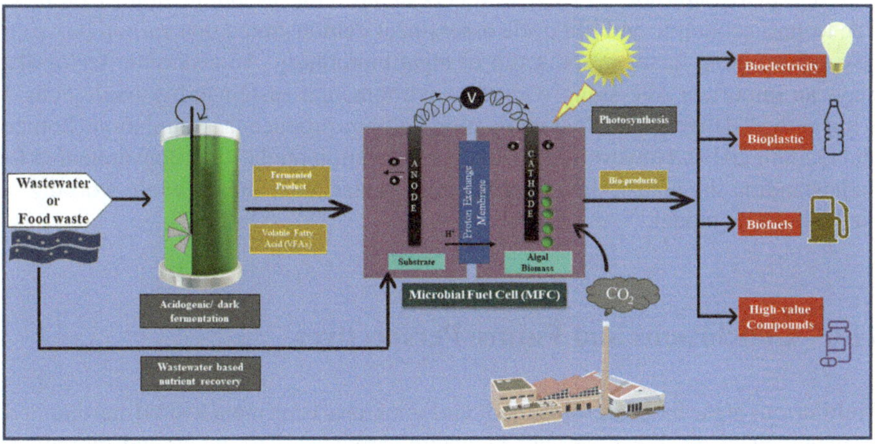

Fig. 11.2 Integrated model for treatment of waste via different biorefinery techniques using microalgae for production of biobased products

Furthermore, algae can survive in the CO_2 saturated surrounding and can grow proficiently under the supplementation of different waste media; biopolymers derived from algae accumulate explicitly under the condition of high CO_2 and waste as carbon substrate. Bioplastics derived there from can be processed into commercial commodities. Still, more research needs to take place on capable algal species at every stage of conversion to biopolymer and even during the end-of-life management of the bioplastic products so as to ensure the sustainability of the process involved and attain circular bioeconomy (Karan et al. 2019; Das et al. 2018).

A study was conducted regarding the feasibility of harnessing biorefinery products derived out from microalga *Scenedesmus obliquus*. The microalgae were cultivated in brewery wastewater followed by extracting biofertilizers, biofuels, and bioactive compounds successfully so as to accomplish the circular bioeconomy. The study postulated that the whole process culminates with the growth of microalgae on waste feedstock followed by the production of biorefinery products, which eventually might turn out to be environmentally friendly and economically feasible to accomplish the circular bioeconomy (Ferreira et al. 2019). The techno-economic analysis and life cycle assessment of the varied integrated biorefinery models have also been shown to postulate the higher feasibility of the product recovery; nevertheless, this field needs to be enquired further so as to get acquainted with the sustainability of the methodology (Rajesh Banu et al. 2020). Besides, the techno-economic assessment of an algal biorefinery that produced β-carotene and fertilizer in different scenarios was performed in another study. The study demonstrated that the usage of photobioreactor on a commercial scale affects the economic value of the process, and the open pond cultivation of *Dunaliella salina* and *Haematococcus pluvialis* for the production of multiple products in a biorefinery setup was economically viable (Thomassen et al. 2016). The significance of carbon dioxide sequestration in the algal biorefinery was considered by researchers along with the production of biofuels. The techno-economic and life cycle assessment demonstrated that carbon proves to be a constraining factor in the setup of algal biorefinery. An economic factor also plays an important role along with the environmental sustainability while considering carbon delivery for algal cultivation and bio-products accumulation (Somers and Quinnn 2019). However, more such investigation ought to be consummated for understanding the plausibility of achieving circular bioeconomy while working on the biorefinery model.

11.6 Conclusions and Future Perspectives

Multifarious algae bioprocessing methodologies have been maneuvered and enquired to obtain manifold bioactive and bioenergy products. The integrated biorefinery approach has been known to aid in attaining a circular bioeconomy in an energy-efficient manner. The feasibility of producing several biochemical compounds on a commercial scale highly depends on the ecological and economical impact of the process involved; therefore, considering the usefulness, more light needs to be shed on the techno-economic analysis and life cycle assessment different approaches possible. Lately, the production of biopolymer derived from different algal species has been known to assist in achieving the biorefinery. However, more such studies need to be effectuated so as to make the process achievable on a large commercial scale in an economically feasible manner without compromising environmental health.

Acknowledgements The authors acknowledge UGC fellowship support (University Grant Commission), India, to Ms. Rimjhim Sangtani. The authors are thankful to DAAD for the LUH-IITI mobility grant and IIT Indore for providing the necessary support. The funding agency has not played any role in the design or decisions regarding the publication of the manuscript.

Conflicts of Interest
There are no conflicts of interest involved with this manuscript.

References

Aaronson S, Ellenbogen BB, Yellen LK, Hutner SH (1967) In vivo differentiation of Euglena cytoplasmic and chloroplast protein synthesis with chloramphenicol and DL-ethionine. Biochem Biophys Res Commun 27:535–538. https://doi.org/10.1016/S0006-291X(67)80020-2

Abdel-Raouf N, Al-Homaidan AA, Ibraheem IBM (2012) Microalgae and wastewater treatment. Saudi J Biol Sci 19:257–275

Abdo SM, Ali GH (2019) Analysis of polyhydroxybutrate and bioplastic production from microalgae. Bull Natl Res Cent 43. https://doi.org/10.1186/s42269-019-0135-5

Ahluwalia SS, Goyal D (2007) Microbial and plant derived biomass for removal of heavy metals from wastewater. Bioresour Technol 98:2243–2257

Aida TM, Nonaka T, Fukuda S, Kujiraoka H, Kumagai Y, Maruta R, Ota M, Suzuki I, Watanabe MM, Inomata H, Smith RL (2016) Nutrient recovery from municipal sludge for microalgae cultivation with two-step hydrothermal liquefaction. Algal Res 18:61–68. https://doi.org/10.1016/j.algal.2016.06.009

Anand V, Kashyap M, Samadhiya K, Ghosh A, Kiran B (2019) Salinity driven stress to enhance lipid production in Scenedesmus vacuolatus: a biodiesel trigger? Biomass Bioenergy 127. https://doi.org/10.1016/j.biombioe.2019.05.021

Asada Y, Miyake M, Miyake J, Kurane R, Tokiwa Y (1999) Photosynthetic accumulation of poly-(hydroxybutyrate) by cyanobacteria—the metabolism and potential for CO_2 recycling. Int J Biol Macromol pp 37–42

Aslan S, Kapdan IK (2006) Batch kinetics of nitrogen and phosphorus removal from synthetic wastewater by algae. Ecol Eng 28:64–70. https://doi.org/10.1016/j.ecoleng.2006.04.003

Azwar MY, Hussain MA, Abdul-Wahab AK (2014) Development of biohydrogen production by photobiological, fermentation and electrochemical processes: a review. Renew Sustain Energy Rev 31:158–173

Bachu S, Brulotte M, Grobe M, Stewart S (2000) Suitability of the alberta subsurface for carbon-dioxide sequestration in geological media

Barbalace MC, Malaguti M, Giusti L, Lucacchini A, Hrelia S, Angeloni C (2019) Anti-inflammatory activities of marine algae in neurodegenerative diseases. Int J Mol Sci 20:3061

Bastiaens L, Van Roy S, Thomassen G, Elst K (2017) Biorefinery of algae: technical and economic considerations. In: Microalgae-based biofuels and bioproducts: from feedstock cultivation to end-products. pp 327–345

Baumann I, Westermann P (2016) Microbial production of short Chain fatty acids from lignocellulosic biomass: current processes and market. Biomed Res Int 2016

Bazdar E, Roshandel R, Yaghmaei S, Mardanpour MM (2018) The effect of different light intensities and light/dark regimes on the performance of photosynthetic microalgae microbial fuel cell. Bioresour Technol 261:350–360. https://doi.org/10.1016/j.biortech.2018.04.026

Beckstrom BD, Wilson MH, Crocker M, Quinn JC (2020) Bioplastic feedstock production from microalgae with fuel co-products: a techno-economic and life cycle impact assessment. Algal Res 46:1–58. https://doi.org/10.1016/j.algal.2019.101769

Belay A, Ota Y, Miyakawa K, Shimamatsu H (1993) Current knowledge on potential health benefits of Spirulina. J Appl Phycol 5:235–241. https://doi.org/10.1007/BF00004024

Bendich A (1989) Carotenoids and the immune system. Carotenoids. Springer, US, pp 323–335

Benemann JR (1992) Microalgae aquaculture feeds. J Appl Phycol 4:233–245. https://doi.org/10.1007/BF02161209

Bethesda (2006) Chlorella-drugs and lactation database (LactMed)

Bhatia SK, Kim SH, Yoon JJ, Yang YH (2017) Current status and strategies for second generation biofuel production using microbial systems. Energy Convers Manag 148:1142–1156

Bini R, Moraes D, Varanda L (2016) Preparation of poly(3-hydroxybutyrate) micro- and nanoparticles as hydrophobic drugs carrier using self-emulsifying nanoemulsion method. Br J Pharm Res 12:1–14. https://doi.org/10.9734/bjpr/2016/28070

Boelee NC, Temmink H, Janssen M, Buisman CJN, Wijffels RH (2012) Scenario analysis of nutrient removal from municipal wastewater by microalgal biofilms. Water (switzerland) 4:460–473. https://doi.org/10.3390/w4020460

Borkhsenious ON, Mason CB, Moroney JV (1998) The intracellular localization of ribulose-1,5-bisphosphate carboxylase/oxygenase in chlamydomonas reinhardtii. Plant Physiol 116:1585–1591. https://doi.org/10.1104/pp.116.4.1585

Borowitzka MA (1998) Algae as food. Microbiology of fermented foods. Springer, US, pp 585–602

Borrelle SB, Ringma J, Lavender Law K, Monnahan CC, Lebreton L, McGivern A, Murphy E, Jambeck J, Leonard GH, Hilleary MA, Eriksen M, Possingham HP, De Frond H, Gerber LR, Polidoro B, Tahir A, Bernard M, Mallos N, Barnes M, Rochman CM (2020) Predicted growth in plastic waste exceeds efforts to mitigate plastic pollution. Science (80-)369:1515–1518. https://doi.org/10.1126/SCIENCE.ABA3656

Brown MR, Mular M, Miller I, Farmer C, Trenerry C (1999) The vitamin content of microalgae used in aquaculture. J Appl Phycol 11:247–255. https://doi.org/10.1023/A:1008075903578

Buono S, Langellotti AL, Martello A, Rinna F, Fogliano V (2014) Functional ingredients from microalgae. Food Funct 5:1669–1685

Cai T, Park SY, Li Y (2013) Nutrient recovery from wastewater streams by microalgae: Status and prospects. Renew Sustain Energy Rev 19:360–369

Carpine R, Raganati F, Olivieri G, Hellingwerf KJ, Pollio A, Salatino P, Marzocchella A (2018) Poly-β-hydroxybutyrate (PHB) production by Synechocystis PCC6803 from CO_2: model development. Algal Res 29:49–60. https://doi.org/10.1016/j.algal.2017.11.011

Carus M, Dammer L (2018) The circular bioeconomy—concepts, opportunities, and limitations. Ind Biotechnol 14:83–91. https://doi.org/10.1089/ind.2018.29121.mca

Chalima A, Oliver L, De Castro LF, Karnaouri A, Dietrich T, Topakas E (2017) Utilization of volatile fatty acids from microalgae for the production of high added value compounds. Fermentation 3

Chandra R, Iqbal HMN, Vishal G, Lee HS, Nagra S (2019) Algal biorefinery: a sustainable approach to valorize algal-based biomass towards multiple product recovery. Bioresour Technol 278:346–359

Chawla J, Kvarnberg D (2014) Hydrosoluble vitamins. In: Handbook of clinical neurology. pp 891–914

Cheng J, Huang Y, Feng J, Sun J, Zhou J, Cen K (2013) Improving CO_2 fixation efficiency by optimizing Chlorella PY-ZU1 culture conditions in sequential bioreactors. Bioresour Technol 144:321–327. https://doi.org/10.1016/j.biortech.2013.06.122

Cheng-Wu Z, Cohen Z, Khozin-Goldberg I, Richmond A (2002) Characterization of growth and arachidonic acid production of Parietochloris incisa comb. nov (Trebouxiophyceae, Chlorophyta). J Appl Phycol 14:453–460. https://doi.org/10.1023/A:1022375110556
Cherubini F (2010) The biorefinery concept: using biomass instead of oil for producing energy and chemicals. Energy Convers Manag 51:1412–1421. https://doi.org/10.1016/j.enconman.2010.01.015
Chiranjeevi P, Patil SA (2020) Microbial fuel cell coupled with microalgae cultivation for wastewater treatment and energy recovery. In: Integrated microbial fuel cells for wastewater treatment. pp 213–227
Chisti Y (2007) Biodiesel from microalgae. Biotechnol Adv 25:294–306
Choi H Il, Hwang S-W, Patel AK, Sim SJ (2021) Microalgal biorefinery: a sustainable technology toward circular bioeconomy and microalgal biomass valorization. Biomass, Biofuels, Biochemicals 323–350
Craggs RJ, Smith VJ, McAuley PJ (1995) Wastewater nutrient removal by marine microalgae cultured under ambient conditions in mini-ponds. Water Sci Technol 31:151–160. https://doi.org/10.1016/0273-1223(95)00502-E
Craigie JS, McLachlan J, Majak W, Ackman RG, Tocher CS (1966) Photosynthesis in algae: ii. Green algae with special reference to Dunaliella spp and tetraselmis Spp. Can J Bot 44:1247–1254. https://doi.org/10.1139/b66-140
Dahiya S, Sarkar O, Swamy YV, Venkata Mohan S (2015) Acidogenic fermentation of food waste for volatile fatty acid production with co-generation of biohydrogen. Bioresour Technol 182:103–113. https://doi.org/10.1016/j.biortech.2015.01.007
Das SK, Sathish A, Stanley J (2018) Production of biofuel and bioplastic from Chlorella pyrenoidosa. In: Materials today: proceedings. pp 16774–16781
De Bhowmick G, Sarmah AK, Sen R (2018) Production and characterization of a value added biochar mix using seaweed, rice husk and pine sawdust: a parametric study. J Clean Prod 200:641–656. https://doi.org/10.1016/j.jclepro.2018.08.002
De Morais MG, Stillings C, Dersch R, Rudisile M, Pranke P, Costa JAV, Wendorff J (2015) Extraction of poly(3-hydroxybutyrate) from Spirulina LEB 18 for developing nanofibers. Polimeros 25:161–167. https://doi.org/10.1590/0104-1428.1686
De Morais MG, De Morais EG, Da Silva VB, Goncalves CF, Lisboa C, Costa JAV (2016) Nanoencapsulation of the bioactive compounds of Spirulina with a microalgal biopolymer coating. J Nanosci Nanotechnol 16:81–91
De Roeck-Holtzhauer Y, Quere I, Claire C (1991) Vitamin analysis of five planktonic microalgae and one macroalga. J Appl Phycol 3:259–264. https://doi.org/10.1007/BF00003584
de Souza FP, de Lima ECS, Urrea-Rojas AM, Suphoronski SA, Facimoto CT, da Silva Bezerra J, de Oliveira TES, de Pádua Pereira U, Di Santis GW, de Oliveira CAL, Lopera-Barrero NM (2020) Effects of dietary supplementation with a microalga (Schizochytrium sp.) on the hemato-immunological, and intestinal histological parameters and gut microbiota of Nile tilapia in net cages. PLoS One 15. https://doi.org/10.1371/journal.pone.0226977
Dionisi D, Silva IMO (2016) Production of ethanol, organic acids and hydrogen: an opportunity for mixed culture biotechnology? Rev Environ Sci Biotechnol 15:213–242
Du D, Huang X, Cai J, Zhang A (2007) Comparison of pesticide sensitivity by electrochemical test based on acetylcholinesterase biosensor. Biosens Bioelectron 23:285–289. https://doi.org/10.1016/j.bios.2007.05.002
Edelmann M, Aalto S, Chamlagain B, Kariluoto S, Piironen V (2019) Riboflavin, niacin, folate and vitamin B12 in commercial microalgae powders. J Food Compos Anal 82. https://doi.org/10.1016/j.jfca.2019.05.009
El-baky HA, Baz FEl, Biol GE-B-IJA, 2004 U (2004) Production of antioxidant by the green alga Dunaliella salina. Int J Agri Biol
Fabregas J, Herrero C (1990) Vitamin content of four marine microalgae. Potential use as source of vitamins in nutrition. J Ind Microbiol 5:259–263. https://doi.org/10.1007/BF01569683

Ferreira A, Ribeiro B, Ferreira AF, Tavares MLA, Vladic J, Vidović S, Cvetkovic D, Melkonyan L, Avetisova G, Goginyan V, Gouveia L (2019) Scenedesmus obliquus microalga-based biorefinery—from brewery effluent to bioactive compounds, biofuels and biofertilizers—aiming at a circular bioeconomy. Biofuels, Bioprod Biorefining 13:1169–1186. https://doi.org/10.1002/bbb.2032

Flórez-Fernández N, Domínguez H, Torres MD (2019) A green approach for alginate extraction from Sargassum muticum brown seaweed using ultrasound-assisted technique. Int J Biol Macromol 124:451–459. https://doi.org/10.1016/j.ijbiomac.2018.11.232

Fu Z, Holtzapple MT (2010) Anaerobic mixed-culture fermentation of aqueous ammonia-treated sugarcane bagasse in consolidated bioprocessing. Biotechnol Bioeng 106:216–227. https://doi.org/10.1002/bit.22679

Fulke AB, Mudliar SN, Yadav R, Shekh A, Srinivasan N, Ramanan R, Krishnamurthi K, Devi S, Chakrabarti T (2010) Bio-mitigation of CO_2, calcite formation and simultaneous biodiesel precursors production using Chlorella sp. Bioresour Technol 101:8473–8476. https://doi.org/10.1016/j.biortech.2010.06.012

Gajda I, Greenman J, Melhuish C, Ieropoulos I (2013) Photosynthetic cathodes for microbial fuel cells. Int J Hydrogen Energy. pp 11559–11564

Galasso C, Gentile A, Orefice I, Ianora A, Bruno A, Noonan DM, Sansone C, Albini A, Brunet C (2019) Microalgal derivatives as potential nutraceutical and food supplements for human health: a focus on cancer prevention and interception. Nutrients 11

García G, Sosa-Hernández JE, Rodas-Zuluaga LI, Castillo-Zacarías C, Iqbal H, Parra-Saldívar R (2021) Accumulation of pha in the microalgae scenedesmus sp. under nutrient-deficient conditions. Polymers (basel) 13:1–23. https://doi.org/10.3390/polym13010131

Ghosh S, Das D (2016) Improvement of harvesting technology for algal biomass production. In: Algal biorefinery: an integrated approach. Springer International Publishing, pp 169–193

Ginzberg A, Cohen M, Sod-Moriah UA, Shany S, Rosenshtrauch A, Arad S (2000) Chickens fed with biomass of the red microalga Porphyridium sp. have reduced blood cholesterol level and modified fatty acid composition in egg yolk. J Appl Phycol. Springer Netherlands 325–330

Gonzalez Olias L, Cameron PJ, Di Lorenzo M (2019) Effect of electrode properties on the performance of a photosynthetic microbial fuel cell for atrazine detection. Front Energy Res 7. https://doi.org/10.3389/fenrg.2019.00105

Gouveia L, Batista AP, Raymundo A, Bandarra N (2008) Spirulina maxima and diacronema vlkianum microalgae in vegetable gelled desserts. Nutr Food Sci 38:492–501. https://doi.org/10.1108/00346650810907010

Gouveia L, Neves C, Sebastião D, Nobre BP, Matos CT (2014) Effect of light on the production of bioelectricity and added-value microalgae biomass in a photosynthetic alga microbial fuel cell. Bioresour Technol 154:171–177. https://doi.org/10.1016/j.biortech.2013.12.049

Grace J (2004) Understanding and managing the global carbon cycle. J Ecol. Wiley 189–202

Gruhn M, Frigon JC, Guiot SR (2016) Acidogenic fermentation of Scenedesmus sp.-AMDD: comparison of volatile fatty acids yields between mesophilic and thermophilic conditions. Bioresour Technol 200:624–630. https://doi.org/10.1016/j.biortech.2015.10.087

Guldhe A, Bhola V, Rawat I, Bux F (2015) Carbon dioxide sequestration by microalgae: biorefinery approach for clean energy and environment. In: Algae and environmental sustainability. Springer India, pp 147–154

Habermann W, Pommer EH (1991) Biological fuel cells with sulphide storage capacity. Appl Microbiol Biotechnol 35(1):128–133

Harnisch F, Schröder U (2010) From MFC to MXC: chemical and biological cathodes and their potential for microbial bioelectrochemical systems. Chem Soc Rev 39:4433–4448. https://doi.org/10.1039/c003068f

Hemalatha M, Sravan JS, Min B, Venkata Mohan S (2019) Microalgae-biorefinery with cascading resource recovery design associated to dairy wastewater treatment. Bioresour Technol 284:424–429. https://doi.org/10.1016/j.biortech.2019.03.106

Hoffmann M, Marxen K, Schulz R, Vanselow KH (2010) TFA and EPA productivities of nannochloropsis salina influenced by temperature and nitrate stimuli in turbidostatic controlled experiments. Mar Drugs 8:2526–2545. https://doi.org/10.3390/md8092526

Hou Q, Nie C, Pei H, Hu W, Jiang L, Yang Z (2016) The effect of algae species on the bioelectricity and biodiesel generation through open-air cathode microbial fuel cell with kitchen waste anaerobically digested effluent as substrate. Bioresour Technol 218:902–908. https://doi.org/10.1016/j.biortech.2016.07.035

Huang YL, Wu Z, Zhang L, Ming Cheung C, Yang ST (2002) Production of carboxylic acids from hydrolyzed corn meal by immobilized cell fermentation in a fibrous-bed bioreactor. Bioresour Technol 82:51–59. https://doi.org/10.1016/S0960-8524(01)00151-1

Hughes AD, Black KD, Campbell I, Davidson K, Kelly MS, Stanley MS (2012) Does seaweed offer a solution for bioenergy with biological carbon capture and storage? Greenh Gases Sci Technol 2:402–407. https://doi.org/10.1002/ghg.1319

International energy agency bioenergy task 42. (2021) https://task42.ieabioenergy.com/

Jacob-Lopes E, Mérida LGR, Queiroz MI, Zepka LQ (2015) Microalgal biorefineries. In: Biomass production and uses

Jahan S, Strezov V (2019) Assessment of trace elements pollution in the sea ports of New South Wales (NSW), Australia using oysters as bioindicators. Sci Rep 9. https://doi.org/10.1038/s41598-018-38196-w

Jaiswal KK, Kumar V, Vlaskin MS, Sharma N, Rautela I, Nanda M, Arora N, Singh A, Chauhan PK (2020) Microalgae fuel cell for wastewater treatment: recent advances and challenges. J Water Process Eng 38. https://doi.org/10.1016/j.jwpe.2020.101549

Jendrossek D, Pfeiffer D (2014) New insights in the formation of polyhydroxyalkanoate granules (carbonosomes) and novel functions of poly(3-hydroxybutyrate). Environ Microbiol 16:2357–2373

Jiang R, Linzon Y, Vitkin E, Yakhini Z, Chudnovsky A, Golberg A (2016) Thermochemical hydrolysis of macroalgae Ulva for biorefinery: taguchi robust design method. Sci Rep 6:1–14. https://doi.org/10.1038/srep27761

Jumaidin R, Sapuan Salit M, Jawaid M, Ridzwan Ishak M, Sahari J (2017) Seaweeds as renewable sources for biopolymers and its composites: a review. Curr Anal Chem 13. https://doi.org/10.2174/1573411013666171009164355

Kaewbai-ngam A, Incharoensakdi A, Monshupanee T (2016) Increased accumulation of polyhydroxybutyrate in divergent cyanobacteria under nutrient-deprived photoautotrophy: an efficient conversion of solar energy and carbon dioxide to polyhydroxybutyrate by Calothrix scytonemicola TISTR 8095. Bioresour Technol 212:342–347. https://doi.org/10.1016/j.biortech.2016.04.035

Kakarla R, Kim JR, Jeon BH, Min B (2015) Enhanced performance of an air-cathode microbial fuel cell with oxygen supply from an externally connected algal bioreactor. Bioresour Technol 195:210–216. https://doi.org/10.1016/j.biortech.2015.06.062

Karan H, Funk C, Grabert M, Oey M, Hankamer B (2019) Green bioplastics as part of a circular bioeconomy. Trends Plant Sci 24:237–249

Kassim MA, Meng TK (2017) Carbon dioxide (CO_2) biofixation by microalgae and its potential for biorefinery and biofuel production. Sci Total Environ 584–585:1121–1129. https://doi.org/10.1016/j.scitotenv.2017.01.172

Kay RA, Barton LL (1991) Microalgae as food and supplement. Crit Rev Food Sci Nutr 30:555–573. https://doi.org/10.1080/10408399109527556

Khan M, Yoshida N (2008) Effect of l-glutamic acid on the growth and ammonium removal from ammonium solution and natural wastewater by Chlorella vulgaris NTM06. Bioresour Technol 99:575–582. https://doi.org/10.1016/j.biortech.2006.12.031

Kiran B, Kumar R, Deshmukh D (2014) Perspectives of microalgal biofuels as a renewable source of energy. Energy Convers Manag 88:1228–1244. https://doi.org/10.1016/j.enconman.2014.06.022

Kiran BR, Venkata Mohan S (2021) Microalgal cell biofactory—therapeutic, nutraceutical and functional food applications. Plants 10

Klemeš JJ, Fan Y Van, Tan RR, Jiang P (2020) Minimising the present and future plastic waste, energy and environmental footprints related to COVID-19. Renew Sustain Energy Rev 127. https://doi.org/10.1016/j.rser.2020.109883
Kovač D, Simeunović J, Babić O, Mišan AČ, Milovanović IL (2013) Algae in food and feed. Food Feed Res 40:21–31
Koyande AK, Chew KW, Rambabu K, Tao Y, Chu DT, Show PL (2019) Microalgae: a potential alternative to health supplementation for humans. Food Sci Hum Wellness 8:16–24
Kumar D, Singh B (2019) Algal biorefinery: an integrated approach for sustainable biodiesel production. Biomass Bioenergy 131:105398
Kumar R, Ghosh AK, Pal P (2020) Synergy of biofuel production with waste remediation along with value-added co-products recovery through microalgae cultivation: a review of membrane-integrated green approach. Sci Total Environ 698
Lacis AA, Schmidt GA, Rind D, Ruedy RA (2010) Atmospheric CO_2: principal control knob governing earth's temperature. Science (80-) 330:356–359. https://doi.org/10.1126/science.1190653
Lavens P, Sorgeloos P (1996) Manual on the production and use of live food for aquaculture (No. 361). Food and Agriculture Organization (FAO)
Li Y, Hua D, Zhang J, Zhao Y, Xu H, Liang X, Zhang X (2013) Volatile fatty acids distribution during acidogenesis of algal residues with pH control. World J Microbiol Biotechnol 29:1067–1073. https://doi.org/10.1007/s11274-013-1270-z
Liu CH, Chang CY, Liao Q, Zhu X, Liao CF, Chang JS (2013) Biohydrogen production by a novel integration of dark fermentation and mixotrophic microalgae cultivation. Int J Hydrogen Energy. Pergamon pp 15807–15814
Liu T, Rao L, Yuan Y, Zhuang L (2015) Bioelectricity generation in a microbial fuel cell with a self-sustainable photocathode. Sci World J 2015. https://doi.org/10.1155/2015/864568
Liu H, Logan BE (2004) Electricity generation using an air-cathode single chamber microbial fuel cell in the presence and absence of a proton exchange membrane. Environ Sci Technol 38(14):4040–4046
Logan BE, Hamelers B, Rozendal R, Schröder U, Keller J, Freguia S, Aelterman P, Verstraete W, Rabaey K (2006) Microbial fuel cells: methodology and technology. Environ Sci Technol 40:5181–5192
Logan BE, Rossi R, Ragab A, Saikaly PE (2019) Electroactive microorganisms in bioelectrochemical systems. Nat Rev Microbiol 17:307–319
López Rocha CJ, Álvarez-Castillo E, Estrada Yáñez MR, Bengoechea C, Guerrero A, Orta Ledesma MT (2020) Development of bioplastics from a microalgae consortium from wastewater. J Environ Manage 263. https://doi.org/10.1016/j.jenvman.2020.110353
Madeira MS, Cardoso C, Lopes PA, Coelho D, Afonso C, Bandarra NM, Prates JAM (2017) Microalgae as feed ingredients for livestock production and meat quality: a review. Livest Sci 205:111–121
Magdalena JA, Tomás-Pejó E, Ballesteros M, González-Fernandez C (2018) Volatile fatty acids production from protease pretreated Chlorella biomass via anaerobic digestion. Biotechnol Prog 34:1363–1369. https://doi.org/10.1002/btpr.2696
Magdalena JA, González-Fernández C (2019) Microalgae biomass as a potential feedstock for the carboxylate platform. Molecules 24
Makkar HPS, Tran G, Heuzé V, Giger-Reverdin S, Lessire M, Lebas F, Ankers P (2016) Seaweeds for livestock diets: a review. Anim Feed Sci Technol 212:1–17
McCauley JI, Labeeuw L, Jaramillo-Madrid AC, Nguyen LN, Nghiem LD, Chaves AV, Ralph PJ (2020) Management of enteric methanogenesis in ruminants by algal-derived feed additives. Curr Pollut Rep 6:188–205
McCracken MD, Middaugh RE, Middaugh RS (1980) A chemical characterization of an algal inhibitor obtained from chlamydomonas. Hydrobiologia 70:271–276. https://doi.org/10.1007/BF00016771

Meixner K, Kovalcik A, Sykacek E, Gruber-Brunhumer M, Zeilinger W, Markl K, Haas C, Fritz I, Mundigler N, Stelzer F, Neureiter M, Fuchs W, Drosg B (2018) Cyanobacteria biorefinery—production of poly(3-hydroxybutyrate) with Synechocystis salina and utilisation of residual biomass. J Biotechnol 265:46–53. https://doi.org/10.1016/j.jbiotec.2017.10.020

Mennaa FZ, Arbib Z, Perales JA (2015) Urban wastewater treatment by seven species of microalgae and analgal bloom: biomass production, N and P removal kinetics andharvestability. Water Res 83:42–51. https://doi.org/10.1016/j.watres.2015.06.007

Mirjafari P, Asghari K, Mahinpey N (2007) Investigating the application of enzyme carbonic anhydrase for CO_2 sequestration purposes. Ind Eng Chem Res 46:921–926. https://doi.org/10.1021/ie060287u

Misra A, Vol RS-MA in PS, 2013 U (2013) Algae as drug plants in India. In: vol 1. pp 237–242

Munoz-Cupa C, Hu Y, Xu C, Bassi A (2021) An overview of microbial fuel cell usage in wastewater treatment, resource recovery and energy production. Sci Total Environ 754

Nikhil GN, Venkata Mohan S, Swamy YV (2014) Systematic approach to assess biohydrogen potential of anaerobic sludge and soil rhizobia as biocatalysts: Influence of crucial factors affecting acidogenic fermentation. Bioresour Technol 165:323–331. https://doi.org/10.1016/j.biortech.2014.02.097

Oliver JWK, Machado IMP, Yoneda H, Atsumi S (2014) Combinatorial optimization of cyanobacterial 2,3-butanediol production. Metab Eng 22:76–82. https://doi.org/10.1016/j.ymben.2014.01.001

Otari SV, Jadhav JP (2021) Seaweed-based biodegradable biopolymers, composite, and blends with applications. In: Energy, environment, and sustainability. pp 121–149

Panda B, Mallick N (2007) Enhanced poly-β-hydroxybutyrate accumulation in a unicellular cyanobacterium, Synechocystis sp. PCC 6803. Lett Appl Microbiol 44:194–198. https://doi.org/10.1111/j.1472-765X.2006.02048.x

Parlevliet D, Moheimani NR (2014) Efficient conversion of solar energy to biomass and electricity. Aquat Biosyst 10

Parsaeimehr A, Lutzu GA (2016) Algae as a novel source of antimicrobial compounds: current and future perspectives. Current and future perspectives. In: Antibiotic resistance: mechanisms and new antimicrobial approaches. pp 377–396

Patel A, Mahboubi A, Horváth IS, Taherzadeh MJ, Rova U, Christakopoulos P, Matsakas L (2021) Volatile fatty acids (VFAs) generated by anaerobic digestion serve as feedstock for freshwater and marine oleaginous microorganisms to produce biodiesel and added-value compounds. Front Microbiol 12. https://doi.org/10.3389/fmicb.2021.614612

Patrício Silva AL, Prata JC, Walker TR, Duarte AC, Ouyang W, Barcelò D, Rocha-Santos T (2021) Increased plastic pollution due to COVID-19 pandemic: challenges and recommendations. Chem Eng J 405

Pedersén M, DaSilva EJ (1973) Simple brominated phenols in the bluegreen alga Calothrix brevissima West. Planta 115:83–86. https://doi.org/10.1007/BF00388608

Perez-Garcia O, Escalante FME, De-Bashan LE, Bashan Y (2011) Heterotrophic cultures of microalgae: metabolism and potential products. Water Res 45:11–36

Pesando D, Caram B (1984) Screening of marine algae from the french mediterranean coast for antibacterial and antifungal activity. Bot Mar 27:381–386. https://doi.org/10.1515/botm.1984.27.8.381

Pisciotta JM, Zou Y, Baskakov IV (2010) Light-dependent electrogenic activity of cyanobacteria. PLoS One 5. https://doi.org/10.1371/journal.pone.0010821

Plaza M, Cifuentes A, Ibáñez E (2008) In the search of new functional food ingredients from algae. Trends Food Sci Technol 19:31–39. https://doi.org/10.1016/j.tifs.2007.07.012

Prathima Devi M, Swamy YV, Venkata Mohan S (2013) Nutritional mode influences lipid accumulation in microalgae with the function of carbon sequestration and nutrient supplementation. Bioresour Technol 142:278–286. https://doi.org/10.1016/j.biortech.2013.05.001

Pratt R, Daniels TC, Eiler JJ, Gunnison JB, Kumler WD, Oneto JF, Strait LA, Spoehr HA, Hardin GJ, Milner HW, Smith JHC, Strain HH (1944) Chlorellin, an antibacterial substance from Chlorella. Science (80-) 99:351–352

Pyle DJ, Garcia RA, Wen Z (2008) Producing docosahexaenoic acid (DHA)-rich algae from biodiesel-derived crude glycerol: effects of impurities on DHA production and algal biomass composition. J Agric Food Chem 56:3933–3939. https://doi.org/10.1021/jf800602s

Rabinovitch-Deere CA, Oliver JWK, Rodriguez GM, Atsumi S (2013) Synthetic biology and metabolic engineering approaches to produce biofuels. Chem Rev 113:4611–4632

Rahman A, Putman RJ, Inan K, Sal FA, Sathish A, Smith T, Nielsen C, Sims RC, Miller CD (2015) Polyhydroxybutyrate production using a wastewater microalgae based media. Algal Res 8:95–98. https://doi.org/10.1016/j.algal.2015.01.009

Rajendran N, Puppala S, Raj MS, Angeeleena BR, Rajam C (2012) Seaweeds can be a new source for bioplastics. J Pharm Res 5:1476–1479

Rajesh Banu J, Preethi, Kavitha S, Gunasekaran M, Kumar G (2020) Microalgae based biorefinery promoting circular bioeconomy-techno economic and life-cycle analysis. Bioresour Technol 302

Ramos-Suarez M, Zhang Y, Outram V (2021) Current perspectives on acidogenic fermentation to produce volatile fatty acids from waste. Rev Environ Sci Biotechnol. https://doi.org/10.1007/s11157-021-09566-0

Rashid T, Sher F, Hazafa A, Hashmi RQ, Zafar A, Rasheed T, Hussain S (2021) Design and feasibility study of novel paraboloid graphite based microbial fuel cell for bioelectrogenesis and pharmaceutical wastewater treatment. J Environ Chem Eng 9. https://doi.org/10.1016/j.jece.2020.104502

Remmers IM, Martens DE, Wijffels RH, Lamers PP (2017) Dynamics of triacylglycerol and EPA production in phaeodactylum tricornutum under nitrogen starvation at different light intensities. PLoS ONE 12:e0175630. https://doi.org/10.1371/journal.pone.0175630

Richmond A (ed) (2008) Handbook of microalgal culture: biotechnology and applied phycology. Wiley

Richmond A, Preiss K (1980) The biotechnology of algaculture. Interdiscip Sci Rev 5:60–70. https://doi.org/10.1179/030801880789767891

Rocha CP, Pacheco D, Cotas J, Marques JC, Pereira L, Gonçalves AMM (2021) Seaweeds as valuable sources of essential fatty acids for human nutrition. Int J Environ Res Public Health 18. https://doi.org/10.3390/ijerph18094968

Rosenbaum M, He Z, Angenent LT (2010) Light energy to bioelectricity: photosynthetic microbial fuel cells. Curr Opin Biotechnol 21:259–264

Sangtani R, Ghosh A, Jha HC, Parmar HS, Bala K (2020) Potential of algal metabolites for the development of broad-spectrum antiviral therapeutics: possible implications in COVID-19 therapy. Phyther Res

Santhosh S, Dhandapani R, Hemalatha N (2016) Bioactive compounds from microalgae and its different applications—a review. Adv Appl Sci Res 7:153–158

Sarkar O, Agarwal M, Naresh Kumar A, Venkata Mohan S (2015) Retrofitting hetrotrophically cultivated algae biomass as pyrolytic feedstock for biogas, bio-char and bio-oil production encompassing biorefinery. Bioresour Technol 178:132–138. https://doi.org/10.1016/j.biortech.2014.09.070

Sarkar O, Mohan SV (2016) Deciphering acidogenic process towards biohydrogen, biohythane, and short chain fatty acids production: multi-output optimization strategy. Biofuel Res J 3:458–469. https://doi.org/10.18331/BRJ2016.3.3.5

Sarker PK, Kapuscinski AR, McKuin B, Fitzgerald DS, Nash HM, Greenwood C (2020) Microalgae-blend tilapia feed eliminates fishmeal and fish oil, improves growth, and is cost viable. Sci Rep 10. https://doi.org/10.1038/s41598-020-75289-x

Schubert M, Kluge S, Schmölz L, Wallert M, Galli F, Birringer M, Lorkowski S (2018) Long-chain metabolites of vitamin E: metabolic activation as a general concept for lipid-soluble vitamins? Antioxidants 7:10

Sengupta S, Nawaz T, Beaudry J (2015) Nitrogen and phosphorus recovery from wastewater. Curr Pollut Rep 1:155–166

Sharma N, Nayak RK, Dadhwal VK, Kant Y, Ali MM (2012) Diurnal and seasonal variation of measured atmospheric CO_2 at dehradun during 2009. ISPRS—Int Arch Photogramm Remote Sens Spat Inf Sci XXXVIII-8/:87–90. https://doi.org/10.5194/isprsarchives-xxxviii-8-w20-87-2011

Shields R, Lupatsch I (2013) 5 Algae for aquaculture and animal feeds. In: Microalgal biotechnology: integration and economy

Shim JY, Shin HS, Han JG, Park HS, Lim BL, Chung KW, Om AS (2008) Protective effects of Chlorella vulgaris on liver toxicity in cadmium-administered rats. J Med Food 11:479–485. https://doi.org/10.1089/jmf.2007.0075

Sieburth JMN (1960) Acrylic acid, an "antibiotic" principle in phaeocystis blooms in Antarctic waters. Science (80-) 132:676–677. https://doi.org/10.1126/science.132.3428.676

Silva T, Salomon PS, Hamerski L, Walter J, Menezes RB, Siqueira JE, Santos A, Santos JAM, Ferme N, Guimarães T, Fistarol GO, Hargreaves PI, Thompson C, Thompson F, Souza TM, Siqueira M, Miranda M (2018) Inhibitory effect of microalgae and cyanobacteria extracts on influenza virus replication and neuraminidase activity. PeerJ 6:e5716. https://doi.org/10.7717/peerj.5716

Singh UB, Ahluwalia AS (2013) Microalgae: a promising tool for carbon sequestration. Mitig Adapt Strateg Glob Chang 18:73–95

Singh AK, Sharma L, Mallick N, Mala J (2017) Progress and challenges in producing polyhydroxyalkanoate biopolymers from cyanobacteria. J Appl Phycol 29:1213–1232

Snell TW, Bieberich CJ, Fuerst R (1983) The effects of green and blue-green algal diets on the reproductive rate of the rotifer Brachionus plicatilis. Aquaculture 31:21–30. https://doi.org/10.1016/0044-8486(83)90254-5

Somers MD, Quinn JC (2019) Sustainability of carbon delivery to an algal biorefinery: a techno-economic and life-cycle assessment. J CO_2 Utilization 30:193–204

Spolaore P, Joannis-Cassan C, Duran E, Isambert A (2006) Commercial applications of microalgae. J Biosci Bioeng 101:87–96. https://doi.org/10.1263/jbb.101.87

Steinbruch E, Drabik D, Epstein M, Ghosh S, Prabhu MS, Gozin M, Kribus A, Golberg A (2020) Hydrothermal processing of a green seaweed Ulva sp. for the production of monosaccharides, polyhydroxyalkanoates, and hydrochar. Bioresour Technol 318:124263. https://doi.org/10.1016/j.biortech.2020.124263

Subhadra BG (2010) Sustainability of algal biofuel production using integrated renewable energy park (IREP) and algal biorefinery approach. Energy Policy 38:5892–5901. https://doi.org/10.1016/j.enpol.2010.05.043

Tababa HG, Hirabayashi S, Inubushi K (2012) Media optimization of parietochloris incisa for arachidonic acid accumulation in an outdoor vertical tubular photobioreactor. J Appl Phycol 24:887–895. https://doi.org/10.1007/s10811-011-9709-9

Tanaka T, Morishita Y, Suzui M, Kojima T, Okumura A, Mori H (1994) Chemoprevention of mouse urinary bladder carcinogenesis by the naturally occurring carotenoid astaxanthin. Carcinogenesis 15:15–19. https://doi.org/10.1093/carcin/15.1.15

Tarawat S, Incharoensakdi A, Monshupanee T (2020) Cyanobacterial production of poly(3-hydroxybutyrate-co-3-hydroxyvalerate) from carbon dioxide or a single organic substrate: improved polymer elongation with an extremely high 3-hydroxyvalerate mole proportion. J Appl Phycol 32:1095–1102. https://doi.org/10.1007/s10811-020-02040-4

Thomassen G, Vila UE, Van Dael M, Lemmens B, Van Passel S (2016) A techno- economic assessment of an algal-based biorefinery. Clean Technol Environ Policy 18(6):1849–1862

Toledo-Cervantes A, Morales M, Novelo E, Revah S (2013) Carbon dioxide fixation and lipid storage by Scenedesmus obtusiusculus. Bioresour Technol 130:652–658. https://doi.org/10.1016/j.biortech.2012.12.081

Trevan MD, Mak AL (1988) Immobilized algae and their potential for use as biocatalysts. Trends Biotechnol 6:68–73

Unnikrishnan PS, Jayasri MA (2018) Marine algae as a prospective source for antidiabetic compounds—a brief review. Curr Diabetes Rev 14:237–245. https://doi.org/10.2174/1573399812666161229151407

Uyar B, Eroglu I, Yücel M, Gündüz U (2009) Photofermentative hydrogen production from volatile fatty acids present in dark fermentation effluents. Int J Hydrogen Energy 34:4517–4523. https://doi.org/10.1016/j.ijhydene.2008.07.057

Varanasi JL, Prasad S, Singh H, Das D (2020) Improvement of bioelectricity generation and microalgal productivity with concomitant wastewater treatment in flat-plate microbial carbon capture cell. Fuel 263. https://doi.org/10.1016/j.fuel.2019.116696

Venkata Mohan S, Mohanakrishna G, Reddy BP, Saravanan R, Sarma PN (2008) Bioelectricity generation from chemical wastewater treatment in mediatorless (anode) microbial fuel cell (MFC) using selectively enriched hydrogen producing mixed culture under acidophilic microenvironment. Biochem Eng J 39:121–130. https://doi.org/10.1016/j.bej.2007.08.023

Venkata Mohan S, Velvizhi G, Vamshi Krishna K, Lenin Babu M (2014) Microbial catalyzed electrochemical systems: a bio-factory with multi-facet applications. Bioresour Technol 165:355–364

Venkata Mohan S, Rohit MV, Chiranjeevi P, Chandra R, Navaneeth B (2015) Heterotrophic microalgae cultivation to synergize biodiesel production with waste remediation: progress and perspectives. Bioresour Technol 184:169–178

Venkata Mohan S, Modestra JA, Amulya K, Butti SK, Velvizhi G (2016) A circular bioeconomy with biobased products from CO_2 sequestration. Trends Biotechnol 34:506–519

Venkata Mohan S, Dahiya S, Amulya K, Katakojwala R, Vanitha TK (2019) Can circular bioeconomy be fueled by waste biorefineries—a closer look. Bioresour Technol Reports 7:100277. https://doi.org/10.1016/j.biteb.2019.100277

Wan AHL, Davies SJ, Soler-Vila A, Fitzgerald R, Johnson MP (2019) Macroalgae as a sustainable aquafeed ingredient. Rev Aquac 11:458–492

Wang X, Feng Y, Liu J, Lee H, Li C, Li N, Ren N (2010) Sequestration of CO_2 discharged from anode by algal cathode in microbial carbon capture cells (MCCs). Biosens Bioelectron 25:2639–2643. https://doi.org/10.1016/j.bios.2010.04.036

Wang S, Li Y, White W, Lu J (2014) Extracts from New Zealand undaria pinnatifida containing fucoxanthin as potential functional biomaterials against cancer in vitro. J Funct Biomater 5:29–42. https://doi.org/10.3390/jfb5020029

Watanabe F, Yabuta Y, Tanioka Y, Bito T (2013) Biologically active vitamin B12 compounds in foods for preventing deficiency among vegetarians and elderly subjects. J Agric Food Chem 61:6769–6775

Wells ML, Potin P, Craigie JS, Raven JA, Merchant SS, Helliwell KE, Smith AG, Camire ME, Brawley SH (2017) Algae as nutritional and functional food sources: revisiting our understanding. J Appl Phycol 29:949–982

Wilberforce T, Sayed ET, Abdelkareem MA, Elsaid K, Olabi AG (2021) Value added products from wastewater using bioelectrochemical systems: current trends and perspectives. J Water Process Eng 39. https://doi.org/10.1016/j.jwpe.2020.101737

Wu XY, Song TS, Zhu XJ, Wei P, Zhou CC (2013) Construction and operation of microbial fuel cell with Chlorella vulgaris biocathode for electricity generation. Appl Biochem Biotechnol 171:2082–2092. https://doi.org/10.1007/s12010-013-0476-8

Xu Y, Ramanathan V, Victor DG (2018) Global warming will happen faster than we think. Nature 564:30–32

Yamaguchi K (1996) Recent advances in microalgal bioscience in Japan, with special reference to utilization of biomass and metabolites: a review. J Appl Phycol 8:487–502. https://doi.org/10.1007/BF02186327

Ye Y, Ngo HH, Guo W, Chang SW, Nguyen DD, Zhang X, Zhang J, Liang S (2020) Nutrient recovery from wastewater: from technology to economy. Bioresour Technol Rep 11:100425

You X, Zhang Z, Guo L, Liao Q, Wang Y, Zhao Y, Jin C, Gao M, She Z, Wang G (2021) Integrating acidogenic fermentation and microalgae cultivation of bacterial-algal coupling system for mariculture wastewater treatment. Bioresour Technol 320:124335. https://doi.org/10.1016/j.biortech.2020.124335

Zhao B, Su Y (2014) Process effect of microalgal-carbon dioxide fixation and biomass production: a review. Renew Sustain Energy Rev 31:121–132

Zheng T, Li J, Ji Y, Zhang W, Fang Y, Xin F, Dong W, Wei P, Ma J, Jiang M (2020) Progress and prospects of bioelectrochemical systems: electron transfer and its applications in the microbial metabolism. Front Bioeng Biotechnol 8

Zhou W, Min M, Li Y, Hu B, Ma X, Cheng Y, Liu Y, Chen P, Ruan R (2012) A hetero-photoautotrophic two-stage cultivation process to improve wastewater nutrient removal and enhance algal lipid accumulation. Bioresour Technol 110:448–455. https://doi.org/10.1016/j.biortech.2012.01.063

Chapter 12
Microalgae Coupled Biofuel Production and Carbon Capture from Thermal Power Plant: A Biorefinery Approach

V. Sruthi, P. Jyothirmai, E. Anagha, S. Aishwarya, Abhilash T. Nair, Samarshi Chakraborty, and K. Sivagami

12.1 Introduction

As the world's population grows, the link between the energy, water, and food industries, and how they coincide with the environment, have been becoming increasingly important. These three key features of modern society have emerged as indivisible topics of attention in developing sustainable guidelines (Wilson et al. 2021). Carbon dioxide (CO_2) is one of the most prominent greenhouse gases, and the increase in carbon dioxide emissions is mainly due to global warming and climate change. Negative emission technologies remove effluent carbon dioxide from the air producing a negative CO_2 balance (Fuss et al. 2016). In this context, large-scale microalgae farming has been offered as a potentially sustainable CO_2 emissions collection and recycling system coupled with contaminated water cleanup. Microalgae are fast-growing microorganisms with high growth rates of 0.44 per day and 1720 cells/ml/day and CO_2 fixation potential exceeds terrestrial plants. Diversification and growth of biologically produced goods and commodities may be achieved by utilizing the high photosynthetic efficiency of algae and their capacity to grow in different types of terrains (Wilson et al. 2021). The main components of algae are moisture, ash, and different types of lipids, starch, carbohydrates, proteins, and amino acids.

The composition of algae varies on factors such as the strain, nutrients of the algal culture medium, and the time of harvest. The conversion process determines the type of biorefinery used and the value-added products that we can receive from the conversion process. Biological sequestration uses microbiological algae for capturing

V. Sruthi · P. Jyothirmai · E. Anagha · S. Aishwarya · S. Chakraborty · K. Sivagami (✉)
School of Chemical Engineering, Vellore Institute of Technology, Chennai, Tamil Nadu, India
e-mail: sivagami.k@vit.ac.in

A. T. Nair
Department of Applied Sciences and Humanities, National Institute of Foundry and Forge Technology (NIFFT), Hatia, Ranchi, Jharkhand, India

Y. K. Nandabalan et al. (eds.), *Zero Waste Biorefinery*, Energy, Environment, and Sustainability, https://doi.org/10.1007/978-981-16-8682-5_12

CO_2 produced from thermal power plants. There are many ways to capture CO_2 from the environment, such as amine-based carbon capture and carbon-based carbon capture. (Cheah et al. 2016). During the microalgae carbon capture process, the algae convert sunlight into chemical energy using photo biological processes and later convert this chemical energy and carbon dioxide through a dark reaction to a stable form of chemical energy of organic matter without any other pollution. The selection of the carbon capture process is based on types of the combustion process, such as post-combustion, pre-combustion, and oxyfuel combustion (Blomen et al. 2009). Furthermore, economic aspects like techno-economic analysis (TEA) and life cycle assessment (LCA) have been described in detail to measure the industrial figures and the economic performance of a process, its products, and its services. This chapter also explains how the circular economy concept helps industries approach a greener environment methodology during the production, removal, and utilization of resources. The main concept of the circular economy in algae biorefineries is to utilize the power and the resources used from the starting of the production to the end to improve the economic viability and sustainability of the production as well as of biorefinery (Budzianowski and Postawa 2016). The schematic representation of the carbon capture cycle from different sources are shown in Fig. 12.1

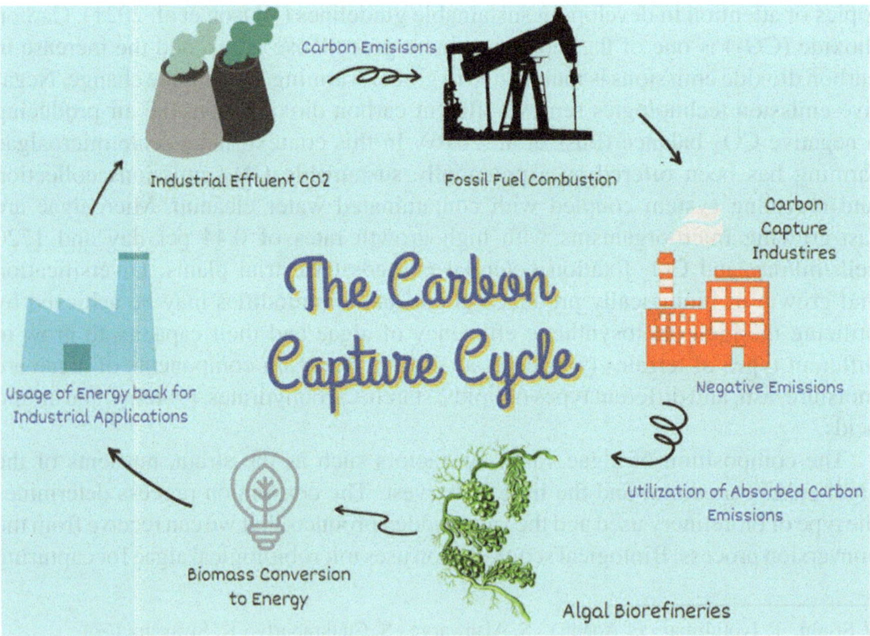

Fig. 12.1 Schematic representation of coupling thermal power plants and CO_2 sequestration

12.2 Microalgae

Microalgae reduce the carbon dioxide concentration in the atmosphere during photosynthesis (Martín-Girela et al. 2020). Microalgae can be co-refined with different species, organisms, and growth to improve wastewater nutrient removal and biomass aggregation because of the synergy between microalgae and different microorganisms (Hena et al. 2015; Mahapatra et al. 2014). In this way, microalgal development for biofuel creation is possible. Based upon the activity of algae, it may or may not be easily maintainable as the fundamental issues lie in the higher capital and activity costs due to countless data sources, including power, supplements, and compound reagents (Cheng et al. 2019).

12.2.1 Algal Biorefinery

The algae biorefinery idea addresses the change of algae biomass through a reasonable preparation approach to deliver biofuels and value-added products (Trivedi et al.

Table 12.1 Overview of conversion routes of plant material to biofuels

Plant material	Conversion route	Primary product	Process	Final product
Ligno cellulosic biomass	Flash pyrolysis	Bio-oil	Hydrotreating and Refining	Hydrocarbon, diesel oil, chemicals, oxygenates and hydrogen
	Gasification	Syngas	Water–gas shift + Separation [Catalyzed synthesis]	Hydrocarbons, methanol, dimethyl ether, FT diesel, SNG (methane) and Bioethanol
	Hydrolysis	Sugar	Fermentation	Hydrocarbons, diesel fuel, chemicals
	Hydrothermal liquefaction	Bio-oil	Hydrotreating and refining purification	SNG (methane)
	Anaerobic digestion	Biogas	–	–
Sugar and starch crops	Milling and hydrolysis	Sugar	Fermentation	Bio-ethanol
Oil plants	Pressing or extraction	Vegetable oil	Esterification pyrolysis	Biodiesel bio-oil, diesel fuel, gasoline

2015). An algae biorefinery includes incorporating upstream, bioreaction, and downstream handling of algae biomass into the finished product. A definite objective for a biomass-based biorefinery industry is to upgrade the use of assets, augment benefit, and limit squanders at the same time (Khoo et al. 2019). Four primary principal types of biorefineries present today are pyrolysis-based, syngas-based treatment facility, aqueous-based processing plant, and aging-based treatment facility. Table 12.1 below provides an outline of the diverse transformation courses of plant material to biofuels (Trivedi et al. 2015).

12.3 Value-Added Products from Algal Biorefinery

Energy demands are increasing worldwide because of industrialization and modernization, triggering the abundant usage of restricted accessible regular fuel holds. The bioenergy production from biomass has acquired attention due to the easy accessibility of biomass, the restricted accessibility of petroleum derivatives, and the increase in the amount of carbon dioxide and other harmful substances, also known as GHGs, in the atmosphere (Kumar et al. 2016; Abdullah et al. 2019). But one of the drawbacks of producing biofuels is the expenditure of its creation, viability, natural maintainability, and minimization of waste (Abdullah et al. 2019; Srivastava 2019). Presently, petroleum derivative industries take up to 70% of the overall energy in the market. Due to the rising pollution effects, many different feedstocks have been used to produce biofuels. This helps in the reduced dependence on petroleum products to a cleaner fuel (Abdullah et al. 2019). The first biofuel production was from plants like soybean, corn, oil palm, and sugarcane (Shuba and Kifle 2018). Although this was a very safe and cleaner method to produce energy, the availability of bagasse was not sufficient for producing biofuel, and was hence later replaced by conventional fuels again. Later, studies of algae and its effects emerged which stated that algae are photosynthetic living beings (Panahi et al. 2019). A study showed that to produce 1 kg of algal biomass, approximately 1.83 kg of carbon dioxide is required (Bhowmick et al. 2019). But algal biomass has numerous advantages like it can absorb water and it can be used as an electron to carry out photosynthesis, has a high biomass efficiency, can produce biomass in any type of water bodies such as sea, hard water, or even in wastewaters.

12.3.1 Biofuels Production from Algae

Algae is considered one of the most significant biofuel sources as the conventional ways of producing biodiesel from fuel oils have raised much concern for everyone (Demirbas and Demirbas 2010). Presently biodiesel formed from microalgae is preferred more than traditional production methods like rapeseed. But if we are going to use algae as biofuel then there is a need for high production of algae which means high usage of fertilizers. Microalgae can create around 10 to 20 times more biodiesel than any other manure which is close to around 8–16 tons/ha/year. (Demirbas and Demirbas 2010). The fast development pace of algae makes it easier to meet the demands of our world, also avoiding any depletion of limited resources. Fundamental methods of thermal conversion are gasification, liquefaction, and pyrolysis. The hydrocarbons in the algae are isolated using the extraction method using any natural dissolving agent after freeze-drying the algae. However, these procedures are not suitable for separation on a large scale because they are costly. The best strategy is by extracting the hydrocarbons using liquefaction of the fuel with the help of high moisture content. Even today, thermochemical liquefaction or pyrolysis is more preferred than algae as algae have a higher moisture content and require a lot of energy for the production of energy. Also, the products formed from the gasification process, such as flammable gases like hydrogen, methane carbon dioxide, and smelling salts pose great damage to the environment (Demirbas and Demirbas 2010). In liquefaction, the microalgae are condensed and separated by CH_2Cl_2. This process occurs at around 302 °C and 10 MPa using steel autoclave with blending, and the oil part is recovered with dichloromethane (Minowa and Sawayama 1999). The hydrothermal liquefaction technique is more effective for the extraction of microalgal biodiesel than using supercritical carbon dioxide and is the most effective option for the production of biodiesel from algae. Nevertheless, due to limited information on the hydrothermal liquefaction of algae, more research in this area would be needed (Demirbas and Demirbas 2010). At very high pressure wet biomass of algae can be converted to pyrolysis oil and gas combined with CO/H_2 or flammable gas (CH4). The process of gasification produces flammable gases like hydrogen, methane, and carbon Gasification is done at elevated temperatures, whereas pyrolysis is done at moderate temperatures. Lipids are a part of microalgae that can be converted into biodiesel. Biodiesel produced from microalgae is preferred over its production from lipids. Numerous specialized and natural issues, for example, land use and manure input actually should be investigated and business creation has still not been attained (Demirbas 2011). Algae produce twice the amount of biodiesel as rapeseed. They need five-ten times more manure that is around 8 to 16 tons hectare per year but these huge amounts of nitrogen and phosphorus could harm the environment. And unsaturated fats are a part of microalgae as film segment stockpiling items which are mainly used as light and energy powerhouses.

The cycle of treating the alkoxy mixture of an ester by liquor is known as transesterification. The oil component of algae is converted to biodiesel using the transesterification process. The content of oil in microalgae is more than 80% of that of the dry algae biomass. Algae can be used as a potential energy crop due to its simple flexibility and development in new or marine waters and keeping away from the utilization of land. Since our earth is covered with 66% water, this would be an extraordinary alternative for the worldwide energy needs (Demirbas 2011). Maturation is utilized industrially for a huge scope in different nations to produce C_2H_5OH (ethanol) yields. The product is made out of the enzymatic reaction of sucrose continued by the maturation of sugars. This process is done using *saccharomyces cerevisiae.* Amylase converts the starch into D-glucose. The enzymatic reaction of sucrose is continued by maturation, refining, and drying to produce bioethanol. Corn as feedstock contains 60–70% starch, which can be converted into bioethanol. The biomass of algae and starch is converted into sugar by a catalytic reaction. The process of producing ethanol using microalgae follows the process below. The starch inside the microalgae is delivered into the cells using a mechanical gear or a compound is the first step. When cells start to shrink, the yeast is added to the biomass, and it starts to age. This aging causes the production of ethanol. It is then transported into a tank and later to a refining unit (Pimentel 2003; Pimentel et al. 2008).

12.3.2 Processes for the Conversion of Algal Biomass to Biofuels and Co-products

There are different pathways for the change of biomass algae into different biofuel and other value-added biochemicals. The quality of biofuel mainly depends on its extraction techniques (Marrone et al. 2018; Khan et al. 2018). Selective extraction towards specific biomaterials reduces the pollution load. Figure 12.2 explains the different pathways for the conversion of biomass into biofuel and value-added products.

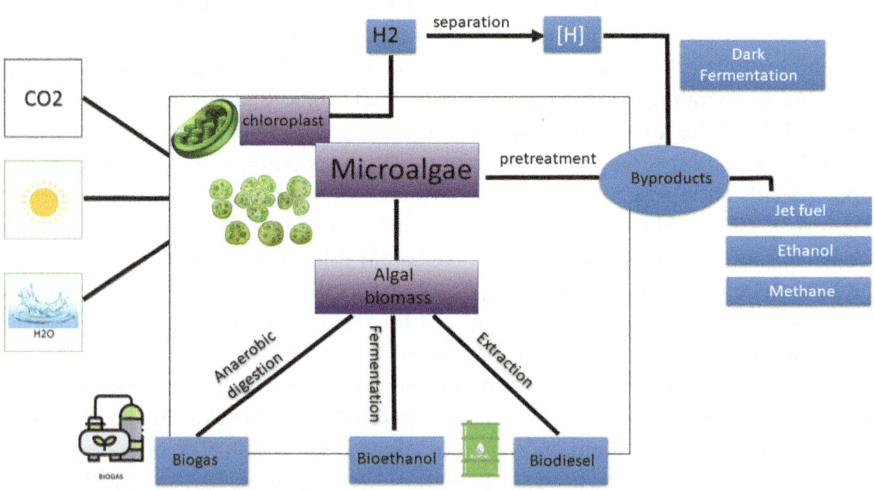

Fig. 12.2 Pathway for conversion of algal biomass into biofuel and value-added products

12.3.2.1 Microbial Fermentation of Algal Biomass

Production of biofuels is essentially done by yeast maturation followed by aging. There are different boundaries included in the creation of the bioethanol production process. For example, screening of powerful microbial strains and choice of substrate play a major role in the production process (Westman et al. 2017). It has been noticed that almost 140,000 L/ha/year of bioethanol is delivered from algae that have high starch content; this is comparatively higher than any other fluid fills. The average amount of sugar present in any microalgae is around 40 units. Even species like *Chlorella* and *Scenedesmus* show up to 50% starch content only (Agwa et al. 2017).

After the pretreatment of algae, the microbial maturation step converts the recuperated sugar into bioethanol. Two methods of enzymatic hydrolysis are used. (i) Separated Hydrolysis and Fermentation (SHF) and (ii) Simultaneous Saccharification and Fermentation (SSF). Difficulties faced in SHF are the partition of the wet algal biomass after hydrolysis since it restricts the yield of the product (Offei et al. 2018). But compared with SHF, SSF is more prudent because of the smaller compound measurement needed alongside the better return of yield.

12.3.2.2 Anaerobic Digestion of Microalgal Biomass

Production of biogas using anaerobic assimilation from microalgal biomass is one of the most feasible methods. The advantages of this method are control of GHG discharges and natural excrement production (Paolini et al. 2018). The four stages associated with absorption of the anaerobic bacteria are hydrolysis, acidogenesis, acetogenesis, and methanogenesis and the deciding stage is methanogenesis (Anukam et al. 2019). The ratio of carbon to nitrogen proportion is an essential part

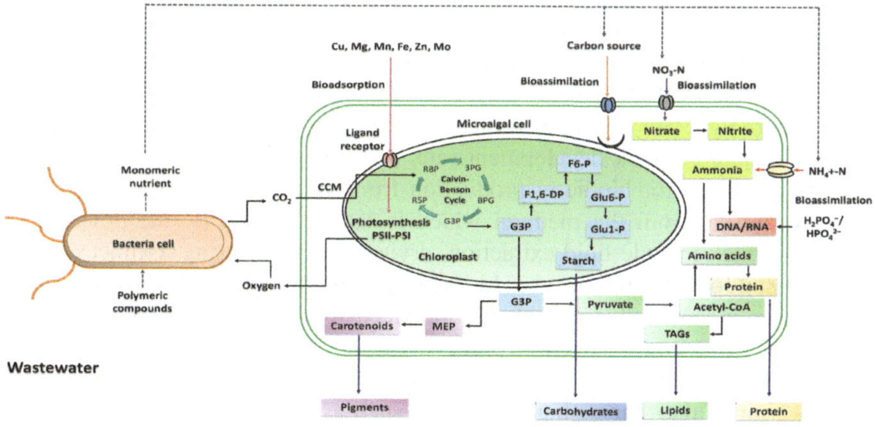

Fig. 12.3 Nutrient and CO_2 metabolism inside microalgae cell (Reproduced with permission from Goswami et al. (2021))

of deciding the algae and microbial interaction. The raw materials required for the production of biodiesel from algae are lipids. After the extraction of lipids, roughly 65% of algae would be obtained which is a significant number for a biofuel business. The microalgal build-up improved mixture using sugar and proteins can be additionally used for the feedstock for the creation of biogas by applying anaerobic assimilation measures (Fig. 12.3).

12.4 Microalgae: Cultivation and Methodologies

Microalgae have a high development rate. The high photosynthesis productivity of algae combined with the capacity to produce a lot of byproducts inside their cells makes them a good possibility to fill in as mechanical crude material (Randrianarison and Ashraf 2017). The development of microalgae doesn't need prolific land, a huge amount of freshwater, or herbicides and pesticides and hence the resources will always be replenished in one or the other method (Khan et al. 2018). Moreover, microalgae only are produced using wastewater. Aside from wastewater treatment, the development of microalgae can likewise assist with the decrease of environmental carbon dioxide through photosynthesis. Regardless of the advantages of microalgae development, its advancements still face a lot of different issues. For instance, the low biomass generation and the small size of cells when produced in fluid medium hinders the harvesting process of microalgae's, making it a very costly method to produce energy. The reaping of microalgae is one of the principal parts of microalgae handling. A few studies have recommended that it makes up to 20 to 30% of the production cost because of high energy interest and capital cost. Various reaping strategies have been utilized to gather biomass, including centrifugation, flotation, flocculation, and filtration (Lardon et al. 2009). In some cases, a mix of at least two methods is utilized for additional expansion collecting effectiveness.

The collection is a process of isolating microalgae from the medium and concentrating the algal biomass. Determination of gathering technologies depends essentially upon what kind of algae is being used in the production process (Cheng et al. 2019). Microalgae are generally preferred in biofuel production due to their less complex structure, quick development rate, and high lipid content. The different types of processes for biodiesel production from algae are the cultivation of the algae, separation from the medium which is later followed by the downstream processes which include lipid extraction, drying, and dewatering. Extracted lipids are processed for biodiesel or other biofuels in similar methods as used by existing technologies and methods for another biofuel feedstock. Algae are usually harvested based on the solid–liquid separation process. The most common harvesting processes are screening, coagulation, flocculation, flotation, sedimentation, filtration, and centrifugation. Other harvesting techniques such as electrophoresis, electro flotation, and ultrasound are used comparatively less. Energy efficiency and cost-effective harvesting pose two significant challenges in the commercialization of biofuels from algae. Life cycle assessment (LCA) studies can be conducted to identify the possible ecological effects and the energy balance for the whole chain of preparation of algae (Lardon et al. 2009). One such LCA study showed that the creation of biodiesel with super low sulfur diesel and canola on outflows and expenses was interpreted that the high yield was the only way to make algal biodiesel monetarily feasible. There has been a great discovery the recent times revealing the use of mineral manures and biogas which are the by-products of anaerobic processing of used oil cakes from algae oil extraction which in turn gives an upper hand in the energy requirement of the process (Lardon et al. 2009). Waste gases produced from existing frameworks like the vent gas (CO_2) obtained as waste products from power plants and also the profluent supplements like nitrogen, phosphorus, and other micronutrients could be changed into a crude material for a large scope of algae development. A thorough techno-economic analysis is important to draw a clear correlation between algal biofuel and other ordinary petroleum products.

12.5 Algae Biofuels and Conversion Process

Algae biomass has a wide variety of applications. Major components of microalgae (i.e., carbohydrates, lipids, and proteins) are easily converted into liquid, gas, and solid fuel (e.g., bioethanol, biodiesel, biogas, biohydrogen, and syngas) through three processes. They are biochemical, chemical, and thermochemical conversion. In thermochemical conversion, the organic matter of algal biomass is broken down to produce value-added liquid, gaseous, solid fuels such as synthesis gas, carbon monoxide, and hydrogen, etc. They are mainly categorized into pyrolysis, liquefaction, gasification, and direct combustion. In the pyrolysis process, the feedstock in the absence of oxygen is heated and decomposed at a definite residence time to produce liquid oil, gas, and solid char. The temperature is around 400–600 °C and the pressure is around 0.1 Mpa. When compared to the gasification process, decomposition occurs at much lower temperatures and helps produce liquid oil. The liquefaction of biomass is a catalytic conversion process carried out at a very low temperature (300–350 °C) and high pressures (5–20 MPa). Excessive moisture content in the feedstock makes the system more suitable. Gasification is a thermochemical transformation of microalgal biomass into useful energy fuels like hydrogen and methane. The reaction is carried out with less oxygen and at high temperatures mainly between 800 °C and 1000 °C. Gasification is mainly classified into four stages: drying, pyrolysis, combustion, and reduction. Direct combustion is a thermochemical conversion technique where the biomass is heated in the open air, or excess air is supplied to produce energy. High moisture content ($< = 50\%$) biomass is not suitable for this process as a lot of energy is needed for drying. The heat generated can be utilized for drying and grinding microalgae reducing additional costs for the process (Yang et al. 2011).

Biochemical conversion is a decomposition process where biomass is broken down using enzymes of bacteria or other microorganisms through anaerobic digestion, fermentation, and photobiological techniques. Anaerobic digestion is suitable for microalgae with high polysaccharides with no lignin and high moisture content. Algal biomass fermentation is one of the promising alternatives solutions to produce bioethanol. Fermentation of microalgae biomass with *Saccharomyces cerevisiae* (yeast) helps to produce bioethanol. Yeast is widely used in ethanol production due to its high ethanol tolerance and it is high in ethanol production. Bioethanol production using fermentation depends on many factors such as inoculum size, sugar concentration, yeast volume, agitation rate, temperature, pH, and fermentation time. Chemical conversion processes taking place during transesterification is a chemical conversion process used for the extraction of lipids. A chemical reaction takes place between lipids and the acylation agents and catalysts. The byproduct glycerol obtained can be used in the pharmaceutical industry and cosmetics. Studies are being carried out to analyze the product yield in ex situ transesterification. Though the ex-situ transesterification was energy-intensive and time-consuming, there was a much higher rate of biodiesel production.

12.6 Flue Gas Clean-Up and Algae Production

Biological sequestration uses algae to capture the CO_2 produced from thermal power plants. The residue can be used to produce solid biomass. Studies show that more than 40 hectares (ha) of algal ponds are needed to fix the carbon released from coal power plants. For ideal CO_2 sequestration, the following characteristics must be followed: high rate of CO_2 absorption and mineralization; results in sequestered carbon forever; trade of high and valuable products to generate revenue; and utilization and consumption of concentrated, anthropogenic CO_2 before atmospheric release. One of the best CO_2 sequestration approaches is algal biomass production using CO_2 in the flue gas. Flue gas from thermal power plants using coal and natural gas as fuel comprises 3–5% and 12–15% of CO_2 by weight (Somers and Quinn 2019). In the present age, although algae cultivation techniques have many limitations, like carbon dioxide feeding efficiency being one of them, two methods have been introduced to consume the flue gas from industrial plants to utilize and capture the carbon dioxide with the help of algae production: amine-based carbon capture and carbonate-based carbon capture. Both the methodologies use algae as a photochemical desorbed of carbon dioxide from an absorption column (Cheah et al. 2016). This helps us scale up the production very easily, increases carbon removal efficiency, and can decrease the capital cost for operation (Singh and Gu 2010). During the microalgae carbon capture process, the algae convert the energy from the sun into chemical energy using photo biological processes and later convert this chemical energy and carbon dioxide through a dark reaction to a stable form of chemical energy of organic matter without any other pollution. Almost 50% of any algal biomass is carbon, and 1 kg of algal biomass can fix almost 1.83 kg of carbon (Thomas et al. 2016). Table 12.2 lists the advantages of carbon capture by algae over conventional techniques.

Table 12.2 Advantages of algae over conventional techniques for CO_2 sequestration (Brennan and Owende 2010; Schipper et al. 2013)

Algal techniques	Conventional techniques		
	Post-combustion capture	Pre-combustion capture	Oxy-combustion
Biomass is produced through the process of carbon dioxide fixation by photosynthesis. This decreases the amount of energy required and the cost of production as compared to conventional techniques	CO_2 absorption, adsorption, cryogenic distillation are the major methods to capture CO_2	CO_2 is removed from the process gas before the steam generation takes place	Pure oxygen is used for combustion to increase the purity of exhaust CO_2
The absorption liquid can be regenerated easily without using large amounts of energy. In addition, the saved energy can be reused for heating, reducing the cost of carbon capture and conversion	Different solvents, adsorbents and membranes are used for carbon capture	Gasifier controls the amount of O_2 inside the combustion chamber which forms partial oxidized products	Purified oxygen eliminates the unwanted by product formation
Utilizing waste CO_2 from power plants can improve algae production effectiveness and decreased energy requirements	Improved adsorbents and membranes reduces the energy cost	Energy requirements are higher than Post combustion capture	Significant energy prerequisites making it not financially feasible for CO_2 reduction
On account of the amine carbon capture method, the absorption liquid is recovered by algae through bicarbonate. When most bicarbonates are depleted, the lean absorption fluid can be reused again for the next cycle. In addition, the utilization of less expensive absorption liquids, such as potassium carbonate and enzymes like carbonic anhydrase, builds up the uptake rate	–	The carbon dioxide capture efficiency is less compared to that of the new modern techniques	CO_2 capture efficiency can be achieved up to 90%, if oxygen production cost is reduced

(continued)

Table 12.2 (continued)

Algal techniques	Conventional techniques		
	Post-combustion capture	Pre-combustion capture	Oxy-combustion
Because of improved CO_2 conversion, the surface area of algae lakes could be reduced by a factor of 1.5–2 times	High volume of gas has to be treated for lesser CO_2 concentration	–	Technology development cost is more costly than other processes

12.7 Techno-economic and Life Cycle Assessment of Algal-Based Biorefineries

Techno-economic analysis (TEA) helps the industry analyze the economic performance of a process, its products, and its services. It uses a wide range of simulations and modeling to estimate certain criteria like cost of production, capital cost, revenue access, etc. Such parameters are calculated to determine the feasibility of the algal biofuel plant. Although biofuel from microalgae can provide a viable substitute for fossil fuels, the high production cost can threaten their financial feasibility. In the USA, the Department of Energy (DOE) also stated that the cost of algae biodiesel was approximate \$2.11/L, which was comparatively higher than that of the biodiesel produced from soya beans (Nagarajan et al. 2013). Hence, to evaluate the feasibility of algal biorefinery, a comprehensive assessment of all economic aspects integrated with the technological process is essential. The assessment includes the valuation of the cultivation and processing, environmental impact, energy cost, and product prices, including the value-added products.

Kovacevic and Wesseler (2010), analyzed the effectiveness of the energy produced from algae biorefinery by comparing it with biodiesel and fossil fuels in transportation, social cost, external costs, and benefits of the fuel. The study concluded that biodiesel produced from algal biomass had the least cost requirements for raw materials utilization. The break-even point of biodiesel formed from algal refineries was also the lowest among all other biodiesel production methods. TEA showed that the price of algae was a major factor determining the final product cost. However, higher lipid content and a lower amount of nutrients in the algae can reduce the cost without damaging the yield capacity of the plant (Kovacevic and Wesseler 2010). Biller et al. (2012), focused on using treated wastewater to help decrease the final cost during cultivation. The results indicated that the efficiency and nutrient utilization increased up to 75% (Biller et al. 2012).

Similarly, Xin and his research group used the TEA method to check the feasibility of algal biofuel production in different types of municipal wastewaters (Xin et al. 2016). The cost of biofuel produced from algae grown in wastewater collected from sludge dewatering plants was \$2.23/gal. Similarly, Sasongko (2018) tried to cultivate algae using wastewater and flue gas. As a result, the cost of algal biomass production reduced from 1605.9 JPY/kg to 160.6 JPY/kg (Biller et al. 2012). Delrue et al. (2012) conducted another study on the development of a model of algae biofuel to produce fuel economically in a PBR cultivation process. The fuel production pathway used was a hybrid raceway/PBR cultivation system, belt filter press for dewatering, wet lipid extraction, oil hydro treating, and anaerobic digestion of residues (Delrue et al. 2012). The fuel cost was between \$2.52 and \$4.35/L which was economical compared to their reference pathway. This was because of the supercritical gasification process which increased the calorific value of fuel and reduced the CO_2 content in the fuel (Xin et al. 2016). Although most TEA studies showed that biofuel production from algal biomass is economical, it is not economically popular. This is due to the high capital investment, high initial production cost, and competitiveness of algal biofuel industries.

Life cycle assessment (LCA) collects and checks the product interaction with output, input, and the environment with the system or surrounding through its lifespan. The lifespan of a product includes extraction of raw material, processing the raw material, manufacturing of the product to the product, and life treatment. Hence it is also known as cradle to grave analysis. It not only focuses on the lifespan of the product but also on the process cycles of input, labor, equipment, and infrastructure of the plant. It gives an excellent qualitative report of the environmental effects that the product could lead to or its production (Delrue et al. 2012). It also provides an excellent quantitative report to make and design an environmentally friendly product. The techno-economic skill assessment helps in designing an excellent LCA design. While TEA tells the manufacturer how much the capital and production cost and also helps in analyzing the different risks that the production of the product would require, LCA takes care of the effect that the product could have after its manufacturing (Campbell et al. 2011).

12.8 Circular Economy Concepts in Biorefineries

A circular economy is a method that approaches a greener environment methodology during the production, removal and utilization of resources. It can be known to be an alternative to the traditional linear economy production, which used to be used by many E-Companies before. Circular economy concepts help in the reduction of pollution and waste created during the industrial production of a product and at the same time help to recycle the wastes or resources back to the environment (Nizami et al. 2017a). Algal biorefineries use the circular economy to its maximum utilization by converting the biomass into different chemicals by-products and produces energy with the least amount of omit it off emissions. The main aim of the circular economy in algae biorefineries is to utilize the power and the resources used from the starting of the production to the end to improve the sustainability and economic viability of the production and a biorefinery (Nizami et al. 2017a). Circular biorefinery is used production of the least amount of waste. For example, the chemical and biological waste formed from the production was later converted into chemical feedstock in

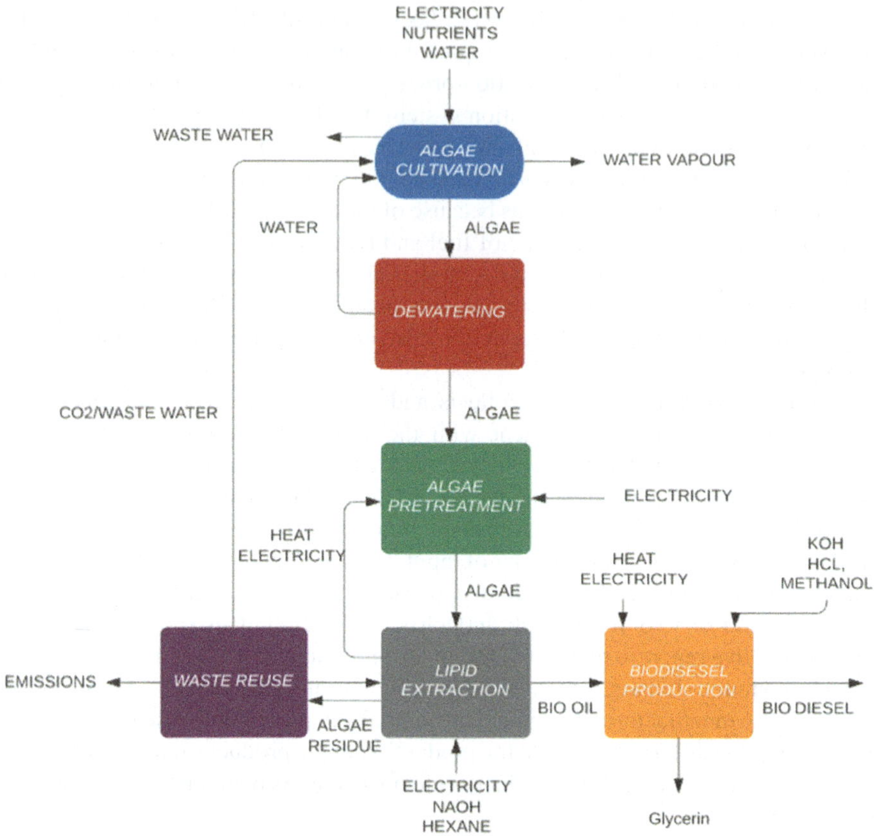

Fig. 12.4 LCA study scope for algal biofuel production

the food and plastic industry (Clark 2017). A recent study calculated that almost $410 billion could be saved up from just recycling community-based waste worldwide (Nizami et al. 2017b). After the techno-economic and life impact analysis, biorefineries can approach practical waste disposal methodologies by cross-country similarities, analysis of consequences on economic maturity, and business model evaluation. The waste on biorefineries can be recycled back into renewable energy, later used as a fuel for the same biorefinery (Fig. 12.4).

12.9 Biorefineries—Their Scenarios and Challenges

The International Energy Agency bioenergy task 42 defined biorefining as "the sustainable processing of biomass into a spectrum of bio-based products (food, feed, chemicals, and materials) and bioenergy (biofuels, power and/or heat)". The major

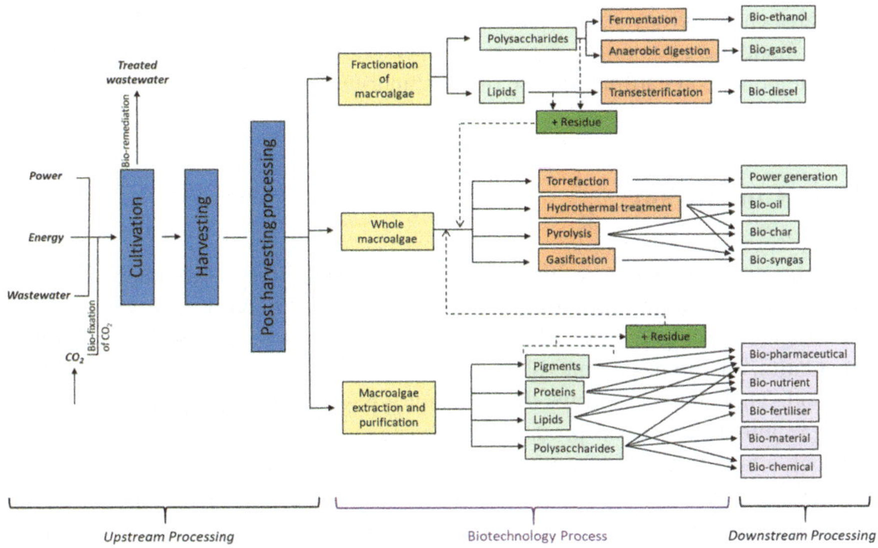

Fig. 12.5 The circular economy concept in biorefinery (adapted from Kuppens et al. (2015))

advantage of the biorefinery is that different value-added products are produced simultaneously, reducing the manufacturing cost and market price. In addition, many process streams are merged to produce valuable products. Biorefineries have been are classified based on the model, status of their technological implementation, size, and type of feedstocks, platforms, processes, and other products (Fig. 12.5).

Biorefineries deal with several technical, environmental, strategic, and commercial issues. The biorefinery concept is applied only on a small scale; its up-gradation and evolution to a higher level seem complicated. Another challenge is the maximum utilization of feedstock and production output. Detailed knowledge of the characteristics and composition of biomass is required to obtain an optimal conversion, optimal availability of biomass, and economic benefit. The main challenge of biorefineries is to convert the biomass residue to value-added products. The main residue of biomass is lignin which requires costly enzymes for its conversion. Alternate cost-effective enzymes need to be explored to reduce the cost of the process. The development of biochemical and chemical catalysts are some of the other challenges researchers are facing. This helps in the production of higher yield desired products. Logistical challenges mainly include feedstock processing, including the conservation and densification techniques and network distribution of products. In the latter case, existing technologies can be used for biofuel transportation. Another challenge is commercial challenges which include investment cost, funding, etc. Deforestation, excessive water, fertilizer, and pesticide consumption to cultivate biofuel feedstock severely impact the environment and biodiversity. Also, the use of agricultural lands to produce energy crops affects the food supply and biodiversity. Another challenge faced is inadequate economic support and technological support, as many countries lack state policies regarding the use of biorefineries.

12.10 Conclusion

Studies have shown that one of the safest ways to reduce carbon dioxide emissions is to utilize the carbon from thermal power plants and feed it to microalgae-based fuel production. Microalgae cultivation and harvesting can later be used for microbial fermentation of biomass, anaerobic digestion, or extraction to produce biodiesel or bio-oil. This chapter reviews the different types of conversion routes to convert the algae into value-added products. Microalgae are considered as green energy feedstock which helps in minimizing industrial emission problems and challenges. Furthermore, it helps to capture carbon dioxide produced from thermal power plants. Carbon dioxide can be captured and used in algal biorefinery through two methods that are amine-based carbon dioxide capture or through carbonate-based carbon dioxide capture system (Algae biofuels, E.E.R.E. U.S. 2012).

The environmental impact of using different methods to capture carbon dioxide from the atmosphere can be recorded and deciphered using the techno-economic assessment and life cycle assessment. Different types of simulations and software are used to evaluate specific criteria like capital cost, revenue, and feasibility of production, which increase the economy of the process and reduce the impact on the environment. The life cycle assessment is used to check the lifespan of a product, its process cycles of input, labor equipment, and infrastructure of the plant. Unlike techno-economic assessment, which gives a quantitative analysis, life cycle assessment is a qualitative analysis that reports on the environmental effects that the product could lead to during the production of the product. Therefore, it helps industries to produce environmentally friendly products. Algal biorefineries use the circular economy concept to maximize the utilization of feedstock with minimum energy input, wastage, and emission.

References

Abdullah B, Muhammad SAFAS, Shokravi Z, Ismail S, Kassim KA, Mahmood AN, Aziz MMA (2019) Fourth generation biofuel: a review on risks and mitigation strategies. Renew Sustain Energy Rev 107:37–50. https://doi.org/10.1016/j.rser.2019.02.018

Agwa OK, Nwosu IG, Abu GO (2017) Bioethanol production from chlorella vulgaris biomass cultivated with plantain (Musa paradisiaca) peels extract. Adv Biosci Biotechnol 8(12):478. https://doi.org/10.4236/abb.2017.812035

Algae biofuels, E.E.R.E. U.S. (2012) Department of energy. In: Growing America's energy future. Alternative fuels data center, engineering news-record, ENR The McGraw-Hill Companies, Inc, Washington, DC, USA

Anukam A, Mohammadi A, Naqvi M, Granström K (2019) A review of the chemistry of anaerobic digestion: methods of accelerating and optimizing process efficiency. Processes 7(8):504. https://doi.org/10.3390/pr7080504

Biller P, Ross AB, Skill SC et al (2012) Nutrient recycling of aqueous phase for microalgae cultivation from the hydrothermal liquefaction process. Algal Res 1:70–76. https://doi.org/10.1016/j.algal.2012.02.002

Blomen E, Hendriks C, Neele F (2009) Capture technologies: improvements and promising developments. Energy Procedia 1:1505–1512. https://doi.org/10.1016/j.egypro.2009.01.197

Brennan L, Owende P (2010) Biofuels from microalgae—a review of technologies for production, processing, and extractions of biofuels and co-products. Renew Sustain Energy Rev 14:557–577. https://doi.org/10.1016/j.rser.2009.10.009

Budzianowski WM, Postawa K (2016) Total Chain integration of sustainable biorefinery systems. Appl Energy 184:1432–1446. https://doi.org/10.1016/j.apenergy.2016.06.050

Campbell SW, Galloway AM, McPherson NA (2011) Techno-economic evaluation on the effects of alternating shielding gases for advanced joining processes. Proc Inst Mech Eng, Part b: J Eng Manuf 225(10):1863–1872. https://doi.org/10.1177/0954405411408353

Cheah WY, Ling TC, Juan JC et al (2016) Biorefineries of carbon dioxide: from carbon capture and storage (CCS) to bioenergies production. Bioresour Technol 215:346–356. https://doi.org/10.1016/j.biortech.2016.04.019

Cheng J, Zhu Y, Zhang Z, Yang W (2019) Modification and improvement of microalgae strains for strengthening CO_2 fixation from coal-fired flue gas in power plants. Bioresour Technol 291:121850. https://doi.org/10.1016/j.biortech.2019.121850

Clark JH (2017) From waste to wealth using green chemistry: the way to long term stability. Curr Opin Green Sustain Chem 8:10–13. https://doi.org/10.1016/j.cogsc.2017.07.008

De Bhowmick G, Sarmah AK, Sen R (2019) Zero-waste algal biorefinery for bioenergy and biochar: a green leap towards achieving energy and environmental sustainability. Sci Total Environ 650:2467–2482. https://doi.org/10.1016/j.scitotenv.2018.10.002

Delrue F, Setier P-A, Sahut C et al (2012) An economic, sustainability, and energetic model of biodiesel production from microalgae. Bioresour Technol 111:191–200. https://doi.org/10.1016/j.biortech.2012.02.020

Demirbas MF (2011) Biofuels from algae for sustainable development. Appl Energy 88(10):3473–3480. https://doi.org/10.1016/j.apenergy.2011.01.059

Demirbas A, Demirbas MF (2010) Algae energy: algae as a new source of biodiesel. Springer Science and Business Media. https://doi.org/10.1016/j.apenergy.2011.01.059

Fuss S, Jones CD, Kraxner F, Peters GP, Smith P, Tavoni M, van Vuuren DP, Canadell JG, Jackson RB, Milne J, Moreira JR (2016) Research priorities for negative emissions. Environ Res Lett 11(11):115007. https://doi.org/10.1088/1748-9326/11/11/115007

Goswami RK, Mehariya S, Verma P, et al. (2021) Microalgae-based biorefineries for sustainable resource recovery from wastewater. J Water Process Eng 40:101747. https://doi.org/10.1016/j.jwpe.2020.101747

Hena S, Fatimah S, Tabassum S (2015) Cultivation of algae consortium in a dairy farm wastewater for biodiesel production. Water Resour Ind 10:1–14. https://doi.org/10.1016/j.wri.2015.02.002

Khan MI, Shin JH, Kim JD (2018) The promising future of microalgae: current status, challenges, and optimization of a sustainable and renewable industry for biofuels, feed, and other products. Microb Cell Fact 17(1):1–21. https://doi.org/10.1016/j.mset.2018.12.007

Khoo KS, Lee SY, Ooi CW et al (2019) Recent advances in biorefinery of astaxanthin from Haematococcus pluvialis. Bioresour Technol 288:121606. https://doi.org/10.1016/j.biortech.2019.121606

Kostas ET, Adams JM, Ruiz HA, Durán-Jiménez G, Lye GJ (2021) Macroalgal biorefinery concepts for the circular bioeconomy: a review on biotechnological developments and future perspectives. Renew Sustain Energy Rev 151:111553. https://doi.org/10.1016/j.rser.2021.111553

Kovacevic V, Wesseler J (2010) Cost-effectiveness analysis of algae energy production in the EU. Energy Policy 38:5749–5757. https://doi.org/10.1016/j.enpol.2010.05.025

Kumar M, Ghosh P, Khosla K, Thakur IS (2016) Biodiesel production from municipal secondary sludge. Bioresour Technol 216:165–171. https://doi.org/10.1016/j.rser.2017.08.042

Kuppens T, Van Dael M, Vanreppelen K et al (2015) Techno-economic assessment of fast pyrolysis for the valorization of short rotation coppice cultivated for phytoextraction. J Clean Prod 88:336–344. https://doi.org/10.1016/j.jclepro.2014.07.023

Lardon L, Hélias A, Sialve B, Steyer JP, Bernard O (2009) Life-cycle assessment of biodiesel production from microalgae. https://doi.org/10.1021/es900705j

Mahapatra DM, Chanakya HN, Ramachandra TV (2014) Bioremediation and lipid synthesis through mixotrophic algal consortia in municipal wastewater. Biores Technol 168:142–150. https://doi.org/10.1016/j.biortech.2014.03.130

Marrone BL, Lacey RE, Anderson DB, Bonner J, Coons J, Dale T, Downes CM, Fernando S, Fuller C, Goodall B, Holladay JE (2018) Review of the harvesting and extraction program within the national alliance for advanced biofuels and bioproducts. Algal Res 33:470–485. https://doi.org/10.1016/j.algal.2017.07.015

Martín-Girela I, Albero B, Tiwari BK, Miguel E, Aznar R (2020) Screening of contaminants of emerging concern in microalgae food supplements. Separations 7(2):28. https://doi.org/10.3390/separations7020028

Mehariya S, Goswami RK, Karthikeysan OP, Verma P (2021) Microalgae for high-value products: a way towards green nutraceutical and pharmaceutical compounds. Chemosphere 280:130553. https://doi.org/10.1016/j.chemosphere.2021.130553

Minowa T, Sawayama S (1999) A novel microalgal system for energy production with nitrogen cycling. Fuel 78:1213–1215. https://doi.org/10.1016/S0016-2361(99)00047-2

Nagarajan S, Chou SK, Cao S et al (2013) An updated comprehensive techno-economic analysis of algae biodiesel. Bioresour Technol 145:150–156. https://doi.org/10.1016/j.biortech.2012.11.108

Nizami AS, Rehan M, Waqas M et al (2017a) Waste biorefineries: enabling circular economies in developing countries. Bioresour Technol 241:1101–1117. https://doi.org/10.1016/j.biortech.2017.05.097

Nizami AS, Shahzad K, Rehan M et al (2017b) Developing waste biorefinery in Makkah: a way forward to convert urban waste into renewable energy. Appl Energy 186:189–196. https://doi.org/10.1016/j.apenergy.2016.04.116

Offei F, Mensah M, Thygesen A, Kemausuor F (2018) Seaweed bioethanol production: a process selection review on hydrolysis and fermentation. Fermentation 4(4):99. https://doi.org/10.3390/fermentation4040099

Panahi Y, Khosroushahi AY, Sahebkar A, Heidari HR (2019) Impact of cultivation condition and media content on Chlorella vulgaris composition. Advan Pharm Bull 9(2):182. https://doi.org/10.15171/apb.2019.022

Paolini V, Petracchini F, Segreto M, Tomassetti L, Naja N, Cecinato A (2018) Environmental impact of biogas: a short review of current knowledge. J Environ Sci Health, Part A 53(10):899–906. https://doi.org/10.1080/10934529.2018.1459076

Pimentel D (2003) Ethanol fuels: energy balance, economics, and environmental impacts are negative. Nat Resour Res 12(2):127–134. https://doi.org/10.1023/A:1024214812527

Pimentel D, Marklein A, Toth MA, Karpoff M, Paul GS, McCormack R, Kyriazis J, Krueger T (2008) Biofuel impacts on world food supply: use of fossil fuel, land and water resources. Energies 1(2):41–78. https://doi.org/10.3390/en1010041

Randrianarison G, Ashraf MA (2017) Microalgae: a potential plant for energy production. Geology, Ecology, Landscapes 1(2):104–120. https://doi.org/10.1080/24749508.2017.1332853

Schipper K, van der Gijp S, van der Stel R, Goetheer E (2013) New methodologies for the integration of power plants with algae ponds. Energy Procedia 37:6687–6695. https://doi.org/10.1016/j.egypro.2013.06.601

Shuba ES, Kifle D (2018) Microalgae to biofuels: 'Promising' alternative and renewable energy, review. Renew Sustain Energy Rev 81:743–755. https://doi.org/10.1016/j.rser.2017.08.042

Singh J, Gu S (2010) Commercialization potential of microalgae for biofuels production. Renew Sustain Energy Rev 14(9):2596–2610. https://doi.org/10.1016/j.rser.2010.06.014

Singh G, Patidar SK (2018) Microalgae harvesting techniques: a review. J Environ Manage 217:499–508. https://doi.org/10.1016/j.jenvman.2018.04.010

Somers MD, Quinn JC (2019) Sustainability of carbon delivery to an algal biorefinery: a techno-economic and life-cycle assessment. J CO2 Util 30:193–204. https://doi.org/10.1016/j.jcou.2019.01.007

Srivastava RK (2019) Bio-Energy production by contribution of effective and suitable microbial system. Mater Sci Energy Technol 2(2):308–318. https://doi.org/10.1016/j.mset.2018.12.007

Thomas DM, Mechery J, Paulose SV (2016) Carbon dioxide capture strategies from flue gas using microalgae: a review. Environ Sci Pollut Res 23:16926–16940. https://doi.org/10.1007/s11356-016-7158-3

Trivedi J, Aila M, Bangwal DP, Kaul S, Garg MO (2015) Algae based biorefinery—how to make sense? Renew Sustain Energy Rev 47:295–307. https://doi.org/10.1016/j.rser.2015.03.052

Westman JO, Wang R, Novy V, Franzén CJ (2017) Sustaining fermentation in high-gravity ethanol production by feeding yeast to a temperature-profiled multifeed simultaneous saccharification and co-fermentation of wheat straw. Biotechnol Biofuels 10(1):1–16. https://doi.org/10.1186/s13068-017-0893-y

Wilson MH, Shea A, Groppo J, Crofcheck C, Quiroz D, Quinn JC, Crocker M (2021) Algae-based beneficial re-use of carbon emissions using a novel photobioreactor: a techno-economic and life cycle analysis. BioEnergy Research 14(1):292–302. https://doi.org/10.1007/s12155-020-10178-9

Xin C, Addy MM, Zhao J et al (2016) Comprehensive techno-economic analysis of wastewater-based algal biofuel production: a case study. Bioresour Technol 211:584–593. https://doi.org/10.1016/j.biortech.2016.03.102

Yang J, Xu M, Zhang X, Hu Q, Sommerfeld M, Chen Y (2011) Life-cycle analysis on biodiesel production from microalgae: water footprint and nutrients balance. Biores Technol 102(1):159–165. https://doi.org/10.1016/j.biortech.2010.07.017

Chapter 13
Seaweed Bioprocessing for Production of Biofuels and Biochemicals

B. Vanavil, P. Ezhilarasi, R. Aanandhalakshmi, P. S. Gowtham, and K. Sundar

13.1 Introduction

Rapid population growth demands increasing energy and resources. Therefore, there is a constant need for alternative resources and processes to satisfy the industrial requirements to produce biofuels and biochemicals. Indian marine environment is appreciated for its rich reserve of structurally unique and prospective bioactive metabolites (Vanavil et al. 2020). Seaweeds are marine macroalgae that comprise approximately 15,000 species. Seaweeds are categorized into red, brown, and green seaweeds on the basis of their color and chemical composition. Indian coast harbours 434, 191 and 216 species of red, brown and green seaweeds respectively (Mantri et al. 2020). According to a 2019 report, the total production of brown seaweeds from Indian waters per year was approximately 600,000 tonnes which include 16,000 tonnes of *Sargassum* and *Turbinaria* (Manickavasagam et al. 2019). Besides human food consumption, seaweeds are extensively utilized for the extraction of phycocolloids such as agar, alginate, and carrageenan that serve for a wide range of commercial applications (Rathour et al. 2021). Marine macroalgae are affluent in bioactive metabolites with a comprehensive spectrum of biological activities (Biris-Dorhoi et al. 2020). As a rich source of organic matter, seaweeds have also been recycled as soil conditioners and fertilizers. Seaweed extracts are extensively employed as plant biostimulants (El Boukhari et al. 2020). Several commercial products extracted from seaweeds finds application in farming and cultivation. Numerous reports uncovered the positive effects of seaweed extracts on plants like early seed germination, improved crop performance and yield, augmented resistance to biotic and abiotic stress, and improved shelf-life of short-lived products (Wang et al. 2016). Extracts from brown algae *Durvillaea potatorum* and *Ascophyllum nodosum* have improved

B. Vanavil (✉) · P. Ezhilarasi · R. Aanandhalakshmi · P. S. Gowtham · K. Sundar
Department of Biotechnology, School of Bio and Chemical Engineering, Kalasalingam Academy of Research and Education, Krishnankoil, Tamil Nadu 626126, India
e-mail: b.vanavil@klu.ac.in

Y. K. Nandabalan et al. (eds.), *Zero Waste Biorefinery*, Energy, Environment, and Sustainability, https://doi.org/10.1007/978-981-16-8682-5_13

growth and productivity in tomato plants. Further, these extracts improved the diversity of the bacterial population and soil nitrogen content (Hussain et al. 2021). Foliar treatment of seaweed extracts ensued in quantitative and qualitative improvement of yield in mango trees (Mohamed and El-Sehrawy 2013). Foliar applications of *Kappaphycus alvarezii* and *Gracilaria edulis* sap enhanced growth, yield and nutritional content in wheat (Shah et al. 2013). Being a wealthy source of functional metabolites, seaweeds extracts can be utilized in the augmentation of safety and quality aspects of food products (Gupta and Abu-Ghannam et al. 2011). *A. nodosum* extract was found to impact the yield and the nutritional value of spinach (Fan et al. 2013).

Apart from the various applications of macroalgae mentioned above, this biomass can potentially be incorporated into bioprocesses under the emerging concept called "Biorefinery framework". By this concept, the seaweed wastes generated during the manufacture of phycocolloids and offshore seaweed waste biomass can be engaged as feedstock for the conversion of biomass into chemicals, energy and value-added products to take full advantage of the value of the generated waste and minimize the wastes through sustainable bioprocessing (D'Este 2017).

Renewable biomasses can be categorized into four main classes: (Cherubini 2010) crops or agro residues, forestry, domestic and industrial wastes and, aquatic (microalgae and seaweeds). Seaweed is an excellent underutilized renewable marine resource due to the lack of insight into its economic potential and hence has not yet been tapped as commercial feedstock for the manufacture of third-generation fuels, biochemicals and value upgraded commodities such as pigments, fatty acids, vitamins, minerals, sterols, terpenoids and functional foods (Enquist-Newman et al. 2013). Macroalgae is of particular interest for "biorefinery" as they do not compete with agricultural land, freshwater, or fertilizers and overcomes the competition with crops for food and feed (Ghadiryanfar et al. 2016).

13.2 Seaweeds

Seaweeds (macroalgae) are macroscopic autotrophs, non-vascular multicellular photosynthetic forms (Pereira 2021) inhabiting over rocks and intertidal regions and belong to a diverse evolutionary lineage. Seaweeds are believed to be originated about 1.5 billion years ago (Martin 2017). Based on pigmentation, seaweeds are categorized into three broader categories namely brown, green and red algae. Brown algae (Phaeophyceae) contain pigments such as fucoxanthin, flavoxanthin, violaxanthin, and lutein. Green algae (Chlorophyceae) have photosynthetic pigments such as chlorophyll a, b, and carotenes (Lobban and Wynne 1981). Red algae (Rhodophyceae) have photosynthetic pigments phycoerythrin and phycocyanin (Brodie and Lewis 2007; McHugh 2002). The phylogeny of three types of macroalgae indicates rich diversity and is distantly related. Seaweeds have a major role in maintaining the biodiversity of the coastal ecosystem and serve as the primary producer in the marine food chain (Satheesh et al. 2017). During photosynthesis, seaweed photosynthetic pigments assimilate CO_2 as bicarbonate in the presence of sunlight, produce its cellular components, and release O_2 (Falkowski and Raven 1997; Thompson et al. 2019). Seaweed hydrocolloids namely agar, alginate, and carrageenan have many industrial uses. Agar, a polysaccharide with two fractions namely 70% agarose made up of alternating entities of D-galactose and 3,6-anhydrogalactose and 30% agaropectin composed of 1,3-glycosidically linked D-galactose units (Syamdidi et al. 2016). Agar is extracted from red seaweeds namely *Acanthopeltis, Ceramium, Fermium, Gelidium, Gelidiella, Gracilaria, Pterocladiella, Pterocladia,* etc. (McHugh 1991; Cardozo et al. 2007). Alginate (algin) consists of 1,4-linked β-D-mannuronic and α-L-guluronic acid units. Alginate is produced from brown seaweeds such as *Ascophyllum, Durvillaea, Ecklonia, Lessonia, Laminaria, Macrocystis, Sargassum and Undaria* (Peteiro et al. 2018). Carrageenan is organized with an alternating units of α (1–4)-3,6-anhydro-D-galactose and β (1–3)-D-galactose. Carrageenan is mined from red algae namely *Eucheuma cottonii, Eucheuma spinosum, Hypnea musciformis,* and *Kappaphycus alvarezii* (Cardozo et al. 2007).

Seaweed hydrocolloids have larger applications in nutraceuticals, pharmaceuticals, agriculture, medicine, and bioenergy. Brown algae namely *Fucus sp., Ascophyllum nodosum, Laminaria sp., Sargassum sp., Turbinaria sp.* are widely used in agriculture (Khan et al. 2009). Seaweed wastes based foliar spray products are produced in the name of Kelp Meal, Maxicrop, Seaspray, etc. (Begum et al. 2018). The seaweed sap has growth-promoting effects in plants, increases photosynthetic rate, reduces senescence, enhances the color and flavor in fruit, imparts disease resistance and stress tolerance (Khan et al. 2009; Crouch and van Staden 1993). The biosorption capability of the seaweed biomass can be used for environmental protection such as wastewater treatment and removal of heavy metals (Sadhukhan et al. 2019; El Boukhari et al. 2020). Figure 13.1 highlights various potential applications of seaweeds. Figure 13.2 shows few types of seaweed collected by the authors from Hare Island, Gulf of Mannar, Tuticorin, Tamilnadu, India.

13.3 Seaweed Waste to Wealth

Tonnes of seaweed wastes get accumulated in the marine environment due to storm-cast and ocean acidification owing to human activities. Similarly, a large magnitude of seaweed waste is created after the extraction of hydrocolloids through industrial processing. The remaining seaweed waste is being processed to get value-added products such as protein, sugars, vitamins, composites, animal and poultry feed, cosmetics, gums, ion exchangers, viscous agents, fertilizers etc. (Mugnai et al. 2008). Instead of disposal, seaweed wastes must be processed through a biorefinery strategy (Wijayanta et al. 2015). Seaweed wastes that have high fiber content can be collected and processed to obtain organic acids, biodiesel, bioalcohol, and biogas. Hence, residual seaweed biomass can be converted into a higher value bioresource with the feasibility of minimal negative environmental impact. The global production of seaweeds has risen to 32.4 million tonnes in 2018 (Chopin and Tacon 2021). In 2015, the United Nations established Sustainable Development Goal 14 (SDG14

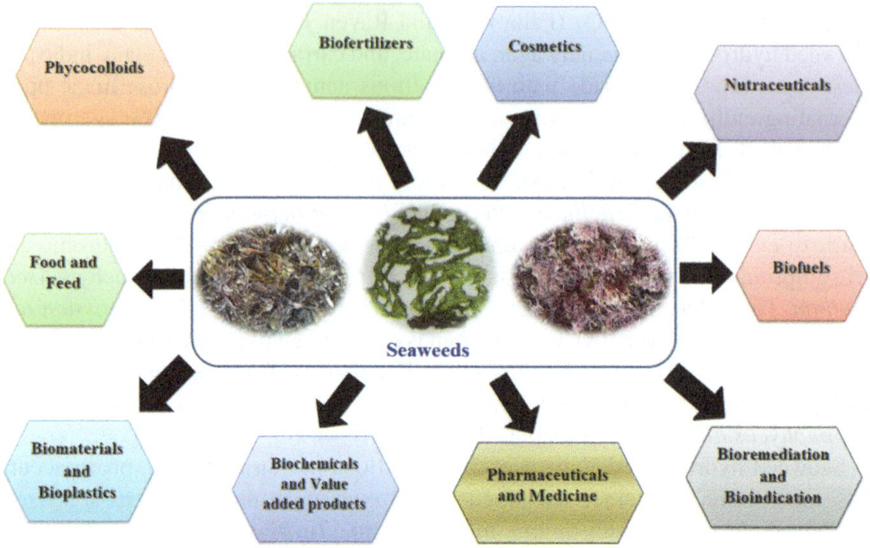

Fig. 13.1 Potential applications of seaweeds

Fig. 13.2 Seaweeds (**a**–**e**) collected by the authors from Hare Island, Gulf of Mannar, Tamilnadu, India

"Life below water") implying "conserve and sustainably use the oceans, seas and marine resources for sustainable development" (Friess et al. 2019). As there is crucial depletion of economy-related to fossil fuels globally, an extensive agenda of research on alternate renewable and sustainable energy is essential. The biorefinery approach by the processing of seaweed wastes is considered as the right path towards renewable green energy (Ragauskas et al. 2006).

13.4 Seaweed Integrated Biorefineries

The International Energy Agency's Bioenergy Task 42 terms biorefining as the "sustainable processing of biomass into a spectrum of bio-based products (food, feed, chemicals) and bioenergy (biofuels, power, and/or heat)" (Taylor 2008; Cherubini 2010). The biorefinery process involves a technical approach in the exploration of biomass for bioenergy production under optimal management practices that include depolymerization of seaweed feedstock. The biorefinery approach is always regarded as sustainable as it completely reduces waste from the environment. Biorefinery preprocessing and its valorization helps in the amplification of multitudinous recovery of products (Brennan and Owende 2010). An integrated biorefinery technology includes handling of biomass, harvesting, physical pre-treatments such as liquefaction, chemical and enzymatic hydrolysis, bioprocessing, fractionation, modification, refining, pyrolysis, and gasification, followed by recovery of the cascade of interesting products to improve blue bio-economy (Lange et al. 2020). Seaweed biorefineries provide a conceptual model to produce high-value products from waste. In this perspective, seaweed integrated biorefineries pave the way for an innovative and cost-effective recovery of the mixture of seaweed-related products such as carbohydrates, protein, organic acids, nutrients, minerals, biocomposites (pollutant absorber) bio-oil, biochar and liquid bio-crude with zero net carbon emissions (Balina et al. 2017). Figure 13.3 illustrates the seaweed biorefinery approach. *Ulva lactuca* is a saline tolerant green alga, which can grow at temperatures with reduced photo intensity, making it a good choice as a biorefinery feedstock (McHugh 2003).

In an integrated biorefinery process using green algae, *Ulva ohnoi* biomass was used to co-produce six products such as salts, starch, lipids, ulvan, protein and cellulose by green extraction methods (Prabhu et al. 2020). In a biorefinery approach, polyhydroxyalkanoates and biochar were produced from seaweed hydrolysate of *Ulva sp.* using extreme halophile *Haloferax mediterranei* at a yield of 0.104 gg^{-1} and 0.194 ± 1.23 gg^{-1} respectively (Ghosh et al. 2021). Carrageenan residues obtained from red algae *Kappaphycus alvarezii* was refined to obtain carbohydrates, proteins, insoluble aromatics, galacturonic acid, hydroxymethylfurfural and lipids (Masarin et al. 2016). Acid and enzymes have been used for saccharification of *Gelidium elegans* polysaccharides (Lobban and Wynne 1981). Saccharification of brown seaweeds *Saccharina sp.* and *Laminaria sp.* have been testified to deliver industrially important fermentable

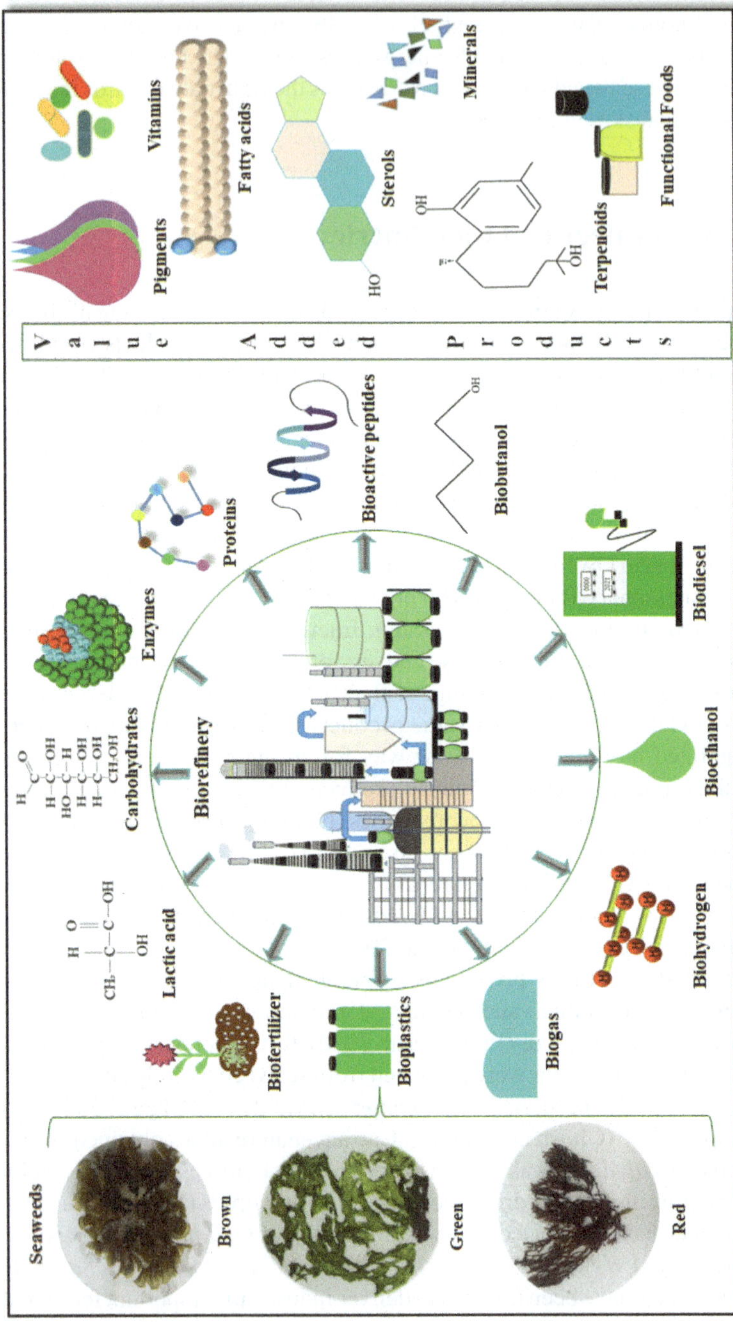

Fig. 13.3 Seaweed biorefinery approach

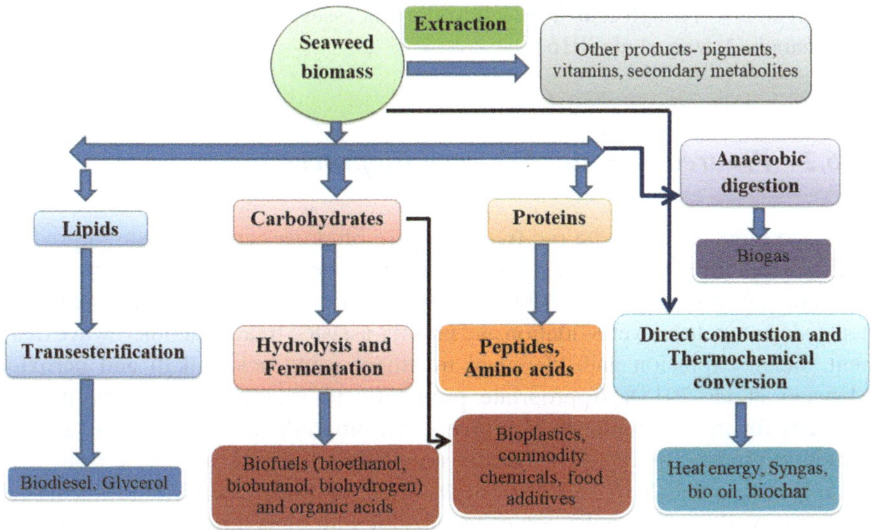

Fig. 13.4 Overview of products from macroalgal biomass and their potential methods of processing

sugars such as L-glucuronate, D-mannuronate, L-fucose, D-glucose, D-galactose and D-xylose (Ge et al. 2011).

13.5 Potential Methods of Seaweed Bioprocessing

Seaweeds can be processed using different conversion technologies for bioenergy, production of biochemicals and other bioproducts. Seaweeds harvested from natural resources or cultivated or waste seaweed biomass generated at agar–agar and alginate manufacturing units can be subjected to thermochemical or biochemical technologies for the production of heat energy, bioethanol, biohydrogen, biobutanol, biogas, bio-oil, biodiesel, biochar, syngas and other commodity chemicals (Michalak 2018) as depicted in Fig. 13.4.

13.6 Sequential Processing of Seaweed Wastes

13.6.1 *Collection of Seaweeds*

Seaweeds are collected or harvested from shore and transported to the biorefinery site and are dehydrated by solar drying, convective air drying, microwave drying, and vacuum drying to obtain dry biomass. Seaweed biomass can also be freeze-dried

(Uribe et al. 2019; Singh et al. 2017). In some processes, wet seaweed is also selected as feedstock (Nazari et al. 2016).

13.6.2 Pre-treatment for Seaweed Bioprocessing

The dried biomass has to be pre-treated before bioprocessing of waste to produce biochemicals. Mechanical pre-treatment includes beating, size reduction, sonication, and washing of seaweed. Other pre-treatment methods such as thermal treatment using autoclave, wet oxidation and plasma-assisted treatment, microwave treatment, steam explosion and, chemical treatment using acid, alkali and peroxide (Maneein et al. 2018). Appropriate pre-treatment is required to fully harness seaweeds for the biosynthesis of biofuels and bioproducts. Sonication is a faster method to produce bioethanol from the simultaneous arrival of glucose from *Ulva rigida* (Korzen et al. 2015). *Saccharina japonica* was pre-treated with low acid such as sulfuric acid for the production of bioethanol with a maximum yield of 6.65 g/L (Lee et al. 2013).

Treatment with dilute acid is the best physicochemical method for treating raw seaweeds especially *Laminaria japonica* for producing various biofuels (Lee et al. 2013). Acetone, butanol, and ethanol are procured from green seaweed *Ulva lactuca* using a mild alkaline pre-treatment (van der Wal et al. 2013). The hyperthermal acid hydrolysis was employed for the manufacture of biofuels such as butanol, acetone, and ethanol from seaweed wastes collected from Gwangalli Beach, Korea using single and sequential fermentation of *Clostridium acetobutylicum* and *Clostridium tyrobutyricum* (Sunwoo et al. 2018). Hydrothermal processing such as hydrothermal liquefaction and hydrothermal gasification is employed for biofuels and bio-oil production from the seaweeds (Smith and Ross 2016). Biodiesel can be produced by pre-treating algal biomass by solvent-based extraction and mechanical press (Behera et al. 2015). Cell disruption is carried out by high-pressure homogenizers, bead mills, autoclaving and hydrolysis with hydrochloric acid, sodium hydroxide and sodium carbonate (Falshaw et al. 2003). Hydrocolloids like alginate and ulvan cannot be metabolized by industrial microbes during bioprocessing. This drawback is overcome by biological pre-treatment with the use of macroalgae-specific enzymes such as β-agarases, fucoidanase, laminarinase to hydrolyze macroalgal carbohydrates (Wegeberg and Selby 2019; Jang et al. 2012).

13.6.3 Thermochemical Conversion of Biomass

Direct combustion is a traditional method of generating heat or steam for domestic and industrial applications or electricity from dry biomass. However, direct combustion is not a preferred method for the production of energy using seaweed as the process results in high ash and sulphur content leading to undesirable emissions, fouling and corrosion of boiler (Milledge and Harvey 2018). Pyrolysis is a thermolytic method carried out in the absence of air for the conversion of biomass to fuels through thermal decay of the organic constituents present in dry biomass. Pyrolysis is categorized as slow, fast and flash on the basis of temperature and processing time (Milledge et al. 2014). Pyrolysis products included pyrolysis oil, char and non-condensable gas (Yanik et al. 2013; Wang et al. 2007). Gasification is the transformation of organic material chiefly into syngas by incomplete oxidation at high temperatures (800–1000 °C). Syngas is utilized to deliver heat or transformed into electricity (Rauch et al. 2014). In the presence of a catalyst and hydrogen, the wet biomass is transformed to bio-oil at high pressure and low temperature through a process known as hydro-liquefaction or hydrothermal liquefaction (Milledge et al. 2014).

13.6.4 Microbial Fermentation

Biochemical methods for the conversion of seaweed biomass are the most promising systems for processing of seaweed into several biofuels and biochemicals. Upon saccharification of biomass, seaweed carbohydrates are hydrolyzed and subjected to fermentation by aerobic or anaerobic microorganisms to produce a cascade of alcohols, hydrogen and biogas.

13.6.5 Extraction Methods

After suitable bioprocessing of seaweeds for energy, the leftover residues can be processed for extraction of proteins, carbohydrates, minerals and other metabolites. Different extraction methods such as solvent-mediated (Prasad and Sharma 2019), supercritical fluid extraction (Gallego et al. 2019), pressurized liquid extraction (Saravana et al. 2016) were carried out to extract the bioactive compounds from the seaweed wastes which is followed by purification by chromatography (Batool and Menaa 2020).

13.7 Role of Enzymes in Seaweed Bioprocessing

As seaweeds constitute complex polysaccharides in their structure, these polysaccharides are to be hydrolyzed into monosaccharides or simple sugars to facilitate the bioconversion or fermentation by different microorganisms towards the production of different biochemicals. Seaweeds constitute diverse carbohydrates such as ulvan, agar, carrageenan, alginate, mannitol, laminarin, fucoidan apart from starch and cellulose (Garcia-Ruiz et al. 2016). Depolymerization of the above carbohydrates requires either microbes that can utilize these seaweed polysaccharides or carbohydrate-degrading enzymes to employ seaweeds as effective biomass for successful bioprocessing. Enzymes such as alginate lyase, laminarase, cellulase, agarase, fucoidanase and carrageenanase, are used to split complex polysaccharides present in seaweed into monomers and are further employed in anaerobic or aerobic microbial fermentation processes to produce organic acids or bio-alcohols (Manns et al. 2016; Goh 2010). Alginate is hydrolysed into unsaturated oligosaccharides using alginate lyase through β-elimination (Wong et al. 2000). Marine algae, marine molluscs, marine, and terrestrial microorganisms are reported to produce alginate lyases (Kim et al. 2011). Fucoidanase cleaves the 1, 4 glycosidic bonds between fucose residues by endolytic action (Kusaykin et al. 2016). Descamps et al. (2006) isolated a bacteria with fucoidanolytic activity belonging to the *Flavobacteriaceae* family that was capable of degrading fucoidans from various brown algae. Laminarinase degrades the β-1, 3- and β-1, 6-glycosidic bond of laminarin (Linton et al. 2020). Laminarinase produced by *Flavobacterium sp.* UMI-01 degrades β-1, 3- glucosyl linkages of laminarin by hydrolytic activity (Qin et al. 2017). Agarase degrades agar by hydrolytic action. Based on the cleaving mechanism, agarase can be classified into alpha and beta agarase (Araki 1937). Alpha-agarase produced by *Alteromonas sp.* degrades agar into agarotetraose and agarotriose residues (Potin et al. 1993). A novel beta-agarase produced by *Alteromonas sp.* E-1 degrades agar into neoagarobiose (Kirimura et al. 1999). Carrageenase was purified from *Pseudomonas carrageenovora* (Mclean et al. 1979). Usage of alginate lyase in the hydrolysis of *Undaria pinnatifida* resulted in the extraction of fucoxanthin and lipids easier (Billakanti et al. 2013). Aqueous pretreated green seaweed *Ulva lactuca* following enzymatic hydrolysis has yielded 225 g protein per kg of dry matter and hydrolysate containing 38.8 g/L sugars. Biofuels such as acetone, butanol, ethanol, and 1, 2-propanediol were produced using this hydrolysate (Bikker et al. 2016).

13.8 Energy from Waste Seaweed Biomass Through Thermochemical Conversion

Energy can be obtained from waste seaweed biomass by direct combustion or thermochemical conversion methods (Demirbaş et al. 2001). Thermochemical conversion of seaweeds (34 MJ/kg) provides higher energy yield when compared to biochemical conversions such as fermentation (7.9 MJ/kg) and anaerobic digestion (8.3 MJ/kg) (Shushpanova and Kapralova 2021). Traditionally seaweed biomass was incinerated to yield soda ash, potash and iodine (Rowbotham et al. 2012). Hydrothermal carbonization was used to produce hydrochar (Patel et al. 2021). *Gracilaria lemaneiformis* waste was processed into levulinic acid and hydrochar with 45–55% energy yield after microwave-assisted low-temperature hydrothermal treatment with mild acid (Cao et al. 2019). The seaweed harvested from polar regions has high potential for biochar production (Kebelmann et al. 2013). Pyrolysis was a promising method for producing bio-oil at a higher yield (Wang et al. 2020). Kositkanawuth et al. (2017) proposed co-pyrolysis of *Sargassum* macroalgae and polystyrene as a prospective energy source that provided a solution for the *Sargassum* disposal problem prevailing in beaches worldwide and recycling plastics. Direct combustion may be necessary before anaerobic digestion for enhancing the output of methane from *U. lactuca* (Nikolaison et al. 2012). Kwon et al. (2012) investigated the influence of CO_2 as co-feed for the pyrolysis or gasification of different macroalgae. The fast pyrolysis method was exploited for producing bio-oil from *S. japonica* (Ly et al. 2015). Biochar derived from kelp is examined for direct carbon solid oxide fuel cells (Wu et al. 2021). *L. digitata* is a favourable feedstock for the generation of bioenergy by thermochemical conversion methods (Adams et al. 2011). Agar–agar industry waste was valorized using conventional and microwave pyrolysis for biofuel production (Ferrera-Lorenzo et al. 2014).

13.9 Production of Biofuels and Biochemicals Using Seaweeds

The richness of carbohydrates that are usually converted into simple sugars through pre-treatment and saccharification process and absence of lignin marks seaweed, a promising feedstock for bioconversion to several biofuels and biochemicals using microbial fermentation.

13.9.1 Bioethanol

Production of ethanol using macroalgae is considered as standard liquid biofuel as studied by several researchers. Though brown, green and red seaweeds can be converted to ethanol through fermentation, brown algae serve as a better substrate due to their maximum content of sugars (Milledge et al. 2014). Tan et al. (2014) optimized bioethanol production using seaweed wastes acquired after the isolation of κ-carrageenan. Floating residue, an excess by-product obtained after the extraction of alginate from *L. japonica* is a potential source for bioethanol production (Ge et al. 2011). Li et al. (2016) examined the practical viability of using hydrogen peroxide pre-treated *U. prolifera* residues acquired after extraction of polysaccharides for the production of bioethanol. The remaining pulp from *Gracilaria verrucosa* after extraction of agar was enzymatically hydrolysed to produce maximum bioethanol production (Shukla et al. 2016). Ethanol concentration of 25.7 mg/mL is achieved with 15% (w/v) autoclave-treated *Gelidium amansii* after 24 h (Kim et al. 2015). Waste seaweed biomass from carrageenan industry with peracetic acid-ionic liquid pre-treatment was reported to be an excellent feedstock for bioethanol production (Wijayanta et al. 2015). *Undaria pinnatifida* is suitable alternate biomass for bioethanol synthesis due to its rapid growth and great productivity. Thermal acid-hydrolyzed *U. pinnatifida* biomass was subjected to bioethanol fermentation using yeast acclimatized to high-salt resulting in 9.42 g/L ethanol (Cho et al. 2013a). Sunwoo et al. (2017) assessed ethanol production using red, brown, and green seaweed waste mixture collected from Gwangalli Beach, Korea. *U. rigida* was studied as a resource for the production of bioethanol in simultaneous saccharification and fermentation supported with sonication pre-treatment (Korzen et al. 2015). *A. nodosum* and *L. digitata* biomass were hydrolyzed with dilute sulphuric acid and enzymes to release glucose and rhamnose. *Scheffersomyces stipitis* and *Kluyveromyces marxianus* were used to ferment the sugars, resulting in a bioethanol concentration of 6 g/L (Obata et al. 2016). *Saccharina latissima* was pre-treated to enhance the hydrolysis of laminarin by altering pH 2–6 using 2 M HCl and 5 M NaOH followed by heat treatment at 65 °C for an hour. Bioethanol yield of 0.45% (v/v) was obtained (Adams et al. 2009).

Red seaweed, *Laurencia obtusa* and brown seaweeds, *Cystoseira compressa, Colpomenia sinuosa* were hydrolysed with acid and alkali at a proportion of 1:10 (w/v) to release sugars such as ribose, galactose and arabinose towards conversion into ethanol using *Saccharomyces cerevisiae* (Hamouda et al. 2015). *Microbulbifer elongatus* HZ11 has the potential to degrade alginate from brown seaweed, *L. japonica* into single-cell detritus particles and are utilized for bioethanol production (Sun et al. 2014). Pre-treated green seaweed, *Ulva fasciata* was subjected to saccharification by cellulase produced by marine fungus, *Cladosporium sphaerospermum,* leading to bioethanol production (Trivedi et al. 2015). Brown seaweed, *A. nodosum,* was directed to saccharification by microwave-assisted acid hydrolysis to release monosaccharides for ethanol production leading to the accumulation of 5.57 g/L bioethanol (Yuan et al. 2015). Brown seaweed, *L. digitata,* rich in glucose was used as the feedstock for a unified production of bioethanol and protein (Hou et al. 2015).

Green seaweed, *Enteromorpha intestinalis,* biomass was hydrolyzed using acid and enzymes and trailed by fermentation using *Saccharomyces cerevisiae* KCTC1126 yielding 8.6 g/L ethanol (Cho et al. 2013b). Red seaweed, *Gracilaria verrucosa,* was pretreated by hyperthermal acid hydrolysis and enzymatic saccharification was supported out through the addition of cellulase and hemicellulase. *S. cerevisiae* was used for bioconversion to bioethanol with a yield of 24.8 g/L (Sukwong et al. 2020). Red seaweed, *Gelidium elegans,* was pre-treated by acid hydrolysis using 2.5% H_2SO_4 at 120 °C for 40 min to release galactose and glucose. *S. cerevisiae* NBRC10217 was cast off for fermentation to obtain an ethanol concentration of 13.27 g/L (Hessami et al. 2019). Hakim et al. (2017) worked on simultaneous saccharification and fermentation of seaweed waste collected from the agar processing industry in Indonesia by *Trichoderma reesei* and *Saccharomyces cerevisiae* respectively for 72 h for bioethanol production.

13.9.2 Biobutanol

n-butanol, acetone and ethanol are produced by acetone–butanol–ethanol (ABE) fermentation of carbohydrates such as starch and glucose (Green 2011). Green seaweed, *Ulva lactuca* hydrolysed using hot-water treatment and cellulase was employed for production of acetone, butanol and ethanol using *Clostridium acetobutylicum* and *Clostridium beijerinckii* (van der Wal et al. 2013). The production of butanol was carried using seaweed wastes recovered from Gwangalli Beach, Korea (Sunwoo et al. 2017). Butanol fermentation was carried out using pre-treated brown seaweed, *Sargassum wightii* with *Clostridium saccharoperbutylacetonicum* resulting in ~15 g/L butanol (Dubey et al. 2015). A high butanol yield of 0.42 g/g was attained by fermentation of *L. digitata* from the Danish North Sea coast as substrate after enzymatic hydrolysis using *Clostridium beijerinckii* DSM-6422 (Hou et al. 2017). Production of butanol attempted with Jamaica bay macroalgae using *Clostridium beijerinckii* and *C. saccharoperbutylacetonicum* resulted in the accumulation of butanol in the fermentation broth to 4 g/L (Potts et al. 2012). Kelp *Saccharina sp.* employed for acetone-butanol fermentation by *C. acetobutylicum* ATCC 824 yielded 0.12 g/g butanol (Huesemann et al. 2012).

13.9.3 Biogas

Seaweeds can also be used predominantly as significant feedstock for methane or biogas production through anaerobic digestion (Milledge et al. 2019). The transformation of macroalgal biomass to biogas through anaerobic digestion is more suitable due to its low lignin and cellulose content, presence of natural sugars and other carbohydrates, lower heating value and high moisture content (Murphy et al. 2013). Edyvean et al. (1988) investigated methane production using waste seaweed residues left after alginate extraction as a substrate resulting in 376 L/kg volatile solids biogas with 63% methane. Residues of *Laminaria* and *Fucus sp.* were used for biogas generation with methane production of 187–195 mL gVS^{-1} (Tedesco et al. 2017). *Sargassum* and food waste were co-digested at a ratio of 25:75 resulting in approximately 292.18 mL gVS^{-1} yield of methane (Thompson et al. 2021). *Saccorhiza polyschides, Ulva sp., Laminaria digitata and Saccharina latissima,* were co-digested independently with bovine slurry ensuing maximum methane yield 255 mL gVS^{-1}, 191 mL gVS^{-1}, 246 mL gVS^{-1} and 335 mL gVS^{-1} respectively (Vanegas et al. 2013a). Vivekanand et al. (2012) investigated co-digestion of *S. latissima* with wheat straw for methane production. Vanegas et al. (2013b) studied the influence of temperature and composition for anaerobic digestion of *L. digitata.* Green seaweed, *U. lactuca* was reported as viable and promising biomass for producing biogas with 49.9% of methane (Sitompul et al. 2012). Nkemka et al. (2010) examined production of biogas using seaweed coupled with heavy metals removal so that the resulting biomass can be effectively used as a biofertilizer. *Sargassum tenerrimum* is anaerobically digested with a mixed culture of methanogenic and algin-degrading bacteria and used for the production of biogas (Anjaneyulu et al. 1989). Irfan et al. (2019) reported the potential of *Sargassum crassifolium* for biogas production.

13.9.4 Biodiesel

Biodiesel is produced by transesterification of oils with methanol in the presence of a catalyst resulting in a mixture of fatty acid alkyl esters (Boro et al. 2011). Vegetable oil as feedstock for biodiesel imposes high production costs and requires large areas for the cultivation of crops and hence lipids extracted from seaweeds can be a better alternative feedstock for biodiesel production (Maceiras et al. 2016). *Ulva lactuca, Padina boryana* and *Ulva intestinalis* have been shown to have substantial fatty acid methyl esters and lipids and hence exploited as promising macroalgal feedstock for biodiesel production (Abomohra et al. 2018). Macroalgae species such as *Gracilaria corticata, Chaetomorpha antennina, Chladophora vagabunda* were compared for biodiesel yields and *Chladophora vagabunda* yielded more biodiesel when compared to other algal species (Sharmila et al. 2012). Green macroalgae, *Caulerpa taxifolia, Chaetomorpha antennina, Chaetomorpha linum, Ulva fasciata,* and *Ulva flexuosa* gathered from the coastline areas of Ghana were examined for their role in biodiesel production (Ameka et al. 2019). Algae oil extracted macroalgae species namely *Gracilaria sp., Spirogyra sp. and Bryopsis pennata* are utilized for biodiesel production by transesterification process (Ahmed et al. 2012). Biodiesel production process using macroalgae, *Chara vulgaris,* were optimized using Box–Behnken design (Siddiqua et al. 2015). Oil from marine macroalgae, *Caulerpa racemosa,* has been chosen as a feedstock for biodiesel generation (Balu et al. 2020). Production of biodiesel was carried out using oil extracted from *Padina boergesenii* (Nageswara Rao et al. 2018). *Ulva lactuca* has been examined as potential renewable biomass for biodiesel production (Bruhn et al. 2011; Abd El Baky et al. 2016).

13.9.5 Biohydrogen

Macroalgae is also a competent biomass for the production of biohydrogen. Biohydrogen (H_2) is a viable and clean fuel produced by the dark fermentation of seaweeds (Kumar et al. 2021). Red algae, *Gelidium amansii,* and the brown algae, *Laminaria japonica,* are probable seaweed biomass resources for biohydrogen generation (Şentürk and Büyükgüngör 2013). Biohydrogen production from *Chaetomorpha antennina* was carried out through pre-treatment with surfactants aided with microwave disintegration (Kumar et al. 2019). Margareta et al. (2020) reported the potential of using green macroalgae *Ulva sp.* as a substrate for H_2 production. Ding et al. (2020) investigated the co-production of biohydrogen and biomethane using pre-treated *Laminaria digitata*. Matsumura et al. (2014) demonstrated hydrogen production with a heterotrophic marine bacterium, *Vibrio tritonius* AM2 using powdered *Saccharina sculpera* as a feedstock, resulting in a yield of 1.6 mol H_2/mol mannitol.

13.9.6 Lactic Acid

Seaweeds also have the potential to be used as substrate for microbial fermentation of lactic acid. At optimal conditions, hydrolysates of *Gracilaria sp.* fermented using lactic acid bacteria provided 19.32 g/L of lactic acid (Lin et al. 2020). Marine single cell detritus, a novel feeding material for shrimp was developed using lactic acid fermentation of *Ulva reticulata* biomass with *Lactobacillus plantarum* (Felix and Pradeepa 2012). Hydrolysed *L. japonica* was used as a carbon source for lactic acid fermentation by *Lactobacillus rhamnosus* with an accumulation of 14.42 g/L lactic acid after 48 h of fermentation (Jang et al. 2011). Hwang et al. (2011) examined the prospect of using seaweed biomass as feedstock for lactic acid production using seven different *Lactobacillus* species. The study concluded that the predicted lactic acid yield for seaweed biomasses is comparable with lignocellulosic biomasses. Hwang et al. (2012) assessed the practicability of utilizing green seaweed, *Enteromorpha prolifera,* as an alternate substrate for biosynthesis of lactic acid and concluded that green seaweed biomass will be economical with lignocellulosic biomass in terms of lactic yield. Jang et al. (2013) used *Gelidium amansii* hydrolysate for lactic acid production using *Lactobacillus rhamnosus.* This process provided a maximum lactic acid yield of 42.03%.

13.9.7 Seaweed as Substrate in Solid-State Fermentation for Enzyme Production

Solid-State Fermentation (SSF) or Solid-State Bioconversion (SSB) enables better production of enzymes and metabolites by consuming less energy (Pandey 2003; Vinigra-Gonzalez et al. 2003). Owing to its advantages such as simplicity, ease of product extraction, high yield, and production of concentrated products (Panesar et al. 2016), SSF can serve as an effective technology for the production of enzymes and a viable alternative for the production of value-added products from seaweed biomass. Rodríguez-Jasso et al. (2013) used algal biomass as substrate for fungal fucoidanase production in a rotating drum bioreactor using SSF. Wang et al. (2016) developed an alternative economical process for reducing alginate content in the *L. japonica* feed through semi-solid fermentation with alginate-degrading *B. amyloliquefaciens* WB1. Pervez et al. (2017) reported utilization of algal biomass such as *Dictyopteris polypodioides, Sargassum wightii, Dictyopteris divaricata, Ulva lactuca* and *Codium tomentosum* for the production of pectinases using SSF. Among all, *U. lactuca* was reported to be the best substrate for pectinase production using *Bacillus licheniformis* KIBGE-IB4 and a higher yield of the enzyme was reported when compared to the usage of conventional agricultural biomass as substrate. Jamal et al. (2017) used solid-state bioconversion of *Ulva* seaweed for protein augmentation using *Phanerochaete chrysosporium*. Fernandes et al. (2019) reported sequential bioprocessing of *Ulva rigida* to synthesize lignocellulolytic enzymes and expand its potential as aquaculture feed in terms of nutritional merit. Production of cellulase production using *Cladosporium sphaerospermum* was achieved by SSF with green macroalgae, *Ulva fasciata* as substrate (Lara et al. 2020). General et al. (2014) assessed the practicality of using *S. japonica* as a raw material for biosynthesis of pigment using *Talaromyces amestolkiae* GT11 in solid-state bioconversion. Table 13.1 summarizes the bioconversion methods, pre-treatment, yield of various products from different seaweeds.

13.9.8 Biofertilizer

In agriculture, seaweeds have been recycled and employed as compost and fertilizers for ages. The advantages of seaweed applications in the agricultural field are incitement of seed germination, improvement in the growth and development of plants, in particular, shoot and root extension, enhanced water and supplement take-up, saline resistance and, biocontrol of phytopathogens (Nabti et al. 2017). Silva et al. (2019) assessed the potential of *Ascophyllum nodosum* and *Sargassum muticum* extracts as fertilizer in agriculture. Nasmia et al. (2021) examined the applicability of organic seaweed-based fertilizer to improve the growth and value of seaweed *Gracilaria verrucosa*. Biofertilizer formulated with seaweed waste improved the bacterial viability in liquid formulations and could increase the development of green bean sprouts (Arfarita et al. 2019). Biofertilizer produced from mixed mackerel and *Undaria* (brown seaweed) wastewater was the promising arrangement for wastewater treatment, and also decreased the usage of chemical fertilizers. In open-flow lettuce hydroponics, it was used as a high-quality biofertilizer (Jung et al. 2020). *Saccorhiza polyschides* extracts have interesting characteristics with all essential nutrients as biofertilizers (Soares et al. 2020). Residual biomass after anaerobic digestion of *Ulva sp.* was evaluated for the growth of *Vigna radiata* (Akila et al. 2019). Hence, the seaweed wastes as such or after bioprocessing steps can be effectively utilized as biofertilizers.

13.9.9 Value-Added Products

Seaweeds are a marvellous reservoir of biologically active phytoconstituents such as pigments, vitamins, minerals, fatty acids, sterols and terpenoids (Dominguez

Table 13.1 Bioconversion methods, pre-treatment, the yield of various biochemical and biofuels using different seaweeds as feedstock

S. No.	Seaweeds	Bioconversion/Processing method	Pre-treatment	Product	Yield	References
1	*Ulva sp.*	Fermentation using *Haloferax mediterranei* ATCC 33500	Subcritical water hydrolysis	Polyhydroxyalkanoate	0.104 g g^{-1}	Ghosh et al. (2021)
2	*Ulva sp.*	Fermentation using *Haloferax mediterranei* ATCC 33500	Subcritical water hydrolysis	Biochar	0.194 ± 1.23 g g^{-1}	Ghosh et al. (2021)
3	*Ulva lactuca*	Acetone-butanol-ethanol fermentation	Hot-water treatment trailed by enzymatic hydrolysis using cellulases	Acetone, butanol, and ethanol	0.35 g ABE/g sugar	van der Wal et al. (2013)
4	*Ascophyllum nodosum*	Fermentation using *Saccharomyces cerevisiae* ATCC no. 200062	Microwave-assisted acid hydrolysis with 0.4 M H_2SO_4	Bioethanol	5.57 g/L	Yuan et al. (2015)
5	*Enteromorpha intestinalis*	Separate hydrolysis and Fermentation and Simultaneous Saccharification and Fermentation	Thermal acid hydrolysis with 75 mM H_2SO_4 followed by hydrolysis using Celluclast and Viscozyme	Bioethanol	8.6 g/L in SHF process and 7.6 g/L in SSF process	Cho et al. (2013a)
6	*Undaria pinnatifida*	Fermentation using *Pichia angophorae* KCTC 17574	Thermal acid hydrolysis using 75 mM H_2SO_4, 121 °C, 60 min	Bioethanol	9.42 g/L	Cho et al. (2013b)

(continued)

Table 13.1 (continued)

S. No.	Seaweeds	Bioconversion/Processing method	Pre-treatment	Product	Yield	References
7	*Gracilaria verrucosa*	Enzymatic saccharification	Hydrolysis using 0.2 M HNO_3, 150 °C, 10 min	Bioethanol	24.8 g/L	Sukwong et al. (2020)
8	*Gelidium elegans*	Fermentation using *Saccharomyces cerevisiae*	Acid hydrolysis 2.5% H_2SO_4, 120 °C, 40 min	Bioethanol	13.27 ± 0.47 g/L	Hessami et al. (2019)
9	*Ulva rigida*	Simultaneous Saccharification and Fermentation	Sonication	Bioethanol	333.3 ± 4.7 mg/ glucose	Korzen et al. (2015)
10	*Ulva prolifera*	Fermentation with *Saccharomyces cerevisiae*	0.2% hydrogen peroxide, 50 °C, pH 4.0, 12 h followed by hydrolysis with commercial cellulase and cellobiase	Bioethanol	13.2 g/100 g dry *Ulva prolifera* residue	Li et al. (2016)
11	*Eucheuma cottonii*	Anaerobic fermentation	Enzymatic hydrolysis using cellulase and fungal β-glucosidase	Bioethanol	5.47 mg/mL	Tan et al. (2014)
12	*Saccharina japonica*	Anaerobic fermentation	Thermal acid hydrolysis with 40 mM H_2SO_4	Bioethanol	7.7 g/L	Jang et al. (2012)
13	*Laminaria digitata*	Anaerobic fermentation	Milling followed by enzymatic hydrolysis using cellulases	Bioethanol	77.7%	Hou et al. (2015)

(continued)

Table 13.1 (continued)

S. No.	Seaweeds	Bioconversion/Processing method	Pre-treatment	Product	Yield	References
14	*Sargassum wightii*	Acetone-butanol-ethanol (ABE) fermentation	2 N H_2SO_4, 120 °C, 30 min	Butanol	~ 15 g/L	Dubey et al. (2015)
15	*Laminaria digitata*	Acetone-butanol-ethanol (ABE) fermentation	Enzymatic hydrolysis using cellulases	Butanol	7.16 g/L	Hou et al. (2017)
16	*Saccharina sp.*	Acetone-butanol fermentation by *Clostridium acetobutylicum*	Enzymatic hydrolysis	Butanol	0.12 g/g sugar	Huesemann et al. (2012)
17	*Ulva lactuca*	Acetone-butanol-ethanol (ABE) fermentation	1% acid hydrolysis at 125 °C for 30 min	Butanol	4 g/L	Potts et al. (2012)
18	*Sargassum sp.*	Anaerobic co-digestion	Hydrothermal pretreatment	Biogas	292.18 ± 8.70 mL/gVS	Thompson et al. (2021)
19	*Laminaria sp.*	Anaerobic digestion	–	Biogas	187–195 mL CH_4 gVS^{-1}	Tedesco et al. (2017)
20	*Fucus sp.*	Anaerobic digestion	–	Biogas	100 mL CH_4 gVS^{-1}	Tedesco et al. (2017)
21	*Ascophyllum sp.*	Anaerobic digestion	–	Biogas	376 L/kg volatile solids at 63% methane	Edyvean et al. (1988)
22	*Saccorhiza polyschides*	Anaerobic digestion	–	Biogas	335 ml gVS^{-1}	Vanegas et al. (2013a)
23	*Saccharina latissima*	Anaerobic digestion	–	Biogas	255 ml gVS^{-1}	Vanegas et al. (2013a)
24	*Laminaria digitata*	Anaerobic digestion	–	Biogas	246 ml gVS^{-1}	Vanegas et al. (2013a)

(continued)

Table 13.1 (continued)

S. No.	Seaweeds	Bioconversion/Processing method	Pre-treatment	Product	Yield	References
25	*Ulva sp.*	Anaerobic digestion	–	Biogas	191 ml gVS^{-1}	Vanegas et al. (2013a)
26	*Saccharina latissima*	Anaerobic digestion and co-digestion with wheat straw	Steam explosion	Biogas	223 to 268 mL g-1 VS	Vivekanand et al. (2012)
27	*Ulva lactuca*	Anaerobic biodegradation	–	Biogas	49.90%	Sitompul et al. (2012)
28	*Caulerpa racemosa*	Transesterification	Ultrasonication	Biodiesel	93%	Balu et al. (2020)
29	*Chara vulgaris*	Transesterification	–	Biodiesel	3.6 ml	Siddiqua et al. (2015)
30	*Padina boergesenii*	Transesterification	–	Biodiesel	18.7 ml/20 ml oil	Nageswara et al. (2018a)
31	*Gelidium amansii*	Dark fermentation	–	Biohydrogen	2.46 L H_2/g VSS/d	Park et al. (2011)
32	*Laminaria japonica*	Dark fermentation	Thermal pretreatment at 170 °C for 20 min	Biohydrogen	109.6 mL H_2/g COD	Jung et al. (2011)
33	*Chaetomorpha antennina*	Dark fermentation	Surfactant aided microwave disintegration	Biohydrogen	74.5 ml/g COD	Kumar et al. (2019)
34	*Laminaria digitata*	Dark fermentation	Hydrothermal pretreatment at 140 °C for 20 min along with 1% H_2SO_4	Biohydrogen	57.4 mL/gVS	Ding et al. (2020)

(continued)

Table 13.1 (continued)

S. No.	Seaweeds	Bioconversion/Processing method	Pre-treatment	Product	Yield	References
35	*Gelidium amansii*	Dark fermentation	Dilute acid treatment with 1% sulfuric acid, 121 °C, 30 min	Biohydrogen	52.8 ± 0.2 mL H_2/g dry biomass	Sivagurunathan et al. (2017)
36	*Gracilaria sp.*	Lactic acid fermentation	Dilute acid pretreatment using 0.4 N HCl for 30 min trailed by enzymatic hydrolysis using cellulase at 37 °C for 48 h	Lactic acid	19.32 g/L	Lin et al. (2020)
37	*Laminaria japonica*	Lactic acid fermentation using *Lactobacillus rhamnosus*	Acid and alkali hydrolysis using 0.5% H_2SO_4 or 1% NH_4OH	Lactic acid	14.42 g/L	Jang et al. (2011)
38	*Enteromorpha prolifera*	Lactic acid fermentation using different species of *Lactobacillus*	Acid hydrolysis with 0.5 M H_2SO_4	Lactic acid	*Lactobacillus salivarius* (68.5%), *Lactobacillus plantarum* (66.0%), *Lactobacillus rhamnosus* (55.8%), *Lactobacillus brevis* (54.5%), and *Lactobacillus casei* (51.4%)	Hwang et al. (2012)
39	*Gelidium amansii*	Lactic acid fermentation using *Lactobacillus rhamnosus*	Hydrolysis using 3% H_2SO_4, 140 °C, 5 min	Lactic acid	42.03%	Jang et al. (2013)
40	*Fucus vesiculosus*	Solid-state fermentation using *Mucor sp.* 3P	Autohydrolysis	Fucoidanase	3.82 U/L	Rodriguez-Jasso et al. (2013)

(continued)

Table 13.1 (continued)

S. No.	Seaweeds	Bioconversion/Processing method	Pre-treatment	Product	Yield	References
41	*Sargassum horneri*	Solid-state fermentation using *Bacillus* M3	–	Alginate lyase	33.74 U/mg	Wang et al. (2016)
42	*Ulva lactuca*	Submerged fermentation using *Bacillus licheniformis* KIBGE-IB4	–	Pectinase	2457 ± 3.31 U/mg	Pervez et al. (2017)
43	*Ulva lactuca*	Solid state fermentation using *Bacillus licheniformis* KIBGE-IB4	–	Pectinase	1432 ± 1.46 U/ mg	Pervez et al. (2017)

2013). The extraction methods have to be holistically designed to recover the highest possible valuable compounds from the seaweed biomass towards achieving zero waste generation (Matos et al. 2021). The nutritional value of wastes of brown seaweed, *Macrocystis pyrifera* was upgraded using the marine fungi *Paradendryphiella salina* to yield mycoproteins that find application as a functional food (Salgado et al. 2021). Seaweeds received a lot of consideration in developing skincare products due to the existence of diverse bioactive chemicals with tyrosinase, collagenase, elastase and hyaluronidase inhibition activities as well as moisture retention and photoprotection activities (Jesumani et al. 2019). Fermentation of *Sargassum sp.* using lactic acid bacteria from the marine environment enhanced the antioxidant and anticoagulation activity of the seaweed thus serving as a novel functional food with therapeutic potential (Shobharani et al. 2013). Rianingsih and Sumardianto (2020) formulated *Sargassum sp.* based fermented drink with the addition of probiotic bacteria such as *Lactobacillus plantarum* FNCC027 and *L. acidophilus* with the extracts of seaweed. Hisaoka (1997) filed a patent for the production of seaweed vinegar through acetic acid fermentation using decomposed raw seaweeds. Sanderson and Wise (1978) have developed an economical process for producing acetic acid using macroalgal biomass. Biomass of brown seaweed, *Saccharina japonica,* was employed as a substrate in a two-stage fermentation system to synthesize the value-added lipids and xanthophylls using *Gluconobacter oxydans* and marine thraustochytrid *Aurantiochytrium sp.* (Arafiles et al. 2014). The nutritional content of *Sargassum binderi* has been improved through fermentation with *Bacillus megaterium* S245 and was used as a feed of laying hens (Dewi et al. 2019). Raw and fermented *Padina tetrastomatica* was demonstrated as a potential feed ingredient for the juveniles of freshwater prawn *Macrobrachium rosenbergii* (Felix and Brindo 2014). Nutritional rich fish feed was developed with *Caulerpa lentillifera, Eucheuma cottonii* and *Sargassum fulvellum* through solid-state fermentation using *Phanerochaete chrysosporium* and *Candida utilis* (Ilias et al. 2015). Fermented *Gracilaria sp.* waste was examined as a feed supplement for improving the productivity of indigenous Indonesian ducks (Santoso et al. 2016). Fermentation of brown seaweed *Eisenia bicyclis* with *Candida utilis* boosted its antibacterial potential against methicillin-resistant *Staphylococcus aureus* and other food-borne pathogens due to intensification of eckol, dieckol, dioxinodehydroeckol, and phlorofucofuroeckol-A (Eom et al. 2013). Fermentation of seaweed using lactic acid bacteria favours the production of γ-aminobutyric acid (Chye et al. 2017).

13.9.9.1 Bioplastics

Seaweeds serve as a resource for the production of eco-friendly bioplastics. *Ulva armoricana* blended with polyvinyl alcohol and starch is used to synthesize bioplastics (Chiellini et al. 2008). Sudhakar et al. (2020) assessed the suitability of red seaweed *Kappaphycus alvarezii* for bioplastic film production. Yusmaniar et al. (2019) prepared *Eucheuma cottonii* based biodegradable plastic in combination with polysaccharides isolated from seeds of tropical fruits. Red algae, *Eucheuma cottonii* is blended with latex of *Artocarpus altilis* and *Calotropis gigantea* for plastic production (Machmud et al. 2013).

13.9.9.2 Recovery of Carbohydrates, Protein and Bioactive Peptides

In addition to several products and biochemicals derived from waste seaweed biomass, these resources can be exploited to extract novel functional proteins (Álvarez-Viñas et al. 2019), bioactive peptides (Admassu et al. 2018a) and carbohydrates for functional food applications. For example, peptides inhibiting α-amylase activity were recovered from proteolytic hydrolysate of *Porphyra sp.* (Admassu et al. 2018b).

13.10 Conclusion and Future Prospects

This chapter discussed seaweed-based bioprocessing in the direction of the synthesis of biofuels and value upgraded biomolecules through waste seaweed biomass. A proper direction for exploiting seaweed biomass as a resource for the biorefinery concept is presented. Though seaweeds are being utilized potentially for the extraction of commercial hydrocolloids, its success concerned with bioprocessing is limited. The main challenge associated with the exploitation of marine macroalgae as feedstock for the processing of biofuels and biochemicals lies in the varied chemical composition respective to the type of species, harvesting season and geographic location. Also, the cost linked with seaweed harvesting together with the seasonal availability of the seaweeds presently makes exploitation of macroalgae too expensive for production of biochemicals and biofuels. Variations in the chemical constituents of the seaweed biomass will affect the growth of microorganisms involved in fermentation leading to a lack of consistency in the quality and yield of the biochemicals. Efforts concerning seaweed cultivation or farming, preservation of seaweeds for continuous feedstock availability and improvement of the biomass conversion process through exploration of novel seaweed polysaccharide degrading enzymes and microorganisms for direct conversion of seaweed carbohydrates into biofuels and biochemicals can revitalize this new stream of biomass for exploitation to its fullest potential in the near future.

References

Abd El Baky HH, El Baroty GS (2016) Potential of macroalgae *Ulva lactuca* as a source feedstock for biodiesel production. Recent Pat Food Nutr Agric 8:199–204. https://doi.org/10.2174/2212798409666170602080725

Abomohra AEF, El-Naggar AH, Baeshen AA (2018) Potential of macroalgae for biodiesel production: screening and evaluation studies. J Biosci Bioeng 125:231–237. https://doi.org/10.1016/j.jbiosc.2017.08.020

Adams JMM, Ross AB, Anastasakis K, Hodgson EM, Gallagher JA, Jones JM, Donnison IS (2011) Seasonal variation in the chemical composition of the bioenergy feedstock *Laminaria digitata* for thermochemical conversion. Bioresour Technol 102:226–234. https://doi.org/10.1016/j.biortech.2010.06.152

Adams JM, Gallagher JA, Donnison IS (2009) Fermentation study on *Saccharina latissima* for bioethanol production considering variable pre-treatments. J Appl Phycol 21:569–574. https://doi.org/10.1007/s10811-008-9384-7

Admassu H, Gasmalla MAA, Yang R, Zhao W (2018a) Bioactive peptides derived from seaweed protein and their health benefits: antihypertensive, antioxidant, and antidiabetic properties. J Food Sci 83:6–16. https://doi.org/10.1111/1750-3841.14011

Admassu H, Gasmalla MA, Yang R, Zhao W (2018b) Identification of bioactive peptides with α-amylase inhibitory potential from enzymatic protein hydrolysates of red seaweed (*Porphyra* spp). J Agric Food Chem 66:4872–4882. https://doi.org/10.1021/acs.jafc.8b00960

Ahmed AS, Khan S, Hamdan S, Rahman R, Kalam A, Masjuki HH, Mahlia TMI (2012) Biodiesel production from macro algae as a green fuel for diesel engine. In: EnCon 2012, 5th engineering conference, "engineering towards change—empowering green solutions" Kuching, Sarawak, pp 393–398

Akila V, Manikandan A, Sukeetha DS, Balakrishnan S, Ayyasamy P M, Rajakumar S (2019) Biogas and biofertilizer production of marine macroalgae: an effective anaerobic digestion of *Ulva sp.* Biocatal Agric Biotechnol 18:101035. https://doi.org/10.1016/j.bcab.2019.101035

Alvarez-Vinas M, Florez-Fernandez N, Torres MD, Dominguez H (2019) Successful approaches for a red seaweed biorefinery. Mar Drugs 17:1–18. https://doi.org/10.3390/md17110620

Ameka GK, Doamekpor LK, Amadu AA, Amamoo AP (2019) Production of biodiesel from marine macroalgae occurring in the Gulf of Guinea, off the Coast of Ghana. Ghana J Agric Sci 60:50–58. https://doi.org/10.4314/gjs.v60i1.5

Anjaneyulu K, Tarwadi SJ, Mehta DJ (1989) Anaerobic digestion of seaweed for biogas: a kinetic evaluation. J Chem Technol Biotechnol 45:5–14. https://doi.org/10.1002/jctb.280450103

Arafiles KHV, Iwasaka H, Eramoto Y, Okamura Y, Tajima T, Matsumura Nakashimada Y, Aki T (2014) Value-added lipid production from brown seaweed biomass by two-stage fermentation using acetic acid bacterium and thraustochytrid. Appl Microbiol Biotechnol 98:9207–9216. https://doi.org/10.1007/s00253-014-5980-4

Araki CH (1937) Acetylation of agar like substance of *Gelidium amansii*. J Chem Soc 58:1338–1350

Arfarita N, Higuchi T, Prayogo C (2019) Effects of seaweed waste on the viability of three bacterial isolates in biological fertilizer liquid formulations to enhance soil aggregation and fertility. J Degrade Min Land Manage 6:1889. https://doi.org/10.15243/jdmlm.2019.064.1889

Balina K, Romagnoli F, Blumberga D (2017) Seaweed biorefinery concept for sustainable use of marine resources. In International scientific conference environmental and climate technologies, CONECT 2017, Energy procedia. Riga, Latvia, pp 504–511. https://doi.org/10.1016/j.egypro.2017.09.067

Balu M, Lingadurai K, Shanmugam P, Raja K, Teja NB, Vijayan V (2020) Biodiesel production from *Caulerpa racemosa* (macroalgae) oil. Indian J Geo Mar Sci 49:616–621

Batool A, Menaa F (2020) Concentration and purification of seaweed components by chromatography methods, In: Torres MD, Kraan S, Dominguez H (eds) Sustainable seaweed technologies, pp 315–370. https://doi.org/10.1016/B978-0-12-817943-7.00013-5

Begum M, Bordoloi BC, Singha DD, Ojha NJP (2018) Role of seaweed extract on growth, yield and quality of some agricultural crops: a review. Agriculture 39:1–6. https://doi.org/10.18805/ag.R-1838

Behera S, Singh R, Arora R, Sharma NK, Shukla M, Kumar S (2015) Scope of algae as third generation biofuels. Front Bioeng Biotechnol 2:90–103. https://doi.org/10.3389/fbioe.2014.00090

Bikker P, van Krimpen MM, van Wikselaar P, Houweling-Tan B, Scaccia N, van Hal JW, López-Contreras AM (2016) Biorefinery of the green seaweed *Ulva lactuca* to produce animal feed, chemicals and biofuels. J Appl Phycol 28:3511–3525. https://doi.org/10.1007/s10811-016-0842-3

Billakanti JM, Catchpole OJ, Fenton TA, Mitchell KA, MacKenzie AD (2013) Enzyme-assisted extraction of fucoxanthin and lipids containing polyunsaturated fatty acids from *Undaria pinnatifida* using dimethyl ether and ethanol. Process Biochem 48:1999–2008. https://doi.org/10.1016/j.procbio.2013.09.015

Biris-Dorhoi ES, Michiu D, Pop CR, Rotar AM, Tofana M, Pop OL, Socaci SA, Farcas AC (2020) Macroalgae—a sustainable source of chemical compounds with biological activities. Nutrients 12:1–23. https://doi.org/10.3390/nu12103085

Boro J, Thakur AJ, Dike D (2011) Solid oxide derived from waste shells of *Turbonilla striatula* as a renewable catalyst for biodiesel production. Fuel Process Technol 92:2061–2067. https://doi.org/10.1016/j.fuproc.2011.06.008

Brennan L, Owende P (2010) Biofuels from microalgae-a review of technologies for production, processing, and extractions of biofuels and co-products. Renew Sustain Energy Rev 14:557–577. https://doi.org/10.1016/j.rser.2009.10.009

Brodie J, Lewis J (2007) Unravelling the algae: the past, present, and future of algal systematics, vol 75. CRC Press, London

Bruhn A, Dahl J, Nielsen HB, Nikolaisen L, Rasmussen MB, Markager S, Olesen B, Arias C, Jensen PD (2011) Bioenergy potential of *Ulva lactuca*: biomass yield, methane production and combustion. Bioresour Technol 102:2595–2604. https://doi.org/10.1016/j.biortech.2010.10.010

Cao L, Iris KM, Cho DW, Wang D, Tsang DC, Zhang S, Ok YS (2019) Microwave-assisted low-temperature hydrothermal treatment of red seaweed (*Gracilaria lemaneiformis*) for production of levulinic acid and algae hydrochar. Bioresour Technol 273:251–258. https://doi.org/10.1016/j.biortech.2018.11.013

Cardozo KH, Guaratini T, Barros MP, Falcão VR, Tonon AP, Lopes NP, Pinto E (2007) Metabolites from algae with economical impact. Comp Biochem Physiol C Toxicol 146:60–78. https://doi.org/10.1016/j.cbpc.2006.05.007

Cherubini F (2010) The biorefinery concept: using biomass instead of oil for producing energy and chemicals. Energy Convers Manag 51:1412–1421. https://doi.org/10.1016/j.enconman.2010.01.015

Chiellini E, Cinelli P, Ilieva VI, Martera M (2008) Biodegradable thermoplastic composites based on polyvinyl alcohol and algae. Biomacromol 9:1007–1013. https://doi.org/10.1021/bm701041e

Cho Y, Kim H, Kim SK (2013a) Bioethanol production from brown seaweed, *Undaria pinnatifida*, using NaCl acclimated yeast. Bioprocess Biosyst Eng 36:713–719. https://doi.org/10.1007/s00449-013-0895-5

Cho Y, Kim MJ, Kim SK (2013b) Ethanol production from seaweed, *Enteromorpha intestinalis*, by separate hydrolysis and fermentation (SHF) and simultaneous saccharification and fermentation (SSF) with *Saccharomyces cerevisiae*. Korean Society Biotechnol Bioeng J 28:366–371. https://doi.org/10.7841/ksbbj.2013.28.6.366

Chopin T, Tacon AGJ (2021) Importance of seaweeds and extractive species in global aquaculture production. Rev Fish Sci Aquac 29:139–148. https://doi.org/10.1080/23308249.2020.1810626

Chye FY, Ooi PW, Ng SY, Sulaiman MR (2017) Fermentation-derived bioactive components from seaweeds: functional properties and potential applications. J Aquat Food Prod Technol 27:144–164. https://doi.org/10.1080/10498850.2017.1412375

Crouch IJ, Van Staden J (1993) Effect of seaweed concentrate from *Ecklonia maxima* (Osbeck) Papenfuss on *Meloidogyne incognita* infestation on tomato. J Appl Phycol 5:37–43. https://doi.org/10.1007/BF02182420

Demirbaş A (2001) Biomass resource facilities and biomass conversion processing for fuels and chemicals. Energy Convers Manag 42:1357–1378. https://doi.org/10.1016/S0196-8904(00)00137-0

Descamps V, Colin S, Lahaye M, Jam M, Richard C, Potin P, Barbeyron T, Yvin JC, Kloareg B (2006) Isolation and culture of a marine bacterium degrading the sulfated fucans from marine brown algae. Mar Biotechnol 8:27–39. https://doi.org/10.1007/s10126-005-5107-0

D'Este M (2017) Algal biomass for bioenergy and bioproducts production in biorefinery concepts. Department of Environmental Engineering, Technical University of Denmark (DTU) Kgs Lyngby, pp 1–58

Dewi YL, Yuniza A, Nuraini SK, Mahata ME (2019) Fermentation of *Sargassum binderi* seaweed for lowering alginate content of feed in Laying Hens. J World's Poult Res 9:147–153. https://doi.org/10.36380/jwpr.2019.18

Ding L, Cheng J, Lin R, Deng C, Zhou J, Murphy JD (2020) Improving biohydrogen and biomethane co-production via two-stage dark fermentation and anaerobic digestion of the pretreated seaweed *Laminaria digitata*. J Clean Prod 251:1–13. https://doi.org/10.1016/j.jclepro.2019.119666

Dominguez H (2013) Algae as a source of biologically active ingredients for the formulation of functional foods and nutraceuticals. In: Functional ingredients from algae for foods and nutraceuticals. Elsevier, Woodhead Publishing, pp 1–19. https://doi.org/10.1533/9780857098689.1

Dubey KK, Dhingra AK, Rana S (2015) Optimisation of process parameters for enhanced biobutanol production from *Sargassum wightii* hydrolysate. Int J Energy Technol Policy 11:303–311. https://doi.org/10.1504/IJETP.2015.073806

Edyvean RGJ, Stanley IM, Stanley SO (1988) Biogas production from seaweed waste following alginate extraction. In: Houghton DR, Smith RN, Eggins HOW (eds) Biodeterioration, vol 7. Springer, Dordrecht, pp 819–824. https://doi.org/10.1007/978-94-009-1363-9_107

El Boukhari M, Barakate M, Bouhia Y, Lyamlouli K (2020a) Trends in seaweed extract based biostimulants: manufacturing process and beneficial effect on soil-plant systems. Plants 9:1–23. https://doi.org/10.3390/plants9030359

El Boukhari ME, Barakate M, Bouhia Y, Lyamlouli K (2020b) Trends in seaweed extract based biostimulants: manufacturing process and beneficial effect on soil-plant systems. Plants 9:359–382. https://doi.org/10.3390/plants9030359

Enquist-Newman M, Faust AME, Bravo DD, Santos CNS, Raisner RM, Hanel A, Sarvabhowman P, Le C, Regitsky DD, Cooper SR, Peereboom L, Clark A, Martinez Y, Goldsmith J, Cho M, Donohoue PD, Luo L, Lamberson B, Tamrakar P, Kim EJ, Villari JL, Gill A, Tripathi SA, Karamchedu P, Paredes CJ, Rajgarhia V, Kotlar HK, Bailey RB, Miller DJ, Ohler NL, Swimmer C, Yoshikuni Y (2013) Efficient ethanol production from brown macroalgae sugars by a synthetic yeast platform. Nature 505:239–243. https://doi.org/10.1038/nature12771

Eom SH, Lee DS, Kang YM, Son KT, Jeon YJ, Kim YM (2013) Application of yeast *Candida utilis* to ferment *Eisenia bicyclis* for enhanced antibacterial effect. Appl Biochem Biotechnol 171:569–582. https://doi.org/10.1007/s12010-013-0288-x

Falkowski PG, Raven JA (1997) Aquatic photosynthesis. Princeton University Press, London

Falshaw R, Bixler HJ, Johndro K (2003) Structure and performance of commercial κ-2 *carrageenan* extracts. Part III. Structure analysis and performance in two dairy applications of extracts from the New Zealand red seaweed Gigartina Atropurpurea. Food Hydrocoll 17:129–139. https://doi.org/10.1016/S0268-005X(02)00045-0

Fan D, Hodges DM, Critchley AT, Prithiviraj BA (2013) Commercial extract of brown macroalga (*Ascophyllum nodosum*) affects yield and the nutritional quality of spinach in vitro. Commun Soil Sci Plant Anal 44:1873–1884. https://doi.org/10.1080/00103624.2013.790404

Felix N, Brindo A (2014) Effects of raw and fermented seaweed, *Padina tetrastomatica* on the growth and food conversion of giant freshwater prawn *Macrobrachium rosenbergii*. Int J Fish Aquat Stud 1:108–113

Felix S, Pradeepa P (2012) Lactic acid fermentation of seaweed (*Ulva reticulata*) for preparing marine single cell detritus (MSCD), Tamilnadu. J Vet Anim Sci 8:76–81

Fernandes H, Salgado JM, Martins N, Peres H, Oliva-Teles A, Belo I (2019) Sequential bioprocessing of *Ulva rigida* to produce lignocellulolytic enzymes and to improve its nutritional value as aquaculture feed. Bioresour Technol 281:277–285. https://doi.org/10.1016/j.biortech.2019.02.068

Ferrera-Lorenzo N, Fuente E, Bermúdez JM, Suárez-Ruiz I, Ruiz B (2014) Conventional and microwave pyrolysis of a macroalgae waste from the agar-agar industry prospects for bio-fuel production. Bioresour Technol 151:199–206. https://doi.org/10.1016/j.biortech.2013.10.047

Friess DA, Aung TT, Huxham M Lovelock C, Mukherjee N, Sasmito S (2019) SDG 14: Life below water—impacts on mangroves. In: Katila P, Colfer CJP, Jong WD, Pacheco P, Winkel G (eds) Sustainable development goals: their impacts on forests and people. Cambridge University Press, pp 445–481. https://doi.org/10.1017/9781108765015.016

Gallego R, Bueno M, Herrero M (2019) Sub-and supercritical fluid extraction of bioactive compounds from plants, food-by-products, seaweeds and microalgae—an update. Trends Anal Chem 116:198–213. https://doi.org/10.1016/j.trac.2019.04.030

Garcia-Ruiz E, Badur A, Rao CV, Zhao H (2016) Characterization and engineering of seaweed degrading enzymes for biofuels and biochemicals production. In: CK Lau P (eds) Quality living through chemurgy and green chemistry. Green chemistry and sustainable technology. Springer, Heidelberg, pp 99–128. https://doi.org/10.1007/978-3-662-53704-6_5

Ge L, Wang P, Mou H (2011) Study on saccharification techniques of seaweed wastes for the transformation of ethanol. Renew Energy 36:84–89. https://doi.org/10.1016/j.renene.2010.06.001

General T, Kim HJ, Prasad B, Ngo HTA, Vadakedath N, Cho MG (2014) Fungal utilization of a known and safe macroalga for pigment production using solid-state fermentation. J Appl Phycol 26:1547–1555. https://doi.org/10.1007/s10811-013-0168-3

Ghadiryanfar M, Rosentrater KA, Keyhani A, Omid M (2016) A review of macroalgae production, with potential applications in biofuels and bioenergy. Renew Sust Energ Rev 54:473–481. https://doi.org/10.1016/j.rser.2015.10.022

Ghosh S, Greiserman S, Chemodanov A, Slegers PM, Belgorodsky B, Epstein M, Golberg A (2021) Polyhydroxyalkanoates and biochar from green macroalgal *Ulva sp.* biomass subcritical hydrolysates: Process optimization and a priori economic and greenhouse emissions break-even analysis. Sci Total Environ 770:145281. https://doi.org/10.1016/j.scitotenv.2021.145281

Goh CS, Lee KT (2010) A visionary and conceptual macroalgae-based third-generation bioethanol (TGB) biorefinery in Sabah, Malaysia as an underlay for renewable and sustainable development. Renew Sustain Energy Rev 14:842–848. https://doi.org/10.1016/j.rser.2009.10.001

Green EM (2011) Fermentative production of butanol-the industrial perspective. Curr Opin Biotechnol 22:337–343. https://doi.org/10.1016/j.copbio.2011.02.004

Gupta S, Abu-Ghannam N (2011) Recent developments in the application of seaweeds or seaweed extracts as a means for enhancing the safety and quality attributes of foods. Innov Food Sci Emerg Technol 12:600–609. https://doi.org/10.1016/j.ifset.2011.07.004

Hakim A, Chasanah E, Uju U, Santoso J (2017) Bioethanol production from seaweed processing waste by simultaneous saccharification and fermentation (SSF). Squalen Bull Mar Fish Postharvest Biotechnol 12:41–47

Hamouda RA, Hussein MH, El-Naggar NEA (2015) Potential value of red and brown seaweed for sustainable bioethanol production, Bangladesh. J Bot 44:565–570. https://doi.org/10.3329/bjb.v44i4.38571

Hessami MJ, Cheng SF, Ambati RR, Yin YH, Phang SM (2019) Bioethanol production from agarophyte red seaweed, *Gelidium elegans*, using a novel sample preparation method for analysing bioethanol content by gas chromatography. 3 Biotech 9:25. https://doi.org/10.1007/s13205-018-1549-8

Hisaoka M Production of seaweed vinegar, Japanese Patent: JPH10304866A

Hou X, Hansen JH, Bjerre AB (2015) Integrated bioethanol and protein production from brown seaweed *Laminaria digitata*. Bioresour Technol 197:310–317. https://doi.org/10.1016/j.biortech.2015.08.091

Hou X, From N, Angelidaki I, Huijge WJ, Bjerre AB (2017) Butanol fermentation of the brown seaweed *Laminaria digitata* by *Clostridium beijerinckii* DSM-6422. Bioresour Technol 238:16–21. https://doi.org/10.1016/j.biortech.2017.04.035

Huesemann MH, Kuo LJ, Urquhart L, Gill GA, Roesijadi G (2012) Acetone-butanol fermentation of marine macroalgae. Bioresour Technol 108:305–309. https://doi.org/10.1016/j.biortech.2011.12.148

Hussain HI, Kasinadhuni N, Arioli T (2021) The effect of seaweed extract on tomato plant growth, productivity and soil. J Appl Phycol 33:1305–1314. https://doi.org/10.1007/s10811-021-02387-2

Hwang HJ, Lee SY, Kim SM, Lee SB (2011) Fermentation of seaweed sugars by *Lactobacillus* species and the potential of seaweed as a biomass feedstock. Biotechnol Bioprocess Eng 16:1231–1239. https://doi.org/10.1007/s12257-011-0278-1

Hwang HJ, Kim SM, Chang JH, Lee SB (2012) Lactic acid production from seaweed hydrolysate of *Enteromorpha prolifera* (Chlorophyta). J Appl Phycol 24:935–940. https://doi.org/10.1007/s10811-011-9714-z

Ilias NN, Jamal P, Jaswir I, Sulaiman S, Zainuddin Z, Azmi AS (2015) Potentiality of selected seaweed for the production of nutritious fish feed using solid state fermentation. In J Eng Sci Technol 30–40

Irfan M, Wahab IH, Subur R, Akbar N (2019) Seaweed *Sargassum* sp. as material for biogas production. Aquac Aquar Conserv Legis 12:2015–2019

Jamal P, Olorunnisola KS, Jaswir I, Tijani IDR, Ansari AH (2017) Bioprocessing of seaweed into protein enriched feedstock: process optimization and validation in reactor. Int Food Res J 24:382–386

Jang SS, Shirai Y, Uchida M, Wakisaka M (2011) Production of L (+)-lactic acid from mixed acid and alkali hydrolysate of brown seaweed. J Food Sci Technol 17:155–160. https://doi.org/10.3136/fstr.17.155

Jang JS, Cho Y, Jeong GT, Kim SK (2012) Optimization of saccharification and ethanol production by simultaneous saccharification and fermentation (SSF) from seaweed, *Saccharina japonica*. Bioprocess Biosyst Eng 35:11–18. https://doi.org/10.1007/s00449-011-0611-2

Jang SS, Shirai Y, Uchida M, Wakisaka M (2013) Potential use of *Gelidium amansii* acid hydrolysate for lactic acid production by *Lactobacillus rhamnosus*. Food Technol. Biotechnol 51:131–136

Jesumani V, Du H, Aslam M, Pei P, Huang N (2019) Potential use of seaweed bioactive compounds in skincare—a review. Mar Drugs 17:1–19. https://doi.org/10.3390/md17120688

Jung KW, Kim DH, Shin HS (2011) Fermentative hydrogen production from *Laminaria japonica* and optimization of thermal pretreatment conditions. Bioresour Technol 102:745–2750. https://doi.org/10.1016/j.biortech.2010.11.042

Jung HY, Kim JK (2020) Complete reutilisation of mixed mackerel and brown seaweed wastewater as a high-quality biofertiliser in open-flow lettuce hydroponics. J Clean Prod 247:119081. https://doi.org/10.1016/j.jclepro.2019.119081

Kebelmann K, Hornung A, Karsten U, Griffiths G (2013) Thermo-chemical behaviour and chemical product formation from Polar seaweeds during intermediate pyrolysis. J Anal Appl Pyrolysis 104:131–138. https://doi.org/10.1016/j.jaap.2013.08.012

Khan W, Rayirath UP, Subramanian S, Jithesh MN, Rayorath P, Mark Hodges D, Critchley AT, Craigie JS, Norrie J, Prithiviraj B (2009) Seaweed extracts as biostimulants of plant growth and development. J Plant Growth Regul 28:386–399. https://doi.org/10.1007/s00344-009-9103-x

Kim HS, Lee CG, Lee EY (2011) Alginate lyase: structure, property, and application. Biotechnol Bioprocess Eng 16:843–851. https://doi.org/10.1007/s12257-011-0352-8

Kim HM, Wi SG, S, Song Y, Bae HJ (2015) Efficient approach for bioethanol production from red seaweed *Gelidium amansii*. Bioresour Technol 175:128–134. https://doi.org/10.1016/j.biortech.2014.10.050

Kirimura K, Masuda N, Iwasaki Y, Nakagawa H, Kobayashi R, Usami S (1999) Purification and characterization of a novel β-agarase from an alkalophilic bacterium, *Alteromonas sp.* E-1. J Biosci Bioeng 87:436–441. https://doi.org/10.1016/S1389-1723(99)80091-7

Korzen L, Pulidindi IN, Israel A, Abelson A, Gedanken A (2015) Single step production of bioethanol from the seaweed *Ulva rigida* using sonication. RSC Adv 5:16223–16229. https://doi.org/10.1039/C4RA14880K

Kositkanawuth K, Bhatt A, Sattler M, Dennis B (2017) Renewable energy from waste: investigation of co-pyrolysis between *Sargassum* macroalgae and polystyrene. Energy Fuels 31:5088–5096. https://doi.org/10.1021/acs.energyfuels.6b03397

Kumar D, Eswari AP, Park JH, Adishkumar S, Banu JR (2019) Biohydrogen generation from macroalgal biomass, *Chaetomorpha antennina* through surfactant aided microwave disintegration. Front Energy Res 7:1–11. https://doi.org/10.3389/fenrg.2019.00078

Kumar MD, Kavitha S, Tyagi VK, Rajkumar M, Bhatia SK, Kumar G, Banu JR (2021) Macroalgae-derived biohydrogen production: biorefinery and circular bioeconomy. Biomass Convers Biorefin 1:23. https://doi.org/10.1007/s13399-020-01187-x

Kusaykin MI, Silchenko AS, Zakharenko AM, Zvyagintseva TN (2016) Fucoidanases. Glycobiology 26:3–12. https://doi.org/10.1093/glycob/cwv072

Kwon EE, Jeon YJ, Yi H (2012) New candidate for biofuel feedstock beyond terrestrial biomass for thermo-chemical process (pyrolysis/gasification) enhanced by carbon dioxide (CO_2). Bioresour Technol 123:673–677. https://doi.org/10.1016/j.biortech.2012.07.035

Lange L, Bak UG, Hansen SC-B, Gregersen O, Harmsen P, Karlsson EN, Meyer A, Mikkelsen MD, Broek LVD, Hreggviðsson GO (2020) Opportunities for seaweed biorefinery, In: Torres MD, Kraan S, Dominguez H (eds) Advances in green chemistry. Elsevier, pp 3–31. https://doi.org/10.1016/B978-0-12-817943-7.00001-9

Lara A, Rodriguez-Jasso, RM, Loredo-Trevino A, Aguilar CN, Meyer AS, Ruiz HA (2020) Enzymes in the third generation biorefinery for macroalgae biomass. In: Singh, SP, Pandey A, Singhania RR, Larroche C, Li Z (eds) Biomass, biofuels, biochemicals. Elsevier, London, pp 363–396. https://doi.org/10.1016/B978-0-12-819820-9.00017-X

Lee JY, Kim YS, Um BH, Oh K (2013) Pretreatment of *Laminaria japonica* for bioethanol production with extremely low acid concentration. Renew Energy 54:196–200. https://doi.org/10.1016/j.renene.2012.08.025

Li Y, Cui J, Zhang G, Liu Z, Guan H, Hwang H, Wang P (2016) Optimization study on the hydrogen peroxide pretreatment and production of bioethanol from seaweed *Ulva prolifera* biomass. Bioresour Technol 214:44–149. https://doi.org/10.1016/j.biortech.2016.04.090

Lin HTV, Huang MY, Kao TY, Lu WJ, Lin HJ, Pan CL (2020) Production of lactic acid from seaweed hydrolysates via lactic acid bacteria fermentation. Ferment 6:1–11. https://doi.org/10.3390/fermentation6010037

Linton SM (2020) The structure and function of cellulase (endo-β-1, 4-glucanase) and hemicellulase (β-1, 3-glucanase and endo-β-1, 4-mannase) enzymes in invertebrates that consume materials ranging from microbes, algae to leaf litter. Comp Biochem Physiol B Biochem Mol Biol, 240:110354. https://doi.org/10.1016/j.cbpb.2019.110354

Lobban CS, Wynne MJ (1981) The biology of seaweeds, vol 17. University of California Press, Oakland

Ly HV, Kim SS, Woo HC, Choi JH, Suh DJ, Kim J (2015) Fast pyrolysis of macroalga *Saccharina japonica* in a bubbling fluidized-bed reactor for bio-oil production. Energy 93:1436–1446. https://doi.org/10.1016/j.energy.2015.10.011

Maceiras R, Cancela A, Sanchez A, Perez L, Alfonsin V (2016) Biofuel and biomass from marine macroalgae waste. Energy Sources Part A Recov Utilization Environ Effects 38:1169–1175. https://doi.org/10.1080/15567036.2013.862584

Machmud MN, Fahmi R, Abdullah R, Kokarkin C (2013) Characteristics of red algae bioplastics/latex blends under tension. Int J Eng Sci 5:81–88. https://doi.org/10.12777/ijse.5.2.81-88

Maneein S, Milledge JJ, Nielsen BV, Harvey PJ (2018) Fermentation a review of seaweed pretreatment methods for enhanced biofuel production by anaerobic digestion or fermentation. Fermentation 4:1–31. https://doi.org/10.3390/fermentation4040100

Manickavasagam S, Bharathi S, Aanand S (2019) Overview of commercial and economic important seaweeds found in Indian coastal waters. Aqua Star 25–28

Manns D, Andersen SK, Saake B, Meyer AS (2016) Brown seaweed processing: enzymatic saccharification of *Laminaria digitata* requires no pre-treatment. J Appl Phycol 28:1287–1294. https://doi.org/10.1007/s10811-015-0663-9

Mantri VA, Kaval MG, Kazi MA (2020) Seaweed biodiversity of India: reviewing current knowledge to identify gaps, challenges, and opportunities. Divers 12:1–22. https://doi.org/10.3390/d12010013

Margareta W, Nagarajan D, Chang JS, Lee DJ (2020) Dark fermentative hydrogen production using macroalgae (*Ulva sp.*) as the renewable feedstock. Appl Energy 262:114574. https://doi.org/10.1016/j.apenergy.2020.114574

Martin WF, Allen JF (2018) An algal greening of land. Cell 174:256–258. https://doi.org/10.1016/j.cell.2018.06.034

Masarin F, Cedeno FRP, Chavez EGS, De Oliveira LE, Gelli VC, Monti R (2016) Chemical analysis and biorefinery of red algae *Kappaphycus alvarezii* for efficient production of glucose from residue of carrageenan extraction process. Biotechnol Biofuels 9:1–12. https://doi.org/10.1186/s13068-016-0535-9

Matos GS, Pereira SG, Genisheva ZA, Gomes AM, Teixeira JA, Rocha CM (2021) Advances in extraction methods to recover added-value compounds from seaweeds: sustainability and functionality. Foods 10:1–20. https://doi.org/10.3390/foods10030516

Matsumura Y, Sato K, Al-saari N, Nakagawa S, Sawabe T (2014) Enhanced hydrogen production by a newly described heterotrophic marine bacterium, *Vibrio tritonius* strain AM2, using seaweed as the feedstock. Int J Hydrog Energy 39:7270–7277. https://doi.org/10.1016/j.ijhydene.2014.02.164

McHugh DJ (1991) Worldwide distribution of commercial resources of seaweeds including *Gelidium*. In: Juanes JA, Santelices B, McLachlan JL (eds.) International workshop on *Gelidium*. Developments in hydrobiology, vol 68. Springer, Dordrecht, pp 19–29. https://doi.org/10.1007/978-94-011-3610-5_2

McHugh DJ (2002) Prospects for seaweed production in developing countries. Food and Agriculture Organization of the United Nations, Rome

McHugh DJ (2003) A guide to the seaweed industry. FAO fisheries technical paper, vol 441. Food and Agriculture Organization, United Nations. Rome

McLean MW, Williamson FB (1979) kappa-Carrageenase from *Pseudomonas carrageenovora*. Eur J Biochem 93:553–558. https://doi.org/10.1111/j.1432-1033.1979.tb12854.x

Michalak I (2018) Experimental processing of seaweeds for biofuels. Wiley Interdiscipl Rev Energy Environ Sci 7:1–28. https://doi.org/10.1002/wene.288

Milledge JJ, Smith B, Dyer PW, Harvey P (2014) Macroalgae-derived biofuel: a review of methods of energy extraction from seaweed biomass. Energies 7:7194–7222. https://doi.org/10.3390/en7117194

Milledge JJ, Nielsen BV, Maneein S, Harvey PJ (2019) A brief review of anaerobic digestion of algae for bioenergy. Energies 12:1–22. https://doi.org/10.3390/en12061166

Milledge JJ, Harvey P (2018) Anaerobic digestion and gasification of seaweed. In: Rampelotto PH, Trincone A (eds) Grand challenges in marine biotechnology. Springer, Cham, Switzerland, pp 237–258. https://doi.org/10.1007/978-3-319-69075-9_7

Mohamed AY, El-Sehrawy OAM (2013) Effect of seaweed extract on fruiting of *Hindy Bisinnara* mango trees. Am J Sci 9:537–544

Mugnai S, Azzarello E, Pandolfi C, Salamagne S, Briand X, Mancuso S (2008) Enhancement of ammonium and potassium root influxes by the application of marine bioactive substances positively affects *Vitis vinifera* plant growth. J Appl Phycol 20:177–182. https://doi.org/10.1007/s10811-007-9203-6

Murphy F, Devlin G, Deverell R, McDonnell K (2013) Biofuel production in Ireland—an approach to 2020 targets with a focus on algal biomass. Energies 6:6391–6412. https://doi.org/10.3390/en6126391

Nabti E, Jha B, Hartmann A (2017) Impact of seaweeds on agricultural crop production as biofertilizer. Int J Environ Sci Technol 14:1119–1134. https://doi.org/10.1007/s13762-016-1202-1

Nageswara Rao L, Al Riyami SSS (2018) Experimental Investigation on Production of Biodiesel from *Padina boergesenii sp.* Macro Algae. Austin Chem Eng 5:1–7

Nasmia RE, Masyahoro A, Putera FHA, Natsir S (2021) The utilization of seaweed-based liquid organic fertilizer to stimulate *Gracilaria verrucosa* growth and quality. Int J Environ Sci Technol 18:1637–1644. https://doi.org/10.1007/s13762-020-02921-8

Nazari L (2016) Hydrothermal liquefaction of high-water content biomass and waste materials for the production of biogas and bio-crude oil. Electronic thesis and dissertation repository 4069

Nikolaison L, Dahl J, Bech KS, Bruhn A, Rasmussen MB, Bjerre AB, Kadar Z, Heiske S, Sander B, Daugbjerg Jensen P (2012) Energy production from macroalgae. In: Proceedings of the 20th European biomass conference, Milan, Italy

Nkemka VN, Murto M (2010) Evaluation of biogas production from seaweed in batch tests and in UASB reactors combined with the removal of heavy metals. J Environ Manage 91:1573–1579. https://doi.org/10.1016/j.jenvman.2010.03.004

Obata O, Akunna J, Bockhorn H, Walker G (2016) Ethanol production from brown seaweed using non-conventional yeasts. Bioethanol 2:134–145. https://doi.org/10.1515/bioeth-2016-0010

Pandey A (2003) Solid-state fermentation. Biochem Eng J 13:81–84. https://doi.org/10.1016/S1369-703X(02)00121-3

Panesar PS, Kaur R, Singla G, Sangwan RS (2016) Bio-processing of agro-industrial wastes for production of food-grade enzymes: progress and prospects. Appl Food Biotechnol 3:208–227

Park JH, Yoon JJ, Park HD, Kim YJ, Lim DJ, Kim SH (2011) Feasibility of biohydrogen production from *Gelidium amansii*. Int J Hydrog Energy 36:13997–14003. https://doi.org/10.1016/j.ijhydene.2011.04.003

Patel N, Acharya B, Basu P (2021) Hydrothermal carbonization (HTC) of seaweed (Macroalgae) for producing hydrochar. Energies 14:1805–1821. https://doi.org/10.3390/en14071805

Pereira L (2021) Macroalgae. Encyclopedia 1:177–188. https://doi.org/10.3390/encyclopedia1010017

Pervez S, Shahid F, Aman A, Qader SAU (2017) Algal biomass: a sustainable, economical and renewable approach for microbial production of pectinolytic enzymes using submerged and solid state fermentation techniques. Biocatal Biotransform 35:442–449. https://doi.org/10.1080/10242422.2017.1364731

Peteiro C (2018) Alginate production from marine macroalgae, with emphasis on kelp farming, In: Rehm BHA, Moradali F (eds) Alginates and their biomedical applications, Springer, Singapore. https://doi.org/10.1007/978-981-10-6910-9_2

Potin P, Richard C, Rochas C, Kloareg B (1993) Purification and characterization of the α-agarase from *Alteromonas agarlyticus* (Cataldi) comb. nov., strain GJ1B. Eur J Biochem 214:599–607. https://doi.org/10.1111/j.1432-1033.1993.tb17959.x

Potts T, Du J, Paul M, May P, Beitle R, Hestekin J (2012) The production of butanol from Jamaica bay macro algae. Environ Prog Sustain Energy 31:29–36. https://doi.org/10.1002/ep.10606

Prabhu MS, Israel A, Palatnik RR, Zilberman D, Golberg A (2020) Integrated biorefinery process for sustainable fractionation of *Ulva ohnoi* (Chlorophyta): process optimization and revenue analysis. In: 23rd international seaweed symposium on journal of applied phycology. Springer Netherlands, Jeju, pp 2271–2282. https://doi.org/10.1007/s10811-020-02044-0

Prasad K, Sharma M (2019) Green solvents for the dissolution and processing of biopolymers. Curr Opin Green Sustain Chem 18:72–78. https://doi.org/10.1016/j.cogsc.2019.02.005

Qin HM, Miyakawa T, Inoue A, Nakamura A, Nishiyama R, Ojima T, Tanokura M (2017) Laminarinase from *Flavobacterium sp.* reveals the structural basis of thermostability and substrate specificity. Sci Rep 7:1–9. https://doi.org/10.1038/s41598-017-11542-0

Ragauskas AJ, Williams CK, Davison BH, Britovsek G, Cairney J, Eckert CA, Tschaplinski T (2006) The path forward for biofuels and biomaterials. Science 311:484–489. https://doi.org/10.1126/science.1114736

Rathour RK, Mehta S, Sharma P, Bhatia RK, Bhatt AK (2021) Seaweed cultivation and its biobusiness status around the world. In: Pant D, Bhatia SK, Patel AK, Giri A (eds) Bioremediation using weeds. Energy, Environment, and Sustainability. Springer, Singapore, pp 151–182. https://doi.org/10.1007/978-981-33-6552-0_7

Rauch R, Hrbek J, Hofbauer H (2014) Biomass gasification for synthesis gas production and applications of the syngas. Wires Energy Environ 3:343–362. https://doi.org/10.1002/wene.97

Rianingsih L, Sumardianto (2020) Antioxidant activity in seaweed (*Sargassum sp.*) extract fermented with *Lactobacillus plantarum* and *Lactobacillus acidophilus*. In: IOP conference series earth environmental science, vol 530. IOP Publishing. https://doi.org/10.1088/1755-1315/530/1/012011

Rodriguez-Jasso RM, Mussatto SI, Sepulveda L, Agrasar AT, Pastrana L, Aguilar CN, Teixeira JA (2013) Fungal fucoidanase production by solid-state fermentation in a rotating drum bioreactor using algal biomass as substrate. Food Bioprod Process 91:587–594. https://doi.org/10.1016/j.fbp.2013.02.004

Rowbotham JS, Dyer PW, Greenwell HC, Theodorou MK (2012) Thermochemical processing of macroalgae: a late bloomer in the development of third-generation biofuels? Biofuels 3:441–461. https://doi.org/10.4155/bfs.12.29

Sadhukhan J, Gadkari S, Martinez-Hernandez E, Ng KS, Shemfe M, Torres-Garcia E, Lynch J (2019) Novel macroalgae (seaweed) biorefinery systems for integrated chemical, protein, salt, nutrient and mineral extractions and environmental protection by green synthesis and life cycle sustainability assessments. Green Chem 21:2635–2655

Salgado CL, Munoz R, Blanco A, Lienqueo ME (2021) Valorization and upgrading of the nutritional value of seaweed and seaweed waste using the marine fungi *Paradendryphiella salina* to produce mycoprotein. Algal Res 53:102135. https://doi.org/10.1016/j.algal.2020.102135

Sanderson JE, Wise DL (1978) Acetic acid production from marine algae. Technical report, Dynatech R/D Co., Cambridge. https://doi.org/10.2172/6620191

Santoso SI, Suprijatna E, Setiadi A, Susanti S (2016) Effect of duck diet supplemented with fermented seaweed wastes on carcass characteristics and production efficiency of indigenous Indonesian ducks. Indian J Anim Res 50:699–704. https://doi.org/10.18805/ijar.11160

Saravana PS, Cho YJ, Park YB, Woo HC, Chun BS (2016) Structural, antioxidant, and emulsifying activities of *fucoidan* from *Saccharina japonica* using pressurized liquid extraction. Carbohydr Polym 153:518–525. https://doi.org/10.1016/j.carbpol.2016.08.014

Satheesh S, Siddik AA, Ba-Akdah MA, Al-Sofyani AA (2017) An Introduction to the ecological significance of seaweeds on coastal ecosystems. In: Nabti E (ed) Biotechnological applications of seaweeds. Nova Science Publishers, New York, pp 1–14

Senturk I, Buyukgungor H (2013) Evaluation of biohydrogen production potential from marine macro algae. In: Veziroglu A, Tsitskishvili M (eds) Black sea energy resource development and hydrogen energy problems. NATO science for peace and security series C: environmental security. Springer, Dordrecht, pp 117–124. https://doi.org/10.1007/978-94-007-6152-0_11

Shah MT, Zodape ST, Chaudhary DR, Eswaran K, Chikara J (2013) Seaweed sap as an alternative liquid fertilizer for yield and quality improvement of wheat. J Plant Nutr 36:192–200. https://doi.org/10.1080/01904167.2012.737886

Sharmila S, Rebecca LJ, Dhanalakshmi V, Susithra G, Mohanta B, Kar M, Patro R (2012) Production of biodiesel from macroalgae. Int J Biotechnol Bioeng 3:11–16

Shobharani P, Halami PM, Sachindra NM (2013) Potential of marine lactic acid bacteria to ferment *Sargassum sp.* for enhanced anticoagulant and antioxidant properties. J Appl Microbiol 114:96–107. https://doi.org/10.1111/jam.12023

Shukla R, Kumar M, Chakraborty S, Gupta R, Kumar S, Sahoo D, Kuhad RC (2016) Process development for the production of bioethanol from waste algal biomass of *Gracilaria verrucosa*. Bioresour Technol 220:584–589. https://doi.org/10.1016/j.biortech.2016.08.096

Shushpanova DV, Viktorovna K (2021) Life-cycle assessment of kelp in biofuel production. In: International science and technology conference. IOP Conf Ser Mater Sci Eng IOP Publishing, pp 1–8. https://doi.org/10.1088/1757-899X/1079/7/072023

Siddiqua S, Al Mamun A, Babar SME (2015) Production of biodiesel from coastal macroalgae (*Chara vulgaris*) and optimization of process parameters using Box-Behnken design. Springerplus 4:1–11. https://doi.org/10.1186/s40064-015-1518-1

Silva LD, Bahcevandziev K, Pereira L (2019) Production of bio-fertilizer from *Ascophyllum nodosum* and *Sargassum muticum* (Phaeophyceae). J Oceanol Limnol 37:918–927. https://doi.org/10.1007/s00343-019-8109-x

Singh S, Gaikwad KK, Park SI, Lee YS (2017) Microwave-assisted step reduced extraction of seaweed (*Gelidiella aceroso*) cellulose nanocrystals. Int J Biol Macromol 99:506–510. https://doi.org/10.1016/j.ijbiomac.2017.03.004

Sitompul JP, Bayu A, Soerawidjaja TH, Lee HW (2012) Studies of biogas production from green seaweeds. Int J Environ Bioenergy 3:132–144

Sivagurunathan P, Kumar G, Kobayashi T, Xu K, Kim SH (2017) Effects of various dilute acid pretreatments on the biochemical hydrogen production potential of marine macroalgal biomass. Int J Hydrog Energy 42:27600–27606. https://doi.org/10.1016/j.ijhydene.2017.05.106

Smith AM, Ross AB (2016) Production of bio-coal, bio-methane and fertilizer from seaweed via hydrothermal carbonisation. Algal Res 16:1–11. https://doi.org/10.1016/j.algal.2016.02.026

Soares C, Svarc-Gajic J, Oliva-Teles MT, Pinto E, Nastic N, Savic S, Almeida A, Delerue-Matos C (2020) Mineral composition of subcritical water extracts of *Saccorhiza polyschides*, a brown Seaweed used as fertilizer in the North of Portugal. J Mar Sci Eng 8:1–11. https://doi.org/10.3390/jmse8040244

Sudhakar MP, Peter DM, Dharani G (2020) Studies on the development and characterization of bioplastic film from the red seaweed (*Kappaphycus alvarezii*). Environ Sci Pollut Res 28:33899–33913. https://doi.org/10.1007/s11356-020-10010-z

Sukwong P, Sunwoo IY, Jeong DY, Kim SR, Jeong GT, Kim SK (2020) Improvement of bioethanol production by *Saccharomyces cerevisiae* through the deletion of GLK1, MIG1 and MIG2 and overexpression of PGM2 using the red seaweed *Gracilaria verrucosa*. Process Biochem 89:34–145. https://doi.org/10.1016/j.procbio.2019.10.030

Sun C, Chen YJ, Zhang XQ, Pan J, Cheng H, Wu M (2014) Draft genome sequence of *Microbulbifer elongatus* strain HZ11, a brown seaweed-degrading bacterium with potential ability to produce bioethanol from alginate. Mar Genomics 18:83–85. https://doi.org/10.1016/j.margen.2014.05.009

Sunwoo IY, Kwon JE, Nguyen TH, Ra CH, Jeong GT, Kim SK (2017) Bioethanol production using waste seaweed obtained from Gwangalli beach, Busan, Korea by co-culture of yeasts with adaptive evolution. Appl Biochem Biotechnol 183:966–979. https://doi.org/10.1007/s12010-017-2476-6

Sunwoo IY, Hau NT, Ra CH, Jeong GT, Kim SK (2018) Acetone–butanol–ethanol production from waste seaweed collected from Gwangalli beach, Busan, Korea, based on pH-controlled and sequential fermentation using two strains. Appl Biochem Biotechnol 185:1075–1087. https://doi.org/10.1007/s12010-018-2711-9

Syamdidi HEI, Giyatmi I (2016) Agar-abundant marine carbohydrate from seaweeds in Indonesia: principles and applications. In: Kim S (ed) Marine glycobiology. CRC Press, Boca Raton, pp 255–262. https://doi.org/10.1201/9781315371399-19

Tan IS, Lee KT (2014) Enzymatic hydrolysis and fermentation of seaweed solid wastes for bioethanol production: an optimization study. Energy 78:53–62. https://doi.org/10.1016/j.energy.2014.04.080

Taylor G (2008) Biofuels and the biorefinery concept. Energy Policy 36:4406–4409. https://doi.org/10.1016/j.enpol.2008.09.069

Tedesco S, Stokes J (2017) Valorisation to biogas of macroalgal waste streams: a circular approach to bioproducts and bioenergy in Ireland. Chem Pap 71:721–728. https://doi.org/10.1007/s11696-016-0005-7

Thompson TM, Young BR, Baroutian S (2019) Advances in the pretreatment of brown macroalgae for biogas production. Fuel Process Technol 195:1–12. https://doi.org/10.1016/j.fuproc.2019.106151

Thompson TM, Young BR, Baroutian S (2021) Enhancing biogas production from Caribbean pelagic *Sargassum* utilising hydrothermal pretreatment and anaerobic co-digestion with food waste. Chemosphere 275:130035. https://doi.org/10.1016/j.chemosphere.2021.130035

Trivedi N, Reddy CRK, Radulovich R, Jha B (2015) Solid state fermentation (SSF)-derived cellulase for saccharification of the green seaweed *Ulva* for bioethanol production. Algal Res 9:48–54. https://doi.org/10.1016/j.algal.2015.02.025

Uribe E, Vega-Gálvez A, García V, Pastén A, López J, Goñi G (2019) Effect of different drying methods on phytochemical content and amino acid and fatty acid profiles of the green seaweed, *Ulva* spp. J Appl Phycol 31:1967–1979. https://doi.org/10.1007/s10811-018-1686-9

van der Wal H, Sperber BL, Houweling-Tan B, Bakker RR, Brandenburg W, López-Contreras AM (2013) Production of acetone, butanol, and ethanol from biomass of the green seaweed *Ulva lactuca*. Bioresour Technol 128:431–437. https://doi.org/10.1016/j.biortech.2012.10.094

Vanavil B, Selvaraj K, Aanandhalakshmi R, Usha Sri K, Arumugam M (2020) Bioactive and thermostable sulphated polysaccharide from *Sargassum swartzii* with drug delivery applications. Int J Biol Macromol 153:190–200. https://doi.org/10.1016/j.ijbiomac.2020.02.332

Vanegas CH, Bartlett J (2013a) Green energy from marine algae: biogas production and composition from the anaerobic digestion of Irish seaweed species. Environ Technol 34:2277–2283. https://doi.org/10.1080/09593330.2013.765922

Vanegas C, Bartlett J (2013b) Anaerobic digestion of *Laminaria digitata*: the effect of temperature on biogas production and composition. Waste Biomass Valorization 4:509–515. https://doi.org/10.1007/s12649-012-9181-z

Viniegra-Gonzalez G, Favela-Torres E, Aguilar CN, De Jesus R-G, Diaz-Godinez G, Augur C (2003) Advantages of fungal enzyme production in solid state over liquid fermentation systems. Biochem Eng J 13:157–167. https://doi.org/10.1016/S1369-703X(02)00128-6

Vivekanand V, Eijsink VG, Horn SJ (2012) Biogas production from the brown seaweed *Saccharina latissima*: thermal pretreatment and codigestion with wheat straw. J Appl Phycol 24:1295–1301. https://doi.org/10.1007/s10811-011-9779-8

Wang S, Jiang XM, Wang N, Yu LJ, Li Z, He PM (2007) Research on pyrolysis characteristics of seaweed. Energy Fuels 21:3723–3729. https://doi.org/10.1021/ef700214w

Wang M, Chen L, Liu Z, Zhang Z, Qin S, Yan P (2016a) Isolation of a novel alginate lyase-producing *Bacillus litoralis* strain and its potential to ferment *Sargassum horneri* for biofertilizer. Microbiol Open 5:1038–1049. https://doi.org/10.1002/mbo3.387

Wang X, Wang L, Li X, Xu Y (2016b) Response surface methodology based optimization for degradation of align in *Laminaria japonica* feedstuff via fermentation by *Bacillus* in *Apostichopus japonicas* farming. Electron J Biotechnol 22:1–8. https://doi.org/10.1016/j.ejbt.2016.04.003

Wang S, Zhao S, Uzoejinwa BB, Zheng A, Wang Q, Huang J, Abomohra AEF (2020) A state-of-the-art review on dual purpose seaweeds utilization for wastewater treatment and crude bio-oil production. Energy Convers Manag 222:1–16. https://doi.org/10.1016/j.enconman.2020.113253

Wegeberg S, Felby C (2019) Algae biomass for bioenergy in Denmark. Biological/technical Challenges and Opportunities, Technical Report, University of Copenhagen, Denmark. https://doi.org/10.13140/RG.2.2.19457.51040

Wijayanta AT, Goto M, Kamiya N (2015) Great potency of seaweed waste biomass from the carrageenan industry for bioethanol production by peracetic acid–ionic liquid pretreatment. Biomass Bioenergy 81:63–69. https://doi.org/10.1016/j.biombioe.2015.05.023

Wong TY, Preston LA, Schiller NL (2000) Alginate lyase: review of major sources and enzyme characteristics, structure-function analysis, biological roles, and applications. Annu Rev Microbiol 54:289–340. https://doi.org/10.1146/annurev.micro.54.1.289

Wu H, Xiao J, Hao S, Yang R, Dong P, Han L, Zhang Y (2021) In-situ catalytic gasification of kelp-derived biochar as a fuel for direct carbon solid oxide fuel cells. J Alloys Compd 865:158922. https://doi.org/10.1016/j.jallcom.2021.158922

Yanik J, Stahl R, Troeger N, Sinag (2013) A Pyrolysis of algal biomass. J Anal Appl Pyrolysis 103:134–141. https://doi.org/10.1016/j.jaap.2012.08.016

Yuan Y, Macquarrie DJ (2015) Microwave assisted acid hydrolysis of brown seaweed *Ascophyllum nodosum* for bioethanol production and characterization of alga residue. ACS Sustain Chem Eng 3:1359–1365. https://doi.org/10.1021/acssuschemeng.5b00094

Yusmaniar Y, Syafei DI, Arum M, Handoko E, Kurniawan C, Asali MR (2019) Preparation and characterization of seaweed based bioplastic blended with polysaccharides derived from various seeds of avocado, jackfruit and durian. J Phys Conf Ser 1402: 055097. https://doi.org/10.1088/1742-6596/1402/5/055097

Part IV
Municipal Solid Waste Biorefinery

Chapter 14
Biochar Pyrolyzed from Municipal Solid Waste—Properties, Activation, Applications and Climate Benefits

S. Sri Shalini, K. Palanivelu, and A. Ramachandran

14.1 Introduction

14.1.1 Municipal Solid Waste Management Scenario

World-wide, the increase in population, urbanisation rate and industrialisation has led to the billion tonnes of municipal solid waste (MSW) generation that are piled up enormously in the environment. In 2016, the World Bank estimated a global MSW generation of 2.01 billion tonnes and anticipated to reach about 2.59 billion tonnes in 2030 and by the year 2050 which could further accumulate upto 3.40 billion tonnes (Silpa et al. 2050). Among them, high-income countries produced 34% i.e. 683 million tonnes of MSW having only 16% of world population. The percentage of waste generation and composition is based upon their socio-economic-cultural status, population density, income, life-style, geographical location and climate (Al-Jarallah and Aleisa 2013; Joseph 2014; Khan et al. 2016). Hence, the per capita waste generation varies in different countries from 0.11 to 4.54 kg waste/capita/day averaging of 0.74 kg waste per capita per day. Whereas, generated MSW disposed off in landfill is around 40% (39% in high income nations; 54% in upper middle income nations), 19% for material recovery (recycling and composting), 11% undergoes modern incineration and remaining 33% end-up in open dumping (Silpa et al. 2050). Most of the developing and under-developing countries majority MSW disposal method is by open dumping (around 93%) due to reasons such as limited number of landfill availability, existing landfill exceeding the capacity, non-availability of space for future landfills and limited or no financial support for the municipal disposal services (World Bank 2018; Ferronato and Torretta 2019; Penteado and Castro 2021). This common practise of open dumping of solid wastes causes air, water and land pollution leading to severe health and environmental complications (Joseph 2014;

S. Sri Shalini (✉) · K. Palanivelu · A. Ramachandran
Centre for Climate Change and Disaster Management, Department of Civil Engineering, College of Engineering Guindy Campus, Anna University, Chennai, Tamil Nadu 600025, India

Y. K. Nandabalan et al. (eds.), *Zero Waste Biorefinery*, Energy, Environment, and Sustainability, https://doi.org/10.1007/978-981-16-8682-5_14

Ferronato and Torretta 2019). MSW disposed in landfills emits methane which is a major contributor of greenhouse gases having 28 fold higher global warming potential when compared to CO_2 (for 100 years) that critically impacts the environment and health. The third largest anthropogenic source of methane are MSW from households, which accounts for about 11% of global methane emissions (Scheehle and Kruger 2006).

Over the years, comprehending the impacts of MSW led to reduction and utilisation of MSW in practice by adopting '3Rs' waste management principle of reduce, reuse and recycle. This progressively paved way for recovering the waste as a resource rather than disposing in open dumps. Furthermore, rather than linear economy aiming to close the loops of the recovered materials from MSW in-flowing into the urban supply chains for accomplishing the circular economy (Priyadarshini and Abhilash 2020). The reuse or recovery methods mainly depends upon the composition of MSW feedstock. MSW is a solid or semi-solid substance consists of food wastes, plastics, yard wastes, paper wastes, cardboard, glass and metal from residential and commercial sources; sometimes the construction and demolition wastes are also included. The composition varies based on the income levels and consumption patterns (Al-Jarallah and Aleisa 2013; Khan et al. 2016). In general, green and food waste contributes to 44% and recyclable MSW fractions of paper, cardboard, plastics, metal and glass accounts to 38%. Generally, MSW contains 50% or higher content of organic waste (exceptional for MSW from Europe, Central Asia and North America having higher percentage of dry waste) (Silpa et al. 2050).

The key challenge is to handle the enormous quantities of waste in a scientific and sustainable way for managing the organic fraction mainly carbonaceous portion of MSW into a useful product or energy. Many MSW conversion technologies are available for waste valorization through (i) physico-chemical methods (extraction and hydrolysis), (ii) thermo-chemical methods (pyrolysis, incineration, gasification and hydrothermal liquefaction), and iii) biochemical methods (enzymatic conversion, fermentation and anaerobic digestion), respectively. Gasification is a clean process than incineration, however the energy production is lower and capital costs are much higher (Seo et al. 2018). Whereas incineration, capital costs are lower, however to control emissions from the polluted gas and fly ash is a problematic issue. Another process is vitrification, where atomizing MSW in a plasma arc method but not cost effective (Gopu et al. 2018). The best method is the pyrolysis process of MSW for "Biochar" production is environmental viable, sustainable, climate benefit and cost effective option than other organic waste conversion methods (Gunarathne et al. 2019). The advantages of biochar from pyrolysis process is that it has higher carbon content, more surface areas, greater calorific values, rich surface functional groups based on their initial feedstock properties (Kumar and Bhattacharya 2021). The biochar production in India is usually made from the residues generated in agricultural practices such as crop wastes, cobs, stovers, bagasse, wheat straw, rice husk, groundnut and walnut shells, leaves, saw dust, etc. (Mankasingh et al. 2011; Mohanty et al. 2013; Parmar et al. 2014; Pawar and Panwar 2020). Recently, few studies focus on biochar production from various MSW components (Borgohain et al. 2020; Nair et al. 2020), as Indian MSW contains high organic content especially more than 50%

(Joshi and Ahmed 2016), which is a potential feedstock for pyrolysis process to yield biochar.

MSW pyrolysis converts the solid waste into products of solid biochar, liquid and gas (Ates et al. 2013). The primary factor of pyrolysis depends on obtaining the solid residue biochar from the process, in addition to capture the syngas and collect bio-oil. This thermo-chemically converted "MSW biochar" have very low amounts of toxic elements (heavy metals and polyaromatic hydrocarbons) which is beneficial to wide applications in heavy metal removal from soil, wastewater and enhancement in plant growth (Gunarathne et al. 2019; Agrafioti et al. 2014; Jin et al. 2014; Taherymoosavi et al. 2017; Jayawardhana et al. 2017; Abedinzadeh et al. 2020).

The different pyrolysis techniques of slow, fast, intermediate and hydrothermal carbonization were adopted for biochar production from MSW in different reactor configurations (Jin et al. 2014; Abedinzadeh et al. 2020; Chen et al. 2014; Benavente et al. 2018). The different catalysts were used for increasing the MSW pyrolysis performance (He et al. 2010; Chen et al. 2018; Li et al. 2020; Wang et al. 2020). The physico-chemical properties of MSW biochar were explored and physical/chemical activation studies were carried out for enhancing the surface area and porosity (Jayawardhana et al. 2017; Benavente et al. 2018; Li et al. 2015; Genuino et al. 2018; Zhang et al. 2019).

MSW biochar has been used for various applications such as organic contaminant removal of dyes and pharmaceutical compounds, inorganic contaminant removal of heavy metals, soil amendment, etc. (Agrafioti et al. 2014; Abedinzadeh et al. 2020; Parshetti et al. 2014; Sumalinog et al. 2018; Hoslett et al. 2020). Biochar, also shown to have climate change mitigation potential. Researchers have shown the biochar can reduce greenhouse gaseous emissions of methane and nitrous oxide, taking up CO_2 in their surface area thereby reducing the CO_2 emissions (Agyarko-Mintah et al. 2017; Sanchez-Monedero et al. 2018).

This chapter critically focuses on the municipal solid waste biochar production methods, pyrolysis reactor configurations with operational conditions, physico-chemical biochar properties, presence of catalysts in pyrolysis, activation/modification, applications for contaminant removal, bioenergy, climate change mitigation and techno-economic aspects of biochar production. It also elaborates the operating conditions and results of current MSW biochar study from kitchen waste that are further compared for their yield and properties with other MSW-biochar.

14.2 MSW Biochar Production Methods

The different thermochemical methods to produce MSW biochar are shown in the Fig. 14.1 along with their process operational details, end-products and various activation or modification methods. Among the various processes, pyrolysis of MSW is a significant thermo-chemical conversion option that produce solid, liquid and gaseous products for direct application in power production, generation of value-added chemicals and other processes (Jin et al. 2014; Chen et al. 2014). The detailed technicalities involved in the various pyrolysis processes are described below.

14.2.1 MSW Pyrolysis Processes

Pyrolysis is a complex thermo-chemical degradation process comprising of multiple reactions taking place at elevated temperatures with no oxygen condition (Lehmann et al. 2003; Bhardwaj et al. 2021) and it is an endothermic process (Jahirul et al. 2012). MSW composed of cellulose, hemicellulose, and lignin content that vary in different MSW feedstock undergoes several reactions during this thermo-chemical conversion processes namely dehydration, degradation, and depolymerisation. Especially in the pyrolysis of thermo-chemical process, it undergoes several concurrent and related mechanisms of (a) depolymerization, (b) isomerization, (c) dehydration, (d) aromatization, (e) decarboxylation, and (f) charring. These reaction mechanisms can be further categorised as primary decomposition/depolymerisation and secondary mechanism. Largely thermal heating aimed at waste decomposition to produce volatile mixtures and biochar through cleaving the chemical bonds. The elevated temperatures and lower vapor condensation conditions favorable for the remaining unstable volatile mixtures and solid complex to produce non-condensable gaseous mixture through secondary mechanisms (Kumar et al. 2020). Pyrolysis processes differ based on their heating rate and residence time such as torrefaction, slow pyrolysis, flash pyrolysis, intermediate pyrolysis and fast pyrolysis. The other types of pyrolysis processes are flash carbonisation, hydrothermal carbonisation (HTC) and microwave assisted pyrolysis. Thermo-catalytic reforming (TCR) process is another process in-use nowadays, where pyrolysis and catalytic reforming are interlinked.

The most common method for the biochar production is slow pyrolysis method (Daful and Chandraratne 2018). Slow pyrolysis has a lesser heating rate of five to seven degree Celsius per minute and higher vapor residence period of more than one hour. Whereas, the fast pyrolysis has an elevated heating rate of greater than 200 °C per minute and shorter vapor residence period of lesser than ten seconds. In the intermediate pyrolysis, it has a lower heating rate of 1–10 °C/min and moderate vapour residence period of 0.5–20 s. The flash pyrolysis has an elevated heating of greater than 1000 °C/min and lower vapor residence period of less than 0.5 s. The end-products from these pyrolysis processes are biochar the solid product, bio-oil the liquid product and syngas the gaseous product. Their yield in each of these processes varied based on their MSW feedstock. Generally, slow pyrolysis yields 35% of char, 30% of bio-oil and 35% gas, where the lesser temperature range and heating conditions generates greater char than other processes (Jahirul et al. 2012). The vapour residence time is very high in this process ranging from five minutes to one hour that makes the vaporous materials to interact more one another that results in biochar and bio-oils. In fast pyrolysis, 60–75% of bio-oils, 15–25% of solids especially biochar content and 10–20% of gases and, in flash pyrolysis it is 12%

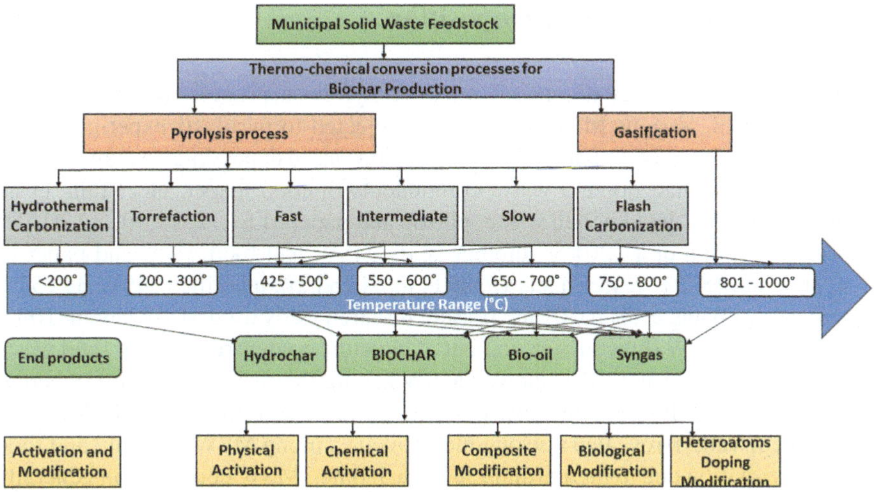

Fig. 14.1 MSW Biochar production methods

char, 75% oil and 13% gas (Jin et al. 2014; Jahirul et al. 2012; Kumar et al. 2020; Sri Shalini et al. 2021).

The other pyrolysis processes are microwave assisted pyrolysis has a rate of heating range from 0.1 to greater than 1000 °C/s under microwave radiation (Li et al. 2016) and flash carbonization process takes place in the elevated pressure conditions of approximately one to two Megapascal pressure (MPa). The flash fire flare up the waste materials at the incidence of air exposure. The process occurs at the temperature range of three hundred to six hundred degree Celsius and at lower residence period of lesser than or equal to thirty minutes. The gasification process occurs at elevated temperature range of 600–900 °C using gasifying agents that predominantly produces syngas/producer gas. Some of the different gasifying agents used in the process are air, carbon dioxide, oxygen, steam, or/and combination of gases. The other hydrothermal carbonization process directly converts high moisture content MSW feedstock to solid residue content product called hydrochar that occurs at subcritical conditions of water at one hundred eighty to two hundred fifty degree Celsius temperature range and two to six megapascal pressure environment (Kumar et al. 2020). In thermo-catalytic reforming process, it integrates pyrolysis between 400 and 500 °C with catalytic reforming of their volatile mixtures. In the intermediate pyrolysis, the products are downstream to post-reforming process primarily for improving yield and quality of syngas. During the reforming stage, products formed in the pyrolysis process are maintained in the temperature range of 500–700 °C. The resultant output of 30–45% of syngas, 7–15% of bio-oil and 25–50% of bio-char and with water as by-product of 15–25% (Wang et al. 2020; Moreno et al. 2020). Hence, for the conversion of MSW to biochar necessitates for careful selection of MSW feedstock, type of pyrolysis process and reactor configuration.

14.2.2 MSW Feedstock Composition

MSW biochar properties are based on their pyrolysis process and MSW feedstock properties. Various MSW feedstocks subjected to pyrolysis experiments are summarised in Table 14.1. The MSW composition varied depending on their sources: residential MSW had organic waste components of food, paper, etc. (Tang et al. 2019); some MSW had kitchen waste, plastic and paper of 57.72, 23.40 and 10.7%, respectively (Song et al. 2018); another study MSW had food, PVC and paper of 46.4, 30.7 and 18.2%, respectively (Fang et al. 2018a, b); landfilled MSW had 57% of yard waste (Wang et al. 2020) and present research study used MSW having 100% kitchen waste. MSW showed high content of moisture content due to their food waste composition. Tang et al. (2019) showed MSW having a moisture of 53.31%, whereas Wang et al. (2020) showed only 5.6% of moisture due to sun-dried after MSW collection. Lu et al. (2020) revealed a strong inter-linkage among MSW moisture and food waste composition due to their affinity with each other. MSW composition generally based on dry-ash free basis it has 40–50% of carbon, 25–35% of oxygen, 5–7% of hydrogen, 0.5–2% of nitrogen and 0.1–0.2% of sulphur with having elevated moisture and high volatile content of 60–80% (Taherymoosavi et al. 2017). Typical mixed MSW contained greater than 80% of carbon and oxygen in their ultimate composition (Song et al. 2018; Fang et al. 2018a) and MSW having pork, rice and kitchen waste composition also gave greater than 70% of carbon and oxygen (Chen et al. 2018). This organic content in MSW is largely converted to carbonaceous biochar product in pyrolysis process. Some MSW biochar studies were carried out in India using urban organic wastes of banana peduncles, fruit market wastes, municipal sewage sludge, food and market waste processing anaerobic sludge (Nair et al. 2020), tea pruning litter (Borgohain et al. 2020), etc. as feedstock. The biochar quality is based on their feedstock, choice of pyrolysis process and reactor configuration.

14.2.3 Pyrolysis Reactors for MSW Biochar

Table 14.1 summarises the various MSW pyrolysis studies conducted in different reactor configurations, varying MSW feedstocks along with their biochar qualities. The pyrolysis reactor configuration is vital for the good yield, quality and its cost-effectiveness. As the MSW feedstock is heterogeneity in nature, the selection of reactors is sometimes intricate. Some of the reactors used for pyrolysis are fluidised bed, fixed bed, tubular, ablative, vacuum reactor, rotating cone, plasma, radiative-convective, auger screw, vortex, microwave, entrained flow, solar, ceramic ball, etc. (Chen et al. 2014; Benavente et al. 2018; Li et al. 2015, 1999; Garcia et al. 1995a, b; Liu et al. 2017; Font et al. 1995; Luo et al. 2010; Zaman et al. 2017). Subsequent paragraphs, however, are indented.

The muffle furnaces have been used for MSW and municipal sewage sludge pyrolysis and yielded biochar that can be used for soil amendment and heavy metals

Table 14.1 MSW pyrolysis studies in different MSW feedstock and reactor configurations

S. No	MSW feedstock	Tem (°C)	Reactor conditions; residence time	Surface area (m^2/g)	Pore volume (cm^3/g)	Country and MSW biochar application	References
Fast pyrolysis							
1	MSW as dry pellets	700–850	Fluidisedsand bed; helium 3.6 cm/s; 20 min	–	–	Spain; kinetics of pyrolysis process assessment	Garcia (1995a, b)
2	MSW (paper, board, plastic, rubber, vegetable, wood, etc.)	850	Externally heated rotary kiln at 3r/min; 7–15 min	–	–	China, manufacturing flue gas from MSW	Li et al. (1999)
3	Municipal sewage sludge	500 600 700 800 900	Horizontal fixed-bed reactor, N_2-0.3 L/min; 20 min	25.424 20.268 32.167 48.499 67.603	0.05608 0.05268 0.06842 0.08989 0.09855	China; pyrolysis temperature and heavy metal adsorption potential	Chen et al. (2014)
Slow pyrolysis							
4	MSW from households (paper, textile, organic waste, plastics, metals, etc.)	400 500 600	Fixed-bed batch Parr pressure with condenser, N_2-100 L/min; 30 min	20.7 29.1 29.8	0.027 0.039 0.038	USA; removal of aqueous As(V) removal	Jin et al. (2014)
5	MSW from residential area	400 600	Fixed bed, N_2-100 mL/min; 2 h	2.0 4.0	0.007 0.029	China; chemical and physical modification for mercury removal	Li et al. (2015)
6	MSW	400 500 700	–	–	–	Spain; Use as soil amendment	Zornoza et al. (2016)

(continued)

Table 14.1 (continued)

S. No	MSW feedstock	Tem (°C)	Reactor conditions; residence time	Surface area (m^2/g)	Pore volume (cm^3/g)	Country and MSW biochar application	References
7	MSW	750 300 500 750 300	Residence time at 2 h; 4 h and 6 h at 300, 500, 750 °C; respectively	10 (2 h) 05 (4 h) 140 (4 h) 155 (4 h) 160 (6 h)	–	Qatar; characterisation of MSW biochar	Rehrah et al. (2016)
8	OFMSW—paper, fiber, plastic, woods and kitchen waste	600 700 800	Vacuum pyrolysis furnace, N_2; 60 min	0.98 28.4 112	0.0003 0.01 0.04	China; priming effects; mineralization of SOC in dry land soil	Liu et al. (2017)
9	Fibrous organic fraction MSW	450	Batch reactor; 30 min and 15 °C /min up to 450 °C	108.47	0.013	Sri Lanka; benzene removal from landfill leachate	Jayawardhana et al. (2017)
10	Composted MSW (domestic waste and kerbsides)	450 550 650	–	–	–	Australia; characterisation of MSW biochar	Taherymoosavi et al. (2017)
11	Municipal sludge	350 500	Muffle furnace, 25 °C/min; 30 min	–	–	Brazil; release of nutrients and contaminants	Figueredo et al. (2017)
12	MSW (paper, yard and textile wastes)	400 - 500	Batch, pressure control, 3 °C/min; 30 min	–	–	USA; humic acid extraction from MSW biochar	Genuino et al. (2017)
13	Composted organic fines MSW & sludge	700	Fixed bed reactor	–	–	Iran; MSW organic fines to gas and char	Agar et al. (2018)

(continued)

Table 14.1 (continued)

S. No	MSW feedstock	Tem (°C)	Reactor conditions; residence time	Surface area (m^2/g)	Pore volume (cm^3/g)	Country and MSW biochar application	References
14	Organic fraction of urban waste	300 500	Tubular carbolite furnace, 10 °C/min; 1 and 5 h	–	–	Spain; on enzymes and plant germination in mediterranean soil	Benavente et al. (2018)
15	Municipal sludge	450	Temp—350, 400, 450, 500, 550 °C. N_2—25 mL/min; 2 h	20.86 (BC450)	0.664 (BC450)	Canada; ammonia removal from wastewater	Tang et al. (2019)
16	Municipal sewage sludge	200 300 500 700	Muffle furnace; 2 h; Heating rate—10 /min	–	–	China; soil amendment and development of turf grass	Tian et al. (2019)
17	Segregated organic fraction of MSW	500	Muffle furnace; 1 h; Heating rate—20 °C /min;	37.47	0.03	Iran; reduction of Pb and Cd in forage maize	Abedinzadeh et al. (2020)
18	MSW (Kitchen waste)	500	1 h in Muffle furnace with a heating rate of 10 °$Cmin^{-1}$	5.54	0.019	India, biochar preparation from kitchen waste	Present study
Hydrothermal carbonization							
19	Urban food waste (Cooked, uncooked and restaurant condiments)	250 350	Parr stirred pressure batch reactor; 20 min	6.07 1.01	0.064 0.001	Singapore; acridine orange and rhodamine 6G dye removal from polluted water	Parshetti et al. (2014)

(continued)

Table 14.1 (continued)

S. No	MSW feedstock	Tem (°C)	Reactor conditions; residence time	Surface area (m^2/g)	Pore volume (cm^3/g)	Country and MSW biochar application	References
20	MSW (Newspaper and Korean Kimchi)	200	Pilot-scale pressures 1.6 Mpa, 200 rpm; 60 min	–	–	Korea; renewable solid fuel	Kim et al. (2017)
Thermo-catalytic reforming							
21	MSW (plastics, biomass, paper, metal, textile, rubber, soil, rubble, fines, concrete, food, electrical item)	450	Continuous pilot scale reactor; 2 kg/h	–	–	UK; conversion of MSW to fuels	Ouadi et al. (2017)
22	MSW from a company (ECOHISPÁNICA S.A.)	550	Two stages of pyrolysis and catalytic cracking; fixed bed, N_2—300 mL/min; 30 min	–	–	Spain; generation of syngas; production of solid refuse derived fuels	Veses et al. (2020)

removal (Abedinzadeh et al. 2020; Figueredo et al. 2017; Tian et al. 2019). Abedinzadeh et al. (2020) studied organic fraction MSW pyrolysis in muffle furnace at 500 °C that yielded a biochar having surface area—37.47 m^2/g and pore volume—0.03 cm^3/g, shown to remove heavy metals from forage maize spiked with heavy-metal contaminated water. The present study conducted pyrolysis of MSW from kitchen waste in a muffle furnace at 500 °C at a residence time one hour yielded a biochar having surface area—5.54 m^2/g and pore volume—0.019 cm^3/g. The common reactors used for slow and fast pyrolysis are fluidised bed, fixed bed, rotary kiln and tubular reactors. The operational details of these reactors are as follows.

14.2.3.1 Fluidised Bed Reactor

The fluidized bed reactor comprises a fluid and solid mixture which behaves as a fluid in nature and accomplished through intrusion of pressurized fluid over a solid particulate that provides high turbulence to the waste sample with greater heat transfer effect. It is prevalent for fast pyrolysis and flash pyrolysis due to (a) faster heat transfer rate, (b) proper regulation of pyrolysis process, (c) control over vapour residence period, (d) bed volume with greater surface area inter-linkage among fluid and solid and (e) prevalence of greater relative velocity among solid and fluid (Jahirul et al. 2012). There are various fluidized bed reactors available such as circulating and bubbling reactors (Zaman et al. 2017). The circulating fluidized reactors have shorter residence time and higher velocity of vapors used for large waste volume compared to the bubbling reactors.

Font et al. (1995) experimented with fluidized sand bed reactor for spherical pellets of MSW mainly composed of paper (38%) operated in an inert atmosphere of helium at a pressure of 1 atm. The temperature of the reactor was kept above 700 °C and MSW samples were sent on the sand bed. The sand bed composed of sand that was calcinated at nine hundred degree Celsius and each sand particle have a diameter of 0.105–0.210 mm, later washed with hydrochloric acid. The primary degradation reaction initiated and secondary reactions conducted above the sand bed, where temperature reduced to lower than 300 °C. At 850 °C, the end-products yielded 47% syngas. Garcia et al. (Garcia et al. 1995a, b) conducted similar experiments in a fluidised sand bed with spherical pellets of MSW having an ash content of 26.7%. The reactor was oven heated kept in an inert helium atmosphere operated at 700, 750, 800 and 850 °C, respectively. Heat transfer coefficient were around 112–559 J/s m^2 K between MSW discharged and fluidized sand bed. At 800 °C, primary tars cracked and gas production increased. At 850 °C, the resultant gas contained methane content of 6.20%, ethane of 0.70–0.32%, ethylene of 5.90–4.50%, propane of 0.034–0.001%, propylene of 1.10–0.17%, acetylene of 0.23–0.19%, butylene of 0.19–0.007%, carbon monoxide of 26.0%, carbon dioxide of 13.90–15.70% and hydrogen of 1.60–2.40%, respectively. Most of the fluidised bed studies conducted for fast pyrolysis of MSW, whereas fixed bed reactors primarily used for fast and slow pyrolysis of MSW.

14.2.3.2 Fixed Bed Reactor

The fixed bed reactor configurations are simpler than other pyrolysis reactors, it is reliable and operates under low gas velocity, long solids residence time, conserve carbon and suitable for small-scale heat or power production. It is predominant for slow pyrolysis process of MSW. The reactors are usually made of steel, firebricks or concrete containing a gas cooling system. MSW samples enter through a vertical shaft in a fixed bed reactor that faces an upward moving counter current gas stream, sometimes difficult for tar removal (Jahirul et al. 2012; Zaman et al. 2017). Different fixed bed type reactors were in use such as horizontal fixed-bed (Chen et al. 2014) and Parr pressure reactor with a condenser (Jin et al. 2014). It is shown to operate in varying temperature ranges (Wang et al. 2020). Several researchers have used fixed bed pyrolysis for municipal sewage sludge (Chen et al. 2014; Agar et al. 2018), MSW from residential areas containing paper, textile, organic wastes, plastics, metals, etc. (Jin et al. 2014; Li et al. 2015) and composted organic fines from MSW (Agar et al. 2018) as shown in Table 14.1. Studies showed that increase in temperatures (range of 400–900 °C) has resulted correspondingly an increase in surface areas and pore volume (from 20.268 to 67.602 m^2/g and 0.05268 to 0.09855 m^3/g (Chen et al. 2014) and 20.7–29.8 m^2/g and 0.027–0.038 m^3/g (Jin et al. 2014), respectively) as presented in Table 14.1. Likewise, Luo et al. (2010) operated fixed bed reactor with a furnace for pyrolysis of MSW samples collected from a transfer station at Wuhan, China. The fixed bed reactor had a screw feeding system to feed MSW into the bed through a pipe. MSW was separated into three different size fractions of below 5 mm, 50–10 mm and above 10 mm. The study exhibited that increasing the temperature of 600 to 900 °C and with smaller MSW particles, yielded an increase in the gas product (H_2 and CO) and reduction in the tar product. Veses et al. (2020) demonstrated fixed bed configuration for thermo-catalytic pyrolysis process for MSW from Spain. The MSW feedstock with uniform size can be used in fixed bed type, whereas MSW with varying mixes of waste can be operated in rotary kiln.

14.2.3.3 Rotary Kiln

Rotary kilns are efficient than fixed bed reactors and suitable for slow pyrolysis to achieve yields (Chen et al. 2015). It is used for MSW of different size can be operated with continuous or batch mode and heated internally or externally. Commonly, internally heated rotary kilns are used but generates producer gas of low heating value, therefore for increasing the heating value the externally heated kilns must be planned. The rotary kiln operated at 3r/min for organic components of MSW having paper, paper board, waste plastic including PVC plastic and PE plastic, rubber, vegetable, wood cloth and orange husk were studied for the fast and slow pyrolysis in batch mode by Li et al. (1999). The rotary kiln reactor was heated externally and operated at 850 °C and sample fed for fast pyrolysis, whereas the sample was heated at five to sixty degree Celsius along with the reactor residence of 7–15 min in slow pyrolysis. The results revealed that under fast heating process, higher conversion of MSW to syngas of 13,000–23,000 kJ/kg obtained. The syngas yield was maximum for PE plastic and minimum for vegetables.

Slow pyrolysis of oil field sludge operated in the rotary kiln at 480–650 °C gave 54–86% of biochar yield (Ma et al. 2014). An indirectly fired rotary kiln was fabricated for biochar production using various feedstocks such as coffee husks, fiberboard, palm date fronds, wood mix, tree bark and olives stone kernels was studied (Haeldermans et al. 2020). The reactor was operated at 450 °C yielded the biochar at the rate of 0.81–1.11 tonne/h of solid fraction. It was demonstrated as a more feasible option for biochar production than microwave pyrolysis. Some pyrolysis studies have been conducted in tubular reactor as described below.

14.2.3.4 Tubular Reactor

The tubular reactors designed with fixed walls structured like tube and MSW samples move inside the reactor through several operating modes and usually heated externally. The different types of tubular configurations are screw pyrolyser, a tubular rectilinear reactor in which the solids pass by vibro-fluidised transport (Chen et al. 2015; Serrano et al. 2001; Aguado et al. 2002). Benavente et al. (2018) fabricated tubular Carbolite furnace for pyrolysis of organic fraction of urban waste. A screw kiln reactor used for thermal and catalytic cracking of low-density polyethylene was studied (Serrano et al. 2001; Aguado et al. 2002). The reactor had a long screw with hopper for sample feed and operated in two temperature zones at an inert nitrogen medium. The inlet plastic material loaded was about 20–41 g/h. The resultant thermal degradation gave a wider product range (Aguado et al. 2002).

As seen earlier, the slow pyrolysis yields higher biochar than other pyrolysis processes. Lu et al. (2020) has studied in detail about the MSW pyrolyzers for co-pyrolyzing the several components of MSW. Ates et al. (2013) and Miskolczi et al. (2013) demonstrated that the co-pyrolysis of individual MSW components and mixture of MSW components yield different products. From the various pyrolysis reactor studies, it was clear that the MSW feedstock, retention time, heating temperature, heating medium, reactor capacity, batch or continuous operation, pressure and inert gas are some of the aspects for the optimum selection of the reactors for pyrolysis process yields desirable quality end-products. The presence of external catalyst in the pyrolysis process can yield higher quality pyrolysis products as described in Table 14.2.

14.2.4 MSW Pyrolysis in Presence of Catalysts

The additives and catalysts usually added to increase the pyrolysis performances of MSW. Some of the catalysts predominantly used for pyrolysis of MSW are MgO, CaO and alkali earth metals as summarized in Table 14.2. Some of the catalytic pyrolysis studies are discussed below.

Table 14.2 The presence of different catalysts used for MSW pyrolysis

S. No.	Feedstock	Catalyst	References
1	MSW and biomass briquettes	alkali and alkaline earth metals (K, Na, Ca, and Mg)	Li et al. (2020)
2	Paper sludge and MSW	MgO	Fang et al. (2018a, b)
3	MSW	MgO. CaO	He et al. (2010); Tursunov (2014); Veses et al. (2020)
4	MSW	CaO, MSW Char	Song et al. (2018)
5	MSW	Char (from pyrolyzed MSW)	Wang et al. (2020)
6	MSW and plastic waste	Y-zeolite, b-zeolite, equilibrium FCC, MoO_3, Ni–Mo-catalyst, HZSM-5 and $Al(OH)_3$	Ates et al. (2013); Miskolczi et al. (2013)
7	Kitchen waste and *Chlorella vulgaris*	$CaCO_3$, CaO, SiO_2, permutit	Chen et al. (2018)
8	Food waste	CO_2	Lee et al. (2020)

The influence of catalyst and non-catalyst in the pyrolysis process was assessed with MSW and municipal plastic waste (MPW) in a pyrolyzer operated at 500, 550 and 600 °C by Ates et al. (2013). The different catalysts used in the study were Y-zeolite, b-zeolite, equilibrium FCC, MoO_3, Ni–Mo-catalyst, HZSM-5 and $Al(OH)_3$. The study used MSW from Hungarian waste recycling plant contained organic waste, paper, plastics, metals, textile, etc. and the plastic components from MSW was considered as municipal plastic waste, both were cut into 10 mm size particles. The results revealed that in the presence of catalyst, the volatile fraction yield was increased, whereas the reaction period decreased, yielded with aromatic and cyclic compounds. The catalyst pore size and Si/Al ratio had a significant impact on catalyst efficiency, where MPW had higher efficiency than MSW. The MSW pyrolysis using Ni–Mo-catalyst had maximum water content and HZSM-5 with lowest water content. In MSW pyrolysis product of bio-oil, the aliphatic hydrocarbons are converted to aromatic and cyclic compounds due to the catalysts productivity and selectivity efficiency. Bio-oil contained less quantity of ketones, alcohols, acids or esters. In MSW pyrolysis of gaseous product contained hydrogen, carbon monoxide and carbon dioxide, whereas MPW pyrolysis gas product contained hydrogen and hydrocarbons. Similar investigations with same set of catalysts and feedstock MPW and MSW components was carried out by Miskolczi et al. (2013) at varying temperatures of 500, 550 and 600 °C. MSW pyrolysis in presence of catalyst revealed coke accumulation, formation of carbon frame and double bound isomerization. The b-zeolite and MoO_3 catalysts decreased the pyrolysis oil's viscosity and removed contaminants.

The effect of zeolite and calcined dolomite catalysts was examined during MSW pyrolysis in fixed bed at 200–750 °C studied by Tursunov (2014). The zeolite catalyst presence revealed that the gas yield (of 49–57 mol %) was increased significantly, whereas oil and char yield was reduced compared to that of the process without the catalyst. At temperatures of 600–750 °C, elevated levels of gas formed containing carbon monoxide that subsequently utilized for energy generation. In another MSW pyrolysis study in fixed bed using calcined dolomite at 750–900 °C exhibited a similar conclusions of increase in syngas yield of 47–67 mol% containing hydrogen and carbon monoxide, with decreased yield of oil and char (He et al. 2010). The syngas produced of 13.87 MJ/N m^3 shall be utilized as a source material for Fischer–Tropsch synthesis for generating transportation fuels. Veses et al. (2020) studied the calcined dolomite catalyst application for longer chain hydrocarbons yield hydrogen and carbon mono-oxide. While using char as catalyst, it produces maximum syngas composition with hydrogen (Wang et al. 2020). Using CO_2 as catalyst, it affects the free radical mechanisms causing obstruction in cyclization processes that reduces the cyclic compounds formation (Lee et al. 2020).

Many co-pyrolysis studies of MSW using catalyst was conducted (Li et al. 2020; Fang et al. 2018a, b). The effect of alkali and alkaline earth metals (AAEMs) of potassium, sodium, calcium and magnesium on the co-pyrolysis of MSW and biomass briquettes (CSBs) under nitrogen and CO_2 atmosphere was investigated by Li et al. (2020). The AAEM catalyst during first pyrolysis stage decreased the initial temperature by inhibiting the volatile release, whereas increased it during subsequent stages. The alkyl and carbonyl emissions varied under different catalysts. The change in N_2 to CO_2 atmosphere caused the reduction in MSW and CSBs residual mass that subsequently lowered by catalyst effect. The CO_2 gas yields were increased by AAEM catalyst under N_2 atmosphere. In the third stage pyrolysis, Ca catalyst had major impact on the carbon monoxide production. Similarly, in another study of CaO catalyst on co-pyrolysis of, biomass, kitchen waste and plastics showed the decreasing trend in tar and oil yield, whereas increasing trend in the syngas yield with composition of hydrogen and carbon mono-oxide (Song et al. 2018). Cao catalyst in co-pyrolysis of *Chlorella vulgaris* and kitchen waste decreased the bio-oil acid values and increased the aliphatic hydrocarbon (Chen et al. 2018). The MgO catalyst decreased the oxygenated substances and increased aliphatic hydrocarbons in bio-oil (Fang et al. 2018b) and decreased biochar product, whereas increased its surface areas and pore volume (Fang et al. 2018a). Further, the activation and modification of the biochar increase the biochar applicability in various fields.

14.3 Activation and Modification of MSW Biochar

The biochar activation is a method for improving the textural properties and chemical functionalities of a biochar (Sajjadi et al. 2019). The biochar activation can be conducted by physical and chemical methods, and biochar modification by biological, composite and hetero-atom doping methods. The physical or chemical activation methods primarily enhance the biochar porosity and develop higher surface area by adding more functional groups on the surface that eventually intensify its adsorptive properties (Genuino et al. 2018; Sajjadi et al. 2019; Liu et al. 2012). The chemical activation combines the carbonization and activation as one process conducted under minimum temperatures that improve the biochar catalytic oxidation efficiency (Liu et al. 2012). The physical activation consists of thermal treatment at temperatures of 300–800 °C (Li et al. 2015; Genuino et al. 2018), CO_2 (Zhang et al. 2019) and steam especially at greater temperatures than 800 °C, sometimes increase the adsorbent cost that may not be realistic for pilot-scale applications. The chemical activation uses chemical reactions with acids, dehydrating/oxidising agents and alkali solutions, it is cost effective as no heat used in the process (Jin et al. 2014; Li et al. 2015; Genuino et al. 2017, 2018; Zhang et al. 2019).

The different chemical agents on biochar has activated several functional groups such as (a) KOH elevates hydroxyl, basicity and condenses organic content and (b) H_2O_2 elevates oxygen-carboxyl groups that leach out organic and ash content. So, washing step is important after the activation and modification treatment (Liu et al. 2012; Wongrod et al. 2018). Physical and chemical activations on various MSW biochar studies are summarised in Table 14.3. Combination of physical and chemical activations were conducted to yield good quality products (Li et al. 2015; Genuino et al. 2018; Zhang et al. 2019). The activated biochar is used for many applications such as wastewater treatment and water contaminant removal especially for heavy metals as described below.

14.4 MSW Biochar Yield and Properties

14.4.1 MSW Biochar Yield

Biochar is a carbonaceous and negatively-charged product has strong affinity to organic/inorganic pollutants that makes it easy for contaminant removal. Several waste biomass can be used for conversion to biochar, but waste-derived biomass such as MSW feedstock is encouraging for production of valuable product at low temperature and no oxygen condition avoids toxic pollutant emissions (Gopu et al. 2018; Sumalinog et al. 2018; Sri Shalini et al. 2021). The MSW biochar yield obtained from different studies is summarised in Table 14.4. The MSW biochar is denoted as MSW-BC and sewage sludge biochar is denoted MS-BC.

The process can be designed for maximising the biochar production by pyrolyzing MSW with higher volatiles under high pressure and with external addition of a carbon source (Song et al. 2018). The extreme high temperature is not favourable for biochar production as described in Table 14.4. The char yield at 900 °C was 15.86% and with presence of catalyst it was 14.92% (He et al. 2010). The present study conducted the slow pyrolysis of kitchen waste at 500 °C yielded a 26.40 ± 1.5%. MSW biochar yield and properties depends upon the MSW pyrolysis heating temperature (Table 14.4).

Table 14.3 Chemical and physical activation of MSW Biochar

S. No.	Activation (physical/chemical)	Conditions	Surface area m^2/g	Pore volume (m^3/g)	Application/removal	Adsorption capacity	References
1	Chemical	500 °C pyrolysed Biochar in 2 M KOH for 1 h	49.1 ± 3.1	0.357 ± 0.002	Arsenic V removal	As (V)—30.98 mg/g	Jin et al. (2014)
2	Chemical	MSW Pyrolysis at 400–500 °C for 30 min at 3 °C/min; Treatment with 0.25–0.75 M KOH for 24 h	–	–	Humic acid (HA) extraction from MSW-BC using artificial neural network	ANN model gave variation of 3.71% among predicted HA yield 180.57 mg/g and experiment 187.52 mg/g	Genuino et al. (2017)
3	Physical and chemical	Physical: 600 °C pyrolysed biochar (C600) with N_2 (40 mL/min) in microwave activation for 4 h (C6W) Chemical: C6W impregnated with 1 wt.% NH_4Cl solution (5 wt.% loading) stirred for 12 h, dried 4 h at 80 °C and 12 h at 90 °C (C6WN5)	C6W: 48.1 C6WN5: 12.4	C6W: 0.072 C6WN5: 0.04	Mercury removal of C6W increased to 22.9% than C600 and C6WN5 increased to 22.9% to 65.9%	Adsorption capacity of C6WN5 was 157.7 μg/g	Li et al. (2015)

(continued)

Table 14.3 (continued)

S. No.	Activation (physical/chemical)	Conditions	Surface area m^2/g	Pore volume (m^3/g)	Application/removal	Adsorption capacity	References
4	Chemical and thermal	Chemical: pyrolysis at 400–500 °C for 30 min; treatment KOH 0.25–0.75 M for 24 h Thermal: KOH activated biochar, tube furnace inert N_2 at 600–800 °C for 30–90 min. Impurities removal by treating with 0.1 M hydrochloric acid for two hours	662.4 ± 5.5	0.577 ± 0.022	Methylene blue with 99% removal efficiency	Methylene blue adsorption on activated biochar with adsorption capacity of 37.0–41.2 mg/g	Genuino et al. (2018)
5	Chemical and physical	Chemical: 0.01 g Municipal sewage sludge biochar at 700 °C for 60 min treatment with two grams of KOH or CH_3COOK for N_2 (400 mL/min) up to 700 °C (10 °C/min for one hour). Physical: SS-BC with CO_2 (100 mL/min) for one hour	C: 81.35 C-KOH 907.95 C-CK: 611.16 C-CO_2: 239.82	C: 0.1150 C-KOH: 0.8341 C-CK: 0.6892 C-CO_2: 0.3874	Sludge biochar with chemical and physical activation	Pb(II) Adsorption capacity: CO_2—22.40 mg/g CH_3COOK—47.59 mg/g KOH—57.48 mg/g	Zhang et al. (2019)

Table 14.4 MSW biochar yield and its properties

S. No.	MSW-biochar	Temp	EC (μS/cm)	pH	C/N	H/C	O/C	Yield	CEC (cmolc/kg)	References
1	Urban waste-BC	300-1 h 300-5 h 500-1 h	450 ± 30 470 ± 50 420 ± 30	7.89 ± 0.16 7.46 ± 0.15 9.35 ± 0.06	12.5 ± 2.0 12.2 ± 1.0 9.3 ± 0.4	–	–	43.1 42.1 30.9	52.35 ± 0.87 73.14 ± 2.02 54.99 ± 8.08	Benavente et al. (2018)
2	MSW-BC	500	3.40 dS/m	9.04	–	0.03	0.18	–	–	Abedinzadeh et al. (2020)
3	MSW-BC	450	0.31 mS/cm	9.7	–	0.04	0.24	–	–	Jayawardhana et al. (2017)
4	MS-BC	500 600 700 800 900	–	8.81 9.54 11.11 12.18 12.15	–	0.09 0.06 0.04 0.04 0.08	0.45 0.30 0.30 0.17 0.12	63.1 ± 0.5 60.3 ± 1.5 58.7 ± 0.7 54.7 ± 1.3 53.3 ± 0.5	76.75 ± 6.53 30.81 ± 2.67 50.34 ± 2.73 126.62 ± 9.16 247.51 ± 7.49	Chen et al. (2014)
5	MS-BC	350 400 450 500 550	–	8.57 8.63 9.47 10.41 10.60	7.07 7.14 7.16 7.36 7.60	0.08 0.07 0.05 0.04 0.04	0.68 0.65 0.64 0.62 0.59	70.3 ± 3.6 64.4 ± 3.6 61.3 ± 3.6 58.4 ± 2.8 57.2 ± 3.0	–	Tang et al. (2019)
6	MS-BC	350 500	–	4.49 5.21	7.68 7.23	0.07 0.04	3.37 3.08	-	9.3 10.3	Figueredo et al. (2017)
7	MS-BC	200 300 500 700	347.0 114.8 73.8 96.2	6.54 7.20 8.70 11.15	7.80 7.61 8.82 15.96	–	–	92.2 81.7 67.8 65.1	–	Tian et al. (2019)

14.4.2 MSW Biochar Properties

MSW biochar physical properties and chemical properties are summarised in Tables 14.4 and 14.5. The MSW biochar properties depends upon pyrolysis conditions of temperature, pressure, heating condition and residence period as given in Tables 14.1, 14.4 and 14.5. MSW converted to biochar mainly composed of organic content of carbon, and other inorganic compounds depending upon the MSW feedstock composition.

The various MSW biochar physical properties are porous structure, pore volume, density, surface area and water holding capacity (WHC). The different MSW biochar chemical properties are pH, electrical conductivity (EC), fixed carbon, volatile content, elemental composition, cation exchange capacity (CEC), and heavy metals (Table 14.4, 14.5 and 14.6). The various heavy metals found in MSW biochar are P, S, K, Ca, Al, Mg, Co (Table 14.6a) and, Cr, Cu, Fe, Mn, Ni, Pb, Zn and Cd (Table 14.6b).

14.4.2.1 Physical Properties of MSW Biochar

The physical properties of MSW biochar are described as follows:

- **Porous structure:** Porous volume and size of MSW biochar occurs based on the temperature and MSW feedstock. Biochar pore structure can exists as nanopores with less than 0.9 nm or micropores with less than 2 nm or macropores as greater than 50 nm. The particle size attained from slow pyrolysis are 5–50 mm, fast pyrolysis are less than 1 mm and flash pyrolysis are less than 0.2 mm (Jahirul et al. 2012). From the literature (Table 14.1), it has been seen that the MSW biochar pore volume varied from 0.0003 (Liu et al. 2017) to 0.064 m^3/g (Parshetti et al. 2014), present study MSW (kitchen waste)—0.019 m^3/g and maximum 0.664 m^3/g for municipal sludge. The increase in residence time upto two hours increase the pore volume. The pore structure gets destructed or deactivated with prolonged time period and eventually reduces the biochar absorption property.
- **Surface Area**: MSW biochar surface area is the vital place where the various reaction mechanisms takes place. Brunauer– Emmett–Teller (BET) surface areas of MSW biochar was in the range of 0.98 (Liu et al. 2017) to 160 m^2/g (Rehrah et al. 2016) and 20.27–67.06 m^2/g for sewage sludge. The present study MSW biochar from kitchen waste gave a BET surface area of 5.54 m^2/g. Temperature has an effect on BET surface area, it increases upto a specific temperature of 200–700 °C as given in Table 14.1, whereas excessive elevated temperatures has a negative impact. It is further supported by Lehmann and Joseph (2009) who have demonstrated the micropores formation and increase in surface area with higher temperatures.
- The elevated pyrolysis pressure shall increase the BET surface area. Residence time affects surface area, the increased residence time up to two hours have a

Table 14.5 Chemical properties of MSW biochar

S. No.	MSW biochar; country	Tem (°C)	Mobile matter (%)	Fixed matter (%)	Ash (%)	C (%)	H (%)	O (%)	N (%)	References
Fast Pyrolysis										
1	MS-BC; China	500 600 700 800 900	–	–	74.21 77.90 81.53 83.93 88.07	17.46 18.40 16.92 16.20 15.92	0.70 0.34 0.21 0.03 0.11	10.5 7.35 6.86 3.64 2.44	1.54 1.38 0.95 0.50 0.53	Chen et al. (2014)
Slow pyrolysis										
2	MSW-BC; USA	400 500 600	22.3 26.4 15.1	65.2 63.8 78.2	6.1 9.2 6.2	48.6 59.5 70.1	12.2 9.1 8.4	31.7 20.8 13.7	0.1 0 0.1	Jin et al. (2014)
3	MSW-BC; China	400 600	26.2 15.1	23.3 30.1	50.1 53.8	80.2 85.0	7.4 2.3	8.5 6.7	2.8 3.6	Li et al. (2015)
4	OFMSW—BC; China	600 700 800	–	–	–	64.4 64.1 64.8	0.22 0.21 0.23	12.4 12.8 12.6	2.4 2.8 2.7	Liu et al. (2017)
5	Fibrous OFMSW-BC; Sri Lanka	450	31.6	46.5	15.6	60.8	2.79	14.6	1.33	Jayawardhana et al. (2017)
6	OFMSW-BC; Iran	500	–	–	30.90	66.46	2.30	12 10	0.68	Abedinzadeh et al. (2020)
7	Composted organic fines MSW-BC; Iran	700	3.9	31.9	64.2	34.96	0.42	3.14	1.13	Agar et al. (2018)

(continued)

Table 14.5 (continued)

S. No.	MSW biochar; country	Tem (°C)	Mobile matter (%)	Fixed matter (%)	Ash (%)	C (%)	H (%)	O (%)	N (%)	References
8	MSW-BC; Spain	400 500 700	–	–	59.9 63.0 68.4	30.1 28.8 28.0	1.6 1.2 0.7	6.8 5.5 1.9	6.57 1.3 1.3	Zornoza et al. (2016)
9	Composted MSW-BC; Australia	450 550 650	11.9 8.5 6.3	13.0 12.6 14.3	72.7 76.2 76.8	68.6 76.7 80.7	4.33 2.84 2.64	20.1 13.6 10.0	6.09 5.79 6.00	Taherymoosavi et al. (2017)
10	MSW-BC; Qatar	3002 h 500 750 3004 h 500 750 3006 h 500 750	–	–	10.1 20.5 20.6 10.3 20.5 30.1 10.5 20.5 30.1	40.7 51.7 61.2 45.1 56.2 62.5 48.4 55.7 60.5	5.2 1.7 0.01 4.6 1.9 0.0 4.5 1.0 0.1	53.6 45.8 38.3 49.7 41.3 37.0 46.6 42.5 38.9	0.3 0.7 0.5 0.6 0.6 0.5 0.6 0.8 0.5	Rehrah et al. (2016)
11	OF urban waste-BC; Spain	300-1 h 300-5 h 500-1 h	48 75 28 60 9 69	22 67 29 02 32 65	28.58 42.38 57.67	–	–	–	–	Benavente et al. (2018)
12	MSW-BC; USA	400–500	61 27	25 93	12.80	59.17	3.65	23 12	1.04	Genuino et al. (2017)

(continued)

Table 14.5 (continued)

S. No.	MSW biochar; country	Tem (°C)	Mobile matter (%)	Fixed matter (%)	Ash (%)	C (%)	H (%)	O (%)	N (%)	References
13	MS-BC; Canada	350 400 450 500 550	–	–	–	22.0 19.4 18.6 17.7 15.7	1.7 1.3 1.0 0.8 0.6	14.9 12.7 11.9 10.9 9.2	3.1 2.7 2.6 2.4 2.1	Tang et al. (2019)
14	MS-BC; Iran	700	3.2	27.4	69.4	28.68	0.09	0.41	1.94	Agar et al. (2018)
15	MS-BC; Brazil	350 500	–	–	51.24 65.81	24.35 20.99	1.73 0.88	75.2 70.8	3.17 2.9	Figueredo et al. (2017)
16	MS-BC; China	200 300 500 700	–	–	68.62 70.14 79.00 85.75	17.09 19.72 15.26 11.33	2.09 1.79 0.73 0.31	10.0 5.76 3.28 1.90	2.19 2.59 1.73 0.71	Tian et al. (2019)
17	MSW-BC	500	–	–	–	32.85	*a	19.4	1.12	Present study
Hydrothermal carbonization										
18	MSW-BC; Korea	200a* 200b*	74.2 75.3	13.3 15.0	12.5 9.7	41.7 43.5	5.3 5.7	40.1 40.0	0.4 1.1	Kim et al. (2017)
19	Food waste-BC; Singapore	250 350	–	–		52.23 56.75	5.70 4.31	38.2 36.4	3.38 2.07	Parshetti et al. (2014)
Thermo-catalytic reforming										
20	MSW-BC	450	–	–	44.9	47.3	0.8	5.7	1.0	Ouadi et al. (2017)
21	MSW-BC	550 °C	13.2	10.8	73.7	22.4	0.4	2.3	0.6	Veses et al. (2020)

*Mixing ratio of waste paper: Kimichi **a** 5:5; **b** 23:7. *a—Not analysed

Table 14.6 **a** Heavy metal content of MSW Biochar (Al, Ca, K, Mg, P, S and Co). **b** Heavy metal content of MSW Biochar (Cr, Cu, Fe, Mn, Ni, Pb, Zn and Cd)

(a)

S.NO	Biochar	Temp (°C)	Al (mg/kg)	Ca (mg/kg)	K (mg/kg)	Mg (mg/kg)	P (mg/kg)	S (mg/kg)	Co (mg/kg)	References
1	OFMSW-BC	300-1 h 300-5 h 500-1 h	–	–	524 ± 33 508 ± 29 597 ± 46	–	2720 ± 280 3810 ± 340 530 ± 20	–	–	Benavente et al. (2018)
2	MSW-BC SS-BC	400 350	14,200 14,300	114,900 29,500	21,800 4400	10,400 8000	–	–	–	Wongrod et al. (2018)
3	SS-BC	500	83,000	10,000	760	1200	96,000	23,000	3700	Figueredo et al. (2017)
4	SS-BC	500 600 700 800 900	28,417.35 28,332.37 32,848.06 34,454.77 35,501.94	59,286 90 62,719 38 64,373 64 65,839 62 69,560 08	8518.52 8498.59 9938.32 9289.42 8684.18	14,736.07 15,457.60 16,369.21 16,573.98 17,523.36	18,185.58 18,760.23 20,350.03 19,348.41 20,238.79	–	–	Chen et al. (2014)
5	MSW-BC	500	2606	17,205	31,738	5982	–	–	–	Present study

(b)

S. No.	MSW biochar	Tem (C)	Cr (mg/kg)	Cu (mg/kg)	Fe (mg/kg)	Mn (mg/kg)	Ni (mg/kg)	Pb (mg/kg)	Zn (mg/kg)	Cd (mg/kg)	References
1	MSW-BC	450	9 27	10.90	1810	305	1.81	2.48	82.80	ND	Jayawardhana et al. (2017)

(continued)

Table 14.6 (continued)

(b)

S. No.	MSW biochar	Tem (C)	Cr (mg/kg)	Cu (mg/kg)	Fe (mg/kg)	Mn (mg/kg)	Ni (mg/kg)	Pb (mg/kg)	Zn (mg/kg)	Cd (mg/kg)	References
2	MSW-BC	450 550 650	35 29 29	163 187 160	–	–	45 23 18	192 193 160	987 791 735	3 <1 2	Taherymoosavi et al. (2017)
3	MSW-BC	400 500 600	43 ± 3 64 ± 2 103 ± 1	63 ± 3 101 ± 1 157 ± 4	–	–	67 ± 16 143 ± 10 599 ± 2	1 ± 0 10 ± 1 38 ± 0	149 ± 17 213 ± 32 356 ± 57	5 ± 1 5 ± 0 6 ± 1	Jin et al. (2014)
4	OFMSW-BC	300-1 h 300-5 h 500-1 h	17 ± 5 12 ± 3 17 ± 2	138 ± 1 150 ± 1 163 ± 2	–	–	82 ± 1 163 ± 1 197 ± 1	183 ± 9 185 ± 13 166 ± 13	326 ± 5 361 ± 19 496 ± 14	7 ± 1 9 ± 1 11 ± 1	Benavente et al. (2018)
5	MSW-BC	300 400 500 700	–	112.5 117.4 143.1 140.9	3853 3915 4004 4283	156 150 176 188	–	–	304 268 289 257	–	Zornoza et al. (2016)
6	SS-BC	500	69 ± 2	375 ± 15	26,113 ± 23	167 ± 5	17 ± 1	30 ± 1	480 ± 7	–	Figueredo et al. (2017)
7	SS-BC	500 600 700 800 900	100.34 100.94 114.58 105.81 112.02	202 44 208 48 242 30 201 49 183 71	31,123.70 33,606.39 35,326.67 35,769.17 37,202.34	749.27 760.43 833.45 799.40 836.45	–	51.52 ND ND ND 5.81	–	3.37 3.70 ND ND ND	Chen et al. (2014)

(continued)

Table 14.6 (continued)

(b)

S. No.	MSW biochar	Tem (C)	Cr (mg/kg)	Cu (mg/kg)	Fe (mg/kg)	Mn (mg/kg)	Ni (mg/kg)	Pb (mg/kg)	Zn (mg/kg)	Cd (mg/kg)	References
8	MSW-BC	500	–	10.4	2657	131.2	–	2.65	–	–	Present study

MSW-BC: MSW biochar; SS-BC: Sewage sludge biochar

positive impact and longer residence time more than two hours shows a negative effect on surface area. Pyrolysis exhibiting high surface areas that can be converted to an activated carbon, a favourable substitute for the marketable activated carbon due to cheap, large BET surface area and good CEC (Siipola et al. 2020). Larger surface area biochar has higher water retention capacity and gas absorption capacity. Lehmann and Joseph (2009) demonstrated that micropores less than 2 nm dia responsible for the surface area of biochars of 750–1360 m^2/g that conduct adsorptive capacities and the macropores less than 50 nm have a smaller surface area of 51–138 m^2/g that conduct soil functions.

- **Density**: Bulk density is the sum of the volume specific weight of a bulk compound along with pores in the solid structure and voids between particles of the bulk. The biochar energy density elevates with rising residence period (Kumar et al. 2020). The pyrolysis of sewage sludge showed that the sewage sludge biochar bulk density increased with increasing pyrolysis temperature (Khanmohammadi et al. 2015). The feedstock and pyrolysis conditions mainly affects the biochar bulk density and their biochar additions on soil decreases the soil bulk density (Ennis et al. 2012).
- **Water Holding Capacity**: The WHC mainly depends on the biochar's porosity and increasing temperature ultimately increases the hydrophobic biochar nature, as it occupies more water in the pores (Weber and Quicker 2018). The pyrolysis of oak saw dust at 400, 500 and 600 °C gave biochars with increasing WHC of 1.16, 1.35 and 1.49 mL/g, respectively (Ulusal and Apayd 2021). The parameters such as surface functional groups, pore volume, pore size and surface area primarily impacts the WHC. During the pyrolysis of tea pruning litter in the range of 250–500 °C showed an increased WHC of the produced biochars with increasing pyrolysis temperatures (Borgohain et al. 2020). The WHC is showed as one of the principal component (PC1) in the factor loadings.

14.4.2.2 Chemical Properties of MSW Biochar

The chemical properties of MSW biochar are described as follows:

- **pH**: pH of the biochar is important for the soil amendment applications. Increasing pH results in increase in biochars alkalinity. From Table 14.2 it is clear that MSW-BC is usually higher than 7 and goes as high as 9.7 (Jayawardhana et al. 2017). Municipal sewage sludge biochar pH was in the range of neutral to 12.15 (Chen et al. 2014), exceptional in some studies (Figueredo et al. 2017). The biochars produced at higher temperatures, resulted in alkaline pH (Chen et al. 2014; Tang et al. 2019) and very high pH are not suitable for soil applications.
- **Ash Content**: MSW feedstock ash content plays a major role on the biochar ash content and usually the ash content of biochar is higher than the MSW feedstock. Similar to pH, WHC and surface area properties, elevating the temperature results in higher ash composition as tabulated in the Table 14.5. The ash content varied as low as 6.1% (Jin et al. 2014) to more than 70% (Veses et al. 2020), based on the initial MSW feedstock characteristics.

- **Electrical Conductivity**: The biochar EC is prospectively associated with ash composition. Biochar EC elevates with elevating temperature and residence period, as studied by Benavente et al. (2018). MSW biochar EC produced at 300 °C with longer residence time of five hours gave higher value (470 ± 50 μS/cm) than biochar EC produced at one hour (450 ± 30 μS/cm).
- **Fixed Carbon and Volatile Matter**: Pyrolysis reaction releases volatile compounds and the left-out carbon is considered as fixed carbon. It shows a lower content of volatiles and a higher amount of fixed carbon than initial MSW feedstock. The lower pyrolysis temperature and slower heating rates produces good fixed carbon content and volatile matter content in biochar (as shown in Table 14.5; Jin et al. 2014; Taherymoosavi et al. 2017; Chen et al. 2014; Li et al. 2015) that is favourable for soil amendment studies. The pyrolysis pressure also impacts biochar carbon, where elevated pressure causes increase in biochar carbon and energy densification of biochar is formed.
- **Elemental Composition**: The main elements of MSW biochar are carbon, hydrogen, oxygen, and nitrogen as tabulated in Table 14.5. The other elements present in biochar with minimum quantities are P, S, K, Ca, Al, Mg, Co, Cr, Cu, Fe, Mn, Ni, Pb, Zn and Cd as given in Table 14.6a, b. The carbon is the main component of biochar, it accumulates with elevating pyrolysis temperature, whereas it decreases with hydrogen, oxygen and nitrogen content (Table 14.5). Hydrogen gives the structure for the biochar contained ionized compounds in it and organic/inorganic oxygen phases are available in biochar. The composition of nitrogen depends upon the MSW feedstock composition of amino acids and proteins.

  MSW biochar carbon content can be as high as 80–85% (Taherymoosavi et al. 2017; Li et al. 2015). Figueredo et al. (2017) produced biochar from municipal sewage sludge at 300 and 500 °C contained organic carbon of 24.31 and 18.86 g/kg, respectively. Biochar produced at 300 °C for a longer residence time of five hours contained higher organic carbon content (223 g/kg) than produced at 300 °C at one hour residence time (Benavente et al. 2018). The ratios of carbon, hydrogen and oxygen decides the aliphatic and aromatic structures. The ratios of C/N, H/C and O/C plays a vital role in biochar application and ratios of various MSW biochar studies are summarised in Table 14.4. The biochar atomic ratios of H/C and O/C are lesser value than the MSW feedstock. The biochar with low H/C ratios had more aromatic structures than lower temperature biochars. The biochar yield, H/C and O/C ratio are inversely proportional to the pyrolysis temperature.
- **Heavy Metals Content**: Composition of heavy metals such as Ca, Al, Mg, Co, Cr, Cu, Fe, Mn, Ni, Pb, Zn and Cd in MSW-biochar are typically based on their feedstock composition and pyrolysis temperature as tabulated in Table 14.6a, b. It demonstrates that lower pyrolysis temperature yield an increase in the heavy metal contents due to metals not easily evaporated causing deposition in the biochar. The municipal sewage sludge biochar produced by Chen et al. (2014) and Figueredo et al. (2017) composed of most of the heavy metals composition (P, S, K, Ca, Al, Mg, Co, Cr, Cu, Fe, Mn, Ni, Pb, Zn and Cd). The present study MSW biochar from

kitchen waste contained 9 heavy metals content of Pb—2.65, Al—2606, Ba—28.3, Ca—17,205, Cu—10.4, Fe—2657, Mg—5982, Mn—131.2, Na—7850 and K—31,738 mg/kg, respectively. Jin et al. (2014) demonstrated the increase in metal content with elevating pyrolysis temperatures. The heavy metal content was higher in MSW biochar pyrolysed at 500 °C (Cd-5, Cr-64, Cu-101, Zn- 213, Ni-143 and Pb-10 mg/kg, respectively) than the raw MSW (Cd-3, Cr-42, Cu-30, Zn-53, Ni-34, and Pb-0 mg/kg, respectively).

- **Cation Exchange Capacity**: It is a quantity of surface charge and it escalates with the age of biochar, as its functional groups accumulates on its surface. MSW biochar produced at longer residence time of five hours had higher CEC (73.14 $\pm$ 2.02 cmolc/kg) than at lower residence of one hour (52.35 $\pm$ 0.87 cmolc/kg) (Benavente et al. 2018). The higher capacity of CEC has higher potential to adsorb nutrients. Municipal sewage sludge biochar's CEC increased by increasing temperature from 600 to 900 °C are from 30.81 $\pm$ 2.67, 50.34 $\pm$ 2.73, 126.62 $\pm$ 9.16 to 247.51 $\pm$ 7.49 cmolc/kg, respectively (Chen et al. 2014). Usually, CEC decreases because of the aromatic carbon oxidation at very high temperature leads to carboxylic group formation.
- **Functional groups**: Biochar contains aliphatic and aromatic compounds. The surface functionalities of the biochar mainly depends on temperature. It has surface functional groups of hydroxyl, epoxy, carboxyl, quinonyl, acyl, carbonyl, ether, and esters. Taherymoosavi et al. (2017) demonstrated the variation in the organic compound concentrations and its interactions with mineral phases of MSW biochar produced at temperatures of 450, 550 and 650 °C, respectively. At 450 °C, MSW biochar has complex organic concentrations and dissolved organic carbon compounds. At 550 °C, biochar surface contained Si/Al/O-rich phases and micropores contained Ca/P/O-rich phases, also dissolved organic carbon compounds present in it. The higher temperature biochar contained higher fixed carbon concentration with more biochar stability. At 650 °C, low molecular weight acids were formed (59.9%). According to the MSW biochar properties, it is widely applied in various fields of contaminant removal, soil amendment, bio-energy and climate change mitigation.

14.5 MSW Biochar Application

Some of the MSW biochar applications are depicted in Fig. 14.2. Many research studies have been carried out using MSW biochar and modified MSW biochar for contaminant removal from wastewaters using adsorption are as follows.

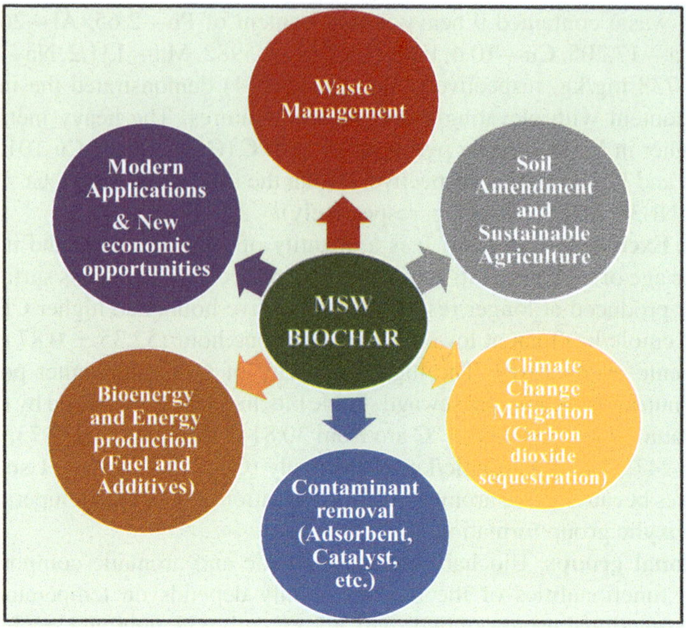

Fig. 14.2 MSW biochar applications

14.5.1 MSW Biochar for Contaminant Removal Using Adsorption

MSW biochar has high adsorption capacity for removing organic contaminants such as dyes and pharmaceuticals compounds (Jayawardhana et al. 2017; Parshetti et al. 2014; Ashiq et al. 2019; Jung et al. 2019) and, inorganic contaminants such as heavy metals of Cr (Agrafioti et al. 2014; Hoslett et al. 2019) As (Agrafioti et al. 2014), Pb, Cd (Abedinzadeh et al. 2020), Zn, Ni (Jin et al. 2014) and nutrients from wastewater.

14.5.1.1 Organic Contaminants Removal by MSW Biochar

The methylene blue dye adsorption using MSW biochar from MSW feedstock in heat pipe pyrolyser at 300 °C for 12 h was studied by Hoslett et al. (2020). The highest adsorption of 7.2 mg/g for 100 mg/l of methylene blue concentration was achieved. The removal of crystal violet and congo red dye removal using two MSW biochar was conducted by Jung et al. (2019). The two MSW biochars was produced from gasification of solid refuse fuel (SRF) from MSW operated at a high steam flow rate (104 kg/h and low air supply of 41% O_2 content; 201 Nm^3/h) and a low steam flow rate (37 kg/h and air supply of 214 Nm^3/h). The MSW biochar produced at low steam rate gave higher congo red and crystal violet removal with the biochar properties of higher BET surface area of 11.4 m^2/g, pore volume of 0.0392 cm^3/g, porosity of 74.7%, FC of 25.8 wt.% and ash of 54.54% than other biochar. The acridine orange and rhodamine 6G dyes removal by MSW biochar produced from urban food waste through hydrothermal carbonisation at 250 °C was investigated by Parshetti et al. (2014). MSW biochar effectively removed the acridine orange and rhodamine 6G dyes concentration of 10—100 mg/L with highest adsorption efficiency of 79.36 mg/g and 71.42 mg/g, respectively. The adsorption mechanisms involved were Van der Waals forces, electrostatic interactions, and hydrogen bonding. The organic contaminant removal of benzene from landfill leachate by MSW biochar was studied by Jayawardhana et al. (Jayawardhana et al. 2017). The fibrous organic fraction MSW pyrolysed at 450 °C was used in the study and its properties are as described in Tables 14.1, 14.4, 14.5, 14.6a, b. The benzene removal was increased with higher pH and highest adsorption efficiency of 576 μg/g was achieved at pH 9 through pseudo-second-order by chemisorption mechanism.

14.5.1.2 Inorganic Contaminants Removal by MSW Biochar

The heavy metal removal of Cr (VI) and As (V) using MSW biochar and municipal sewage sludge biochar was studied by Agrafioti et al. (2014). MSW biochar was prepared by pyrolysis of organic component of MSW from recovery facility and municipal sewage sludge biochar produced from wastewater treatment plant at 300 °C in a muffle furnace for sixty minutes of residence period. The BET surface areas of sewage sludge biochar and MSW biochar was 51 and 5 m^2/g respectively and its pore volumes were 0.153, 0.0584 and 0.0291 cm^3/g, respectively. Sewage sludge biochar showed higher removal of 89% of Cr (VI) and 53% of As (V), than MSW biohcar removal 44% of Cr (VI) and 55% of As (V). Both removed 99% of Cr (III). Copper removal using MSW biochar was studied by Hoslett et al. (2019). The mixed municipal discarded material composed of plastic, meat, paper and food waste pyrolysed in a heat-pipe reactor at 300 °C under N_2 atmosphere for 12 h. The produced MSW biochar had the highest adsorption of 4–5 mg/g at pH >4.5, where initial Cu content was about 50–100 mg/L. The heavy metals removal of Cd and Pb contaminated water irrigated on forage maize using MSW biochar and *E. cloacae R7* strain was explored by Abedinzadeh et al. (2020). It resulted in adsorption capacity of Pb was 25.23 mg/kg and Cd was 23.41 mg/kg with initial concentrations of 100 and 5 mg/L, respectively.

14.5.2 Chemically/Physically Activated MSW Biochar for Contaminant Removal

Several physical and chemical activation of MSW biochar studies were carried out to increase the surface area and porosity for higher adsorption efficiency as tabulated in the Table 14.3. The As (V) removal using chemically activated MSW biochar was studied by Jin et al. (2014) and its preparation using KOH is given in Table 14.3. The activated biochar had increased the surface area, porous structure with functional groups gave the highest As (V) adsorption capacity of 30.98 mg/g more than the pristine biochar (24.49 mg/g). Similar chemical activation was used for Cr removal was studied by Paranaque et al. (2020). The chemical activation increased the surface area to 6.5 m^2/g, whereas pristine MSW biochar was 2.6 m^2/g. This increased surface area for chemically activated MSW biochar at pH 12 increased the maximum Cr total removal to 98.97% containing Cr concentration of 10 mg/mL. Also, nitrate removal studies were carried out using KOH activated MSW biochar by Revilla et al. (2020). The highest nitrate removal at pH 2 with initial 30 mg/L and ten milligram per milliliter of adsorbent was 66.97%. KOH activated MSW biochar shown to extract humic acid compounds studied by Genuino et al. (2017). MSW biochar produced by pyrolysing the MSW feedstock (procedure as tabulated in Table 14.1) was experimented for the effect of different concentrations of KOH (0.25–0.75 M), extractant dose (10–30 g/L), contact time (1–12 h) and precipitant volume (0.5–2.5 mL) on the humic acid yield and optimized using Artificial Neural Network (ANN). Humic acid yield from ANN model had a 3.71% variation from the predicted humic acid (180.57 mg/g) and experimented (187.52 mg/g). Similarly, another KOH activation MSW biochar study for the removal of acetaminophen and methylene blue was conducted by Sumalinog et al. (2018). MSW biochar thermally activated at 683 °C in N_2 atm for fifty three minutes and subsequently soaked in 0.1 M hydrochloric acid and washed by water. The methylene blue adsorption was effective than acetaminophen with maximum removal of 99.9%. The pH impacted the acetaminophen adsorption where pH change from 2 to 12 reduced the adsorption capacity by 89%. Whereas, methylene blue not affected when pH changed from two to ten with adsorption of 32.9–33.3 mg/g and removal efficiency greater than 98.8%. In another study, CO_2 activation of MSW biochar produced at 408 °C for methylene blue was studied by Gopu et al. (Gopu et al. 2018). Biochar CO_2 activation at 600 and 900 °C for 30 min, revealed surface areas over 300 m^2/g. Methylene blue adsorption capacity was 250 mg/g for activated biochars and followed Langmuir isotherm.

The experiments with potassium hydroxide, hydrogen peroxide and MSW-biochar washing on lead adsorption were investigated by Wongrod et al. (2018). The biochars were produced from sewage sludge digestate and MSW digestate at 350 °C, further were chemically modified using 2 M KOH or 10% H_2O_2 and subsequently washed by semi-continuous or continuous operation with ultrapure water. The BET specific surface areas of sewage sludge semi-continuous washing biochar treated with H_2O_2 and KOH gave 3.6 ± 0.1 and 3.0 ± 0.1 m^2/g, respectively. The BET surface area of sewage sludge continuous washing biochar with H_2O_2 and KOH gave 5.7 ± 0.1 and 7.9 ± 0.1 m^2/g, respectively. Pb adsorption of MSW biochar gave 73 mg/g, whereas chemically activated biochar by H_2O_2 was 90 mg/g and KOH was 106 mg/g, respectively. The sewage sludge biochar gave 25 mg/g after H_2O_2 activation. In another study, the Pb adsorption by biochar from sludge pre-deashed by HCl or HF and subsequent activation with potassium acetate was investigated by Zhang et al. (2020). The pre-dashing with HCl or HF acid increase the biochar pore size thereby further activation by potassium acetate. The adsorption capacity increased after hydrofluoric acid treatment (16.70 mg/g) and potassium acetate activation (49.47 mg/g) than untreated biochar (7.56 and 38.49 mg/g).

The elemental mercury (Hg^0) removal using chemical and physically activated MSW biochar was studied by Li et al. (2015). The preparation of the biochar activation is summarised in the Table 14.3, where physical activation by microwave steam and chemical activation by NH_4Cl. Both activation processes together and separately were compared for removal performance. The combined physical and chemically activated MSW biochar showed better removal of mercury than other sorbents and especially that pyrolyzed at 600 °C gave Hg0 better removal than pyrolyzed at 400 °C, adsorption by chemisorption process.

The ciproflaxin (CPX) removal using MSW biochar and modification of MSW biochar composite of bentonite clay was studied by Ashiq et al. (2019). MSW from dumpsite was pyrolysed at 450 °C in a muffle furnace with 30 min residence time. Subsequently, MSW biochar-clay composite was prepared by bentonite-MSW slurry (1:5 ratio) similar to earlier conditions. The modification increased the functional groups on the MSW biochar clay composite. The MSW biochar-clay composite showed higher adsorption for CPX at pH 6 of mildly acidic condition with adsorption capacity of 190 mg/g with 40% higher efficiency than MSW biochar. Similarly, a composite of montmorillonite (MMT) with MSW biochar for CPX removal was conducted (Ashiq et al. 2019a). The specific surface area of the adsorbents of MSW biochar, MMT and MSW biochar-MMT were 4.33, 35.75 and 6.51 m^2/g, respectively. The maximum CPX removal was achieved in MSW biochar-MMT composite with π-induced electrostatic interactions as primary mechanism in the adsorption.

14.5.3 MSW Biochar for Soil Amendment

Several soil amendment studies were conducted using MSW biochar and sewage sludge biochars (Abedinzadeh et al. 2020; Tian et al. 2019). MSW biochar pyrolysed at 500 °C from organic fraction of MSW along with bacteria E. cloacae R7 have profound effect on the forge maize plant development was demonstrated by Abedinzadeh et al. (2020). The characteristics of the produced biochar is given in the Tables 14.1 and 14.4. The MSW biochar and strain gave a profound impact on elevating the root and shoot dry weight from 29 to 33% and 32 to 43%, respectively. Another study showed that the municipal sewage sludge biochar application to soil has good turf grass growth explored by Tian et al. (2019). The sewage sludge biochar onto turf grass resulted in increased soil organic carbon from 3 to 8 times, black carbon by 7–25 times, total nitrogen by 2–9 times, available phosphorus by 10–19 times and potassium by 1.4 to two folds. Also, grass dry weight elevated from 43 to 147% after biochar soil amendment, whereas 4–70% lesser heavy metals were observed in soil.

Biochar production from tea (*Camellia sinensis L.*) pruning litter from tea growing region of Assam, India at varying pyrolysis temperatures in the range of 250–500 °C was studied for soil applications (Borgohain et al. 2020). The study showed biochars having appropriate properties required for agricultural applications such as H:C and O:C ratios less than 0.6 and 0.4, respectively. In another study, slow pyrolysis wood biochar was applied in Indian soil conditions for the crop productivity showed beneficial agronomic outcomes (Deb et al. 2016). The MSW biochar also has climate change mitigation potential by reducing the gaseous emissions.

14.5.4 MSW Biochar for Climate Change Mitigation

The carbon imbalance prevails in the environment between carbon release and uptake that leads to atmospheric CO_2 equivalent (CO_2eq) to 9.5 PgC/yr (Peters et al. 2012). Therefore, a viable option for storing the carbon from the atmosphere for prolonged period is important. Biochar can store carbons for several years and influence of biochar on gaseous emission reduction is detailed elsewhere (Sri Shalini et al. 2021). Several studies has described the advantages of biochar for reduction in GHGs and improving the soil fertility and agriculture (Ashiq et al. 2019; Lehmann 2007; Woolf et al. 2010). Still more research gaps to be addressed with respect MSW biochar to soil health improvement and greenhouse gaseous reduction mechanisms for achieving the climate change mitigation potential. The benefits of biochar systems are way different from other energy processes, as biochar a carbon-negative process removes atmospheric CO_2 and store as soil carbon (Lehmann et al. 2006). The soil biomass mineralization and microbial respiration emits CO_2, which could be trapped by the biochar added to the soil (Qambrani et al. 2017). The biochar is recalcitrant in nature hence it can present for longer duration in the soil.

An emission factor for biochar technology was studied by Roberts et al. (2010). By biochar technology, a net GHGs emissions possibly lowered to extent of 1.8 Pg CO_2eq per year (i.e. 12% of current anthropogenic CO_2eq emissions) and 130 Pg CO_2eq in a century. Biochar with greater climate change mitigation potential when compared to combustion of bioenergy (Woolf et al. 2010). Biochar largely reduces the atmospheric carbon dioxide levels. Biochar from prevailing biomass flows shall eliminate 0.77–0.87 PgC/yr (Lenton 2010). The potential of the carbon dioxide removal (CDR) by biochar production is at 0.56 PgC/yr and it is 5.5–9.5 PgC/yr for 180–310 EJ/yr all by pyrolysis process (Lehmann et al. 2006). The CDR by all residues converting to biochar is around 0.16–0.34 PgC yr-1 (Lenton 2010). All biomass energy by pyrolysis then CDR will be 0.18 PgC/yr (Lehmann et al. 2006). The sewage sludge life cycle of energy system was compared with combined anaerobic digestion and fast pyrolysis process and, a fast pyrolysis process for GHG emissions (of CO_2eq) (Cao and Pawłowski 2013). The study showed that net GHG emissions of combined system gave—15.8 t of CO_2eq, whereas fast pyrolysis gave—11.8 t CO_2eq. Net negative GHG emissions have an optimistic impact on climate change as abundant GHG is consumed than emitted. Biochar is a carbon neutral or carbon negative product (Hasan et al. 2021). The climate neutrality can be achieved by this biochar carbon negative system.

14.5.5 MSW Biochar for Bioenergy and Energy Production

MSW organic waste has organic compounds prospective for energy storage. Pyrolysis produces bioenergy substances of bio-oil and syngas (Parañaque et al. 2020). The syngas for power production and bio-oil as alternate for transport fuel (Woolf et al. 2010). The syngas containing combustible substances between 36 and 54% for sewage sludge with their lower heating value of 11.8–19.1 MJ/m^3 and pyrolysis gases for organic fraction of MSW was 62–72% and their lower heating value was 18.2–21.0 MJ/m^3 (Agar et al. 2018). Fast pyrolysis produces more bio-oil than slow pyrolysis. Biochar containing lesser ash have greater heating condition apt for fuels. Fang et al. (2018a) demonstrated the pyrolysis with ultrasound lowered the oxygenated substances in bio-oil thereby increasing its fuel quality. Roberts et al. (2010) considered four feedstock stover, early stover, switchgrass, and yard waste for demonstrating the energy balance during pyrolysis. Biochar shown to have higher energy production than utilised energy for the feedstock and positive net energies for dry late stover was 4116 MJ/t, early stover was 3044 MJ/t, switch grass was 4899, and yard waste was 4043 MJ/t, respectively. The energy yield is inversely proportional to the ash content present in the produced biochar (Kumar et al. 2020). Moreno et al. (2020) studied in detail about the biomass conversion by TCR process and its techno-economic aspects. The other techno-economic aspects in biochar production is as follows.

14.6 Techno-economic Analysis of Biochar Production

Techno-economic analysis (TEA) is vital for discovering feasibility and viability of the biochar for its commercial production. TEA not only consider the operational cost, it's inclusive of cost associated with feedstock collection, operation and production costs. The assessment of biochar value for fast and slow pyrolysis gave 2.85 $ and 10.98 $/t feedstock, respectively (Kung et al. 2013) and net loss of 26.9 and 20.5 $/t feedstock, respectively. This revealed that fast pyrolysis attributable for higher power production with lower economic and environmental profitability (Kumar et al. 2020). The biochar price might also be higher due to the lack of production and high demand for it (Vochozka et al. 2016). The life cycle analysis shows the breakeven price for biochar production at only US$7 Mg/C from yard waste, but US$147 Mg/C from crop residues and US$227 Mg/C from bioenergy crops (Roberts et al. 2010). If biochar is produced from waste residues, then it is cheaper rather than using high-tech equipment's for its production (Vochozka et al. 2016).

This study revealed the MSW feedstock to be a feasible and cost-effective option for biochar production and its wide application makes it market profitable in the long-run. In India, the legal framework for waste management is the Solid waste management rules, 2016. Yet policies related to biochar production is to be framed. The outcomes of this study could help policy-makers for designing a policy or legal framework on the biochar technologies.

14.7 Conclusion

This chapter covered in detail about the scientific and technical aspects of MSW biochar production, reactor conditions, activation and modification studies with their wide application in different fields. The different MSW feedstock was explored for MSW biochar production by various pyrolysis methods. MSW with high organic fraction components shown to be effective for carbon-rich biochar production. Kitchen waste pyrolysis produced good biochar yield and properties. The feedstock properties and pyrolysis conditions are the deciding parameters for the biochar quality. MSW activated and modified biochar shown to have wider applications due to their improved catalytic properties, higher surface area and porosity. The MSW biochar demonstrated efficient removal for organic and inorganic contaminants, potential for soil applications and bio-energy generation. Carbon-negative system of biochar production a significant prospect for the climate change mitigation. The study outcomes help the policy-makers for a legal framework on biochar technologies. Still some research gaps in MSW biochar applications for climate change potential in real-time scenario has to be explored.

Acknowledgements The corresponding author (Dr. S. Sri Shalini) gratefully acknowledges the financial support provided by the Department of Science and Technology (DST), Ministry of Science and Technology, Government of India under Women Scientists Scheme-A (WOS-A) (Grant no. SR/WOS-A/EA-37/2018).

References

Abedinzadeh M, Etesami H, Alikhani HA, Shafiei S (2020) Combined use of municipal solid waste biochar and bacterial biosorbent synergistically decreases Cd (II) and Pb (II) concentration in edible tissue of forage maize irrigated with heavy metal–spiked water. Heliyon 6(8):e04688

Agar DA, Kwapinska M, Leahy JJ (2018) Pyrolysis of wastewater sludge and composted organic fines from municipal solid waste: laboratory reactor characterisation and product distribution. Environ Sci Pollut Res 25:35874–35882

Agrafioti E, Kalderis D, Diamadopoulos E (2014) Arsenic and chromium removal from water using biochars derived from rice husk, organic solid wastes and sewage sludge. J Environ Manag 133:309–314

Aguado J, Serrano DP, Escola JM, Garagorri E (2002) Catalytic conversion of low density polyethylene using a continuous screw kiln reactor. Catal Today 75:257–262

Agyarko-Mintah E, Cowie A, Singh BP, Joseph S, Van Zwieten L, Cowie A, Harden S, Smillie R (2017) Biochar increases nitrogen retention and lowers greenhouse gas emissions when added to composting poultry litter. Waste Manage 61:138–149

Al-Jarallah R, Aleisa E (2013) Investigating causes contributing to increased municipal solid waste in Kuwait: a national survey. J Eng Res 1:123–143

Ashiq A, Adassooriya NM, Sarkar B, Rajapaksha AU, Ok YS, Vithanage M (2019) Municipal solid waste biochar-bentonite composite for the removal of antibiotic ciprofloxacin from aqueous media. J Environ Manage 236:428–435

Ashiq A, Sarkar B, Adassooriya N, Walpita J, Rajapaksha AU, Ok YS, Vithanage M (2019a) Sorption process of municipal solid waste biochar-montmorillonite composite for ciprofloxacin removal in aqueous media. Chemosphere 236:124384

Ates F, Miskolczi N, Borsodi N (2013) Comparision of real waste (MSW and MPW) pyrolysis in batch reactor over different catalysts. Part I: product yields, gas and pyrolysis oil properties. Bioresour Technol 133:443–454

Benavente I, Gasco G, Plaza C, Paz-Ferreiro J, Mendez A (2018) Choice of pyrolysis parameters for urban wastes affects soil enzymes and plant germination in a mediterranean soil. Sci Total Environ 634:1308–1314

Bhardwaj G, Kumar M, Mishra PK, Upadhyay SN (2021) Kinetic analysis of the slow pyrolysis of paper wastes. Biomass Convers Biorefin

Borgohain A, Konwar K, Buragohain D, Varghese S (2020) Bioresource technology temperature effect on biochar produced from tea (Camellia sinensis L.) pruning litters : a comprehensive treatise on physico-chemical and statistical approaches. Bioresour Technol 318:124023

Cao Y, Pawłowski A (2013) Life cycle assessment of two emerging sewage sludge-to- energy systems: evaluating energy and greenhouse gas emissions implications. Bioresour Technol 127:81–91

Chen T, Zhang Y, Wang H, Lu W, Zhou Z, Zhang Y, Ren L (2014) Influence of pyrolysis temperature on characteristics and heavy metal adsorptive performance of biochar derived from municipal sewage sludge. Bioresour Technol 164:47–54

Chen D, Yin L, Wang H, He P (2015) Reprint of: pyrolysis technologies for municipal solid waste: a review. Waste Manag 37:116–136

Chen L, Yu Z, Liang J, Liao Y, Ma X (2018) Co-pyrolysis of chlorella vulgaris and kitchen waste with different additives using TG-FTIR and Py-GC/MS. Energ Convers Manage 177:582–591

Daful AG, Chandraratne M (2018) Biochar production from biomass waste-derived material. Reference Module in Materials Science and Materials Engineering, Elsevier Inc. (S&T Books and Cell Press), United Kingdom, pp 11249–11258–9

Deb D, Kloft M, Lässig J, Walsh S (2016) Agroecology and sustainable food systems variable effects of biochar and P solubilizing microbes on crop productivity in different soil conditions. Agroecol Sustain Food Syst 40(2):145–168

Ennis CJ, Evans AG, Islam M, Ralebitso-Senior TK, Senior E (2012) Biochar: carbon sequestration, land remediation, and impacts on soil microbiology. Crit Rev Environ Sci Technol 42(22):2311–2364

Fang S, Gu W, Chen L, Yu Z, Dai M, Lin Y, Liao Y, Ma X (2018a) Ultrasonic pretreatment effects on the co-pyrolysis of municipal solid waste and paper sludge through orthogonal test. Bioresour Technol 258:5–11

Fang S, Yu Z, Ma X, Lin Y, Chen L, Liao Y (2018b) Analysis of catalytic pyrolysis of municipal solid waste and paper sludge using TG-FTIR, Py-GC/MS and DAEM (distributed activation energy model). Energy 143:517–532

Ferronato N, Torretta V (2019) Waste mismanagement in developing countries: a review of global issues. Int J Environ Res Public Health 16(6):1060

Figueredo NA, Costa LM, Melo LCA, Siebeneichlerd EA, Tronto J (2017) Characterization of biochars from different sources and evaluation of release of nutrients and contaminants. Rev Cienc Agron 48(3):395–403

Font R, Marcilla A, Garcia AN, Caballero JA, Conesa JA (1995) Comparison between the pyrolysis products obtained from different organic wastes at high temperatures. J Anal Appl Pyrol 32:41–49
Garcia AN, Font R, Marcilla A (1995a) Kinetic study of the flash pyrolysis of municipal solid waste in a fluidized bed reactor at high temperature. J Anal Appl Pyrol 31:101–121
Garcia AN, Font R, Marcilla A (1995b) Gas production by pyrolysis of municipal solid waste at high temperature in a fluidized bed reactor. Energy Fuels 9:648–658
Genuino DAD, Bataller BG, Capareda SC, de Luna MDG (2017) Application of artificial neural network in the modeling and optimization of humic acid extraction from municipal solid waste biochar. J Environ Chem Eng 5(4):4101–4107
Genuino DAD, de Luna MDG, Capareda SC (2018) Improving the surface properties of municipal solid waste-derived pyrolysis biochar by chemical and thermal activation: Optimization of process parameters and environmental application. Waste Manag 72:255–264
Gopu C, Gao L, Volpe M, Fiori L, Goldfarb JL (2018) Valorizing municipal solid waste: waste to energy and activated carbons for water treatment via pyrolysis. J Anal Appl Phys 133:48–58
Gunarathne V, Ashiq A, Ramanayaka S, Wijekoon P, Vithanage M (2019) Biochar from municipal solid waste for resource recovery and pollution remediation. Environ Chem Lett 17:1225–1235
Haeldermans T, Campion L, Kuppens T, Vanreppelen K, Cuypers A, Schreurs S (2020) A comparative techno-economic assessment of biochar production from different residue streams using conventional and microwave pyrolysis. Bioresour Technol 318:124083
Hasan S, Girindran R, Zacharia PU, Jaya H, Kooren P, Sayooj P, Benny S, Joseph D, Hussain SV (2021) Climate resilient products development through valorization of *Eichhornia crassipes* to biofuel and biochar. Int J Environ Sci Technol
He M, Xiao B, Liu S, Hu Z, Guo X, Luo S, Yang F (2010) Syngas production from pyrolysis of municipal solid waste (MSW) with dolomite as downstream catalysts. J Anal Appl Pyrol 87:181–187
Hoslett J, Ghazal H, Ahmad D, Jouhara H (2019) Removal of copper ions from aqueous solution using low temperature biochar derived from the pyrolysis of municipal solid waste. Sci Total Environ 673:777–789
Hoslett J, Ghazal H, Mohamad N, Jouhara H (2020) Removal of methylene blue from aqueous solutions by biochar prepared from the pyrolysis of mixed municipal discarded material. Sci Total Environ 714:136832
Jahirul MI, Rasul MG, Chowdhury AA, Ashwath N (2012) Biofuels production through biomass pyrolysis—a technological review. Energies 5:4952–5001
Jayawardhana Y, Mayakaduwa SS, Kumarathilaka P, Gamage S, Vithanage M (2017) Municipal solid waste-derived biochar for the removal of benzene from landfill leachate. Environ Geochem Health 41(4):1739–1753
Jin H, Capareda S, Chang Z, Gao J, Xu Y, Zhang J (2014) Biochar pyrolytically produced from municipal solid wastes for aqueous As(V) removal: adsorption property and its improvement with KOH activation. Bioresour Technol 169:622–629
Joseph K (2014) Municipal solid waste management in India. In: Municipal solid waste management in Asia and the Pacific Islands. In: Pariatamby A, Tanaka M (eds) Environmental Science and Engineering. Springer-Verlag, Singapore, pp 113–138. ISBN 978-981-4451-72-7
Joshi R, Ahmed S (2016) Status and challenges of municipal solid waste management in India: a review. Cogent Environ Sci 2:1139434
Jung H, Sewu DD, Ohemeng-Boahen G, Lee DS, Woo SH (2019) Characterization and adsorption performance evaluation of waste char by-product from industrial gasification of solid refuse fuel from municipal solid waste. Waste Manag 91:33–41
Khan D, Kumar A, Samadder SR (2016) Impact of socioeconomic status on municipal solid waste generation rate. Waste Manag 49:15–25
Khanmohammadi Z, Afyuni M, Mosaddeghi MR (2015) Effect of pyrolysis temperature on chemical and physical properties of sewage sludge biochar. Waste Manag Res 33(3):275–283
Kim D, Park KY, Yoshikawa K (2017) Conversion of municipal solid wastes into biochar through hydrothermal carbonization. In: Huang W-H (ed) Engineering applications of biochar, IntechOpen

Kumar A, Bhattacharya T (2021) Biochar: a sustainable solution. Environ Dev Sustain 23(5):6642–6680

Kumar A, Saini K, Bhaskar T (2020) Hydochar and biochar: production, physicochemical properties and techno-economic analysis. Bioresour Technol 310:123442

Kung CC, McCarl BA, Cao X (2013) Economics of pyrolysis-based energy production and biochar utilization: a case study in Taiwan. Energy Policy 60:317–323

Lee Y, Kim S, Kwon EE, Lee J (2020) Effect of carbon dioxide on thermal treatment of food waste as a sustainable disposal method. J CO_2 Util 36:76–81

Lehmann J (2007) A handful of carbon. Nature 447:143–144

Lehmann J, Gaunt J, Rondon M (2006) Bio-char sequestration in terrestrial ecosystems—a review. Mitig Adapt Strateg Glob Chang 11:403–427

Lehmann J, Joseph S (2009) Biochar for environmental management: science and technology, 1st edn. Earthscan

Lehmann J, Da Silva JP Jr, Steiner C, Nehls T, Zech W, Glaser B (2003) Nutrient availability and leaching in an archaeological Anthrosol and a Ferralsol of the Central Amazon basin: fertilizer, manure and charcoal amendments. Plant Soil 249:343–357

Lenton TM (2010) The potential for land-based biological CO_2 removal to lower future atmospheric CO_2 concentration. Carbon Manag. 1(1):145–160

Li AM, Li XD, Li SQ, Yan JH, Cen KF (1999) Experiment on manufacture medium heating value flue gas by pyrolyzing municipal refuse in a rotary kiln. J Chem Ind Eng (china) 50:101–107

Li G, Shen B, Li F, Tian L, Singh S, Wang F (2015) Elemental mercury removal using biochar pyrolyzed from municipal solid waste. Fuel Process Technol 133:43–50

Li J, Dai J, Liu G, Zhang H, Gao Z, Fu J, He Y (2016) Biomass and bioenergy biochar from microwave pyrolysis of biomass: a review. Biomass Bioenerg 94:228–244

Li Y, Xing X, Ma P, Zhang X, Wu Y, Huang L (2020) Effect of alkali and alkaline earth metals on co-pyrolysis characteristics of municipal solid waste and biomass briquettes. J Therm Anal Calorim 139:489–498

Liu P, Liu WJ, Jiang H, Chen JJ, Li WW, Yu HQ (2012) Modification of bio-char derived from fast pyrolysis of biomass and its application in removal of tetracycline from aqueous solution. Bioresour Technol 121:235–240

Liu G, Xie M, Zhang S (2017) Effect of organic fraction of municipal solid waste (OFMSW)-based biochar on organic carbon mineralization in a dry land soil. J Mater Cycle Waste Manag 19:473–482

Lu JS, Chang YJ, Poon CS, Lee DJ (2020) Slow pyrolysis of municipal solid waste (MSW): a review. Bioresour Technol 123615

Luo SY, Xiao B, Hu ZQ, Liu SM, Guan YW, Cai L (2010) Influence of particle size on pyrolysis and gasification performance of municipal solid waste in a fixed bed reactor. Bioresour Technol 101:6517–6520

Ma Z, Gao N, Xie L, Li A (2014) Study of the fast pyrolysis of oil- field sludge with solid heat carrier in a rotary kiln for pyrolytic oil production. J Anal Appl Pyrol 105:183–190

Mankasingh U, Choi P, Ragnarsdottir V (2011) Biochar application in a tropical, agricultural region: a plot scale study in Tamil Nadu India. J Appl Geochem 26:218–221

Miskolczi N, Ateş F, Borsodi N (2013) Comparison of real waste (MSW and MPW) pyrolysis in batch reactor over different catalysts. Part II: contaminants, char and pyrolysis oil properties. Bioresour Technol 144:370–379

Mohanty P, Nanda S, Pant KK, Naik S, Kozinski JA, Dalai AK (2013) Evaluation of the physiochemical development of biochars obtained from pyrolysis of wheat straw, timothy grass and pinewood: effects of heating rate. J Anal Appl Pyrol 104:485–493

Moreno VC, Iervolino G, Tugnoli A, Cozzani V (2020) Techno-economic and environmental sustainability of biomass waste conversion based on thermocatalytic reforming. Waste Manag 101:106–115

Nair RR, Mondal MM, Weichgrebe D (2020) Biochar from co-pyrolysis of urban organic wastes—investigation of carbon sink potential using ATR-FTIR and TGA. Biomass Convers Biorefin

Ouadi M, Jaeger N, Greenhalf C, Santos J, Conti R, Hornung A (2017) Thermo-catalytic reforming of municipal solid waste. Waste Manag 68:198–206
Parañaque JEL, Maguyon-Detras MC, Migo VP, Alfafara CG (2020) Chromium removal from chrome-tannery effluent after alkaline precipitation by adsorption using municipal solid waste-derived activated Biochar. IOP Conf Ser Mater Sci Eng 778:012134
Parmar A, Nema P, Agarwal T (2014) Biochar production from agro-food industry residues: a sustainable approach for soil and environmental management. Curr Sci 107(10):1673–1682
Parshetti GK, Chowdhury S, Balasubramanian R (2014) Hydrothermal conversion of urban food waste to chars for removal of textile dyes from contaminated waters. Bioresour Technol 161:310–319
Pawar A, Panwar NL (2020) Experimental investigation on biochar from groundnut shell in a continuous production system. Biomass Convers Biorefin
Penteado SC, Castro MAS de (2021) Covid-19 effects on municipal solid waste management : what can effectively be done in the Brazilian scenario. Resour Conserv Recycl 164:105152
Peters G, Andrew R, Boden T, Canadell J, Ciais P, Le Quere C, Marland G, Raupach M, Wilson C (2012) The challenge to keep global warming below 2 °C. Nat Clim Change 3:4–6
Priyadarshini P, Abhilash PC (2020) Circular economy practices within energy and waste management sectors of India: a meta-analysis. Bioresour Technol 304:123018
Qambrani NA, Rahman MM, Won S, Shim S, Ra C (2017) Biochar properties and eco-friendly applications for climate change mitigation, waste management, and wastewater treatment: a review. Renew Sustain Energy Rev 79:255–273
Rehrah D, Bansode RR, Hassan O, Ahmedna M (2016) Physico-chemical characterization of biochars from solid municipal waste for use in soil amendment. J Anal Appl Pyrol 118:42–53
Revilla PND, Maguyon-Detras MC, Migo VP, Alfafara CG (2020) Nitrate removal from aqueous solution by adsorption using municipal solid waste-derived activated biochar. IOP Conf Ser Mater Sci Eng 778:012135
Roberts KG, Gloy BA, Joseph S, Scott NR, Lehmann J (2010) Life cycle assessment of biochar systems: estimating the energetic, economic, and climate change potential. Environ Sci Technol 44(2):827–833
Sajjadi B, Zubatiuk T, Leszczynska D, Leszczynski J, Chen WY (2019) Chemical activation of biochar for energy and environmental applications: a comprehensive review. Rev Chem Eng 35:777–815
Sanchez-Monedero MA, Cayuela ML, Roig A, Jindo K, Mondini C, Bolan N (2018) Role of biochar as an additive in organic waste composting. Bioresour Technol 247:1155–1164
Scheehle E, Kruger D (2006) Global anthropogenic methane and nitrous oxide emissions. In: Yatchew A (ed) Multi-greenhouse gas mitigation and climate policy, Special issue-3 Energy Journal. International Association for Energy Economics, Cleveland, Ohio
Seo Y-C, Alam MT, Yang W-S (2018) Gasification of municipal solid waste, gasification for low-grade feedstock, Yun Y (ed), IntechOpen, Rijeka
Serrano DP, Aguado J, Escola JM, Garagorri E (2001) Conversion of low density polyethylene into petrochemical feedstocks using a continuous screw kiln reactor. J Anal Appl Pyrol 789:58–59
Siipola V, Pflugmacher S, Romar H, Wendling L, Koukkari P (2020) Low-cost biochar adsorbents for water purification including microplastics removal. Appl Sci 10:788
Silpa K, Lisa CY, Perinaz BT, Frank VW (2018) What a waste 2.0: a global snapshot of solid waste management to 2050. Urban Development. World Bank, Washington, DC
Song Q, Zhao HY, Xing WL, Song LH, Yang L, Yang D, Shu X (2018) Effects of various additives on the pyrolysis characteristics of municipal solid waste. Waste Manage 78:621–629
Sri Shalini S, Palanivelu K, Ramachandran A, Raghavan V (2021) Biochar from biomass waste as a renewable carbon material for climate change mitigation in reducing greenhouse gas emissions—a review. Biomass Convers Biorefin 11:2247–2267
Sumalinog DAG, Capareda SC, de Luna MDG (2018) Evaluation of the effectiveness and mechanisms of acetaminophen and methylene blue dye adsorption on activated biochar derived from municipal solid wastes. J Environ Manage 210:255–262

Taherymoosavi S, Verheyen V, Munroe P, Joseph S, Reynolds A (2017) Characterization of organic compounds in biochars derived from municipal solid waste. Waste Manag 67:131–142

Tang Y, Alam MS, Konhauser KO, Alessi DS, Xu S, Tian WJ, Liu Y (2019) Influence of pyrolysis temperature on production of digested sludge biochar and its application for ammonium removal from municipal wastewater. J Clean Prod 209:927–936

Tian Y, Cui L, Lin Q, Li G, Zhao X (2019) The sewage sludge biochar at low pyrolysis temperature had better improvement in urban soil and turf grass. J Agron 9:156

Tursunov O (2014) A comparison of catalysts zeolite and calcined dolomite for gas production from pyrolysis of municipal solid waste (MSW). Ecol Eng 69:237–243

Ulusal A, Apayd E (2021) Opportunity for sustainable biomass valorization to produce biochar for improving soil characteristics. Biomass Convers Biorefin 1041–1051

Veses A, Sanahuja-Parejo O, Callén MS, Murillo R, García T (2020) A combined two-stage process of pyrolysis and catalytic cracking of municipal solid waste for the production of syngas and solid refuse-derived fuels. Waste Manag 101:171–179

Vochozka M, Marouskova A, Vachal J, Strakova J (2016) Biochar pricing hampers biochar farming. Clean Techn Environ Policy 18:1225–1231

Wang N, Qian K, Chen D, Zhao H, Yin L (2020) Upgrading gas and oil products of the municipal solid waste pyrolysis process by exploiting in-situ interactions between the volatile compounds and the char. Waste Manage 102:380–390

Weber K, Quicker P (2018) Properties of biochar. Fuel 217:240–261

Wongrod S, Simon S, Guibaud G, Lens PNL, Pechaud Y, Huguenot D, van Hullebusch ED (2018) Lead sorption by biochar produced from digestates: consequences of chemical modification and washing. J Environ Manage 219:277–284

Woolf D, Amonette JE, Street-Perrott A, Lehmann J, Joseph S (2010) Sustainable biochar to mitigate global climate change. Nat Commun 1:56–64

World Bank (2018) Integrated solid waste management and carbon finance project. Independent Evaluation Group (IEG). Report No 123798. https://ieg.worldbank.org/sites/default/files/Data/reports/ppar-brazilintsolidwastemgmt.pdf. Accessed on 20 Sept 2021

Zaman CZ, Pal K, Yehye WA, Sagadevan S, Shah ST, Adebisi GA, Marliana E, Rafique RF, Johan RB (2017) Pyrolysis : a sustainable way to generate energy from waste. Intechopen, pp 3–36

Zhang J, Shao J, Jin Q, Li Z, Zhang X, Chen Y, Zhang S, Chen H (2019) Sludge-based biochar activation to enhance Pb(II) adsorption. Fuel 252:101–108

Zhang J, Shao J, Jin Q, Zhang X, Yang H, Chen Y, Zhang S, Chen H (2020) Effect of deashing on activation process and lead adsorption capacities of sludge-based biochar. Sci Total Environ 716:137016

Zornoza R, Moreno-Barriga F, Acosta JA, Munoz MA, Faz A (2016) Stability, nutrient availability and hydrophobicity of biochars derived from manure, crop residues, and municipal solid waste for their use as soil amendments. Chemosphere 144:122–130

Chapter 15
Municipal Solid Waste for Sustainable Production of Biofuels and Value-Added Products from Biorefinery

Vishnu Manirethan, Justin Joy, Rijin Thomas Varghese, and Priyanka Uddandarao

15.1 Introduction

With the increase in cosmopolitan population and industrialization there is a significant increase in the utilization of the energy assets and the waste generation. In this scenario, a significant increase in the Municipal solid waste (MSW) generation on a global scale is observed. As of now, fossil fuels are the most dependable resources of energy accounting for 84% of the worldwide energy interest (Stephenson et al. 2021). Fossil fuel assets are rapidly depleting and the burning of non-renewable energy resources such as coal, natural gas, oil represent an immediate adverse consequence on the climate that poses a serious threat to our ecosystem. India's population is about 1.27 billion contributing to 17.6% of the total world population. The global annual MSW generation was 2.01 billion tons in 2018 and it is expected that the increase up to 2.59 billion tons by 2030 further to 3.40 billion tons by 2050 (Ellis 2018). Municipal solid waste refers to the materials discarded in urban areas, predominantly including household waste with sometimes the addition of commercial wastes, collected and disposed by the municipalities. The U.S. Environmental Protection Agency considers MSW as a renewable energy resource because the waste would otherwise be sent to landfills. In India, rapid and uncontrolled urbanization has resulted in a lack of sewerage and proper solid waste management (SWM) systems in a majority of the cities (Yaashikaa et al. 2020).

MSW fractions are generally divided into two main waste streams, viz., biodegradable and non-biodegradable or inorganic fractions. The typical MSW classifications are accordingly with the source generation, as follows: (i) Residential waste: the

V. Manirethan · J. Joy · R. T. Varghese
Department of Chemical Engineering, Amal Jyothi College of Engineering Kanjirappally, Kottayam, India

P. Uddandarao (✉)
School of Nature Sciences, National University of Ireland, Galway, Republic of Ireland

Y. K. Nandabalan et al. (eds.), *Zero Waste Biorefinery*, Energy, Environment, and Sustainability, https://doi.org/10.1007/978-981-16-8682-5_15

waste generated in dwellings, either houses or apartments which includes biodegradable food waste, yard waste, biodegradable plastics, recyclable plastic, paper, metals, bottles, aluminum cans (Abdel-Shafy et al. 2018), electrical appliances such as light bulb, TV screens, batteries (Forti et al. 2018) and composite clothing, tetra pack, cartons (Krauklis et al. 2021) (ii) Commercial waste: the waste generated in commercial facilities, department stores, supermarkets, restaurants, marketplaces, manure, slaughterhouse waste, sewage (Zheng et al. 2013), hazardous paint, chemicals, tires, batteries, aerosol and spray cans (iii) Institutional and service waste: the waste generated in governmental, private offices, education centers and recreation centers (iv) Construction and demolition waste: the waste generated in construction and demolition sites. (v) Special waste: the waste which needs special techniques for control, either because of being relatively hazardous, because of its condition or state or because control is enforced by standing the environmental regulations. This waste is generated in sectors such as scientific research, health that includes biomedical pharmaceutical waste, syringes and medicine bottles (Datta et al. 2018), industrial and automobile maintenance shops, human and veterinarian drugstores, airports, terrestrial transportation terminals (vi) Industrial waste: the waste generated in any process of extraction, benefit, transformation, production of goods it further includes demolition waste, construction waste, debris, dirt and rocks (vii) Agricultural and animal husbandry waste: the waste produced in rural and animal cultivation exercises comprising of harmful material pesticides, herbicides, fungicides and mechanical cleaning specialists (Devi et al. 2018).

MSW has been exceptionally viewed as a sustainable power source. The usage of MSW as an energy asset can diminish the exceptional ecological effects of inappropriate waste administration practices and fossil fuel-based electric power generation. Another vital element of waste components is the energy content or calorific value. The energy content or calorific qualities of MSW is the heating value of combustible MSW which falls within the range of 7–12 MJ/kg. The time has come to understand that the capability of waste to energy (WTE) is a possibility for maintainable strong waste administration and as perhaps the main future sustainable power sources (Lorenzo and Efstratios 2019). WTE plants can change over this modest and promptly accessible sustainable power asset into helpful energy. The term WTE refers to the treatment of waste for energy recovery in the form of heat and electricity or other alternative fuels in gaseous, liquid and solid forms. A huge scope of WTE innovations is accessible to deliver a particularly diverse stream of end products from the complexly composed feedstock, i.e., MSW (Farooq et al. 2021).

This scenario eventuates to a zero-waste system. The bio-refinery bio-transform MSW into energy and value added products in a sustainable approach. It is an integrated close to zero waste system exploiting a sequential process including extraction, followed by a combination of biochemical and thermal processing, with internal recycling of energy and waste gases. A sustainable process that can reduce, recycle and reuse these wastes is essential for an energy-efficient biorefinery. Because of its wide range production strategy, it has several advantages over conventional approaches.

It minimizes the dependency on fossil fuels, diversifies the bio-based resources, reduces the emission of greenhouse gases, and protects the natural environment by stimulating greener development of rural and regional areas.

15.2 MSW to Energy Conversion Processes

Waste-to-energy (WTE) recovers the energy from the waste. This technology converts waste into value added products by various treatment methods/processing. There are three main WTE conversion pathways, viz., thermochemical, biochemical and physicochemical. All types of WTE process technologies follow one of the three conversion pathways. Either direct combustion (e.g., incineration, pyrolysis and gasification) or production of combustible fuels in the forms of methane, hydrogen and other synthetic fuels (e.g., anaerobic digestion, mechanical, biological treatment and waste-derived fuel). Then again, the advances which follow the thermochemical transformation can use the majority of the waste parts in MSW waste streams and can straightforwardly change over this waste into combined heat and power; moreover, few innovations additionally produce solid, liquid or gaseous fuels as value-added products. Table 15.1 gives the insights on biofuel/bioenergy production from MSW in biorefinery concept. Strategies for bioenergy and biofuel production from MSW biomass are given below.

15.2.1 Thermochemical

Thermochemical conversion needs thermal energy to break down the molecular structure of MSW components and convert larger molecules into smaller molecules. The advanced technologies which adapt thermochemical conversion pathways, use very high temperature to convert different fractions of MSW into heat, electricity and other value-added products. These innovations incorporate incineration, pyrolysis, gasification and plasma arc gasification. This innovation changes over the carbonaceous parts of MSW into gaseous, liquid, and solid fuels (char). The significant segments of syngas are carbon monoxide (CO) and hydrogen (H_2). Alongside CO and H_2, more modest measures of carbon dioxide (CO_2), water vapor (H_2O), nitrogen (N_2) and methane (CH_4) were additionally discovered (Farooq et al. 2021; Chanthakett et al. 2021). The heat and hot flue gases produced in the incineration process are utilized to deliver high-pressure steam which is utilized in a steam turbine to produce power. For applications in which MSW is prepared, the intense heat breaks up the molecular structure of the organic material to produce such simpler gaseous molecules as CO_2, CO and H_2. Incineration is capable of totally combusting the organic components present in MSW to reduce its volume and to convert it into heat and power (Fig. 15.1). An incinerator plant works at a high temperature of around 850–1100 °C. The principle segments of an incineration plant are the

Table 15.1 Biofuel production and bioenergy from MSW in biorefinery concept

Type of MSW	Energy conversion process (method)	Product/Energy generated	References
Waste newspaper	Hydrolysis and then fermentation	Bio-ethanol	Byadgi and Kalburgi (2016)
Kitchen waste	Hydrolysis and then fermentation	Bio-ethanol	Hafid et al. (2017), Li et al. (2007)
Waste office paper and cardboard	Dilute acid and oxidative lime pre-treatment and then fermentation	Bio-ethanol	Wang et al. (2013)
Ligno-cellulosic biomass (wood, straw, grasses)	Feedstock gasification and fermentation (or) Hydrolysis and fermentation	Bio-ethanol, bio-ethylene	Irena/Iea-Etsap (2013)
Carbohydrate food wastes	Anaerobic digestion using human excreta as inoculum	Biogas, biofertilizer, electricity from burning biogas	Owamah et al. (2014)
Municipal food wastes	Anaerobic digestion	Biogas/biomethane, biofertilizer	Paul et al. (2018)
Fruit, food, yard, paper and mixed waste	Anaerobic digestion	Biogas/biomethane	Getahun et al. (2014)
Organic fraction of MSW	Acidogenic fermentation, digestate consumption by PHA producing bacterial strains	PHA (Polyhydroxy alkanoate) containing bioplastic	Ivanov et al. (2015)
Organic fraction of MSW including food wastes	Acidogenic fermentation	Volatile fatty acids (VFA)	Cheah et al. (2019)
Urban organic waste	Alkaline pH fermentation	Volatile fatty acids (VFA)	Moretto et al. (2019)
Food wastes	Anaerobic fermentation	Volatile fatty acids (VFA), biohydrogen	Strazzera et al. (2018)
Woody biomass, agricultural residues	Dark fermentation	Biohydrogen	Venkata Mohan (2009)

feeding system, incinerator (combustion chamber), exhaust gas system and remaining removal framework.

Pyrolysis is a thermal process that is performed without oxygen. Pyrolysis is heating biomass in anaerobic conditions bringing about the creation of solid charcoal, liquid bio-oil and fuel gases (Guo et al. 2015) (Fig. 15.2). Pyrolysis is a promising innovation and is now used in numerous regions of the world for MSW disposal and energy generation. The pyrolysis process can recover up to 80% of the energy from the

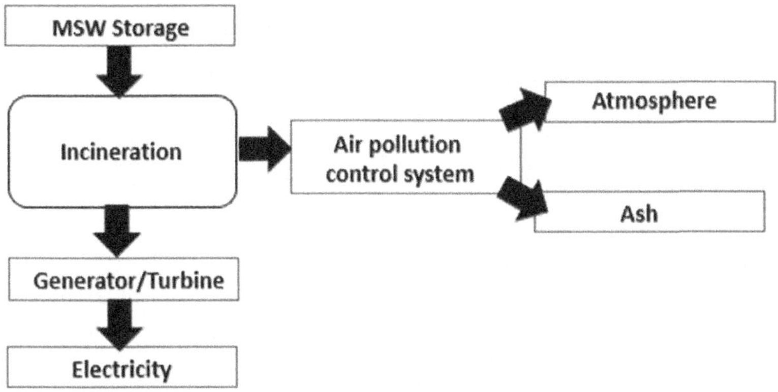

Fig. 15.1 Schematic of incineration process (Beyene et al. 2018)

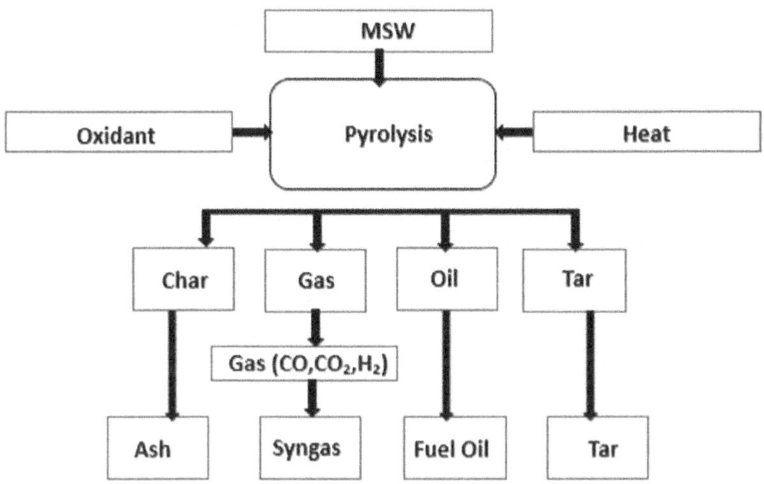

Fig. 15.2 Schematic diagram of pyrolysis (Beyene et al. 2018)

carbonaceous fraction of MSW. The most usually utilized innovation for technology for pyrolysis reaction is the fluidized bed reactors. Pyrolysis of MSW yields around 43% bio-oil, 27% biochar and 25% syngas (Hasan et al. 2021). Gasification is an adjusted pyrolysis measure where the response occurs within the sight of a restricted measure of oxygen or steam (Fig. 15.3). Gasification changes over the biomass into syngas which can be utilized for the creation of biofuel, energy and synthetic compounds (Molino et al. 2018). The gasification reaction is an indirect combustion process that results in an exothermic process. It releases thermal energy because of the carbon and oxygen response. The entire interaction happens at a high temperature of around 800 °C. Ordinarily, the yield of gasification items is exceptionally high (up to 85%). The temperature in the gasifier ought to be in the range of 500–1000 °C, the

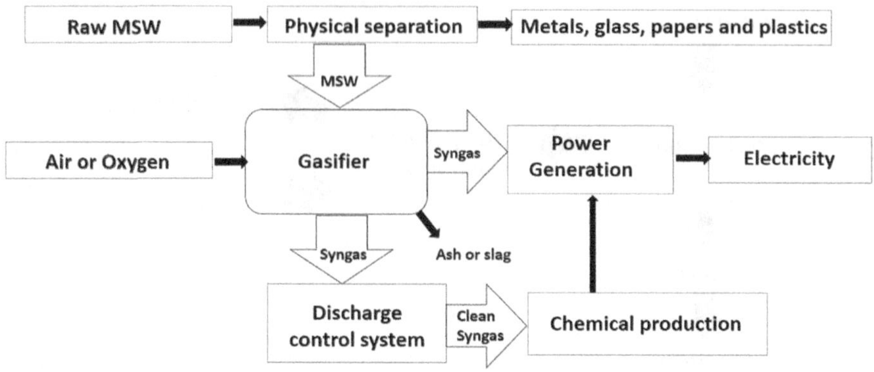

Fig. 15.3 Schematic diagram of gasification (Beyene et al. 2018)

feedstock particles ought to be little and uniform in size, this dampness substance ought to be underneath 15%. Gasification strategies have an energy yield of 547–571 kWh/ton of MSW.

The plasma arc technology is a heating method technique that can be utilized in both pyrolysis and gasification systems. Plasma arc technology innovation utilizes exceptionally high temperatures (7000 °F) to separate the feedstock into elemental by-products. Plasma arc gasification can be a possible innovation for MSW the executives because of the greater energy yield of 816–1000 kWh per ton of waste (Mlonka-Mędrala et al. 2021). Plasma gasification is thermal plasma that offers the chance of disintegration of biomass by unadulterated pyrolysis without oxygen or gasification to create top notch syngas (Hrabovsky et al. 2011).

15.2.2 Biochemical

Biochemical transformation of MSW to energy includes biological agents or microorganisms like yeast to change the organic fraction into biofuels. Anaerobic processing and maturation are the WTE innovations that follow the biochemical transformation pathway (Ofori-Boateng et al. 2013). Anaerobic digestion is one of the biochemical transformation methods that decay the organic fraction of MSW with the assistance of microorganisms without free oxygen (Fig. 15.4). It is an interaction that can change organic matter into a bioenergy source i.e., biogas (Lesteur et al. 2010). The organisms associated with various stages of the anaerobic digestion measure are exceptionally delicate to pH level and need explicit conditions to develop and support the yield of end products. Consequently, this cycle happens in unique reactors that work at explicit conditions that incorporate optimum temperature and pH level. The organic feedstock is blended well and kept in the digester for 5–10 days and during this time the anaerobic digestion process happens in four distinct stages, viz., hydrolysis, acidogenesis, acetogenesis, and methanogenesis. The productivity

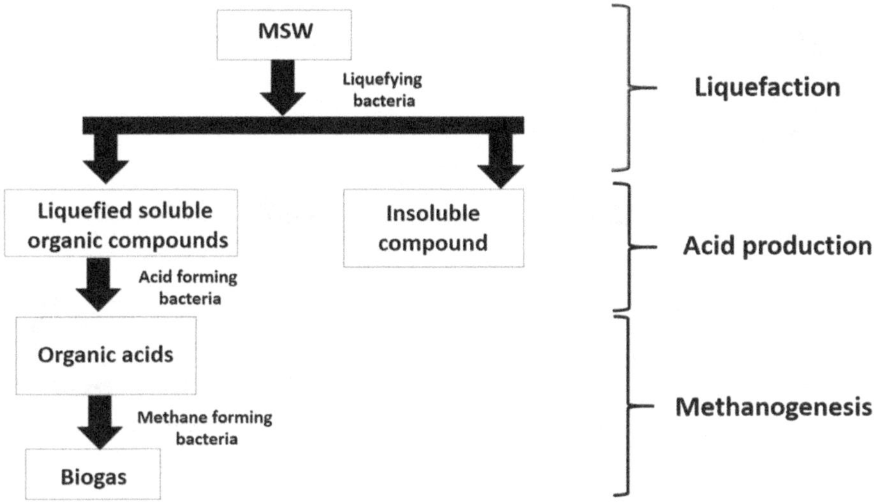

Fig. 15.4 Schematic diagram of anaerobic digestion (Saini et al. 2021)

of heat and power production from biogas is around 5.5–7.5 kWh/m^3 which is less when contrasted with flammable gas (Mlonka-Mędrala et al. 2021).

15.2.3 Physicochemical

Transesterification, also known as alcoholysis, is the displacement of alcohol from an ester by another alcohol (Thangaraj et al. 2019). Biodiesel synthesis from MSW is done by the transesterification of fat fractions of food waste. Transesterification of processed fat is done using monoalkyl alcohols such as ethyl or methyl alcohol in the presence of an acid or base catalyst. Ultrasound-assisted processes have been widely used in biodiesel production. The main advantage is increasing the conversion of esters at reduced reaction times (acoustic cavitation), with significantly lower production costs. The parameters such as frequency and mode of operation plays a key role in the efficiency of the process (Oliveira et al. 2018). Supercritical fluid technologies are used to integrate simultaneous extraction, transesterification and valorisation into a single step (Bernal et al. 2012). The feedstock preparation, extraction and transesterification in the biodiesel production process are conducted using microwaves (Nomanbhay et al. 2017). The transesterification process yields biodiesel as the main product with glycerol as the by-product. The products generated by transesterification, biodiesel and glycerol with high heating values of 40.17 MJ/kg (megajoule/kilogram) and 19 MJ/kg, respectively can be employed for electricity generation.

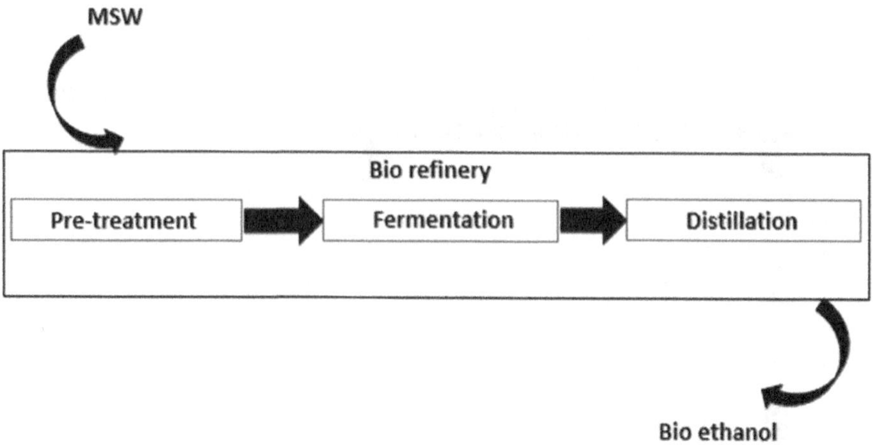

Fig. 15.5 Schematic of the fermentation organic matter (Yaashikaa et al. 2020)

The fermentation cycle additionally utilizes microorganisms for the disintegration of organic materials in an oxygen-free environment. The fermentation process likewise incorporates practically every one of the stages like anaerobic processing, with the exception of the methanogenesis stage. Hence, the final result of fermentation is biofuel instead of biogas. The fermentation process ends at acetogenesis where diluted liquor is isolated from the matured digestate by playing out an extra advance known as distillation. Fermentation is finished utilizing liquor or catalysts to converts the biomass into valuable products. Ethanol production utilizing yeast for maturation at 20–35 °C is quite possibly the most utilized biofuel production technique (Azhar et al. 2017) (Fig. 15.5).

Sanitary landfills are areas for the controlled removal of waste, to diminish its natural negative effect and for the control of lixiviate material. A few landfills create power from the biogas delivered. Landfill gas (LFG) is formed when natural squanders break down anaerobically in a landfill. In spite of the fact that LFG gas is created under aerobic and anaerobic conditions, the underlying vigorous stage is brief and delivers a gas with a much lower energy content than that drawn out of anaerobic stage (De Souza et al. 2018) (Fig. 15.6). The enzymatic response has several advantages over the different strategies because of its gentle response conditions, simple product recovery, no wastewater age, no saponification and higher quality of items (Noraini et al. 2014).

The primary item is biogas which contains 50–80% methane (CH_4), 20–50% carbon dioxide (CO_2), small traces of sulfide and ammonia. The other two items that are produced alongside biogas, are fiber and fluid digestate. The fundamental yield from the aging interaction is ethanol. Other than ethanol, CO_2 refined dried grains (DDGs) and stillage (wastewater) are likewise acquired as side-effects from the aging cycle. The ethanol created by the aging interaction can be utilized to supplant gas as a transportation fuel and an anaerobic absorption plant to deliver biogas (Farooq et al. 2021).

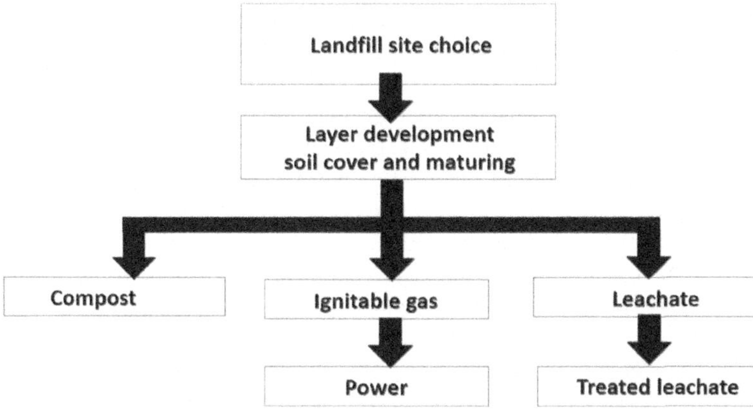

Fig. 15.6 Schematic of landfill process with its process (Reinhart and Townsend 2018)

15.3 Value-Added Products from MSW Biorefineries

The term biorefinery was first identified in the year 1981 which described the concept as a process for the production of acids, liquid fuels and chemicals by combining the three step process of biomass conversion into organic acids by fermentation and its subsequent electrolysis. A biorefinery is a refinery that converts biomass into energy and other useful products. A major part of the energy production in our world is from fossil fuel refineries. The use of fossil fuels for energy production has a great impact on our environment. An increase in the price as well as the uncertainty of the availability of fossil fuels has led to the biorefinery concept (Suhag and Sharma 2015). Hence, replacing fossil fuels with biomass is in the advancing stage. Biorefineries are promising industries for the future because of their potential to derive high value-added products. This in turn also increases the demand for the construction of biorefineries as well. The main focus of these biorefineries will be to produce alternative fuels in the future (Menon and Rao 2012).

The biorefinery concept uses the advancement in technology to separate the components of the biomass into their original constituents like carbohydrates, proteins etc. that are later transformed into value-added products, biofuel, and other chemicals. So a biorefinery is a place where this process of biomass conversion takes place to produce bioenergy. The efficient production of biofuel for transportation is one of the most encouraging factor for construction of new biorefineries around the world (Taylor 2008). Currently, most of the biofuel in the world is not produced in a biorefinery but a single supply stream. For effective use of biorefinery, we have to cut down the use of nonrenewable resources and its effect on the environment.

Unlike oil refineries which are large plants, biorefineries will comparatively be smaller. The byproducts from the biorefineries can in turn be used as raw materials for other industries. The vast availability of biomass is the main driving force for improvement in bioenergy production. The important technical movement within a biorefinery is to depolymerize and deoxygenate the biomass feedstock or components and to convert the feedstock into valuable products (Cherubini 2010). An efficient method to reduce the demand for fossil fuels and to improve the economy as well as help prevent global climate changes are biorefineries (Bauer et al. 2017). The leaders in the distribution and diffusion of renewable energy technologies are Switzerland in Biogas, Netherlands in biomass gasification and Ukraine in agro-bioenergy as well as Sweden and Denmark. European directives on renewable energy and fuel quality, US renewable fuel standard policies that were initiated for biofuel has played a major role in developing this region. The vision for the USA up to 2030 regarding technologies based on biomass have identified targets for biofuels and bioproducts. This includes major progression in research areas such as biomass characteristics, production, conversion and processing. This also emphasized the need to create regulations and a market environment for bio-based products. In countries obtaining good yield from biorefineries, the biofuel production is prioritized in the markets and simultaneously the tax schemes are made in order to support this change and it is prioritized by the policymakers as well as mandatory blending requirements in standard fuel products are implemented (Bauer et al. 2017). The following are some of the examples of value added products from biorefineries.

15.3.1 Biofuel Category

Biofuels are the gaseous or liquid fuels obtained by the biological treatment of the organic matter. They can be used in addition to conventional fuel sources or can be used to replace them for transportation, stationary or portable applications. The second generation of biofuel is produced from different sources such as cellulosic biomass, agricultural residue, and industrial waste. They also include fuels produced from mixed paper waste segregated from MSW (Byadgi and Kalburgi 2016).

15.3.1.1 Bioethanol

Bioethanol can be derived from multiple sources containing carbohydrates and sugars. They are blended with conventional gasoline with characteristic blend ratios of E10 (10% ethanol with 90% gasoline) and E15 (15% ethanol with 85% petrol). Bioethanol is also a raw material for ethylene, acetaldehyde, butadiene, ethyl acetate, and acetic acid production (Byadgi and Kalburgi 2016).

Cellulose is a good contender to produce bioethanol as it is widespread, ample, and not a part of the human food chain and hence is inexpensive. Lignocellulosic materials (i.e., agricultural residues, woods, residues from pulp and paper industry, urban lignocellulosic wastes) consist of long-chained simple sugars such as hexose sugars (e.g.: Glucose) and pentose sugars (e.g. Ribose). Newspapers are a good source of cellulosic materials and are commonly found in MSWs. They are potential sources of bioethanol due to the relatively high levels of carbohydrates they contain and can hence be effectively utilized apart from recycling. They are then hydrolyzed using bacteria such as *Cytophaga hutchnisonni* to reduce the carbohydrate to simple sugars. After doing a neutralizing wash to retain the pH at 7 and to remove inhibitors of yeast metabolism, it is then fermented using common yeast. This yielded bioethanol although the yeast was only able to ferment the hexose sugar. Grass, agricultural and garden wastes, scrap paper, and cards are the other sources used to produce bioethanol. Here also pre-treatment is done to obtain the reduced sugars needed for fermentation. The pre-treatment process is similar to the one done for the waste newspapers (Li et al. 2007).

Kitchen waste is another element of MSW that can be used in the production of bioethanol. Kitchen wastes also generate leachate in landfills due to high moisture and require secondary wastewater treatment. Kitchen waste generally contains around 65% carbohydrates which can then be reduced into fermentable sugars. This is achieved by pretreatment methods such as acid treatment and enzymatic saccharification. Fermentable sugars are obtained at a low cost from waste after an effective pre-treatment. A minimum of 10% concentration of sugars should be available after pre-treatment techniques to manage the costs of the subsequent steps in the process. Yeast (*Saccharomyces cerevisiae*) or *Zymomonas mobilis* (acid-tolerant-ethanol bacterium) converts the newly formed sugars into bioethanol. The crude bioethanol is further purified from the fermentation broth by distillation, rectification, and dehydration to obtain high-quality ethanol. It is of particular importance that the waste is segregated at the source to improve recyclability and to obtain proper feedstock for bioethanol production (Cekmecelioglu and Uncu 2013; Hafid et al. 2017).

15.3.1.2 Bioethylene

Ethylene is one of the major raw materials for the production of plastics, textiles, solvents, and many other industries. It is highly active and is used as a raw material for many chemicals and polymers. The process to convert ethylene to liquid transportation fuels is being developed. Ethylene is found in plants as a hormone and it modulates the growth and development of the plant and functions as a defence mechanism against biotic and abiotic stress. Ethylene is usually produced by the cracking of Naphtha (IEA-ETSAP and IRENA 2013). With the advent of strict environmental regulations and to reduce the effects of fossil fuels, ethylene production by biological means or as a derivative of bio-products is a viable option (Eckert et al. 2014). Ethylene can also be sourced from bioethanol by its catalytic dehydration. The newly formed bio-ethylene is chemically and structurally similar to the ethylene derived from fossil fuels (Mohsenzadeh et al. 2017).

Pseudomonas syringae and *Penicillium digitatum* can be employed for the production of ethylene from simple sugars. As a result, there have been studies conducted to assess if commercial production of bioethylene is viable. The discovery of the ethylene-forming-enzyme (EFE) in *P. syringae* has allowed for the extraction of the gene expressing EFE and introducing it to several common hosts such as *S. cervisiae* and *E. coli*. These hosts can utilize various carbon sources like lignocellulosic materials and carbohydrates from kitchen wastes to produce bioethylene. These materials can be found from MSWs with ease. Ethylene easily separates from the cultures as it is a gas and hence it is also nontoxic. The bioethylene can be harvested while mitigating the risk of the presence of O_2 that is formed along with the ethylene (Eckert et al. 2014).

15.3.2 Value-Added Commodities

15.3.2.1 Biogas

Biogas is an important alternative to traditional energy sources and for the reduction of greenhouse gases. The anaerobic digestion of domestic sewage and MSWs results in the formation of biogas or landfill gas (LFG). Anaerobic digestion occurs in anoxic conditions in anaerobic digesters where bacteria are the primary organisms. The organic matter in the MSW consisting of nitrogen and phosphorous satisfies the required amounts for cell growth of the anaerobes that digest the waste. Other factors that influence the yield of methane, a major constituent of biogas are the organic loading rate, retention time, moisture content (optimum 90%) and temperature. The C/N ratio or the level of carbon and nitrogen in the raw feed influences the biogas production rates. The C/N ratio needs to be in the 25:1 and 30:1 range. Methane constitutes around 50–70% of the generated biogas.

The efficiency of the anaerobic digestion process and the production of biogas can be enhanced by the pre-treatment of waste. The different pre-treatment techniques include mechanical, thermal and biochemical methods. Size diminution to increase the surface area of the same thereby increasing the anaerobic digestion efficiency is the mechanical pre-treatment step. It also helps to reduce the digester volume without affecting the efficiency of production. Thermal pre-treatment has the benefits of pathogen eradication, enhances dewatering benefits, and helps decrease the sample viscosity. Chemical pre-treatment is done to maintain pH, improve digestibility and hydrolysis of lignocellulosic materials.

Domestic sewage is added to MSW to be used not only as a source of nutrition for the anaerobes that produce biogas but also as a diluent to get the required concentration of total solids (TS) in the feedstock. Chicken manure was another component similar to domestic waste, which was also mixed with the MSW. The non-degradable components of the MSW such as wood, plastics, sand, metals, glass, etc., were segregated and the leftover fermentable components were homogenized by mixing. The mixed waste showed the maximum yield of methane compared to the individual fractions of segregated wastes and forms slurry that can be better handled and easily transported in pipes. The sewage/manure also contains methanogens that can digest the MSW in an anaerobic digester. The digester is operated at temperatures between 26 and 36 °C while maintaining a pH range of 6.5–7.5. Mixing is done in regular periods to avoid the development of dry and dormant layers of flotates as well as to vent the biogas produced in the digester. It also gives high calorific value to the gas while saving time that would be spent for segregating the waste components. The maximum yield of methane obtained from the digestion process is 0.36 m^3/kg of VS (volatile solids) added per day (Getahun et al. 2014). Concentrating methane in biogas yields biomethane. The process involves washing with water, membrane separation technology, amine gas treating, pressure swing adsorption and Selexol adsorption to remove the CO_2 and H_2S in biogas. Biomethane is similar in characteristics to natural gas and can be stored in tanks and hence it can be used in gas engines and vehicle fuels (Matheri et al. 2017).

Nutrient-rich fertilizer is obtained from the digestate after the completion of the digestion process. The biogas obtained from MSW digestion is pre-treated and then utilized for power-producing activities in internal combustion engines, turbines and micro-turbines. The potential decrease in the amount of CO_2 generated due to the utilization of biogas is immense. The bioconversion of MSW to biogas which is then used as a power source reduces the generation of CO_2 many folds (Aguilar-Virgen et al. 2014; Elango et al. 2007).

15.3.2.2 Biopolymers

A biopolymer is a natural polymer that is synthesized by microbes or living cells of plants and animals. Hence they are completely biodegradable and have great potential for utilization in our daily life. Bioplastics are produced from renewable sources with improved technologies and materials that contribute to environmental sustainability. As the cost of producing bioplastics is higher compared to conventional petroleum-derived plastics, moving towards using raw materials that are cheaper and applying innovative technologies is necessary. Organic fractions of MSW (OFMSW) like carbohydrates can be fermented acidogenically to produce Polyhydroxyalkanoates (PHAs) and then use the organic acids for the biosynthesis of the PHAs. The acidogenic fermentation process releases volatile fatty acids (VFA) and hydrogen from carbohydrates which are the most favourable substances to produce PHAs.

Different types of bacteria can be used for acidogenic fermentation like *Acetobacter, Clostridium, Bacteroides, Propionibacterium, Enterobacter,* etc. Anaerobic conditions are sustained while maintaining pH between the ranges of 5.5–8 to obtain maximum fermentation productivity from MSW. An economic and effective technique of maintaining the pH of the MSW substrate is to mix the carbohydrate and protein-rich solid wastes. Hydrolysis of proteins releases ammonia and hydroxide ions that balance the pH during acidogenic fermentation of MSW (Ivanov et al. 2015).

A biopolymer that is produced by using restructured starch with specific amount of plasticizers at particular extrusion conditions are called Thermoplastic starch (TPS). The starch is obtained from the agricultural and food-processing waste fractions of MSW. The degradation rate of the biopolymer depends on its monomer and its physicochemical properties such as molecular weight, chemical structure, etc. Other biopolymers that can be formed from MSW are polylactic acid (PLA), polycaprolactones (PCL) and polyethylene 2,5-furandicarboxylate (PEF). Their sources include carbohydrate-rich crops and vegetable-derived fructose. They find specific applications in bottling, textiles, food packaging, carpeting, electronics and automotive industries. The PLA biopolymer can be reinforced with cellulose to produce high value biopolymers that are economical, safe and can be used for food packaging while contributing to sustainable development and waste reduction (Kabir et al. 2020). Biopolymers at the end of their life cycle can be returned to the MSW where they can act as a substrate to produce other bioproducts (Wojnowska-Baryła et al. 2020).

15.3.2.3 Biofertilizers

Biofertilizers assure enhanced food security by increasing the agricultural yield. They are prepared by anaerobic digestion by using a variety of raw materials and techniques. After biogas is produced from the anaerobic digestion of the MSW, remaining is the nutrient-rich digestate. The constituents of the digestate are microbial biomass, inorganic compounds and semi-degraded organic matter. As biofertilizers originate through organic means, unlike conventional fertilizers they do not persist in the soil. This also improves the quality of the soil and the food, also ensures the health and safety of humans and animals as well as improve the overall quality of the environment. Hence they break up into simple molecules and thus the contamination of groundwater sources is avoided. Digestate is classified based on different factors such as raw materials used, forms of utilization, the source of microorganisms and digester configurations, etc. The microbes in the biofertilizer include nitrogen-fixing free-living bacteria such as *Clostridium* and *Klebsiella spp.* At the same time *Bacillus* and *Pseudomonas spp.* are used as phosphate solubilizing biofertilizers (Owamah et al. 2014).

The main components for producing digestate/biofertilizers are the abundant wastes found locally in the region, although carbohydrate food wastes form the major component. Meanwhile, the inoculum is collected from the anaerobic digestion plant and introduced into the slurry mixture. Similar to the biogas production process, here also the pH is maintained at 6.5–8 and the temperature is set between 26 and 36 °C. This temperature range is most suitable for the mesophilic microbes while the pH facilitates the growth of methanogens for biogas production (Paul et al. 2018).

15.3.2.4 Volatile Fatty Acids

Volatile fatty acids (VFAs) are short-chain fatty acids composed mainly of aliphatic monocarboxylate compounds that serve as starting materials to synthesise various products like biopolymers, reduced chemicals and derivatives (esters, ketones, alkanes, alcohols). They are widely used in chemical industries. Conventionally, they are produced by the oxidation or carboxylation of chemical precursors, such as aldehydes and alkenes, derived from petroleum processing. By applying the biorefinery concept, where wastes are converted into value-added products, VFAs from MSW can be produced (Strazzera et al. 2018).

Food wastes (FW) are an ideal source to generate VFA through the anaerobic digestion process as they are abundant in very large amounts throughout all time and seasons. The characteristic features of FWs are the high concentrations of nitrogen and phosphorous, presence of key nutrients and high levels of organic matter. The fermentation process is employed to convert the waste into VFA.

Production of VFAs are influenced by various parameters such as temperature, pH, hydraulic retention time (HRT) and the organic loading rate (OLR). The maximum amount of VFAs are generated at pH maintained around 7, while the product distributions are different at different pH. At pH 10 acetic acid production is favoured; butyric acid is favoured at pH 5 and production of propionic acid is maximum under all tested pH ranges. Overall, the maximum yield of VFAs are obtained at mesophilic temperature (30 °C). This is in part due to acidogenic microbes having an optimal growth rate around the mesophilic temperature range. The HRT should be kept long to allow the formation of acidogens and maximum VFA. Different VFAs are formed with maximum concentration at varying retention times. Shorter HRT mainly formed acetic acid while longer ones gave propionic acid. Increasing the OLR increased the production of VFA. The maximum yield of VFA is obtained by employing the two-stage anaerobic digestion process. The first stage consists of hydrolysis and acidogenesis while the second stage involves methanogenesis. VFAs form during acidogenesis step and the excess volatile acids are sent to a second stage where it is used to produce biogas.

VFAs can be used to produce biopolymers such as polyhydroxyalkanoates (PHAs), biogas, biohydrogen and to remove biological nutrients from wastewater. Biological nutrients are removed to aid the denitrification step. By adding VFAs to the wastewater, COD (chemical oxygen demand) of the wastewater is reduced (Elefsiniotis et al. 2004). VFAs can also be used to generate electricity by employing microbial fuel cells (MFC) in which microbes oxidize the substrate in the anode to produce electrons and transport the electrons to the cathode through an external circuit and here the electrons, protons, neutrons combine to form water (Luo et al. 2019).

15.4 Case Studies in Indian Scenario

Thapa et al. (2019) studied the conversion of MSW into ethanol which is a commercially important product to be used as a blend with gasoline in the ratio of 1:9. The process involved the collection of MSW from the household and then pretreatment which later was followed by treatment with yeast (*Saccharomyces cerevisiae*). Hydrolysis of the treated MSW at a temperature range of 110–150 °C and sulphuric acid concentration of 1–10% was followed by the fermentation step.

Hydrolysis results showed an increased reducing sugar production at 7.5% sulphuric acid and 135 °C. The highest ethanol production obtained from the pilot study was 13.78 g/L. A projected study using 11,558 tons of waste per day in the national capital of Delhi using the pilot study results revealed that 28.53 L of ethanol could be generated from 1 ton of wet waste.

Another case is the Zero Waste Electoral Ward Initiative at Katraj Gaon region in Pune, Maharashtra, India. This public-sector organization generates electricity by converting organic waste from the Pune Municipal Corporation (PMC) to biogas. Streets lights of 4 km stretch Katraj–Kondhwa road in PMC is lit up by the power generated from biogas since 2009. PMC generates 2550 tons/day of MSW out of which 40–60% is organic matter. The feed wastes to the plant include digestible food and kitchen waste from houses, paper waste which includes waste from households and institutions, garden trimmings from home and parks, market dumping including vegetable residues, and discarded food and catering businesses, outdated food from supermarkets, etc. An NGO, Solid Waste Collection and Handling (SWaCH), has trained workers who collect the segregated waste and feeds the biogas plant. They collect nine tons of waste each day.

The process involves segregation at the source which is the best practice for the treatment of MSW. MSW on shredding or crushing is fed to the anaerobic digestion chamber with an ample quantity of water added. Anaerobic digestion can be single-stage, multi-stage, or in the batch process depending upon the total solids content in the MSW. The biogas generated will be stored in a balloon-type holder and the sludge rich in nutrients will be removed and later dried and used as manure. Biogas is further used in a gas engine generator to generate electricity. The electricity generated is used to light up the street lights of a certain area in PMC.

PMC has a population of 31 million with an area of 243 km^2 which is the feeder area of the unit. Land use of the unit is 0.03 ha, capital investment is 180,000 USD with labor of 4 persons. The output is 300–325 m^3/day processing 5 tons/day, electricity generation of 144 MWh/year, and 180 tons/year of bio-sludge used manure.

15.5 Biorefinery Concept for VFA Production from MSW

To avoid the loss of potential energy and to reduce the amount of garbage going to landfill, biodegradable waste has been converted into biogas using AD. Anaerobic consortia carry out a sequence of syntrophic chemical events known as hydrolysis, acidogenesis, acetogenesis, and methanogenesis during the process. The disposal of the organic fraction of MSW face environmental and economic issues. When the generation of methane is prioritized in a series of anaerobic reactions, the potential to obtain a more valuable product is lost, because VFAs produced during the acetogenesis stage have a higher market cost. Using the effluent from acidogenic AD directly in downstream processes, eliminating expensive purification operations as food is used as a carbon source for biological processes for the generation of both medium chain fatty acids and PHA. Modern biorefineries continue to use AD as a core process technology, with the potential for waste stabilization, chemical building block recovery for the carboxylate platform, renewable energy, and nutrient recovery. To convert Organic Fraction of Municipal Solid Waste (OFMSW) into chemical building blocks, a two-phase AD system with a hydrolytic digester and a methanogenic digester is used (VFAs and biogas). Additionally, digestate is also used to generate solid and liquid biofertilizers, in order to valorize the whole OFMSW.

In order to boost biological VFA synthesis, the population of methanogens in the first hydrolytic digester of the two-phase AD system must be inhibited. This can be accomplished by using thermal pretreatment, pH shocks, or inhibitor addition as inoculation techniques. Inhibiting methane production by overfeeding the methanogenic population is a good way to cut operational costs and get the hydrolytic digester up and running quickly.

In the complicated metabolic pathways present in mixed culture fermentative systems, anaerobic microbes compete for substrate. When the operational pH is not in the ideal range, the activity of the enzymes engaged in the hydrolytic phase is susceptible to decline. In the anaerobic breakdown of complex substrates like OFMSW to VFA, this is the first and most important stage. In order to reduce CH_4 production, the operating pH value must be kept outside of the optimum range for methanogenic archaea (6.8–8) at all times. As a result, the most beneficial pH range for maximizing acidification, with or without pH regulation, is 5–6. Several parameters, including hydraulic retention time (HRT), organic loading rate (OLR), pH, temperature, and trace element content, can be adjusted to change the VFA yield and spectrum of individual VFAs produced (Fig. 15.7).

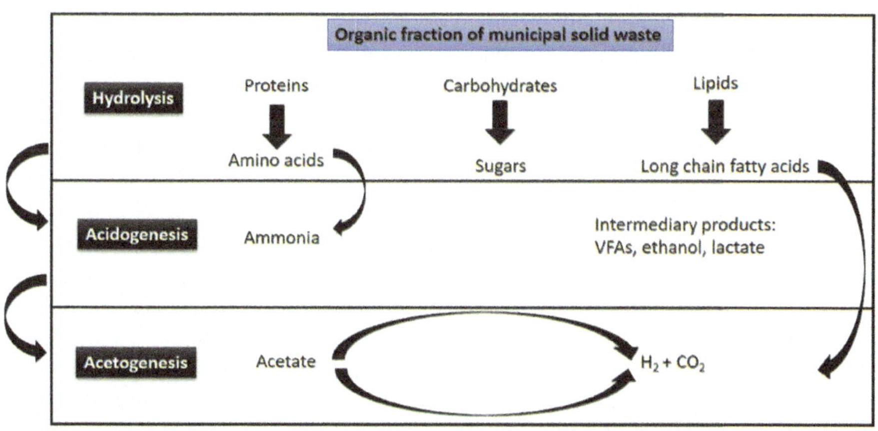

Fig. 15.7 MSW biorefinery for VFA production (Pérez et al. 2020)

15.6 Future Work

MSW is a good resource to substitute for crops in the production of bioproducts. But there are certain challenges to use MSW as they are complex and have non-standardized composition. The different food habits followed by the people in a region and the sites of waste generation are two elements which add to the complexity in the composition of the waste. Moreover, an increment in the cost of processing is seen during segregation of wastes into their individual constituents.

A way to deal with this system is to encourage source segregation of the organic wastes from recyclables at domestic households. Bringing awareness to the public and educating them about the importance of kitchen wastes and their separation at source is important to change the perception of throwing away the wastes. This can be done through social campaigns by both governmental and non-governmental organizations. Another measure involves the erection and operation of waste recycling facilities through the allocation of specific budgets by municipalities and local authorities. Governments can encourage the utilization of separated bins by introducing laws. These bins can be used to segregate household wastes into organic and recyclable components (Hafid et al. 2017).

Pre-treatment and fermentation processes are the steps that can be further improved by introducing cost-effective strategies and new technologies. The commercial scale-up of the digestion process to make it economically viable is another important step. The development of new fermentation processes to decrease the digestion time and the introduction of microbes that can perform both hydrolysis and fermentation are contested areas of research (IEA-ETSAP and IRENA 2013; Wang et al. 2013). Genetically modified organisms can be used on the MSW feedstock to produce a plethora of bioproducts from it. Through genetic engineering, the organisms can be tailored to produce specific products from the MSW at a commercial scale.

Current challenges faced by the MSW biorefinery sector include integration aspects, techno-economic sustainability aspects and environmental sustainability aspects. Biorefineries, on the other hand, should not rely on a single conversion and should instead produce fuels, energy, or biochemicals from a variety of sources. MSW has the advantage of being diverse as a substrate, allowing for the creation of a wide range of bio-based products. However, in order to make proper and full use of the substrate, the biorefinery operations in any MSW biorefinery sector should be interconnected. Using a combination of technology platforms, an integrated biorefinery strategy enables for efficient synthesis of both energy and biochemicals/products. Many of these biorefineries have been built around the world, but feedstock availability and transportation have still remained a problem.

Any biorefinery should be sustainable not just from a technological standpoint but also from a socioeconomic standpoint when it comes to sustainability. The substrate, power or energy, non-recurring and recurrent costs, utilization, and commercial obtainability of the output should all be addressed in this study, just as they are in petroleum refineries. Pretreatment and recovery procedures are extremely

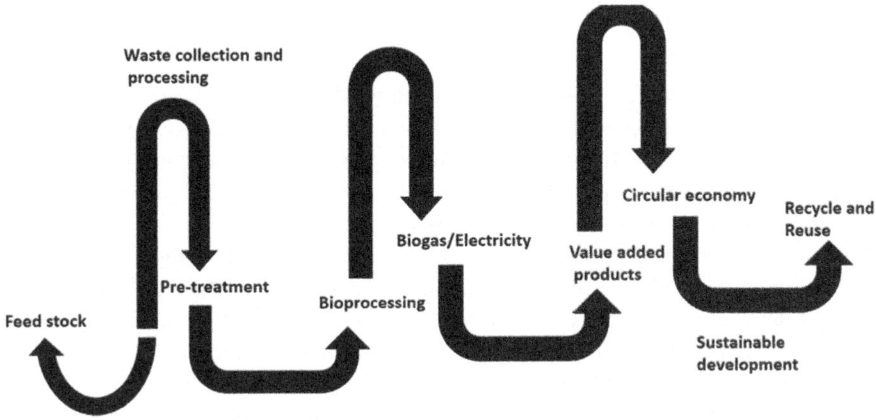

Fig. 15.8 Circular economy in MSW (Yaashikaa et al. 2020)

energy/power input dependent, forcing any biorefinery industry to use a portion of the heat or bioenergy created during its operation, which also includes a portion of the heat or bioenergy generated during its operation. To understand the specific techno-enviro-economic elements of MSW-based biorefinery, a thorough picture of the cost–benefit analysis as well as the life-cycle assessment (LCA) technique is required. In MSW, life cycle analysis (LCA) by implementation of circular economy in MSW is a critical technique for identifying societal, environmental, technical, and economic constraints in waste treatment, particularly in the context of an integrated solid waste management (ISWM) approach (Fig. 15.8). LCA provides a clearer picture of the environmental and economic impacts of trash collection, treatment, and disposal before creating electricity from MSW, allowing for the most efficient energy generation.

Introducing financial subsidies for renewable energy sources, removing import duties, conducting lifecycle assessments (LCAs) of different technology pathways, studying the spatial distribution of waste feedstock and the local conditions under which waste feedstock will be processed are the other ways through which production of bioproducts from organic fractions of MSW (OFMSW) can be made economically feasible (Liu and Rajagopal 2019). Adopting centralized and decentralized strategies while organizing the informal sector (such as waste pickers) along with government and private sectors in solid waste management will help in facing the MSW management problems (Kendall 2012).

Other advanced management techniques of MSW involve the development of plastic-eating worms, use of Sal saw wood and coffee waste to build electrode materials for Vanadium redox batteries and the recycling of construction/demolition waste that forms a fraction of MSW for road construction, drainpipes and to build gravel and bricks (Kumar and Agrawal 2020).

15.7 Conclusions

There are considerable challenges concerning the legitimacy of biorefineries. Perceived slow development of technologies, significant expenses, vulnerability with respect to future sustainability requirements and the ruined standing of certain biomass feedstock are the prominent ones. The assortment of MSW is the bottleneck in biorefinery operations since it's a disorganized process. Decreasing the collection time while expanding the simplicity of waste assortment will help to relieve the challenge of a reliable and predictable feedstock supply. Season, climate and location are factors that establish the heterogeneous composition of MSW, which should be taken care of by the waste recycling facilities that are under development. Biofuels sourced from MSW face certain challenges such as recovery, purification, and quality maintenance during its scale-up production process. The low price of oil, high investment costs, project financing and lack of policy supports are some of the reasons for the sluggish growth of this sector in most of the developing countries. Other reasons include the variable accessibility to feedstock and uncertainty in the pricing of biomass. These reasons in overall affect the public perceptions of these bio-derived products and biorefineries and create mixed messages about their production and utilization. Furthermore, critical development of biorefineries depends on its integration with existing technical systems. In conclusion, MSW has tremendous potential as it is a cheap and abundant resource for bioproduct formation. Hence it needs more research and development to realise its potential. Developing nations are the greatest benefactors of utilizing biorefineries that process MSW. These biorefineries help these nations to handle wastes and finally attain sustainability with socio-economic and environmental advantages.

References

Abdel-Shafy HI, Mansour MS (2018) Solid waste issue: sources, composition, disposal, recycling, and valorization. Egypt J Pet 27(4):1275–1290

Aguilar-Virgen Q, Taboada-González P, Ojeda-Benítez S, Cruz-Sotelo S (2014) Power generation with biogas from municipal solid waste: prediction of gas generation with in situ parameters. Renew Sustain Energy Rev 30:412–419

Azhar SHM, Abdulla R, Jambo SA, Marbawi H, Gansau JA, Faik AAM, Rodrigues KF (2017) Yeasts in sustainable bioethanol production: a review. Biochem Biophys Rep 10:52–61

Bauer F, Coenen L, Hansen T, McCormick K, Palgan YV (2017) Technological innovation systems for biorefineries: a review of the literature. Biofuels Bioprod Biorefin 11(3):534–548

Bernal JM, Lozano P, García-Verdugo E, Burguete MI, Sánchez-Gómez G, López-López G, Pucheault M, Vaultier M, Luis SV (2012) Supercritical synthesis of biodiesel. Molecules 17(7):8696–8719

Beyene HD, Werkneh AA, Ambaye TG (2018) Current updates on waste to energy (WtE) technologies: a review. Renew Energy Focus 24:1–11

Byadgi SA, Kalburgi PB (2016) Production of bioethanol from waste newspaper. Procedia Environ Sci 35:555–562

Cekmecelioglu D, Uncu ON (2013) Kinetic modeling of enzymatic hydrolysis of pretreated kitchen wastes for enhancing bioethanol production. Waste Manage 33(3):735–739
Chanthakett A, Arif MT, Khan MMK, Oo AM (2021) Performance assessment of gasification reactors for sustainable management of municipal solid waste. J Environ Manage 291:112661
Cheah YK, Vidal-Antich C, Dosta J, Mata-Álvarez J (2019) Volatile fatty acid production from mesophilic acidogenic fermentation of organic fraction of municipal solid waste and food waste under acidic and alkaline pH. Environ Sci Pollut Res 26(35):35509–35522
Cherubini F (2010) The biorefinery concept: using biomass instead of oil for producing energy and chemicals. Energy Convers Manage 51(7):1412–1421
Datta P, Mohi G, Chander J (2018) Biomedical waste management in India: critical appraisal. J Lab Phys 10(01):006–014
De Souza LLP, Lora EES, Palacio JCE, Rocha MH, Renó MLG, Venturini OJ (2018) Comparative environmental life cycle assessment of conventional vehicles with different fuel options, plug-in hybrid and electric vehicles for a sustainable transportation system in Brazil. J Clean Prod 203:444–468
Devi KS, Sujana O, Singh TC (2018) Hazardous waste management in India—a review. Int J Creative Res Thoughts 6(1):1547–1555
Eckert C, Xu W, Xiong W, Lynch S, Ungerer J, Tao L, Gill R, Maness PC, Yu J (2014) Ethylene-forming enzyme and bioethylene production. Biotechnol Biofuels 7(1):1–11
Elango D, Pulikesi M, Baskaralingam P, Ramamurthi V, Sivanesan S (2007) Production of biogas from municipal solid waste with domestic sewage. J Hazard Mater 141(1):301–304
Elefsiniotis P, Wareham DG, Smith MO (2004) Use of volatile fatty acids from an acid-phase digester for denitrification. J Biotechnol 114(3):289–297
Ellis C (2018) World Bank: global waste generation could increase 70% by 2050. Published Sept 23
Farooq A, Haputta P, Silalertruksa T, Gheewala SH (2021) A framework for the selection of suitable waste to energy technologies for a sustainable municipal solid waste management system. Frontiers Sustain 2:27
Forti V, Baldé K, Kuehr R (2018) E-waste statistics: guidelines on classifications, reporting and indicators
Getahun T, Gebrehiwot M, Ambelu A, Van Gerven T, Van der Bruggen B (2014) The potential of biogas production from municipal solid waste in a tropical climate. Environ Monit Assess 186(7):4637–4646
Guo M, Song W, Buhain J (2015) Bioenergy and biofuels: history, status, and perspective. Renew Sustain Energy Rev 42:712–725
Hafid HS, Shah UKM, Baharuddin AS, Ariff AB (2017) Feasibility of using kitchen waste as future substrate for bioethanol production: a review. Renew Sustain Energy Rev 74:671–686
Hasan MM, Rasul MG, Khan MMK, Ashwath N, Jahirul MI (2021) Energy recovery from municipal solid waste using pyrolysis technology: a review on current status and developments. Renew Sustain Energy Rev 145:111073
Hrabovsky M (2011) Thermal plasma gasification of biomass. Progress Biomass Bioenergy Prod 39
IEA-ETSAP, I.R.E.N.A. (2013) Production of bio-ethylene: technology brief. International Renewable Energy Agency, Abu Dhabi, UAE
Ivanov V, Stabnikov V, Ahmed Z, Dobrenko S, Saliuk A (2015) Production and applications of crude polyhydroxyalkanoate-containing bioplastic from the organic fraction of municipal solid waste. Int J Environ Sci Technol 12(2):725–738
Kabir E, Kaur R, Lee J, Kim KH, Kwon EE (2020) Prospects of biopolymer technology as an alternative option for non-degradable plastics and sustainable management of plastic wastes. J Clean Prod 258:120536
Kendall A (2012) A life cycle assessment of biopolymer production from material recovery facility residuals. Resour Conserv Recycl 61:69–74

Krauklis AE, Karl CW, Gagani AI, Jørgensen JK (2021) Composite material recycling technology—state-of-the-art and sustainable development for the 2020s. J Composites Sci 5(1):28

Kumar A, Agrawal A (2020) Recent trends in solid waste management status, challenges, and potential for the future Indian cities—a review. Current Res Environ Sustain 2:100011

Lesteur M, Bellon-Maurel V, Gonzalez C, Latrille E, Roger JM, Junqua G, Steyer JP (2010) Alternative methods for determining anaerobic biodegradability: a review. Process Biochem 45(4):431–440

Li A, Antizar-Ladislao B, Khraisheh M (2007) Bioconversion of municipal solid waste to glucose for bio-ethanol production. Bioprocess Biosyst Eng 30(3):189–196

Liu B, Rajagopal D (2019) Life-cycle energy and climate benefits of energy recovery from wastes and biomass residues in the United States. Nat Energy 4(8):700–708

Lorenzo Llanes J, Kalogirou E (2019) Waste-to-energy conversion in Havana: technical and economic analysis. Social Sci 8(4):119

Luo K, Pang Y, Yang Q, Wang D, Li X, Lei M, Huang Q (2019) A critical review of volatile fatty acids produced from waste activated sludge: enhanced strategies and its applications. Environ Sci Pollut Res 26(14):13984–13998

Matheri AN, Ndiweni SN, Belaid M, Muzenda E, Hubert R (2017) Optimising biogas production from anaerobic co-digestion of chicken manure and organic fraction of municipal solid waste. Renew Sustain Energy Rev 80:756–764

Menon V, Rao M (2012) Trends in bioconversion of lignocellulose: biofuels, platform chemicals & biorefinery concept. Prog Energy Combust Sci 38(4):522–550

Mlonka-Mędrala A, Evangelopoulos P, Sieradzka M, Zajemska M, Magdziarz A (2021) Pyrolysis of agricultural waste biomass towards production of gas fuel and high-quality char: experimental and numerical investigations. Fuel 296:120611

Mohan SV (2009) Harnessing of biohydrogen from wastewater treatment using mixed fermentative consortia: process evaluation towards optimization. Int J Hydrogen Energy 34(17):7460–7474

Mohsenzadeh A, Zamani A, Taherzadeh MJ (2017) Bioethylene production from ethanol: a review and techno-economical evaluation. ChemBioEng Reviews 4(2):75–91

Molino A, Larocca V, Chianese S, Musmarra D (2018) Biofuels production by biomass gasification: a review. Energies 11(4):811

Moretto G, Valentino F, Pavan P, Majone M, Bolzonella D (2019) Optimization of urban waste fermentation for volatile fatty acids production. Waste Manage 92:21–29

Nomanbhay S, Ong MY (2017) A review of microwave-assisted reactions for biodiesel production. Bioengineering 4(2):57

Noraini MY, Ong HC, Badrul MJ, Chong WT (2014) A review on potential enzymatic reaction for biofuel production from algae. Renew Sustain Energy Rev 39:24–34

Ofori-Boateng C, Lee KT, Mensah M (2013) The prospects of electricity generation from municipal solid waste (MSW) in Ghana: a better waste management option. Fuel Process Technol 110:94–102

Oliveira PA, Baesso RM, Moraes GC, Alvarenga AV, Costa-Félix RP (2018) Ultrasound methods for biodiesel production and analysis. Biofuels-State of Development; IntechOpen, London, UK, pp 121–148

Owamah HI, Dahunsi SO, Oranusi US, Alfa MI (2014) Fertilizer and sanitary quality of digestate biofertilizer from the co-digestion of food waste and human excreta. Waste Manage 34(4):747–752

Paul S, Dutta A, Defersha F, Dubey B (2018) Municipal food waste to biomethane and biofertilizer: a circular economy concept. Waste Biomass Valorization 9(4):601–611

Pérez V, Pascual A, Rodrigo A, Torreiro MG, Latorre-Sánchez M, Lozano CC, David-Moreno A, Oliva-Dominguez JM, Serna-Maza A, García NH, Granados IG (2020) Integrated innovative biorefinery for the transformation of municipal solid waste into biobased products. In: Waste biorefinery, pp 41–80

Reinhart DR, Townsend TG (2018) Landfill bioreactor design and operation. Routledge

Saini JK, Kumar G, Singh S, Kuhad RC (2021) Municipal solid waste biorefinery for sustainable production of bioenergy. In: Waste biorefinery, pp 207–233

Stephenson JR, Sovacool BK, Inderberg THJ (2021) Energy cultures and national decarbonisation pathways. Renew Sustain Energy Rev 137:110592

Strazzera G, Battista F, Garcia NH, Frison N, Bolzonella D (2018) Volatile fatty acids production from food wastes for biorefinery platforms: a review. J Environ Manage 226:278–288

Suhag M, Sharma HR (2015) Biorefinery concept: an overview of producing energy, fuels and materials from biomass feedstocks. Int Adv Res J Sci Eng Technol 2(12):103–109

Taylor G (2008) Biofuels and the biorefinery concept. Energy Policy 36(12):4406–4409

Thangaraj B, Solomon PR, Muniyandi B, Ranganathan S, Lin L (2019) Catalysis in biodiesel production—a review. Clean Energy 3(1):2–23

Thapa B, Patidar SK, Khatiwada NR, Kc AK, Ghimire A (2019) Production of ethanol from municipal solid waste of India and Nepal. In: Waste valorisation and recycling. Springer, Singapore, pp 47–58

Wang L, Sharifzadeh M, Templer R, Murphy RJ (2013) Bioethanol production from various waste papers: economic feasibility and sensitivity analysis. Appl Energy 111:1172–1182

Wojnowska-Baryła I, Kulikowska D, Bernat K (2020) Effect of bio-based products on waste management. Sustainability 12(5):2088

Yaashikaa PR, Kumar PS, Saravanan A, Varjani S, Ramamurthy R (2020) Bioconversion of municipal solid waste into bio-based products: a review on valorisation and sustainable approach for circular bioeconomy. Sci Total Environ 748:141312

Zheng W, Phoungthong K, Lü F, Shao LM, He PJ (2013) Evaluation of a classification method for biodegradable solid wastes using anaerobic degradation parameters. Waste Manage 33(12):2632–2640

Chapter 16
Recent Advances in Biorefineries for Energy and Nutrient Recovery from Food Waste

Srisowmeya Guruchandran, Muthahar Fazila Amanullah, Chakravarthy Muninathan, and Nandhini Devi Ganesan

16.1 Introduction

Sustainable food waste management approaches are considered significant to assure comprehensive resource utilization and ameliorate its impact on the environment. Food waste management (FWM) has been recognized as a global challenge because the collection and retrieval cost is higher than their overall economic worth. However, the increase in food waste (FW) generation over years has become an issue of public concern and demonstrates the need for technologies for appropriate waste management enabling maximum recovery of resources. Food waste is defined as the wastes or byproducts generated before, after or during the processing of foods broadly classified based on their origin as food supply chain waste and domestic waste. The global food waste statistics according to Food and Agricultural Organization (FAO), indicate that one-third of the food produced for human consumption is wasted every year which is approximately 1.3 billion tonnes with an estimated cost of about 0.65 trillion USD. In the EU 27, about 89 million tons of food waste is generated annually, of the 71.2 million tons recorded it is tacit that FW is generated at each phase in a food supply chain (Secondi et al. 2015). In Asian countries, the food waste statistics demonstrate about 278 million tons in 2020 and it is further predicted to increase to 416 million tons by 2025 (Melikoglu et al. 2013). Lack of adequate facilities, awareness, planning and advanced technologies to combat the challenges associated with climate changes, unstable markets, and over production are the main causes of food waste generation leading to adverse economic and environmental effects (Joardder and Masud 2019).

S. Guruchandran · M. F. Amanullah · C. Muninathan · N. D. Ganesan (✉)
Centre for Food Technology, Department of Biotechnology, Anna University, Chennai, India

Y. K. Nandabalan et al. (eds.), *Zero Waste Biorefinery*, Energy, Environment, and Sustainability, https://doi.org/10.1007/978-981-16-8682-5_16

16.1.1 Sources of Food Waste

The food supply chain (FSC) encompasses all the processes involved in the conversion of raw material into a product. The various stages of a food supply chain can be broadly classified as production, processing, distribution, consumption and disposal. The management of food waste generated during various stages of an FSC is very challenging especially in the case of developing countries and countries with dense population. This is mainly because the rate of food waste generation is very high when compared to its rate of management with the existing technologies. Food wastage in an FSC can be broadly classified as food loss and food waste. Food loss is defined as an accidental deterioration of the quality and quantity of the food that was essentially intended for human consumption but ultimately not consumed by humans due to the decline in their nutritional, economic value or safety. In an FSC food loss is particular to production and processing while food waste is associated with distribution, retail, restaurant, catering and domestic consumption. Food waste is defined as the disposal of food products appropriate for human consumption due to improper behaviour/handling of the seller and consumer of FSC ultimately resulting in the spoilage or deterioration of the essential characteristics of food making it unsuitable for human consumption. The FW generation is reported to be more than 40% during the post-harvest and processing stages in developed countries while in the case of developing countries it is during the marketing and consumption stage (Gustavsson et al. 2011). It is therefore considered to be a prerequisite to figure the stage or source of FW generated in the FSC to decide on the appropriate food waste management approach.

According to European Commission in 2014 FW has been broadly classified into three categories in the FSC. 1. Food loss: loss of food during the production stage, 2. Inevitable FW: foodstuff that is lost during the consumption stage which is not suitable for human consumption. 3. Unnecessary FW: foods appropriate for human consumption lost due to improper management and behaviour. Inevitable FW cannot be reduced or eliminated but can be managed and valorized to value-added products. Few examples of inevitable FW valorizations include dietary fibre from banana peels, chitosan from crustacean shells etc., (Dima et al. 2017; Budhalakoti 2019). Based on the post-consumer phase, the FW is further classified as avoidable FW, probably avoidable FW and unavoidable FW. Avoidable FW includes the food that is wasted as it not required or is left unnoticed to expiry (Lipinski et al. 2021). This type of FW can be reduced or eliminated with appropriate food waste management systems with awareness for self-realized monitoring and maintenance. Probably avoidable FW is the food that will be eaten by some people and not by others. This can be reduced with person-specific/ preferred food or meal type. In the case of unavoidable FW, the food wasted during the processing or preparation is not suitable for consumption but inevitable. The understanding on the type and the source of FW is necessary to address the challenge of food waste management appropriately.

16.1.2 Characteristics of Food Waste

Food waste is the most generated biodegradable waste and is majorly comprised of carbohydrates such as starch, cellulose, and hemicellulose, protein, lipids, organic acids, inorganic mineral content. Though inherent characteristics of FW such as high moisture content, high biodegradability, rich in organic load and fewer toxic substances is considered to be a challenging aspect of FW management, it remains to be the positive rationale for most of the successful sustainable techniques of FW management. The high moisture content (80–90%) of the FW remains the major limitation for incineration methods. The pH of the FW varies majorly between 4 and 8 depending on the composition. The nature and the composition of the FW decide its appropriateness for the choice of biorefinery.

16.1.3 Food Waste Biorefineries

Waste to energy technologies (WTE) are recognized to be a sustainable approach for resolving the challenges associated with waste management and chronic energy demands. WTE technologies and their advancements are obligatory for the sustainable development of a Country. Numerous technologies have been investigated and adopted for nutrient, energy and resource recovery from FW including biogas, liquid biofuels, chemicals, biopolymers etc., (Dahiya et al. 2018). FW biorefineries can be broadl classified into three categories (1) biological process that involves enzymes or microorganisms for the conversion of FW to value added products, examples: anaerobc digestion, photo and electro fermentation (2) chemical process that involves the application of chemicals as catalyst for FW valorisation (3) thermochemical process involves the treatment of FW with chemicals at higher temperature, this includes gasification, pyrolysis and liquefaction (Tsegaye et al. 2021). Considering the various WTE technologies anaerobic digestion is regarded as the most potent and widely accepted sustainable technology for treating food waste mainly due to its environmental and economic advantages over the other WTE technologies. The high moisture content, organic load, biodegradability, nutrient balance are the few inherent characteristics of FW that makes anaerobic digestion the most appropriate and promising biorefinery for food waste management.

16.2 Food Waste Anaerobic Digestion

16.2.1 Anaerobic Digestion

Anaerobic digestion (AD) is a bioconversion process in which the complex organic molecules are converted into biogas. During anaerobic digestion, the complex organic matter undergoes four subsequent stages—hydrolysis, acidogenesis, acetogenesis and methanogenesis to produce biogas (Fig. 16.1). Biogas is composed of 50–85% methane (CH_4), 20–35% Carbon dioxide (CO_2) and residual H_2S and water vapour (Srisowmeya et al. 2020). Biogas is upgraded by removing impurities particularly carbon dioxide to improve its suitability as renewable low carbon fuel for electricity generation and transportation. The calorific value of the biogas ranges between 4800 and 6900 kcal/m^3 and pure biomethane at 15.5 °C and 1 atm is 9100 kcal/m^3 respectively (Charcosset 2013). AD of FW demonstrates a typical theoretical methane potential of 0.4–0.5 LCH_4 gVS^{-1} (Li et al. 2018a). Therefore, anaerobic digestion is proposed as a cost-effective technology for renewable energy production and waste management.

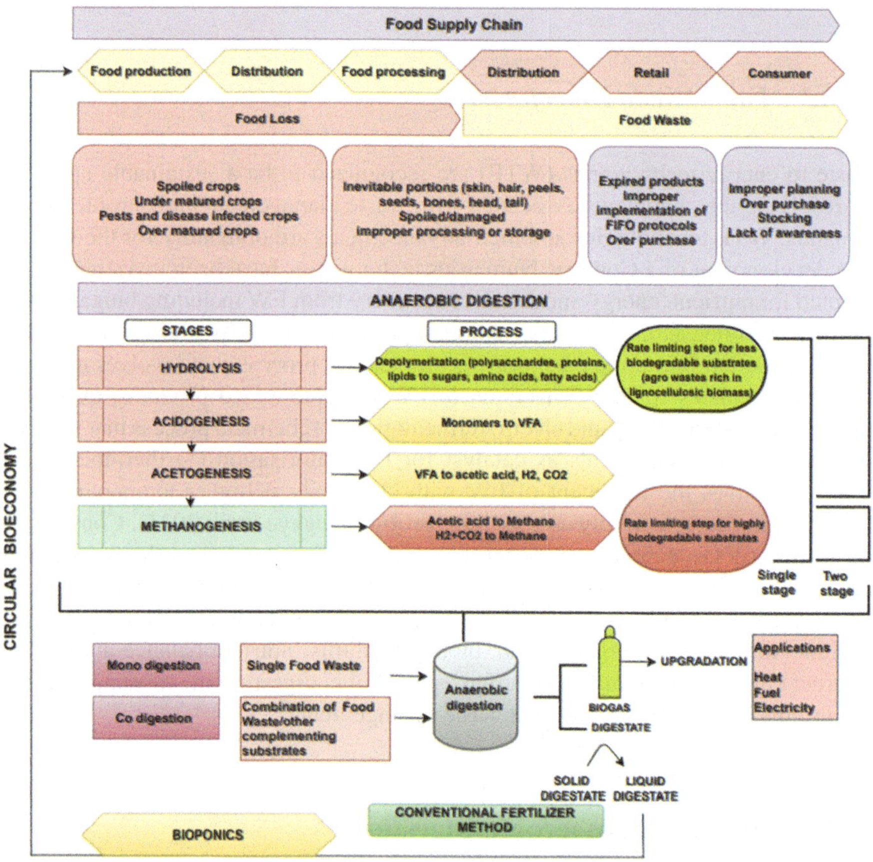

Fig. 16.1 Anaerobic digestion food waste biorefinery

16.2.2 Stages of Anaerobic Digestion

Anaerobic digestion is a biological process mediated by consortia of microorganism performed under an oxygen-free environment. AD initiates with hydrolysis, the depolymerization step where the complex food components such as carbohydrates, proteins and lipids are broken down into simpler molecules such as sugars, amino acids and fatty acids. Hydrolysis is a vital process mediated by microbial extracellular enzymes usually surface-bound or secreted into the medium. Several microbial communities are involved in the hydrolysis process depending on the nature of the substrate, the most predominant ones include *Clostridium, Bacteroides, Staphylococcus, Bacillus, Micrococcus* (Menzel et al. 2020; Angelidaki et al. 2011). Followed by acidogenesis where simpler monomers are further converted into volatile fatty acids (VFA). In the subsequent stage of acetogenesis, the VFA is biotransformed to acetate, hydrogen (H_2) and Carbon dioxide (CO_2). The final stage of methanogenesis involves the conversion of acetogenic components into CH_4 and CO_2. Methanogenesis has two pathways namely acetoclastic methanogenesis which uses acetate as the substrate for methane production and hydrogenotrophic methanogenesis which utilizes hydrogen and CO_2 for methane production. About 64% of the methane produced is by utilizing acetate as substrate and the rest by from hydrogen (Batstone et al. 2002). The microbial community and the microbial enzymes are the key players in the overall bioconversion of waste to biogas. Each stage is unique and its optimal operation is essential for the overall process stability and performance. Similarly, coordination among the various microbial community responsible for various stages of anaerobic digestion is necessary to achieve optimal yield. The rate-limiting stages differ with respect to the nature of the substrate. Less biodegradable substrates like agro-waste rich in cellulose and hemicellulose have hydrolysis as the rate-limiting step while methanogenesis is for highly biodegradable substrates like food waste generated in the distribution and consumption sector (Srisowmeya et al. 2020).

16.2.3 Crucial Parameters

Anaerobic digestion is a comprehensive biological process that requires ideal operating conditions for optimal process stability and performance. Several factors influence the performance of an AD process impacting the quality and quantity of the biogas produced. The influence of crucial influencing parameters is covered in the following subsections.

16.2.3.1 Substrate Characteristics

Food waste comprises several substrates that are originated during various stages of an FSC such as agricultural waste, food processing industrial wastes, spoiled/expired foods, foods wasted during consumption etc., The nature and composition of FW is wide and diverse. Typical FW has a total solid content of 18.1–30.9% (w/w) and volatile solids (VS) content of 17.1–26.35% respectively. The composition of FW is usually 41–62% carbohydrates, 13–30% lipids, 15–25% protein and 70–80% moisture content (Braguglia et al. 2018). The biochemical methane potential (BMP) defined as the maximum volume of methane produced per g of VS substrate demonstrating the ability of the substrate to yield methane on anaerobic digestion (Jingura and Kamusoko 2017). The BMP is highest for lipid rich substrates followed by protein and carbohydrate-rich substrates. However, at higher concentrations lipid, protein and carbohydrate-rich substrates suffer inhibitions due to long-chain fatty acids, ammonia and VFA accumulations (Yenigün and Demirel 2013; Dasa et al. 2016). Therefore, the composition of the substrate is a crucial factor influencing the performance and yield of AD. Anaerobic digestion of various food waste is given in Table 16.1. The optimal composition of FW recommended for AD is in the ratio of 40:40:20 parts of carbohydrates, protein and lipids. Elemental nutrient balance represented as C/N (carbon: nitrogen) is another significant factor that provides insights into the suitability of the substrate for AD. The optimal C/N for AD is in the range of 15–30. Higher C/N depicts the chances of VFA inhibition while lower C/N indicates the probability of ammonia inhibition (Srisowmeya et al. 2020).

Substrate biodegradability is another significant factor that describes the ability of the substrate to be broken down for further bioconversions. The degree of biodegradability differs from the substrate to substrate and not all substrates are readily biodegradable. The degree of biodegradability is in the following order: VFA, alcohols > hemicellulose, fats and proteins > cellulose > lignin. Less biodegradable substrates oppose microbial hydrolysis and therefore remain recalcitrant for AD. The degree of biodegradability of the substrates can be simply understood from their C/N. Highly biodegradable substrates have a C/N of 20–25 and less biodegradable substrates have C/N > 40 (Zhang et al. 2014). Such substrates require pretreatments to enhance microbial accessibility and thereby bioconversion. Several methods have been adopted for FW pretreatments including thermal (Saragih et al. 2019), mechanical (Izumi et al. 2010), Alkali (Kim et al. 2003; Goud and Mohan 2012), Ultrasound (Elbeshbishy and Nakhla 2011), Ozonation (Yeom et al. 2002), biological (Moon and Song 2011) and combination of these methods (Monlau et al. 2012; Jin et al. 2009). Therefore, understanding the substrate characteristics and composition is obligatory to enable the optimal operation of AD.

16.2.3.2 Inoculum Characteristics

The inoculum is considered a key player of an anaerobic digestion system. AD is a complex process in which the substrate undergoes series of biochemical transformations mediated by consortia of microorganisms. The microbial community is specific for each process in an AD and therefore, a pivotal balance in the metabolism and the growth of all the microbial communities is essential for overall performance. Hydrolysis is a rate-limiting step performed predominantly by anaerobes of the

Table 16.1 Production of biogas from different food waste via anaerobic digestion

Type of waste	Type of digestion/digester	Co-digestion	Codigestate	Operating conditions					Removal efficiency	Methane yield	References
				Temperature	pH	Retention time	Organic loading rates	Reactor volume			
Food waste from refectory	Single-stage	Yes	Waste activated sludge	Mesophilic: 35 °C ± 2, and thermophilic: 45 ± 2, 55 ± 2 °C	–	188 days	Mesophilic: <5 g VS/L-d Thermophilic: 7 g VS/L-d	2 L	75% (55 ± 2 °C)	0.40 L CH_4/g VS	Gou et al. (2014)
Tomato pomace	Multiple fed-batch	Yes	Dry bovine manure and mature green waste compost	Mesophilic: 30 °C and thermophilic: 55 °C	–	Thermophilic: 15–20 days mesophilic: 60–75 days	Mesophilic: 5% (w/w) thermophilic: 1% (w/w)	–	–	Mesophilic 5% (w/w): 201 ml CH_4/g dry tomato pomace; thermophilic 1% (w/w); 91.29 CH_4/g dry tomato pomace	Achmon et al. (2019)
Grape pomace	Multiple fed-batch	Yes	Dry bovine manure and mature green waste compost	Mesophilic: 30 °C and thermophilic: 55 °C	–	Thermophilic: 15- 20 days mesophilic: 60–75 days	Mesophilic: 5% (w/w) thermophilic: 1% (w/w)	–	–	Mesophilic 5% (w/w): 132 ml CH_4/g dry grape pomace; thermophilic1% (w/w): 131 CH_4/g dry grape pomace	Achmon et al. (2019)

(continued)

Table 16.1 (continued)

Type of waste	Type of digestion/digester	Co-digestion	Codigestate	Operating conditions					Removal efficiency	Methane yield	References
				Temperature	pH	Retention time	Organic loading rates	Reactor volume			
Cheese whey	Two-stage	No	–	Mesophilic (35 °C)	5.2,7.9	1,4.4	47.4 g COD L-1 d-1	3,15 L	94.2% COD	5.04 L CH_4/L-d	Antonopoulou et al. (2008)
Citrus waste	Two-stage with recirculation	No	–	Thermophilic (55 °C)	5–6	15, 120 days	5 g VS/L d	2 L STR 0.4 L UASB	89–91% COD	203 N mL/g VS	Lukitawesa et al. (2018)
Food waste	Single stage batch	Yes	Green waste (40:60)	Mesophilic (37 °C)	7.3–8.0	50 days	5–20% TS	0.5 L	–	272.1 ml/g VS	Chen et al. (2014)
Mixed Food waste (meat, rice, noodles, leafy vegetables, fish, eggs)	Single-stage fed-batch	No	–	Mesophilic (37 °C) and thermophilic (55 °C)	5.02 ± 0.03	57 days	2.5 g VS/ L-d [thermophilic], 1.5 g VS/ L-d [mesophilic]	0.4 L working volume	–	541 mL/g VS [thermophilic], 371 mL/g VS [mesophilic]	Liu et al. (2017b)
food and vegetable waste	Two-phase	No	–	Mesophilic (35°)	6.70–7.36	37 days	3.1 g VS/dm^3. d	1.3 dm^3	72.4% VS	1.5 dm^3 CH_4/ dm^3 d	Mtz.-Viturtia et al. (1995)
Food waste	Semi-continuous	Yes	Dairy manure	Mesophilic (36 °C)	–	178 days	3 g VS/L d	2 L (1.8 L working volume)	–	0.51 L CH_4/g VS	Agyeman and Tao (2014)
Food waste (mixture of noodle, rice and vegetables)	Two-stage	Yes	Chicken manure and grass (3:5:5) VS basis	Mesophilic temperature (35 ± 1 °C)	7.62	30 days (each stage)	4 g VS/L d	250 mL	–	113.4 mL CH_4/g VS	Li et al. (2018b)

(continued)

Table 16.1 (continued)

Type of waste	Type of digestion/digester	Co-digestion	Codigestate	Operating conditions					Removal efficiency	Methane yield	References
				Temperature	pH	Retention time	Organic loading rates	Reactor volume			
Food waste (rice, potato, breads, boiled egg yolks, cabbage, cauliflower, lentils, noodles, etc.), peeled material of raw vegetables (cucumber, tomato, carrot, etc.) and, stale food items)	Single-stage batch	Yes	Cow dung, granular sludge and waste activated sludge (1:1:1:1)	Mesophilic temperature (35 ± 1 °C)	6.19–7.11	30 days	–	450 ml	38.4% VS	109 mL CH_4/g VS	Gaur and Suthar (2017)
Extruded food waste	Single-stage batch	–	–	Mesophilic temperature (35 °C)	–	31 days	Inoculum to substrate ratio (ISR) 0.5	0.5 L	70% VS	505 mL CH_4/g VS	Kong et al. (2016)
Food waste	Batch type	Yes	Green waste (1:1)	Mesophilic (35 ± 2°C) and thermophilic (50 ± 2 °C)	7.3–7.6	28 days	6.5 g VS/L Substrate to inoculum ratio 4.0	1 L	90.8% VS	430 mL CH_4/g VS	Liu et al. (2009a)

(continued)

Table 16.1 (continued)

Type of waste	Type of digestion/digester	Co-digestion	Codigestate	Operating conditions					Removal efficiency	Methane yield	References
				Temperature	pH	Retention time	Organic loading rates	Reactor volume			
Food waste (rice, noodles, meat, vegetables and condiments)	Three-stage	Yes	Horse manure	Mesophilic (35 ± 2 °C)	7.3–7.9	60 days	12.5 g VS/L	20 L	71% VS	0.37 L CH4/g VS	Zhang et al. (2017)
Food waste	Semi-continuous	Yes	Cattle manure (2:1)	Mesophilic (35 ± 1 °C)	7–7.8	20–28 days	10 g VS/L	1 L	–	317 ml CH_4/g VS	Zhang et al. (2013)
Synthetic food waste	Two-stage	–	–	Thermophilic condition	–	127 days	–	2 l, 8 l, 4 l (three reactors)	82.02% VS	0.459 ± 0.115 L CH_4/ g VS	Xiao et al. (2018)

(continued)

Table 16.1 (continued)

Type of waste	Type of digestion/digester	Co-digestion	Codigestate	Operating conditions					Removal efficiency	Methane yield	References
				Temperature	pH	Retention time	Organic loading rates	Reactor volume			
Banana peels	Batch	Yes	Cow dung, water hyacinth (Mixing ratio 1.5)	–	5.2–5.8	50 days	–	700 mL	44% VS	253 ± 3 mL/g VS	Barua et al. (2019)
Food waste (mainly composed of rice, noodles, meat, vegetables and condiments)	Multi-phase-baffled reactor	No	–	Mesophilic 35 ± 1 °C	–	175 days	0.5–1.0 g VS/L d	–	86.4% VS	215.57 mL CH_4/g VS d	Ahamed et al. (2015)

genera *Streptococcus, Lactobacillus* and *Enterobacterium* mediated by hydrolytic enzymes such as amylases, proteases and lipases (Odnell et al. 2016). Acidogenesis is primarily performed by the microorganism of the genera *Clostridium, Bacillus, Pseudomonas, Micrococcus* or *Flavobacterium*. The key players of acetogenesis include *Syntrophomas* and *Syntrophobacter* for the conversion of acidogenic components to acetates and H_2. The higher VFA conversion is carried out by the specific groups of acetogens namely *Methanobacterium suboxydans* (pentanoic acid to propionic acid) and *Methanobacterium propionicum* (Propionic acid to acetic acid) respectively (Nguyen et al. 2019). Methanogens can be broadly classified as acetoclastic methanogens and hydrogenotrophic methanogens. Acetoclastic methanogens convert acetate to CH_4 and CO_2 and hydrogenotrophic methanogens convert H_2 and CO_2 to CH_4. More than 65% of the methane produced is via the acetoclastic methanogenesis pathway and therefore acetates are considered as a key substrate for methane production. *Methanosarcina* and *Methanosaeta* are the acetoclastic methanogens and *Methanomicrobiales* and *Methanobacteriales* are the hydrogenotrophic methanogens that remain as the key players of methanogenesis. The factors influencing the efficiency of the inoculum includes the substrate to inoculum ratio (S/I), pH, temperature, the concentration of VFA, ammonia, alkalinity, particle size etc., S/I is a crucial factor deciding the stability and the overall yield of AD. A Higher S/I ratio can result in longer hydraulic retention time (HRT) whereas a lower S/I can result in process instability and low yield. S/I of less than 0.5 is usually recommended for optimal operation (Pellera and Gidarakos 2016; Yoon et al. 2014). The characteristics of the inoculum can be recognized from the source digester conditions. The ideal digester characteristics that can be a good source of inoculum was reported by Holliger et al. (2016) and provided the indicative operational values as follows pH: 7–8.5; VFA less than 1.0 g acetic acid/L; alkalinity greater than 3 g $CaCO_3$/L; and Ammonia less than 2.5 g/L. The right of choice of inoculum in combination with the optimal operating parameters is obligatory to achieve anticipated process stability, performance and yield.

16.2.3.3 pH, VFA and alkalinity

pH is a crucial parameter that plays a significant role in maintaining process stability as well as an indicator of approaching process failure. The pH sensitivity of the distinctive microbial communities mediating acidification (acidogens) and methanogenesis (methanogens) varies significantly. The optimal pH for the effective performance of acidogens and methanogens are 4–8 and 6.5–7.5 respectively. Acidogens perform effectively over a wide range of pH whereas methanogens are considered to be highly sensitive to pH changes. Therefore, the optimal pH recommended for AD of food waste is 6.8–7.2 (Cioabla et al. 2012). Neglection in maintaining the optimal digester operating parameters can result in process instability. The most reported cause of process failure in the AD of FW is the accumulation of VFA (Lee et al. 2014; Wang et al. 1999).

Volatile fatty acids are intermediate compounds in an AD process produced during acidogenesis and acetogenesis. The pH of a system and ultimately the overall process stability and performance is greatly influenced by the concentration of VFA. Increasing concentration of VFA results in pH drop, this pH change/fluctuation affects the performance of sensitive methanogens. VFA accumulation is mainly due to the fact that the rate of production of VFA (rate of acidification) is very much high when compared to the rate of consumption (methanogenesis). This imbalance is acceptable to a particular threshold level and when this difference exceeds the prescribed limit it results in process failure. VFA in an anaerobic digestion system includes the following predominant acids: acetic acid, propionic, butyric, valeric and caproic. An ideally operated anaerobic digester is expected to have a VFA concentration below 500 mg/L acetic acid. An increase in VFA concentration of more than 1500–2000 mg/L can be recognized as a clear indication of approaching process failure (Labatut and Gooch 2014). Among the various VFA, the concentration of propionic acid is highly influential. Propionic acid concentration higher than 1 g/L is extremely toxic to methanogens. Indicative factors can be monitored to predict the chances of process instability. VFA to alkalinity ratio (T_{VFA}/T_{ALK}) greater than 2.5 is considered as a threat for process stability (Chakraborty et al. 2018), propionic acid to acetic acid ratio exceeding 1.4, acetic acid concentration exceeding 1.8 g/L (Zhang et al. 2014). On the other hand, the composition of VFA produced during AD is influenced by the pH. Low pH favours the production of acetic and butyric acid while high pH (8.0) favours the production of acetic acid and propionic acid (Morales-Polo et al. 1804). Monitoring and maintaining the VFA concentration and pH is essential to assure process stability and optimal biogas yield.

Alkalinity is another highly influential factor that decides the stability of AD. Alkalinity is defined as the buffering capacity of the reactor demonstrating its ability to manage pH fluctuations. Adequate buffering capacity is required for the process to be stable during the series of reactions that causes production and consumption of intermediate products that are probably capable of affecting the system pH, most importantly to withstand the shock loads of VFA. Bicarbonates (HCO_3^-) are the key components that contribute to the buffering capacity of the system. Bicarbonate alkalinity of about 5500 mg/L $CaCO_3$ is considered to be adequate (Labatut and Pronto 2018a). Continuously monitoring the crucial parameters can aid in the smooth and stable operation of AD as well as understanding the indicative factors can enable strategies to overcome the approaching failure at the earliest possible.

16.2.3.4 Temperature

Temperature plays a significant role in the growth and metabolism of microorganism as well as the activity of enzymes and coenzymes. This demonstrates the influence of temperature on AD. The anaerobic microbes can be distinguished into three groups based on their growth temperature namely psychrophiles (4–20 °C), mesophiles (20–40 °C) and thermophiles (>40 °C). However, mesophilic or thermophilic conditions are usually recommended for effective AD and biogas yield. AD is a process that requires reaction specific operating conditions to achieve optimal performance and yield. Endergonic reactions of AD such as hydrolysis, acidogenesis and acetogenesis perform effectively in a thermophilic condition while the sensitive methanogens involved in the exergonic reaction of methanogenesis perform effectively in a mesophilic condition (Gallert and Winter 1997). Thermophilic operation is advantageous in the case of hydrolysis and acidification reaction as it increases the activity of microbes and microbial enzymes responsible for hydrolysis thereby increasing the substrate availability for bioconversion to VFA. Thermophilic operation is beneficial while considering low retention time, better degradability and high organic loading rates (OLR). However, the limitations of thermophilic operation include increased probability of process instability, ammonia inhibition and high energy requirement. The mesophilic operation ensures better stability and low energy requirement. Thermophilic operations can be adopted for hydrolysis and acidification of less biodegradable substrates and mesophilic for highly biodegradable substrates in order to synchronize the efficiency of various stages in an AD system. The sensitivity of methanogenesis to thermophilic operation is to be considered critically to evade process failure. The rate of temperature changes greater than 1 °C/day can be detrimental and is usually recommended to be less than 0.6 °C/day (Srisowmeya et al. 2020). Considering the temperature-dependent performance of acidogens and methanogens, the concept of temperature phased anaerobic digestion (TPAD) was established. In TPAD, the acidification reactions are performed under thermophilic conditions and methanogenesis under the mesophilic condition to establish stage separated process-specific operations for enhanced performance, biogas yield and digestate quality.

16.2.3.5 Organic Loading Rate and Hydraulic Retention Time

Organic loading rate (OLR) is defined as the quantity of substate entering the digester per unit time and is expressed as g/L-d. Hydraulic retention time is defined as the average time (days), the substrate spends inside the digester until maximum conversion to an intended product. OLR can be calculated as a ratio of influent substrate concentration in g/L (VS or COD basis) to HRT (days). Though the biogas production increases with increasing OLR, beyond a threshold limit it can result in VFA accumulation thereby triggering process failure. While in the case of low OLR, the system suffers nutrient deficiency resulting in downgraded performance. The optimal OLR for food waste AD is recommended to be in the range of 1.5–2.5 g VS/L-d (Jiang et al. 2020). The efficiency of handling various OLR is influenced by substrate characteristics, temperature, HRT, and inoculum characteristics. Thermophilic digesters can effectively handle higher OLR than mesophilic digesters (Liu et al. 2017a). Similarly, high OLR requires adequate HRT for complete and stable bioconversion. High OLR and shorter HRT is a noxious combination that can certainly result in process instability and failure.

The HRT in an AD process depends on the specific growth rate of the particular microbial community. Acidogens exhibit a higher specific growth rate of 0.172 h^{-1}

demonstrating lesser HRT when compared to that of methanogens with a specific growth rate of 0.0167–0.02 h^{-1} (Lo et al. 2009). The recommended HRT of acidogens and methanogens are typically 2–8 days and 20–30 days. Operating the methanogenic phase at lower HRT can result in process failure. HRT depends on several factors such as substrate characteristics, temperature, and OLR. Easily biodegradable substrates such as cooked food waste require shorter HRT whereas less biodegradable substrates such as plant matter require longer HRT. Temperature is a significant influencer of the HRT; the higher the temperature lower is the HRT. HRT of a mesophilic digester is usually 14–40 days while that of a thermophilic digester is 14–20 days (Srisowmeya et al. 2020). However, lower HRT than optimal can result in incomplete substrate utilization and product formation. Shorter HRT can cause stress for the microbial community ultimately favouring the growth and metabolism of hydrogen-producing microbes.

16.2.3.6 Ammonia Inhibition

Ammonia is majorly a hydrolytic product from nitrogen/protein-rich substrates. On depolymerization of protein-rich substrates, ammonia is released as ammonium ion (NH_4^+) or free ammonia (NH_3) which is collectively refereed as total ammonia nitrogen (TAN). Though ammonia aids in improving the buffering capacity (Morales-Polo et al. 1804), it is considered to exhibit an inhibitory response in most conditions. Among the two forms, free ammonia is regarded as more toxic in AD. The inhibitory concentrations of free ammonia and TAN are >650 mg/L and ≥1200 mg/L respectively (Hartmann and Ahring 2005). However, the intensity of inhibition depends majorly on temperature and pH. The ammonia inhibitory effect increased with increasing pH and temperature (Fernandes et al. 2014). With respect to the microbial community, methanogens particularly acetoclastic groups are more sensitive to ammonia inhibition than acidogens. Additionally, thermophilic methanogens exhibit better tolerance to ammonia concentration than mesophilic methanogens. Therefore, maintaining a free ammonia concentration of less than 0.2 g/L is highly recommended. Several strategies can be adopted to remove excess ammonia and maintain it within the optimal levels (Krakat et al. 2017; Ghyselbrecht et al. 2018; Walker et al. 2011).

Process specific operation enables optimal process performance, stability and yield. Understanding the relationship and the influence of the crucial parameters is essential to ensure stable and effective anaerobic digestion. Monitoring and maintaining the critical limits of the crucial parameters are obligatory to achieve maximum resource utilization and energy recovery.

16.2.4 Advanced Strategies

Anaerobic digestion of food waste is a potential approach that anticipates optimal operations and continuous monitoring to achieve maximum biogas production and good quality of digestate. Since AD of FW confronts limitations due to VFA accumulation, ammonia inhibition, imbalanced nutrients, insufficient buffering capacity, and high biodegradability, advanced strategies such as codigestion, stage separation and recirculation have been adopted to minimize these limitations to a greater extent.

16.2.4.1 Codigestion

Mono digestion of FW suffers process instability due to imbalanced nutrients such as improper C/N ratio, lack of micronutrients, excess of macro nutrients, lack of buffering components, the unhealthy composition of carbohydrate, lipid, and protein. Such imbalances result in process failure due to VFA accumulation picking the digester, pH fluctuations affecting the performance of methanogens driving the overall process to instability and hence breakdown. Codigestion is an effective approach particularly to resolve the problems associated with the nutrient imbalance and buffering potential. During Codigestion FW is digested along with a substrate preferably with complementing characteristics. Organic wastes especially animal manures and sewage sludge are recognized to be a potential codigestate for FW. Animal manures (AM) rich in ammonia aids in improving the buffering capacity of the digester attributing to improve process stability (Zhang et al. 2013). AM also contributes to improving the C/N ratio resolving the most limitations of mono-digestion. Sewage sludge (SS) as a codigestate is recognized to be significant considering the economic and environmental aspects. FW codigestion with SS can aid in resolving the limitations of FW as well as the toxicity of SS (Mehariya et al. 2018). Codigestion of FW and SS is advantageous in various aspects including (i) balanced C/N ratio: SS has low C/N and FW has high C/N therefore codigestion aids in bringing the C/N to optimal levels (ii) reducing or diluting the toxicity of SS (iii) nutrient balance: SS provides the necessary micronutrients (iv) improved rate of hydrolysis. The other advantages of codigestion over mono-digestion of FW includes methane enrichment, economic feasibility, yield enhancement, treating two different complementing wastes at once is beneficial on environmental aspects, higher energy recovery, reducing toxicity and greenhouse emissions. Codigestion has been demonstrated to be considerably potential in increasing the biogas production from 35 to 400% (Cavinato et al. 2010; Shah et al. 2015). However, the choice of codigestate, organic loading rates, complementing characteristics of the substrates, the synergy between the substrates, the biodegradability of the substrates are to be carefully regarded to avoid process saturation and ultimately failure. The appropriate choice of co-substrate is essential to achieve the desired process performance and biogas yield.

16.2.4.2 Stage Separation

Anaerobic digestion takes place in four stages namely hydrolysis, acidogenesis, acetogenesis and methanogenesis. In single-stage AD (SSAD), all four stages take place in a single compartment which suffers limitations due to VFA accumulations, pH drop, ammonia inhibition, compromised stability and performance (Srisowmeya et al. 2020). In order to overcome these glitches, the stage separation concept was established. Two-stage anaerobic digestion (TSAD) is where acidification and methanogenesis are performed in different compartments providing optimal operating conditions specific for each stage. The operating parameters of acidification and methanogenesis are different and hence stage separation aids in establishing conditions for maximum performance and efficiency. Acidogens are recognized to be more tolerant and perform effectively over a wide range of conditions whereas methanogens are highly sensitive and hence requires precise conditions for existence. The optimal pH for methanogenesis is 6.0–8.0 and acidification is 5.2–6.5 respectively (Matheri et al. 2016; Li et al. 2013) The HRT of the acidification stage is shorter than the methanogenesis stage and this is due to the difference in the specific growth rates and rate of bioconversion (rate of acidification > rate of methanogenesis) (Srisowmeya et al. 2020; Cremonez et al. 2021). TSAD can effectively handle higher OLR and a wide range of total solid content (2–40%) when compared to SSAD. OLR has a greater impact on the performance of methanogens than acidogens. Acidification reactions are favourable under both mesophilic and thermophilic conditions however, the rate of bioconversion increased with increasing temperature. While the optimal temperature for methanogens is strictly mesophilic as they are sensitive to higher temperatures (Lay et al. 2003). The major advantage of TSAD is biohythane production. Biohythane production is the simultaneous production of biohydrogen in the acidification stage and biomethane in the methanogenesis stage. This increases the overall energy efficiency of the process (Srisowmeya et al. 2021). Stage separated process, therefore, enables the chance of providing process-specific operating conditions such as pH, temperature, HRT, OLR, and C/N. Though, TSAD is advantageous over SSAD as it provides better stability, shorter HRT, efficiency to handle high OLR, improved removal efficiencies, reduces VFA accumulation and ammonia inhibition, it exhibits few practical shortcomings such as the need for skilled labour, complexity in continuous monitoring and maintenance (Srisowmeya et al. 2020). Advanced automation is anticipated for constant monitoring and optimal system operations. However, TSAD is a potential approach that can significantly resolve most challenges associated with FW AD and assure higher bioenergy yield and good quality digestate.

16.2.4.3 Recirculation

Recirculation is an approach usually integrated with the TSAD system where the effluent from the second stage (methanogenesis) is recirculated into the first stage (acidification). Recirculation is advantageous as it improves biomethane yield, provides higher degradability, stable operation and higher VS removal efficiency (Xue et al. 2020). The TSAD with recirculation established a stable operating pH of 5.0–6.0 while the other counterpart suffered a pH drop of 3.7–4.7 demonstrating the efficiency of recirculation to improve process stability (Lukitawesa et al. 2018). Similarly, VS efficiency of the recirculating system was 33–35% higher (Xing et al. 2020). Recirculation shows better efficiency in handling digesters with high OLR and shorter HRT. Recirculation enabled semi-continuous TSAD showed efficiency in handling OLR up to 4 g VS/L-d for cotton and 10 g VS/L-d for starch respectively (Aslanzadeh et al. 2013). At higher OLR, recirculation aids in increasing the system alkalinity providing adequate buffering capacity for maintaining the optimum pH for methanogens (Wu et al. 2018). Recirculation ratio (RR) is defined as the portion of the effluent recirculated. Ideal RR is necessary to provide maximum efficiency which differs with respect to the substrates, digester volume, OLR, HRT and temperature. Usually, an RR of 25–40% is recommended for optimal efficiency and maximum biogas yield which can however differ based on the biogas digester or plant characteristics (Müller et al. 2017; Chen et al. 2020).

16.3 Biogas Upgradation and Utilization

16.3.1 Biogas Upgradation

The composition of biogas and the proportion of CH_4 and CO_2 varies depending on the substrate and the operating conditions. Biogas upgradation is a process of removing CO_2 and other trace gases to improve the purity of methane ($\geq$90%). Biogas upgradation increases the energy value of the resultant gas. The average calorific value of biogas is 20–28 MJ/m^3 and natural gas is 32.60–36.20 MJ/m^3 (Charcosset 2013; Carvill 1993). The difference in the calorific value is due to the incombustible counterpart which is majorly CO_2 in the case of biogas. High-grade biomethane has an upper calorific value of 38 MJ/m^3 demonstrating the significance of biogas upgradation. Upgraded biomethane application on transportation can reduce greenhouse gas (GHG) emission by 60–80%. Several methods such as absorption, adsorption, permeation, cryogenic separation and in situ methane enrichment have been adopted for biogas upgradation. The highest purity of methane is obtained from cryogenic separation and membrane permeation. Cryogenic separation is advantageous as it provides the highest methane purity (99%), lowest methane loss (<0.1%), no chemicals involved and most importantly the pressure of the upgraded biogas is sufficient for vehicular fuel purpose and requires no additional compression.

Membrane permeation is where silicon, cellulose polymers are used as a membrane for biogas upgradation. Membrane permeation exhibits high purity, low mechanical wear, efficient design and operation (Kárászová et al. 2015). The high capital cost, energy requirement and operating cost are the major limitations of both methods. Absorption methods (physical and chemical) have lower maintenance cost, simple operation, economical and less methane loss (Abdeen et al. 2016; Sahota et al. 2018). Several methods such as industrial lung and supersonic separation are anticipated biogas upgradation methods under development (Dhanya et al. 2020). Though several upgradation methods are available, the appropriate choice will depend on the composition of the biogas, capacity requirement, economic feasibility, expected final biogas purity.

16.3.2 Biogas Utilization

Biogas plants are considered potential for energy generation and the application of biogas for fuel, heat and electricity is increasing progressively. The choice of biogas application largely depends on the biogas plant capacity. Small scale digesters are usually recommended for domestic direct combustion for heat using specially designed stoves. Large scale biogas plants are used for electricity, heat or fuel. The energy generation technologies combined with AD to convert biogas to fuel, heat or electricity includes fuel cells, combined heat and power (CHP), turbines and reciprocating engines (Labatut and Pronto 2018b). Fuel cell techniques have the advantage of least emissions and greater fuel flexibility. Compressed biomethane can be used as a vehicular fuel and is also known as bio compressed natural gas (Bio-CNG). The energy efficiency of biogas for electricity generation is 8–54%, Transport 4–18% and heat 16–83% respectively. The electricity efficiency of biogas utilization technologies are as follows: fuel cells (40–60%), reciprocating engines (22–45%), microturbines (25–35%) and gas turbines (22–36%). Similarly, the capital cost of the technologies is highest for fuel cells followed by microturbines > reciprocating engines > gas turbines (Labatut and Pronto 2018b). Biogas technologies are recognized to exhibit higher efficiency over conventional methods. Boilers burned with methane showed 10% higher efficiency than coal (Stafford et al. 1981). The efficient utilization of biogas for versatile applications requires further advancements considering efficiency and economic feasibility. Biogas technologies are recognized to be potential and are standardized in few countries while it is still under consideration in most developing countries. Further approaches for standardizing biogas technologies with guidelines for purity levels and lucrative applications are anticipated.

16.4 Anaerobic Digestate Utilization

16.4.1 Conventional Methods

A byproduct of biogas production is the digestate also known as effluent/slurry. Anaerobic digestates are considered to exhibit potential properties of quality fertilizer. Digestates contain adequate nutrient requirements to be employed as a fertilizer. The major nutrients of digestates include nitrogen, phosphorous and potassium encompassed as NPK. The cost efficiency and environmentally friendly aspects of anaerobic digestates such as low levels of heavy metals, reduced pathogens, ability to eliminate weeds are significantly beneficial over conventional fertilizers (Mukhuba et al. 2018; Furukawa and Hasegawa 2006). Conventional methods demonstrate the direct application of digestates into farmlands to enrich the soil quality and nutrients. However, the digestate quality, quantity, nutrient availability, fertilizer dose calculations are prerequisites to this approach. The nutrient requirement of crops differs with respect to their type, stage, flowering or fruiting nature. In such cases, micronutrient fortification of anaerobic digestates can be adopted to provide complete nutrient requirements for the crop. Anaerobic digestates ensured its application as a potential fertilizer by improving the crop yield of tomato (Abdelhamid et al. 2011; Ferdous et al. 2018), xiao bai cai (*Brassica rapa*) (Cheong et al. 2020), chillies, brinjal (Bhatnagar and Mutnuri 2015), and others (Pastorelli et al. 2021; Baştabak and Koçar 2020; Koszel and Lorencowicz 2015). The conventional method of digestate utilization confronts limitations due to incomplete anaerobic digestion which can result in GHG emissions, imbalanced nutrients that can disturb the innate quality of the soil, and excessive nutrients in plant unexploitable form. Therefore, adequate investigations and approaches are anticipated to improve the suitability of anaerobic digestate for effective crop production and maximum nutrient recovery.

16.4.2 Bioponics

Bioponics is an advancing technology that is an integration of biogas production component and hydroponic component. The anaerobic digestate from the biogas plant is used as a nutrient medium for hydroponics (water-based cultivation). Bioponics system enables a sustainable approach with thoughtful consideration to circular bioeconomy. Several studies have revealed the potential of bioponics using anaerobic digestate for the production of high-quality crops (Liu et al. 2009b). Conventional methods demonstrate the direct application of the entire digestates into farmlands to enrich the soil quality and nutrients. However, the digestate quality, quantity, nutrient availability, fertilizer dose calculations are prerequisites to this approach. The conventional method of digestate utilization confronts limitations due to incomplete anaerobic digestion which can result in GHG emissions, imbalanced nutrients that can disturb the innate quality of the soil, and excessive nutrients in plant unexploitable form. Adequate nutrients in plant exploitable form with the optimized environment (pH, temperature, humidity) are the major advantage of the bioponics system. Bioponics is a vertical farming approach that conserves space, time, nutrients and water. Crops grow at a rate of 30–50% faster and healthier in a bioponics system. Moreover, bioponics is appropriate to be regarded as organic farming (Shubha et al. 2019). Bioponics systems showed efficiency for a variety of crops such as cucumber (Endo et al. 2016), bok choy (Pelayo Lind et al. 2021; Bergstrand et al. 2020), kai choy (Jamison et al. 2021), lettuce (Ronga et al. 2019), and tomatoes (Neal and Wilkie 2014). However, the efficiency of bioponics depends on several factors such as the composition of the digestate, levels of plant inhibitory factors, the concentration of essential nutrients, availability of nutrients (in appropriate forms), nutrient adequacies and deficiencies (micronutrients). The most significant factor deciding the success of a bioponics system is the relationship between the residual nutrient availability of the anaerobic digestate and the nutrient requirement of the crop of choice, when these two factors are in line the system can be a potential approach for sustainable bioconversion. Bioponics systems require further investigations on the standardization of the digestate quality, methods to increase the plant exploitable nutrients, safety protocols, monitoring and maintenance practices. Bioponics is an advancing approach considered sustainable to simultaneously address waste management and food supply demands especially in the case of developing countries like India with exponentially growing population. Moreover, utilizing anaerobic digestate for crop production can be a prospective approach to increase agricultural sustainability. Global availability of food waste is subjected to geographical features. The rate of food waste generation is high in the European Union followed by India thus presenting promising opportunities to recover nutrients from the waste for the purpose of food generation (Ranganathan et al. 2020). The safety concerns of the bioponics produce requires consideration as the increasing concern on health and awareness regarding the loss of significant nutrients during the process has increased the demand for fresh, minimally processed foods. Less processed foods have higher chance of the pathogen outbreak and dissemination. Therefore, finest approach to maintain compliance with the safety of fresh produce is to adhere strictly to the guidelines proposed by the eminent organizations predominantly USDA and Global GAPs. Further technical advancements can ensure large scale implementation of these novel systems to progress towards sustainability.

16.5 Other Products from Food Waste Biorefineries

16.5.1 Bioalcohols

Alcohols have high demands as industrial solvents and renewable fuel. Though a range of alcohols can be produced by fermentation of FW, ethanol and 1-butanol are often studied and preferred for their applicability as fuels by blending with conventional fuels. Methanol is often produced by chemical conversion methods and does not recognize as an important fuel. The need for crop production or sourcing raw materials at a higher cost for alcohol production can be avoided by utilizing FW based on the composition and characteristics related to microbial alcoholic fermentation. Fuel alcohols can be produced economically from carbohydrate-rich biomass than using lignocellulosic biomass and molasses (Hegde and Trabold 2018). The sources of compatible FW from alcohol generation can be found in all stages of food lifecycle i.e., from crop processing to post food-processing however, appropriate pre-treatment methods can help achieve optimum production. The physical nature of the waste is an important concern in the process and classified between solid and liquid waste. Solid wastes are often rich in starch, lignin, cellulose and hemicellulose. The major source is inedible dough, spent coffee grounds, waste bakery products, mashed fruit residues, vegetable residues etc. Common liquid effluents are dairy whey, yoghurt, tofu whey, cooked rice water, brewery effluents, vegetable processing waste water, etc.

The organic composition of FW plays an important role in alcoholic fermentation. For higher microbial metabolism, high reducing sugar content (20–10%) is desired but very high free sugar levels can suppress alcohol production by feedback mechanism (Tahir et al. 2010). For solids and substrates with complex structure, appropriate pretreatment technologies like acid hydrolysis, alkali treatment, heat treatment or enzymatic hydrolysis are performed prior to the fermentation step. The next important factor for good alcohol production is the optimum carbon to nitrogen ratio (C/N ratio). For biomass like plantain pith, banana and other fruit peels containing high polymerized carbohydrate chains require simultaneous saccharification and fermentation for continuous alcohol production. In most cases, *Aspergillus niger* was used for saccharification and *Saccharomyces cerevisiae* was used for bioethanol conversion (Itelima et al. 2013). Altered stimulatory effects on yeast ethanol tolerance can be found in yeast in the presence of certain lipid molecules while some fatty acids were known to function as a growth stimulant for yeast and anti-foam agent in fermentation (Ghareib et al. 1988). In alcoholic fermentation like bioethanol fermentation or ABE (Acetone, butanol, ethanol) fermentation, fed-batch fermentation was preferred over batch or continuous fermentation (Zabed et al. 2016). For microbial metabolite production, fed-batch fermentation provides good process stability and works with lower substrate concentration and tolerates higher product concentration. Allowing control of crucial parameters like temperature, pH and dissolved oxygen while engaging a variety of food waste becomes possible by feedback control and the productivity of metabolite is limited to substrate concentration and microbial cell growth in the fed-batch process (Chandel et al. 2013). Ethanol and butanol are usually recovered by fractional distillation or distillation combined with adsorption (Mosier et al. 2005). The spent fermentation media is treated in wastewater treatment plants or converted into various other co-products.

While a good understanding of the composition of bio-molecules and their role in optimum fermentation are continuously acquired, the main challenge in consistent alcohol conversion is interrupted by highly unstable feedstock composition and characteristics. For a successful FW to alcohol conversion unit, a quality check on the physical and chemical consistency of the feedstock should be performed regularly since it may vary due to the origin of feedstock, time since collection and storage methods. Other barriers to commercialization of alcohol production from food waste are low butanol yield, lower alcohol tolerance by microbes, storage and transportation issues with feedstock, lack of decentralized design in waste processing and lack of policy and work plan towards utilizing FW for alcohol production (Hegde et al. 2018) (Table 16.2).

Table 16.2 Products obtained from the alcoholic fermentation of food wastes

Food waste	Organism employed	Product	Yield (g/L)	References
Potato starch	*Saccharomyces cerevisiae*	Ethanol	19	Abouzied and Reddy (1986)
	Clostridium acetobutylicum	1-butanol	15.3	Kheyrandish et al. (2015)
Waste potato	*Saccharomyces cerevisiae*	Ethanol	31	Izmirlioglu and Demirci (2010)
	Clostridium beijerinckii CCM 6218	1-butanol	4.73	Patáková et al. (2009)
Whey permeate	*Candida pseudotropicalis*	Ethanol	120	Shabtai and Mandel (1993)
	Kluyveromyces fragilis	2-methyl 1-butanol	80	Parrondo et al. (2000)
	Kluyveromyces fragilis	2-methyl 1-butanol	0.05	
Deproteinated cheese whey	*Clostridium acetobutylicum DSM 792*	1-butanol	8.9	Raganati et al. (2013)
Carrot pomace	*Kluyveromyces marxianus*	Ethanol	37	Yu et al. (2013)
Grape pomace	*Pichia rhodanensis*	Ethanol	18.1	Korkie et al. (2017)
Carbonated drinks	*Clostridium beijerinckii NCIMB 8052*	1-butanol	8.7	Dwidar et al. (2012)
	Zymomonas mobilis ZM4	Ethanol	25	
Apple pomace	*Saccharomyces cerevisiae*	2-methyl 1-butanol	120	Hang et al. (1981)
	Saccharomyces cerevisiae	Ethanol	43	Hang et al. (1981)
	Clostridium acetobutylicum	1-butanol	22	Voget et al. (1985)

16.5.2 Biodiesel

Biodiesel is fatty acid alkyl esters (FAME) derived from transesterification of oils with alcohols in the presence of alkali, acid or enzyme as catalyst (Table 16.3). Biodiesel can be produced from waste cooking/frying oil, rendered fats from animal, fish and plant sources including direct transesterification by using a catalyst or producing oil by microalgal fermentation for biodiesel conversion. FW can be of dual-use in the bio-diesel conversion process. Oil-rich FW are often exploited and recognized as feedstock for biodiesel production while some food wastes have the characteristics to act as a catalyst in the process of esterifying fats to FAME. Heterogenous catalyst from FW includes bone, animal shells, ash, eggshells etc. (Li and Yang 2016). Calcinated shells perform better than bones or mollusc shells since eggshell is porous and provides larger surface area. FW derived catalyst are cheaper than

Table 16.3 Biodiesel production from food waste feedstocks

Food waste	Catalyst	Acyl acceptor	Biodiesel yield (%)	References
Fish waste	Potassium hydroxide	Methanol	72	Jung et al. (2019)
	$SO_4^{2-}/SnO_2\text{-}ZrO_2$	Ethanol	82.58	Enascuta et al. (2018)
	Novozyme 435	Ethanol	82.91	Marín-Suárez et al. (2019)
Kitchen waste cooking oil	Sodium methoxide	Methanol	80.9	Wang et al. (2017)
	Potassium hydroxide	Methanol	94	Sahar Sadaf (2018)
	Sulfonated carbon microspheres	Methanol	89.6	Tran et al. (2016)
	Immobilized *Pseudomonas mendocina*	Methanol	91.8	Chen et al. (2018)
Bitter apple waste	2-(4-Sulfobutyl) pyrazolium hydrogen sulfate	Methanol	89.5	Elsheikh (2014)
Melon seeds	Potassium hydroxide	Methanol	93.16	Fadhil (2013)
Olive pomace waste	Sulfuric acid	Methanol	90.8	Ouachab and Tsoutsos (2013)
Waste sardine oil	Immobilized lipase from *Aspergillus niger*	Methanol	94.55	Arumugam and Ponnusami (2017)
Waste Manilkara zapota seeds	Immobilized lipase B from *Candida antarctica*	2-propanol	92	Cruz-Izquierdo et al. (2014)
Palm oil	Immobilized Burkholderia *Cepacian lipase*	Methanol	100	Jegannathan et al. (2010)

conventional catalysts and offer more reusability than the highly efficient yet costly enzymes. Biodiesel from waste fats is successfully converted to biodiesel in UK and EU. Few commercial establishments like SENECA green catalysts are also demonstrating pilot-scale biodiesel production plants from food waste (Lin et al. 2013). The usual process for successful biodiesel production from FW involves two steps: (1) conversion of free fatty acids to FAME using homogenous catalyst (2) conversion of triacylglycerols to biodiesel using enzyme-mediated reaction. Though many successful methods have been demonstrated for conversion of oil to FAME using different sources, catalyst and acyl acceptors, the real challenge lies in the extraction of oil from oil-rich food waste fraction, segregation of oil-rich food fraction, channelizing the feedstock supply, storage and transport. The technical problems involve oil extraction by economic methods. A combined approach by Brocklesby Ltd, US involved mincing and thermal treatment to recover oils entrapped in food waste followed by continuous oil separation using tricanter centrifuge resulting in 98% oil recovery (Woodgate and Veen 2004). Such approaches should be economical from the grounds of geography where the technology is demanded for successful establishments of biodiesel bio-refinery. The carbohydrate-rich residues can further be used to produce bio-oils and biochar by thermal treatments. An upcoming yet potential challenge with biodiesel production plants is glycerol removal (Gholami et al. 2014). Increasing bio-diesel plants have driven the value of glycerol to near-zero resulting in the need for technologies for glycerol utilization (Quispe et al. 2013). Crude glycerol has potential contaminants and hence the need for purification also arises in processing however the alcohols and catalysts used in this process are abundant and the prices remain unchanged in the recent past (Mohammadshirazi et al. 2014). The cost of biodiesel production from waste cooking oil has always been more expensive than conventional fuels for which sustainable glycerol utilization and value addition may help reducing the overall economics.

16.5.3 Bioactive Component, Biochemicals and Bioproducts

The strength of biorefinery lies in its potential of handling waste to generate bioproducts and reuse them as feedstock, energy commodities or any useful bioproducts (Table 16.4). A more interesting technology for the production of bioproducts and biochemicals from waste is solid-state fermentation. Though submerged fermentation techniques have reported success for the production of various bioproducts from waste, the limitation to treat liquid fraction is overcome by solid-state fermentation (SSF). The most common bioproducts produced through SSF on waste food biomass are enzymes with high market value thus helping an integrated biorefinery keeping the economics on the positive side. In SSF fermentation, the microbial process happens in absence of free water or near the absence of free water on the solid substrates (Thomas et al. 2013). This condition helps to establish microbial populations that can produce hydrolytic enzymes often required by bio-fuel industries in the pretreatment processes (Farinas 2015). Notable enzymes produced by utilizing food waste

Table 16.4 Various biochemicals and bio-products obtained from food waste

Enzymes	Food waste	Organism employed	Product	Yield (U/g dry weight)	References
	Soy fibre waste	*Thermomyces lanuginosus*	Amylase	228,000	Cerda et al. (2016)
	Soybean meal	*Aspergillus niger*	Xylanase	365	Khanahmadi et al. (2018)
	Wheat bran	*Aspergillus niger*	Xylanase	1137	
	Wheat bran	*Aspergillus niger*	Xylanase	2919	
	Corn cob	*Aspergillus niger*	Xylanase	380	
	Coffee husk	Compost	Cellulase	10 FPU/g dry weight	Cerda et al. (2017a)
	Coffee husk	Special consortia	Xylanase	48	Cerda et al. (2017b)
Biochemicals	Food waste	Method employed	Product	Category	References
	Grape pomace	Ultrasound assisted extraction	Pectin	Biopolymer/dietary fibre	He et al. (2016)
	Grape pomace	Pulsed electric field	Polyphenol	Nutraceutical	Kantar et al. (2018)
	Grape pomace	Pulsed electric field	Anthocyanin	Nutraceutical	Parniakov et al. (2014)
	Orange peel	Pulsed electric field	Hesperidin	Nutraceutical	
	Onion skin	Subcritical water extraction	Quercetin	Nutraceutical	Bleve et al. (2008)
	Grapeseed	Supercritical fluid extraction	Resveratrol	Nutraceutical	Casas et al. (2010)
	Tomato skin	Enzyme assisted extraction	Lycopene	Pigment/nutraceutical	Oroian and Escriche (2015)
	Flax seeds	Solvent extraction	Lignans	Antioxidant/nutraceutical	Todaro et al. (2009)

are amylases, xylanases, cellulases and hemicellulases. Other bioproduct incudes bio-surfactants, bio-plastics, bio-pesticides, bio-fertilizers, organic acids and antibiotics. Bio-surfactants are less toxic, highly degradable and sustainably producible used in cosmetics and bio-polymer industries (Singh et al. 2019).

Apart from these bioproducts, bio-actives are also produced. Bioactive compounds include a wide range of non-nutritive products yet has significant effects on human health. Vitamins, minerals, dietary fibres, antioxidants and phenolics form a major portion of a complete and balanced human diet with functional properties (Martins and Ferreira 2017). Olive waste, grape pomace, pear seeds are a rich source of polyphenols and other antioxidants that can be extracted to be functional additive in any food (Akhtar et al. 2015; Chougui et al. 2015). Sour acid peel, pomegranate peels and gac fruit peel extracts are rich in ferulic acids and sinapic acids (Sharma et al. 2017). Glycosylated flavones and polymethoxylated flavones extracted from citrus fruit peels can exhibit antioxidant, anti-inflammatory and anti-carcinogenic effect upon consumption in food (Chen et al. 2012). Wasted cabbage, beet molasses, cheese whey, peach peel, waste oil, watermelon husk can also be used to produce carotenoids using heterothallic *Blakeslea trispora* (Papadaki and Mantzouridou 2021). Astaxanthin extracted from shrimp shells exhibits blood pressure and cholesterol regulating properties (Ambigaipalan and Shahidi 2017). Most bio-chemicals extracted from FW are important ingredients in cosmetic and functional food and nutraceutical industries as additive products with significant physical, chemical and biological functions.

16.6 Current Challenges, Future Perspective and Research Needs of Food Waste-Based Biorefineries

The importance of food waste management and the consequence of its negligence has been very well established in the advancing world. Despite the clear potential of FW management via anaerobic digestion, the challenges encountered during biogas plant establishment is obvious. Lack of profound policy and funding support remain the main challenges that remain unaddressed over decades. Encouraging decentralized/standalone biogas plants with fundamental investment support or subsidiary by the government can be beneficial and appreciable. Further insights and awareness on the consequence of improper food waste management and the potential of FW for resource recovery are required to enable self-initiated response/attempt for food waste management and valorization. Food waste valorization can be a potential platform for entrepreneur development. The complications associated with food waste anaerobic digestion can be overcome by further investigations on the (i) process automation to continuously monitor and control the system; (ii) strategies to improve process stability concerning the digester design; (iii) approaches to enable management of higher OLR. Bioponics is a developing technique currently under investigation. Further studies and intense investigations are required to provide insights into the potential of bioponics. Advanced strategies for converting the digestate nutrients into plant exploitable forms are anticipated to demonstrate the prospects of FWM via combined AD and bioponics for a circular bioeconomy. Safety studies to assure the safety of the produce is obligatory to ensure the success of the system/approach. Advanced technologies for exploring food waste valorization effectively considering the aspects of sustainability and circular bioeconomy are the promises for a better future.

16.7 Conclusion

Food waste is a sustainable source of nutrients and energy that requires viable approaches and technologies to resolve the management glitches. Several biorefineries have been adopted to ensure maximum resource recovery from food waste. The organic richness of food waste with an adequate composition of most essential nutrients promotes its potential for recovering substantial resources via promising technologies. Food waste biorefineries are the future prospects to simultaneously unravel the complexity of food waste management and energy crisis. Anaerobic digestion systems coupled with bioponics system are considered to be foreseen technologies that encourage the aspects of sustainability and circular bioeconomy. Coupling these systems can enable maximum resource recovery postulating towards zero waste biorefinery. The successful establishment of these technologies necessitates significant contributions from the scientific community to further achieve economic and environmental benefits. Further advancements in anaerobic digestion coupled bioponics are anticipated to attain the vision of sustainable development addressing the constraints of circular bioeconomy.

References

Abdeen FRH, Mel M, Jami MS, Ihsan SI, Ismail AF (2016) A review of chemical absorption of carbon dioxide for biogas upgrading. Chinese J Chem Eng 24:693–702. https://doi.org/10.1016/j.cjche.2016.05.006

Abdelhamid MT, Selim EM, EL-Ghamry AM (2011) Integrated effects of bio and mineral fertilizers and humic substances on growth, yield and nutrient contents of fertigated cowpea (Vigna unguiculata L.) grown on sandy soils. J Agron 10:34–9. https://doi.org/10.3923/ja.2011.34.39

Abouzied MM, Reddy CA (1986) Direct fermentation of potato starch to ethanol by cocultures of Aspergillus niger and Saccharomyces cerevisiae. Appl Environ Microbiol 52:1055–1059. https://doi.org/10.1128/aem.52.5.1055-1059.1986

Achmon Y, Claypool JT, Pace S, Simmons BA, Singer SW, Simmons CW (2019) Assessment of biogas production and microbial ecology in a high solid anaerobic digestion of major California food processing residues. Bioresour Technol Reports 5:1–11. https://doi.org/10.1016/j.biteb.2018.11.007

Agyeman FO, Tao W (2014) Anaerobic co-digestion of food waste and dairy manure: effects of food waste particle size and organic loading rate. J Environ Manage 133:268–274. https://doi.org/10.1016/j.jenvman.2013.12.016

Ahamed A, Chen CL, Rajagopal R, Wu D, Mao Y, Ho IJR, Lim JW, Wang JY (2015) Multi-phased anaerobic baffled reactor treating food waste. Bioresour Technol 182:239–244. https://doi.org/10.1016/j.biortech.2015.01.117

Akhtar S, Ismail T, Fraternale D, Sestili P (2015) Pomegranate peel and peel extracts: chemistry and food features. Food Chem 174:417–425. https://doi.org/10.1016/j.foodchem.2014.11.035

Ambigaipalan P, Shahidi F (2017) Bioactive peptides from shrimp shell processing discards: antioxidant and biological activities. J Funct Foods 34:7–17. https://doi.org/10.1016/j.jff.2017.04.013

Angelidaki I, Karakashev D, Batstone DJ, Plugge CM, Stams AJM (2011) Biomethanation and its potential. Methods Enzymol 494:327–351. https://doi.org/10.1016/B978-0-12-385112-3.00016-0

Antonopoulou G, Stamatelatou K, Venetsaneas N, Kornaros M, Lyberatos G (2008) Biohydrogen and methane production from cheese whey in a two-stage anaerobic process. Ind Eng Chem Res 47(15):5227–5233. https://doi.org/10.1021/ie071622x

Arumugam A, Ponnusami V (2017) Production of biodiesel by enzymatic transesterification of waste sardine oil and evaluation of its engine performance. Heliyon 3:e00486. https://doi.org/10.1016/j.heliyon.2017.e00486

Aslanzadeh S, Rajendran K, Jeihanipour A, Taherzadeh MJ (2013) The effect of effluent recirculation in a semi-continuous two-stage anaerobic digestion system. Energies 6:2966–2981. https://doi.org/10.3390/en6062966

Barua VB, Rathore V, Kalamdhad AS (2019) Anaerobic co-digestion of water hyacinth and banana peels with and without thermal pretreatment. Renew Energy 134:103–112. https://doi.org/10.1016/j.renene.2018.11.018

Baştabak B, Koçar G (2020) A review of the biogas digestate in agricultural framework. J Mater Cycles Waste Manag 22:1318–1327. https://doi.org/10.1007/s10163-020-01056-9

Batstone DJ, Keller J, Angelidaki I, Kalyuzhnyi SV, Pavlostathis SG, Rozzi A et al (2002) The IWA anaerobic digestion model no 1 (ADM1). Water Sci Technol 45:65–73. https://doi.org/10.2166/wst.2002.0292

Bergstrand KJ, Asp H, Hultberg M (2020) Utilizing anaerobic digestates as nutrient solutions in hydroponic production systems. Sustain 12:1–12. https://doi.org/10.3390/su122310076

Bhatnagar N, Mutnuri S (2015) Digestate from anaerobic reactor as a potential fertilizer. Carbon Sci Technol 7(2):17–24

Bleve M, Ciurlia L, Erroi E, Lionetto G, Longo L, Rescio L, Schettino T, Vasapollo G (2008) An innovative method for the purification of anthocyanins from grape skin extracts by using liquid and sub-critical carbon dioxide. Sep Purif Technol 64:192–197. https://doi.org/10.1016/j.seppur.2008.10.012

Braguglia CM, Gallipoli A, Gianico A, Pagliaccia P (2018) Anaerobic bioconversion of food waste into energy: a critical review. Bioresour Technol 248:37–56. https://doi.org/10.1016/j.biortech.2017.06.145

Budhalakoti N (2019) Formulation and standardisation of banana peel extracted insoluble dietary fibre based buns. Curr J Appl Sci Technol 32:1–9. https://doi.org/10.9734/cjast/2019/45832

Carvill J (1993) Thermodynamics and heat transfer. In: Carvill J (ed) Mechanical engineering data handbook. Butterworth Heineman, Elsevier, Oxford, United Kingdom, pp 102–145. https://doi.org/10.1016/B978-0-08-051135-1.50008-X

Casas L, Mantell C, Rodríguez M, de la Ossa EJM, Roldán A, De OI, Blandino ICA (2010) Extraction of resveratrol from the pomace of Palomino fino grapes by supercritical carbon dioxide. J Food Eng 96:304–308. https://doi.org/10.1016/j.jfoodeng.2009.08.002

Cavinato C, Fatone F, Bolzonella D, Pavan P (2010) Thermophilic anaerobic co-digestion of cattle manure with agro-wastes and energy crops: comparison of pilot and full scale experiences. Bioresour Technol 101:545–550. https://doi.org/10.1016/j.biortech.2009.08.043

Cerda A, Gea T, Vargas-García MC, Sánchez A (2017a) Towards a competitive solid state fermentation: cellulases production from coffee husk by sequential batch operation and role of microbial diversity. Sci Total Environ 589:56–65. https://doi.org/10.1016/j.scitotenv.2017.02.184

Cerda A, Mejías L, Gea T, Sánchez A (2017b) Cellulase and xylanase production at pilot scale by solid-state fermentation from coffee husk using specialized consortia: the consistency of the process and the microbial communities involved. Bioresour Technol 243:1059–1068. https://doi.org/10.1016/j.biortech.2017.07.076

Cerda A, El-Bakry M, Gea T, Sánchez A (2016) Long term enhanced solid-state fermentation: inoculation strategies for amylase production from soy and bread wastes by Thermomyces sp. in a sequential batch operation. J Environ Chem Eng 4:2394–2401. https://doi.org/10.1016/j.jece.2016.04.022

Chakraborty D, Karthikeyan OP, Selvam A, Wong JWC (2018) Co-digestion of food waste and chemically enhanced primary treated sludge in a continuous stirred tank reactor. Biomass Bioenerg 111:232–240. https://doi.org/10.1016/j.biombioe.2017.06.002

Chandel AK, da Silva SS, Singh OV (2013) Detoxification of lignocellulose hydrolysates: biochemical and metabolic engineering toward white biotechnology. Bioenergy Res 6:388–401. https://doi.org/10.1007/s12155-012-9241-z

Charcosset C (2013) Combining membrane processes with renewable energy technologies: erspectives on membrane desalination, biofuels and biogas production, and microbial fuel cells. In: Gugliuzza A, Basile A (eds) Membranes for clean and renewable power applications, pp 44–62. Woodhead Publishing Ltd, Elsevier, Sawston, Cambridge, United Kingdom. https://doi.org/10.1533/9780857098658.1.44

Chen ZT, Chu HL, Chyau CC, Chu CC, Der DuhP (2012) Protective effects of sweet orange (Citrus sinensis) peel and their bioactive compounds on oxidative stress. Food Chem 135:2119–2127. https://doi.org/10.1016/j.foodchem.2012.07.041

Chen X, Yan W, Sheng K, Sanati M (2014) Comparison of high-solids to liquid anaerobic co-digestion of food waste and green waste. Bioresour Technol 154:215–221. https://doi.org/10.1016/j.biortech.2013.12.054

Chen G, Liu J, Yao J, Qi Y, Yan B (2018) Biodiesel production from waste cooking oil in a magnetically fluidized bed reactor using whole-cell biocatalysts. Energy Convers Manag 138:556–564. https://doi.org/10.1016/j.enconman.2017.02.036

Chen R, Li Z, Feng J, Zhao L, Yu J (2020) Effects of digestate recirculation ratios on biogas production and methane yield of continuous dry anaerobic digestion. Bioresour Technol 316:123963. https://doi.org/10.1016/j.biortech.2020.123963

Cheong JC, Lee JTE, Lim JW, Song S, Tan JKN, Chiam ZY, et al (2020) Closing the food waste loop: Food waste anaerobic digestate as fertilizer for the cultivation of the leafy vegetable, xiao bai cai (Brassica rapa). Sci Total Environ 715:136789. https://doi.org/10.1016/j.scitotenv.2020.136789

Chougui N, Djerroud N, Naraoui F, Hadjal S, Aliane K, Zeroual B, Larbat R (2015) Physicochemical properties and storage stability of margarine containing Opuntia ficus-indica peel extract as antioxidant. Food Chem 173:382–390. https://doi.org/10.1016/j.foodchem.2014.10.025

Cioabla AE, Ionel I, Dumitrel GA, Popescu F (2012) Comparative study on factors affecting anaerobic digestion of agricultural vegetal residues. Biotechnol Biofuels 5:39. https://doi.org/10.1186/1754-6834-5-39

Cremonez PA, Teleken JG, Weiser Meier TR, Alves HJ (2021) Two-stage anaerobic digestion in agroindustrial waste treatment: a review. J Environ Manage 281:111854. https://doi.org/10.1016/J.JENVMAN.2020.111854

Cruz-Izquierdo Á, Picó EA, López C, Serra JL, Llama MJ (2014) Magnetic cross-linked enzyme aggregates (mCLEAs) of Candida antarctica lipase: an efficient and stable biocatalyst for biodiesel synthesis. PLoS One 9. https://doi.org/10.1371/journal.pone.0115202

Dahiya S, Kumar AN, Shanthi Sravan J, Chatterjee S, Sarkar O, Mohan SV (2018) Food waste biorefinery: sustainable strategy for circular bioeconomy, vol 248. https://doi.org/10.1016/j.biortech.2017.07.176

Dasa KT, Westman SY, Millati R, Cahyanto MN, Taherzadeh MJ, Niklasson C (2016) Inhibitory effect of long-chain fatty acids on biogas production and the protective effect of membrane bioreactor. Biomed Res Int 5727203. https://doi.org/10.1155/2016/7263974

Dhanya BS, Singh D, Jana AK, Dwiwedi AK, Sharma AK, Chamundeeswari M, Verma ML (2020) Recent trends in biogas upgrading technologies for biomethane production. In: Srivastava N, Srivastava M, Mishra P, Gupta V (eds) Biofuel production technologies: critical analysis for sustainability. Clean Energy Production Technologies. Springer, Singapore, pp 239–281. https://doi.org/10.1007/978-981-13-8637-4_9

Dima JB, Sequeiros C, Zaritzky N (2017) Chitosan from marine crustaceans: production, characterization and applications. In: Shalaby E (ed) Biological activities and application of marine polysaccharides. InTech. London, United Kingdom. https://doi.org/10.5772/65258

Dwidar M, Lee S, Mitchell RJ (2012) The production of biofuels from carbonated beverages. Appl Energy 100:47–51. https://doi.org/10.1016/j.apenergy.2012.02.054

El Kantar S, Boussetta N, Lebovka N, Foucart F, Rajha HN, Maroun RG, Louka N, Vorobiev E (2018) Pulsed electric field treatment of citrus fruits: improvement of juice and polyphenols extraction. Innov Food Sci Emerg Technol 46:153–161. https://doi.org/10.1016/j.ifset.2017.09.024

Elbeshbishy E, Nakhla G (2011) Comparative study of the effect of ultrasonication on the anaerobic biodegradability of food waste in single and two-stage systems. Bioresour Technol 102:6449–6457. https://doi.org/10.1016/j.biortech.2011.03.082

Elsheikh YA (2014) Optimization of novel pyrazolium ionic liquid catalysts for transesterification of bitter apple oil. Process Saf Environ Prot 92:828–834. https://doi.org/10.1016/j.psep.2013.05.001

Enascuta CE, Stepan E, Bolocan I, Bombos D, Calin C, Oprescu EE, Lavric V (2018) Simultaneous production of oil enriched in ω-3 polyunsaturated fatty acids and biodiesel from fish wastes. Waste Manag 75:205–214. https://doi.org/10.1016/j.wasman.2018.01.022

Endo R, Yamashita K, Shibuya T, Kitaya Y (2016) Use of methane fermentation digestate for hydroponic culture: analysis of potential inhibitors in digestate to cucumber seedling. Eco-Eng 28(3):67–72. https://doi.org/10.11450/seitaikogaku.28.67

Fadhil AB (2013) Optimization of transesterification parameters of melon seed oil. Int J Green Energy 10:763–774. https://doi.org/10.1080/15435075.2012.727200

Farinas CS (2015) Developments in solid-state fermentation for the production of biomass-degrading enzymes for the bioenergy sector. Renew Sustain Energy Rev 52:179–188. https://doi.org/10.1016/j.rser.2015.07.092

Ferdous Z, Ullah H, Datta A, Anwar M, Ali A (2018) Yield and profitability of tomato as influenced by integrated application of synthetic fertilizer and biogas slurry. Int J Veg Sci 24:445–455. https://doi.org/10.1080/19315260.2018.1434585

Fernandes TV, Keesman KJ, Zeeman G, van Lier JB (2014) Effect of ammonia on the anaerobic hydrolysis of cellulose and tributyrin. Biomass Bioenerg 47:316–323. https://doi.org/10.1016/j.biombioe.2012.09.029

Furukawa Y, Hasegawa H (2006) Response of Spinach and Komatsuna to biogas effluent made from source-separated kitchen garbage. J Environ Qual 35:1939–1947. https://doi.org/10.2134/jeq2005.0482

Gallert C, Winter J (1997) Mesophilic and thermophilic anaerobic digestion of source-sorted organic wastes: effect of ammonia on glucose degradation and methane production. Appl Microbiol Biotechnol 48:405–410. https://doi.org/10.1007/s002530051071

Gaur RZ, Suthar S (2017) Anaerobic digestion of activated sludge, anaerobic granular sludge and cow dung with food waste for enhanced methane production. J Clean Prod 164:557–566. https://doi.org/10.1016/j.jclepro.2017.06.201

Ghareib M, Youssef KA, Khalil AA (1988) Ethanol tolerance of Saccharomyces cerevisiae and its relationship to lipid content and composition. Folia Microbiol (praha) 33:447–452. https://doi.org/10.1007/BF02925769

Gholami Z, Abdullah AZ, Lee KT (2014) Dealing with the surplus of glycerol production from biodiesel industry through catalytic upgrading to polyglycerols and other value-added products. Renew Sustain Energy Rev 39:327–341. https://doi.org/10.1016/j.rser.2014.07.092

Ghyselbrecht K, Monballiu A, Somers MH, Sigurnjak I, Meers E, Appels L, Meesschaert B (2018) Stripping and scrubbing of ammonium using common fractionating columns to prove ammonium inhibition during anaerobic digestion. Int J Energy Environ Eng 9:447–455. https://doi.org/10.1007/s40095-018-0283-7

Gou C, Yang Z, Huang J, Wang H, Xu H, Wang L (2014) Effects of temperature and organic loading rate on the performance and microbial community of anaerobic co-digestion of waste activated sludge and food waste. Chemosphere 105:146–151. https://doi.org/10.1016/j.chemosphere.2014.01.018

Goud RK, Mohan SV (2012) Acidic and alkaline shock pretreatment to enrich acidogenic biohydrogen producing mixed culture: Long term synergetic evaluation of microbial inventory, dehydrogenase activity and bio-electro kinetics. RSC Adv 2(15):6336–6353. https://doi.org/10.1039/c2ra20526b

Gustavsson J, Cederberg C, Sonesson U, Van Otterdijk R, Meybeck A (2011) Global food losses and food waste: extent, causes and prevention. Food and agricultural organization. http://www.fao.org/3/mb060e/mb060e.pdf

Hang YD, Lee CY, Woodams EE, Cooley HJ (1981) Production of alcohol from apple pomace. Appl Environ Microbiol 42:1128–1129. https://doi.org/10.1128/aem.42.6.1128-1129.1981

Hartmann H, Ahring BK (2005) A novel process configuration for anaerobic digestion of source-sorted household waste using hyper-thermophilic post-treatment. Biotechnol Bioeng 90:830–837. https://doi.org/10.1002/bit.20485

He B, Zhang LL, Yue XY, Liang J, Jiang J, Gao XL, Yue PX (2016) Optimization of ultrasound-assisted extraction of phenolic compounds and anthocyanins from blueberry (Vaccinium ashei) wine pomace. Food Chem 204:70–76. https://doi.org/10.1016/j.foodchem.2016.02.094

Hegde S, Lodge JS, Trabold TA (2018) Characteristics of food processing wastes and their use in sustainable alcohol production. Renew Sustain Energy Rev 81:510–523. https://doi.org/10.1016/j.rser.2017.07.012

Hegde S, Trabold TA (2018) Sustainable waste-to-energy technologies: fermentation. Sustain. In: Trabold TA, Babbitt CW (eds) Food waste-to-energy systems, pp 69–88. Academic press Elsevier, London, United Kingdom. https://doi.org/10.1016/B978-0-12-811157-4.00005-X

Holliger C, Alves M, Andrade D, Angelidaki I, Astals S, Baier U et al (2016) Towards a standardization of biomethane potential tests. Water Sci Technol 74:2515–2522. https://doi.org/10.2166/wst.2016.336

Itelima J, Onwuliri F, Onwuliri E, Onyimba I, Oforji S (2013) Bio-ethanol production from banana, plantain and pineapple peels by simultaneous saccharification and fermentation process. Int J Environ Sci Dev 4:213–216. https://doi.org/10.7763/ijesd.2013.v4.337

Izmirlioglu G, Demirci A (2010) Ethanol production from waste potato mash by using Saccharomyces cerevisiae. Appl Sci 2(4):738–753. https://doi.org/10.13031/2013.29692

Izumi K, Okishio Y ki, Nagao N, Niwa C, Yamamoto S, Toda T (2010) Effects of particle size on anaerobic digestion of food waste. Int Biodeterior Biodegrad 64:601–608. https://doi.org/10.1016/j.ibiod.2010.06.013

Jamison J, Khanal SK, Nguyen NH, Deenik JL (2021) Assessing the effects of digestates and combinations of digestates and fertilizer on yield and nutrient use of Brassica juncea (Kai Choy). Agronomy 11:509. https://doi.org/10.3390/agronomy11030509

Jegannathan KR, Jun-Yee L, Chan ES, Ravindra P (2010) Production of biodiesel from palm oil using liquid core lipase encapsulated in κ-carrageenan. Fuel 89:2272–2277. https://doi.org/10.1016/j.fuel.2010.03.016

Jiang J, He S, Kang X, Sun Y, Yuan Z, Xing T et al (2020) Effect of organic loading rate and temperature on the anaerobic digestion of municipal solid waste: process performance and energy recovery. Front Energy Res 8:89. https://doi.org/10.3389/fenrg.2020.00089

Jin Y, Li H, Mahar Rb, Wang Z, Nie Y (2009) Combined alkaline and ultrasonic pretreatment of sludge before aerobic digestion. J Environ Sci 21:279–84. https://doi.org/10.1016/S1001-0742(08)62264-0

Jingura RM, Kamusoko R (2017) Methods for determination of biomethane potential of feedstocks: a review. Biofuel Res J 4:573–86. https://doi.org/10.18331/BRJ2017.4.2.3

.Joardder MUH, Masud MH (2019) Causes of food waste. In: Joardder MUH, Masud MH (eds) Food preservation in developing countries: Challenges and solutions. Springer, Cham, Switzerland, pp 27–55. https://doi.org/10.1007/978-3-030-11530-2_2

Jung JM, Oh JI, Park YK, Lee J, Kwon EE (2019) Biodiesel synthesis from fish waste via thermally-induced transesterification using clay as porous material. J Hazard Mater 371:27–32. https://doi.org/10.1016/j.jhazmat.2019.02.109

Kárászová M, Sedláková Z, Izák P (2015) Review: Gas permeation processes in biogas upgrading: a short review. Chem Pap 69:1277–1283. https://doi.org/10.1515/chempap-2015-0141

Khanahmadi M, Arezi I, Amiri M sadat, Miranzadeh M (2018) Bioprocessing of agro-industrial residues for optimization of xylanase production by solid- state fermentation in flask and tray bioreactor. Biocatal Agric Biotechnol 13:272–282. https://doi.org/10.1016/j.bcab.2018.01.005

Kheyrandish M, Asadollahi MA, Jeihanipour A, Doostmohammadi M, Rismani-Yazdi H, Karimi K (2015) Direct production of acetone-butanol-ethanol from waste starch by free and immobilized Clostridium acetobutylicum. Fuel 142:129–133. https://doi.org/10.1016/j.fuel.2014.11.017

Kim J, Park C, Kim T-H, Lee M, Kim S, Kim S-W et al (2003) Effects of various pretreatments for enhanced anaerobic digestion with waste activated sludge. J Biosci Bioeng 95:271–275. https://doi.org/10.1016/s1389-1723(03)80028-2

Kong X, Xu S, Liu J, Li H, Zhao K, He L (2016) Enhancing anaerobic digestion of high-pressure extruded food waste by inoculum optimization. J Environ Manage 166:31–37. https://doi.org/10.1016/j.jenvman.2015.10.002

Korkie LJ, Janse BJH, Viljoen-Bloom M (2017) Utilising grape pomace for ethanol production. South African J Enol Vitic 23. https://doi.org/10.21548/23-1-2152

Koszel M, Lorencowicz E (2015) Agricultural use of biogas digestate as a replacement fertilizers. Agric Agric Sci Procedia 7:119–124. https://doi.org/10.1016/j.aaspro.2015.12.004

Krakat N, Anjum R, Dietz D, Demirel B (2017) Methods of ammonia removal in anaerobic digestion: a review. Water Sci Technol 76:1925–1938. https://doi.org/10.2166/wst.2017.406

Labatut RA, Gooch C (2014) Monitoring of anaerobic digestion process to optimize performance and prevent system failure. Proc Got Manure? Enhancing Environ Econ Sustain 209–225

Labatut RA, Pronto JL (2018) Sustainable waste-to-energy technologies: anaerobic digestion. Sustainable Food Waste-to-Energy System. In: Trabold TA, Babbitt CW (eds) Sustainable food waste-to-energy systems, pp 47–67. Academic press Elsevier, London, United Kingdom Elsevier. https://doi.org/10.1016/B978-0-12-811157-4.00004-8

Labatut RA, Pronto JL (2018) Sustainable waste-to-energy technologies: anaerobic digestion. In: Thomas A, Trabold, Babbitt CW (eds) Sustainable Food waste-to-energy systems. Academic press Elsevier, London, United Kingdom Elsevier, pp 47–67. https://doi.org/10.1016/b978-0-12-811157-4.00004-8

Lay JJ, Fan KS, Chang IJ, Ku CH (2003) Influence of chemical nature of organic wastes on their conversion to hydrogen by heat-shock digested sludge. Int J Hydrogen Energy 28:1361–1367. https://doi.org/10.1016/S0360-3199(03)00027-2

Lee WS, Chua ASM, Yeoh HK, Ngoh GC (2014) A review of the production and applications of waste-derived volatile fatty acids. Chem Eng J 235:83–99. https://doi.org/10.1016/j.cej.2013.09.002

Li Y, Zhang R, Chen C, Liu G, He Y, Liu X (2013) Biogas production from co-digestion of corn stover and chicken manure under anaerobic wet, hemi-solid, and solid state conditions. Bioresour Technol 149:406–412. https://doi.org/10.1016/j.biortech.2013.09.091

Li L, Peng X, Wang X, Wu D (2018a) Anaerobic digestion of food waste: a review focusing on process stability. Bioresour Technol 248:20–28. https://doi.org/10.1016/j.biortech.2017.07.012

Li W, Loh KC, Zhang J, Tong YW, Dai Y (2018b) Two-stage anaerobic digestion of food waste and horticultural waste in high-solid system. Appl Energy 209:400–408. https://doi.org/10.1016/j.apenergy.2017.05.042

Li S, Yang X (2016) Biofuel production from food wastes. In: Luque R, Lin C, Wilson K, Clark J (eds) Handbook of biofuels production: processes and technologies, 2nd edn. Elsevier, Cambridge, United Kingdom, pp 617–653. https://doi.org/10.1016/B978-0-08-100455-5.00020-5

Lin CSK, Pfaltzgraff LA, Herrero-Davila L, Mubofu EB, Abderrahim S, Clark JH et al (2013) Food waste as a valuable resource for the production of chemicals, materials and fuels. Current situation and global perspective. Energy Environ Sci 6:426–464. https://doi.org/10.1039/c2ee23440h

Lipinski B, Hanson C, Waite R, Searchinger T, Lomax J (2021) Reducing food loss and waste, pp 1–40. World Resource Institute, Washington DC. Available at: https://www.wri.org/research/reducing-food-loss-and-waste

Liu G, Zhang R, El-Mashad HM, Dong R (2009a) Effect of feed to inoculum ratios on biogas yields of food and green wastes. Bioresour Technol 100:5103–5108. https://doi.org/10.1016/j.biortech.2009.03.081

Liu WK, Yang QC, Du L (2009b) Soilless cultivation for high-quality vegetables with biogas manure in China: feasibility and benefit analysis. Renew Agric Food Syst 24:300–307. https://doi.org/10.1017/S1742170509990081

Liu C, Wang W, Anwar N, Ma Z, Liu G, Zhang R (2017a) Effect of organic loading rate on anaerobic digestion of food waste under mesophilic and thermophilic conditions. Energy Fuels 31:2976–2984. https://doi.org/10.1021/acs.energyfuels.7b00018

Liu C, Wang W, Anwar N, Ma Z, Liu G, Zhang R (2017b) Eff ect of organic loading rate on anaerobic digestion of food waste under mesophilic and thermophilic conditions. Energy Fuels 31(3):2976–2984. https://doi.org/10.1021/acs.energyfuels.7b00018

Lo YC, Su YC, Chen CY, Chen WM, Lee KS, Chang JS (2009) Biohydrogen production from cellulosic hydrolysate produced via temperature-shift-enhanced bacterial cellulose hydrolysis. Bioresour Technol 100:5802–5807. https://doi.org/10.1016/j.biortech.2009.06.066

Lukitawesa, Wikandari R, Millati R, Taherzadeh MJ, Niklasson C (2018) Effect of effluent recirculation on biogas production using two-stage anaerobic digestion of citrus waste. Molecules 23(12):3380. https://doi.org/10.3390/molecules23123380

Marín-Suárez M, Méndez-Mateos D, Guadix A, Guadix EM (2019) Reuse of immobilized lipases in the transesterification of waste fish oil for the production of biodiesel. Renew Energy 140:1–8. https://doi.org/10.1016/j.renene.2019.03.035

Martins N, Ferreira ICFR (2017) Wastes and by-products: upcoming sources of carotenoids for biotechnological purposes and health-related applications. Trends Food Sci Technol 62:33–48. https://doi.org/10.1016/j.tifs.2017.01.014

Matheri AN, Belaid M, Seodigeng T, Ngila CJ (2016) Modelling the kinetic of biogas production from co-digestion of pig waste and grass clippings. In: Proceedings of the world congress on engineering 2016, vol II (WCE 2016). Retrieved from http://www.iaeng.org/publication/WCE2016/WCE2016_pp813-820.pdf. Accessed 8 Nov 2018

Mehariya S, Patel AK, Obulisamy PK, Punniyakotti E, Wong JWC (2018) Co-digestion of food waste and sewage sludge for methane production: current status and perspective. Bioresour Technol 265:519–531. https://doi.org/10.1016/j.biortech.2018.04.030

Melikoglu M, Lin CSK, Webb C (2013) Analysing global food waste problem: pinpointing the facts and estimating the energy content. Cent Eur J Eng 32(3):157–164. https://doi.org/10.2478/S13531-012-0058-5

Menzel T, Neubauer P, Junne S (2020) Role of microbial hydrolysis in anaerobic digestion. Energies 13(21):5555. https://doi.org/10.3390/en13215555

Mohammadshirazi A, Akram A, Rafiee S, Bagheri KE (2014) Energy and cost analyses of biodiesel production from waste cooking oil. Renew Sustain Energy Rev 33:44–49. https://doi.org/10.1016/j.rser.2014.01.067

Monlau F, Barakat A, Steyer JP, Carrere H (2012) Comparison of seven types of thermo-chemical pretreatments on the structural features and anaerobic digestion of sunflower stalks. Bioresour Technol 120:241–247. https://doi.org/10.1016/j.biortech.2012.06.040

Moon HC, Song IS (2011) Enzymatic hydrolysis of foodwaste and methane production using UASB bioreactor. Int J Green Energy 8:361–371. https://doi.org/10.1080/15435075.2011.557845

Morales-Polo C, del Mar Cledera-Castro M, Yolanda Moratilla Soria B (2009) Reviewing the anaerobic digestion of food waste: from waste generation and anaerobic process to its perspectives. Appl Sci 8(10):1804. https://doi.org/10.3390/app8101804

Mosier N, Wyman C, Dale B, Elander R, Lee YY, Holtzapple M, Ladisch M (2005) Features of promising technologies for pretreatment of lignocellulosic biomass. Bioresour Technol 96:673–686. https://doi.org/10.1016/j.biortech.2004.06.025

Mtz.-Viturtia A, Mata-Alvarez J, Cecchi F (1995) Two-phase continuous anaerobic digestion of fruit and vegetable wastes. Resour Conserv Recycl 13:257–67. https://doi.org/10.1016/0921-3449(94)00048-A

Mukhuba M, Roopnarain A, Adeleke R, Moeletsi M, Makofane R (2018) Comparative assessment of bio-fertiliser quality of cow dung and anaerobic digestion effluent. Cogent Food Agric 4:1435019. https://doi.org/10.1080/23311932.2018.1435019

Müller FPC, Maack GC, Buescher W (2017) Effects of biogas substrate recirculation on methane yield and efficiency of a liquid-manure-based biogas plant. Energies 10:1–11. https://doi.org/10.3390/en10030325

Neal J, Wilkie AC (2014) Anaerobic digester effluent as fertilizer for hydroponically grown tomatoes. J Undergraduate Res 15(3):1–5

Nguyen LN, Nguyen AQ, Nghiem LD (2019) Microbial community in anaerobic digestion system: progression in microbial ecology. In: Bui XT, Chiemchaisri C, Fujioka T, Varjani S (eds) Water and wastewater treatment technologies. Energy, environment, and sustainability, pp 331–355. Springer, Singapore. https://doi.org/10.1007/978-981-13-3259-3_15

Odnell A, Recktenwald M, Stensén K, Jonsson BH, Karlsson M (2016) Activity, life time and effect of hydrolytic enzymes for enhanced biogas production from sludge anaerobic digestion. Water Res 103:462–471. https://doi.org/10.1016/j.watres.2016.07.064

Oroian M, Escriche I (2015) Antioxidants: Characterization, natural sources, extraction and analysis. Food Res Int 74:10–36. https://doi.org/10.1016/j.foodres.2015.04.018

Ouachab N, Tsoutsos T (2013) Study of the acid pretreatment and biodiesel production from olive pomace oil. J Chem Technol Biotechnol 88:1175–1181. https://doi.org/10.1002/jctb.3940

Papadaki E, Mantzouridou FT (2021) Natural β-carotene production by blakeslea trispora cultivated in spanish-style green olive processing wastewaters. Foods 10. https://doi.org/10.3390/foods10020327

Parniakov O, Barba FJ, Grimi N, Lebovka N, Vorobiev E (2014) Impact of pulsed electric fields and high voltage electrical discharges on extraction of high-added value compounds from papaya peels. Food Res Int 65:337–343. https://doi.org/10.1016/j.foodres.2014.09.015

Parrondo J, Garcia LA, Diaz M (2000) Production of an alcoholic beverage by fermentation of whey permeate with Kluyveromyces fragilis II: aroma composition. J Inst Brew 106:377–382. https://doi.org/10.1002/j.2050-0416.2000.tb00528.x

Pastorelli R, Valboa G, Lagomarsino A, Fabiani A, Simoncini S, Zaghi M et al (2021) Recycling biogas digestate from energy crops: effects on soil properties and crop productivity. Appl Sci 11:1–20. https://doi.org/10.3390/app11020750

Patáková P, Lipovský J, Čížková H, Fořtová J, Rychtera M, Melzoch K (2009) Exploitation of food feedstock and waste for production of biobutanol. Czech J Food Sci 27:276–283. https://doi.org/10.17221/106/2009-cjfs

Pelayo Lind O, Hultberg M, Bergstrand KJ, Larsson-Jönsson H, Caspersen S, Asp H (2021) Biogas digestate in vegetable hydroponic production: pH dynamics and ph management by controlled nitrification. Waste Biomass Valorization 12:123–133. https://doi.org/10.1007/s12649-020-00965-y

Pellera FM, Gidarakos E (2016) Effect of substrate to inoculum ratio and inoculum type on the biochemical methane potential of solid agroindustrial waste. J Environ Chem Eng 4:3217–3229. https://doi.org/10.1016/j.jece.2016.05.026

Quispe CAG, Coronado CJR, Carvalho JA (2013) Glycerol: production, consumption, prices, characterization and new trends in combustion. Renew Sustain Energy Rev 27:475–493. https://doi.org/10.1016/j.rser.2013.06.017

Raganati F, Olivieri G, Procentese A, Russo ME, Salatino P, Marzocchella A (2013) Butanol production by bioconversion of cheese whey in a continuous packed bed reactor. Bioresour Technol 138:259–265. https://doi.org/10.1016/j.biortech.2013.03.180

Ranganathan S, Dutta S, Moses JA, Anandharamakrishnan C (2020) Utilization of food waste streams for the production of biopolymers. Heliyon 6:e04891. https://doi.org/10.1016/j.heliyon.2020.e04891

Ronga D, Setti L, Salvarani C, De Leo R, Bedin E, Pulvirenti A et al (2019) Effects of solid and liquid digestate for hydroponic baby leaf lettuce (Lactuca sativa L.) cultivation. Sci Hortic (Amsterdam) 244:172–181. https://doi.org/10.1016/j.scienta.2018.09.037

Sahota S, Shah G, Ghosh P, Kapoor R, Sengupta S, Singh P, Vijay V, Sahay A, Vijay VK, Thakur IS (2018) Review of trends in biogas upgradation technologies and future perspectives. Bioresour Technol Reports 1:79–88. https://doi.org/10.1016/j.biteb.2018.01.002

Sahar Sadaf S, Iqbal J, Ullah I, Bhatti HN, Nouren S, Rehman H, Nisar J, Iqbal M (2018) Biodiesel production from waste cooking oil: an efficient technique to convert waste into biodiesel. Sustain Cities Soc 41:220–226. https://doi.org/10.1016/j.scs.2018.05.037

Saragih FNA, Priadi CR, Adityosulindro S, Abdillah A, Islami BB (2019) The effectiveness of anaerobic digestion process by thermal pre-treatment on food waste as a substrate. IOP Conf Ser Earth Environ Sci 251:012014. https://doi.org/10.1088/1755-1315/251/1/012014

Secondi L, Principato L, Laureti T (2015) Household food waste behaviour in EU-27 countries: a multilevel analysis. Food Policy 56:25–40. https://doi.org/10.1016/J.FOODPOL.2015.07.007

Shabtai Y, Mandel C (1993) Control of ethanol production and monitoring of membrane performance by mass-spectrometric gas analysis in the coupled fermentation-pervaporation of whey permeate. Appl Microbiol Biotechnol 40:470–476. https://doi.org/10.1007/BF00175733

Shah FA, Mahmood Q, Rashid N, Pervez A, Raja IA, Shah MM (2015) Co-digestion, pretreatment and digester design for enhanced methanogenesis. Renew Sustain Energy Rev 42:627–642. https://doi.org/10.1016/j.rser.2014.10.053

Sharma K, Mahato N, Cho MH, Lee YR (2017) Converting citrus wastes into value-added products: economic and environmently friendly approaches. Nutrition 34:29–46. https://doi.org/10.1016/j.nut.2016.09.006

Shubha K, Mukherjee A, Dubey A, Koley TK (2019) Bioponics—a new way to grow soilless vegetable cultivation. Agric Food Mag 1. https://doi.org/10.13140/RG.2.2.21868.31368

Singh P, Patil Y, Rale V (2019) Biosurfactant production: emerging trends and promising strategies. J Appl Microbiol 126:2–13. https://doi.org/10.1111/jam.14057

Srisowmeya G, Chakravarthy M, Nandhini Devi G (2020) Critical considerations in two-stage anaerobic digestion of food waste—a review. Renew Sustain Energy Rev 119:109587 https://doi.org/10.1016/j.rser.2019.109587

Srisowmeya G, Chakravarthy M, Bakshi A, Nandhini Devi G (2021) Improving process stability, biogas production and energy recovery using two-stage mesophilic anaerobic codigestion of rice wastewater with cow dung slurry. Biomass Bioenergy 152:106184. https://doi.org/10.1016/J.BIOMBIOE.2021.106184

Stafford DA, Hawkes DL, Horton R (1981) Methane production from waste organic matter. CRC Press Inc., Boca Raton, United states FL

Tahir A, Aftab M, Farasat T (2010) Effect of cultural conditions on ethanol production by locally isolated saccharomyces cerevisiae bio-07. J Appl Pharm 2:72–78. https://doi.org/10.21065/19204159.2.72

Thomas L, Larroche C, Pandey A (2013) Current developments in solid-state fermentation. Biochem Eng J 81:146–161. https://doi.org/10.1016/j.bej.2013.10.013

Todaro A, Cimino F, Rapisarda P, Catalano AE, Barbagallo RN, Spagna G (2009) Recovery of anthocyanins from eggplant peel. Food Chem 114:434–439. https://doi.org/10.1016/j.foodchem.2008.09.102

Tran TTV, Kaiprommarat S, Kongparakul S, Reubroycharoen P, Guan G, Nguyen MH et al (2016) Green biodiesel production from waste cooking oil using an environmentally benign acid catalyst. Waste Manag 52:367–374. https://doi.org/10.1016/j.wasman.2016.03.053

Tsegaye B, Jaiswal S, Jaiswal AK (2021) Food waste biorefinery: pathway towards circular bioeconomy. Foods 10:1–21. https://doi.org/10.3390/foods10061174

Voget CE, Mignone CF, Ertola RJ (1985) Butanol production from apple pomace. Biotechnol Lett 7:43–46. https://doi.org/10.1007/BF01032418

Walker M, Iyer K, Heaven S, Banks CJ (2011) Ammonia removal in anaerobic digestion by biogas stripping: an evaluation of process alternatives using a first order rate model based on experimental findings. Chem Eng J 178:138–145. https://doi.org/10.1016/j.cej.2011.10.027

Wang Q, Kuninobu M, Ogawa HI, Kato Y (1999) Degradation of volatile fatty acids in highly efficient anaerobic digestion. Biomass Bioenerg 16:407–416. https://doi.org/10.1016/S0961-9534(99)00016-1

Wang X, Qin X, Li D, Yang B, Wang Y (2017) One-step synthesis of high-yield biodiesel from waste cooking oils by a novel and highly methanol-tolerant immobilized lipase. Bioresour Technol 235:18–24. https://doi.org/10.1016/j.biortech.2017.03.086

Woodgate S, Veen JV (2004) The role of fat processing and rendering in the European Union animal production industry. Biotechnol Agron Société Environ 8:283–294

Wu C, Huang Q, Yu M, Ren Y, Wang Q, Sakai K. Effects of digestate recirculation on a two-stage anaerobic digestion system , particularly focusing on metabolite correlation analysis. Bioresour Technol 251:40–8 (2018). https://doi.org/10.1016/j.biortech.2017.12.020

Xiao B, Qin Y, Wu J, Chen H, Yu P, Liu J, Li YY (2018) Comparison of single-stage and two-stage thermophilic anaerobic digestion of food waste: Performance, energy balance and reaction process. Energy Convers Manag 156:215–223. https://doi.org/10.1016/j.enconman.2017.10.092

Xing T, Kong X, Dong P, Zhen F, Sun Y (2020) Leachate recirculation effects on solid-state anaerobic digestion of Pennisetum hybrid and microbial community analysis. J Chem Technol Biotechnol 95:1216–1224. https://doi.org/10.1002/jctb.6310

Xue S, Qiu L, Guo X, Yao Y (2020) Effect of liquid digestate recirculation on biogas production and enzyme activities for anaerobic digestion of corn straw. Water Sci Technol 82:144–156. https://doi.org/10.2166/wst.2020.338

Yenigün O, Demirel B (2013) Ammonia inhibition in anaerobic digestion: a review. Process Biochem 48:901–911. https://doi.org/10.1016/j.procbio.2013.04.012

Yeom IT, Lee KR, Lee YH, Ahn KH, Lee SH (2002) Effects of ozone treatment on the biodegradability of sludge from municipal wastewater treatment plants. Water Sci Technol 46:421–425. https://doi.org/10.2166/wst.2002.0641

Yoon YM, Kim SH, Shin KS, Kim CH (2014) Effects of substrate to inoculum ratio on the biochemical methane potential of piggery slaughterhouse wastes. Asian-Australasian J Anim Sci 27:600–607. https://doi.org/10.5713/ajas.2013.13537

Yu C-Y, Jiang B-H, Duan K-J (2013) Production of bioethanol from carrot pomace using the thermotolerant yeast Kluyveromyces marxianus. Energies 6:1–8

Zabed H, Sahu JN, Boyce AN, Faruq G (2016) Fuel ethanol production from lignocellulosic biomass: an overview on feedstocks and technological approaches. Renew Sustain Energy Rev 66:751–774. https://doi.org/10.1016/j.rser.2016.08.038

Zhang C, Xiao G, Peng L, Su H, Tan T (2013) The anaerobic co-digestion of food waste and cattle manure. Bioresour Technol 129:170–176. https://doi.org/10.1016/j.biortech.2012.10.138

Zhang C, Su H, Baeyens J, Tan T (2014) Reviewing the anaerobic digestion of food waste for biogas production. Renew Sustain Energy Rev 38:383–392. https://doi.org/10.1016/j.rser.2014.05.038

Zhang J, Loh KC, Lee J, Wang CH, Dai Y, Wah TY (2017) Three-stage anaerobic co-digestion of food waste and horse manure. Sci Rep 7:1–10. https://doi.org/10.1038/s41598-017-01408-w

Part V
Conversion Technologies

Chapter 17
Commercial or Pilot-Scale Pyrolysis Units for Conversion of Biomass to Bio-Oils: State of the Art

Ravneet Kaur and Simar Preet Singh

17.1 Introduction

The depletion of fossil fuels reserves, environmental issues, and expansion in industrialization have stimulated interest in producing renewable liquid fuels. The use of renewable and sustainable energy alternatives became the positive approach to replace fossil fuels and their harmful effects. Considerable research has been conducted to discover renewable energy alternatives to replace gasoline (Dupuis et al. 2019; Zupko 2019; Gamliel et al. 2018), jet fuel (Zhang et al. 2015; Yang et al. 2018), and diesel (Kumar et al. 2019; Zhang and Brown 2019; Lopez et al. 2019; Sun et al. 2018). Lignocellulosic biomass is abundant, widely distributed, and made up of cellulose, hemicellulose, and lignin (Soltanian et al. 2020). The primary application of biomass is used as heat in many rural and developing countries.

Figure 17.1 shows the different conversion processes, products, and applications of various obtained products. The conversion of biomass to biofuels includes thermochemical and biochemical techniques.

The various products of thermochemical conversion of biomass are gases from gasification, liquids from pyrolysis, and solids from combustion. The upgrading of obtained products is required to liquid fuels and valuable chemicals. Ethanol and biogas are the products obtained from biochemical conversion methods. Combustion of biomass yields solid char, and its commercialization is done in most industrialized and developing countries. Fuel gas is a final product of gasification and is used to generate heat and electricity (Lepage et al. 2021). There are different types of gasifiers used in the demonstration and pre-commercial levels, and it was well-reviewed by Bridgewater and Maniatis (Bridgwater and Maniatis 2004). The cost of the gasification technologies is too high (Beenackers 2001). Pyrolysis is the breakdown of

R. Kaur (✉)
Sant Longowal Institute of Engineering and Technology, Sangrur, Longowal, India

S. P. Singh
Bennett University, Greater Noida, Uttar Pradesh, India

Y. K. Nandabalan et al. (eds.), *Zero Waste Biorefinery*, Energy, Environment, and Sustainability, https://doi.org/10.1007/978-981-16-8682-5_17

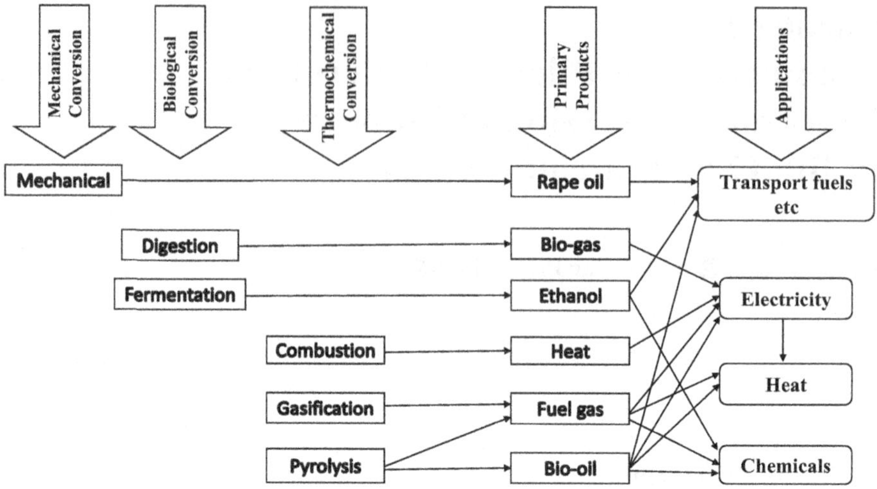

Fig. 17.1 Conversion routes, products and their applications

biomass molecules in the presence of heat at a temperature of 300–650 °C. The products obtained from pyrolysis are gas, liquid, and char used for various purposes (Fig. 17.1). Different pyrolysis processes have been categorized based on their operating conditions and attained products and are tabulated in Table 17.1 (Vamvuka 2011; Kaur et al. 2016).

This chapter aims to emphasize the emerging technology of pyrolysis used to produce liquid fuels. This chapter discusses the different commercial pyrolysis plants available worldwide. A commercial process comprises of three stages:

(a) Feed preparation, pre-treatment, storage and handling;
(b) Biomass conversion to valuable liquid fuels, i.e., bio-oil by pyrolysis;
(c) Conversion of bio-oil to marketable final product such as bio-fuels or chemicals, electricity and heat.

Table 17.1 Pyrolysis types, processing parameters and major products

Technology	Temperature, °C	Residence time, s	Heating rate	Major products
Fast pyrolysis	650–900	<1 s	High	Liquids, gases
	1000–3000	<1 s	Very high	Gases
Flash pyrolysis	400–650	0.1–2 s	High	Liquids
Vacuum pyrolysis	350–450	2–30 s	Medium	Liquids
Pressurized hydro pyrolysis	<500	<10 s	High	Liquids
Slow pyrolysis	700–900	5–30 min	Medium	Charcoal, gases
	400–600	Hours	Low	Charcoal, liquids, Gases
	300–500	Hours-days	Very low	Charcoal

17.2 Fast Pyrolysis

In this type of pyrolysis, the decomposition of biomass is done very rapidly, i.e., few seconds and produces the maximum amount of liquid fuel. The obtained liquid is dark brown and it has heating value just half of the non-renewable fuel. The product distribution based on a dry-feed basis is 75 wt.% of liquid and the remaining 25 wt.% of char and gas. The different parameters that affect the product yields are the type of biomass, temperature, vapor residence time, ash content of biomass and catalyst effect (Kaur et al. 2016). More liquid yield is attained at moderate temperature and in short residence time, while maximum solid product yield is obtained at a lower temperature range and longer residence time (Amenaghawon et al. 2021). An increase in the gas yield was acquired at an elevated temperature and residence time.

The vital features of a fast pyrolysis process for obtaining a high yield of oils are:

(i) Very high heat transfer and heating rates by which heat must be transferred rapidly from heating medium to the sample even to the particle is less than 3 mm;
(ii) Temperature control should be reasonable to obtain the maximum yield of liquid products as the vapor phase temperature range is 400–450 °C;
(iii) Vapor residence time should be fewer to lessen the secondary reactions;
(iv) Quick cooling of vapors to yield the maximum amount of liquid product (bio-oil);
(v) Speedy removal of char to reduce cracking of vapors.

Fast pyrolysis is an operative method for densifying bulk biomass for decentralized densification or centralized conversion platform models (Amenaghawon et al. 2021; Sims et al. 2008). Figure 17.2 represents the fast pyrolysis process. The fast pyrolysis technology using straw as raw material was first developed by Forschungszentrum Karlsruhe and reported by Henrich et al. (2002). Fine particles of straw are fed in a twin-screw reactor similar to the LR reactor developed by Lurgi in the 1950s. The heating of straw is done with the help of recirculated sand. Char and oil were obtained as a final product.

Figure 17.3 represents the different stages involved in the technological expansion of fast pyrolysis of biomass. Considering the biorefinery process, bio-oil can be used to produce various value-added products. Due to some undesirable fuel properties, the bio-oil cannot be used directly in engines. For large-scale applications, the upgradation of bio-oil is the best way to use it as a substitute for transport fuel.

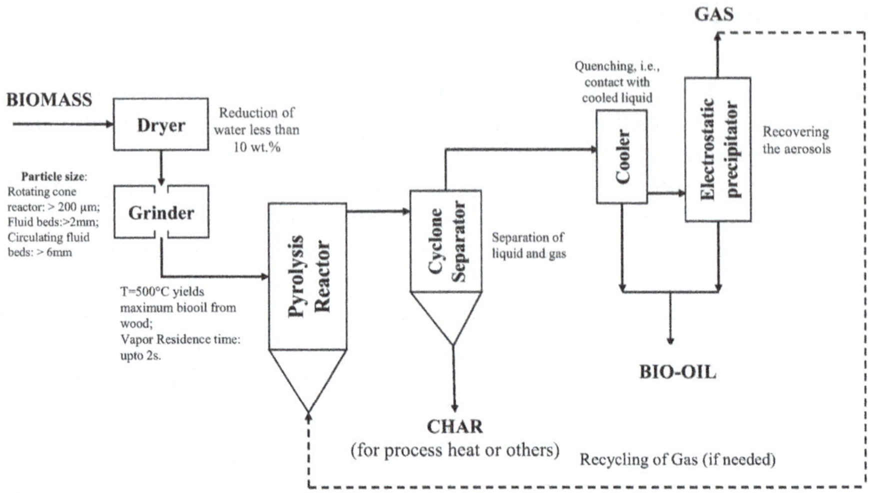

Fig. 17.2 Fast pyrolysis process

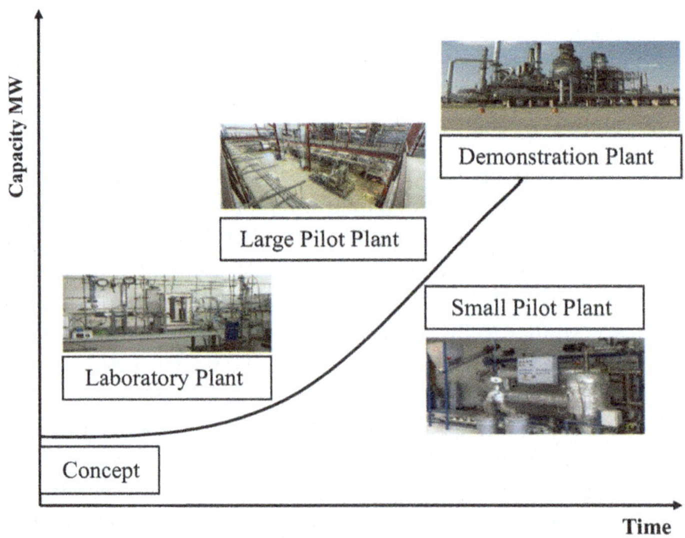

Fig. 17.3 Stages involved in the technological development of the fast pyrolysis process

17.3 Pyrolysis Process Technologies

17.3.1 An Overview

The thermal decomposition of different organic materials has been developed for hundreds or thousands of years, but the growth of fast pyrolysis occurred less than 30 years ago. During the 1980-early 1990, the most emergent research area was developing special reactors like rotating blades, rotating cone, vortex, cyclone, vacuum, entrained flow, circulating fluid-bed (CFB) and bubbling fluid-bed reactor (BFB). In fast pyrolysis, the reactor is considered the heart of the process. The investment in the reactor is approx. 10–15% of the entire capital cost of an integrated system. Numerous studies have been conducted on the development and testing of various reactor configurations using multiple feedstocks. In the later 1990, countries like Spain (Union Fenosa), Italy (Enel), UK (Wellman), Canada (Pyrovac, Dynamotive), Finland (Fortum) and Netherlands (BTG) started the construction work of the pilot plant. The advancement of pyrolysis reactors includes fixed beds, fluidized beds, suspended beds, moving beds, horizontal shaft kilns, stationary vertical shafts, and single and multi-hearth reactors. Bubbling fluidised beds (BFBs) and circulating fluidised beds (CFBs) are the reactors used on the commercial scale for the manufacture of bio-oil (Briens et al. 2008). Table 17.2 represents the reactor selection criteria based on different parameters. Ensyn's circulating fluidised bed process has been applied commercially in the USA and Canada used this technology to produce a food ingredient termed "liquid smoke." Commercial pyrolysis plants Dynamotive, Canada and BTG, Netherlands was operated with a capacity of 2–4 tons biomass/hour throughput to produce biofuels. The technology strength vs. market attractiveness of various fast pyrolysis technologies is shown in Fig. 17.4. Pyrolysis reactors and their applications on commercial and pilot scale are elaborated in this chapter. The additional area of research is regulating and improving liquid quality and upgrading liquid

Table 17.2 Reactor selection criteria

Reactor type	Overall Technology	Capital expense	Operational expense	Low temperature	Low gas/solid ratio	Easy Scale-up
BFB	✔✔	✔✔✔	✔✔	✔✔	✔	✔✔✔
CFB	■	✔	■	✔✔	□	✔✔
Ablative	■	✔✔	✔	✔✔	✔✔✔	□□
Rotating cone	■	✔	■	✔✔	✔✔	□
Auger	✔✔	✔✔✔	✔	✔✔	✔✔✔	□
Vacuum	■	✔✔✔	□	✔✔✔	✔✔✔	□

□□ :Very Poor □:Poor ■: Fair ✔: Good ✔✔:Very Good, ✔✔✔:Excellent

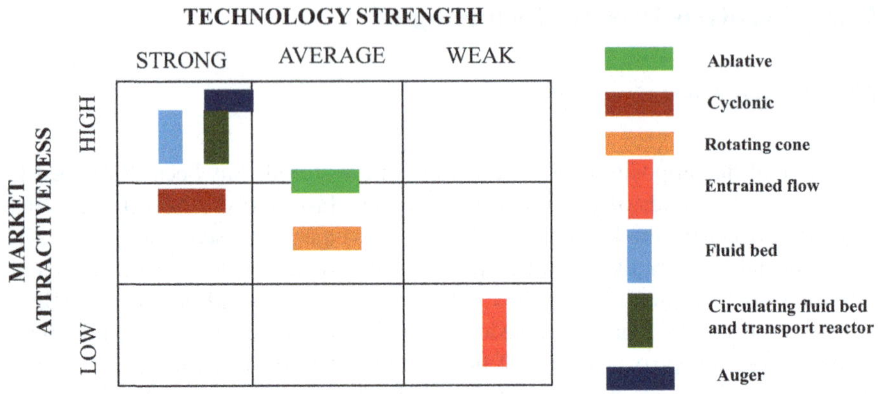

Fig. 17.4 The commercial potential of pyrolysis technologies

collection systems. Many research organizations are also involved in the upgradation of bio-oil using different processes (Bridgwater 2012).

17.3.2 Historical Background

In the early seventies, USA and Japan launched several initiatives in pyrolysis. The main activities in Western Europe are pyrolysis of household refuse and industrial waste using a rotary kiln. Developing a pyrolysis process to recuperate chemicals from rubber and metals from composite wastes.

The Danish *Destrugas* process was the first development done by European Commission, 2000 Proposal for a Directive on the Promotion of Electricity from Renewable Energy Sources in the Internal Electricity market, COM 279, of 10.05.00, Brussels. This process involves vertical shaft furnace heats externally upto a temperature of 1050 °C. In the Destrugas reactor, feed is settled by gravity and char is collected over a star value and quenched. The process having a capacity of 5 t day^{-1} was designed but never established. Low volumetric capacity, low rates of heat transfer, huge amount of heat required and output of the produced gas is low and has containments were the significant drawbacks of this system. The limitations of vertical shaft reactor were documented by Warren Spring Laboratory (Stevenage, England). They established a cross-flow reactor where dried wastes are fed to the vertical furnace then goes downward in the extraction system. This pyrolysis system contains vertical grate bars and gases to heat them (Buekens and Schoeters 1985).

Foster Wheeler conducted further experiments in Hartlepool to develop a pyrolysis plant where scrap tyres, refuse and forest residues were used as raw material. In this process, some produced gas is recycled to attain better heat transfer to the raw material. Organic and aqueous fractions are condensed and char is removed from the pyrolysis reactor. The commercial-scale tyre pyrolysis plant capacity of 7 t h^{-1}

was commissioned at Tyrolysis Ltd (A daughter company of Foster Wheeler Power Products Ltd.). The produced oil was distilled and commercialized. The produced gas was used to heat the reactor and char was used for the boiler fuel (Buekens and Schoeters 1985).

The Babcock Krauss Maffei Industrieanlagen (BKMI) is an engineering firm, is a member of the Deutsche Babcock group and its environmental division developed the pyrolysis processes of municipal refuse. A demonstration plant capacity of 2 * 3 t h^{-1} was established at Gunzbueg (Bavaria) in June 1983. The plant has operated erratically for a total period of 100 days with a maximum capacity of 4.2 t h^{-1}.

The BKMI pyrolysis process to obtain product gas and char is published in literature (Buekens and Schoeters 1986). In this process, all the refuse is collected into a storage pit. The shredded refuse is then mixed with limestone ($CaCO_3$) provided on a horizontal belt conveyor. A second belt conveyor controls the depth of layer, overcoming the first one and thus, levelling the flow rate brought to a chute. Feeding system is then sealed by chute from the atmosphere. Refuse is then put into a barrel and then to the kiln. The kiln's heating is done with six separate heating zones and is coupled to a standard flute. The preheating of the feed produces light heating oil and then product gas (heating value of 7500 kJ m^{-3}). The carbonized char is then collected from the Martin residue quenching tub. The addition of limestone in feed leads to small amounts of impurities like HCl and SO_2.

The Kiener process was designed and established by a German Engineer and Industrialist named Mr. Kiener (Nowak 1978). This process is suitable for the processing of domestic, commercial, industrial refuse, and sewage sludge. The applications of the product gas are as fuel and as feed to produce plastics. A research group did the first phase of this program at Stuttgart University, including the system's scientific and technical evaluation. The second phase was completed by the end of 1977 involved optimization of the process. The third phase comprises constructing a demonstration plant with a refuse capacity of 3 tons hr^{-1}. The final operation of the plant is started in the year 1978. This process was patented in the United States on June 7, 1977, under Patent no. 4028068 (Buekens and Schoeters 1986).

The refuse with a maximum particle size of 3–4 inches is placed inside a rotating drum having special vanes and these vanes are served as a heat exchanger. The additional function of the vanes is mixing and transporting the materials in the drum. The refuse present in the drum is heated to 750–950 F in an air-free atmosphere. The gases are then passed through cyclone gas cleaner, where fly ash and soot are removed. The gas flow then enters to gas cracking and generating unit at a temperature of 650 F. Air is added in this section and the upper heating temperature is done at a temperature of 1100–1200 °C. In this section, the cracking of hydrocarbons produces methane, hydrogen (H_2), carbon dioxide (CO_2), carbon monoxide (CO) and minor quantities of simple bonded hydrocarbons. The gases are again passed through the second cyclone gas cleaner to remove dust particles. Cleaning of gas is done in a gas cleaner with the help of freshwater. Cooling of gas is performed using a gas cooler of about 72 F. Gas is then passed through an Induced draft fan and then to the storage tank. Gas from storage is used to produce electricity. Different alternative methods were also used for the power generation using Kiener pyrolysis process.

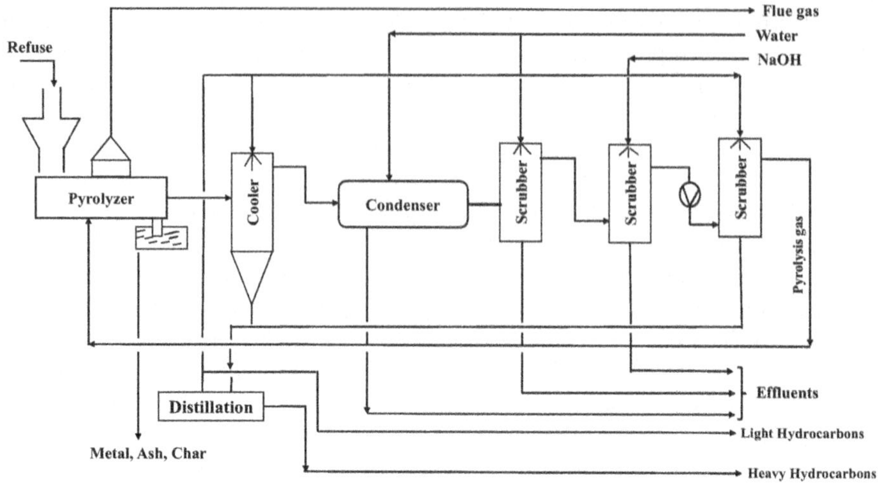

Fig. 17.5 Rotopyr pyrolysis process

The Odapyr process was invented by Dr. C. Otto and Co., a leading company in Coking Furnace Technology. This process involves heating at a temperature of 650–700 °C using an externally heated rotary kiln. The other steps involved in gas purification are quenching, separation of products (oil and water), scrubbing, low temperature condensation. The maximum amount of product gas is achieved while oil and char are in lesser amounts.

Rotopyr process was developed by a grouping of three films (1) Eisen und Metall A.G. (Gelsenkirchen); (2) Mannesmann Veba Umwelttechnik GmbH (Herne); (3) Rutgerswerk A.G. (Frankfurt) (Buekens and Schoeters 1986). This process is specially designed to obtain oil fractions and recover metals present in industrial wastes like cable scrap, rubber and plastic wastes, etc. Figure 17.5 shows the schematic overview of the Rotopyr process. The reaction at a temperature of 700 °C is done in the Pyrolyzer. The cooling of the gases is done in two steps. In the first step, gas is cooled at 150 °C and heavy oil is collected. Light oil comprising 30–40 wt.% of Benzene Toluene Xylene (BTX) compounds was found in the second cooling step. The oil yield of 18% was obtained, with a maximum of 30% for a feedstock of plastics. The char yield varies from 15 to 50% of the charged feedstock.

17.4 Pyrolysis Reactors and Their Implementation

17.4.1 Fluidized Bed Reactors

Prof. Kaminsky developed the fluidized bed pyrolysis process at Hamburg University. The objective of this development is to convert waste plastic and tires into pyrolysis products. The Deutsche Reifen- und Kunststoff Pyrolyse GmbH (DRP) process is based on this technology. The suggested heat transfer approach in a fluidized bed reactor is 90% conduction, 9% convection, and 1% radiation. Numerous studies were reported in the literature using fluidized bed reactors and different biomass as feedstock (Luo et al. 2005; Eri et al. 2017; Kersten et al. 2005; Pielsticker et al. 2019; Ku et al. 2019). A bench-scale fluidized bed reactor capacity of 1–5 kg/h was constructed, which consists of three sections: (i) feeding section, (ii) reactor section, and (iii) product collection section. A total bio-oil yield of 41.5 wt.% was obtained. The maximum product yield of gas 43.3 wt.% and char yield was 15.2 wt.% was obtained (Chen et al. 2010). Commercialization of pyrolysis technology was done using a plant size of 5–120 tons per day (Meier et al. 2013) and 7.2–400 tons per day (Butler et al. 2011). A commercial-scale pyrolysis plant capacity of 1–3 tons per hour was developed on downdraft fluidized bed technology. The process comprises of six parts: (i) a feeding system, (ii) a heat carrier system, (iii) a reactor, (iv) a cyclone system, (v) a condensation system, (vi) a carbon separating system. A maximum bio-oil yield of 48.1 wt.% was gained at 550 °C. The maximum amount of phenolics compounds of 14.92 wt.% were formed (Cai and Liu 2016). The commercialization of biomass fast pyrolysis technology was done by a detailed understanding of operation status and its product characteristics.

17.4.2 Bubbling Fluidized Beds

Figure 17.6 shows the typical configuration of a bubbling fluid bed reactor. The various advantages of bubbling fluidized beds (BFB) are (i) simple in operation and construction, (ii) reasonable temperature control, (iii) easy scaling, (iv) well-understood technology, (v) high liquid yield (70–75 wt.%) and (vi) rapid and efficient heating of biomass (Hasan et al. 2021). In this system, gas is injected vertically skyward through a bed of coarse material like sand at adequate velocity, by which gas and solid form an emulsion that looks like a fluid. The heating of the bed is done by externally combusting the produced gas and/or char and transferring this heat via direct heat transfer through hot solids or indirect heat transfer through steam and hot gas through tubes (Bridgwater 2003). The particle size of >2–3 mm is generally preferred for better heat transfer. The fluidizing gas flowrate controls the residence time of solids and vapors and is higher for char than for vapors. Generally, the vapor residence time of 0.2–5 s is preferred in a bubbling fluidized bed and it depends on the reactor size. Specially designed cyclones are used to capture the fine char. The large particles of char are separated by making segregation zones in the reactor and it is removed continuously with high purity.

Numerous installations of BFB technology are in innumerable universities and commercial facilities around the world (Bridgwater 2010). Examples are Waterloo (Scott and Piskorz 1984), Dynamotive, Union Fenosa (Cuevas et al. 1995), and Wellman. In 1980s, the early research on the science of fast pyrolysis was started at University of Waterloo, Canada. Union Fenosa did further developments on BFB

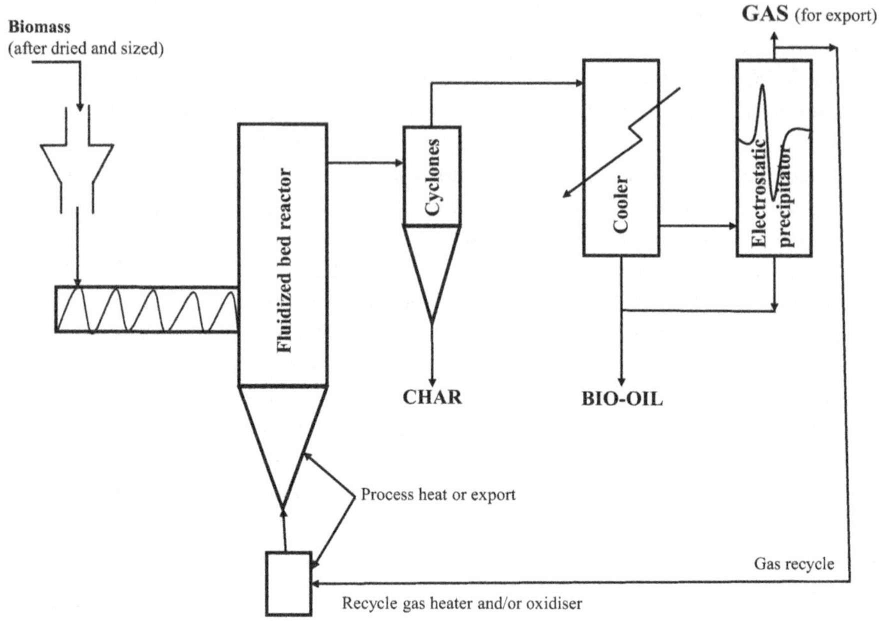

Fig. 17.6 Bubbling fluidized bed reactor

technology in 1990. They built a pilot plant in Spain and worked on a 200 kg/h unit to generate, transmit, and distribute electricity. Dynamotive was united in 1991, which is situated at British Columbia research in Vancouver. This company is involved in various advanced technologies, especially fast pyrolysis of wastes to produce value-added products. Plants with a capacity of 14–24 tpd have been installed in China based on BFB technology (Wu et al. 2010; Lu et al. 2007; Zhang et al. 2011). The scaling problems associated with Dynamotive design are modified by Biomass Engineering Ltd (UK), with a plant capacity of 250 kg/h. The direct heat transfer into the reactor can overcome the confines of indirect heat transfer. Using a twin fluidized bed may overcome the heating problem as solids are heated by the combustion of gas and solids in one bed, whereas the second bed is used for pyrolysis. This twin fluidized bed was designed and built by Wellman Process Engineering under an EU-sponsored project coordinated by Aston University in Birmingham. The construction of the pilot plant was completed in 1999, but it was never started due to permit problems. The novel-design pyrolyzer offered by Agri-Therm (Canada) is used to convert agricultural residue to bio-oil. Agri-Therm is a branch of Western University Institute for Chemicals and Fuels from Alternative Resources (ICFAR).

University of Waterloo, Canada, started a project to convert forest residue into liquid fuels. The research work done on modern fast/flash pyrolysis is highly publicized by University of Waterloo. A bench-scale continuous flash pyrolysis unit of following reactor conditions was installed: feed rate of 50 g/h, particle size 140 and 60 mesh, atmosphere nitrogen, temperature range 400–650 °C. The liquid yields of 60–70% were obtained from hardwoods like aspen, maple and poplar at 0.5 s of retention time. Around 40–60% of organic liquids were found from agricultural wastes like bagasse, corn stover and wheat straw.

Bridgwater and Peacocke (2000) represent the process designed at Waterloo. Sand is used to made the bed material of the reactor. The recycled product gas is consisting of CO-CO_2-CH_4 and used as fluidizing agent. Electric heaters are used to preheat the gas, while extra heat is provided through heating coils after wrapping the reactor. The fluid bed design is such that the char was entrained easily without affecting the sand particles from the bed. The reactor is operated at 425–625 °C of temperature, 1.25 bars of pressure and a feed rate of 1.5–3 kg h^{-1}. Feed as wood with moisture content ~7% and particle size of −595 μm is conveyed from a hopper to reactor by passing through twin screw feeder. After the reaction, the char is separated with the help of a cyclone separator. The vapours and gaseous products were then passed through hot water (60 °C) and ice water (0 °C) condenser. The final product, i.e., bio-oil, is obtained at the bottom of the condenser. The gases are then passed through the filter to eliminate impurities like tar mist, etc., and then pass through the compressor. The compressed gas is then used to fluidize the reactor bed while the excess gas is vented out. The product gas (CO and CO_2) is then analyzed using an infra-red gas analyzer -recorder and gas chromatography (GC). The higher heating value of liquid product obtained after pyrolysis of Brockville poplar, White spruce and Red maple was 23.20, 22.70, 22.40 MJ/kg, respectively. The high heating value of produced gas was about 14.4 MJ/kg. After successful results from this plant, a pilot plant based on the same technology of capacity 200 kg/hr was designed and built in Spain by Union Fenosa.

Canadian company Dynamotive has built and examined a 75 kg h^{-1} and 400 kg h^{-1} of pilot plants based on RTI design in Canada. These fluidized bed pyrolyzers are also known as BioTherm™. The laboratory plant was started its operation in 1996. Further, they constructed four installations, out of which the largest capacities of 100 tpd and 200 tpd are in West Lorne and Guelph, respectively. In West Lorne plant waste sawdust with a moisture content of 15% or less is used as feedstock to produce bio-oil. This plant started operation in early February of 2005.100 tonnes of biomass per day produced around 70 t of bio-oil, 20 t of char, and 10 t of noncondensable gases. But at the commencement of 2008, the plant was not in sound production and did not grasp the calculated bio-oil production capacity.

The obtained pyrolysis liquid is a mixture of compounds as well as value-added products. BioLime® is produced from the reaction between pyrolysis oil and slaked lime. It has applications for the regulation of oxides of sulfur and nitrogen in coal combustion processes.

The making of pyrolysis liquids for fuels was after 1989 is done by Union Fenosa. Waterloo Flash Pyrolysis Process used this technology and settled by the University of Waterloo in Canada. A plant capacity of 200 kg h^{-1} was commissioned in October 1992. By 1993, the plant has begun some problems with dry wood throughput of 160 kg/h. In the past year, several modifications were made to improve the process technology.

In the Union Fenosa plant, feed (eucalyptus) was first dried, screened and then used in the reactor. Eucalyptus was used as a feedstock due to its abundance, while oak and pine were also used in the other feedstocks. After drying and screening, the feed is transferred to the pyrolyzer using a belt conveyor. Propane-fired heat exchanger is used to heat the fluidizing gas. The separation of char is done by a cyclone separator. Quenching of the vapours is done in two water-cooled heat exchangers and then goes to demister and then finally goes to cooler so that maximum amount of bio-oil is obtained. Oil yields upto 70% and char yields upto 20% wt. on dry feed were reported from this plant (Williams and Horne 1994).

Wellman worked on the thermal conversion of biomass and coal. A fast pyrolysis fluidized bed pilot plant capacity of 250 kg h^{-1} was constructed by the European Commission (Bridgwater and Peacocke 2000). The Wellman process Engineering pilot plant process is continuous, which makes it a long-term operation. The heating of the fluidized bed reactor is done by the combustion of char in an annular fluid bed combustor. The obtained vapours are then passed through cyclone separators and then an electrostatic precipitator. The higher heating value of obtained liquid is 17 MJ kg^{-1}.

Agri-Therm, Canada, invented a mobile fast pyrolysis unit to produce bio-oil and biochar from bio-residue. The advantage of mobile pyrolysis is that pyrolysis is done directly into the agricultural and forestry operation, which leads to reduce the transportation cost. The unit comprises of a fluidized bed reactor where pyrolysis of biomass takes place. All the non-condensable gases generated in the reactor is burned in a provided fluidized bed combustor. Lift tubes are used to increase ablation from reactor to combustor and transfers the heat to solids. This technology is ideal for agricultural residue pyrolysis. The first-generation demonstration unit (MPS100) was ready for complete testing in early 2008 with capacity of 10 t/day. The second-generation unit (MPS200) has been designed and engineered. Comparing the MPS200 to MPS100 all improvements like safety, maintenance, reactor design, condensation system was made in MPS200 design. The capital cost of a commercial mobile pyrolysis unit is approximately $1.5 million.

In the year 1992, the former personnels of the University of Waterloo founded Resource Transforms International, Canada. Prof. D. Scott from the Department of Chemical Engineering exploits potentials of production of value-added chemicals by pyrolysis (Bridgwater and Peacocke 2000). In 1996, Resource Transforms International (RTI) and Dynamotive decided work together in the field of "biomass refinery". The pyrolysis plant capacity of 15 kg h^{-1} has been started by RTI. The sawdust or agri and wood waste and the lignocellulosic part of municipal solid waste (MSW) were used as feedstock. The pyrolysis plant with a capacity of 25 tons per day has been designed. Biomass was dried upto moisture content of 15% and used in the reactor. Bio-oil with 65–72%, biochar 15–20% and 12–18% of non-condensable gases are obtained as products. BioSol is a mixture of pyrolysis gas and mists, which is when condensed forms bio-oil. The char and ash are also used as activated carbon. The upgradation of bio-oil forms a variety of chemical products.

The Institute of Technical and Macromolecular Chemistry under the University of Hamburg, Germany, worked on pyrolysis at laboratory and pilot plants of different capacities. Plastics, rubber, bio-polymers, sewage sludge and oil shale are the feedstocks. This work includes the production of benzene, toluene and xylene (BTX) at temperatures more than 700 °C. The reactor made up of quartz sand bed and having an inner dia. of 450 mm is used. The feed enters the reactor using a double flap gate where depolymerization of the polymers is done at 450–900 °C (Lu et al. 2007). 50% of the feed is recovered as a liquid and contains 95% of aromatics.

Bioware Technologia (scientifically maintained via the University of Campinas, Brazil) intends to grow and produce value-added products from agro-industrial and forest waste. Elephant grass, cane trash and bagasse are used as raw materials. The fully automated process works on a temperature range of 480–500 °C, capacity: 300 kg/h, gas/feedstock mass ratio: 0.4 on a dry basis. The obtained bio-oil is utilized as fuel for energy generation, emulsifying agent for heavy petroleum, a phenol substitute in PF resin formulations, and an additive for cellular concrete. Biochar is used as fuel in boiler ovens, as an activated carbon, a pre-reducer for iron ore pellets, and catalytic substrate.

Researchers from University of Leeds, UK, and VTT, A Technical Research Centre of Finland was also involved in the research on fluidized units. They have installed plants having capacity of <1 kg/h which are easy to operate and give good results (Oasmaa et al. 2021).

17.4.3 Ablative Pyrolysis

This process involves the contact between biomass particles and carrier gas at velocities more than 100 m s^{-1}. Solid particles are then centrifuged to hot wall of the vortex reactor, and an immediate heat transfer happens to the surface of the particles. This pyrolysis process is also known as surface pyrolysis. Surface pyrolysis is done at the temperature range of 550–600 °C, in the presence of nitrogen and steam and independent of particle size. Ablative pyrolysis works on the same mode of reaction as melting butter in a frying pan. In the place of butter, wood as feedstock is used and heat is transferred through the reactor walls by which biomass gets melted and the product's vaporization process takes place. Large particles of char and other impurities are removed and recycled again. In this process, the reaction rate is inclined by the relative velocity of wood, pressure, heat-exchange surface and temperature (Bridgwater 2003). Figure 17.7 represents the Ablative reactor. The various advantages of Ablative reactor include: (i) Large particle sizes particles can be used, (ii) controllable residence time, (iii) Good heat transfer, (iv) Inert gas is not essential as it makes processing equipment smaller, (v) reaction system is more intensive, (vi) wall temperature of the reactor should be >600 °C (Cai et al. 2021). Drawbacks of Ablative reactor are: (i) Process is costly, (ii) High gas flow and product dilution, (iii) Process is driven mechanically makes the reactor more complex.

The first plant-based on ablative flash pyrolysis was BBC, Canada, where disposed tires are used as the raw material capacity of 10–25 kg/h. The technology was sold to Castle Capital Inc. Canada. They installed a plant capacity of 1500–2000 kg h^{-1} in

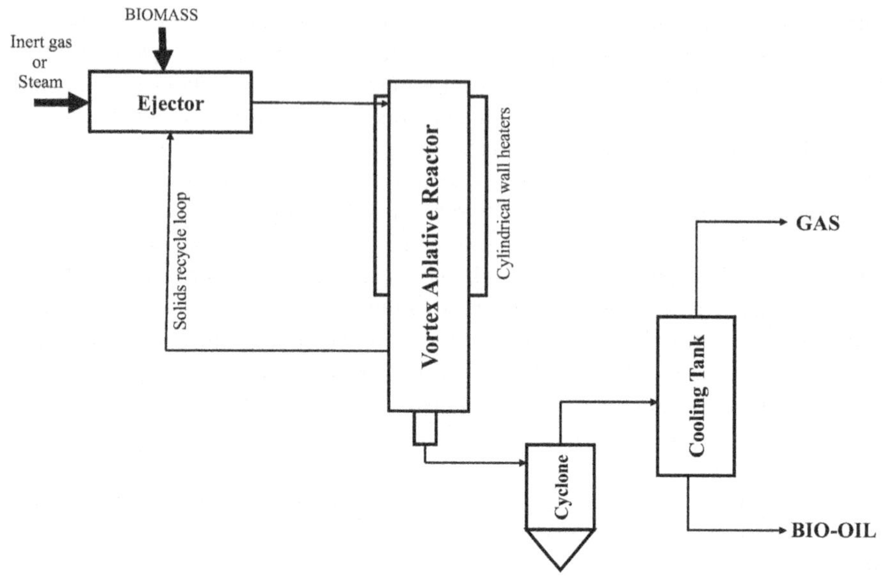

Fig. 17.7 Ablative reactor

Halifax, Nova Scotia, where solid waste is used. PyTec (Germany) installed a plant with a capacity of 6 tpd based on ablative technology. This plant aims to produce bio-oil and use this product in a CHP unit to run a diesel engine (Faix et al. 2010).

In the BBC ablative process, Tyre rubber as feed at a throughput of 10–25 kg h^{-1} with 1–3 mm particle size has been used. Rubber particles are then fed into the feeder, which then feeds the reactor. External heating is provided to the reactor and nitrogen is used to regulate reaction time. Sump tank is used to collect the liquid product. Two packed columns are used with exit temperature of vapour/gas at 100 °C (1st) and 40 °C (2nd). The particle size of product char is >10 μm.

Castle Capital Inc. Canada is based on Continuous Ablative Regenerator (C.A.R.). This process converts the liquid and solid organic wastes to fuels and chemical feedstocks. The feedstock is first fed into a loading screw then inserted into the preheated gas in reactor with adequate volume. The reactor is heated through indirect heating. The applications of product gas involve as a burner fuel, dry feedstock, and produce electricity from a gas turbine. The obtained vapours are then condensed to get liquids and used as a feedstock for chemicals.

The Aston University Bioenergy Research Group (BERG) has worked on pyrolysis for more than 25 years. A novel ablative plate reactor was used to get liquid fuels from woody biomass. The reactor capacity is 3 kg h^{-1} to get liquid yields upto 80 wt.% based on dry feedstock. The biomass particles size of 0.635 cm is fed into the reactor. The system consists of four asymmetric blades that rotate upto 200 rpm, generating mechanical pressure goes to the reactor having temperature of 600 °C. The mechanical act of the blades forms biomass particles to pyrolyze under high reactive motion on the reactor surface. Dilution of the products with nitrogen was performed to separate char in a cyclone.

Aston University BERG group is emphases the preparation of biomass and pre-treatment (Scott et al. 2010), the effect of growing conditions (Hodgson et al. 2010, 2011), fast pyrolysis in fluid bed systems and ablative pyrolysis reactor (Peacocke et al. 1994), supervision of thermal decomposition and secondary reactions, design and expansion of liquid collection (Sitzmann and Bridgwater 2007), significant developments in CFD modeling (Papadikis et al. 2010, 2008, 2009a, 2009b, 2009c; Gerhauser and Bridgwater 2003; Gerhauser et al. 2008), upgradation of bio-oils. The other group of Aston University European Bioenergy Research Institute was just formed. They are working on innovative research and knowledge transfer in various fields of bioenergy, including intermediate pyrolysis, Application of aqueous pyrolysis fractions, catalytic conversion of intermediate products, Micro kinetic studies of lignin degradation as well as tar formation, Direct use of intermediate pyrolysis bio-oil.

The CNRS laboratories did the fundamental work on ablative pyrolysis in Nancy in France. The first aim of this work is to produce non-equilibrium olefinic gases from biomass. The modified purpose is to produce liquid fuels using a zeolite catalyst reactor system. The produced hydrocarbons are BTX, gasoline, and chemical fractions. The biomass is fed into the hopper, then passes through the screw feeder and enters a steam ejector. The used ablative reactor is placed longitudinally at a wall temperature of 625 °C. The heating of the reactor is done superficially using three-zone electric furnaces. Recycling of partially pyrolyzed feedstock and char particles was done using an insulated recycle loop at the reactor exit. The vortex reactor has capacity of 50 kg h^{-1} (Bridgwater and Peacocke 2000; Diebold and Scahill 1987, 1984; Diebold 1980).

In the modified version of the NREL plant biomass having particle size of 5 mm is used as feedstock. Biomass is then mixed with the recycle stream and a gas, passes the vortex reactor at speed upto 400 m/s tangentially. The product from the reactor leaves by the axial exit, extends partway to the reactor. The used reactors have high capacity and high rate of heat transfer. Cyclone with a diameter of 4 in and temperature of 475–500 °C is used to separate char from the product stream. The vapours are passed through a first heat exchanger at 80 °C. The cooled gas stream is collected after passing through series of equipments. The liquids and water go to receiver. The oil yield of 55 wt.% and 13 wt.% char based on dry feed basis was obtained (Bridgwater and Peacocke 2000).

17.4.4 Circulating Fluid Bed (CFB) and Transported Bed

Circulating Fluid Beds and Transported beds have high heat transfer rates and short residence times. Similar to BFB technology, the residence time of char is nearly the same as vapour and gas. The difference between BFBs and CFBs is the quantity of gas used to fluidize the bed. The heat transfer medium is made up of the bed of particles of sand and catalyst etc. The reactor runs at high superficial gas velocity by which char is more attrited. The char particles are then mixed with collected bio-oil. The significant advantages of CFB are: (i) High heating rates, (ii) Controllable residence time, (iii) well-understood technology, (iv) suitable for very large throughputs, (v) Good heat and mass transfer. Disadvantages of the technology include: (i) Smaller size of feed is required, (ii) High rate of gas flow and product dilution, (iii) Char comprises of sand, (iv) High separation and quenching required, (v) Challenging to operate/separation.

Figure 17.8 represents a typical layout of a CFB reactor. The first Circulating Fluidized-bed was developed in the 1970s and 1980s at the University of Western Ontario (Bridgwater 2003). In this process, prepared biomass (drying and grinding) is fed in the riser of the CFB reactor. Hot sand as a bed material is added on the other side of the reactor. Biomass and sand get mixed in the riser section of the CFB reactor. The separation of char and sand is done at the exit of the riser by a cyclone separator. The particulate matter is then burned in the combustor in the presence of air, heating the bed media, and returning to the riser's bottom. The recovery system of the process consists of the cooler and electrostatic precipitator. The product gas is passed through the recovery system for bio-oil recovery from noncondensable gases.

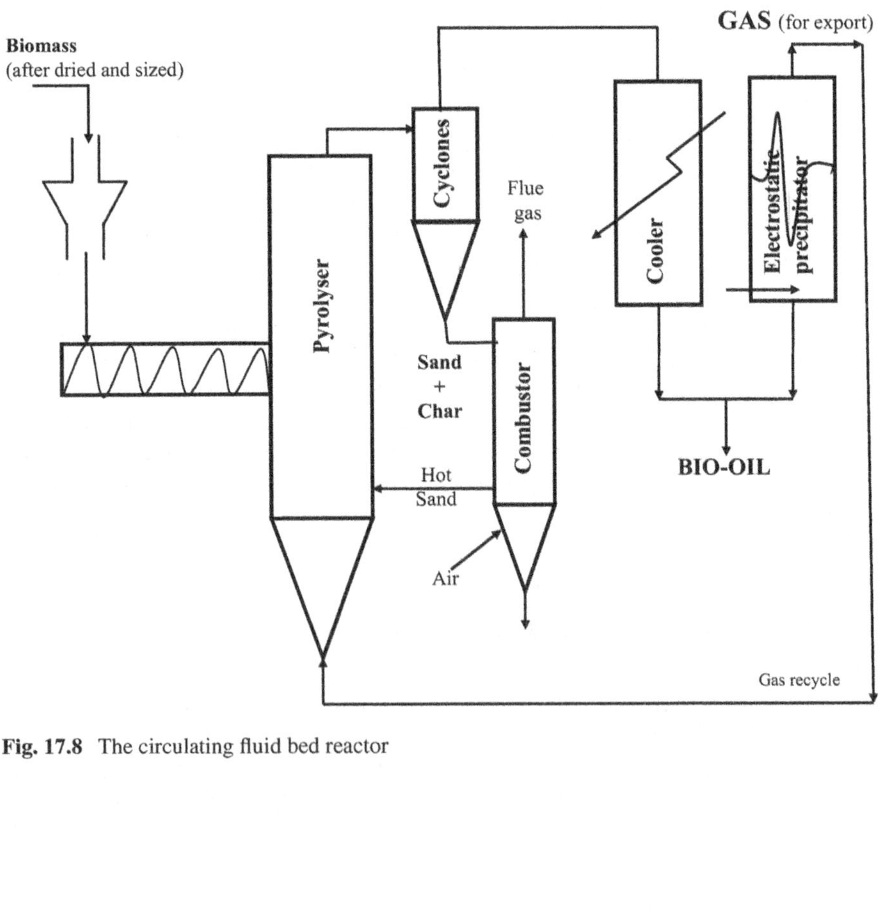

Fig. 17.8 The circulating fluid bed reactor

Enysn Technologies Inc., Canada, is the CFB technology leader based on Rapid Thermal Processing (RTP). Commercialization of this technology enabled in 1990 after approving Red Arrow Food Products Company Ltd of Wisconsin. The final products are liquid smoke and browning agents for the food industry. The plants based on the same technology are Red Arrow in Wisconsin, ENEL in Italy, VTT in Finland and Ensyn in Ottawa. In this process, biomass particle size ~6 mm and moisture content not more than 10% is used in transported bed reactor. After reaction, the separation of products is done using two cyclones. As solids get separated, the vapour is cooled using a multiple-stage system. The hot vapours get freezes in hundred milliseconds. Woody biomass yields 83% of the liquid by weight on a dry feed basis. The products (Oil and char) obtained from Ensyn Technology at Red Arrow is used to manufacture chemicals and as boiler fuel respectively (Underwood 1992).

The joint venture of Ensyn includes (i) UOP, a Honeywell firm for engineering, RTP equipment supply and upgrading of RFO to electricity generation and transportation fuels, (ii) For RFO refinery applications and upgrading with Chevron Technology Ventures, (iii) Fibria Celulose, and (iv) Felda Global Ventures- leading oil palm producer in Malaysia. The ASTM D7544-10 standard was developed by Ensyn, which is used for Pyrolysis Liquid Biofuel (ASTM standards). It confirms that pyrolysis oil is a liquid biofuel driven from biomass. It has applications as fuel in industrial burners to produce heat and power.

Ensyn is working on energy conversion projects in collaboration with Canmet ENERGY, Manitoba Hydro, and BMA.

17.4.5 Rotating Cone Pyrolysis

Rotating cone pyrolysis was settled by the Biomass Technology Group of Twente University, Netherlands (Wagenaar and Prins 1996). This technology is grounded on the fast heat transfer from the surface to small wood particles, using sand or catalytically active material. The feed and sand are introduced at the down of the rotating cone, where solids are mixed and then forced for reaction. The solids are collected out from the rotating cone. The vapours are then passed through condensation train. Advantages of Rotating cone reactor are: (i) No carrier gas required, (ii) Easy quenching, (iii) Centrifugal forces heats the sand and biomass. The various disadvantages of a Rotating cone reactor are (i) Difficult to scale up, (ii) Complex process, (iii) Capital cost is much higher, and (iv) Small size of particles. The Malaysian plant based on this technology was commissioned in 2006 with 50 t d^{-1}, but the plant is not in operation (Faix et al. 2010; Venderbosch and Prins 2010) EMPRYO European Project demonstrates this expertise in Europe on a scale of 120 t d^{-1}, targeting to produce 20,000 and 25,000 t y^{-1} of pyrolysis oil, process steam, electricity and organic acids. Figure 17.9 represents the Rotating cone reactor technology.

Biomass Technology Group B.V., (BTG) Netherlands, a self-governing film that includes consultants, engineers, researchers which has background in the Department

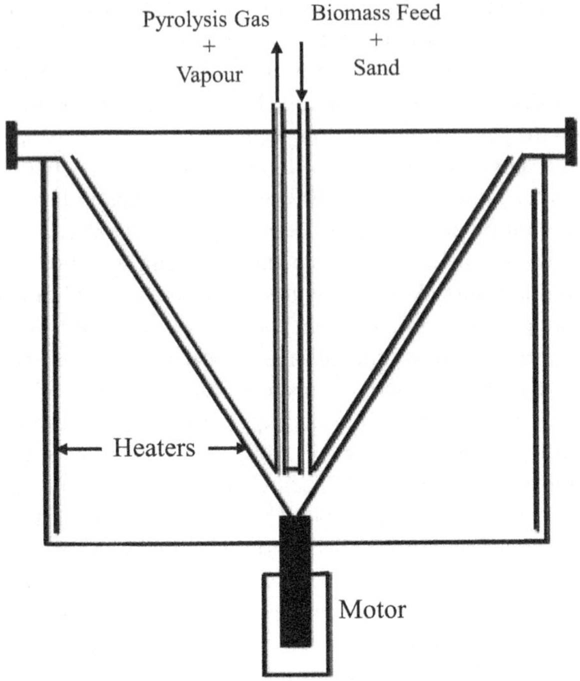

Fig. 17.9 Rotating cone pyrolyzer

of Chemical Engineering University of Twente, Netherlands. In 1989, the expansion of rotating cone technology leads to a 10 kg/h prototype reactor (Wagenaar et al. 1994). The work continued due to another project developing a fully heat-integrated lab. Plant and catalytic pyrolysis (Wagenaar and Prins 1996). BTG has two plants, one has capacity of 1–5 kg h^{-1} unit and another is a pilot plant where 200 kg h-1 is used to produce rich quantity products along with the complete information. BTG is involved in various funded projects for the progress of best technology for various types of biomass conversion and its applications.

University of Twente, Netherlands, worked on the rotating cone pyrolysis process where cone temperature is 600 °C with speed of 900 rpm. The outside the reactor has a cone which is filled with char and sand, limiting the experiment from performing more than 10 min. The modification was done where sand and char are removed from the reactor.

The commercialization of this technology results in a 50 kg h^{-1} unit sent to the Shenyang Agricultural University, China for two years in 1994. In this process, feed of 260 kg h^{-1} is fed to the reactor over a lock hopper while hot sand is added in a char combustor. The rotation of the cone is 600 rpm and surrounded within an electric oven. All the gaseous products were cooled in a condenser are used to produce heat. The obtained char contains some amount of sand that goes to the char combustor. The highest bio-oil yield of 75 wt.% based on feed (dry basis) was obtained using biomass like wood, wheat straw, rice husks, and organic waste.

17.4.6 Vacuum Pyrolysis

Vacuum pyrolysis requires temperature (T) ~450 °C and pressure of 15 kPa. The rate of heating is low as compared to other systems. The liquid yield of 55–60% wt. on feed (dry basis) was found. In this process, the extent of decomposition reactions is less, as quick volatilization of fragmented products takes place (Mohan et al. 2006; Marshall et al. 2014). The advantages of vacuum reactor are flexibility of feed particle size, bio-oil free from char, fewer aerosols formed, No extra requirement of carrier gas and dilution of product. Disadvantages of vacuum reactor are Low heating efficiency, Low Bio-oil yield, High capital and maintenance cost, More pyrolytic water generation and high sealing requirements.

Dr. Christian Roy along his team members were researched Pyrocycling™ at the University de Sherbrooke in the year 1981–1985. Pyrovac Institute Inc., a research centre working and refining the products attained from various industrial wastes started by Dr. Roy in 1988 and then get shaped into Pyrovac International Inc., in 1990. Pyrovac Group Inc. is the farm of Pyrovac Institute Inc. and Pyro System Inc. The head office is situated in Quebec City, Canada (Bridgwater and Peacocke 2000).

The Pyrocycling™ vacuum pyrolysis process includes converting organic materials to valuable products at 450 °C and 15 kPa. The pyrolysis reactor operates under continuous feed mode. The feedstock is introduced to Pyrocycler under vacuum. The feedstock is then pass over horizontal plates where it is heated at 530 °C. The heating of salts was done with a burner and a supply of non-condensable was provided in the burner. After heating, feedstock decomposition takes place and the produced vapours are separate out from the reactor instantaneously with the help of a vacuum pump. These vapours are then passed through condensers where heavy and light fractions of oils and an aqueous fraction are obtained. The solid product was also collected from the reactor after cooling.

17.4.7 Entrained Downflow

Entrained downflow fast pyrolysis is a simple technology where the 1–5 mm biomass particle size was fed into a stream of hot, inert gas. The temperature of the reactor tube was 700–800 °C and the residence time was of few seconds. This technology has some disadvantages like poor heat transfer, lower liquid yields, High gas flow rates and large plant size and has not been so successful. The technology was first developed at the Georgia Institute of Technology and then transferred to Egemin in Belgium. The plant was dismantled in 1993 due to its disadvantages.

The work is based on a project which was initiated in 1980 using a rotating tube furnace. The oil yield of 28 wt.% was obtained from it. Entrained flow reactor was then designed and operated successfully until 1989. Egemin built another plant based on entrained flow process. GTRI unit was constructed and completed in 1983 (Bridgwater and Peacocke 2000). To optimize the bio-oil yield, several modifications were made in 1985, out of which two were: (i) replacement of reactor tube from 8 in. to 6 in., (ii) adding of quenching pot and a demister.

The feed of particle size of 1.5 mm was fed to reactor via a rotary valve. A vertical reactor made up of stainless steel and has an inner diameter of 6 inch was used. The feed particles are then entrained in a stream of hot gas. In the reactor tube, the wood and gas flow co-currently where pyrolysis takes place. The obtained products were: gases, oil vapour and, char. The hot effluent is then entered into a water sprayed quench vessel, then an air cooler condenser. Accumulation of tarry materials in an air-cooled condenser becomes a big problem. The gas product is then goes through two demisters which are connected in series to remove aerosols. The remaining effluent is then entered into a flare where the burning takes place and combustible products are then exhausted to the atmosphere.

The liquid yield of 60 wt.% was achieved based on feed used, while modeling and optimization studies resulted that 70 wt.% of liquid product yield was obtained by using a modified reactor and system.

The Egemin flash pyrolysis plant was commissioned in 1991 (Maniatis et al. 1993). This research work aims to obtain a high bio-oil yield at a short residence time by providing better heat transfer. The wood particle size of 1–5 mm was used as feedstock. Feed is transferred to the reactor using a screw auger. Nitrogen gas is purging into it. Then it was entrained to the down-flowing reactor along with the hot gas at a temperature of 700–800 °C. At 490 °C, products left the reactor and passed through a cyclone separator. Separation of liquid production was done using a venturi scrubber. The inlet temperature of vapours in venture scrubber was 400 °C and outlet was 55 °C. The oil yield of 39.9%, char 16.2%, water 14.9% and gas (by difference) 29.0% were obtained as final products.

17.5 Current Scenario and Future Recommendations

The quality of feedstock is an essential parameter for pyrolysis operations. Wood is considered a traditional feedstock for the pyrolysis process. High ash content in biomass is a significant problem using biomass in the fast pyrolysis process (Venderbosch and Prins 2010). Realistic solutions for dealing with high ash content biomass will be done. Several studies are reported in literature where different catalysts, including zeolites and noble metals, are used (Xia et al. 2021; Nanduri et al. 2021; Miranda et al. 2021). High selectivity is the primary objective of the catalyst. Noble catalyst results in increased activity and selectivity. The drawbacks of noble catalysts are metal site poisoning, environmental issues and less availability. The catalyst preparation, its evaluation, optimization, understanding of its behavior in products at the molecular level, characterization of the catalyst including physical and chemical techniques, kinetic and so on are the main challenges. The application of catalysts needs to develop and optimization. The catalytic fast pyrolysis process termed Bio-TCat™ was developed by Anellotech, New York. The final products of this process are BTX aromatics and are used as fuels and petrochemicals. ZSM-5 catalysts were developed and tested using a fluidized bed reactor to improve the selectivity to BTX (Bond et al. 2014; Huber et al. 2006; Carlson et al. 2010). The reaction is done at a temperature of 600 °C in the presence of ZSM-5 catalyst, and around 10–15 wt.% of aromatic products were formed using woody biomass as feedstock (Carlson et al. 2011; Jae et al. 2014). Significant production of coke upto 35 wt.% was formed and is a disadvantage of the process (Foster et al. 2012). They have recently constructed a plant of 1–2 tpd capacity to authenticate the working of the Bio-TCat™ process.

The Research Triangle Institute is developing an integrated catalytic pyrolysis process with hydroprocessing, almost similar to the Ensyn RTP process. The catalyst used in this process is RTI-A9 in the range of 10–20 wt.%, based on metal oxide and tungsten oxide. On a bench scale fluidized bed was used and the reduction of oxygen was observed from 37.7 to 19.9 wt.%. The yield of bio-oil was increased with the yield of char and gases) A pilot plant of capacity 1 tpd was designed and installed for the better understanding of the integrated process and assess the performance of the catalyst (Dayton et al. 2015). Pine was used as feedstock after drying with the particle size of 6 mm and it has less ash content, nitrogen and sulfur content. The feed is then heated to 450–600 °C and nitrogen is used as a fluidized agent. The bio-oil yield of 65 wt.% was attained with 26 wt.% of gas yield and 2 wt.% of char yield. The optimization studies of different parameters, i.e., Temperature, residence time, regenerator temperature, catalyst circulation rate, catalyst regeneration effectiveness, coking and catalyst deactivation on pilot runs were studied. The effect of catalytic as well as non-catalytic hydro-pyrolysis of biomass has also been studied by RTI (Dayton et al. 2016). The use of hydrogen suppresses the yield of char/coke formation. A minor impact on the yield of gas and oil was observed, while catalytic hydro-pyrolysis reduces the oxygen content in liquid products (Dayton et al. 2013).

The process developed by Gas Technology Institute (GTI) is IH^2. It integrates catalytic hydro-pyrolysis and hydroconversion to yield gasoline and diesel fraction liquid from biomass (Marker et al. 2012, 2014). In 2010–11, U.S. Department of Energy funded this project and give rights to Criterion Inc., for the commercialization of this technology.

In this process, biomass reacts with hydrogen in a fluidized bed reactor at 400–450 °C and 100–500 psi (hydrogen). The total pressure of the reaction should be 3000–4000 kPa. In the IH^2 process, the hydro pyrolysis step is exothermic and balanced in nature. The initial research on IH^2 at the pilot scale was done using various biomasses as feedstock with 5 g/min of feed rates and the process is well explained (Perkins et al. 2018). The difference in oxygen content was observed using catalytic hydro pyrolysis (<3 wt.%) and non-catalytic hydro pyrolysis (14 wt.%). A variety of bench and mini- pilot reactors were used to investigate the effect of biomass process parameters. The various projects on biomass are supported by EPSRC—fluidized bed gasification and the Engineering Doctorate Centre on Efficient Fossil Energy Technologies. Conversion and Resource Evaluation (CARE) Ltd, U.K. is working on designing, building, and commissioning a slow pyrolysis reactor having capacity of 50 kgh^{-1} for char production with environmental obedience, upgrading liquid products, and pyrolysis of high ash material. Future blends, U.K. is working on the upgradation of bio-oils to transport fuels. Canmet ENERGY, Canada has collaboration with Ensyn and ABRI- Tech in the expansion of fast pyrolysis. They are working on catalytic hydro-processing of biomass-includes exploration and evaluation of chemicals for fast pyrolysis oils. Finland researchers are working in fast pyrolysis. VTT supports industrial projects consortium of Metso Power, UPM Kymmene and Fortum. There main is to validate pyrolysis oil production and long-run use (Miranda et al. 2021; Oasmaa et al. 2010; Lehto et al. 2014, 2010; Solantausta et al. 2012). Energy Research Centre, Netherlands, is working on the thermochemical conversion of lignin to phenols by bubbling fluidized bed.

A very less research work has been directed on the complete structure, working principles, characteristics and stability of bioproducts by the commercial-scale pyrolysis plant (Cai and Liu 2016). Handling and storage issues are barriers to developing a market for bio-oil (Kargbo et al. 2021; Radlein 2010). The novel pyrolysis technologies i.e., IH^2 by GTI and CHP co-production model by Metso (now Valmet)/UPM/VTT can be well known technologies if they can be successfully demonstrated and commercialized.

17.6 Conclusions

Pyrolysis is a well-known technology and the research on it is still going on. This chapter reviews the different technologies used in the pyrolysis for the conversion of biomass to bio-oils and bio-char. Different types of reactors along with commercial as well as pilot plants used for the production of value-added products are elaborate in this chapter. Some technologies such as rotating cone and fluidized bed is used on the commercial scale. The advancement in the technology under optimized parameters may lead to enhance the product yield. Further research is required to make the process cost effective on large scale and to overcome the use of fossil fuels.

References

Amenaghawon AN, Anyalewechi CL, Okieimen CO, Kusuma HS (2021) Biomass pyrolysis technologies for value-added products: A state-of-the-art review. Environ Dev Sustain 1–55

Beenackers A (2001) Bio-energy implementation: constraints for large scale commercialisation, Conclusions of Workshop. In: Proceedings

Bond JQ, Upadhye AA, Olcay H, Tompsett GA, Jae J, Xing R, Alonso DM, Wang D, Zhang T, Kumar R (2014) Production of renewable jet fuel range alkanes and commodity chemicals from integrated catalytic processing of biomass. Energy Environ Sci 7:1500–1523

Bridgwater AV (2003) Renewable fuels and chemicals by thermal processing of biomass. Chem Eng J 91:87–102

Bridgwater AV (2012) Review of fast pyrolysis of biomass and product upgrading. Biomass Bioenerg 38:68–94. https://doi.org/10.1016/j.biombioe.2011.01.048

Bridgwater AV, Peacocke GVC (2000) Fast pyrolysis processes for biomass. Renew Sustain Energy Rev 4:1–73

Bridgwater AV (2010) Fast pyrolysis of biomass for energy and fuels. Thermochem Convers biomass to Liq fuels Chem 146–191
Bridgwater AV (2003) Renewable fuels and chemicals by thermal processing of biomass. Chem Eng J 91:87–102.https://doi.org/10.1016/S1385-8947(02)00142-0
Bridgwater AV, Maniatis K (2004) The production of biofuels by the thermochemical processing of biomass, in molecular to global photosynthesis
Briens C, Piskorz J, Berruti F (2008) Biomass valorization for fuel and chemicals production—a review. Int J Chem React Eng 6
Buekens AG, Schoeters JG (1986) European experience in the pyrolysis and gasification of solid wastes. Conserv Recycl 9:253–269.https://doi.org/10.1016/0361-3658(86)90016-0
Buekens A, Schoeters J (1985) European experience in the pyrolysis and gasification of solid wastes. American Institute of Chemical Engineers, New York, NY
Butler E, Devlin G, Meier D, McDonnell K (2011) A review of recent laboratory research and commercial developments in fast pyrolysis and upgrading. Renew Sustain Energy Rev 15:4171–4186
Cai W, Luo Z, Zhou J, Wang Q (2021) A review on the selection of raw materials and reactors for biomass fast pyrolysis in China. Fuel Process Technol 221:106919
Cai W, Liu R (2016) Performance of a commercial-scale biomass fast pyrolysis plant for bio-oil production. Fuel 182:677–686
Carlson TR, Jae J, Lin Y-C, Tompsett GA, Huber GW (2010) Catalytic fast pyrolysis of glucose with HZSM-5: the combined homogeneous and heterogeneous reactions. J Catal 270:110–124
Carlson TR, Cheng Y-T, Jae J, Huber GW (2011) Production of green aromatics and olefins by catalytic fast pyrolysis of wood sawdust. Energy Environ Sci 4:145–161
Chen T, Deng C, Liu R (2010) Effect of selective condensation on the characterization of bio-oil from pine sawdust fast pyrolysis using a fluidized-bed reactor. Energy Fuels 24:6616–6623
Cuevas A, Reinoso C, Scott DS (1995) Pyrolysis oil production and its perspectives. Proc Power Prod from biomass II, Espoo
Dayton DC, Carpenter J, Farmer J, Turk B, Gupta R (2013) Biomass hydropyrolysis in a pressurized fluidized bed reactor. Energy Fuels 27:3778–3785
Dayton DC, Carpenter JR, Kataria A, Peters JE, Barbee D, Mante OD, Gupta R (2015) Design and operation of a pilot-scale catalytic biomass pyrolysis unit. Green Chem 17:4680–4689
Dayton DC, Hlebak J, Carpenter JR, Wang K, Mante OD, Peters JE (2016) Biomass hydropyrolysis in a fluidized bed reactor. Energy Fuels 30:4879–4887
Diebold JP (1980) Ablative pyrolysis of macroparticles of biomass. In: Proceedings of the specialists workshop on the fast pyrolysis of biomass, Copper Mountain. Co, p 237
Diebold JP, Scahill JW (1984) Ablative, entrained-flow fast pyrolysis of biomass. In: Proceedings 16th biomass thermochemical conversion contractors meeting. Portland, OR, May 1984
Diebold J, Scahill J (1987) Production of primary pyrolysis oils in a vortex reactor. Prepr Pap, Am Chem Soc, Div Fuel Chem (United States) 32
Dupuis DP, Grim RG, Nelson E, Tan ECD, Ruddy DA, Hernandez S, Westover T, Hensley JE, Carpenter D (2019) High-octane gasoline from biomass: experimental, economic, and environmental assessment. Appl Energy 241:25–33
Eri Q, Zhao X, Ranganathan P, Gu S (2017) Numerical simulations on the effect of potassium on the biomass fast pyrolysis in fluidized bed reactor. Fuel 197:290–297
Faix A, Schweinle J, Scholl S, Becker G, Meier D (2010) Life-cycle assessment of the BTO (R)-process (biomass-to-oil) with combined heat and power generation. Environ Prog Sustain Energy 29:193–202
Foster AJ, Jae J, Cheng Y-T, Huber GW, Lobo RF (2012) Optimizing the aromatic yield and distribution from catalytic fast pyrolysis of biomass over ZSM-5. Appl Catal A Gen 423:154–161
Gamliel DP, Bollas GM, Valla JA (2018) Two-stage catalytic fast hydropyrolysis of biomass for the production of drop-in biofuel. Fuel 216:160–170

Gerhauser H, Bridgwater AV (2003) Cfd modelling of biomass pyrolysis for better fluidised bed design. In: Proceedings of pyrolysis Gasification of biomass waste, Strasbourg, France. CPL Sci Press, Newbury, UK

Gerhauser H, Generalis SC, Hague RA, Bridgwater AV (2008) CFD for the modelling of entrainment in fluidised bed fast pyrolysis of biomass. In: Progress in thermochemical biomass conversion. Wiley-Blackwell, pp 1281–1295

Hasan MM, Rasul MG, Khan MMK, Ashwath N, Jahirul MI (2021) Energy recovery from municipal solid waste using pyrolysis technology: A review on current status and developments. Renew Sustain Energy Rev 145:111073

Henrich E, Dinjus E, Weirich F (2002) A new concept for biomass gasification at high pressure. In: Proceedings of the 12th European conference and technology exhibition on biomass for energy, industry and climate protection, Amsterdam ETA, Florence and WIP, Munich. pp 628–632

Hodgson EM, Fahmi R, Yates N, Barraclough T, Shield I, Allison G, Bridgwater AV, Donnison IS (2010) Miscanthus as a feedstock for fast-pyrolysis: does agronomic treatment affect quality? Bioresour Technol 101:6185–6191

Hodgson EM, Nowakowski DJ, Shield I, Riche A, Bridgwater AV, Clifton-Brown JC, Donnison IS (2011) Variation in Miscanthus chemical composition and implications for conversion by pyrolysis and thermo-chemical bio-refining for fuels and chemicals. Bioresour Technol 102:3411–3418

Huber GW, Iborra S, Corma A (2006) Synthesis of transportation fuels from biomass : chemistry, catalysts, and engineering. 2:4044–4098

Jae J, Coolman R, Mountziaris TJ, Huber GW (2014) Catalytic fast pyrolysis of lignocellulosic biomass in a process development unit with continual catalyst addition and removal. Chem Eng Sci 108:33–46

Kargbo H, Harris JS, Phan AN (2021) "Drop-in" fuel production from biomass: Critical review on techno-economic feasibility and sustainability. Renew Sustain Energy Rev 135:110168

Kaur R, Gera P, Jha MK (2016) Study on effects of different operating parameters on the pyrolysis of biomass: a review. J Biofuels Bioenergy 1:135. https://doi.org/10.5958/2454-8618.2015.00015.2

Kersten SRA, Wang X, Prins W, van Swaaij WPM (2005) Biomass pyrolysis in a fluidized bed reactor. Part 1: literature review and model simulations. Ind Eng Chem Res 44:8773–8785

Ku X, Shen F, Jin H, Lin J, Li H (2019) Simulation of biomass pyrolysis in a fluidized bed reactor using thermally thick treatment. Ind Eng Chem Res 58:1720–1731

Kumar P, Varkolu M, Mailaram S, Kunamalla A, Maity SK (2019) Biorefinery polyutilization systems: production of green transportation fuels from biomass. In: Polygeneration with polystorage for chemical and energy hubs. Elsevier, pp 373–407

Lehto J, Oasmaa A, Solantausta Y, Kytö M, Chiaramonti D (2014) Review of fuel oil quality and combustion of fast pyrolysis bio-oils from lignocellulosic biomass. Appl Energy 116:178–190. https://doi.org/10.1016/j.apenergy.2013.11.040

Lehto J, Jokela P, Alin J, Solantausta Y (2010) Bio-oil production integrated with a fluidized bed boiler-experiences from a pilot project. In: Best Papers awards 2010: PEI power engineering. pp 182–187

Lepage T, Kammoun M, Schmetz Q, Richel A (2021) Biomass-to-hydrogen: A review of main routes production, processes evaluation and techno-economical assessment. Biomass Bioenergy 144:105920

Lopez JS, Dagle RA, Dagle VL, Smith C, Albrecht KO (2019) Oligomerization of ethanol-derived propene and isobutene mixtures to transportation fuels: catalyst and process considerations. Catal Sci Technol 9:1117–1131

Lu Q, Zhu X, Li Q, Guo Q, Zhu Q (2007) Biomass fast pyrolysis for liquid fuels. Prog Chem 19:1064

Luo Z, Wang S, Cen K (2005) A model of wood flash pyrolysis in fluidized bed reactor. Renew Energy 30:377–392

Maniatis K, Baeyens J, Peeters H, Roggeman G (1993) The Egemin flash pyrolysis process: commissioning and initial results. In: Advances in thermochemical biomass conversion. Springer, pp 1257–1264

Marker TL, Felix LG, Linck MB, Roberts MJ (2012) Integrated hydropyrolysis and hydroconversion (IH^2) for the direct production of gasoline and diesel fuels or blending components from biomass, part 1: proof of principle testing. Environ Prog Sustain Energy 31:191–199

Marker TL, Felix LG, Linck MB, Roberts MJ, Ortiz Toral P, Wangerow J (2014) Integrated hydropyrolysis and hydroconversion (IH^2) for the direct production of gasoline and diesel fuels or blending components from biomass, PART 2: continuous testing. Environ Prog Sustain Energy 33:762–768

Marshall ASJ, Wu PF, Mun SH, Lalonde C (2014) Commercial application of pyrolysis technology in agriculture. In: 2014 Montreal, Quebec Canada July 13-July 16, 2014. American society of agricultural and biological engineers, p 1

Meier D, Van De Beld B, Bridgwater AV, Elliott DC, Oasmaa A, Preto F (2013) State-of-the-art of fast pyrolysis in IEA bioenergy member countries. Renew Sustain Energy Rev 20:619–641

Miranda NT, Motta IL, Maciel Filho R, Maciel MRW (2021) Sugarcane bagasse pyrolysis: a review of operating conditions and products properties. Renew Sustain Energy Rev 149:111394

Mohan D, Pittman CU, Steele PH (2006) Pyrolysis of wood/biomass for bio-oil : a critical review. 848–889

Nanduri A, Kulkarni SS, Mills PL (2021) Experimental techniques to gain mechanistic insight into fast pyrolysis of lignocellulosic biomass: a state-of-the-art review. Renew Sustain Energy Rev 148:111262

Nowak F (1978) Development of the Kiener pyrolysis system for environmental protection, energy recovery and recycling. Energy Conserv through waste Util 1978

Oasmaa A, Solantausta Y, Arpiainen V, Kuoppala E, Sipila K (2010) Fast pyrolysis bio-oils from wood and agricultural residues. Energy Fuels 24:1380–1388

Oasmaa A, Lehto J, Solantausta Y, Kallio S (2021) Historical review on VTT fast pyrolysis bio-oil production and upgrading. Energy Fuels 35:5683–5695

Papadikis K, Bridgwater AV, Gu S (2008) CFD modelling of the fast pyrolysis of biomass in fluidised bed reactors, Part A: Eulerian computation of momentum transport in bubbling fluidised beds. Chem Eng Sci 63:4218–4227

Papadikis K, Gu S, Bridgwater AV, Gerhauser H (2009a) Application of CFD to model fast pyrolysis of biomass. Fuel Process Technol 90:504–512

Papadikis K, Gu S, Bridgwater AV (2009b) CFD modelling of the fast pyrolysis of biomass in fluidised bed reactors: modelling the impact of biomass shrinkage. Chem Eng J 149:417–427

Papadikis K, Gu S, Bridgwater AV (2009c) CFD modelling of the fast pyrolysis of biomass in fluidised bed reactors. Part B: heat, momentum and mass transport in bubbling fluidised beds. Chem Eng Sci 64:1036–1045

Papadikis K, Gu S, Bridgwater AV (2010) Computational modelling of the impact of particle size to the heat transfer coefficient between biomass particles and a fluidised bed. Fuel Process Technol 91:68–79

Peacocke GVC, Madrali ES, Li C-Z, Guell AJ, Wu F, Kandiyoti R, Bridgwater AV (1994) Effect of reactor configuration on the yields and structures of pine-wood derived pyrolysis liquids: a comparison between ablative and wire-mesh pyrolysis. Biomass Bioenerg 7:155–167

Perkins G, Bhaskar T, Konarova M (2018) Process development status of fast pyrolysis technologies for the manufacture of renewable transport fuels from biomass. Renew Sustain Energy Rev 90:292–315

Pielsticker S, Schlogel K, Kreitzberg T, Hatzfeld O, Kneer R (2019) Biomass pyrolysis kinetics in a fluidized bed reactor: measurements and plausibility verification for reaction conditions. Fuel 254:115589

Radlein D (2010) The past and future of fast pyrolysis for the production of bio-oil. In: Symposium on thermal and catalytic sciences for biofuels and biobased products. USA

Scott SW, Nowakowski DJ, Bridgwater A V, Jones JM (2010) Assess the impact of alkali metals in biomass on the physical and chemical properties of fast pyrolysis bio-oil. In: Proceedings of the bioten conference on biomass bioenergy and biofuels, Newbury, UK, CPL Press
Scott DS, Piskorz J (1984) The continuous flash pyrolysis of biomass. Can J Chem Eng 62:404–412
Sims R, Taylor M, Saddler J, Mabee W (2008) From 1st-to 2nd-generation biofuel technologies. Paris Int Energy Agency Organ Econ Co-Operation Dev 16–20
Sitzmann J, Bridgwater AV (2007) Upgrading fast pyrolysis oils by hot vapour filtration. In: Proceeding of 15th European energy from biomass conference, Berlin; 7e11 May
Solantausta Y, Oasmaa A, Sipila K, Lindfors C, Lehto J, Autio J, Jokela P, Alin J, Heiskanen J (2012) Bio-oil production from biomass: steps toward demonstration. Energy Fuels 26:233–240
Soltanian S, Aghbashlo M, Almasi F, Hosseinzadeh-Bandbafha H, Nizami A-S, Ok YS, Lam SS, Tabatabaei M (2020) A critical review of the effects of pretreatment methods on the exergetic aspects of lignocellulosic biofuels. Energy Convers Manag 212:112792
Sun S, Yang R, Wang X, Yan S (2018) Hydrogenation and hydrodeoxygenation of biomass-derived oxygenates to liquid alkanes for transportation fuels. Data Br 17:638–646
Underwood G (1992) Commercialisation of fast pyrolysis products. Biomass Therm Process 226–228
Vamvuka D (2011) Bio oil, solid and gaseous biofuels from biomass pyrolysis processes-an overview. Int J Energy Res 35:835–862
Venderbosch RH, Prins W (2010) Fast pyrolysis technology development. Biofuels Bioprod Biorefining 4:178–208
Wagenaar BM, Prins W, Van Swaaij WPM (1994) Pyrolysis of biomass in the rotating cone reactor: modelling and experimental justification. Chem Eng Sci 49:5109–5126
Wagenaar BM, Prins W (1996) The pyrolysis pilot plant, a characterisation program. Prep NOVEM, Utrecht, Netherlands, Rep 355200:2070
Williams PT, Horne PA (1994) The role of metal salts in the pyrolysis of biomass. Renew Energy 4:1–13
Wu CZ, Yin XL, Yuan ZH, Zhou ZQ, Zhuang XS (2010) The development of bioenergy technology in China. Energy 35:4445–4450
Xia C, Cai L, Zhang H, Zuo L, Shi SQ, Lam SS (2021) A review on the modeling and validation of biomass pyrolysis with a focus on product yield and composition
Yang Z, Qian K, Zhang X, Lei H, Xin C, Zhang Y, Qian M, Villota E (2018) Process design and economics for the conversion of lignocellulosic biomass into jet fuel range cycloalkanes. Energy 154:289–297
Zhang C, Zhang R, Li X, Li Y, Shi W, Ren X, Xu X (2011) Bench scale fluidized bed fast pyrolysis of peanut shell for bio oil production. Environ Prog Sustain Energy 30:11–18
Zhang Y, Bi P, Wang J, Jiang P, Wu X, Xue H, Liu J, Zhou X, Li Q (2015) Production of jet and diesel biofuels from renewable lignocellulosic biomass. Appl Energy 150:128–137
Zhang X, Brown RC (2019) Introduction to thermochemical processing of biomass into fuels, chemicals, and power. Thermochem Process Biomass Convers Fuels Chem Power 1–16
Zupko R (2019) Life cycle assessment of the production of gasoline and diesel from forest residues using integrated hydropyrolysis and hydroconversion. Int J Life Cycle Assess 24:1793–1804

Chapter 18
An In-Depth Evaluation of Feedstock, Production Process, Catalyst for Biodiesel Production

Deepak Kumar Yadav, Narsi R. Bishnoi, Somvir Bajar, and Anita Singh

18.1 Introduction

Energy demand is expanding persistently with increment in the world population. Worldwide petroleum product demand is increasing persistently because its consumption rate is 105 times faster in comparison to its formation naturally (Thoai et al. 2019). The major contribution to temperature rise is greenhouse gas (GHGs) emission from anthropogenic activates such as the burning of fossil fuel in four sectors of our economy viz. transportation, industrial, residential and commercial sector (Karmakar 2019). Intergovernmental Panel on Climate Change (IPCC) published a report that conclude that if the average global temperature increased a minimal margin of 1.5 °C, up to one million species could become extinct (Pasha et al. 2021). The fluctuating price of petrol in the global market, uncertainties concerning petroleum availability, emission of GHGs, and increased environmental concerns from using conventional sources of energy have stimulated the search for alternative sources (Sajid et al. 2016). In the past few years, biodiesel received ample attention. In the last decade, dramatic growth is observed in the manufacturing of biodiesel. The characteristics of biodiesel are low GHGs emission, eco-friendly and biodegradable molecular structure with minimal combustion of toxicity (Christopher et al. 2014; Thoai et al. 2019). For biodiesel production there are wide range of feedstock are available like edible, non-edible oil crop, microalgae as well as waste cooking oils. In biodiesel production the uses of separation conditions and diverse reaction, and

D. K. Yadav · N. R. Bishnoi
Department of Environmental Science and Engineering, GJUS&T, Hisar, India

S. Bajar
Department of Environmental Science and Engineering, J.C. Bose University of Science and Technology, YMCA, Faridabad, Haryana 121006, India

A. Singh (✉)
Department of Environmental Sciences, Central University of Jammu, Jammu, Jammu and Kashmir 180011, India

Y. K. Nandabalan et al. (eds.), *Zero Waste Biorefinery*, Energy, Environment, and Sustainability, https://doi.org/10.1007/978-981-16-8682-5_18

different types of catalyst make manufacturing of biodiesel is a complex process (Change CS-NC 2019).

Bio-diesel is commonly produced by transesterification of lipids (triglyceride) and is treated with alcohols like ethanol or methanol which produces methyl esters (Biodiesel) and glycerol as a by-product. It has higher oxygen content, so better combustion quality. It has a higher cetane number, high viscosity, and very low emissions of greenhouse gases compared to other petroleum products (Mathew et al. 2021). Biodiesel production can be done by using various feedstocks such as edible, non-edible oil crop, microalgae as well as waste cooking oils. The biodiesel fuel properties significantly depend on the composition of fatty acid and physico-chemical properties of using feedstock. Bio-diesel becomes crucial for several nations as it provides opportunities like getting rid of oil dependency by using renewable resources, reducing CO_2 production as it is emitted by traditional fossil fuels and world is running out of fossil fuel resources which leads to conflict among nations. There are several processes present for the production of bio-diesel fuel, among which transesterification is the best process for the production of bio-diesel (Kirubakaran et al. 2018). It uses alkali catalysis and gives high level of conversion of triglycerides to their corresponding methyl esters in short reaction time. The aim of this book chapter is to present a broad review evaluation of feedstock, production process, catalyst for biodiesel production.

18.2 Bio-diesel Production Status

From vegetable oil to coal dust and then kerosene, various types of fuels were used to operate diesel engine within few years of its invention in 1890 by Rudolph Diesel. In 1900 for the first time at World's Fair, demonstration of diesel engine based on vegetable oil was done by the Government of French for the public. They ordered the auto company to make an engine that runs on peanut oil. The Government of France was concerned about using vegetable oils for remote areas as a fuel (Mishra and Goswami 2018). Dr. Diesel did extensive research work on vegetable oils to promote an idea, believing that farmers could get benefit from producing their fuel. In 1913, after the demise of Rudolph Diesel, petroleum becomes easily accessible at cheap rates (Oliveira and Coelho 2017). The modifications were made into diesel engine to coordinate the properties of petroleum-based fuel. The result was an engine fuel which was extremely incredible, productive and turned into a standard where force, economy, and quality are required (Khanal and Shah 2021). G. Chavanne in 1937 used palm oil as a feedstock for an ethyl-ester and was granted a Belgian patent. In 1985, an agriculture college of Austria started the first commercial manufacturing unit of biodiesel. Commercial manufacturing of bio-diesel started in Europe in 1992, and Germany becomes the major producer. Yellowstone National Park was offered bio-diesel by University of Idaho in the year 1995 for their vehicle which provides no harm to its engine even after several years of use. Other national parks also adopted similar practices for their automobiles (Oliveira and Coelho 2017). The leading

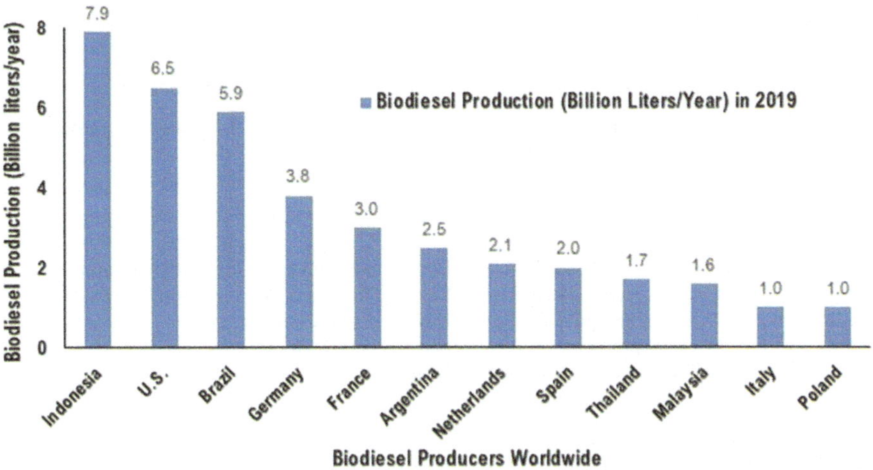

Fig. 18.1 Leading biodiesel producers worldwide in 2019 (*Source* https://www.statista.com/statistics/271472/biodiesel-production-in-selected-countries/

nations in biodiesel production are Indonesia, United State and Brazil (Fig. 18.1). They are the largest producers of biodiesel in the world. The U.S is projected to reach one billion gallon of biodiesel production by 2025. The energy policy act of 2005 help in increasing biodiesel production in U.S by providing tax incentive for different types of energy. Currently the volumetric ethanol excise tax credit is the main financial support for the production of biofuels. U.S shows the highest bioenergy capacity in the world totalling 12,450 megawatts (Statista 2021).

India initiated "Ethanol Blending Programme" in 2002, and mandated a 5% (E5) blending of ethanol with petrol in 9 states and 4 Union Territories with effect from January 2003 (Usmani 2020). In July 2002, Planning Commission of India constitute a committee on development of biofuels. In year 2009, Government of India launched the National Biodiesel Mission and identified Jatropha as most suitable oil plant for biodiesel production and projected 20% blending in conventional diesel by 2017. Jatropha plants are cultivated on unused land and sides of railway tracks in India. The National Biofuel Policy 2018 approved by Government of India set a goal of 20% (E20) blending of ethanol in petrol and 5% blending of bio-diesel in diesel by 2030 (Das 2020).

18.3 Feedstock for Biodiesel

There are different kinds of feedstock have been used for the generation of biofuels. The wide range of resources include cottonseed, soyabean, jatropha, palm oil, jatropa, mahua, waste cooking oils, animal fat, microalgae, etc. The resources are classified into four categories or generations (Knothe et al. 1962) as shown in Table 18.1. First

Table 18.1 Classifications of biofuels (biodiesel) depending on feedstock, their advantages and disadvantages

Generation	Feedstock's	Advantages	Disadvantages
First generation	Edible feedstocks: Oil (Rapeseed, groundnut, canola Corn, olive, rice, coconut, soybean, Palm)	Crop availability Easy conversion process	Food vs feed problem
Second generation	Non-edible feedstocks: Oil (Neem, Jatropha, Nagchampa, Karanja, Calophyllum inophyllum, Rubber seed, Mahua indica,, Jojoba)	Eco-friendly in nature, Less production cost, Less land requirement and can be grown on unimportant land	Plant yield might be low for some feedstock
Third generation	Algae, animal fat, and waste oil	Higher growth rate and productivity, Less land requirement	Investment is high, Requirement of sunlight, expense of oil extraction process from algae, Focused research required
Fourth generation	Photobiological solar fuels and electro-fuels	Feedstock widely available, inexhaustible and cheap	Initial investment high and much focused research required

generation biofuel are produced from edible feedstock's while 2nd generation biofuel are produced from non-food crops and biomass. The "third-generation" biofuels are usually manufactured through microalgal species which contain approximately 70% of lipid (Banerjee et al. 2019; Kang et al. 2014).Genetic modifications are practiced on algal biomasses to produce fourth-generation biofuels to achieve higher yield. Improving light penetration, photosynthetic efficiency and high lipid content are common strategies used in fourth generation bio-fuel or GM of microalgae (Abdullah et al. 2019). These sophisticated biofuels must, fulfil sustainability demands like reducing greenhouse gas emissions with greater proportion, more affordable, and have many environmental advantages compared to conventional fuels (Banerjee et al. 2019; Ziolkowska 2020).

18.4 Biodiesel Production Chain

In a simplified way, the biodiesel production chain can be represented as in Fig. 18.2 including the following steps.

(1) Feedstocks production: Biodiesel is produced by different feedstocks including vegetable oils, animal fats including lard and microalgae.

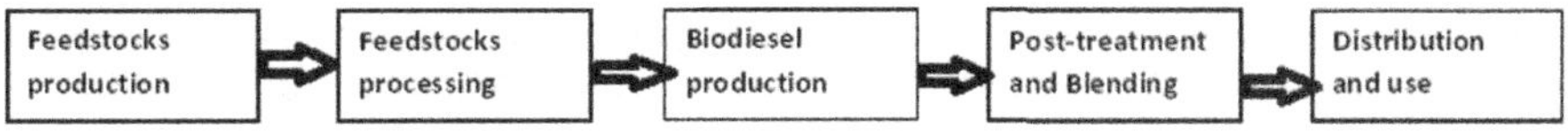

Fig. 18.2 Bio-diesel production steps

(2) Feedstocks processing: Including pre-treatment process which removes impurities from the feedstocks and oil is extracted from feedstocks by using conventional methods like Folch method and Bligh and Dyer method.

(3) Biodiesel production: It is usually made by transesterification of lipids such as vegetable oil or animal fat is treated with alcohol, producing a methyl ester (Biodiesel).

(4) Bio-diesel post-treatment and blending: Bio-diesel is most commonly used as a blend with petroleum diesel. In post treatment of biodiesel the settling process is penultimate step in which biodiesel and glycerine mixture is settled by using a decanter or a centrifuge to separate the biodiesel from glycerine.

(5) Distribution and final use: The distribution of bio-diesel from point of production to fuel terminal or wholesalers by trucks or railcars. It should not be contaminated and trucks or railcars should be washed from previous load to prevent mixing with leftover residuals or water.

18.5 Biodiesel Production from First Generation Feedstock's

The first generation feedstock's for biodiesel production includes oil from edible feedstock's like rapeseed, groundnut, canola, corn, olive, rice, coconut, soybean. The advantages and disadvantages of 1st generation feedstock's are mentioned in Table 18.1. Biodiesel production from few feedstock from 1st generation are discussed below.

18.5.1 Bio-diesel Production Using Palm Oil

The palm oil is appropriate feedstock for biodiesel production and it is commercially available in equatorial regions. Free fatty acids (FFA) present in palm oil esterified with glycerol like any other fat. A large quantity of saturated fatty acids present in palm oil which leads to its solidification at low temperature. Palm oil contains 16-carbon saturated fatty acid which is known as palmitic acid and oil named as palm oil (Mamilla et al. 2012). In transesterification process, NaOH and methanol catalyst are used, and the conversion rate is 92% at 60 °C (Abdullah et al. 2019). Any kind of modification in diesel engine is not required to use palm biodiesel directly. The palm oil can be used directly or mixed with petrol/diesel at any ratio (Zahan and Energies 2018). Environmental sustainability could be taken to a step further by using biodiesel

Table 18.2 Comparison of physico-chemical properties of various feedstocks for biodiesel production (*Source* (Mathew et al. 2021; Lamichhane et al. 2020; Kumar and Sharma 2014))

SI No	Fuel properties	Unit	Microalgae	Jatropha	Palm oil	Waste cooking oil
1	Density	Kg m^{-3}	919	940	918	926
2	Flash point	°C	65–115	242	40	–
3	Cetane no	–	51	38	–	–
4	Acid value	Mg KoH g^{-1}	0.13	28	5.3	2.1
5	Heating value	MJ Kg^{-1}	41	38	–	–
6	Kinematic viscosity at 40 °C	mm^2 S^{-1}	33.06	24.5	40	40.2

and use of biodiesel should be initiated at the national level. Fluidic property of fuel is a determining factor for engine's effective performance. The fluidity of fuel should be helpful in starting the engine which helps in moving its mechanical parts freely. Otherwise, it may deteriorate the engine in long run (Harahap et al. 2019). The major issue in regular usage of palm biodiesel as alternative biodiesel fuel is its cold flow properties (Table 18.2).

18.5.2 Biodiesel Production from Coconut Oil

Coconut (*Cocos nucifera*) oil can be used as a feedstock for biodiesel production due to high oil content and inherent properties like specific energy, high content of saturated fatty acid, cetane number solubility (Singh et al. 2020). Coconut oil as feedstock for biodiesel production is famous in Philippines, second largest producer of coconut worldwide (Jr et al. 2021). Bambase et al. (2021) studied biodiesel production from coconut oil using OH-impregnated CaO as heterogeneous catalyst method and obtained 66.36% biodiesel conversion from coconut oil within 10 min. The authors found addition of cosolvent (tetrahydrofuran) enhanced biodiesel conversion process by 81.70%. Lugo-Méndez et al. (2021) used coconut oil for biodiesel production using homogeneous basic catalysis method with NaOH as catalyst, and also characterise various blends of biodiesel with diesel (Lugo-Méndez et al. 2021).

18.5.3 Biodiesel Production from Soybean Oil

Soybean (Glycine max) oil as feedstock for biodiesel production is used worldwide. Soybean oil is having high content of linoleic acid, oleic acid which make this suitable for biodiesel production (Singh et al. 2020). Colombo et al. (2019) produced biodiesel from soybean oil using methanol via heterogenous catalyst (calcium oxide) in a recycle reactor, and concluded that the process is cheaper and continues (Colombo et al. 2017). Chen and Lin (2017) used soybean oil for biodiesel production using lithium metasilicate catalyst and obtained a yield of more than 95% within an hour (Chen and Lin 2017). The efficiency of simletaneous supercritical transesterification and partial hydrogenation process was studied for conversion of soybean oil into biodiesel with Pd/Al_2O_3 catalyst by Lee et al. (2021).

18.6 2nd Generation Feedstock's

The second generation feedstock's for biodiesel production solves the problem raised by first generation i.e. feed vs fuel. In this generation, non-food feedstock's are used for biodiesel production as mentioned in Table 18.1 along with their advantages and disadvantages. Biodiesel production from few feedstock from 2nd generation are discussed below.

18.6.1 Biodiesel Production from Jatropha Plant

Jatropha plants have high potential and multipurpose attributes. Its' leaves and latex have medicinal uses, and biodiesel can be extracted from its seeds which contains considerable amount of oil. Its' plant helps in eradicating soil erosion as a living fence. Its' easy availability and sustainable property apparently helps in producing biodiesel (Baral et al. 2020). Also, it is a passive source of income for the rural farmers, and proved to be a main source of renewable energy at local and international level. Biodiesel can be extracted from its seeds which prove them as most valuable resource. The oil was extracted from seeds by either mechanical process or chemical process by using solvent like hexane (Kamel et al. 2018). The base catalysed transesterification is best process to make biodiesel from jatropha oil. The final products formed during transesterification were jatropha oil methyl ester and glycerine.

18.6.2 Biodiesel Production from Jojoba

The main producers of jojoba (*Simmondsia Chinensis*) worldwide are Mexico, India, Arizona, California due to its adaptability in harsh environmental conditions, and the oil obtained from its seeds can be used as feedstock for biodiesel production (Singh et al. 2020). Shah et al. (2014) produced biodiesel from tranesterification of jojoba oil using various catalyst and found dibutyltin diacetate gave 92.6% conversion efficiency of oil into biodiesel (Shah et al. 2014). Abdulrahman et al. (2021) synthesised biodiesel form jojoba oil and maximum yield was obtained under conditions of molar ratio of methanol to jojoba 15.99, NaOH concentration 1.5wt% in 15 min of reaction time (Abdulrahman et al. 2021). The effect of nanoparticles like CuO (Rastogi et al. 2021), La_2O_3 (Arunprasad et al. 2021) on performance and emission characteristics of using jojoba biodiesel on diesel engine were also evaluated by many researchers.

18.6.3 Biodiesel Production from Karanja

Karanja (*Millettia pinnata/Pongamia pinnata*) having a seed production capacity of 9–90 kg per plant and grown in South-east Asia, Australia, China, and US (Singh et al. 2020). Much attention is given to this feedstock due to its ability to survive in heat, drought, and salinity conditions (Patel et al. 2017). Alhassan and Kumar (2016) used deep eutectic solvent catalyst for biodiesel production from karanja seed oil in a single step process and obtained 97.53% conversion efficiency (Alhassan and Kumar 2016). Dhingra et al. (2013) studied on improvement and optimization of conditions for biodiesel production from karanja oil using genetic algorithm and response surface methodology and obtained 90% conversion efficiency (Dhingra et al. 2013). Karanja oil methyl ester (KOME) process can be used for biodiesel production from Karanja oil (KO) via transesterification process with an efficiency of 97% (Patel et al. 2017).

18.6.4 Biodiesel Production from Mahua

India is the largest producer of mahua and its oil can be converted into biodiesel using transesterification process. Mahua oil was converted to biodiesel using manganese doped zinc oxide nanocatalyst via transesetrification process and obtained 97% yield (Baskar et al. 2017). Senthil et al. (2016) compared two catalyst KOH and activated red mud for biodiesel production from mahua oil and studied engine performance with blended diesel (Senthil et al. 2016). The authors found 7.5% higher NO_x emission in KOH biodiesel as compared to red mud biodiesel.

18.7 3rd Generation Feedstock's

The third generation feedstock's for biodiesel production includes algae, animal fat and waste cooking oil as mentioned in Table 18.1. Biodiesel production process from microalgae and waste cooking oil is discussed below.

18.7.1 Production of Biodiesel from Waste Cooking Oil

Biodiesel manufacturing plants use cooking oils as their feedstock because cooking oil contains free fatty acids (FFA). When FFA reacted with alcohol and via transesterification becomes biodiesel. Biodiesel production by transesterification reaction can be catalysed with alkali, acidic, or enzymatic catalysts. Used cooking oil has different characteristics depending upon numerous variables, including the source of oil, duration of use, frying temperature, and food ingredients fried in it and the main characteristics of utilised cooking oil includes free fatty acids, viscosity and density (Tangy et al. 2017; Sahar et al. 2018). Density is defined as mass of biodiesel per unit of volume at particular temperature. The density and viscosity of used cooking oil is about 10% and 10 times respectively higher than normal diesel. These properties play a crucial role in combustion; therefore modification is necessary prior to use of used cooking oils in the engine. There are many techniques present to reduce the specific gravity and kinematic viscosity of used cooking oil, which include emulsification, pyrolysis and transesterification. Transesterification is best process because it is carried out at normal conditions, relatively easy and gives the best quality of the converted fuel and high conversion efficiency (Sahar et al. 2018).

18.7.2 Microalgae as a Bio-diesel Producer

Microalgae have anticipated latency for biofuels production compared with the various feedstocks. Microalgae are fast-growing organisms, and having crucial amount of lipids in intervals of their cells. Algal-based bio-fuel does not produce clashes among food and fuel because they don't require fertile land for their cultivation (Ananthi et al. 2021). The utilization of wastewater in their cultivation as a nutrient source leads to management of wastewater. They also assist in CO_2 mitigation biologically, which increases the worth of algal biofuel (Yin et al. 2020). The algal species that have been broadly calculated for bio-diesel production are *Chlorella vulgaris, Nannochloropis sp., Botryococcusbraunii, Dunaliellasalina, Neochlorisoleoabundans, and haematococcuspluvialis* (Singh and Olsen 2011; Mata and Martins 2010). Cultivating algae at commercial scale still, face various technical problems that hamper the commercialization of bio-diesel production using algae as a feedstock. Challenges can be classified into two classes (Lam et al. 2019) (Fig. 18.3).

(1) **Upstream procedure**—In this procedure, the steps included are algal species selection and cultivation methods, supplement sources, energy contribution for working in closed photobioreactor, and reusability of water.
(2) **Downstream procedure**—In this procedure, the steps included are drying and harvesting methods for lipid extraction, transesterification process, and improves biodiesel quality

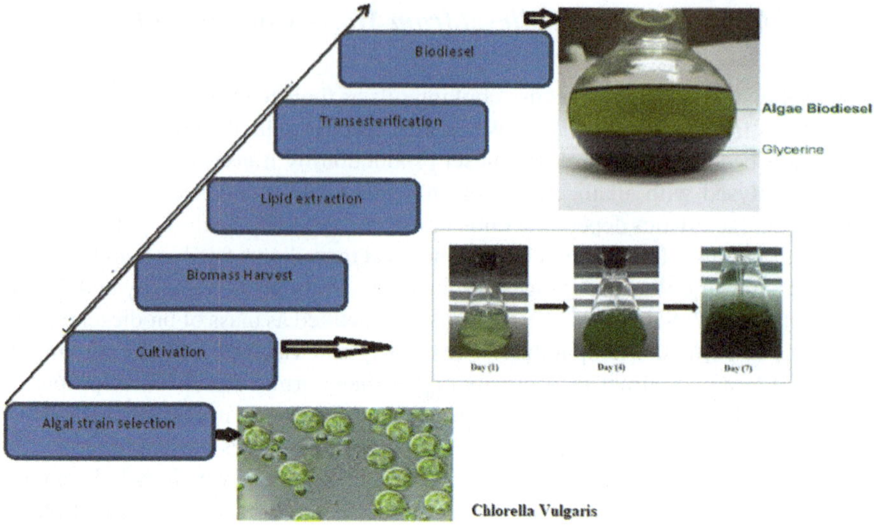

Fig. 18.3 General processes of biodiesel production from microalgae

18.7.2.1 Algae Cultivation System

Microalgae cultivation is one of the major steps for biodiesel production. Based on scientific literature a microalga cultivation system is divided into 3 main categories that are open pond, closed system and hybrid cultivation.

A. **Open pond cultivation system:** Open pond cultivation system has been used from long time for mass cultivation of microalgae because of its simple and easily operating design. It is not effective for the cultivation of algal biomass because it has high impurities level of contaminated by fungus and bacteria, which adversely influenced algal growth.

B. **Closed system for algae cultivation**: Photo bio-reactor is the well-known examples of closed system. It provides a controlled environment to cultivate algae. The closed photobioreactor will maintain axenic cultures with great production yield.

C. **Hybrid cultivation system**: It joined the benefits of both exposed ponds as well as closed photobioreactor (Duran et al. 2021). The utilization of a hybrid cultivation system was discovered by Brennan and Owende (Brennan 2010). A hybrid cultivation system is a two-stage process. In the primary phase to achieve high biomass, by provided controlled development conditions in a closed photobioreactor. In the next stage, permitted environmental stress (nutrients adequate conditions) applied to the algal culture in an open pond that to improve the lipid production in algae. Hybrid cultivation system (integrated cultivation system) is responsible for higher operational and capital expenses in the algal biofuel commercialization than any single bioreactor system (Singh and Olsen 2011) (Table 18.3).

Table 18.3 Different feedstocks and their oil contents and biodiesel productivity

Raw material	Oil content (% in biomass dry weight)	land used		Biodiesel productivity
		Oil yield (L/ha/year)	(m^2/year/kg biodiesel)	(kg bioiesel/ha/year
Sunflower	40	1070	11	946
Soybeans	18	636	18	562
Castor	48	1307	9	1156
Palm oil	36	5366	2	4747
Rapeseed	41	974	12	862
Microalgae	70	1,36,900	0.1	1,21,104

Source (Mata and Martins 2010)

18.7.2.2 Methods for Harvesting and Drying of Algal Biomass

The production of algae biomass is relating to the segregation of algae biomass from water for further biofuel production, the method involves two distinct steps:

(A) In bulk, algae are isolated from bulk suspension during harvesting using flotation, flocculation and gravity sedimentation process.
(B) In the second step thickening, the concentration of microalgae slurries, after a bulk suspension by the utilizing of methods like centrifugation and filtration (Kligerman and Bouwer 2015).

Harvesting algal biomass proved to be a very difficult task because of their small size (1–20 μm) and suspension in water (Rastogi et al. 2018). The ratio between mass of algal biomass and water is observed as remarkably low, despite the fact that the algae are developed in an airtight photobioreactor. During large scale production of algal biomass algal cultivation system (usually closed photobioreactor) requires a normal 73 tons of water when harvesting a single ton biomass of algae. Amount of water used in this process is very significant and therefore evolving effective harvesting techniques especially to support the possibility of commercializing algal biofuels generation (Konur 2021).

18.7.2.3 Lipid Extraction

The major idea or fundamental thought for the cultivation of microalgae is the transformation of algal lipids into biodiesel (Mubarak et al. 2015). Biodiesel is a substitute of regular diesel oil that compromises of numerous benefits to nature and environment; it's eco-accommodating and non-toxic, further has incredible lubricant along with low CO as well as low SO_x emission (Mubarak et al. 2015) (Marella et al. 2019). For the production of biodiesel, firstly, the algal biomass was subjected to drying followed by lipids extraction. Dissimilar to conventional harvests, lipids extraction through algae as biomass is comparatively difficult as the thickness of the cell wall obstructs inter lipid discharge. A mechanical press is nominal for separation of oil from traditional (oil-bearing) harvests and commonly not applicable for algal biomass. The extraction of lipid innovation must demonstrate a higher level of selectivity along with specificity only to algal lipids to stay away from the co-extraction of different mixes, for example, starches, protein, carotenes that can't be changed over to biodiesel (Hena et al. 2015). Chemical solvents like, methanol, blended polar or nonpolar synthetic solvents, n-hexane and ethanol are powerful enough for extracting the lipids from the cells of algae. However, the extraction effectiveness is especially subject to microalgae strains (Lam and Lee 2012). For lowering the energy utilization, extracting lipids through wet algal feedstock would eliminate the energy utilized for algal biomass dewatering procedure. In this way, one of a kind recovery or extraction procedure was developed that utilized the SPS (Switchable polarity solvents to separate lipids from the algal culture) in the presence of CO_2 (Samorì et al. 2013). The cell wall of algae is made of complex carbohydrates and glycoprotein which make cell wall protection from a wide scope of chemicals and high mechanical strength. Extraction procedure, for example, ultrasonication, irradiation, and homogenization, could influence total lipid yield (Naghdi et al. 2016). Nanotechnology is an alternate for lipid extraction from cell mass of microalgae without hurting the algal cell. Lipid extraction from microalgae cells is possible by using nanospheres, which are utilized consistently for the accumulation of lipids, consequently declining re-cultivation cost of algae.

18.8 Biodiesel Production Process

Various suitable catalysts are helpful in production of biodiesel through transesterification which is an established core technology for this purpose by using different feedstock. Biodiesel has higher oxygen content and holds a better combustion quality; however significant industrial production of biodiesels is still a challenging task. The general processes used in biodiesel production are as discussed below.

18.8.1 Pre-treatment

For effective conversion of cellulosic compounds into bioethanol, pre-treatment is required which deal with recalcitrant nature of cellulosic biomass. The cellulosic biomass is composed of complex structure like cellulose, hemicellulose and lignin. By pre-treatment, various physical and chemical changes take place in the complex cell wall structure. The main aim of pre-treatment practice is to split carbohydrate from the lignin compounds (Lamichhane et al. 2020). The various pre-treatment methods are physical, chemical and biological methods. These include concentrated acid, hydrogen peroxide, steam, wet oxidation, organic solvent, CO_2 explosion treatment and etc. are used (Kadir et al. 2018). Feedstock that contain high FFA greater than about 1% (w/w) must be pre-treated to either convert the FFA to esters or remove the FFA before carrying out the biodiesel generation reaction or transesterification (Ananthi et al. 2021). The FFA can't be transformed to biodiesel, forming soap in its place that confines the mass transfer between phases, reduces the reaction rate and its selectivity towards biodiesel, and then complicated in the separation of phases after the biodiesel generation (Yew et al. 2019).

18.8.2 Homogeneous Catalyst

In homogeneous catalyst the catalyst and reactant have same physical state. In biodiesel production the most common homogenous catalysts used is sodium methylate in transesterification and sulphuric during esterification (Syakirah Talha and Sulaiman 2016). During transesterification process, the amount of sulphuric acid added is based on free fatty acid (FFA) content of the feedstocks being processed (Lam et al. 2019).

18.8.3 Heterogeneous Catalyst

In heterogeneous catalyst the catalyst and reactant are in different phases. In liquid or gaseous reaction media mainly solid heterogeneous catalysts acts on the substrate. Alkaline earth metal carbonates ($CaCO_3$), alkaline metals carbonates (Na_2CO_3, K_2), alkaline earth metal oxides (CaO, MgO, SrO, BaO) and other oxides as ZnO are basic heterogeneous catalysts which are usually used (Lima et al. 2016). Heterogeneous catalysts can be utilised in a fixed-bed reactor and shows few noble properties like, an environment-friendly operation which are harmless and inexpensive. Biodiesel production will generate about 10% (w/w) glycerol as the main by-product. The by-product glycerol is neutralization by heterogeneous catalysts and can be extracted out through simple filtration process (Tang et al. 2018) (Fig. 18.4).

18.8.4 Enzymatic Catalysis Process

The enzymatic approach for the conversion of oils and fats gives high purity products and reduce the post processing operations and costs. By using enzymatic approach waste is minimized and by-products are more easily purified. Free lipase, conventionally immobilized lipase which immobilized mainly on nonmagnetic material, and lipase restrained on magnetic nanoparticles can be used as biocatalysts. As enzymes are biocompatible, environmentally accepted and biodegradable,

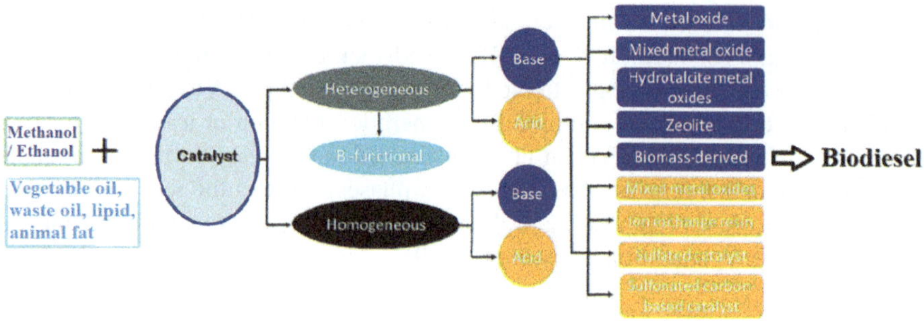

Fig. 18.4 Catalyst types used in biodiesel production (*Source* Changmai et al. 2020)

they are proved superior to homogeneous catalysts (Ullah et al. 2016). In non-aqueous media, Lipases shows remarkable stability and catalytic activity. *Aspergillus niger, Bacillus thermoleovorans, Penicillium cyclopium, Rhizopus thermosus, Pseudomonas fluorescens,* etc. are the major sources of microbial lipases (Thangaraj et al. 2019).

18.8.5 Transesterification

After lipid extraction, they are converted into biodiesel by transesterification, however, the lipids and short-chain alcohols (for example methyl alcohol) requires a catalyst to react (Norjannah et al. 2016). A reversible reaction with alcohol to oil at the reflux temperature of 60–70 °C along with homogeneous base (NaOH/KOH) as a catalyst and two clear layers are developed as a result of gravity and partition coefficient and top layer indicates biodiesel whereas, glycerol at the base layer as a by-product (Fig. 18.5).

Numerous purification steps like washing (water), filtration as well as evaporation are employed to enhance the purity of biodiesel. A recent study, *Chlorella vulgaris* based algal biodiesel yield was obtained at 43 °C (response temperature), along with methanol to oil molar proportion of 14, NaOH (0.42wt%) and reaction time (90 min) (Meher et al. 2006).

Moreover, with the presence of a large amount of FFA in algal lipids (>0.5% w/w), a base catalyst which is homogeneous must not be used for the process of transesterification due to the reaction taking place between FFA and base catalyst to create soap (Meher et al. 2006). This reaction brings about a low yield of biodiesel along with causing an imperative issue in separation and filtration of products. Alternatively, a superior acid catalyst such as H_2SO_4 is not sensitive towards FFA present in oil and subsequently, transesterification along with esterification process will happen at the same time (Go et al. 2016).

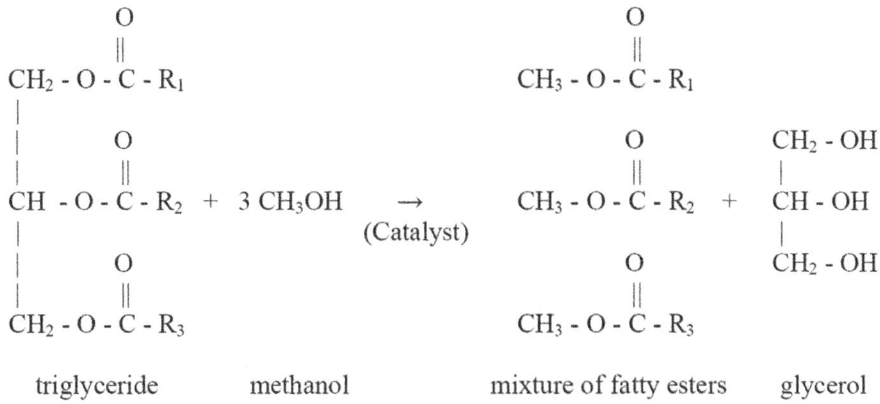

Fig. 18.5 Bio-diesel conversion reaction (*Source* File: Transesterification reaction.png 2021)

18.9 Environmental Impacts of Biodiesel Use

Recently, due to its environmental aid's biodiesel becomes fancier. In the upcoming decades due to its environmental virtues, the segment of bio-fuel in the automobile industry will ascend Merits of biodiesel are:

I. They can be effortlessly obtained from biomass sources.
II. There is a significant combustion of them in carbon dioxide cycle.
III. They are environment friendly.
IV. They hold the benefits for consumer, nature and economy.
V. They are sustainable and biodegradable.

Compared to biodiesel, mainly sulphur is present in petroleum along with other chemical molecules. The burning of gasoline and diesel fuels releases host pollutants and heavy metals that affect local and regional air quality, and these are well linked with global warming issues (Menetrez 2012).

18.10 Economic Feasibility

The most important factor for evaluating the sustainability of biodiesel production is economic performance which plays vital role in industrialization of any process. In biodiesel production the major challenge is its higher production cost at industrial scale (Das 2020). In last decades extensive research has been conducting for concerning the advanced technology and reduces the production cost. Production cost and fixed capital investment are analysed for the total investment required in biodiesel production. The flowsheet process are used to estimate the cost and it is affected by the technology used, construction material and energy balance (Menetrez 2012). The key factors that determine the economic performance of biodiesel plant are production capacity, type of feedstock, and technology used in biodiesel production.

18.11 Conclusion

Biodiesel production from renewable sources has become more attractive in the previous years, extending its possibility from conventional to advance feedstock with less environmental impact and greater energy efficiency. This process revealed numerous key challenges, especially in economic feasibility of biodiesel: feedstock availability, extraction process of lipids; uses of different catalysts and impurities in biodiesel that face economic challenges. Continued help from innovation advancement, policymakers or legislative issues and approval from the community are the steering forces to make an economic feasibility of biodiesel production and demonstrated an innovative drift in the renewable energy sector.

References

Abdullah B, Muhammad S, ZS-... and sustainable energy, 2019 undefined. Fourth generation biofuel: a review on risks and mitigation strategies

Abdulrahman A, Ali A, Alfazazi A (2021) Synthesis and process parameter optimization of biodiesel from Jojoba oil using response surface methodology. Arab J Sci Eng 46(7):6609–6617

Alhassan Y, Kumar N (2016) Single step biodiesel production from Pongamia pinnata (Karanja) seed oil using deep eutectic solvent (DESs) catalysts. Waste Biomass Valorization 7(5):1055–1065

Ananthi V, Raja R, Carvalho IS, Brindhadevi K, Pugazhendhi A, Arun A (2021) A realistic scenario on microalgae based biodiesel production: third generation biofuel. Fuel. 284:118965

Arunprasad J, ... RT-MT, 2021 undefined. Experimental investigation of performance and emission characteristics of jojoba biodiesel with La_2O_3 nanoparticles

Banerjee R, Kumar SPJ, Mehendale N, Sevda S, Garlapati VK (2019) Intervention of microfluidics in biofuel and bioenergy sectors: technological considerations and future prospects. Renew Sustain Energy Rev 1(101):548–558

Bambase Jr, ME, Almazan RAR., Demafelis RB, Sobremisana MJ, Dizon LSH (2021) Biodiesel production from refined coconut oil using hydroxide-impregnated calcium oxide by cosolvent method. Renew Energy, 163, pp. 571–578.

Baral NR, Neupane P, Ale BB, Quiroz-Arita C, Manandhar S, Bradley TH (2020) Stochastic economic and environmental footprints of biodiesel production from Jatropha curcas Linnaeus in the different federal states of Nepal. Renew Sustain Energy Rev 120:109619

Baskar G, Gurugulladevi A, Nishanthini T, Energy RA-R, 2017 undefined. Optimization and kinetics of biodiesel production from Mahua oil using manganese doped zinc oxide nanocatalyst

Brennan L, reviews PO-R and sustainable energy, 2010 undefined. Biofuels from microalgae—a review of technologies for production, processing, and extractions of biofuels and co-products. Elsevier

Changmai B, Vanlalveni C, Ingle A, Advances RB-R, 2020 undefined. Widely used catalysts in biodiesel production: a review. pubs.rsc.org [Internet]. [cited 2021 Sep 26]

Change CS-NC, 2019 undefined. Wasting less electricity before use. https://www.nature.com

Chen Y, Lin D, Chemical BC-J of the TI of, 2017 undefined. Transesterification of acid soybean oil for biodiesel production using lithium metasilicate catalyst prepared from diatomite

Christopher L, Kumar H, Energy VZ-A, undefined (2014) Enzymatic biodiesel: challenges and opportunities

Colombo K, Ender L, Barros AAC (2017) The study of biodiesel production using CaO as a heterogeneous catalytic reaction. Egypt J Pet 26(2):341–349

Colombo K, Santos MM, Ender L and Barros AC (2019). Production of biodiesel from soybean oil and methanol, catalyzed by calcium oxide in a recycle reactor. South African J Chem Eng, 28(1):19–25.

Das S (2020) The national policy of biofuels of India—a perspective. Energy Policy 143:111595

De Oliveira FC, Coelho ST (2017) History, evolution, and environmental impact of biodiesel in Brazil: a review. Renew Sustain Energy Rev 1(75):168–179

de Lima AL, Ronconi CM, Mota CJA (2016) Heterogeneous basic catalysts for biodiesel production. Catal Sci Technol [Internet]. [cited 27 Sep 2021], 6(9):2877–2891

Dhingra S, Bhushan G, Energy KD-F in, 2013 undefined. Development of a combined approach for improvement and optimization of karanja biodiesel using response surface methodology and genetic algorithm 7(4):495–505

Duran SK, Kumar P, Sandhu SS (2021) A review on microalgae strains, cultivation, harvesting, biodiesel conversion and engine implementation 12(1):91–102

File: Transesterification reaction.png—Wikimedia commons [Internet]. [cited 30 Sep 2021]. Available from: https://commons.wikimedia.org/wiki/File:Transesterification_reaction.png

Go AW, Sutanto S, Ong LK, Tran-Nguyen PL, Ismadji S, Ju YH (2016) Developments in in-situ (trans) esterification for biodiesel production: a critical review. Renew Sustain Energy Rev 1(60):284–305

Harahap F, Silveira S, Khatiwada D (2019) Cost competitiveness of palm oil biodiesel production in Indonesia. Energy 1(170):62–72

Hena S, Fatimah S, Tabassum S (2015) Cultivation of algae consortium in a dairy farm wastewater for biodiesel production. Water Resour Ind 1(10):1–14

Jr MB, Almazan R, Energy RD-R, 2021 undefined. Biodiesel production from refined coconut oil using hydroxide-impregnated calcium oxide by cosolvent method

Kadir WNA, Lam MK, Uemura Y, Lim JW, Lee KT (2018) Harvesting and pre-treatment of microalgae cultivated in wastewater for biodiesel production: a review. Energy Convers Manag 1(171):1416–1429

Kamel DA, Farag HA, Amin NK, Zatout AA, Ali RM (2018) Smart utilization of jatropha (Jatropha curcas Linnaeus) seeds for biodiesel production: optimization and mechanism. Ind Crops Prod 1(111):407–413

Kang NK, Lee B, Choi G-G, Moon M, Park MS, Lim J et al (2014) Enhancing lipid productivity of Chlorella vulgaris using oxidative stress by TiO_2 nanoparticles. Korean J Chem Eng 31(5):861–867

Karmakar B, management GH-E conversion and, 2019 undefined. Progress and future of biodiesel synthesis: advancements in oil extraction and conversion technologies

Khanal A, Shah A (2021) Oilseeds to biodiesel and renewable jet fuel: an overview of feedstock production, logistics, and conversion. Biofuels Bioprod Biorefining 15(3):913–30

Kirubakaran M, Arul Mozhi Selvan V (2018) A comprehensive review of low cost biodiesel production from waste chicken fat. Renew Sustain Energy Rev 82:390–401

Kligerman DC, Bouwer EJ (2015) Prospects for biodiesel production from algae-based wastewater treatment in Brazil: a review. Renew Sustain Energy Rev 1(52):1834–1846

Knothe G, Krahl J, Van Gerpen J, Knapp H, Kodali D (1962) The biodiesel handbook second edition library of congress cataloging-in-publication data. I Knothe, Gerhard II Krahl, Jürgen 669–22

Konur O (2021) Algal biomass production for biodiesel production : a review of the research. Biodiesel Fuels Based edible nonedible feed wastes, Algae [Internet]. [cited 27 Sep 2021]: 695–717

Kumar M, Sharma MP (2014) Potential assessment of microalgal oils for biodiesel production: A review. J Mater Environ Sci 5(3):757–766.
Lam MK, Lee KT (2012) Microalgae biofuels: a critical review of issues, problems and the way forward. Biotechnol Adv 30(3):673–690
Lam MK, Khoo CG, Lee KT (2019) Scale-up and commercialization of algal cultivation and biofuels production. Biofuels Algae 1:475–506
Lamichhane G, Khadka S, Adhikari S, Koirala N, Poudyal DP (2020) Biofuel production from waste cooking oils and its physicochemical properties in comparison to petrodiesel. Nepal J Biotechnol 8(3):87–94
Lee D, Choi J, Lee Y, and JL-CER, 2021 undefined. Design and economic analysis of biodiesel production process of simultaneous supercritical transesterification and partial hydrogenation using soybean oil with Pd
Lugo-Méndez H, Fuel MS-D-, 2021 undefined. Synthesis of biodiesel from coconut oil and characterization of its blends
Major biodiesel producing countries 2019|Statista [Internet]. [cited 30 Sep 2021]. Available from: https://www.statista.com/statistics/271472/biodiesel-production-in-selected-countries/
Mamilla V, Mallikarjun M, Res GR-IJC, 2012 undefined. Biodiesel production from palm oil by transesterification method
Marella TK, Datta A, Patil MD, Dixit S, Tiwari A (2019) Biodiesel production through algal cultivation in urban wastewater using algal floway. Bioresour Technol 1(280):222–228
Mata T, Martins A, energy NC-R and sustainable, 2010 undefined. Microalgae for biodiesel production and other applications: a review. Elsevier
Mathew GM, Raina D, Narisetty V, Kumar V, Saran S, Pugazhendi A, et al (2021) Recent advances in biodiesel production: challenges and solutions. Sci Total Environ 794:148751
Meher LC, Vidya Sagar D, Naik SN (2006) Technical aspects of biodiesel production by transesterification—a review. Renew Sustain Energy Rev 10(3):248–268
Menetrez MY (2012) An Overview of Algae Biofuel Production and Potential Environmental Impact. Environ Sci Technol [Internet] [cited 2021 Sep 27], 46(13):7073–85
Mishra VK, Goswami R (2018) A review of production, properties and advantages of biodiesel 9(2):273–89
Mubarak M, Shaija A, Suchithra TV (2015) A review on the extraction of lipid from microalgae for biodiesel production. Algal Res 1(7):117–123
Naghdi FG, González LMG, Chan W, Schenk PM (2016) Progress on lipid extraction from wet algal biomass for biodiesel production. Microb Biotechnol 9(6):718–726
Norjannah B, Chyuan Ong H, Masjuki HH, Juan JC, Chong WT (2016) Enzymatic transesterification for biodiesel production: a comprehensive review. RSC Adv [Internet] [cited 2021 Sep 27], 6(65):60034–60055
Pasha MK, Dai L, Liu D, Guo M, Du W (2021) An overview to process design, simulation and sustainability evaluation of biodiesel production. Biotechnol Biofuels 1:14(1)
Patel R, Reviews CS-R and SE, 2017 undefined. Biodiesel production from Karanja oil and its use in diesel engine: a review
Rastogi RP, Pandey A, Larroche C, Madamwar D (2018) Algal green energy—R&D and technological perspectives for biodiesel production. Renew Sustain Energy Rev 1(82):2946–2969
Rastogi PM, Sharma A, Kumar N (2021) Effect of CuO nanoparticles concentration on the performance and emission characteristics of the diesel engine running on jojoba (Simmondsia Chinensis) biodiesel. Fuel 286:119358
Sahar, Sadaf S, Iqbal J, Ullah I, Bhatti HN, Nouren S, et al (2018) Biodiesel production from waste cooking oil: an efficient technique to convert waste into biodiesel. Sustain Cities Soc 41:220–6
Sajid Z, Khan F, Zhang Y (2016) Process simulation and life cycle analysis of biodiesel production. Renew Energy 1(85):945–952
Samorì C, Barreiro DL, Vet R, Pezzolesi L, Brilman DWF, Galletti P, et al (2013). Effective lipid extraction from algae cultures using switchable solvents. Green Chem [Internet]. [cited 27 Sep 2021], 15(2):353–356

Senthil M, Visagavel K, ... CS-FP, 2016 undefined. Investigations of red mud as a catalyst in Mahua oil biodiesel production and its engine performance

Shah M, Ali S, Tariq M, Khalid N, Ahmad F, Fuel MK-, et al (2014) Catalytic conversion of jojoba oil into biodiesel by organotin catalysts, spectroscopic and chromatographic characterization

Singh D, Sharma D, Soni S, Sharma S, Fuel PS-, 2020 undefined. A review on feedstocks, production processes, and yield for different generations of biodiesel

Singh A, Olsen SI (2011) A critical review of biochemical conversion, sustainability and life cycle assessment of algal biofuels. Appl Energy 88(10):3548–3555

Statista—The statistics portal for market data, market research and market studies [Internet]. [cited 30 Sep 2021]. Available from: https://www.statista.com/

Syakirah Talha N, Sulaiman S (2016) Overview of catalysts in biodiesel production. [cited 27 Sep 2021], 11(1)

Tang ZE, Lim S, Pang YL, Ong HC, Lee KT (2018) Synthesis of biomass as heterogeneous catalyst for application in biodiesel production: state of the art and fundamental review. Renew Sustain Energy Rev 1(92):235–253

Tangy A, Pulidindi IN, Perkas N, Gedanken A (2017) Continuous flow through a microwave oven for the large-scale production of biodiesel from waste cooking oil. Bioresour Technol 1(224):333–341

Thangaraj B, Solomon PR, Muniyandi B, Ranganathan S, Lin L (2019) Catalysis in biodiesel production—a review. Clean Energy [Internet] [cited 2021 Sep 27], 3(1):2–23

Thoai D, Tongurai C, Prasertsit K, Agricultural AK-B and, 2019 undefined. Review on biodiesel production by two-step catalytic conversion

Ullah F, Dong L, Bano A, Peng Q, Huang J (2016) Current advances in catalysis toward sustainable biodiesel production. J Energy Inst 89(2):282–292

Usmani RA (2020) Potential for energy and biofuel from biomass in India. Renew Energy 1(155):921–930

Yew GY, Lee SY, Show PL, Tao Y, Law CL, Nguyen TTC, et al (2019) Recent advances in algae biodiesel production: from upstream cultivation to downstream processing. Bioresour Technol Reports7:100227

Yin Z, Zhu L, Li S, Hu T, Chu R, Mo F, et al (2020) A comprehensive review on cultivation and harvesting of microalgae for biodiesel production: environmental pollution control and future directions. Bioresour Technol 301:122804.

Zahan K, Energies MK-, 2018 undefined. Biodiesel production from palm oil, its by-products, and mill effluent: a review. https://www.mdpi.com

Ziolkowska JR (2020) Biofuels technologies: an overview of feedstocks, processes, and technologies. Biofuels a More Sustain Futur Life Cycle Sustain Assess Multi-Criteria Decis Mak 1:1–19

Chapter 19
Techniques Used in the Process of Biodiesel Production and Its Merits and Demerits from a Historical Perspective

Iram Gul, Shujaul Mulk Khan, Umar Nawaz, Zahoor Ul Haq, Abdullah, Zeeshan Ahmad, and Majid Iqbal

19.1 Introduction

Non-renewable energy resources are depleting at an alarming rate. The modern world cannot remain operational without such energy resources. Therefore, scientists are trying to explore new sources and ways of renewable and non-renewable energy. Bioenergy or green energy is one of such modern-day explorations. Biodiesel can be made from renewable resources and are more acceptable due to environmental advantages. It is the end product obtained from renewable feedstuff like animal fats and vegetable oils. Biofuel is also gaining much and more important because of depleting reservoirs of fossil fuel. Climate change and limited fuel reserves have led to the realization that eco-friendly biofuel should be a substitute. Liquid fuel is expected to drive planes, trucks, and ships in the near time (Alonso et al. 2010). The possible solution is biodiesel. Biofuel production raised to 83 billion liters in 2008, a small share of the fuel consumption market (Florini and Sovacool 2009). The 1st generation biofuel was questioned about climatic changes and the growth of the economy. There was hope for 2nd generation biofuel to generate under commercially feasible situation between 2015 and 2020 (Sims et al. 2008). There is a need for some advanced techniques to produce easily accessible biodiesel.

I. Gul · S. M. Khan (✉) · Z. U. Haq (✉) · Abdullah · Z. Ahmad · M. Iqbal
Department of Plant Sciences, Quaid-i-Azam University Islamabad, Islamabad, Pakistan

Z. U. Haq
Department of Botany, Shaheed Benazir Bhutto University, Wari Campus, Wari, Dir Upper, Pakistan

U. Nawaz
Department of Nanoscience and Technology, NCP, Quaid-i-Azam University Campus, Islamabad, Pakistan

Y. K. Nandabalan et al. (eds.), *Zero Waste Biorefinery*, Energy, Environment, and Sustainability, https://doi.org/10.1007/978-981-16-8682-5_19

19.1.1 Brief History of Biodiesel

Wood was mostly used as fuel for heating and cooking before the nineteenth century. The world >80% of energy requirements are fulfilled by fossils fuel (Ezeonu and Ezeonu 2016). About a hundred years ago, Rudolf used vegetable oil on a trial basis for the engine (Shay 1993). Vegetable oils were used only on an emergency basis from 1929 to 1941. It was used in Europe from 1920 to 1940 in European countries. The energy crises in the 1970 and 1980s compelled people to think about alternative energy sources to cope with power crises. The sunflower was considered an alternative source of energy in 1980 (Knothe 2010). The global demand for energy in 2010 was 5.5×10^{20} J, and it may increase up to 5.5×10^{20} in 2040 (Chen 2011). Human energy is 3–4 times less than bioenergy captured by land plants each year. The potential of bioenergy for the whole world is estimated to be 190×10^{18} J yr^{-1}, making 35% of total energy demand (Guo et al. 2015). There are many sources of biodiesel production as few sources for biodiesel are edible and non-edible sources. The non-edible sources are, *Jatropha* spp, *Raphanus sativus, Melia azedarach, Salvadora oleoides, Nicotiana tabacum, Simmondsia chinensis, Carthamus oxyacantha* (Mariod et al. 2006), *Madhuca indica, Azadirachta indica, Eruca sativa,* Rubber seed, *Melia azedarach, Linum usitatissimum, Salvadora oleoides, Raphanus sativus, Simmondsia chinensis, Carthamus oxyacantha, Nicotiana tabacum, Cannabis sativa, Pistacia chinensis, Thevetia peruviana, Raphanus* spp, *Xanthium sibiricum* (Azad et al. 2014) and *Pongamia pinnata*—(Kibazohi and Sangwan 2011). The edible plant examples are soya bean (Singh and Singh 2010), palm (Hill 2007) and rapeseed (Pramanik 2003), etc. Organic products are used to solve the problem of oil prices and environmental problems. Global warming can only be reduced by using environment-friendly fuel like biodiesel (Kildiran et al. 1996). The benefit of biodiesel is that it minimizes pollution and eliminates all cancer-causing materials, reduces the emission of CO_2, unburned hydrocarbon, particulate matter, and sulfates (Krawczyk 1996).

19.1.2 Techniques Used for Biodiesel Production

Numbers of techniques are currently used for the formation of biodiesel. It comprehended the modification of materials with the help of catalyst accompanied by heat or mixing are pyrolysis and thermal cracking (Sonntag 1979). The large biochemical bonds are broken down into small molecules with the help of pyrolysis (Weisz et al. 1979), give heat without oxygen or air. Direct use and blending include the mixtures of different fuels, i.e., 20% oil and 80% diesel fuel. The oil obtained from the vegetable is used as a fuel and there was substantial conversation start in 1980. The idea that food plants are used as fuel was presented by Bartholomew (1981). Microemulsions is a technique that has been intended to solve the high viscosity difficulties (Peterson et al. 1983) (Strayer et al. 1983). All those difficulties occur in different oils obtained from vegetables with mixing different solvents i.e., ethanol, 1-butanol, and methanol. Microemulsions were explained as a colloid stability dispersal of optic isotropic liquid plus microstructure within proportion usually formed from 2 to149 nm impulsively two ordinarily liquids and many ionic or non-ionic solvents (Schwab et al. 1988). Transesterification is the technique in which the triglyceride is changed to an ester and glycerine by adding alcohol, also known as alcoholysis (Srivastava and Prasad 2000). The biox co-solvent method is used for inert co-solvents selection in which triglycerides are converted to esters that produce a rich system of oil at one stage (Boocock et al. 1996). A distinctive fluid plus steam stage no longer occur, under such condition (Gerpen et al. 2004). Production of biodiesel from oil through in situ process is a new methodology. This chapter is all about different techniques and their trends decade-wise.

19.1.3 Advancement in Catalysts for Biodiesel Production

Biodiesel production is a multistep process that requires catalysts for the reaction to complete. Catalysts increase the conversion rate of biodiesel production (Hossain and Mazen 2010). Few catalysts for biodiesel production are biocatalyst, photocatalyst, homogeneous and heterogeneous catalysts. Various catalysts studied for biodiesel production are acid, base, or biocatalysts (Hossain and Mazen 2010). Liquid catalysts show high activity among homogeneous reactions. The biodiesel production from the solution is too costly that's why various catalysts used to reduce the cost and enhance the quality are discussed below.

19.1.3.1 Homogeneous Catalysts

Homogeneous acid catalysis is used on a priority basis where the yield of free fatty acid is in ample amount and that's why the same catalyst is used (Ullah et al. 2016; Goff 2017). Few examples of homogeneous catalysts are BF_3, HCl, H_2SO_4, $FeSO_4$, H_3PO_4 (Ullah et al. 2016). Factors that affect the catalytic process are catalyst nature, catalyst amount, reaction time, and temperature for biodiesel production through transesterification reaction (Bhatti et al. 2008). After the esterification process, the extracted fatty acids from the palm had a high amount of FFA (free fatty acids), which reduced from 93 weight % to less than two weight % (Oh et al. 2012). The catalyzed reaction speeds the reaction and eliminates the reaction's intermediate steps (Lotero et al. 2005). The common reaction and corrosive nature compel the researchers to use Brønsted acids instead of metal Lewis acids. Homogeneous base catalysts demand is increasing because it is faster, less corrosive, and reliable compared to homogeneous acid catalysts (Lin et al. 2011). Common homogeneous catalysts are potassium hydroxide, sodium hydroxide, potassium carbonate, and barium hydroxide (Arzamendi et al. 2008). Renowned scholars (Helwani et al. 2009) prefer the use of homogeneous base catalysts to extract biodiesel. Due to their effectiveness and efficiency in converting fatty acid into alkyl esters compared to homogeneous acid catalysts. However, biodiesel production may be affected after the production of soap as free fatty acids react with catalysts or as the production of water molecules while NaOH or KOH reacts to form methoxide. To resolve this issue, scientists introduce two steps reaction (Lotero et al. 2005). The first step acid catalyst adds to triglycerides to form methyl esters of free fatty acids in vegetable oil while the same base reaction has to proceed with same base catalysts. To reduce cast we have to remove the homogeneous acid catalyst before base-catalyzed transesterification (Ullah et al. 2016).

19.1.3.2 Biocatalysts

Biocatalysts enzymes lipase reduce the shortcomings in base catalysts and homogeneous acids and make biodiesel more sustainable and economical (Abbaszaadeh et al. 2012). The alcoholysis procedure is complete in two steps wherein step 1 ester bond hydrolyzed to release alcohol moiety and in step 2 esterification of the second substrate take place (Luna et al. 2016). The cast and deactivation by impurities is a hurdle in the commercialization. But recovery to utilize it will further ensure its sustainability (Ullah et al. 2016).

19.1.3.3 Heterogeneous Catalysts

Heterogeneous catalysts are usually solids and can easily be separated. Fewer waste production properties compelled scientists to use classical heterogeneous solid catalysts than homogeneous catalysts. The few demerits of heterogeneous catalysts are that the production of yield methyl ester is less than homogeneous reaction and deactivation due to leaching, cooking, and poisoning (Marchetti et al. 2007). That's why heterogeneous catalysts are recommended due to their activity at low temperatures and pressure during the reaction (Ullah et al. 2016). Solid acid catalysts are more recommendable in heterogeneous catalysis due to non-saponification during transesterification (Lopez et al. 2005), but temperature exceeds 170°C in yielding biodiesel from vegetable oil. Heterogeneous solid base catalysts activity is the same as homogeneous basic catalysts in various biodiesel production reactions (Kim et al. 2004). It is recommended for vegetable oils with low free fatty acids content. Commonly used heterogeneous solid base catalysts are hydrotalcite, basic zeolites and many other compounds with high basic properties (Encinar et al. 2005). One of the limitations in heterogeneous solid catalysts is the need for large surface areas and pores for active species (Semwal et al. 2011). In a nutshell, heterogeneous catalysts on a commercial basis are economical and eco-friendly compared to homogeneous catalysts (Ullah et al. 2016). Further studies on heterogeneous catalysts will reduce costs and make the biodiesel industry more eco-friendly and sustainable in the long run (Table 19.1).

Table 19.1 Various types of catalysts and their merits and demerits

S. No.	Types of catalysts	Examples	Merits	Demerits	References
1	Homogeneous acid	H_2SO_4, HF, HCL, H_3PO_4	Insensitive to water and free fatty acids in oil catalyzed synchronized transesterification and esterification reactions, dodge soap formation	Reaction rate low and long reaction time, cause equipment corrosion, high reaction pressure and temperature, high demand of alcohol, non-recyclable catalysts	Mondala et al. (2009), Fattah (2020)
2	Homogeneous base	NaOH, KOH etc.	Faster reaction, high catalytic activity, favorable kinetics, low cost	Highly sensitive to free fatty acids and water, Soap formation, Saponification occurs as a side reaction, High volume of wastewater, catalysts non-recyclable, Equipment corrosion	(Mondala et al. (2009), Fattah (2020)
3	Heterogeneous acid	ZnO, MoO, ZrO, TiO, Zeolites etc.	Insensitive to water and free fatty acids in oil catalyzed synchronized transesterification and esterification reactions, dodge soap formation, non-corrosive, recyclable	Long time reactions, slow rate of reaction, weak catalytic potential, leaching	Ali et al. (2017), Macario et al. (2010)
4	Heterogenous base	MgO, CaO, SrO, hydrotalcite and mixed oxides	Environmentally friendly, non-corrosive, recyclable, longer catalysts life, recyclable	Highly sensitive to water, slow reaction rate, saponification occurs as a side reaction, soap formation, high cost of catalysts synthesis	Ali et al. (2017)

(continued)

Table 19.1 (continued)

S. No.	Types of catalysts	Examples	Merits	Demerits	References
5	Biocatalysts	Lipases	Required low temperature, Methy ester formation, easy recovery of glycerol	Low rate of the reaction, expensive	Ali et al. (2017)
6	Nanocatalysts	$Ni_{0.5}Zn_{0.5}Fe_2O_4$, Cao, MgO, Fe_3O_4, KF/Al_2O_3, $Ca/Al/Fe_3O_4$	High Surface, highly stable, resistive to saponification, reusable	Require more severe conditions, product recovery after neutralization and generation of residue	Ingle et al. (2020), Mondala et al. (2009)

19.1.3.4 Photocatalysts

The photoreaction in the presence of a catalyst is photocatalysis divided into homogeneous photocatalysis and heterogeneous photocatalysis (Wu and Chang 2006). In homogeneous photocatalysis, the catalysts and reactants exist in the same phase. The photo- Fenton system and ozone are common photocatalysts. The reactive species is •OH ion which is used for various purposes. The •OH production by ozone follows two steps.

$$H_2O_2 + h\nu \rightarrow \bullet OH + \bullet OH$$

The Fenton system produces hydroxyl radicals by the below process (Peternel et al. 2007).

$$Fe^{2+} + HO\bullet \rightarrow Fe^{3+} + OH^-$$

This method utilizes sunlight having 450 nm wavelength, which avoids costly methods of energy. The disadvantage of this procedure is the low pH values needed during the reaction (Peternel et al. 2007). In heterogeneous photocatalysis, the reactants and catalysts don't exist in the same phase. It includes a variety of reactions. The most common heterogeneous photocatalysts are transition metal oxide and semiconductor. The energy needed for the reaction should be in the bandgap. The main disadvantage in heterogeneous photocatalysis is, increase or decrease in the energy level affects the function of photocatalysts. The energy level or bandgap is normalized by various techniques i.e., phase hetero-junctions noble-metal nanoparticles and substitutional cation doping. The oxidative reaction to produce the ultimate •OH is stated in the given reaction (Daneshvar et al. 2004) recently (Guo et al. 2021) prepared a reusable and highly active photocatalyst by using the sol–gel method. In this study, a highly active and reusable La^{3+}/ZnO-TiO_2 photocatalyst was prepared by Sol–Gel method. They introduced a two-step method to produce biodiesel from waste cooking oil which is important progress in biodiesel-related techniques. In the following section, we have given few photocatalytic Oxidative reactions.

$$UV + MO \rightarrow MO\ (h + e^-)$$

Here MO stands for metal oxide.

$$H_2O_2 \rightarrow 2 \bullet OH$$

The reductive reaction due to photocatalytic effect (Linsebigler et al. 1995).

$$HOOH \rightarrow HO \bullet + \bullet OH$$

The literature related to biodiesel and its techniques is scattered. People who relate to this field face difficulties in collecting relevant data required to them. This attempt will compile the available information in a single sheet.

19.1.3.5 Nanocatalysts

Scientists are trying to bring progressions in catalytical chemistry. In the current era, several breakthroughs have been done in nanocatalyst technology. According to Ingle et al. (2020) there is a symmetrical correlation between cost-competitive biodiesel production and nanocatalyst technology. Nanocatalysts have exceptional properties such as high activation energy, easy recovery, controlled rate of reaction, selective reactivity and recyclability. MgO, CaO, Fe_3O_4, KF/Al_2O_3, Ca/Al/Fe_3O_4, SnO_2, etc. are important metal oxides used in the production of biodiesel (Zuliani et al. 2017). On the other hand, zirconia, zeolites, hydrotalcite, sulfated oxides are the important nanocatalysts used for biodiesel production (Mallesham et al. 2014) (Fig. 19.1).

19.1.4 Data Collection

The present chapter was developed from the literature available on two online bibliographical databases i.e., Google Scholar and Science Direct Navigator. The survey was conducted for 53 years, from 1971 to 2021. There were more than 174 journals that mentioned seven techniques published i.e. thermal cracking (pyrolysis), direct use and blending, microemulsions and transesterification, Biox co-solvent process, In in situ biodiesel process, Supercritical alcohol process either written in the title, abstract or keywords. Forty-seven journals have repeated those techniques more than a hundred times. Nineteen journals have mentioned those techniques more than fifty

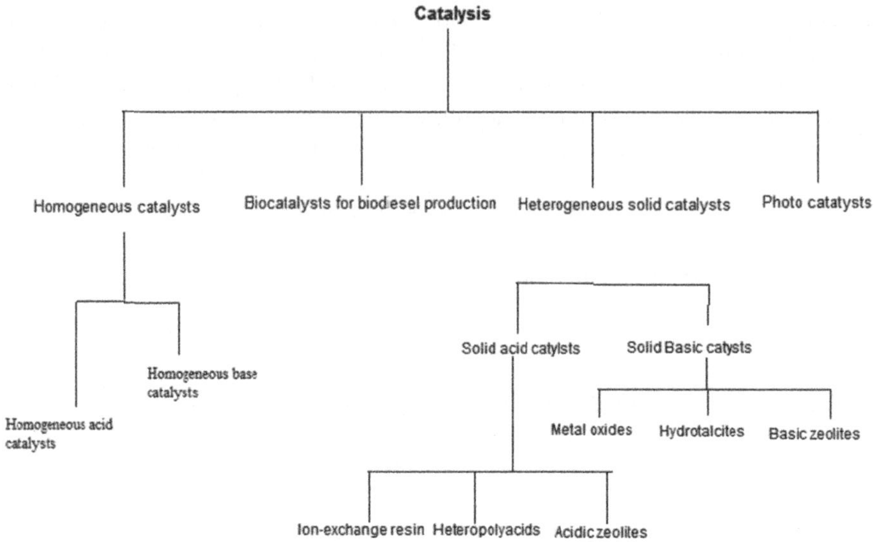

Fig. 19.1 Classification of catalysts used in biodiesel production

times. The remaining one hundred and eight journals have repeated those techniques in less than fifty times. Data were organized and estimation of techniques used maximum time in the last 53 years were recorded.

19.1.5 Method of Preparation of Biodiesel

The four main techniques used for the production of biodiesel are: thermal cracking (pyrolysis), direct use and blending, transesterification, and microemulsions. Some other processes are also used for biodiesel production but at a low scale, such as Biox co-solvent process, In situ biodiesel process, Supercritical alcohol process. The transesterification technique is commonly used for animal fats and vegetable oils to convert it into biodiesel as this technique is well-established and becoming increasingly important (Refaat 2010). The trend of use of techniques decade wise: use of pyrolysis increase (1971–2021) from 4.7 to 36.5%, direct use and blending from 0.8 to 57.2%, microemulsion from 2.3 to 23.5%, transesterification from 0.9 to 51.1 Biox co-solvent are not in practice in situ-biodiesel process from 0 to 94.2%, Supercritical process from 0 to 53.2%. The data from the Elsevier group of the journal shows that in situ-biodiesel technique is more used (94.2%) after that direct use and blending (57.2%), supercritical process (53.2%), transesterification (51.1%), pyrolysis (36.4%), microemulsion (23.4%) and bio-co solvent is least in practice for the production of biodiesel (Fig. 19.2).

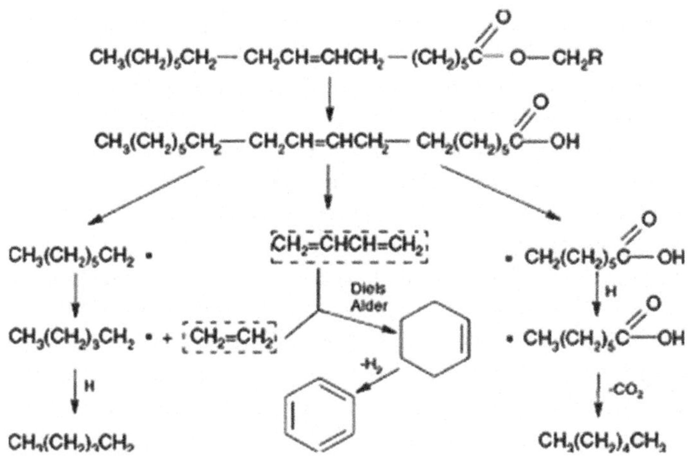

Fig. 19.2 Thermal decomposition of triglycerides mechanism (Abbaszaadeh et al. 2012)

19.1.5.1 Thermal Cracking (Pyrolysis)

The word pyrolysis means the breakdown of some things with the help of heat (Sonntag 1979). The large biochemical bonds are broken down into small molecules with the help of pyrolysis (Weisz et al. 1979). Pyrolytic reaction contains the variation of reaction and many products of this reaction. The chemistry of pyrolytic reaction is tough to describe. The pyrolysis material includes natural fatty acids, animal fat, methyl esters of fatty acids and natural fatty acids. The pyrolysis method for fats was explored in the area where there was no deposition of petroleum for more than 100 years (Sonntag 1979). The first pyrolysis material was vegetable oil which is used to produce petroleum from these oils. Since world war-1, vegetable oils pyrolysis has been studied by several detectives to obtain yields proper for fuel. Tung oil calcium soap thermal cracking was stated in 1947 at a large scale. (Chang and Wan 1947) The Tung oil was primarily saponified with lime after being thermally broken, then polished to produce diesel fuel and minor amounts of kerosene and gasoline. The soap obtained from the process of saponification of Tung oil was 68kgs and yields 50 L of crude oil. (Crossley et al. 1962) noticed the result of temperature on the variety of yields achieved from heated glycerides. In numerous studies, catalysts have been used especially metallic salts to get olefins and paraffin similar to those existing in petroleum sources.

In a standard ASTM distillation tool, the decomposition of soybean oil was done thermally, and the distillation was done in nitrogen and air sparged (Schwab et al. 1988). According to Schwab et al. (1988), safflower oil is used as an elevation to control oleic oil. The recognized hydrocarbons were 80–88% and 73–77% from the extraction of high oleic safflower oils and soybean, respectively. The oils obtained from vegetable produce biofuel have been studied in catalytic cracking (Pioch et al. 1993). The lower molecular weights products such as gases, solids and liquids, the oil of palm and the oil of copra stearin are break on the normal catalyst of petroleum SiO_2/Al_2O_3 in 450 °C. The yield of biogasoline plus biodiesel fuels in the shortened carbon-based stage was fractionated. The chemical configurations of these biodiesel portions were just like the fuels of fossils. The apparatus used for pyrolysis and thermal cracking is very costly nowadays. The yield obtained from pyrolysis is chemically just like diesel fuel and petroleum-derived gasoline. The oxygen is eliminated through the process of pyrolysis and eliminates any ecological welfare using an oxygenated fuel. The process formed more or less material with low value, and gasoline is more than the biodiesel fuel. The Pyrolysis technique trend year wise is as follow; from 1971 to 1980 the percentage is 4.7, from 1981 to1990 is 12.61, from 1991 to 2000 is 20.48, from 2001 to 2011 is 25.82 from 2011 to 2021 is 36.47. The main advantage of pyrolysis is that the fuel standard chemically is similar to gasoline and petroleum (Singh and Singh 2010), The pyrolysis method effectively densifies biomass (Butler et al. 2011). Moreover, the disadvantages are energy-intensive and costly (Alonso et al. 2010).

19.1.5.2 Direct Use and Blending

This method includes the mixtures of different fuels such as 20% oil and 80% diesel fuel. The oil obtained from vegetables is used as a fuel and there was substantial conversation start in 1980. The idea that food plants are used as fuel was presented by Bartholomew (1981). According to him, oil is obtained from vegetables and alcohol is the alternative fuel to that of petroleum and also, the energy that is renewable needs initiate to takings the place of the non-renewable means. In South Africa, because of the oil restriction, the most progressive work with sunflower oil arose. The blend of 10% oil obtained from vegetables to withstand the total power within chamber pre-combustion engine, the engine in 1980 by Brazil. The oil obtained from the vegetable is not 100% useful for diesel fuel. However, 20% and 80% diesel fuel extracted from vegetables were fruitful. But in some cases, the experiment was carried out in the ratio of 50/50. The worldwide conference in North Dakota on oils of vegetable and plants as diesel in August 1982 perceived. The principal concerns conferred were the fuel preparation, specifications and additives, durability, fuel costs and the vegetable oils fuels effect on the engine's performance. In this meeting, different techniques used for the production of oils, treating the oilseeds and oil taking out from seeds were also well-thought-out (ASAE 1982). They used frying oil to make diesel and then ran through sieved to clean it (Anon 1982). A mixture of diesel fuel which is 5% and used cooking oil which is 95%, were used and recycled cooking oil. For instance, necessary to recompense for cooler air temperature, mixing or preheating was used. The problem of carbon accumulation and other element presence has not occurred. The mere problem testified was lubricating oil impurity and the key was recommended to the (polyunsaturated vegetable oils polymerization increase the viscosity). The greasing oil must be transformed to 4000–4500 miles.

The benefits of the oils obtained from vegetables and used as a fuel are the content of heat, which is diesel fuel about 80%, liquid nature-portability, renewability, and prepared availability. Furthermore, some of its drawbacks are lower volatility, higher viscosity and the unsaturated hydrocarbon chains reactivity (Pryde 1983). The oil is obtained from vegetables and put into the engine and the engine operates well for a longer time. The difficulties seem only after that, mainly in those engines in which oil is directly injected. The main difficulties contain other elements accumulation and sound off nearby creation taking place the injectors to a range that energy reduce a liquid to a very fine mist does not happen accurately or as uniform prohibited as a consequence of preserved outlets. Accumulation of carbon, oil disc penetrating and clotting plus crystalizing of the greasing oil due to impurity by the oil obtained from vegetables. The 1:2 and 1:1 of soya bean and diesel fuel were experienced for engine action and grease crankcase viscosity for a total of 600 h in a direct-injection, 6.6 L displacement, John Deere 6-cylinder and turbocharged engine (Adams et al. 1983). The combination with 1:1 greasing oil also possible crystallizing occurred, but with the combination of 1:2, it did not occur. Throughout times of diesel fuel lacks before division's consequences showed that combination of 1:2 would be proper as energy for cultivated tools. Two severe obstacles i.e., half burning and oil deterioration, were related when vegetable oil was used as a fuels source (Peterson et al. 1983). Through storing polymerization and glue production process by oxidation or at the pressure and elevated temperature of ignition through thermal polymerization and tough oxidative the polyunsaturated fatty acids were very liable, because of incombustible gum the greasing fat solidifying and also resultant pledges of carbon. For the abundant production, the rapeseed of winter variety contains 45% oil content and 46.7% high content of the erucic acid used as diesel fuel has been studied (Peterson et al. 1983). The linoleic acid content of about 75–85% was five times higher than the gum formation rate in winter rapeseed oil. The combination either was 50/50 or 70/30 of the diesel and oil obtained from winter rapeseed. The viscosities of the oil obtained from winter rapeseed were 6–18 times much higher than No. 2 diesel. The oil obtained from winter rapeseed plus No. 1 diesel in the combination of 70/30 was used fruitfully to influence the engine run by diesel and it was single-cylinder for 850 h.

No special effect on the yield and greasing oil were renowned. Compared to vegetable oil, the oil obtained from canola is greatly stickier and similar to all other fluids; its viscosity depends on temperature. The viscosity of the oil obtained from canola at 10 °C was 100 CST (centistokes); a mixture such as 75/25 of oil obtained from canola and diesel is forty (40) CST; and if mixture fifty out of fifty is nineteen CST; and 4 CST was fuel density (Strayer et al. 1983). At a similar pressure and at −4 °C the flow rate of diesel is greater than the flow rate of canola. When pure methanol or ethanol was added, they dropped their viscosity. On mixing 10% methanol or ethanol with oil obtained from canola, its viscosity was 21.15 CST, at 37 °C, whereas 37.82 CST was straight canola oil. For testing, various oils were obtained from sunflower and cottonseed. This oil was in the form of a crude, dewaxed-degummed mixture and degummed using a pre-combustion chamber engine in the form of a single-cylinder (Engler et al. 1983). However, negative consequences were shown. The oils modified through various processes were to point well than the oils which are crude nearly and these were not suitable for use as substitute fuels, although, for the time being, they achieved acceptably. Due to the accumulation of carbon plus greasing oil polluting the oils were not proper. Seventy-five parts of diesel and 25 parts of sunflower oil were mixed as a fuel used for transport purposes (Ziejewski et al. 1984). At 40 °C, density is 4.88 CST, whereas above 40 °C extreme definite stage of ASTM is 4.0 CST. However, it was well-thought-out that these were not proper for use in engines directly. A mixture of oil was obtained from safflower and diesel (25/75) at 40 °C. The viscosity was 4.92 CST. The combination of 50/50 Stoddard solvent and oil obtained from soybean and union oil (52% naphthenes and 48% paraffin's) at 38 °C had 5.12 cSt viscosity (Goering and Fry 1984). The EMA (Engine Manufactures' Association) 200 h test passed with the combination of soybean oil plus safflower.

The crude oil obtained from soybean form plus ethyl ester destroy the engine if used directly (Pryor et al. 1983). The use of this crude oil decreases thermal efficiency and also causes carbon accumulation. Various experiments were conducted longer using different fuel mixtures, such as crude oil obtained from soybean is 75% and diesel fuel is 25% (Schlautman et al. 1986). The engine's fuel mixture was burned for 159 h because the engine's constant load would not be apprehended. Rise in the greasing oil viscosity up to 670%, the test let-down occurred into the screening test after 90 h. It assessed that 2.59 L crude oil into the engine and checked their performance. The engine comprises 3 cylinders and 2600 chains shallow diesel working automatically on crude oil obtained from soybean and oil obtained from sunflower and these were mixed in the ratio 25:75 v/v basis with fuel such as No. 2 diesel (Schlick et al. 1988). Throughout 200 h of the process, the power keeps on constant. The excessive deposition of carbon due to different crude oil on burning parts of the chamber, the different fuel mixtures were excluded from usage only used at definite EMA working situations. Either indirect or direct engines run by diesel, the consumption of fuel extracted from vegetables or mixture of the fuel has commonly been well-thought-out to be unworkable and unsuitable. The problems are oxidation during combustion, acid deposition, clotting of lubricant oil, carbon deposition and high viscosity. The direct use and blending trend year wise is as follow; from 1971 to 1980 the percentage is 0.73, from 1981 to 1990 is 2.64, from 1991 to 2000 is 60.02, from 2001 to 2011 is 33.44 from 2011 to 2021 is 57.19.

The direct use and blending process main advantages are the content of heat (diesel fuel is 80%), Liquid nature-portability, freely obtainable, renewability (Kaya et al. 2009) and the disadvantages are higher viscosity as compared to diesel fuel 11–17 times higher.

19.1.5.3 Microemulsions

The microemulsion is a technique that has been intended to solve the high viscosity difficulties. All those difficulties occur in different oils obtained from vegetables by mixing different solvents, including ethanol, methanol, and 1-butanol. The microemulsions have been well-explained as colloid stability dispersal of isotropic optic liquid plus microstructure within proportion usually from 1 to 150 nm impulsively 2 ordinarily liquids un-mixable and many ionic or non-ionic solvents (Ziejewski et al. 1984).

The volatile evaporation of the ingredients having low boiling in the spray features can recover (Pryde 1983). The microemulsions of oil obtained from soybean in aqueous ethanol were closely as good as that fuel such as No. 2 diesel, in both non-ionic and ionic solvents in the meanness of the energy content and the centane number (Goering 1985). Their permanence was not secure. (Ziejewski et al. 1984) prepared winterized sunflower oil and a mixture of 1-butanol, which was 33.4% (vol.), 190-proof ethanol, which was 13.3% (vol.) and alkali-refined, which was 33.4% (vol.). This non-ionic emulsion has less than 0.01% ash content, at 40 °C 6.31 CST viscosity and 25 were of centane number. With an increase of 1-butanol, the viscosity was lesser and detected patterns of well spray. Transmission durability test in a 200 h test center, definitely not weighty depreciations in the presentation were detected, but deposition of carbon occur heavily, incomplete combustion, rough injector pointer stick and viscosity of the greasing oil were increase and stated. The fuel such as Shipp non-ionic contained, 190-proof ethanol was 5%, No. 2 diesel fuel was 50%, 1-butanol was 20% and oil obtained from soybean was 25% in the form of alkali-refined and degummed was estimated in EMA transmission experiment in the 200 h (Goering and Fry 1984). On the tips of the injector due to deposition of varnish and carbon, the cylinder liners in-take valves even though in EMA test the fuel passed 200 h but these were the main difficulties. A mixture of 25% oil obtained from sunflower seeds in diesel oil performed poor than SNI fuel. In microemulsions, the oil obtained from sunflower which was 53% and the mixture of oil obtained from sunflower seeds in diesel which was 25% the engine presentation were the same (Ziejewski et al. 1984). A microemulsion set through mixing oil obtained from soybean, methanol, 2-octanol and centane reformer at the proportion of 52.7:13.3:33.3:1.0 to approved EMA test 200 h (Goering and Fry 1984). In the microemulsions technique octanol, hexanol and butanol were used to lower down the density in No. 2 fuel. In triolein and Glycine max oil, an operative amphiphile was 2-octanol in methanol micellar solubilization. Frequently used solvent was methanol because it has cost-effective benefits as compared to ethanol. The microemulsion trend year wise is as follow; from 1971 to 1980 the percentage is 2.43, from 1981 to 1990 is 12.46, from 1991 to 2000 is 20.98, from 2001 to 2011 is 40.74 from 2011 to 2021 is 32.41. The main advantages of microemulsions process are patterns of spray is better in combustion, viscosities of lower fuel and the disadvantages of this technique are mainly lower centane number, Lower energy content (Kaya et al. 2009), and the process of Microemulsions containing three main components phase of an oil, a surfactant and phase of an aqueous and this process are pure, steady isotropic liquids. The main problem of using this in engines cause, sticking of injector needle irregularly, combustion is partial, deposition of carbon heavily and viscosity of greasing oil increased (Zubr 1997).

19.1.5.4 Transesterification

Transesterification is a technique in which the triglyceride is converted into ester and glycerine in the presence of alcohol (Srivastava and Prasad 2000). The reaction in transesterification occurs in the addition of catalyst and is complete in 3 steps and is also reversible. To decrease triglycerides viscosity, which is high, the transesterification procedure has been extensively used. The reaction for transesterification is given in Fig. 19.3.

In transesterification reaction, different alcohol is used but mainly used alcohol is methanol and then this reaction is called methanolysis. The reaction of transesterification is reversible if the reactant is adding continuously. To speed up the conversion reaction, the catalyst occurrence is necessary, either base or strong acid.

Glycerol and fatty acids alkyls esters are the product of triglycerides obtain through the transesterification process. At the base of the container, the glycerol layer settles down. The intermediate product was the monoglycerides as well as diglycerides in this method. The reaction occurs step by step in the presence of alcohol and these reactions are reversible. The first-order reaction is a forward reaction and the second-order reaction is the reversible reaction in being there of additional alcohol. When catalyzed by alkali it was also detected that trans esterification is faster (Freedman et al. 1986). Tetrahedral intermediate formation consequences when the first step consists of the triglyceride molecule that contains carbonyl carbon and the alkoxide ion attack on these. In the second step, the alkoxide ion produces in the reaction of tetrahedral in-between in the presence of an alcohol (Fig. 19.4). An

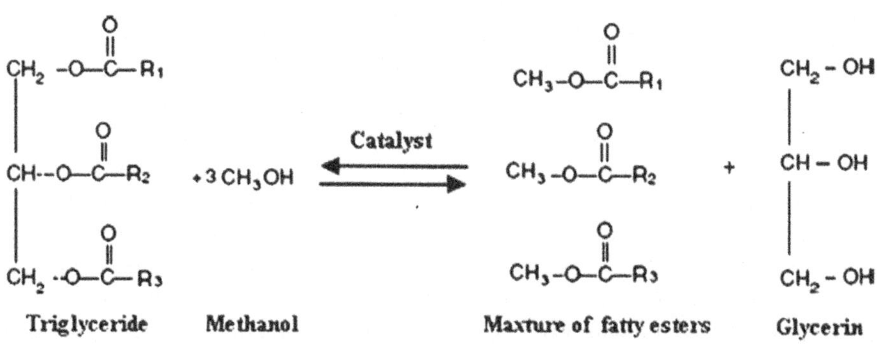

Fig. 19.3 Transesterification of triglycerides with methanol (Srivastava and Prasad 2000)

$$\mathrm{RCOOR^1} + \mathrm{R^2OH} \xrightleftharpoons{\text{Catalyst}} \mathrm{RCOOR^2} + \mathrm{R^1OH}$$

Ester Alcohol Ester Alcohol

Fig. 19.4 Transesterification general equation

ester and diglyceride formation occur in the last step during tetrahedral intermediate rearrangement (Zubr 1997).

The reaction of transesterification can be catalyzed by sulfonic and sulphuric acids, which are Brownsted acids. The reactions are slow, but in alkyl esters, the very high yields are due to the catalysts necessitating usually temperature above 100 °C and for completing the transformation 3 h or more than 3 h would be taken (Freedman et al. 1986). Though, it could be stretch to di-glycerides and triglycerides. The ester-containing carbonyl group carbocation is resulting; afterward alcohol nucleophile attack yields intermediate of tetrahedral. The tetrahedral intermediate reinforces the catalyst and to form new esters after removal of glycerol. In 20 h change was detected to be completed, at a reaction temperature of 65 °C while Ethanolysis at 78 °C and butanolysis at 117 °C having alcohol and catalyst with equal magnitudes taking 18 h and 3 h., respectively (Freedman et al. 1986).

The catalysts such as heterogeneous and homogeneous in the reaction of trans-esterification are being used. Alkalis plus acids are included in homogeneous catalysts. KoH, NaoH, CH_3 NaO were abundantly using alkali catalytic agent. In the background literature, in this sense many references could be (Schwab et al. 1988; Fröhlich and Rice 1995; Vicente et al. 1998, 2004). The usual catalysts are HCl, H_2SO_4 and sulfonic acid. Though, these catalysts have been less studied (Kildiran et al. 1996; Freedman et al. 1986; Harrington and D'Arcy-Evans 1985; Canakci et al. 1999). Enzymes are included in heterogeneous catalysts (Mittelbach 1990; Shimada et al. 1999; Nelson et al. 1996) alkaline-earth metal compounds (Gryglewicz 1999), titanium-silicates (Bayense 1994), guanadines heterogenized on carbon-based polymers (Guo et al. 2021; Ingle et al. 2020; Schuchardt et al. 1998; Fröhlich and Rice 1995; Ali et al. 2017; Said et al. 2015).

The most commonly used catalysts are the basic catalysts, the procedure is more rapid and their response situation is moderate (Freedman et al. 1986). The application of basic catalysts in transesterification of oil obtained from vegetable yields soaps to reduce the effect of oil containing free fatty acid and saponification of triglyceride. Formations of soap somewhat consume the catalyst due to which yield of biodiesel decrease and the separating and refining stages is problematic so, the formations of soap are undesirable side-reaction. To use the oil obtained from vegetables containing low content of free fatty acid (>0.5%), the neutralization of free fatty acid could be avoided. Though, the furthermost commercial raw constituents (e.g. fats, low-value fats and waste product cooking extracts) are generally free-fatty- acid having high content. Catalysts such as NaOH and KOH containing hydroxyl group (OH). Due to this, the saponification side-reaction takes place. Though, hydroxide ions are present as an impurity in the basic methoxides. In this sense, they do not yield soap (Fröhlich and Rice 1995).

By using an acid catalyst, the soap formation can be avoided. To enhance the yield of fuel catalyst is added. Additional pressure conditions and extreme temperatures are required for the transesterification and acid catalysts mixing is slower than trans-esterification of basic-catalyzed reaction (Fröhlich and Rice 1995). The progress of catalysts such as heterogeneous increase the methyl ester yield in recent times because they significantly make simpler and save the product post-treatment, such as separation and purification when used in the transesterification reaction. Further neutralizing free fatty acid and saponification of triglyceride use catalysts such as heterogeneous does not produce soaps. However, extreme reaction conditions are required for the heterogeneously catalyzed reaction compared to the alkali catalysts. The yield of methyl ester and time of the reaction is unfavourable (Vicente et al. 1998).

Mostly studied, rate of conversion of triglyceride, during reaction the product composition changes or calculation of some biodiesel parameters quality. Though, firstly Rice and (Vicente et al. 2004).given out additional related features of the method such as stage of biodiesel produce the phase treatment afterward before and the effect of reaction on the production of biofuel.

Agreeing to Frohlich, two sources are the saponification of the triglyceride and the dissolution of the methyl ester in the glycerol phase and yield loss in triglyceride transesterification. Through the material stability of the method, the production and the production losses can be evaluated. The transesterification trend year wise is as follow; from 1971 to 1980 the percentage is 0.92, from 1981 to 1990 is 1.64, from 1991 to 2000 is 14.70, from 2001 to 2011 is 31.76 from 2011 to 2021 is 51.01. The advantage of the transesterification process is that they are renewable, maximum centane no. higher burning efficiency and lower emissions (Fröhlich et al. 2000). The main disadvantage is by-product (Wastewater and glycerol) disposal (Schinas et al. 2009).

19.1.5.5 Biox Co-solvent Process

In this method, due to inert co-solvents selection, triglycerides are converted to esters that produce a rich system of oil one stage (Goodrum et al. 2003). Boocock et al. (1996) developed Biox co-solvent process. On the way to overwhelmed the time of the reaction which is slow make happen by the alcohol solubility which is very low in the triglyceride stage co-solvent options exist. To make the methanol soluble. A co-solvent such as a tetrahydrofuran (THF) is used. The biodiesel and glycerol separation stage is unpolluted and the co-solvents such as tetrahydrofuran and alcohol in excess form could be recovering at the end of the reaction (Goodrum et al. 2003). Though, for a reason that of the toxic phenomenon and potential threat of the co-solvent, the phase of glycerol and the phase of biodiesel essentially to be removed totally so that product obtained should be free from water (Boocock et al. 1996). The bio- co-solvent trend year wise is as follow; from 1971 to 1980 the percentage is 0, from 1981 to 1990 is 0, from 1991 to 2000 is 0, from 2001 to 2011 is 100 from 2011 to 2021 is 0.

Biox co-solvent process has a distinct advantage that it used passive, co-solvent in reaction at suitable pressure and temperature that takes only a few seconds and no remains of catalysts are to be seen in each, the phase of glycerol or the phase of biodiesel. In this process, feedstocks such as grain-based cannot only be used but other fats and oil such as animal fats and cooking oil waste can also be used (Gerpen et al. 2004) disadvantage is that by this method the extra alcohol regaining is so difficult for the reason that boiling point of methanol is nearly similar to that of THF co-solvent.

19.1.5.6 Supercritical Alcohol Process

The distinctive fluid plus steam stage no longer occur, under such condition (Boocock et al. 1996). The supercritical methanol process has established this technique in producing biofuel through non-catalytic (Demirbas 2008). In the supercritical state, as the dielectric constant value is lower due to methanol addition, this methodology is thought to resolve the difficulties related to using methanol to triglyceride to make a mixture and the short time taken during reaction completion (Saka and Kusdiana 2001).

In a high-pressure reactor, supercritical transesterification is carried out through heat. Gas gives vent to and in the container where the product is tipped into a vessel used for accumulating after the reaction is complete. The left-behind contents in the container are removed (Han et al. 2005). Numerous variables such as temperature and pressure affected the yield of different biodiesel products and the main production could be achieved under the optimum conditions during the whole process. The Supercritical process trend year wise is as follow; from 1971 to 1980 the percentage is 0, from 1981 to 1990 is 1.9, from 1991 to 2000 is 6.32, from 2001 to 2011 is 38.61 from 2011 to 2021 is 53.17.

The supercritical alcohol process, when compared with the alkali-catalyzed method, the several benefits have been seen such as time of reaction and purification but the supercritical process requires high energy, high pressure and temperature (Goodrum et al. 2003; Bunyakiat et al. 2006).

19.1.5.7 In Situ Biodiesel Process

In situ process, to attain acylglycerols transesterification, at suitable pressure and temperature with a solution of methanol, the oilseeds are directly treated in which the catalyst has been up to that time dissolved. This means that earlier to fatty acid esters transesterification the seeds containing oil is not to be isolating (Demirbaş 2002). Converting oil to biodiesel and production of biodiesel through in situ process is a new methodology (Haas et al. 2004).

During in situ transesterification for high effectiveness alcohol necessity is to be decrease and the seed containing oil need to be fully dried before the reaction occurs (Demirbaş 2002). The crude oil and the alcohol are mixed and reflux for about 1–5 h and also heated and they also contain catalyst dissolved in it. Two layers are formed after completion. The phase of the alcohol could be recovered which is formed at the lower layer. The biodiesel, which is in the crude form, formed the upper layer and these are continuously washed away with water to remove pollutants until neutral the solution through washing. After completing the washing steps the biodiesel is filtered and the anhydrous sodium sulfate is dried over (Demirbaş 2002; Harrington and D'Arcy-Evans 1985) found that the transformation of the oilseed is very high (about 98%) and the biodiesel can be conformed to the ASTM standard. The in situ-biodiesel process trend year wise is as follows; from 1971 to 1980 the percentage is 0, from 1981 to 1990 is 0, from 1991 to 2000 is 0, from 2001 to 2011 is 5.89 from 2011 to 2021 is 94.12.

This process removes the requirement for separation and may be used to purify the oilseed containing fatty acid. The method can decrease biodiesel charges; decrease the lengthy size of the making arrangement related by degumming, pre-extraction, and increase biodiesel production. The disadvantage of this technique is that this is not favorable for animal fats and waste of cooking extracts, which decrease feedstock charges (Qian et al. 2008).

19.1.6 Conclusion

Number of studies on biodiesel production point of view has been carried out all over the globe. Numerous techniques have been used for biodiesel production. In current research work, we have tried to combine scatter datasets from 1971 to 2021 on all these techniques in a single chapter. Based on our search we have observed that the in situ-biodiesel technique is the highly used technique, followed by direct use and blending, Supercritical process, Transesterification, Pyrolysis, Microemulsion and Biox co-solvent. In the current era, more breakthroughs have been made and more advanced techniques have been introduced i.e. Nano-catalytic technology. The use of nan-catalytic techniques in the production of biodiesel is user-friendly and has exceptional properties such as high activation energy, easy recovery, controlled rate of reaction, selective reactivity and recyclability. The development of inexpensive and more effective catalysts for biodiesel production is essential to produce biodiesel at large scales and overcome the load of petroleum-based fuels. The current study may provide a baseline to researchers, scientists and environmentalists who are interested to work on biodiesel production and related techniques advancements.

References

Abbaszaadeh A, Ghobadian B, Omidkhah MR, Najafi G (2012) Current biodiesel production technologies: a comparative review. Energy Convers Manage 63:138–148

Adams C, Peters JF, Rand MC, Schroer BJ, Ziemke MC (1983) Investigation of soybean oil as a diesel fuel extender: endurance tests. J Am Oil Chem Soc 60(8):1574–1579

Ali OM, Mamat R, Rasul MG, Najafi G (2017) Potential of biodiesel as fuel for diesel engine. In: Clean energy for sustainable development. Academic Press, pp 557–590

Alonso DM, Bond JQ, Dumesic JA (2010) Catalytic conversion of biomass to biofuels. Green Chem 12(9):1493–1513

Anon M (1982) Filtered used frying fat powers diesel fleet. J Am Oil Chem Soc 59:780A-781A

Arzamendi G, Arguinarena E, Campo I, Zabala S, Gandia LM (2008) Alkaline and alkaline-earth metals compounds as catalysts for the methanolysis of sunflower oil. Catal Today 133:305–313
ASAE (1982) Vegetable oil fuels. In: Proceedings of the international conference on plant and vegetable oils as fuels. In: Backers L (ed) ASAE. Elsevier, St Joseph, MI
Azad AK, Rasul MG, Khan MMK, Sharma SC (2014) Review of non-edible biofuel resources in Australia for second generation (2G) biofuel conversion. In: International green energy conference, Tainjin, China, pp. 867–878
Bartholomew D (1981) Vegetable oil fuel. J Am Oil Chem Soc (United States) 4:58
Bayense CR (1994) Esterification process, European patent no. 0 623 581 A2
Bhatti HN, Hanif MA, Faruq U, Ahmad SM (2008) Acid and base catalyzed transesterification of animal fats to biodiesel
Boocock DG, Konar SK, Mao V, Sidi H (1996) Fast one-phase oil-rich processes for the preparation of vegetable oil methyl esters. Biomass Bioenerg 11(1):43–50
Bunyakiat K, Makmee S, Sawangkeaw R, Ngamprasertsith S (2006) Continuous production of biodiesel via transesterification from vegetable oils in supercritical methanol. Energy Fuels 20(2):812–817
Butler E, Devlin G, Meier D, McDonnell K (2011) A review of recent laboratory research and commercial developments in fast pyrolysis and upgrading. Renew Sustain Energy Rev 15(8):4171–4186
Canakci M, Monyem A, Van Gerpen J (1999) Accelerated oxidation processes in biodiesel. Trans ASAE 42(6):1565
Chang CC, Wan SW (1947) China's motor fuels from Tung oil. Ind Eng Chem 39(12):1543–1548
Chen CJ (2011) Physics of solar energy. John Wiley & Sons
Crossley A, Heyes TD, Hudson BJF (1962) The effect of heat on pure triglycerides. J Am Oil Chem Soc 39(1):9–14
Daneshvar N, Salari D, Khataee AR (2004) Photocatalytic degradation of azo dye acid red 14 in water on ZnO as an alternative catalyst to TiO_2. J Photochem Photobiol A Chem 162(2–3):317–322
Demirbaş A (2002) Biodiesel from vegetable oils via transesterification in supercritical methanol. Energy Convers Manage 43(17):2349–2356
Demirbas A (2008) Fuel properties of biodiesels. In: Biodiesel: a realistic fuel alternative for diesel engines, pp.141–160
Encinar JM, Gonzalez JF, Rodríguez-Reinares A (2005) Biodiesel from used frying oil. Variables affecting the yields and characteristics of the biodiesel. Ind Eng Chem Res 44(15):5491–5499
Engler CR, Johnson LA, Lepori WA, Yarbrough CM (1983) Effects of processing and chemical characteristics of plant oils on performance of an indirect-injection diesel engine. J Am Oil Chem Soc 60(8):1592–1596
Ezeonu CS, Ezeonu NC (2016) Alternative sources of petrochemicals from readily available biomass and agro-products in Africa: a review. J Pet Environ Biotechnol 7(5):12–23
Fadhel AZ, Pollet P, Liotta CL, Eckert CA (2010) Combining the benefits of homogeneous and heterogeneous catalysis with tunable solvents and nearcritical water. Molecules 15(11):8400–8424
Fattah IM, Ong HC, Mahlia TMI, Mofijur M, Silitonga AS, Rahman SM, Ahmad A (2020) State of the art of catalysts for biodiesel production. Front Energy Res 8:101
Florini A, Sovacool BK (2009) Who governs energy? The challenges facing global energy governance. Energy Policy 37(12):5239–5248
Freedman B, Butterfield RO, Pryde EH (1986) Transesterification kinetics of soybean oil 1. J Am Oil Chem Soc 63(10):1375–1380
Fröhlich A, Rice B (1995) The preparation and properties of biodiesel grade methyl ester from waste cooking oil. In: Minutes of the activity meeting of the IEA, vol 18. Vienna
Fröhlich A, Rice B, Vicente G (2000) The conversion of waste tallow into biodiesel grade methyl ester. In: 1st world conference and exhibition on biomass for energy and industry. Proceedings of the conference held in Seville (Spain), vol 697. 2000, June

Gerpen J, Shanks B, Pruszko R, Clements D, Knothe G (2004) Biodiesel Analytical Methods: August 2002--January 2004 (No. NREL/SR-510-36240). National Renewable Energy Lab., Golden, CO (US)

Goering CE (1985) Effect of nonpetroleum fuels on durability of direct-injection diesel engines: Illinois agricultural experiment station

Goering CE, Fry B (1984) Engine durability screening test of a diesel oil/soy oil/alcohol microemulsion fuel. J Am Oil Chem Soc 61(10):1627–1632

Goff J (2017) Carcinogenicity of petrogenic PAHs. Petrogenic polycyclic aromatic hydrocarbons in the aquatic environment: analysis, synthesis, Toxicity and environmental impact. Bentham Science Publishers, Sharjah, UAE, pp 65–110

Goodrum JW, Geller DP, Adams TT (2003) Rheological characterization of animal fats and their mixtures with# 2 fuel oil. Biomass Bioenerg 24(3):249–256

Gryglewicz S (1999) Rapeseed oil methyl esters preparation using heterogeneous catalysts. Biores Technol 70(3):249–253

Guo M, Song W, Buhain J (2015) Bioenergy and biofuels: history, status, and perspective. Renew Sustain Energy Rev 42:712–725

Guo M, Jiang W, Chen C, Qu S, Lu J, Yi W, Ding J (2021) Process optimization of biodiesel production from waste cooking oil by esterification of free fatty acids using La^{3+}/ZnO-TiO_2 photocatalyst. Energy Conver Manag 229:113745

Haas MJ, Scott KM, Marmer WN, Foglia TA (2004) In situ alkaline transesterification: an effective method for the production of fatty acid esters from vegetable oils. J Am Oil Chem Soc 81(1):83–89

Han H, Cao W, Zhang J (2005) Preparation of biodiesel from soybean oil using supercritical methanol and CO_2 as co-solvent. Process Biochem 40(9):3148–3151

Harrington KJ, D'Arcy-Evans C (1985) A comparison of conventional and in situ methods of transesterification of seed oil from a series of sunflower cultivars. J Am Oil Chem Soc 62(6):1009–1013

Harrington KJ, D'Arcy-Evans C (1985) Transesterification in situ of sunflower seed oil. Ind Eng Chem Prod Res Dev 24(2):314–318

Helwani Z, Othman MR, Aziz N, Kim J, Fernando WJN (2009) Solid heterogeneous catalysts for transesterification of triglycerides with methanol: a review. Appl Catal A 363(1–2):1–10

Hill J (2007) Environmental costs and benefits of transportation biofuel production from food-and lignocellulose-based energy crops A Review. Agron Sustain Develop 27(1):1–12

Hossain ABMS, Mazen M (2010) Aust J Crop Sci 4:550–555

Ingle AP, Chandel AK, Philippini R, Martiniano SE, da Silva SS (2020) Advances in nanocatalysts mediated biodiesel production: a critical appraisal. Symmetry 12(2):256

Kaya C, Hamamci C, Baysal A, Akba O, Erdogan S, Saydut A (2009). Methyl ester of peanut (Arachis hypogea L.) seed oil as a potential feedstock for biodiesel production. Renew Energy 34(5):1257–1260

Kibazohi O, Sangwan RS (2011) Vegetable oil production potential from Jatropha curcas, Croton megalocarpus, Aleurites moluccana, Moringa oleifera and Pachira glabra: assessment of renewable energy resources for bio-energy production in Africa. Biomass Bioenerg 35(3):1352–1356

Kildiran G, Yücel SÖ, Türkay S (1996) In-situ alcoholysis of soybean oil. J Am Oil Chem Soc 73(2):225–228

Kim HJ, Kang BS, Kim MJ, Park YM, Kim DK, Lee JS, Lee KY (2004) Transesterification of vegetable oil to biodiesel using heterogeneous base catalyst. Catal Today 93:315–320

Knothe G (2010) Biodiesel and renewable diesel: a comparison. Prog Energy Combust Sci 36(3):364–373

Krawczyk T (1996) Biodiesel. Alternative fuel makes inroads but hurdles remain

Lin L, Cunshan Z, Vittayapadung S, Xiangqian S, Mingdong D (2011) Opportunities and challenges for biodiesel fuel. Appl Energy 88(4):1020–1031

Linsebigler AL, Lu G, Yates JT Jr (1995) Photocatalysis on TiO_2 surfaces: principles, mechanisms, and selected results. Chem Rev 95(3):735–758

Lopez DE, Goodwin JG Jr, Bruce DA, Lotero E (2005) Transesterification of triacetin with methanol on solid acid and base catalysts. Appl Catal A 295(2):97–105
Lotero E, Liu Y, Lopez DE, Suwannakarn K, Bruce DA, Goodwin JG (2005) Synthesis of biodiesel via acid catalysis. Ind Eng Chem Res 44(14):5353–5363
Luna C, Luna D, Calero J, Bautista FM, Romero AA, Posadillo A, Verdugo-Escamilla C (2016) Biochemical catalytic production of biodiesel. In: Handbook of biofuels production. Woodhead Publishing, pp 165–199
Macario A, Giordano G, Onida B, Cocina D, Tagarelli A, Giuffrè AM (2010) Biodiesel production process by homogeneous/heterogeneous catalytic system using an acid–base catalyst. Appl Catal A 378(2):160–168
Mallesham B, Sudarsanam P, Reddy BM (2014) Production of biofuel additives from esterification and acetalization of bioglycerol over SnO2-based solid acids. Ind Eng Chem Res 53(49):18775–18785
Marchetti JM, Miguel VU, Errazu AF (2007) Possible methods for biodiesel production. Renew Sustain Energy Rev 11(6):1300–1311
Mariod A, Klupsch S, Hussein IH, Ondruschka B (2006) Synthesis of alkyl esters from three unconventional Sudanese oils for their use as biodiesel. Energy Fuels 20(5):2249–2252
Mittelbach M (1990) Lipase catalyzed alcoholysis of sunflower oil. J Am Oil Chem Soc 67(3):168–170
Mondala A, Liang K, Toghiani H, Hernandez R, French T (2009) Biodiesel production by in situ transesterification of municipal primary and secondary sludges. Biores Technol 100(3):1203–1210
Nelson LA, Foglia TA, Marmer WN (1996) Lipase-catalyzed production of biodiesel. J Am Oil Chem Soc 73(9):1191–1195
Oh PP, Lau HLN, Chen J, Chong MF, Choo YM (2012) A review on conventional technologies and emerging process intensification (PI) methods for biodiesel production. Renew Sustain Energy Rev 16(7):5131–5145
Peternel IT, Koprivanac N, Božić AML, Kušić HM (2007) Comparative study of UV/TiO_2, UV/ZnO and photo-Fenton processes for the organic reactive dye degradation in aqueous solution. J Hazard Mater 148(1–2):477–484
Peterson CL, Auld DL, Korus RA (1983) Winter rape oil fuel for diesel engines: recovery and utilization. J Am Oil Chem Soc 60(8):1579–1587
Pioch D, Lozano P, Rasoanantoandro M, Graille J, Geneste P, Guida A (1993) Biofuels from catalytic cracking of tropical vegetable oils. Oleagineux 48(6):289–292
Pramanik K (2003) Properties and use of Jatropha curcas oil and diesel fuel blends in compression ignition engine. Renewable Energy 28(2):239–248
Pryde EH (1983) Vegetable oils as diesel fuels: overview. J Am Oil Chem Soc 60(8):1557–1558
Pryor RW, Hanna MA, Schinstock JL, Bashford LL (1983) Soybean oil fuel in a small diesel engine
Qian J, Wang F, Liu S, Yun Z (2008) In situ alkaline transesterification of cottonseed oil for production of biodiesel and nontoxic cottonseed meal. Biores Technol 99(18):9009–9012
Refaat AA (2010) Different techniques for the production of biodiesel from waste vegetable oil. Int J Environ Sci Technol 7(1):183–213
Sahoo PK, Das LM (2009) Process optimization for biodiesel production from Jatropha Karanja and Polanga oils. Fuel 88(9):1588–1594
Said NH, Ani FN, Said MFM (2015) Review of the production of biodiesel from waste cooking oil using solid catalysts. J Mech Eng Sci 8(unknown):1302–1311
Saka S, Kusdiana D (2001) Biodiesel fuel from rapeseed oil as prepared in supercritical methanol. Fuel 80(2):225–231
Schinas P, Karavalakis G, Davaris C, Anastopoulos G, Karonis D, Zannikos F, Stournas S, Lois E (2009) Pumpkin (Cucurbita pepo L.) seed oil as an alternative feedstock for the production of biodiesel in Greece. Biomass Bioenergy 33(1):44–49
Schlautman NJ, Schinstock JL, Hanna MA (1986) Unrefined expelled soybean oil performance in a diesel engine. Trans ASAE 29(1):70–0073

Schlick ML, Hanna MA, Schinstock JL (1988) Soybean and sunflower oil performance in a diesel engine. Trans ASAE 31(5):1345–1349

Schuchardt U, Sercheli R, Vargas RM (1998) Transesterification of vegetable oils: a review. J Braz Chem Soc 9:199–210

Schwab AW, Dykstra GJ, Selke E, Sorenson SC, Pryde EH (1988) Diesel fuel from thermal decomposition of soybean oil. J Am Oil Chem Soc 65(11):1781–1786

Semwal S, Arora AK, Badoni RP, Tuli DK (2011) Biodiesel production using heterogeneous catalysts. Biores Technol 102(3):2151–2161

Shay EG (1993) Diesel fuel from vegetable oils: status and opportunities. Biomass Bioenerg 4(4):227–242

Shimada Y, Watanabe Y, Samukawa T, Sugihara A, Noda H, Fukuda H, Tominaga Y (1999) Conversion of vegetable oil to biodiesel using immobilized Candida Antarctica lipase. J Am Oil Chem Soc 76(7):789–793

Sims R, Taylor M, Saddler J, Mabee W (2008) From 1st-to 2nd-generation biofuel technologies. International Energy Agency (IEA) and Organisation for Economic Co-Operation and Development, Paris, pp 16–20

Singh SP, Singh D (2010) Biodiesel production through the use of different sources and characterization of oils and their esters as the substitute of diesel: a review. Renew Sustain Energy Rev 14(1):200–216

Sonntag N (1979) Reactions of fats and fatty acids. Bailey's industrial oil and fat products, vol. Swern, D

Srivastava A, Prasad R (2000) Triglycerides-based diesel fuels. Renew Sustain Energy Rev 4(2):111–133

Strayer RC, Blake JA, Craig WK (1983) Canola and high erucic rapeseed oil as substitutes for diesel fuel: preliminary tests. J Am Oil Chem Soc 60(8):1587–1592

Thangaraj B, Solomon PR, Muniyandi B, Ranganathan S, Lin L (2019) Catalysis in biodiesel production—a review. Clean Energy 3(1):2–23

Ullah F, Dong L, Bano A, Peng Q, Huang J (2016) Current advances in catalysis toward sustainable biodiesel production. J Energy Inst 89(2):282–292

Vicente G, Coteron A, Martinez M, Aracil J (1998) Application of the factorial design of experiments and response surface methodology to optimize biodiesel production. Ind Crops Prod 8(1):29–35

Vicente G, Martınez M, Aracil J (2004) Integrated biodiesel production: a comparison of different homogeneous catalysts systems. Biores Technol 92(3):297–305

Weisz PB, Haag WO, Rodewald PG (1979) Catalytic production of high-grade fuel (gasoline) from biomass compounds by shape-selective catalysis. Science 206(4414):57–58

Wu CH, Chang CL (2006) Decolorization of reactive Red 2 by advanced oxidation processes: comparative studies of homogeneous and heterogeneous systems. J Hazard Mater 128(2–3):265–272

Ziejewski M, Kaufman KR, Schwab AW, Pryde EH (1984) Diesel engine evaluation of a nonionic sunflower oil-aqueous ethanol microemulsion. J Am Oil Chem Soc 61(10):1620–1626

Zubr J (1997) Oil-seed crop: Camelina sativa. Ind Crops Prod 6(2):113–119

Zuliani A, Ivars F, Luque R (2017) Advances in nanocatalysts design for biofuels production

Chapter 20
Prospect of Metabolic Engineering for Biochemical Production

Rohit Ruhal and Rashmi Kataria

20.1 Introduction

Several strategies have been developed for the large scale production of medically relevant biochemicals. Metabolic engineering comprises strategies including genetic modification to produce relevant biochemicals. It deals with the amendment and optimization of metabolic pathways in living organisms. These objectives are not limited to only higher yield but maximum substrate conversion too. In general, to produce novel compounds with industrial and medical interest, indeed need metabolic pathway changes (Woolston et al. 2013; García-Granados et al. 2019). Any process involving engineered bacteria for the desired metabolite is considered successful after achieving a high yield with a low-cost investment. To reduce the cost of the process, a strategy for efficient conversion of renewable feedstocks is desirable (Kim et al. 2019). In the early days, metabolic engineering practised bottleneck analysis of pathways using metabolic flux calculations. But today, we have a volume of data accumulated at gene and metabolite levels, and in addition, numerous precise techniques are available. All these developed techniques can manipulate one pathway or even a complete cell. These cell factories have higher efficiencies in converting even waste feedstocks. The most straightforward strategy is creating a waste utilizing process based on metabolic engineering shown in Fig. 20.1. Since metabolic engineering aims to augment the production of desired biochemicals, several parameters must be considered (Woolston et al. 2013; García-Granados et al. 2019). In any process involved in biochemical production, one has to take care of factors like nutrient uptake and limited toxic accumulation. Nowadays, several biochemicals are considered desirable, including amino acids, biofuels, or any other industrially relevant fermentation products. The parameters to determine high efficient processes

---

R. Ruhal (✉) · R. Kataria
Department of Biotechnology, Delhi Technological University, Delhi, India

R. Kataria
e-mail: rashmikataria@dtu.ac.in

Y. K. Nandabalan et al. (eds.), *Zero Waste Biorefinery*, Energy, Environment, and Sustainability, https://doi.org/10.1007/978-981-16-8682-5_20

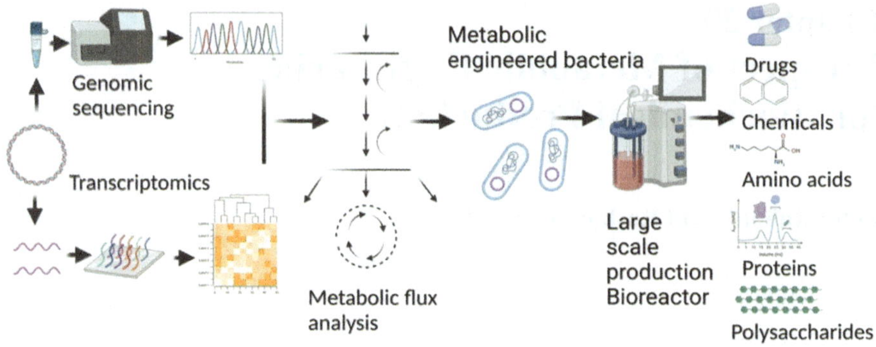

Fig. 20.1 Schematic representation of metabolic engineering steps. The desired pathway is first explored at genomic level (genome sequencing, transcriptomics) followed by metabolic flux analysis. On the basis of all information collected, desired bacteria is engineered for high production of biochemicals. Lab scale production is taken to pilot or large scale production in bioreactor. Metabolic engineering is used for production of range of value-added metabolites drugs, chemicals, amino acids, proteins and polysaccharides. we acknowldge biorender for figure

are product yield and productivity (Khan et al. 2018). Therefore, understanding the biological network needs to visualize how the desired biochemical is synthesized even at the transcriptional level. With the advent of vast genomic and proteomics data, it is now feasible to develop models.

In the current situation, the zero-waste biorefinery is necessary for a safer environment. The microbes can ferment wastes as substrate and produce value-added metabolites, and natural or non-engineered bacteria cannot wholly utilise biomass due to toxic produced during biorefinery. To make microbes more efficient so that the total biomass is the substrate, one has to study the detail of the metabolic pathway for substrate uptake. Thus, microbial engineering pathways may pave the way for rewiring metabolism for zero waste biorefinery. This book chapter displays in-depth detail of metabolic engineering and how it can support efficient production of biochemicals and absolute waste consumption.

20.2 Prerequisite for Metabolic Engineering

The importance of metabolic engineering may be realized by the example of bioethanol production from yeast (Selim et al. 2018). Bioethanol is a top product of biorefinery produced from biomass. During biorefinery, toxic by-products are produced as side products of sugars and yeast fermentation of sugars is hindered due to these toxics. Thus, to make the process efficient and achieve a higher yield of ethanol and complete substrate conversion, an approach of engineering yeast metabolism is established (Hasunuma and Kondo 2012; Ndukwe et al. 2020).

The choice of biochemicals for large-scale production are based on applications in pharmaceutical, industrial, medical, agricultural, health, vaccine, nutraceuticals, etc. In general, biochemicals demand and output are calculated based on market demand and economy. Sometimes economy of industrial production of biochemicals may be reduced using different bio-based materials as a carbon source for fermentation. For instance, lignocellulose (bio-based) material may be utilized to produce chemicals, nutraceuticals, bioplastic, etc. (Kim et al. 2019; Su et al. 2020). Similarly, glycerol from biodiesel waste is considered ideal for using carbon sources and engineering bacteria accordingly (Westbrook et al. 2019; Chen and Liu 2016). But there are limitations for using waste, e.g. the incomplete conversion of debris due to toxins produced during biorefinery. In that case, either detoxification is done, or toxic tolerant strains are developed (Christopher et al. 2016; Bhatia et al. 2020a).

There are specific parameters that determine if engineering bacteria for any process is a success. These parameters are measurements of biochemical production like yield, concentration, or productivity. The yield is defined as the quantity of product formed from the substrate (Ruhal et al. 2011). Another measurement used to quantify biochemical production is concentration, e.g. organic acids like propionic acid measured in the supernatant (Ruhal and Choudhury 2012). Thus engineering any microbe for high production needs to have a high yield or concentration compared to wild strains. There are methods we can calculate theoretical yield, which is defined as a maximum flux for the biosynthesis of metabolite in a metabolic network (Shastri and Morgan 2004). It is difficult to go beyond theoretical yield as microbe must also survive for living and form biomass. The process with high yield have an advantage for complete carbon substrate utilization and thus fulfil the mandate of zero waste biorefinery. However, a high concentration of product may help downstream processing less complicated. Therefore, we need to increase the yield or concentration of our product along with limited side product formation. Developing recombinant strain to produce a high product concentration or high yield may reduce cost and simplify the techniques involved. Another term known as productivity is defined as biochemical produced per unit time and can reduce overall bioprocess economy.

Metabolic engineering is now a day applied for numerous applications. These include nutraceuticals and high value-added metabolites with many applications (Yuan and Alper 2019). In addition, many health-promoting metabolites, pharmaceutical and agricultural applications are also produced. Bio-based or renewable-based products are preferred in comparison to chemically synthesized. Even little cost could be afforded if preformed products have similar properties as chemically synthesized (Moleirinho et al. 2019). Another crucial aspect of developing a metabolic engineering strategy could be bioremediation and biodegradation. It involves the goal of degradation of environmental pollutants, and thus efforts are made to create strains with efficient degradation ability of target ecological pollutants.

20.3 Synthetic Biology Tools for Modifying Metabolic Pathways Towards Biochemical Biosynthesis

As mentioned above, metabolic engineering strategies identify target metabolites that we want to develop for large-scale production. The second step is to recognize natural metabolic pathways available in microorganisms or eukaryotes, and complete knowledge of its biosynthesis pathway is essential. With the advancement of genome sequencing and other computational software, large databases are available. Thus, one can start using the online or offline database for pathways linked to desired metabolites' biosynthesis.

20.3.1 Tools for Mining Genes and Enzymes Involved in Biochemicals Biosynthesis

With the advent of technology, numerous software and databases are available to determine the desired product's biosynthesis (Albarano et al. 2020). For instance, the most accessible approach is the KEGG pathway (Kanehisa and Goto 2000) and BRENDA (Placzek et al. 2017). Several DNA manipulating tools are available, and the synthetic biological system may be developed after optimizing several components, including genetic parts and procedures. To design a natural system controlling gene expression is also one of the challenges for developing an efficient method. Thus there is a need to understand different *omic* technologies with other tools and strategies (Jensen et al. 2014; Niu 2018). For instance, genomics is a branch of science that describes the genetic structure of an organism and finally leads to the identification of possible genetic changes. In genomics, whole genomic sequencing is done identifying genes from the known database. The gene expression data can be analyzed by transcriptomics which may help understand the cell's metabolism under specific conditions (Lowe et al. 2017). Transcriptomic may be done by microarray or RNA-sequencing-based methods (Tachibana 2015). In both scenarios, RNA is isolated from bacterial cells, one needs to be careful as RNA are rapidly degraded, and hence RNA of a particular stage must be fridged immediately. While microarray is a chip method, in general, all gene sequences are placed on a glass slide or chip, and RNA isolated is converted into DNA and allowed to bind on the chip (Tachibana 2015). Microarray does face certain limitations, and nowadays, RNA-sequencing-based transcriptomic is popular. Primarily RNA-seq based may be used to identify unknown gene expression whose sequence is not available.

Further, proteomics may help determine protein expression profile, and data may be interpreted in correlation to transcriptomic data (Wen et al. 2020). Metabolomics may be applied by using GC–MS, HPLC, liquid chromatography. All these analyses may help in determining rate-limiting steps based on the quantitative data of pathway metabolites. In addition, several computational tools are available for mining the genome for biosynthetic gene clusters. One of them is antiSMASH software that may help in rapid genome-wide identification and annotation. This software can further aid in analyzing metabolite biosynthesis gene clusters (Blin et al. 2019). One can identify not only bacteria but fungal and plants metabolites as well. So one can add gene sequences of targeted metabolites, and the website gives information about the pathways. Another computational software PRISM can identify biosynthetic genes and, based on sequences it can predict chemical structures. In 2020, a published paper mentioned PRISM4 software, which can predict antibiotics chemical structure (Skinnider et al. 2020). Similarly, genome mining tool RODEO can identify biosynthetic gene clusters and have the capability of predicting modified peptides (Skinnider et al. 2020).

20.3.2 Metabolic Flux Optimization

Metabolic pathway analysis and metabolic fluxe are considered critical steps in metabolic engineering (Antoniewicz 2021). It includes the passage of metabolites over a reaction in a particular time and conditions and the need to analyze detailed flux maps of microbial metabolism. Metabolic flux analysis for quantitative data into the flow of carbon and energy especially measuring fluxes under various perturbed conditions (generic manipulation of the pathway).

In the last few years, substantial progress for accurate quantification of *in-vivo* fluxes has been achieved. One can tabulate stoichiometric coefficients for all possible metabolic reactions in the biosynthesis of the desired product mathematically. During the development of such metabolic models, it is assumed that intracellular metabolites levels are constant, and calculation is based on the addition of all fluxes towards metabolite biosynthesis and must be equal to the sum of all changes. The advantage of this approach is that one can calculate theoretical yield. For example, to our interest, if we set up the substrate uptake to one mol and need to enhance the desired product output, considering subject to the stoichiometric constraints. Another approach is metabolic flux analysis is calculated by doing a different experiment, and rates are measured for substrate and oxygen uptake rates, growth rates and quantity of product formed. ^{13}C metabolic flux analyses have garnered attention in the past. The advantage was accurate and precise quantification.

^{13}C- labelled carbons are incorporated into intracellular metabolites, and then labelling patterns are quantified as various cellular metabolites with techniques like mass spectrometry, NMR. The data collected may be fitted in experimental data, external rates and metabolic models. Metabolic flux analysis needs a very detailed analysis of pathway enzymes of desired biochemical. For example, the pathway of trehalose synthesis in bacteria when given different substrates like starch, glucose sucrose, or glycerol (Ruhal et al. 2013). This calculation of the amount of carbon flow in each step helps to realize the bottleneck steps for maximum theoretical yield, which is the main objective of metabolic flux. Metabolic flux analysis determines the fate of a molecular substrate and the flow of carbon within different pathways. For calculating metabolic changes, perturbations are given, and the rate of each product formed is calculated, which helps in deciding for deletion or overexpression of a particular gene accordingly.

Several computational techniques were proposed to predict pathways. One of them includes RetroPath RL method; this method explored bioretrosynthesis space by artificial intelligence-based approach (Koch et al. 2020). This method was validated for a dataset of 20 experimental pathways while, experimentally, pinocembrin was produced. AGORA resource software works based on a genome-scale model (Magnúsdóttir et al. 2017). AGORA may be defined as the assembly of all the gut organisms using reconstruction and analysis of sequencing data. Similarly, the Merlin framework can help in the genome-scale metabolic model. The RAVEN is another software that uses metabolic model reconstruction using genetic information (Wang et al. 2018).

20.3.3 Designing Tools for Controlling Gene Expression at the Transcriptional Level

To start with metabolic engineering techniques, we need to understand the central dogma of the concept (Crick 1970). The DNA, RNA and proteins are an essential part of this dogma (Morange 2009). Especially it is important to understand transcriptional, post-transcriptional and translational regulations—all the factors involved in the expression of desired genes. Although all this information is beyond the limit of this book chapter, we need to know few important points. The DNA consist of all the information for proteins to be formed and flows this message with the help of messenger RNA and is broadly termed as transcription (Pérez-Ortín et al. 2019). In today's world, DNA sequences may give extreme information about all the proteins. Translation refers to ribosomal factories which translate the message in mRNA to proteins. All the detailed information on transcription and translation regulation can help develop strategies for expressing our desired biochemicals.

One of the first strategies to control at the transcriptional level is to create a promoter library with variation in efficiency (Kotopka and Smolke 2020). In general, the sequence of the promoter of a particular gene may vary in different species. The most efficient promoter belongs to the category with the most consensus sequences. Further different approaches were made to change the efficiency of the promoter (Sanches-Medeiros et al. 2018; Dong et al. 2012; Peng et al. 2015). For instance, the spacer region between consensus sequences was modified. Even random mutation could be generated and screened for best production. In yeast, which also includes enhancers (eukaryotic transcription), synthetic hybrid promoters may be developed with a combination of promoters and enhancer elements (Peng et al. 2015).

After modifying consensus sequences between −35 and −10 boxes and spacer regions, a library of promoters may be constructed. Transcription is controlled by optimizing and creating promoters with various strengths or copy numbers (Xu et al. 2019; Vogl et al. 2018). In eukaryotes, after a detailed analysis of transcription machinery, elements known as transcription factors (TFs) may be engineered for a global transcription network (Peng et al. 2015). Different tools for controlling genes at a post-transcriptional level were designed (Niu et al. 2018). The 5-UTR can be planned accordingly for particular expression levels thermodynamically or through mathematical models. Further CRISPR-based technologies are advancing in regulating both at transcriptional and translational levels (Lino et al. 2018). For post-translational control, one can harmonize codon preference of gene corresponding to native host and lead to proper folding of polypeptides.

20.3.4 Techniques Used for Genome Editing to Produce Large Scale Biochemicals

Numerous technologies are available for genome engineering to get the desired product. Several genome techniques can be applied to manipulate the genome for our desired result, as shown in Fig. 20.2. BioBrick assembly from NewEngland is a readily available method for changing the genetic system (Sleight et al. 2010). In this approach, all the materials required are first standardized, e.g. DNA fragments encoding proteins, promoters, ribosome binding sites. Further, they are contained in a parts registry of plasmids with identical restriction sites flanking the payload of the part. Gibson assembly discovered by Daniel Gibson from Craig institute described a robust method that involves a single reaction with five èxonuclease

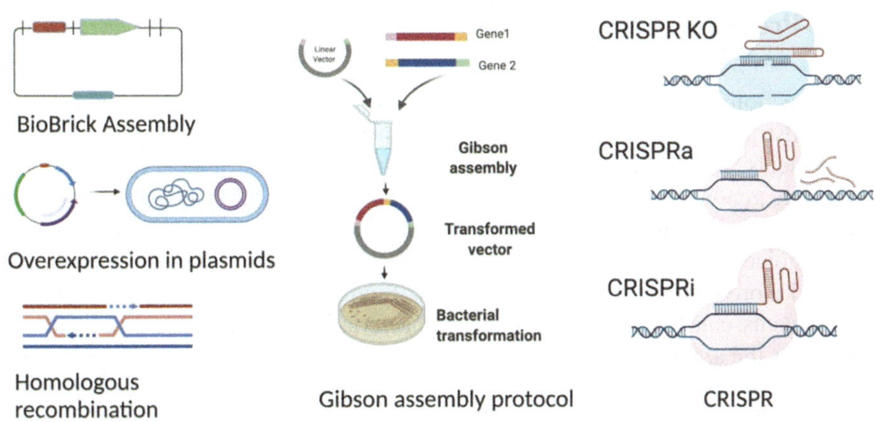

Fig. 20.2 Schematic representation of different genome editing methods used for engineering bacteria for specific biochemical pathway. BioBrick assembly is available for large length of gene expression or operon for biosynthesis of biochemical. Overexpression in plasmid is traditional approach. Gibson assembly protocol utilize one step method where we do not need to do restriction digestion or ligation seperatly. CRISPR technology is now very popular for efficiency and chromosomal genome manipulation. CRISPR KO may help in gene deletion while CRISPRa can help in gene activation and CRISPRi for blocking gene expression. we acknowldge biorender for figure

generates overhangs, polymerase, and DNA ligase. This method is adopted as a significant workhorse for different synthetic biology methods (Gibson 2017). The transformation associated with recombination is another cloning approach that allows entire genes and large chromosomal regions. Especially, homologous recombination during transformation in *S cerevisiae* is feasible. Further chromosomal engineering is also important as plasmids are not much stable, chances of loss. Identification of pathways or relevant genes is possible by transposon mutagenesis. Transposons are jumping genes to a different region of chromosomes, and their mutant libraries are used for screening genes linked to the biosynthesis of relevant metabolites (Tyo et al. 2009). It can be followed by appropriate knockout and overexpression of genes. Site-specific integration is possible by integrating the desired gene in the chromosome. While these can be possible by homologous recombination in which certain specific plasmids transform cells and allow an exchange. Although nowadays CRISPR/Cas technology is more often used. It is an RNA-guided target specific DNA cleavage system which originated from the bacterial adaptive immune system. Day by Day, this technology is improved for making it more efficient (Rainha et al. 2021; Zhao et al. 2020). CRISPRi can help gene expression modulation; for example, effector protein Cas are catalytically inactivated along with guide RNAs, and transcription of the target gene is blocked. During CRISPRa, the gene overexpression tool utilizes an inactive effector Cas proteins and leads to overexpression.

20.4 Examples of Metabolic Engineering of Different Bacteria and Their Applications for Biochemical Production from Wastes

In this section, we will discuss different strains used for metabolic engineering. There exist pervasive literature on the use of other bacteria for producing desired biochemicals. The choice of host-microbe remains based on available extensive data concerning genomics, proteomics and metabolomics. In addition, how readily adaptable is microbe for all the molecular techniques required. One public concern that makes choices limited is safety since microorganisms under GRAS for humans are preferred more than pathogens (Liong 2008). Some metabolites produced naturally by pathogens can be engineered in these bacteria, which are considered GRAS. Sometimes the argument is given to use thermophilic bacteria to prevent contamination while halophiles are manipulated to save freshwater. Therefore we will provide examples of bacteria like *E. coli*, LAB, *Corynebacterium* and *Pseudomonas putida,* which are the top choices.

20.4.1 Escherichia Coli

Escherchia coli is the model organism with several advantages as a microbial cell factory and is employed to produce relevant biochemicals (Yang et al. 2020). A few of *E. coli's* benefits are the availability of complex physiology, fast growth, high cell density, and availability of detailed genes and their sequences. Further, numerous genome engineering tools are available. Due to its long history, it is well established in academia and the industrial environment. It has the advantage to use in all fields, including industrial, environmental, medical. With a feexamplesle of using *E. coli* as a host strain, we can understand how to apply metabolic engineering to improve the overall process. The 1,4 butanediol is one of the important commodities in various industrial production. *E. coli* host cells converted renewable feedstock and achieved a final concentration of 18 g/l (Yim et al. 2011). The strategy of metabolic engineering starts with exploring metabolic pathway analysis and flux calculation of 1,4 Butanediol. For this, a pathway identification algorithm was used. In addition, the genome-scale model was used as guidance for target genes. It was found that BDO was produced from Glucose, xylose, sucrose, which make it advantageous to use biomass-derived mixed sugar streams. So this research utilized SimPheny Biopathway predictor software and elucidated all potential pathways from *E. coli* to BDO (Yim et al. 2011). The rankings based on maximum theoretical yield and thermodynamic feasibility of metabolite produced were used in consequence. This process included deleting genes involved in aerobic respiration and carbon flux moving towards the TCA cycle.

In another study, CRISPRi technologies as an effective gene-editing tool were used to pinpoint *gltA* gene heterologous replacement of *lpknock-in* of six genes involved in BDO biosynthesis (Wu et al. 2017). The first step led to 0.9 g/l of BDO while it was increased to 1.8 g/l using CRISPRi technology to suppress competing genes and hence diverted towards biosynthesis of BDO. This study shows the potential of CRISPR and CRISPRi technology. Several other works for metabolic engineering was done, e.g. Vitamin B12 yield of a recombinant *E. coli* was increased to 307 ug/g biomass (Fang et al. 2018). In this work, different enzymes playing a role in the biosynthesis of Vitamin B12 were selected from other microbes, and a strain of *E. coli* was engineered by introducing those genes with respective plasmids. The expression of the product was determined parallelly.

The lignocellulose derived Glucose and xylose to biofuels is hindered as microbe consume first glucose, and then xylose (Kim et al. 2019). *E. coli* strains may be developed to utilize Glucose and xylose simultaneously. Whole-genome sequencing of evolved strain finds relevant mutations in metabolic and regulatory genes. This information details genes used for improved developing strains with an appropriate regulatory gene mutation and sugar co-utilization. In another study, an engineered strain was made with the capability of utilizing both Glucose and xylose. This strain was obtained by evolving a mutant with araC deletion. All the genes involved in pentose metabolism were consistently expressed. The strain does have an extra deletion in genes which enhance xylose utilization. Further, in another study, dark fermentation was used for hydrogen and ethanol co-production. A genetically engineered *E. coli* was used, and lignocellulose feedstock was obtained from wheat straw and corn stover as feedstock. The engineered strain was capable of increased 30% co-production of hydrogen and ethanol.

20.4.2 Lactic Acid Bacteria (LAB)

Dairy bacteria are commonly used due to their safety. Lactic acid bacteria (LAB) are gaining attention worldwide, especial their fermentation pharmaceutical and GRAS status. Bacteria belonging to LAB are generally found in food and milk and are considered safe. Therefore, *Lactobacillus* becomes the choice of organism considering its safety value. With growing time, genetics control of central metabolism makes them a potential candidate for industrial applications. Lactic acid bacteria received attention due to their ability to produce many relevant metabolites, such as exopolysaccharides, which give probiotic activity (Bajpai et al. 2016). Exopolysaccharides made from LAB have also shown immunomodulatory properties. LAB bacteria are known for their property of producing lactic acid as a significant fermentation product.

LAB are Gram-positive bacteria, and their primary product is lactic acid. Several genera important for fermentation is *Lactobacillus, Lactococcus, Leuconostoc, Streptococcus*. LAB has the potential for lactic acid production as well as a probiotic supplement. Lactic acid is an organic acid and is natural. Although it had s applications in foomedicineine, chemicals, and materials in recent years, its application for polylactic acid (PLA) is considered attainable plastic material, thus generating demand for lactic acid monomer. For this, we need to create high quality monomeric and one optical isomer with chemical purity. Therefore, metabolic engineering aspects are exploited for the production of lactic acid. Initially, the production of pure lactic acid is primary interest as there is increased demand for plant-based plastics.

In one study, a *Lactobacillus Plantarum* strain was engineered by deleting lactate dehydrogenase encoding D-LDH. But even this not led to a reduction of D-LA. Therefore, lactate racemose interconvert D-lactic acid, and L-lactic acid was also deleted, leading to L-LA purity up to 99% (Okano et al. 2018). The production was made using corn starch. The commonly used gene editing methods used in LAB are plasmid-based homologous recombination, double-stranded DNA recombination. Now CRISPR/Cas9 gene-editing techniques are applied easily on LAB, it was applied to *Lactococcus lactis* for deleting the *ldh* gene for one of the optical isomers. Further, to make lactic acid production more economical, efforts were made to utilize lignocellulosic feedstock.

The limitation is using both hexoses and pentoses to produce lactic acid from renewable sources such as agricultural residues. Lignocellulose feedstock hydrolysate generally obtained after acid or alkali pre-treatment consists of a mixture of sugars, including pentoses and hexoses. Therefore, in general, metabolic engineering strategies are applied to gain this advantage of all sugars. In the study, two copies of the *xylAB* operon were introduced in the *LactobacilPlantarumarum*, and the genes were taken from *Lactobacillus pentosus*. The resulting strains could ferment 75 g/l glucose and produced D-Lactic acid and a very high yield of 0.8 g/g of the substrate. In another similar study, the yield of lactic acid was obtained for arabinose fermentation by deleting the phosphoketolase gene and changing (Guo et al. 2019) metabolic flux.

Similarly, lactic acid and other biochemicals may be produced by LAB, especially for the biochemicals whose natural production happens in pathogenic bacteria. Since LAB comes under the GRAS category, therefore, it is safe for display. One such compound is 2,3-butanediol used as antifreeze, plastics, solvents, and other chemical industries naturally produced by *Klebsiella pneumonia* (pathogen). A patent US Patent was filed for production of 2,3-BDO by LAB and was able to get 49% of 2,3 BDO in the strain *Lactobacillus plantarum*. The metabolic engineering was applied for lactate dehydrogenase activity and heterologous expression of butanediol dehydrogenase. They further engineered NAD + cofactor recycling and redirected glucose fermentation towards 2,3 BDO for yield up to 67% (maximum theoretical work). Similarly, using renewable waste like glycerol from biodiesel waste to produce 1,3 PDO has been extensively studied. The 1,3 PDO biosynthetic pathway convert glycerol to 3-hydroxypropioaldehyde and consequently to 1,3-PDO. It is also made naturally in pathogens and hence may be LAB used for production.

Lactobacillus diolivorans can yield 1,3-PDO up to 85 g/g, in the range of other native producers. Many LAB bacteria can become hosts for biofuel production from lignocellulosic biomass. The advantage of using these strains is their ability to metabolize Glucose and xylose, which are majorly present in biomass hydrolysates. In addition, ethanol is also a fermentation product in LAB, biosynthesized by the reduction of acetaldehyde. Thus, efforts are made using metabolic engineering to increase ethanol production in LAB by manipulating pyruvate decarboxylase and alcohol dehydrogenase from yeast *Z. mobilis*. The advantage remains as *Z. mobilis* is the highest producer of ethanol. In addition, LAB is famous for producing many food supplements like diacetyl, alanine and low-calorie polyols. Thus, much application focus has been placed on using LAB to treat many gastrointestinal disorders. LAB have shown an immense future for producing bacteriocins. Bacteriocins are peptides synthesized by bacteria that inhibit or kill other related or unrelated microorganisms (Soltani et al. 2021). Bacteriocins have shown characteristics of the narrow spectrum as they usually inhibit taxonomically close or broad-spectrum by inhibiting a wide variety of bacteria. The first important applications of bacteriocins are safe food preservatives as bacteriocins cannot harm the human gastrointestinal tract and easily get digested. Bacteriocins do have the capability of use as an alternative to chemical preservatives.

20.4.3 Corynebacterium, Pseudomonas Putida

Corynebacterium glutamicum is a facultative anaerobic Gram-positive bacteria and is considered GRAS (generally recognised safe) status. It is a choice of bacteria for producing many organic acids, vitamins and nutraceuticals—the deprivation energy and carbon isareluxare towards desired product formation. Now *C. glutamicum* is used to utilize lignocellulose waste hydrolysate, the limitation was xylose utilization as it is considered a cornerstone for efficient utilization (Lee et al. 2016; Jo et al. 2017). *C. glutamicum* lacks a xylose isomerase pathway and cannot use xylose, but introducing xylose isomerase from *E. coli* led to improved xylose utilization (Lee et al. 2016). Further engineering *C. glutamicum* for sustainable production of chemicals is due to flexible cellular metabolism, high-stress tolerance to carbon source and resistance to fermentation inhibitors. It has genetic stability due to the lack of a recombination repair system and a limited restriction-modification system.

Corynebacterium is also known for producing the amino acid-like L-arginine (Park et al. 2014). This amino acid numerous applications in food, health and pharmaceuticals. The choice of *Corynebacterium* for L-arginine production is based on solid flux towards L-glutamate formation. This bacteria do not have the arginine degrading gene found in other bacteria like *E. coli* leading to the accumulation of arginine intracellularly. One can also give a range of different substrates to bacteria. Different metabolic engineering strategies were adapted for large scale production, e.g., repression of the L-arginine biosynthesis by regulator argR and feedback inhibition by a enzyme kinase (argB) were removed to make flow towards L arginine. Similarly, engineering of *Corynebacterium* for L-serine production was done, which is another essential amino acid (Zhang et al. 2019; Lai et al. 2012). This research first identified genes *serC* and *SerB,* which codes for aminotransferase and phosphatase. After careful experiments with different combinations of deletion and overexpression of genes in the pathway of serine biosynthesis, they found genes responsible for overexpression. They developed a strain overexpressing *users, serB* and *serA* (another pathway gene), deleted dehydratase and hydroxymethyl transferase, and achieved maximum amino acid concentration. Considering the importance of *Corynebacterium* in the production of amino acids, CRISPR technology was also used. The advantage of using CRISPR is more efficient than conventional homologous recombination. Primarily CRISPRi was utilized, which facilitate the mapping of gene expression levels to metabolic outputs, deactivated Cas9 to repress genes in *C. glutamicum* (Cleto et al. 2016). The effects of repression was analyzed based on amino acid titers. The reduced expression of *pgi, pck* and *pyk* have led to titre enhancement up to 98%. This was done in 3 days and was efficient for metabolic pathway remodelling.

One recent article utilized *Pseudomonas putida* for metabolic engineering for PHAs production (Elmore et al. 2021). PHA is polyhydroxyalkanoates and, in general known as bioplastic. This research primarily utilized carbon sources from lignocellulosic biorefineries. This research is fascinating to utilize lignin for producing biochemicals. In this study *P. putida* was engineered for high yield production of tricarboxylic acid cycle derived building blocks chemical itaconic acid and source used was aromatic compounds derived from lignin. In this study two distinct itaconic acid pathways and dynamic regulation were studied, gene expression tuning of TCA cycle and deletion of genes from competing pathways led to desired results. A maximum yield of 56% and titre of around 1.4 g/l obtained from alkali-treated lignin. This work represents the best objective of metabolic engineering, where metabolic control was used to reroute carbon and produce valuable chemicals from lignin. The robust itaconic acid production was limited to fungi, but bacterial strain like *P. putida* was engineered to produce itaconic acid from Glucose, acetate and glycerol.

20.5 Scale-Up of Biochemicals Using Engineered Microbes

If the metabolic engineering seems successful at lab scale and yield desired biochemical produced is nearby theoretical yield, then an overproduced strain needs to show high performance in scaled-up pilot scale (Lu et al. 2020; Bhatia et al. 2020b). One of the important parameters remains controlled conditions: constant temperature, oxygen control, and similar pH during all fermentation. One can also try continuous fermentation, which can be defined as stage freeze for continuous production of biochemical. The input and output are constant at this phase. So we try to scale up because of some limitations, such as higher chances of local substrate concentration differences due to poor mixing. However, one can scale up processes and design projects and strains more resistant to changes in fermentation conditions. System metabolic engineering helps for the scale-up by manipulating metabolic pathways accordingly. In addition, systems metabolic engineering may also help reengineer the strain based on feedstock provided to strain, primarily if desired product is formed at a large scale there are chances of lower cost for downstream processing. In addition, if we are using plasmids for overexpression of genes, then a risk of plasmid instability and loss is possible during large scale-up. At the same timresolved be sresolved by genomes intthe egration of desired gene. In addition, we have to be careful of using antibiotics as markers of genetic changes.

20.6 Conclusion

The industrial production of metabolites from the microbial origin with industrial, pharmaceutical and medical applications have ancient origins. Metabolic engineering has increased production efficiency at a much higher level and has improved downstream processing costs. This achievement was due to identification of bottleneck of metabolic pathways and engineering them accordingly. As we see metabolic engineering for bacterial strain included a number of molecular methods but CRISPR methodology is revolutionary. The advantage of CRISPR is direct manipulation in chromosomes, including deletion activation and gene suppression. The success of metabolic engineering relies on the fact how it can be successfully used for biorefinery. Globally industrialization has increased the wastes, amongst them biodiesel and lignocellulose waste are most popular. Biorefinery is the field of science that includes different treatments of waste and leads to the generation of toxic molecules. These toxic molecules makes bacteria to difficult to adapt in media and grow. Further range of sugars are released during pretreatment which is not utilized by different bacteria. In such cases, metabolic engineering must be great help as a bacteria can be engineered to adapt to waste extracted carbon sources and utilize all kinds of sugars released. Further, a process may be developed with simultaneous saccharification and fermentation. Although metabolic engineering may be developed to engineer plants with week lignin it may raise ethical issues. Thus engineering bacteria for efficient lignin degradation is a much better notion.

Acknowledgements The authors would like to thank CSIR-Govt of India for providing salary under the scheme of CSIR Scientist Pool Scheme to Rohit Ruhal and DBT-Govt of India to provide Ramalingaswamy fellowship Rashmi Kataria. We thanks www.Biorender.com for all figures.

References

Albarano L, Esposito R, Ruocco N, Costantini M (2020) Genome mining as new challenge in natural products discovery

Antoniewicz MR (2021) A guide to metabolic flux analysis in metabolic engineering: methods, tools and applications

Bajpai VK, Rather IA, Majumder R, Shukla S, Aeron A, Kim K, Kang SC, Dubey RC, Maheshwari DK, Lim J, Park YH (2016) Exopolysaccharide and lactic acid bacteria: perception, functionality and prospects

Bhatia SK, Jagtap SS, Bedekar AA, Bhatia RK, Patel AK, Pant D, Rajesh Banu J, Rao CV, Kim YG, Yang YH (2020a) Recent developments in pretreatment technologies on lignocellulosic biomass: effect of key parameters, technological improvements, and challenges

Bhatia R, Winters A, Bryant DN, Bosch M, Clifton-Brown J, Leak D, Gallagher J (2020b) Pilot-scale production of xylo-oligosaccharides and fermentable sugars from Miscanthus using steam explosion pretreatment. Bioresour Technol 296. https://doi.org/10.1016/j.biortech.2019.122285

Blin K, Shaw S, Steinke K, Villebro R, Ziemert N, Lee SY, Medema MH, Weber T (2019) Anti-SMASH 5.0: updates to the secondary metabolite genome mining pipeline. Nucleic Acids Res 47. https://doi.org/10.1093/nar/gkz310

Chen Z, Liu D (2016) Toward glycerol biorefinery: metabolic engineering for the production of biofuels and chemicals from glycerol

Christopher M, Anusree M, Mathew AK, Nampoothiri KM, Sukumaran RK, Pandey A (2016) Detoxification of acidic biorefinery waste liquor for production of high value amino acid. Bioresour Technol 213. https://doi.org/10.1016/j.biortech.2016.03.054

Cleto S, Jensen JVK, Wendisch VF, Lu TK (2016) Corynebacterium glutamicum metabolic engineering with CRISPR Interference (CRISPRi). ACS Synth Biol 5. https://doi.org/10.1021/acssynbio.5b00216

Crick F (1970) Central dogma of molecular biology. Nature 227. https://doi.org/10.1038/227561a0

Dong H, Tao W, Zhang Y, Li Y (2012) Development of an anhydrotetracycline-inducible gene expression system for solvent-producing Clostridium acetobutylicum: a useful tool for strain engineering. Metab Eng 14. https://doi.org/10.1016/j.ymben.2011.10.004

Elmore JR, Dexter GN, Salvachúa D, Martinez-Baird J, Hatmaker EA, Huenemann JD, Klingeman DM, Peabody GL, Peterson DJ, Singer C, Beckham GT, Guss AM (2021) Production of itaconic acid from alkali pretreated lignin by dynamic two stage bioconversion. Nat Commun 12. https://doi.org/10.1038/s41467-021-22556-8

Fang H, Li D, Kang J, Jiang P, Sun J, Zhang D (2018) Metabolic engineering of Escherichia coli for de novo biosynthesis of vitamin B12. Nat Commun 9. https://doi.org/10.1038/s41467-018-07412-6

García-Granados R, Lerma-Escalera JA, Morones-Ramírez JR (2019) Metabolic engineering and synthetic biology: synergies, future, and challenges. Front Bioeng Biotechnol 7. https://doi.org/10.3389/fbioe.2019.00036

Gibson DG (2017) Gibson assembly cloning guide. Sgi-Dna (2017)

Guo T, Xin Y, Zhang Y, Gu X, Kong J (2019) A rapid and versatile tool for genomic engineering in Lactococcus lactis. Microb Cell Fact 18. https://doi.org/10.1186/s12934-019-1075-3

Hasunuma T, Kondo A (2012) Development of yeast cell factories for consolidated bioprocessing of lignocellulose to bioethanol through cell surface engineering

Jensen PR, Chavarria KL, Fenical W, Moore BS, Ziemert N (2014) Challenges and triumphs to genomics-based natural product discovery

Jo S, Yoon J, Lee SM, Um Y, Han SO, Woo HM (2017) Modular pathway engineering of Corynebacterium glutamicum to improve xylose utilization and succinate production. J Biotechnol 258. https://doi.org/10.1016/j.jbiotec.2017.01.015

Kanehisa M, Goto S (2000) KEGG: kyoto encyclopedia of genes and genomes

Khan AZ, Bilal M, Rasheed T, Iqbal HMN (2018) Advancements in biocatalysis: from computational to metabolic engineering. Cuihua Xuebao/Chin J Catal 39. https://doi.org/10.1016/S1872-2067(18)63144-4

Kim J, Tremaine M, Grass JA, Purdy HM, Landick R, Kiley PJ, Reed JL (2019) Systems metabolic engineering of escherichia coli improves coconversion of lignocellulose-derived sugars. Biotechnol J 14. https://doi.org/10.1002/biot.201800441

Koch M, Duigou T, Faulon JL (2020) Reinforcement learning for bioretrosynthesis. ACS Synth Biol 9. https://doi.org/10.1021/acssynbio.9b00447

Kotopka BJ, Smolke CD (2020) Model-driven generation of artificial yeast promoters. Nat Commun 11. https://doi.org/10.1038/s41467-020-15977-4

Lai SJ, Zhang Y, Liu SW, Liang Y, Shang XL, Chai X, Wen TY (2012) Metabolic engineering and flux analysis of corynebacterium glutamicum for L-serine production. Sci China Life Sci 55. https://doi.org/10.1007/s11427-012-4304-0

Lee J, Saddler JN, Um Y, Woo HM (2016) Adaptive evolution and metabolic engineering of a cellobiose- and xylose- negative corynebacterium glutamicum that co-utilizes cellobiose and xylose. Microb Cell Fact. 15. https://doi.org/10.1186/s12934-016-0420-z

Lino CA, Harper JC, Carney JP, Timlin JA (2018) Delivering crispr: a review of the challenges and approaches

Liong MT (2008) Safety of probiotics: translocation and infection

Lowe R, Shirley N, Bleackley M, Dolan S, Shafee T (2017) Transcriptomics technologies. PLoS Computat Biol 13. https://doi.org/10.1371/journal.pcbi.1005457

Lu C, Zhang H, Zhang Q, Chu CY, Tahir N, Ge X, Jing Y, Hu J, Li Y, Zhang Y, Zhang T (2020) An automated control system for pilot-scale biohydrogen production: Design, operation and validation. International J Hydrogen Energy 45. https://doi.org/10.1016/j.ijhydene.2019.04.288

Magnúsdóttir S, Heinken A, Kutt L, Ravcheev DA, Bauer E, Noronha A, Greenhalgh K, Jäger C, Baginska J, Wilmes P, Fleming RMT, Thiele I (2017) Generation of genome-scale metabolic reconstructions for 773 members of the human gut microbiota. Nat Biotechnol 35. https://doi.org/10.1038/nbt.3703

Moleirinho MG, Silva RJS, Alves PM, Carrondo MJT, Peixoto C (2019) Current challenges in biotherapeutic particles manufacturing

Morange M (2009) The central dogma of molecular biology. Resonance. 14. https://doi.org/10.1007/s12045-009-0024-6

Ndukwe JK, Aliyu GO, Onwosi CO, Chukwu KO, Ezugworie FN (2020) Mechanisms of weak acid-induced stress tolerance in yeasts: prospects for improved bioethanol production from lignocellulosic biomass

Niu G (2018) Genomics-driven natural product discovery in actinomycetes

Niu FX, He X, Wu YQ, Liu JZ (2018) Enhancing production of pinene in escherichia coli by using a combination of tolerance, evolution, and modular co-culture engineering. Front Microbiol 9. https://doi.org/10.3389/fmicb.2018.01623

Okano K, Uematsu G, Hama S, Tanaka T, Noda H, Kondo A, Honda K (2018) Metabolic engineering of lactobacillus plantarum for direct l-lactic acid production from raw corn starch. Biotechnol J 13. https://doi.org/10.1002/biot.201700517

Park SH, Kim HU, Kim TY, Park JS, Kim SS, Lee SY (2014) Metabolic engineering of corynebacterium glutamicum for L-arginine production. Nat Commun 5. https://doi.org/10.1038/ncomms5618

Peng B, Williams TC, Henry M, Nielsen LK, Vickers CE (2015) Controlling heterologous gene expression in yeast cell factories on different carbon substrates and across the diauxic shift: a comparison of yeast promoter activities. Microb Cell Fact 14. https://doi.org/10.1186/s12934-015-0278-5

Pérez-Ortín JE, Tordera V, Chávez S (2019) Homeostasis in the central dogma of molecular biology: the importance of mRNA instability. RNA Biol 16. https://doi.org/10.1080/15476286.2019.1655352

Placzek S, Schomburg I, Chang A, Jeske L, Ulbrich M, Tillack J, Schomburg D (2017) BRENDA in 2017: new perspectives and new tools in BRENDA. Nucleic Acids Res 45. https://doi.org/10.1093/nar/gkw952

Rainha J, Rodrigues JL, Rodrigues LR (2021) CRISPR-Cas9: a powerful tool to efficiently engineer saccharomyces cerevisiae

Ruhal R, Choudhury B (2012) Use of an osmotically sensitive mutant of Propionibacterium freudenreichii subspp. shermanii for the simultaneous productions of organic acids and trehalose from biodiesel waste based crude glycerol. Bioresour Technol 109. https://doi.org/10.1016/j.biortech.2012.01.039

Ruhal R, Aggarwal S, Choudhury B (2011) Suitability of crude glycerol obtained from biodiesel waste for the production of trehalose and propionic acid. Green Chem 13. https://doi.org/10.1039/c1gc15847c

Ruhal R, Kataria R, Choudhury B (2013) Trends in bacterial trehalose metabolism and significant nodes of metabolic pathway in the direction of trehalose accumulation

Sanches-Medeiros A, Monteiro LMO, Silva-Rocha R (2018) Calibrating transcriptional activity using constitutive synthetic promoters in mutants for global regulators in Escherichia coli. Int J Genomics. https://doi.org/10.1155/2018/9235605

Selim KA, El-Ghwas DE, Easa SM, Abdelwahab Hassan MI (2018) Bioethanol a microbial biofuel metabolite; New insights of yeasts metabolic engineering

Shastri A, Morgan J (2004) Calculation of theoretical yields in metabolic networks. Biochem Mol Biol Educ 32. https://doi.org/10.1002/bmb.2004.494032050379

Skinnider MA, Johnston CW, Gunabalasingam M, Merwin NJ, Kieliszek AM, MacLellan RJ, Li H, Ranieri MRM, Webster ALH, Cao MPT, Pfeifle A, Spencer N, To QH, Wallace DP, Dejong CA, Magarvey NA (2020) Comprehensive prediction of secondary metabolite structure and biological activity from microbial genome sequences. Nat Commun 11. https://doi.org/10.1038/s41467-020-19986-1

Sleight SC, Bartley BA, Lieviant JA, Sauro HM (2010) In-fusion biobrick assembly and re-engineering. Nucleic Acids Res 38. https://doi.org/10.1093/nar/gkq179

Soltani S, Hammami R, Cotter PD, Rebuffat S, Said L ben, Gaudreau H, Bédard F, Biron E, Drider D, Fliss I (2021) Bacteriocins as a new generation of antimicrobials: toxicity aspects and regulations

Su Y, Zhang W, Zhang A, Shao W (2020) Biorefinery: the production of isobutanol from biomass feedstocks

Tachibana C (2015) Transcriptomics today: microarrays, RNA-seq, and more. Science. https://doi.org/10.1126/science.opms.p1500095

Tyo KEJ, Espinoza FA, Stephanopoulos G, Jin YS (2009) Identification of gene disruptions for increased poly-3-hydroxybutyrate accumulation in synechocystis PCC 6803. Biotechnol Prog 25. https://doi.org/10.1002/btpr.228

Vogl T, Kickenweiz T, Pitzer J, Sturmberger L, Weninger A, Biggs BW, Köhler EM, Baumschlager A, Fischer JE, Hyden P, Wagner M, Baumann M, Borth N, Geier M, Ajikumar PK, Glieder A (2018) Engineered bidirectional promoters enable rapid multi-gene co-expression optimization. Nat Commun 9. https://doi.org/10.1038/s41467-018-05915-w

Wang H, Marcišauskas S, Sánchez BJ, Domenzain I, Hermansson D, Agren R, Nielsen J, Kerkhoven EJ (2018) RAVEN 2.0: a versatile toolbox for metabolic network reconstruction and a case study on Streptomyces coelicolor. PLoS Comput Biol 14. https://doi.org/10.1371/journal.pcbi.1006541

Wen B, Zeng WF, Liao Y, Shi Z, Savage SR, Jiang W, Zhang B (2020) Deep learning in proteomics

Westbrook AW, Miscevic D, Kilpatrick S, Bruder MR, Moo-Young M, Chou CP (2019) Strain engineering for microbial production of value-added chemicals and fuels from glycerol

Query ID="Q1" Text="" Woolston BM, Edgar S, Stephanopoulos G (2013) Metabolic engineering: Past and future. Ann Rev Chem Biomol Eng 4. https://doi.org/10.1146/annurev-chembioeng-061312-103312

Wu MY, Sung LY, Li H, Huang CH, Hu YC (2017) Combining CRISPR and CRISPRi systems for metabolic engineering of E. coli and 1,4-BDO biosynthesis. ACS Synth Biol 6:2350–2361. https://doi.org/10.1021/acssynbio.7b00251

Xu N, Wei L, Liu J (2019) Recent advances in the applications of promoter engineering for the optimization of metabolite biosynthesis

Yang D, Park SY, Park YS, Eun H, Lee SY (2020) Metabolic engineering of escherichia coli for natural product biosynthesis

Yim H, Haselbeck R, Niu W, Pujol-Baxley C, Burgard A, Boldt J, Khandurina J, Trawick JD, Osterhout RE, Stephen R, Estadilla J, Teisan S, Schreyer HB, Andrae S, Yang TH, Lee SY, Burk MJ, van Dien S (2011) Metabolic engineering of Escherichia coli for direct production of 1,4-butanediol. Nature Chem Biol 7. https://doi.org/10.1038/nchembio.580

Yuan SF, Alper HS (2019) Metabolic engineering of microbial cell factories for production of nutraceuticals

Zhang X, Lai L, Xu G, Zhang X, Shi J, Koffas MAG, Xu Z (2019) Rewiring the central metabolic pathway for high-yield l-serine production in corynebacterium glutamicum by using glucose. Biotechnol J 14. https://doi.org/10.1002/biot.201800497

Zhao J, Fang H, Zhang D (2020) Expanding application of CRISPR-Cas9 system in microorganisms. Synth Syst Biotechnol 5. https://doi.org/10.1016/j.synbio.2020.08.001

Chapter 21
Mathematical Models for Optimization of Anaerobic Digestion and Biogas Production

Venkata Naga Surya Gunasri Appala, Nitin Naresh Pandhare, and Shailendra Bajpai

21.1 Introduction

The management of waste and generation of energy at the same time have captivated global attention. Research has centered on developing and implementing long-term bioprocess technologies for resource recovery from complicated waste (Dhanya et al. 2020). Biogas production is a low-cost, easy-to-set-up and-operate technique for treating organic waste (Stazi and Tomei 2018). Two of the most pressing scientific and social concerns today are creating energy from renewable sources and effectively managing trash. Anaerobic digestion, which combines the recycling of various waste products with the generation of biogas, has been demonstrated to be a viable approach capable of solving both of these concerns during the previous two decades. The goal of current anaerobic digestion systems is to convert organic waste into biogas (Zhang et al. 2016).

Biogas plants, which create the renewable energy source "biogas," continue to pique people's curiosity. The average plant size is growing along with the number of plants. The demand for an utterly simulated design and optimized maneuver grows as investment and running expenses rise to execute biogas plants successfully (Balat and Balat 2009). There are various intricate interactions in the biogas manufacturing process. A critical initial step in this method is the creation of a complete model for the entire process. Detailed models for anaerobic digestion are essential components of a complete process model. Several models have been created in recent decades (Gerber and Span 2008). An overview of current models based on a biological and physicochemical backdrop is discussed in this chapter.

V. N. S. G. Appala · N. N. Pandhare (✉) · S. Bajpai
Department of Chemical Engineering, Dr. B. R. Ambedkar National Institute of Technology, Jalandhar, Punjab 144011, India
e-mail: pandharenn@nitj.ac.in

Y. K. Nandabalan et al. (eds.), *Zero Waste Biorefinery*, Energy, Environment, and Sustainability, https://doi.org/10.1007/978-981-16-8682-5_21

21.2 Biochemical Process of Anaerobic Digestion

Organic matter conversion to biogas involves successive sub-processes, including hydrolysis, acidogenesis, acetogenesis, and methanogenesis.

21.2.1 Hydrolysis

Carbohydrates, lipids, and proteins are the primary components of organic matter. These complex biopolymers are not metabolized by microorganisms directly. Biopolymers must first be decomposed into soluble constituents to cross the cell wall of acidogenic bacteria (Shin and Song 1995). Further, to hydrolyze biopolymers, acidogenic bacteria release enzymes into the outer environment. The biopolymer, substrate content, particle size, pH, and temperature all influence the rate of hydrolysis (Veeken and Hamelers 1999). The kinetic expression for hydrolysis is as follows (Loganath and Mazumder 2020):

$$\frac{dS_H}{dt} = -\frac{K_H S_H X_H}{K_{S_H} + S_H} \tag{21.1}$$

where: K_H = maximum specific rate of hydrolysis $(\text{day})^{-1}$, S_H = hydrolysable substrate concentration (mg/L), X_H = concentration of hydrolytic microorganisms (mg/L), K_{S_H} = half velocity constant for hydrolysis (mg/L).

21.2.2 Acidogenesis

Acidogenesis is the digestion of amino acid residues and saccharides by acid-forming bacteria and the oxidation of long-chain carboxylic acids and alcohols (Batstone et al. 2003). Short-chain volatile fatty acids such as acetic, propionic, butyric, and valeric acid will accumulate during this process. Acid-forming bacteria multiply quickly, with a least doubling time of 30 min (Mosey 1983). Therefore, they favor degradation to acetic acid because it provides them with the most energy for development. The kinetic expression for acidogenesis is as follows (Loganath and Mazumder 2020):

$$\frac{dS_{\text{LCFA}}}{dt} = \frac{K_H S_H X_H}{K_{S_H} + S_H} - \frac{K_A S_{\text{LCFA}} X_A}{K_{S_A} + S_{\text{LCFA}}} \tag{21.2}$$

where: K_A = maximum specific rate of acidogenesis (day^{-1}), S_{LCFA} = concentration of long chain fatty-acids (mg/L), X_A = concentration of acidogens (mg/L), K_{S_A} = half velocity constant for acidogenesis (mg/L).

21.2.3 Acetogenesis

Acetogenesis is the oxidation of volatile fatty acids (mainly C_3 and C_4 acids) by acetogenic bacteria into acetic acid and hydrogen. Because hydrogen inhibits this sub-process, a build-up of hydrogen must be avoided. As a result, hydrogen-utilizing and acetogenic bacteria coexist in agglomerates (Mosey 1983). Acetogenic microorganisms multiply gradually under optimal conditions, such as a low amount of dissolved hydrogen, multiplying every 1.5–4 days (Lawrence and McCarty 1969).

21.2.4 Methanogenesis

Methanogenesis is methane generation by methanogens from acetate, hydrogen, and carbon dioxide with a minimum time for doubling of approximately six hours (Bryant 1979). As they govern the synthesis of short-chain fatty acids, they are regarded as the "autopilot" of the anaerobic process. Ambient factors like temperature, pH, alkalinity, inhibitors, hazardous substances, etc., can impact all sub-processes. In addition, all sub-processes are interconnected and impacted by one another. All sub-processes are included in most biogas production models; however, only the rate-limiting phase is significant for modeling the process. There is considerable debate as to which phase is the rate-limiting stage. According to Andrews (1969), acetic acid breakdown to methane is rate-limiting. Hydrolysis is the rate-limiting phase, according to Veeken and Hamelers (1999). The kinetic expression for methane production and methane accumulation is as follows (Loganath and Mazumder 2020):

Methane production

$$\frac{dS_{\text{SCFA}}}{dt} = \frac{K_A S_{LCFA} X_A}{K_{S_A} + S_{\text{LCFA}}} - \frac{K_M S_{\text{SCFA}} X_M}{K_{S_M} + S_{\text{SCFA}}} \tag{21.3}$$

where: K_M = maximum specific-rate of methanogenesis $(\text{day})^{-1}$, S_{SCFA} = concentration of short-chain fatty acids (mg/L), X_M = concentration of methanogens (mg/L), K_{S_M} = half-velocity constant for methanogenesis (mg/L).

Methane accumulation

$$\frac{dS_M}{dt} = \frac{K_M S_{\text{SCFA}} X_M}{K_{S_M} + S_{\text{SCFA}}} \tag{21.4}$$

In anaerobic digestion, methane's theoretical or stoichiometric generation can be estimated as follows (Andriani et al. 2014). The methane levels generated from the breakdown of various substrates may be calculated using those stoichiometric equations.

$$C_\alpha H_\beta O_\delta N_\gamma S_\varepsilon + yH_2O \rightarrow xCH_4 + \gamma NH_3 + \varepsilon H_2S + (\alpha - x)CO_2$$

where: $x = (4\alpha + \beta - 2\delta - 3\gamma - 2\varepsilon)/8$ and $y = (4\alpha - \beta - 2\delta + 3\gamma + 2\varepsilon)/4$.

Methane production from fat breakdown was calculated to be over 70%; protein decomposition resulted in 63%, and cellulose decomposition resulted in 50% (Ward 2010). Methane levels from the breakdown of various substrates ranged from 50–70%, whereas CO_2, the second most crucial element of biogas after methane (Tortora et al. 2004), was about 30–50%.

21.3 Mathematical Modeling

Models of physicochemical and biological processes can include both qualitative and quantitative aspects. Examples include flow dynamics and transport phenomena to microbial cells in various bioreactor designs and operational factors (Xie et al. 2016). However, because anaerobic digestion is a complicated multi-stage continuous system that necessitates consortia of microbial groups, determining the kinetic constants is difficult. In addition, changes in media composition, intermediate metabolites enzyme/metabolite inhibitions, substrate loading rate, solid/fluid retention time, temperature, and reactor architecture can all affect the microbial community structure (Markowski et al. 2014). As a result, anaerobic digestion mathematical process models may be classified into five categories.

21.3.1 Fundamental Kinetics Models

The entire anaerobic process must be examined when investigating the kinetics of biogas generation. Processes can be classified as discontinuous or continuous, depending on the availability of substrate (Gerber and Span 2008). The reaction rate must be proportionate to product generation and depending on cell concentration. The anaerobic digestion process is based on bacterial growth kinetics, which is highly dependent on growth needs and the medium (Kythreotou et al. 2014). Like any other living thing, the life-cycle of bacterial cultures is marked by several development stages (Monod 1949). Due to continuously changing nutrients and inhibitors, bacteria cultures go through a stationary phase with visible, dynamic cellular metabolism or death, unlike steady-state processes. Minor temporal delays develop for discontinuous processes due to this continual adaptation, resulting in observable kinetic parameter variations (Xie et al. 2016). Many factors influence the growth curve, including ambient circumstances, substrate type and concentration, bacterium type, inoculum physiological parameters, and beginning bacterium concentration (Hagos et al. 2017).

The Monod model was the first to establish the notion of growth-limiting or regulating substrate (Monod 1949). The semi-empirical model has dominated the area of microbial growth kinetics in recent decades. The linear relationship between bacterial rate of growth and concentrations is calculated using Monod's model, with particular rates of growth expressed as exponential percent coefficient (Velázquez-Martí et al. 2018):

$$\frac{dX}{dt} = \mu X \tag{21.5}$$

$$\mu = \mu_{\max} \frac{S}{K_M + S} \tag{21.6}$$

where: parameter μ is the specific growth rate, $\mu_{\max}$ can be defined as the growth in biomass per unit time (no limiting nutrients) under optimum feeding circumstances, and K_M is the substrate concentration at which organisms' development rate is limited to half of its current maximum value.

In recent years, many alternative models for forecasting anaerobic digestion have been suggested. Equations (21.5) and (21.6) describe the relationship between microbial growth and substrate consumption as a result of mass formation (21.7).

$$\frac{dS}{dt} = \frac{1}{Y_{X/S}} \frac{dX}{dt} \tag{21.7}$$

where, yield coefficient $Y_{X/S}$ is defined as follows and is considered to be constant.

$$Y_{X/S} = \frac{dX}{dS} \tag{21.8}$$

Biogas is the end product of anaerobic digestion. The kinetics of product generation may be predicted using the kinetics of substrate breakdown and bacterial growth. Several biogas production models are based on Gaden's idea that the product is mainly created through primary energy metabolism and substrate degradation (Gaden 1959). As a result, kinetic Eq. (21.5) may be utilized to explain the production process.

$$\frac{dP}{dt} = Y_{P/X} \frac{dX}{dt} \tag{21.9}$$

where, yield coefficient $Y_{P/X}$ is defined as:

$$Y_{P/X} = \frac{dP}{dX} \tag{21.10}$$

The process of microbial growth produces heat that can be described as:

Table 21.1 Models of cell growth rate (Xie et al. 2016)

Type	Equation	Model Name and Year
Kinetic models without inhibition	$\mu = \mu_{\max}\left(1 - e^{-S/K_s}\right)$	Tessier (1942)
	$\mu = \mu_{\max}\frac{S^n}{K_s a + S^n}$	Moser (1958)
	$\mu = \mu_{\max}\frac{S}{BX+S}$	Contois (1959)
Kinetic models with inhibition	$\mu = \mu_{\max}\frac{1}{K_s+S+\frac{S^2}{K_{is}}}$	Andrews and Noak (1968)
	$\mu = \mu_{\max}\frac{S\left(1+\frac{\beta S}{K_{is}}\right)}{K_s+S+\frac{S^2}{K_{is}}}$	Webb (1963)
	$\mu = \mu_{\max}\left[e^{-\frac{S}{K_{si}}} - e^{-\frac{S}{K_s}}\right]$	Teissier (1942)
	$\mu = \mu_{max}\frac{S}{K_s+S} - K_{si}(s - s_c)$	Wymann and Tseng (1976)

$$\frac{dE}{dt} = Y_{E/X}\frac{dX}{dt} \tag{21.11}$$

$$Y_{E/X} = \frac{dE}{dX} \tag{21.12}$$

Several additional common anaerobic digestion kinetics equations based on microbial growth and substrate consumption rates restricted by a growth-limiting substrate concentration are also available (Table 21.1).

Modified Gompertz (Eq. 21.13) and Dual-pooled first order kinetics models are two more kinetics models used to predict methane production during batch digestion (Eq. 21.14).

$$M(t) = P.\exp\{-\exp\left[\frac{R_{\max}.e}{B_o}(\lambda - t) + 1\right]\} \tag{21.13}$$

$$M(t) = P.\left[1 - \alpha.\exp\left(-K_f t\right) - (1 - \alpha).\exp(-K_L t)\right] \tag{21.14}$$

where: M indicates the total amount of methane produced (mL/g VS); P stands for the maximum amount of methane produced (mL/g VS) [VS: volatile solids]; t refers to the amount of time it takes for a substrate to break down (d); $R_{\max}$ is the maximum rate of methane generation (mL/g VS/d); λ is the time for lag-phase (d); e is the constant 2.7183; K_f is the rate constant for a substrate that degrades quickly (d^{-1}), K_L is the rate constant for a substrate that degrades slowly (d^{-1}), and α is the proportion of quickly degradable substrate to a total biodegradable substrate that is rapidly degradable.

The dynamics of rate-limiting stages may be oversimplified in these kinetics models. In other words, because operational conditions might impact the rate-limiting step, it is unlikely to remain constant. Furthermore, intermediate inhibition cannot be used as a predictor of digester steadiness in these models since it is complicated to quantify. At the same time, only one kind of microbe (e.g., acetoclastic methanogens) is taken into account (Xie et al. 2016). Fundamental kinetics models (typically employed in lab-scale batch research) cannot give direct practical information for industrial-scale application. As a result of these problems, further practical use of such models is hampered, prompting more sophisticated mechanistic models like Anaerobic Digestion Model 1(ADM1) (Kythreotou et al. 2014).

21.3.2 Anaerobic Digestion Model 1

To compute biodegradability and gas flow at a steady state, the aforementioned early models relied on single-stage kinetics. Anaerobic Digestion Model No. 1 was developed later as a consequence of various researchers' efforts to create a common platform for modelling. The ADM1 model is a typical model for anaerobic digestion that includes both biochemical processes (including live organisms) and physico-chemical processes (not involving living organisms) (Thorin et al. 2012). It has five stages: decomposition, hydrolysis, acidogenesis, acetogenesis, and methanogenesis. ADM1 is a complex model representing (i) four equations for particle degradation, eight equations for soluble degradation, and seven equations for biomass concentrations are among the nineteen biochemical processes; (ii) six equations for acid/base equilibrium together with pH calculation; (iii) three equations for gas–liquid mass transfer (CH_4, CO_2, H_2), (iv) inhibitors, (v) several other factors (Markowski et al. 2014). Several researchers have used this model to simulate and evaluate real-world applications. The ADM1 model was also tweaked to accommodate various biogas production methods and substrates. Co-digestion of organic waste with activated sludge (Derbal et al. 2009), cow manure (Lübken et al. 2007), energy crops (Zhou et al. 2011), and dry digestion of the organic component municipal rubbish (Bollon et al. 2011) are examples of two-step digesters with a thermophilic pre-treatment stage and a mesophilic main treatment phase (Blumensaat and Keller 2005). On the other side, the ADM1 model has some challenges as any other model in that it must optimize energy output while recovering resources at the same time. Furthermore, the intricate relationships between substrates that might lead to synergistic or antagonistic effects are largely unproven (Haugen et al. 2013). Other shortcomings in ADM1 have been identified, including stoichiometry errors, solid retention time issues, and a lack of constraints for the thermodynamic limits (Kleerebezem and Loosdrecht 2006).

21.3.3 Statistical Models

The composition of media, which includes nutrient sources, besides culture conditions, has a significant impact on metabolite synthesis and microbial development.

A key necessity for bioprocess engineering is optimizing these variables, which may be accomplished using either traditional one factor at a time (OFAT) or statistical approaches (Breig and Luti 2021). The OFAT approach is a time-consuming strategy that fails to explain the interaction between factors due to the large number of trials required. In practice, applying this approach to get the genuine best value of these variables is difficult (Li et al. 2002). On the other hand, a statistical optimization strategy is recommended to eliminate the requirement for many runs to optimize all variables that may be improved and therefore maximize production (Chang et al. 2011). RSM (response surface methodology) is a statistical and mathematical technique for model building that evaluates the effect of numerous independent factors to determine the best value of variables to achieve desired results (Montgomery et al. 2021). RSM has been effectively utilized to optimize and model fermentation factors in biochemical and biotechnological processes (Ramírez et al. 2000). Optimization using central composite design (CCD) is largely explored (Wang et al. 2013). In statistical models, the relationship between critical variables (e.g., organic loading rate, the substrate to co-substrate, temperature, and carbon to nitrogen ratio) and the outputs is highlighted (e.g., methane yield and volatile solids reduction). Based on experimental data, the following second-order polynomial model coefficients may be estimated to represent functional connections between responses (Y) and a set of variables (X_1 and X_2) (McLeod et al. 2015).

$$Y = \beta_o + \beta_1 X_1 + \beta_2 X_2 + \beta_{11} X_{1^2} + \beta_{22} X_{2^2} + \beta_{12} X_1 X_2 \tag{21.15}$$

where: Y reflects the anticipated methane potential response, X_1 is the ratio of M_1/M_2; X_2 is the ratio of carbon/nitrogen; β_0 is a constant; β_1 and β_2 are linear coefficients; β_{12} is an interaction coefficient and β_{11} and β_{22} are quadratic coefficients.

Statistical models can help formulate beginning circumstances and variables for the anaerobic digestion system's maximal capacity functioning. The prognosis may vary depending on the sensitivity of the different variables listed in each study. The accuracy of a full-scale prognosis depends on the similarity between reaction kinetics, sensory and laboratory scales, and full-scale inhibition (McLeod et al. 2015).

21.3.4 Computational Models

Many researchers have proposed the computational Fluid Dynamics (CFD) approach as an alternative to some of the costly measurements performed on- or off-line. The CFD is based on the numerical solution of mass, momentum, turbulence, and energy equations using finite volumes , finite differences, or finite element discretization techniques, among others. In this simulation method, the biogas reactor's main design, geometrical details, and particularities can be taken into account. It is also possible to properly represent the operating characteristics as boundary and initial conditions (Thorin et al. 2012). The main modeling possibilities for the multiphase mixture in the bioreactor are:

(1) Eulerian/Eulerian where the injected gas is assumed to be a continua similar to the liquid phase. The phases transport equations are coupled via inter-phase exchange terms that account for all forces, mass, and energy between the gas and the liquid.
(2) Eulerian/Lagrangian, where the injected gas is treated as dispersed bubbles which are individually tracked in the liquid during the simulation. Meroney and Colorado (Meroney and Colorado 2009) successfully used a combination VOF/Lagrangian where the top gas in the reactor is considered continuous, having a distinct interface with the liquid. The injected gas is assumed dispersed and tracked with a Lagrangian method.
(3) Volume Of Fluid (VOF), which is an interface tracking method where the interface between gas and liquid is reconstructed, and the bubbles and/or gas pockets are automatically generated under the influence of the fluids and flow characteristics as well as the surface tension (Lindmark et al. 2009, 2011). This method needs a very refined mesh and can be more costly in CPU time than the other methods.

Computational models provide a dynamic framework for studying stream and velocity fields, shear, particle paths, heat transfer rates, transport of dissolved constituents. Digester shape, feed locations, and operating parameters calculate volumes of high mixing intensity and stagnant regions. Mixing allows active microorganisms and feeds substrates to come into close contact resulting in the better mass transfer of intermediate by-products inside digesters, resulting in more effective anaerobic digestion (Yu et al. 2013). The following procedures can be used to implement these models: i. Create the digester geometry in a simulation program; ii. Split the entire domain into smaller cells by placing a mesh; iii. Entrances, exits, and barriers should all be established as boundary conditions; iv. Define the distinct phases' attributes (gas, liquid, and solid); v. Use various resolution models and turbulence models to compute how geometry and boundary conditions affect the phase/phases in each cell specified by the mesh. Once confirmed, the CFD model may simulate, evaluate, and optimize anaerobic fermentation (Lindmark et al. 2014). The computerized model shows dissolved constituents' flow pattern and movement in a simulated anaerobic digester with intuitive illustrative data analysis. Although CFD has been widely used in the hydraulic engineering of anaerobic digestion systems, the basic link between mixing and methane output has yet to be thoroughly understood. The primary reason is that when biological rate equations are linked, numerical simulation and model stability are difficult to achieve (Wu 2012).

21.3.5 *Artificial Intelligence-Based Models*

Artificial intelligence (AI), known as natural-inspired computing (NIC), is relatively recent. Nature maintains equilibrium through optimum searching, which is the basis for developing algorithms for process engineering optimization issues. Iterative techniques for giving computations or suggestions step-by-step suited for specific purposes are known as algorithms (Ramachandran et al. 2019). Developing techniques for planning, executing, and evaluating optimization problems is what computational optimization are all about (Banos et al. 2011). Performance, efficiency, profit maximization, and energy and cost reduction are all examples of optimization. Any problem could be addressed if limitless time were available. However, this is not the case in real life. Intelligent approaches are necessary when time and resources are limited. Computer simulation becomes a must-have tool when dealing with non-linear systems like anaerobic digestion (Walid et al. 2021).

Machine learning (ML) has emerged as a fresh approach for data mining and model creation, thanks to recent advances in computational algorithms and increased computer capacity (Krogh 2008). Computer learning is a type of artificial intelligence in which a system learns from prior data rather than creating it from scratch. It has the potential to reveal hidden interactions between a variety of input characteristics and output findings, allowing for output prediction. This method is unaffected by the intricate interconnections utilized in mathematical models, resulting in better prediction accuracy (Portugal et al. 2018). Artificial neural networks (ANN), support vector machines (SVM), random forests (RF), logistic regression multiclass (GLMNET), and k-nearest neighbors (KNN) are examples of recently developed methods that may be used to make such predictions in regression and classification models.

Machine learning approaches have proven their ability to forecast utilizing traditional experimental data and/or microbial community information in various bioreactors. Machine learning algorithms have also been used to try to forecast biogas output. ANN and SVM were used to quantitatively forecast methane generation from digestion (Qdais et al. 2010; Wang et al. 2020). A web-based application was created to predict biogas output and income, mainly utilizing regression models (Clercq et al.

2020). Wang et al. (2020) used different machine learning approaches (including RF, GLMNET, SVM, and KNN) in regression and classification models to inspect the possiblity of forecasting production of methane from experimental data. Kusiak and Wei (2014) compared with different data-mining techniques and found that adaptive neuro-fuzzy inference system surpassed the others for methane forecast with the lowest error percent and zeroed fractional bias, representing high forecast precision and consensus between simulated and experimental values. To mimic the up-flow anaerobic sludge blanket operation, which is extremely susceptible to load fluctuations, the organic content of the effluent was calculated using ANN (Mendes et al. 2015).

Back-propagation with feed-forward Nair et al. (2016) employed ANN to explore the effects of different organic loading rates and substrates on methane generation in order to reach a target of 60–70% CH_4 from domestic waste. The study helped determine the best parameters for coefficient methane production; training and validation correlation coefficients of 0.88 or higher suggested that the model could learn and adapt successfully. Multi-layer regression network approaches were utilized by Dach et al. (2016) to describe the anaerobic fermentation process of slurry from agricultural and animal waste. ANN was shown to be effective for quantitative and qualitative methane prediction in various configurations with varied input/hidden/output layers. ANN has shown to be an excellent solution for non-linear multivariable systems in all of this study. It is perfect for black-box activities where identifying the impact of input elements on outputs is challenging. Once built, the model can only work with input data identical to the data used for training, suggesting that it lacks extrapolation capabilities. Furthermore, ANN calibration demands significant data, and the technique does not correctly manage missing data. ANN is not suited for describing the reasons for algorithm convergence failure in issues needing process explanation.

Genetic Algorithms (Kamalinasab and Vakili 2014) were used to maximize methane production in research that established the ideal operating parameters for best outcomes. The model was developed to enhance biogas for these kinds of applications, and biogas was used in a combustion engine. Barik and Murugan (2015) created an ANN–GA model to improve biogas quality and slurry fertilizer value, which discovered the best combination of cow dung and Karanja seeds, yielding an 89.8% biogas production with 73% methane content. A key advantage of GA is its ability to solve multi-objective optimization problems (Yan et al. 2016). To build a model for biomethane optimization, three competing goals were used: biomethane production, green energy maximisation, and energy reduction. A multi-layer perceptron neural network (MLPNN) and a PSO were combined to obtain the maximum methane percentage in biogas, biogas quantity, and biogas quality (Akbaş et al. 2015). Model prediction was made with MLPNN, which had a regression coefficient of 0.91, indicating a strong prediction of modeled outputs. Model optimization was done with PSO, which helped employ biogas production at maximum output levels.

Researchers have utilized ANN and GA to estimate methane generation from a combination of organic carbon sources due to the complexity of anaerobic processes (Kusiak and Wei 2014). The non-linear process was efficiently modeled, and a technique for optimum methane synthesis employing those organic substrates was created. As a result, this artificial intelligence-based method can cut the time it takes for anaerobic digestion to occur in half. These algorithm approaches to profit from the fact that they require a rudimentary understanding of reaction mechanisms and monitoring a few parameters during anaerobic digestion. The lack of flexibility in bioreactor design and scale-up and the necessity for diverse and complex input–output connections to train the technique for the realistic situation is also disadvantages (Lauwers et al. 2013).

21.4 Biogas Production from Organic Wastes

Anaerobic digestion involves the microbial decomposition of organic waste to create methane-rich biogas. The digestate generated might be utilised as a residue or as a side-product as fertiliser to improve soil quality (Chiu and Lo 2016; Morales-Polo et al. 1804). The nutritional composition of organic waste, particularly carbohydrates, proteins, and fat, might impact biogas generation. Fats have been proven to produce the greatest biogas in studies (Esposito et al. 2012). Nevertheless, because of their poor biodegradability, the process will take longer to complete. Carbohydrates and proteins, on the other hand, convert at a quicker pace; nevertheless, biogas output is minimal (Esposito et al. 2012). Organic wastes are municipal wastes, kitchen wastes, agricultural wastes, animal wastes, and specific fractions of industrial waste (bakery, food-processing, leather, etc.). Co-digestion of these organic wastes is proposed as the best way to take advantage of process synergies (Table 21.2).

21.5 Conclusions

This chapter summarized different mathematical modeling approaches for understanding the bioprocess kinetics of anaerobic digestion and biogas production. Besides, emphasis on process optimization techniques, computational methods such as computational fluid dynamics, machine learning and other related algorithm-based methods were discussed. Research review analysis deciphered that batch modeling studies carried out in lab-scale is helpful for fundamental process understanding. However, for biogas plants at a commercial scale, continuous mode process understanding and control studies are yet to be carried out. Therefore, future research can explore modern techniques such as artificial intelligence and block-chain for optimized anaerobic fermentative process control.

Table 21.2 Biogas production from different organic wastes

Substrate	Biogas observations	References
Animal dung	The cow dung slurry produced 19.20 mL of gas and when co digested with food waste amount of gas produced is increased to 30.58 mL as food wastes contain more nutrients than the dung	Chibueze et al. (2017)
Kitchen waste	When 75% organic kitchen waste is co-digested with 25% cow manure additions, the maximum methane production of 14,653.5 mL/g-VS is produced	Aragaw and Gessesse (2013)
Tannery waste	Leather shavings and tannery sludge are combined in which biogas yielded between 21 and 30 mL/kgVSS with maximum methane content of 59% v/v was observed	Agustini et al. (2018)
Garden waste	When garden waste is co-digested with pig manure, which is excellent for energy generation, maximum biogas production of 0.650 m^3/m^3.d with 65% methane concentration was recorded for 22 days	Zagorskis et al. (2012)
Withered flowers	Bioreactor with 300 L capacity resulted in optimum biogas production when the substrate concentration was adjusted at 7%	Lakshmi and Vijayalakshmi (2017)
Cotton wastes	Cotton stalks, cotton seed hull, and cotton oil cake were used for methane production which resulted in 65, 86, and 78 mL per gram were obtained in 23 days	Isci and Demirer (2007)

Acknowledgements Ms. Venkata Naga Surya Gunasri Appala acknowledges Dr. B.R. Ambedkar National Institute of Technology Jalandhar for providing financial support from the Ministry of Education (MoE), Govt. of India.

References

Agustini CB, Spier F, da Costa M, Gutterres M (2018) Biogas production for anaerobic co-digestion of tannery solid wastes under presence and absence of the tanning agent. Resour Conserv Recycl 130:51–59

Akbaş H, Bilgen B, Turhan AM (2015) An integrated prediction and optimization model of biogas production system at a wastewater treatment facility. Bioresour Technolol 196:566–576

Andrews JF (1969) Dynamic model of the anaerobic digestion process. J Sanit Eng Div 95(1):95–116

Andriani D, Wresta A, Atmaja TD, Saepudin A (2014) A review on optimization production and upgrading biogas through CO_2 removal using various techniques. Appl Biochem Biotechnol 172:1909–1928

Aragaw T, Gessesse A (2013) Co-digestion of cattle manure with organic kitchen waste to increase biogas production using rumen fluid as inoculums. Inter J Hysical Sci 8:443–450

Balat M, Balat H (2009) Biogas as a renewable energy source—a review. Energy Sour Part A 31(14):1280–1293

Banos R, Manzano-Agugliaro F, Montoya F, Gil C, Alcayde A, Gómez J (2011) Optimization methods applied to renewable and sustainable energy: a review. Renew Sustain Energy Rev 15:1753–1766

Barik D, Murugan S (2015) An artificial neural network and genetic algorithm optimized model for biogas production from co-digestion of seed cake of karanja and cattle dung. Waste Biomass Valorization 6:1015–1027

Batstone DJ, Pind PF, Angelidaki I (2003) Kinetics of thermophilic, anaerobic oxidation of straight and branched chain butyrate and valerate. Biotechnol Bioeng 84(2):195–204

Blumensaat F, Keller J (2005) Modelling of two-stage anaerobic digestion using the IWA Anaerobic Digestion Model No. 1 (ADM1). Wat Res 39:171–183

Bollon J, Le-Hyaric R, Benbelkacem H, Buffiere P (2011) Development of a kinetic model for anaerobic dry digestion processes: focus on acetate degradation and moisture content. Biochem Eng J 56:212–218

Breig SJM, Luti KJK (2021) Response surface methodology: a review on its applications and challenges in microbial cultures. Mater Today: Proc (2021)

Bryant M (1979) Microbial methane production—theoretical aspects. J Anim Sci 48(1):193–201

Chang C-C, Chen W-C, Ho T-F, Wu H-S, Wei Y-H (2011) Development of natural anti-tumor drugs by microorganisms. J Biosci Bioeng 111:501–511

Chibueze U, Okorie N, Oriaku O, Isu J, Peters E (2017) The production of biogas using cow dung and food waste. Inter J Mater Chem 7:21–24

Chiu SL, Lo IM (2016) Reviewing the anaerobic digestion and co-digestion process of food waste from the perspectives on biogas production performance and environmental impacts. Environ Sci Pollut Res 23:24435–24450

De Clercq D, Wen Z, Fei F, Caicedo L, Yuan K, Shang R (2020) Interpretable machine learning for predicting biomethane production in industrial-scale anaerobic co-digestion. Sci Total Environ 712:134574

Dach J, Koszela K, Boniecki P, Zaborowicz M, Lewicki A, Czekała W, Skwarcz J, Qiao W, Piekarska-Boniecka H, Białobrzewski I (2016) The use of neural modelling to estimate the methane production from slurry fermentation processes. Renew Sustain Energy Rev 56:603–610

Derbal K, Bencheikh-Lehocine M, Cecchi F, Meniai A-H, Pavan P (2009) Application of the IWA ADM1 model to simulate anaerobic co-digestion of organic waste with waste activated sludge in mesophilic condition. Biores Technol 100:1539–1543

Dhanya B, Mishra A, Chandel AK, Verma ML (2020) Development of sustainable approaches for converting the organic waste to bioenergy. Sci Total Environ 723:138109

Esposito G, Frunzo L, Giordano A, Liotta F, Panico A, Pirozzi F (2012) Anaerobic co-digestion of organic wastes. Rev Environ Sci Bio/techno 11:325–341

Gaden EL Jr (1959) Fermentation process kinetics. J Biochem Microbiol Technol Eng 1(4):413–429

Gerber M, Span R (2008) An analysis of available mathematical models for anaerobic digestion of organic substances for production of biogas. Proc. IGRC, Paris

Hagos K, Zong J, Li D, Liu C, Lu X (2017) Anaerobic co-digestion process for biogas production: Progress, challenges and perspectives. Renew Sustain Energy Rev 76:1485–1496

Haugen F, Bakke R, Lie B (2013) Adapting dynamic mathematical models to a pilot anaerobic digestion reactor

Isci A, Demirer GN (2007) Biogas production potential from cotton wastes. Renew Energy 32:750–757

Kamalinasab M, Vakili A (2014) Utilization of genetic algorithm to optimize biogas production from livestock waste to use in a CHP Plant in agricultural farms. Biotechnol J Int 1149–1164s
Kiely G, Tayfur G, Dolan C, Tanji K (1997) Physical and mathematical modelling of anaerobic digestion of organic wastes. Water Res 31(3):534–540
Kleerebezem R, Van Loosdrecht M (2006) Waste characterization for implementation in ADM1. Water Sci Technol 54:167–174
Krogh A (2008) What are artificial neural networks? Nat Biotechnol 26:195–197
Kusiak A, Wei X (2014) Prediction of methane production in wastewater treatment facility: a data-mining approach. Ann Oper Res 216:71–81
Kythreotou N, Florides G, Tassou SA (2014) A review of simple to scientific models for anaerobic digestion. Renew Energy 71:701–714
Lakshmi C, Vijayalakshmi S (2017) Studies on biogas production using withered flowers as a substrate. Res J Pharm Technol 10:4223–4226
Lauwers J, Appels L, Thompson IP, Degrève J, Van Impe JF, Dewil R (2013) Mathematical modelling of anaerobic digestion of biomass and waste: power and limitations. Prog Energy Combust Sci 39:383–402
Lawrence AW, McCarty PL (1969) Kinetics of methane fermentation in anaerobic treatment. J Water Pollut Control Fed R1-R17
Li C, Bai J, Cai Z, Ouyang F (2002) Optimization of a cultural medium for bacteriocin production by Lactococcus lactis using response surface methodology. J Biotechnol 93:27–34
Lindmark J, Thorin E, Fdhila RB, Dahlquist E (2014) Effects of mixing on the result of anaerobic digestion. Renew Sustain Energy Rev 40:1030–1047
Lindmark J, Bel Fdhila R, Thorin E (2009) On modelling the mixing in a digester for biogas production. In: MATHMOD 09, 6th international conference on mathematical modelling. Vienna, Austria
Lindmark J, Bel Fdhila R, Thorin E (2011) Retention time and nutrient tracking inside a digester for biogas production. In: 52nd international conference of scandinavian simulation society ScanSIMS, pp 29–30
Loganath R, Mazumder D (2020) Development of a simplified mathematical model for anaerobic digestion. In: Sustain Waste Manage Policies Case Stud, Springer, pp 571–578
Lübken M, Wichern M, Schlattmann M, Gronauer A, Horn H (2007) Modelling the energy balance of an anaerobic digester fed with cattle manure and renewable energy crops. Water Res 41:4085–4096
Markowski M, Białobrzewski I, Zieliński M, Dębowski M, Krzemieniewski M (2014) Optimizing low-temperature biogas production from biomass by anaerobic digestion. Renew Energy 69:219–225
McLeod JD, Othman MZ, Beale DJ, Joshi D (2015) The use of laboratory scale reactors to predict sensitivity to changes in operating conditions for full-scale anaerobic digestion treating municipal sewage sludge. Biores Technol 189:384–390
Mendes C, da Silva Magalhes R, Esquerre K, Queiroz LM (2015) Artificial neural network modeling for predicting organic matter in a full-scale up-flow anaerobic sludge blanket (UASB) reactor. Environ Model Assess 20:625–635
Meroney RN, Colorado P (2009) CFD simulation of mechanical draft tube mixing in anaerobic digester tanks. Water Res 43:1040–1050
Monod J (1949) The growth of bacterial cultures: annual reviews in microbiology 3
Montgomery DC, Peck EA, Vining GG (2021) Introduction to linear regression analysis. Wiley, New York
Morales-Polo C, Cledera-Castro MDM, Moratilla Soria BY (2018) Reviewing the anaerobic digestion of food waste: from waste generation and anaerobic process to its perspectives. Appl Sci 8:1804
Mosey F (1983) Mathematical modelling of the anaerobic digestion process: regulatory mechanisms for the formation of short-chain volatile acids from glucose. Water Sci Technol 15(8–9):209–232

Nair VV, Dhar H, Kumar S, Thalla AK, Mukherjee S, Wong JW (2016) Artificial neural network based modeling to evaluate methane yield from biogas in a laboratory-scale anaerobic bioreactor. Biores Technol 217:90–99

Portugal I, Alencar P, Cowan D (2018) The use of machine learning algorithms in recommender systems: a systematic review. Expert Syst Appl 97:205–227

Qdais HA, Hani KB, Shatnawi N (2010) Modeling and optimization of biogas production from a waste digester using artificial neural network and genetic algorithm. Resour Conserv Recycl 54:359–363

Ramachandran A, Rustum R, Adeloye AJ (2019) Review of anaerobic digestion modeling and optimization using nature-inspired techniques. Processes 7:953

Ramírez J, Santos I, Morales O, Morrissey M, Vázquez M (2000) Application of microbial transglutaminase to improve mechanical properties of surimi from silver carp utilizacióndé transglutaminasa microbiana paramellora-laspropiedades mecánicas de surimide carpaplateada utilizacióndé transglutaminasamicrobiana paramejorarlaspropiedades mecánicas de surimide carpaplateada. CYTA-J Food 3:21–28

Shin H-S, Song Y-C (1995) A model for evaluation of anaerobic degradation characteristics of organic waste: focusing on kinetics, rate-limiting step. Environ Technol 16(8):775–784

Stazi V, Tomei MC (2018) Enhancing anaerobic treatment of domestic wastewater: state of the art, innovative technologies and future perspectives. Sci Total Environ 635:78–91

Thorin E, Nordlander E, Lindmark J, Dahlquist E, Yan J, Bel Fdhila R (2012) Modeling of the biogas production process-a review. In: International conference on applied energy ICAE 2012, Suzhou, China, 5–8 Jul 2012

Tortora GJ, Funke BR, Case CL, Weber D, Bair W (2004) Microbiology: an introduction. Benjamin Cummings San Francisco, CA

Veeken A, Hamelers B (1999) Effect of temperature on hydrolysis rates of selected biowaste components. Biores Technol 69(3):249–254

Velázquez-Martí B, Meneses-Quelal OW, Gaibor-Chavez J, Niño-Ruiz Z (2018) Review of mathematical models for the anaerobic digestion process. In: Anaerobic digestion, IntechOpen

Walid F, El Fkihi S, Benbrahim H, Tagemouati H (2021) Modeling and optimization of anaerobic digestion: a review. In: E3S web of conferences. EDP Sciences, p 01022

Wang X, Yang G, Li F, Feng Y, Ren G, Han X (2013) Evaluation of two statistical methods for optimizing the feeding composition in anaerobic co-digestion: Mixture design and central composite design. Biores Technol 131:172–178

Wang L, Long F, Liao W, Liu H (2020) Prediction of anaerobic digestion performance and identification of critical operational parameters using machine learning algorithms. Bioresour Technol 298:122495

Ward A (2010) Biogas potential of fish wax with cattle manure. Department of Biosystems Engineering, Fakulty of Agricultural Sciences, University of Aarhus, Internal Report-Animal Science

Wu B (2012) CFD simulation of mixing for high-solids anaerobic digestion. Biotechnol Bioeng 109:2116–2126

Xie S, Hai FI, Zhan X, Guo W, Ngo HH, Price WE, Nghiem LD (2016) Anaerobic co-digestion: a critical review of mathematical modelling for performance optimization. Biores Technol 222:498–512

Yan N, Ren B, Wu B, Bao D, Zhang X, Wang J (2016) Multi-objective optimization of biomass to biomethane system. Green Energy Environ 1:156–165

Yu L, Wensel PC, Ma J, Chen S (2013) Mathematical modeling in anaerobic digestion (AD). J Bioremed Biodegradation S 4

Zagorskis A, Baltrėnas P, Misevičius A, Baltrėnaitė E (2012) Biogas production by anaerobic treatment of waste mixture consisting of cattle manure and vegetable remains. Environ Eng Manag J 11:849–856

Zhang Q, Hu J, Lee D-J (2016) Biogas from anaerobic digestion processes: research updates. Renew Energy 98:108–119

Zhou H, Löffler D, Kranert M (2011) Model-based predictions of anaerobic digestion of agricultural substrates for biogas production. Biores Technol 102:10819–10828

The manufacturer's authorised representative in the EU is Springer Nature Customer Service Centre GmbH, Europaplatz 3, 69115 Heidelberg, Germany. If you have any concerns regarding our products, please contact ProductSafety@springernature.com

Printed and bound by CPI Group (UK) Ltd, Croydon, CR0 4YY
30/09/2026
02225332-0018